Introduction to Engineering Electromagnetics

Yeon Ho Lee

Introduction to Engineering Electromagnetics

Second Edition

 Springer

Yeon Ho Lee
College of Information and Communication
Engineering
Sungkyunkwan University
Seoul, Korea (Republic of)

ISBN 978-3-031-28661-2   ISBN 978-3-031-28659-9 (eBook)
https://doi.org/10.1007/978-3-031-28659-9

This Springer imprint is published by the registered company Springer Nature Switzerland AG
The registered company address is: Gewerbestrasse 11, 6330 Cham, Switzerland

Preface

This textbook on engineering electromagnetics is designed for undergraduate sophomore- or junior-level courses. The book covers two semesters. The first part begins with the vector algebra and coordinate systems covered in Chap. 1 and the vector calculus covered in Chap. 2. These two chapters usually take approximately half a semester of their full coverage. Although the instructor may skip some materials in these chapters, students may use them as references. Chapter 3 discusses electrostatics, and Chap. 4 deals with steady currents. Chapter 5 is on magnetostatics. The second part of the book begins with the time-varying fields and Maxwell's equations presented in Chap. 6, and the concept of wave motion is discussed in Chap. 7. In these two chapters, students learn about the interrelationship between time-varying electric and magnetic fields, and the concept of plane waves. Chapter 8 addresses the propagation of electromagnetic waves in various materials, Chap. 9 discusses the transmission lines, and Chap. 10 focuses on waveguides. In addition, the newly added Chap. 11 of the second edition deals with the antennas.

Although electromagnetics is one of the most fundamental subjects in electrical engineering and has attracted many students, some students feel that it is difficult to master electromagnetics. Electromagnetics covers a wide range of topics that not only deal with various physical concepts but also involve many different mathematical concepts, such as vector functions, coordinate systems, integrals, derivatives, complex numbers, and phasors. Confusion arises not from the number of mathematical theorems and formulas but from the lack of thorough knowledge of the mathematical rules and rigorous application of the rules to electromagnetic problems. Moreover, confusion worsens owing to the lack of consistency in the notation used to denote various physical quantities and constants in electromagnetics.

The main objective of this book is to present electromagnetic concepts more consistently and rigorously. This goal can be achieved through the demonstration of elaborate reasoning and strict application of mathematical concepts. However, this does not necessarily imply that lengthy mathematical steps are necessary. In contrast, I encourage students to obtain solutions to electromagnetic problems in an intuitive manner by considering the symmetry of configurations and applying the uniqueness theorem.

The book contains detailed accounts of the following:

1. Students face difficulties with the concept of vector fields at the beginning of class, because they are only familiar with discrete vectors that represent, for example, the force acting on a rigid body. These vectors are closely related to the displacement of the object, which may occur subsequently. However, the constituent vector of a vector field does not necessarily imply the displacement of an object in space; it is a quantity specific to a point in space, and in most cases, is not allowed to move to another point in space.

2. The cylindrical and spherical coordinate systems are meaningful only if the geometry under consideration has cylindrical or spherical symmetry. When a position vector is expressed as $R\,\mathbf{a}_R$ in spherical coordinates, the unit vector in the radial direction $\mathbf{a}_R$ is generally a function of position but is treated as a constant in the presence of spherical symmetry. Base vectors in these coordinates are generally functions of position, and are therefore differentiable and integrable.

3. Symmetry is an integral part of Gauss's and Ampere's laws. The final form of the electric flux density or magnetic field intensity of a given problem should be predicted from symmetry considerations, so that a Gaussian surface or an Amperian path can be constructed. Typical symmetries in electromagnetics, including cylindrical, spherical, translational, and two-fold rotational symmetries, are discussed in detail in the text. Symmetry considerations are useful for intuitively solving the boundary value problems and the problems of solenoidal and toroidal coils.

4. The inconsistency in notation between different books is a less attractive aspect of electromagnetics. For some authors, the meaning of the notation V_{12} is the potential difference between points 1 and 2 ($V_1 - V_2$), yet for others, it signifies the work done in moving a unit charge from point 1 to point 2 ($V_2 - V_1$). It is very confusing if the electric force acting on charge q_2 due to charge q_1 is denoted by $\mathbf{F}_{21}$ and expressed as $\mathbf{F}_{21} = |\mathbf{F}_{21}|\,\mathbf{a}_{12}$, with a unit vector $\mathbf{a}_{12}$ directed from q_1 to q_2. This book adopts new notation to avoid confusion. In our notation, the potential difference is denoted as $V_{1-2} = V_1 - V_2$, in which the subscript $1-2$ mimics the subtraction on the right-hand side, while the hyphen implies a sense of backward direction, such as "from 2 to 1," or the effect at point 1 due to a cause at point 2. Accordingly, the electric force on q_1 due to q_2, which is in the direction of a unit vector pointing from q_2 to q_1, can be expressed as $\mathbf{F}_{1-2} = |\mathbf{F}_{1-2}|\,\mathbf{a}_{1-2}$. Subscript 12 always represents something "from 1 to 2". For example, Ψ_{12} represents the magnetic flux through loop 2 due to the current in loop 1.

5. The electromagnetic quantities can take different forms. The static field quantities are denoted by boldface letters, such as $\mathbf{E}$, for a static electric field, whereas the time-varying fields are denoted by script letters, such as $\mathscr{E}$, for a time-varying electric field. Scalars are denoted by regular letters, such as E, for the magnitude of the electric field intensity. The complex quantities are denoted by the caret at the top, such as $\hat{E}_o$, for the complex amplitude of the electric field intensity. Because an electric field phasor is independent of time, it is denoted as $\tilde{\mathbf{E}}$ with a tilde at the top.

In the second edition, the number of chapters has been expanded to 11 to discuss the principles and applications of antennas, and more figures and problems have been included and updated. Moreover, almost all of the text has been rewritten for better readability, and the English grammar has been checked with Paperpal.

In electromagnetics, a graphical description of a given problem can facilitate solutions. Therefore, many figures are drawn in three dimensions to explain the physical concepts and describe the problems in the examples and practice problems given at the end of the chapters. In addition, the numerical data are drawn graphically to scale for an accurate description of the results. The second edition contains 332 figures. Each subsection includes several examples that are elaborated upon by putting the aforementioned concepts and relations into use and illustrating rigorous approaches in steps. Each subsection ends with exercises and answers that can be solved in a few simple steps and used to check whether students correctly understood the concepts. At the end of each section, several review questions are provided to highlight the key concepts and relations discussed in the section. As it has been found that open-ended questions are ignored by many students, review questions are provided with hints referring to related equations and figures. This edition contains 466 end-of-chapter problems.

I would like to thank the professors and students who provided valuable comments and suggestions and corrected the errors in the examples and exercises. I also thank my wife Hyunjoo for her patience, inspiration, and encouragement while writing this book. I would also like to thank my daughters Sehkyong and Yoonkyong for their understanding and patience.

Yeon Ho Lee
Professor, College of Information
and Communication Engineering
Sungkyunkwan University
Seoul, Korea (Republic of)
pfyeonlee@skku.edu

Contents

Chapter 1
Vector Algebra and Coordinate System

Electromagnetics is primarily concerned with electric and magnetic phenomena occurring in free space and material media. It comprises three core subjects: electrostatics, which concerns static electric fields; magnetostatics, which relates to static magnetic fields; and electrodynamics, which deals with time-varying electric and magnetic fields. Electromagnetic theories are based on electromagnetic models that consist of sources such as electric charges and currents, basic quantities such as electric and magnetic field intensities, rules of operations such as vector algebra and vector calculus, and fundamental laws such as Coulomb's law and Biot-Savart law. The first two chapters of the text discuss the rules of operations, including vector algebra, the coordinate system, and vector calculus.

The basic quantities of electromagnetics are scalar and vector quantities. A scalar is completely specified by its magnitude, as with an electric potential, whereas a vector is uniquely defined by its magnitude and direction, as with an electric force. Under sinusoidal steady-state conditions, in which all electromagnetic quantities vary sinusoidally with time, complex numbers can be conveniently used to represent the harmonic time-dependence of scalar and vector quantities. Electrodynamic quantities are generally continuous functions of position and time, whereas the quantities of electrostatics and magnetostatics may vary only with the space coordinates. Therefore, in addition to discrete scalars and vectors defined at separate points in space, the concept of continuous scalar and vector fields is essential for describing the electromagnetic phenomena.

Mathematical rules and techniques are required not only to express electromagnetic concepts in a compact form but also to construct an electromagnetic model of a problem for a solution. Vector algebra, vector calculus, and coordinate systems are three basic mathematical tools for electromagnetics. Vector algebra concerns vector operations such as vector addition, scaling, scalar product, and vector product. Vector calculus involves the derivatives and integrals of scalar and vector fields, some of which are formulated into special operators called *gradient*, *divergence*, and *curl*. Coordinate systems allow expressions of geometric elements such as points, lines, surfaces, and volumes in terms of mathematical equations. Because the physical

© The Author(s), under exclusive license to Springer Nature Switzerland AG 2024
Y. H. Lee, *Introduction to Engineering Electromagnetics*,
https://doi.org/10.1007/978-3-031-28659-9_1

quantities and laws are independent of the coordinate system, we can select the coordinate system that best fits the geometry under consideration, thereby facilitating the construction of an electromagnetic model of the problem.

1.1   Vector and Vector Field

The vector represents the ***magnitude*** and ***direction***. An arrow can be conveniently used to represent a vector in visual form. The length of the arrow represents the magnitude and the arrowhead indicates the direction. The tail of the arrow is called the ***initial point***, and the tip is called the ***terminal point***. It is customary to use a letter in boldface, such as $\mathbf{A}$, or a letter with an arrow at the top, such as $\vec{A}$, to denote a vector. The vector can take the following forms:

$$\mathbf{A} = A\,\mathbf{a}_A = |\mathbf{A}|\,\mathbf{a}_A = A\frac{\mathbf{A}}{|\mathbf{A}|} \tag{1.1}$$

where A and $|\mathbf{A}|$ are the magnitudes, and $\mathbf{a}_A$ and $\mathbf{A}/|\mathbf{A}|$ are the unit vectors. The magnitude is a positive real number, and carries the unit of the vector. The unit vector has a magnitude of unity, $|\mathbf{a}_A| = 1$, and is dimensionless. A unit vector may be displaced to another point in space to show the direction with respect to the other vectors defined at the point.

If a vector is defined at each point in a region of space, it is said that a ***vector field*** exists in that region. A ***vector function*** is a mathematical expression for a vector field, and is generally a continuous function of position and time. In other words, vector field is defined as a vector function.

In the presence of a vector field in a three-dimensional space, as shown in Fig. 1.1, point p_1 can be defined by Cartesian coordinates x_1, y_1, and z_1, denoted by $p_1:(x_1, y_1, z_1)$. In our notation, colon is used to distinguish a point from scalar function $p_1(x, y, z)$. A point can also be specified by the ***position vector*** drawn from the origin to the point. The position vectors always begin at the origin. The magnitude of the position vector $\mathbf{r}_1$ represents the distance between the origin and point p_1, and its unit vector indicates the direction from the origin to the point.

As shown in Fig. 1.1, a vector quantity $\mathbf{A}$ is observed at point p_1 such that it has magnitude A and points in the direction of $\mathbf{a}_A$. It is important to note that both A and $\mathbf{a}_A$ are quantities specific to point p_1 or the initial point of the arrow representing $\mathbf{A}$. Let us assume that the vector field shown in Fig. 1.1 represents the wind velocity in the air observed at time $t = t_o$. Referring to vector $\mathbf{A}'$, if we follow position vector $\mathbf{r}_2$ and arrive at point p_2, we can detect the wind blowing in the direction of $\mathbf{a}_{A'}$ at speed A' [m/s].

Let the wind velocity be defined by a vector function $\mathbf{F}(\mathbf{r}, t)$ with position vector $\mathbf{r}$ and time t. As the wind velocity is a real physical quantity, $\mathbf{F}(\mathbf{r}, t)$ must be a smooth function of position and time. A constituent vector $\mathbf{A}$ is then specified as

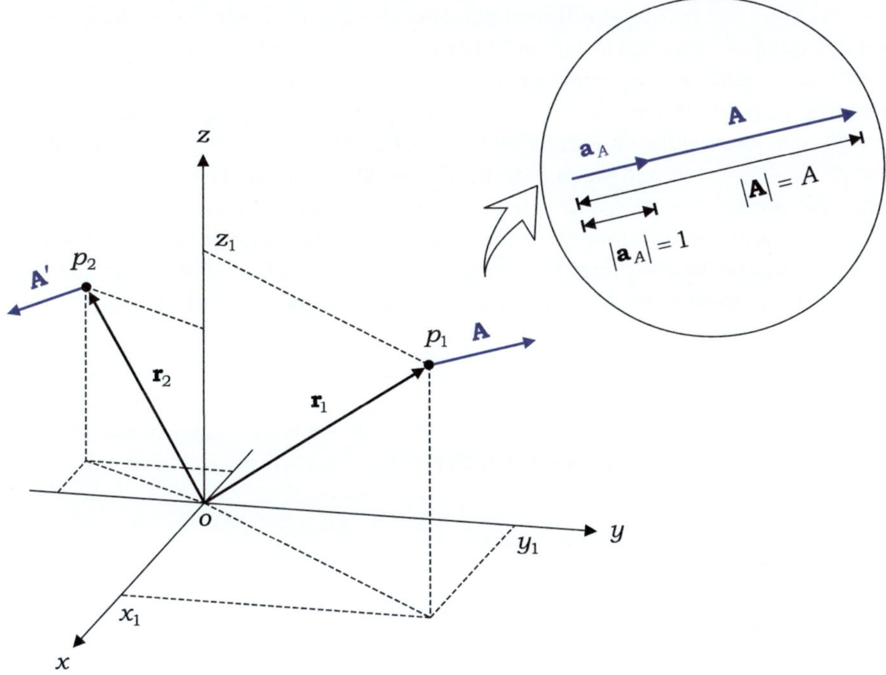

Fig. 1.1 Vector field in three-dimensional space

$$\mathbf{F}(\mathbf{r}_1, t_o) = \mathbf{F}(x_1, y_1, z_1; t_o) = \mathbf{A} \tag{1.2}$$

Note that point p_1 can be specified either by the position vector $\mathbf{r}_1$ or Cartesian coordinates x_1, y_1, and z_1. Similarly,

$$\mathbf{F}(\mathbf{r}_2, t_o) = \mathbf{A}' \tag{1.3}$$

The constant t_o indicates that the two wind velocities in Eqs. (1.2) and (1.3) were measured simultaneously.

If $\mathbf{A}'$ has the same magnitude and direction as $\mathbf{A}$, the two vectors are equal, but not identical, because they are positioned at different points.

Although the definition of a vector is straightforward, care should be taken when dealing with vector fields that typically involve different types of vectors. The position vector always starts at the origin and ends at a point in space. Meanwhile, the ***distance vector*** starts at one point in space, not at the origin, and ends at another. Both are useful for specifying the distance between two points and the direction from the initial point to the terminal point of the arrow. As the terminal point of these vectors is always a point of interest, the magnitude and direction of these vectors are used to describe the electromagnetic phenomena observed at the terminal point. If a vector field is defined in a region of space, its constituent vectors, for example $\mathbf{A}$ and $\mathbf{A}'$, as shown in Fig. 1.1, belong to different points in space in the sense that each represents a vector quantity observed at a point. When an arrow is used to graphically represent a constituent vector, its magnitude and direction define the quantity measured at its tail, which is the point of interest. In this case, the tip of the arrow does not correspond to any point in space if the unit of the vector is not meter. Consequently, the constituent vectors of the vector field are not arbitrarily displaced. However, this is not the case with unit vectors that show only the directions. The **base vectors** are three mutually orthogonal unit vectors along the three coordinate axes at a point in space. The base vectors point in the directions of increase of their respective coordinates and may change their orientations as functions of position in the cylindrical and spherical coordinate systems.

Exercise 1.1
Determine which of the following expressions represents the vector field:
(a) $\mathbf{A}(x)$, (b) $B(\mathbf{r})$, (c) $\mathbf{C}$ at point p, (d) $\mathbf{D}(\mathbf{r} - \mathbf{r}_1)$, (e) $\mathbf{r}$, (f) $\mathbf{r}_1$, and (g) $\mathbf{r} - \mathbf{r}_1$.

Ans. (a), (d), (e), (g).

Exercise 1.2
Which of the following statements does not make sense? (a) $\mathbf{A}$ is the negative of $\mathbf{B}$, (b) $\mathbf{A}$ is negative, and (c) the magnitude of $\mathbf{A}$ is negative.

Ans. (b), (c).

Review Questions

RQ 1.1   What are the differences between vector and vector field?   [Fig. 1.1]

RQ 1.2   What are the differences between position and distance vectors?   [Fig. 1.4]

1.2   Vector Algebra

1.2.1   Vector Addition and Subtraction

Two vectors, $\mathbf{A}$ and $\mathbf{B}$, can be added. The result is another vector $\mathbf{C}$. The **vector addition** is expressed as follows:

$$\mathbf{A} + \mathbf{B} = \mathbf{C} \tag{1.4}$$

Vector addition can be performed graphically using either the **parallelogram rule** or **head-to-tail rule**. According to the parallelogram rule, the diagonal vector of the parallelogram formed by $\mathbf{A}$ and $\mathbf{B}$ corresponds to the resultant vector, $\mathbf{C}$, as shown in Fig. 1.2a. Meanwhile, under the head-to-tail rule, vector $\mathbf{B}$ is first displaced, such that its initial point touches the terminal point of $\mathbf{A}$. The resultant vector $\mathbf{C}$ corresponds to an arrow drawn from the initial point of $\mathbf{A}$ to the terminal point of $\mathbf{B}$, as shown

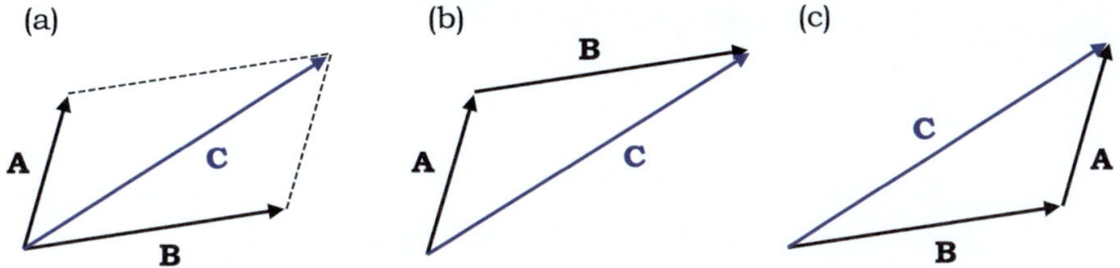

Fig. 1.2 Vector addition $\mathbf{A} + \mathbf{B}$: **a** Parallelogram rule, **b** and **c** Head-to-tail rule

in Fig. 1.2b. The same result is obtained when the roles of $\mathbf{A}$ and $\mathbf{B}$ are reversed, as shown in Fig. 1.2c.

Vector addition obeys the ***associative*** and ***commutative*** laws of algebra:

$$\mathbf{A} + (\mathbf{B} + \mathbf{C}) = (\mathbf{A} + \mathbf{B}) + \mathbf{C} \quad \text{(associative-law)} \tag{1.5a}$$

$$\mathbf{A} + \mathbf{B} = \mathbf{B} + \mathbf{A} \quad \text{(commutative-law)} \tag{1.5b}$$

The subtraction of vector $\mathbf{B}$ from vector $\mathbf{A}$ is equal to the addition of $\mathbf{A}$ to the negative of $\mathbf{B}$, that is,

$$\mathbf{A} - \mathbf{B} = \mathbf{A} + (-\mathbf{B}) = \mathbf{A} + B(-\mathbf{a}_B) \tag{1.6}$$

The negative of $\mathbf{B}$, denoted by $-\mathbf{B}$, has the same magnitude as $\mathbf{B}$, but points in the direction opposite to that of $\mathbf{B}$. Vector subtraction can be performed graphically in the same way as was followed to perform vector addition (see Figs. 1.3 and 1.4).

The rules in Eqs. (1.4)–(1.6) are directly applicable to the vector fields. The addition of two vector fields, $\mathbf{A}(\mathbf{r}) + \mathbf{B}(\mathbf{r})$, is defined such that a constituent vector of vector field $\mathbf{A}$ is added to its counterpart in vector field $\mathbf{B}$ at a given point in space. The addition of two vector fields results in another vector field.

Although the addition of two vectors $\mathbf{A}$ and $\mathbf{A}'$, as shown in Fig. 1.1, is possible from a mathematical point of view, the resultant vector may have no physical significance. This is because $\mathbf{A}$ and $\mathbf{A}'$ are two constituent vectors defined at different points in space and the final quantity is unrelated to any point in space. For example, if a velocity of 2 [m/s] in the northerly direction results from the addition of two wind velocities measured at two different points in the air, what is its use?

Fig. 1.3 Vector subtraction
$\mathbf{A} - \mathbf{B}$

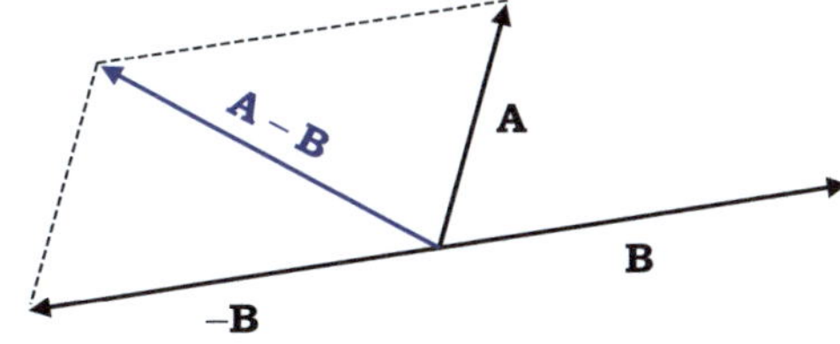

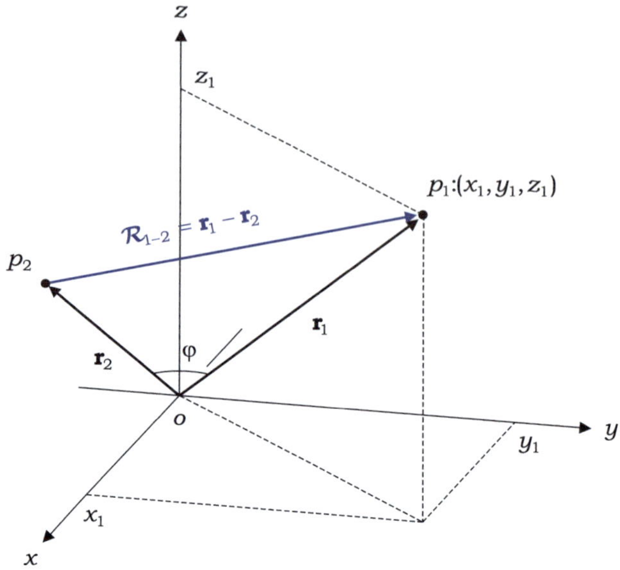

Fig. 1.4 The distance vector from point 2 to point 1 is denoted by $\mathcal{R}_{1-2}$

The subtraction of position vector $\mathbf{r}_2$ from another position vector $\mathbf{r}_1$ is expressed as $\mathcal{R} = \mathbf{r}_1 - \mathbf{r}_2$, which is a ***distance vector*** that represents the directed distance from p_2 with position vector $\mathbf{r}_2$ to p_1 with position vector $\mathbf{r}_1$ (Fig. 1.4). The distance vector is particularly useful for expressing the physical quantity observed at one point (***field point***), caused by a source at another point (***source point***). Here, we introduce a new type of subscript that is used consistently throughout the text. Hereafter, the distance vector is denoted by $\mathcal{R}_{1-2} = \mathbf{r}_1 - \mathbf{r}_2$. The subscript "1–2" mimics subtraction on the right-hand side and denotes a certain effect propagated "from 2 to 1". However, the conventional subscript "12" denotes something propagated "from 1 to 2".

Exercise 1.3
Which of the following expressions does not make sense?
(a) $-(-\mathbf{A}) = \mathbf{A}$, (b) $-(\mathbf{A} + \mathbf{B}) = -\mathbf{A} - \mathbf{B}$, and (c) $\mathbf{A} - (\mathbf{B} - \mathbf{C}) = (\mathbf{A} - \mathbf{B}) + \mathbf{C}$.

Ans. None.

1.2.2 Vector Scaling

The multiplication of vector $\mathbf{A}$ and positive scalar k is equivalent to lengthening or shortening the vector by a factor of k, without changing its direction. Expressed mathematically,

$$k\mathbf{A} = kA\,\mathbf{a}_A \tag{1.7}$$

In view of Eq. (1.7), a negative scalar k reverses the direction of $\mathbf{A}$, in addition to changing the magnitude by a factor of $|k|$.

Vector scaling obeys the associative, commutative, and distributive laws.

$$k\,(l\mathbf{A}) = l\,(k\mathbf{A}) \quad \text{(associative-law)} \tag{1.8a}$$

$$k\mathbf{A} = \mathbf{A}k \quad \text{(commutative-law)} \tag{1.8b}$$

$$(k+l)\,\mathbf{A} = k\mathbf{A} + l\mathbf{A} \quad \text{(distributive-law)} \tag{1.8c}$$

Exercise 1.4
Which of the following expressions is not defined?
(a) $k(\mathbf{A} + \mathbf{B}) = k\mathbf{A} + k\mathbf{B}$, (b) $-k\mathbf{A} = k(-\mathbf{A})$, and (c) $k(\mathbf{A} + c) = k\mathbf{A} + kc$.

Ans. (c).

1.2.3   Scalar or Dot Product

The scalar and vector products are unique vector operations, each involving two vectors. These are very different from the simple multiplication of vectors using scalars. As implied by its name, the scalar product yields scalar, whereas the vector product yields vector. The scalar product involves the cosine of the angle between the two vectors, whereas the vector product involves the sine of the angle. Furthermore, the scalar and vector products allow us to write mathematical expressions containing cosine or sine in vector form, where the equation is separated into two parts and is thus much easier to work with.

The **scalar product** (or **dot product**) of $\mathbf{A}$ and $\mathbf{B}$ is denoted by $\mathbf{A} \cdot \mathbf{B}$ (read "A dot B") and is defined as

$$\boxed{\mathbf{A} \cdot \mathbf{B} = AB \cos \theta_{AB}} \tag{1.9}$$

where A and B are the magnitudes of $\mathbf{A}$ and $\mathbf{B}$, respectively, and θ_{AB} is the smaller angle between $\mathbf{A}$ and $\mathbf{B}$. The dot product results in a positive real number for $0 \leq \theta_{AB} < 90°$, negative real number for $90° < \theta_{AB} \leq 180°$, and zero for $\theta_{AB} = 90°$, indicating that the two vectors are mutually orthogonal (Fig. 1.5).

Term $B \cos \theta_{AB}$ on the right-hand side of Eq. (1.9) represents the **projection** of $\mathbf{B}$ in the direction of $\mathbf{A}$, or simply the projection of $\mathbf{B}$ onto $\mathbf{A}$ (see Fig. 1.6). It is also called the **scalar component** of $\mathbf{B}$ in the direction of $\mathbf{A}$. Thus, the dot product $\mathbf{A} \cdot \mathbf{B}$ is viewed as the product of A and the projection of $\mathbf{B}$ onto $\mathbf{A}$ or the product of B and the projection of $\mathbf{A}$ onto $\mathbf{B}$.

Projection of one vector onto another is useful for resolving a vector into its components. The **vector component** of vector $\mathbf{C}$ in the direction of $\mathbf{A}$ is obtained by simply attaching unit vector $\mathbf{a}_A$ to the scalar component, that is, $(\mathbf{C} \cdot \mathbf{a}_A)\,\mathbf{a}_A$.

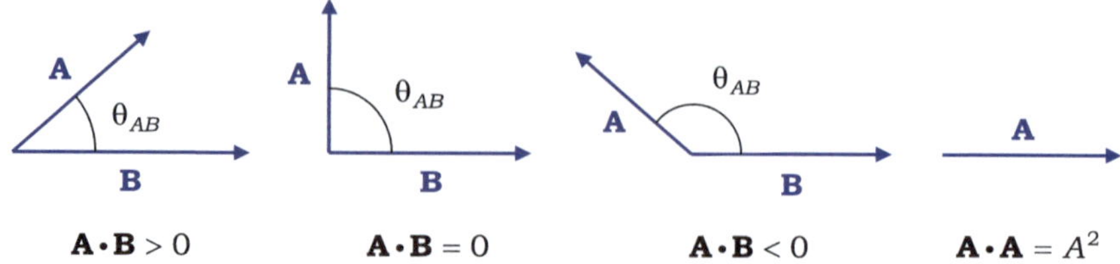

$$\mathbf{A} \cdot \mathbf{B} > 0 \qquad \mathbf{A} \cdot \mathbf{B} = 0 \qquad \mathbf{A} \cdot \mathbf{B} < 0 \qquad \mathbf{A} \cdot \mathbf{A} = A^2$$

Fig. 1.5 Dot product of **A** and **B**

Fig. 1.6 Projection of **B** onto **A**

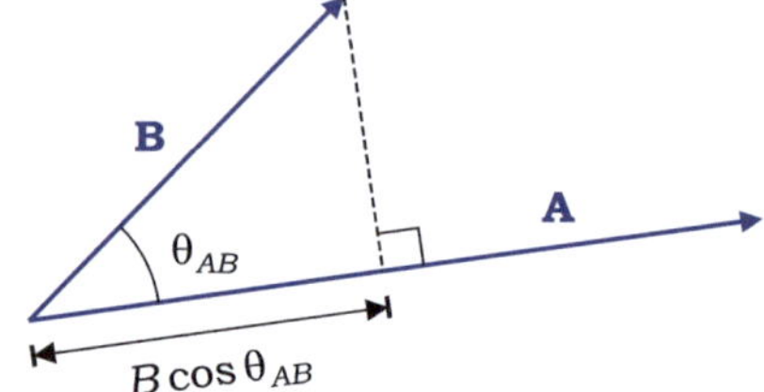

Resolving a vector into its vector components can be considered a reversal of vector addition. For example, vector addition $\mathbf{A} + \mathbf{B} = \mathbf{C}$ directly leads to the resolution of $\mathbf{C}$ into $\mathbf{A}$ and $\mathbf{B}$. A vector can also be projected onto a plane, yielding a vector component that is tangential to the plane.

It is obvious from Eq. (1.9) that $\mathbf{A} \cdot \mathbf{B}$ cannot be greater than the product of the two magnitudes, that is,

$$\mathbf{A} \cdot \mathbf{B} \leq AB \tag{1.10}$$

The dot product of a vector with itself gives the squared magnitude:

$$\mathbf{A} \cdot \mathbf{A} = |\mathbf{A}|^2 = A^2 \tag{1.11}$$

Thus, the magnitude of $\mathbf{A}$ can be obtained using Eq. (1.11) as follows:

$$A = |\mathbf{A}| = \sqrt{\mathbf{A} \cdot \mathbf{A}} \tag{1.12}$$

The magnitude of $\mathbf{A}$ is equal to the positive square root of $\mathbf{A} \cdot \mathbf{A}$.

The dot product obeys commutative and distributive laws.

$$\mathbf{A} \cdot \mathbf{B} = \mathbf{B} \cdot \mathbf{A} \quad \text{(commutative-law)} \tag{1.13a}$$

$$\mathbf{A} \cdot (\mathbf{B} + \mathbf{C}) = \mathbf{A} \cdot \mathbf{B} + \mathbf{A} \cdot \mathbf{C} \quad \text{(distributive-law)} \tag{1.13b}$$

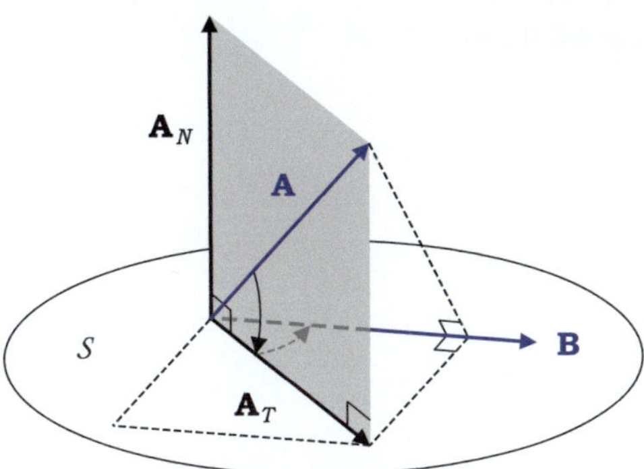

Fig. 1.7 Two-step projection

The commutative law follows directly from the relationship $\theta_{AB} = \theta_{BA}$, and the distributive law can be proven graphically, as shown in Example 1.2.

Next, we investigate an indirect method to determine the projection of **A** onto **B**, called the ***two-step projection***. This method is particularly useful when the angle between two vectors is not explicitly known. First, the projection of **A** onto a plane containing **B** is performed to obtain the vector component tangential to the plane, $\mathbf{A}_T$. Subsequently, the projection of $\mathbf{A}_T$ onto **B** is performed for $\mathbf{A} \cdot \mathbf{a}_B$. This method can be readily proven in a few steps. Because **A** can be resolved into normal and tangential components $\mathbf{A}_N$ and $\mathbf{A}_T$, as shown in Fig. 1.7, the projection of **A** onto **B** can be expressed as

$$\mathbf{A} \cdot \mathbf{a}_B = (\mathbf{A}_T + \mathbf{A}_N) \cdot \mathbf{a}_B = \mathbf{A}_T \cdot \mathbf{a}_B \tag{1.14}$$

where we have used the distributive law of the dot product and relation $\mathbf{A}_N \cdot \mathbf{a}_B = 0$. Thus, the two-step projection is verified.

Example 1.1 Express the magnitude of the distance vector $\mathcal{R}$, as shown in Fig. 1.4, in terms of $|\mathbf{r}_1|$, $|\mathbf{r}_2|$, and φ.

Solution

The dot product of $\mathcal{R}$ with itself yields

$$\begin{aligned}
\mathcal{R} \cdot \mathcal{R} = \mathcal{R}^2 &= (\mathbf{r}_1 - \mathbf{r}_2) \cdot (\mathbf{r}_1 - \mathbf{r}_2) \\
&= r_1^2 + r_2^2 - \mathbf{r}_1 \cdot \mathbf{r}_2 - \mathbf{r}_2 \cdot \mathbf{r}_1
\end{aligned} \tag{1.15a}$$

Inserting $\mathbf{r}_1 \cdot \mathbf{r}_2 = \mathbf{r}_2 \cdot \mathbf{r}_1 = r_1 r_2 \cos \varphi$ into Eq. (1.15a) yields

$$\mathcal{R}^2 = r_1^2 + r_2^2 - 2\,r_1 r_2 \cos \varphi \tag{1.15b}$$

This is known as the ***law of cosines***. Thus, the magnitude of $\mathcal{R}$ is obtained as

$$\mathcal{R} = \left[r_1^2 + r_2^2 - 2\,r_1 r_2 \cos \varphi \right]^{1/2}$$

Fig. 1.8 Law of cosines

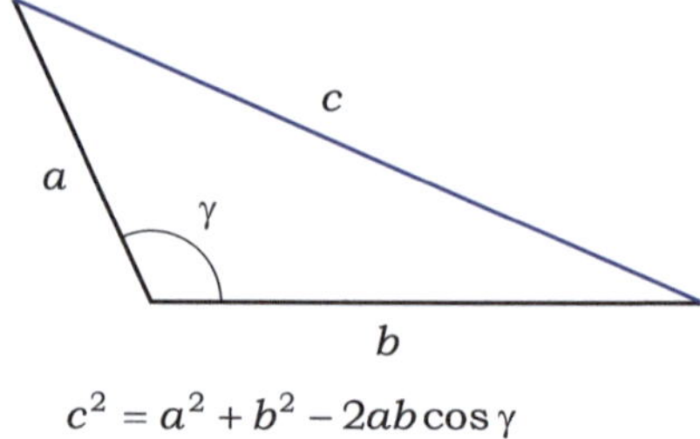

$$c^2 = a^2 + b^2 - 2ab\cos\gamma$$

For future reference, the law of cosines is illustrated in Fig. 1.8.

Example 1.2 Verify the distributive law of the dot product on three coplanar vectors.

Solution

Consider three vectors $\mathbf{A}$, $\mathbf{B}$, and $\mathbf{C}$ lying in a plane, as shown in Fig. 1.9. Line segments b, c, and d are projections of $\mathbf{B}$, $\mathbf{C}$, and $\mathbf{B} + \mathbf{C}$, respectively, onto $\mathbf{A}$. Obviously,

$$d = b + c \tag{1.16a}$$

Multiplying both sides of Eq. (1.16a) by A yields

$$A\,d = A\,b + A\,c \tag{1.16b}$$

By inserting the reversals of b, c, and d into Eq. (1.16b), we obtain

$$A\,(\mathbf{B} + \mathbf{C}) \cdot \mathbf{a}_A = A\,\mathbf{B} \cdot \mathbf{a}_A + A\,\mathbf{C} \cdot \mathbf{a}_A \tag{1.16c}$$

Using $A\,\mathbf{a}_A = \mathbf{A}$, Eq. (1.16c) can be rewritten as

$$\mathbf{A} \cdot (\mathbf{B} + \mathbf{C}) = \mathbf{A} \cdot \mathbf{B} + \mathbf{A} \cdot \mathbf{C} \tag{1.16d}$$

Fig. 1.9 Distributive law of dot product on three coplanar vectors

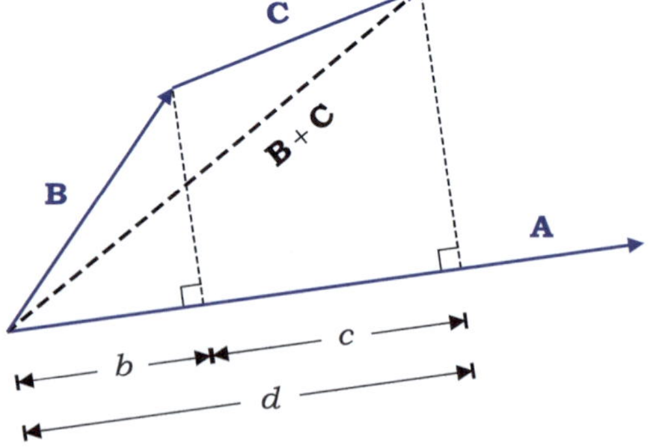

Thus, the distributive law of the dot product is verified in a plane.

Exercise 1.5
Expand dot product $(\mathbf{A} + \mathbf{B}) \cdot (\mathbf{C} + \mathbf{D})$ and name the law used.

Ans. $\mathbf{A} \cdot \mathbf{C} + \mathbf{A} \cdot \mathbf{D} + \mathbf{B} \cdot \mathbf{C} + \mathbf{B} \cdot \mathbf{D}$. The distributive law was applied twice.

Exercise 1.6
Which of the following expressions does not make sense?
(a) $(-\mathbf{A}) \cdot (-\mathbf{B}) = \mathbf{A} \cdot \mathbf{B}$, (b) $(-\mathbf{A}) \cdot \mathbf{B} = -(\mathbf{A} \cdot \mathbf{B})$, and (c) $(\mathbf{A} \cdot \mathbf{B}) \cdot \mathbf{C} = \mathbf{A} \cdot (\mathbf{B} \cdot \mathbf{C})$.

Ans. (c).

Exercise 1.7
What can be said about the projection of $\mathbf{A}$ onto $\mathbf{B}$, if $\mathbf{A} \cdot \mathbf{B} < 0$?

Ans. The projection falls onto $-\mathbf{B}$.

1.2.4 Vector or Cross Product

The ***vector product*** (or ***cross product***) of $\mathbf{A}$ and $\mathbf{B}$ is denoted by $\mathbf{A} \times \mathbf{B}$ (read "A cross B") and is defined as

$$\boxed{\mathbf{A} \times \mathbf{B} = AB \sin \theta_{AB}\, \mathbf{a}_N} \tag{1.17}$$

where θ_{AB} is the smaller angle between $\mathbf{A}$ and $\mathbf{B}$, and $\mathbf{a}_N$ is a unit vector normal to the plane containing both $\mathbf{A}$ and $\mathbf{B}$. The direction of $\mathbf{a}_N$ is governed by the ***right-hand rule***; the right thumb points in the direction of $\mathbf{a}_N$ when the fingers rotate from $\mathbf{A}$ to $\mathbf{B}$ through angle θ_{AB} (see Fig. 1.10).

The direction of $\mathbf{A} \times \mathbf{B}$ remains unchanged when θ_{AB} increases from zero to $180°$ (Fig. 1.11).

The magnitude of the cross product is equivalent to the area of the parallelogram formed by the two vectors. It is evident from Fig. 1.12 that magnitude A is the base of the parallelogram, and $B \sin \theta_{AB}$ is the height of the parallelogram.

Fig. 1.10 Right-hand rule

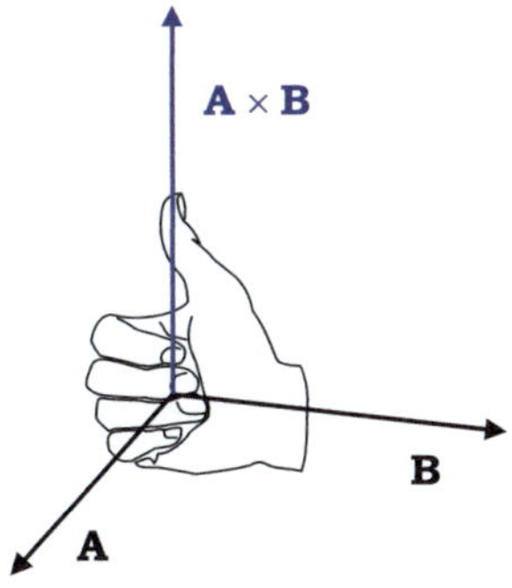

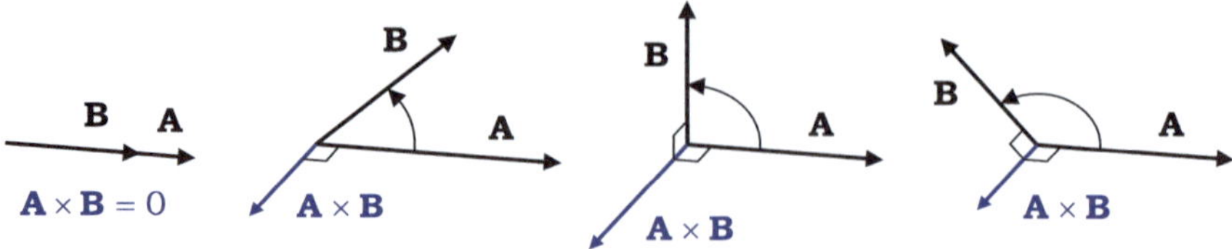

Fig. 1.11 Cross product of **A** and **B**

Fig. 1.12 Parallelogram formed by **A** and **B**

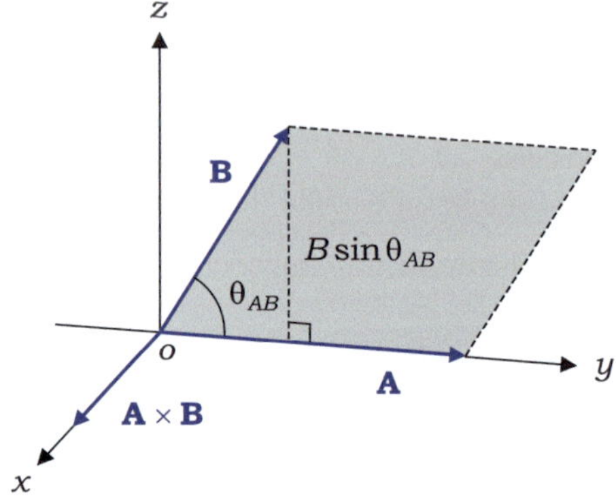

Furthermore, the cross product is useful for measuring the perpendicular distance from a point to a straight line. From Fig. 1.12, the perpendicular distance from the tip of **B** to vector **A** is given by $|\mathbf{B} \times \mathbf{a}_A|$.

Mathematical expressions containing sine can be conveniently written in vector form using a cross product. For example, when a circular cylinder rotates about the z-axis at an angular velocity of ω [rad/s], as shown in Fig. 1.13, the tangential velocity at position **r** is given by $v = \omega r \sin \theta$ [m/s]. This velocity can be expressed in a vector form as $\boldsymbol{v} = \boldsymbol{\omega} \times \mathbf{r}$ with the aid of the definition $\boldsymbol{\omega} \equiv \omega \, \mathbf{a}_z$.

The cross product obeys distributive and anticommutative laws, but not associative law.

$$\mathbf{A} \times (\mathbf{B} \times \mathbf{C}) \neq (\mathbf{A} \times \mathbf{B}) \times \mathbf{C} \quad \text{(not associative)} \tag{1.18a}$$

$$\mathbf{A} \times \mathbf{B} = -(\mathbf{B} \times \mathbf{A}) \quad \text{(anticommutative-law)} \tag{1.18b}$$

$$\mathbf{A} \times (\mathbf{B} + \mathbf{C}) = \mathbf{A} \times \mathbf{B} + \mathbf{A} \times \mathbf{C} \quad \text{(distributive-law)} \tag{1.18c}$$

The anticommutative law follows directly from the right-hand rule: if the right fingers rotate from **B** to **A**, instead of rotating from **A** to **B**, the thumb should point in the opposite direction.

Fig. 1.13 Rotating cylinder with angular velocity ω and tangential velocity v

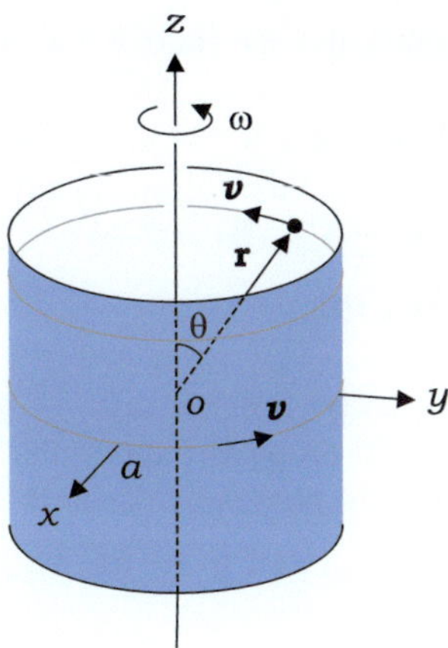

Example 1.3 Verify the distributive law $\mathbf{A} \times (\mathbf{B} + \mathbf{C}) = \mathbf{A} \times \mathbf{B} + \mathbf{A} \times \mathbf{C}$ if $\mathbf{A}$ is perpendicular to both $\mathbf{B}$ and $\mathbf{C}$.

Solution

We start with the parallelogram P formed by $\mathbf{B}$ and $\mathbf{C}$ in plane $\mathcal{S}$, as shown in Fig. 1.14. The two vectors $\mathbf{A} \times \mathbf{B}$ and $\mathbf{A} \times \mathbf{C}$ form another parallelogram, P', in $\mathcal{S}$. The four sides of P' are scaled by a factor of $|\mathbf{A}|$ and rotated by 90° with respect to their counterparts in P.

In exactly the same manner as above, the cross product of $\mathbf{A}$ and $\mathbf{B} + \mathbf{C}$ causes the diagonal of P to undergo the same scaling and rotation to coincide with that of P'. That is,

$$\mathbf{A} \times (\mathbf{B} + \mathbf{C}) = (\mathbf{A} \times \mathbf{B}) + (\mathbf{A} \times \mathbf{C}) \tag{1.19}$$

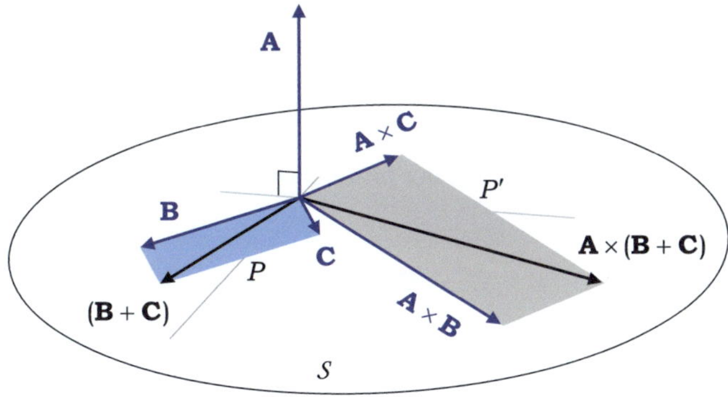

Fig. 1.14 Distributive law of cross product

Thus, the distributive law is verified.

Exercise 1.8
Which of the following expressions makes sense? (a) $-(\mathbf{A} \times \mathbf{B}) = (-\mathbf{A}) \times (-\mathbf{B})$, (b) $-(\mathbf{A} \times \mathbf{B}) = (-\mathbf{A}) \times \mathbf{B}$, and (c) $\mathbf{A} \times (\mathbf{B} \times \mathbf{C}) = -\mathbf{C} \times (\mathbf{A} \times \mathbf{B})$.

Ans. (b).

Exercise 1.9
Given $\mathbf{A}$ and $\mathbf{B}$, express the vector component of $\mathbf{A}$ in the direction (a) parallel to $\mathbf{B}$, and (b) perpendicular to $\mathbf{B}$.

Ans. (a) $(\mathbf{A} \cdot \mathbf{B})\,\mathbf{B}/B^2$, (b) $(\mathbf{B} \times \mathbf{A}) \times \mathbf{B}/B^2$.

Exercise 1.10
Sketch $\mathbf{A}$, $\mathbf{B}$, and $\mathbf{C}$, which all lie in a plane and satisfy (a) $\mathbf{A} \cdot \mathbf{B} = \mathbf{A} \cdot \mathbf{C}$, but $\mathbf{B} \neq \mathbf{C}$, and (b) $\mathbf{A} \times \mathbf{B} = \mathbf{A} \times \mathbf{C}$, but $\mathbf{B} \neq \mathbf{C}$.

Ans.

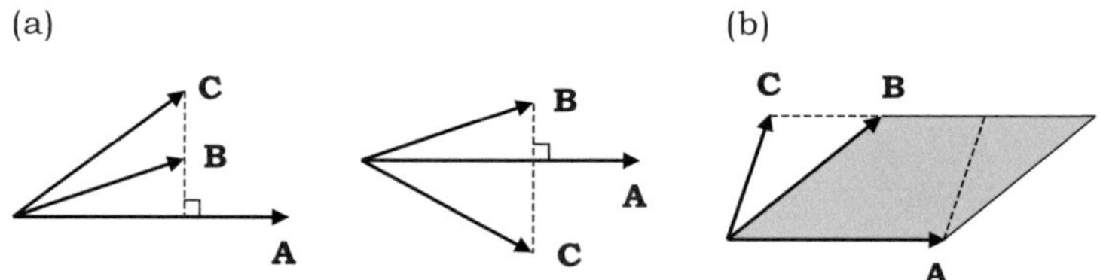

1.2.5   Scalar and Vector Triple Products

The dot and cross products can be extended to triple products that involve three vectors. **Scalar triple product** and **vector triple product** are frequently encountered in electromagnetics. They are so named because the former results in a scalar and the latter results in a vector.

(1)   The scalar triple product is the dot product of a vector and the cross product of two other vectors:

$$\mathbf{A} \cdot (\mathbf{B} \times \mathbf{C}) = \mathbf{B} \cdot (\mathbf{C} \times \mathbf{A}) = \mathbf{C} \cdot (\mathbf{A} \times \mathbf{B}) \qquad (1.20)$$

The three scalar-triple-products in Eq. (1.20) may be viewed as three possible cyclic permutations of the three vectors, that is, ABC-BCA-CAB. Note that the dot and cross symbols can be interchanged such that $\mathbf{A} \cdot \mathbf{B} \times \mathbf{C} = \mathbf{A} \times \mathbf{B} \cdot \mathbf{C}$. There is no need for parentheses because the cross product always precedes the dot product; otherwise, it is completely nonsense.

Fig. 1.15 Scalar triple product $\mathbf{A} \cdot (\mathbf{B} \times \mathbf{C})$

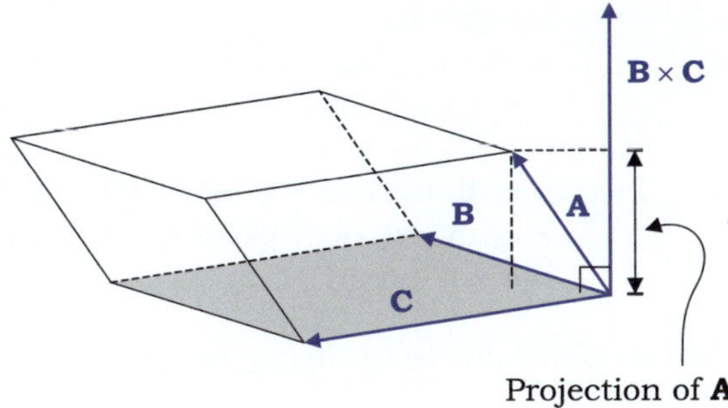

The scalar triple product $\mathbf{A} \cdot (\mathbf{B} \times \mathbf{C})$ is equivalent to the volume of the parallelepiped defined by $\mathbf{A}$, $\mathbf{B}$, and $\mathbf{C}$. It is evident from Fig. 1.15 that magnitude $|\mathbf{B} \times \mathbf{C}|$ is equal to the area of the base, and the projection of $\mathbf{A}$ onto $\mathbf{B} \times \mathbf{C}$ corresponds to the height of the parallelepiped.

(2) The vector triple product is the cross product of a vector and the cross product of two other vectors:

$$\mathbf{A} \times (\mathbf{B} \times \mathbf{C}) = \mathbf{B}\,(\mathbf{A} \cdot \mathbf{C}) - \mathbf{C}\,(\mathbf{A} \cdot \mathbf{B}) \qquad (1.21)$$

It is also known as "BAC-CAB" rule. The vector triple product can be easily verified by expanding the vectors in component form and directly evaluating the dot and cross products (see Sect. 1.3). Parentheses are required for the vector triple product to indicate the cross product to be conducted first. Note that associative law cannot be applied to a vector triple product, that is,

$$\mathbf{A} \times (\mathbf{B} \times \mathbf{C}) \neq (\mathbf{A} \times \mathbf{B}) \times \mathbf{C} \qquad (1.22)$$

It should be noted that the scalar and vector triple products are just mathematical identities. For example, the right-hand side of Eq. (1.21), rather than the left-hand side, can greatly facilitate the calculation, or vice versa.

Example 1.4 Verify the distributive law of the cross product on three coplanar vectors, $\mathbf{A}$, $\mathbf{B}$, and $\mathbf{C}$.

Solution

Let $\mathbf{a}_n$ be a unit normal to the plane containing the three vectors, and let a coplanar vector $\mathbf{D}$ be defined by $\mathbf{A} \equiv \mathbf{D} \times \mathbf{a}_n$. Then,

$$
\begin{aligned}
\mathbf{A} \times (\mathbf{B} + \mathbf{C}) & \qquad : \text{Three coplanar vectors} \\
= (\mathbf{B} + \mathbf{C}) \times (\mathbf{a}_n \times \mathbf{D}) & \qquad : \mathbf{A} \equiv \mathbf{D} \times \mathbf{a}_n \\
= \mathbf{a}_n[(\mathbf{B} + \mathbf{C}) \cdot \mathbf{D}] - \mathbf{D}[(\mathbf{B} + \mathbf{C}) \cdot \mathbf{a}_n] & \qquad : \text{Vector triple product} \\
= \mathbf{a}_n[\mathbf{B} \cdot \mathbf{D} + \mathbf{C} \cdot \mathbf{D}] & \qquad : \text{Distributive law of dot product} \\
= \mathbf{a}_n[\mathbf{B} \cdot (\mathbf{a}_n \times \mathbf{A}) + \mathbf{C} \cdot (\mathbf{a}_n \times \mathbf{A})] & \qquad : \mathbf{D} = \mathbf{a}_n \times \mathbf{A} \\
= \mathbf{a}_n[(\mathbf{A} \times \mathbf{B}) \cdot \mathbf{a}_n + (\mathbf{A} \times \mathbf{C}) \cdot \mathbf{a}_n] & \qquad : \text{Scalar triple product} \\
= \mathbf{A} \times \mathbf{B} + \mathbf{A} \times \mathbf{C} & \qquad : (\mathbf{A} \times \mathbf{B}) \| \mathbf{a}_n, \ (\mathbf{A} \times \mathbf{C}) \| \mathbf{a}_n
\end{aligned}
$$

Thus, the distributive law of the cross product on the three coplanar vectors is verified.

Exercise 1.11

Under what conditions does $\mathbf{A} \cdot (\mathbf{B} \times \mathbf{C})$ become zero?

Ans. $\mathbf{B} \| \mathbf{C}$, or the three vectors lie in the same plane.

Exercise 1.12

Under what condition is $(\mathbf{A} \times \mathbf{B}) \times \mathbf{A}$ equal to $\mathbf{B}$?

Ans. $\mathbf{A} \perp \mathbf{B}$ and $|\mathbf{A}| = 1$.

Exercise 1.13

Is the expression $(-\mathbf{A}) \cdot \mathbf{B} \times \mathbf{C} = \mathbf{A} \cdot (-\mathbf{B}) \times \mathbf{C}$ always true?

Ans. Yes.

Review Questions

RQ 1.3	What are the vector field and function?	[Fig. 1.1,(1.2)]
RQ 1.4	What carries the unit of vector $\mathbf{A}$?	[Fig. 1.1]
RQ 1.5	What is the physical meaning of the negative sign of $-\mathbf{B}$?	[Fig. 1.3]
RQ 1.6	Define the dot product.	[(1.9)]
RQ 1.7	Define scalar and vector components.	[Fig. 1.6]
RQ 1.8	State the distributive law of the dot product.	[(1.13b)]
RQ 1.9	Explain the two-step projection.	[Fig. 1.7]
RQ 1.10	Define the cross product.	[(1.17)]
RQ 1.11	State the distributive law of the cross product.	[(1.18c)]
RQ 1.12	Can an equation with a cosine or sine be expressed in vector form?	[(1.9)(1.17)]

1.3   Orthogonal Coordinate Systems

A coordinate system uses three numbers, called ***coordinates***, to uniquely define a point in a three-dimensional space. By extending the coordinates of a point to the lines, surfaces, and volumes, the coordinate system leads to a mathematical formulation of these geometric elements. A coordinate may be the distance from the origin to a point, or the angle measured from the coordinate axis. In the ***orthogonal coordinate system***, a point is defined by the intersection of three mutually perpendicular surfaces, called ***surfaces of constant coordinate***, over which one coordinate is held constant, whereas the others are allowed to vary. The ***right-handed coordinate system*** enumerates the three coordinates of a point in an order compatible with the right-hand rule. That is, when the right fingers rotate from the direction of increase of the first coordinate to that of the second coordinate, the thumb points in the direction of increase of the third coordinate. The Cartesian (or rectangular), cylindrical, and spherical coordinate systems are the three most common orthogonal coordinate systems used in electromagnetics.

In nature, the physical quantities and laws are independent of the coordinate system. Therefore, any coordinate system can be selected for an electromagnetic problem. However, a particular coordinate system may be advantageous if the surfaces of constant coordinate fit the geometry under consideration. The cylindrical coordinate system is particularly useful for problems with ***cylindrical symmetry***, whereas the spherical coordinate system is useful for those with ***spherical symmetry***. Most electromagnetic problems involve source and conceptual observer, whose locations are two separate and mutually independent entities. Thus, it would be useful to adopt two independent coordinate systems: one defines the position of the source, and the other specifies the position of the observer. We refer to this system as the ***mixed-coordinate system***. Subsequently, it may be necessary to express the final results in certain coordinates by transforming the coordinates of a point and the components of a vector from one coordinate system to another. This is referred to as the ***coordinate transformation***, which is discussed in the last section of this chapter.

1.3.1 Cartesian Coordinate System

The Cartesian coordinate system uses three variables, x, y, and z, to define a point in a three-dimensional space. The position of point $p_1:(x_1, y_1, z_1)$ is defined by the intersection of three planes: the $x = x_1$, $y = y_1$, and $z = z_1$ planes. At p_1, three unit vectors $\mathbf{a}_x$, $\mathbf{a}_y$, and $\mathbf{a}_z$ are defined in such a way that they are perpendicular to the $x = x_1$, $y = y_1$, and $z = z_1$ planes and point in the directions of increase of x, y, and z, respectively. These are called the ***base vectors***. Although the base vectors are constant in the Cartesian coordinate system, they are generally functions of position in other coordinate systems.

The base vectors obey ***orthonormality***:

$$\mathbf{a}_x \cdot \mathbf{a}_y = \mathbf{a}_y \cdot \mathbf{a}_z = \mathbf{a}_z \cdot \mathbf{a}_x = 0 \qquad (1.23a)$$

$$\mathbf{a}_x \cdot \mathbf{a}_x = \mathbf{a}_y \cdot \mathbf{a}_y = \mathbf{a}_z \cdot \mathbf{a}_z = 1 \qquad (1.23b)$$

This indicates that the base vectors are unit vectors and mutually perpendicular.

In the right-handed Cartesian coordinate system, the base vectors follow *cyclic relations*:

$$\mathbf{a}_x \times \mathbf{a}_y = \mathbf{a}_z \tag{1.24a}$$

$$\mathbf{a}_y \times \mathbf{a}_z = \mathbf{a}_x \tag{1.24b}$$

$$\mathbf{a}_z \times \mathbf{a}_x = \mathbf{a}_y \tag{1.24c}$$

The cyclic permutations in Eq. (1.24) follow the right-hand rule: when the right fingers rotate from the first to the second vector, the thumb points in the direction of the third vector.

The position vector always starts at the origin and ends at a point in space. For example, $\mathbf{r}_1$ in Fig. 1.16 is the position vector of point $p_1:(x_1, y_1, z_1)$. Its magnitude represents the distance between the origin and p_1, and its unit vector shows the direction from the origin to p_1. Because p_1, or the terminal point of $\mathbf{r}_1$, is the point of interest at which physical events can be observed and measured, all quantities related to p_1, including the position vector, should be expressed in terms of the coordinates and base vectors given at p_1. As discussed earlier, the vector components of $\mathbf{r}_1$ in the x-, y-, and z-directions are $x_1\mathbf{a}_x$, $y_1\mathbf{a}_y$, and $z_1\mathbf{a}_z$, respectively. By omitting subscript 1 for generalization, the position vector in Cartesian coordinates can be expressed in the *component form* as follows:

$$\boxed{\mathbf{r} = x\,\mathbf{a}_x + y\,\mathbf{a}_y + z\,\mathbf{a}_z} \tag{1.25}$$

Note that the scalar components of the position vector are the Cartesian coordinates of the point.

When vector $\mathbf{A}$ is defined at point p_1 in the Cartesian coordinate system as shown in Fig. 1.17, it represents a vector quantity specific to a point; thus, its magnitude and direction should be expressed in terms of the coordinates and base vectors given at p_1. This can be achieved by simply expanding $\mathbf{A}$ in terms of the base vectors given at p_1. That is,

$$\boxed{\mathbf{A} = A_x\,\mathbf{a}_x + A_y\,\mathbf{a}_y + A_z\,\mathbf{a}_z} \tag{1.26}$$

The scalar components A_x, A_y, and A_z are the projections of $\mathbf{A}$ onto base vectors $\mathbf{a}_x$, $\mathbf{a}_y$, and $\mathbf{a}_z$, respectively. If the unit of $\mathbf{A}$ is not meter, the arrowhead of $\mathbf{A}$ is not a spatial point, but is only used to graphically show the direction and relative magnitude of $\mathbf{A}$.

The expression in Eq. (1.26) can be extended to the vector field $\mathbf{B}(\mathbf{r})$ to express it in component form, such as $\mathbf{B}(\mathbf{r}) = B_x(\mathbf{r})\,\mathbf{a}_x + B_y(\mathbf{r})\,\mathbf{a}_y + B_z(\mathbf{r})\,\mathbf{a}_z$. In this case, the scalar components are generally functions of space coordinates, but the base

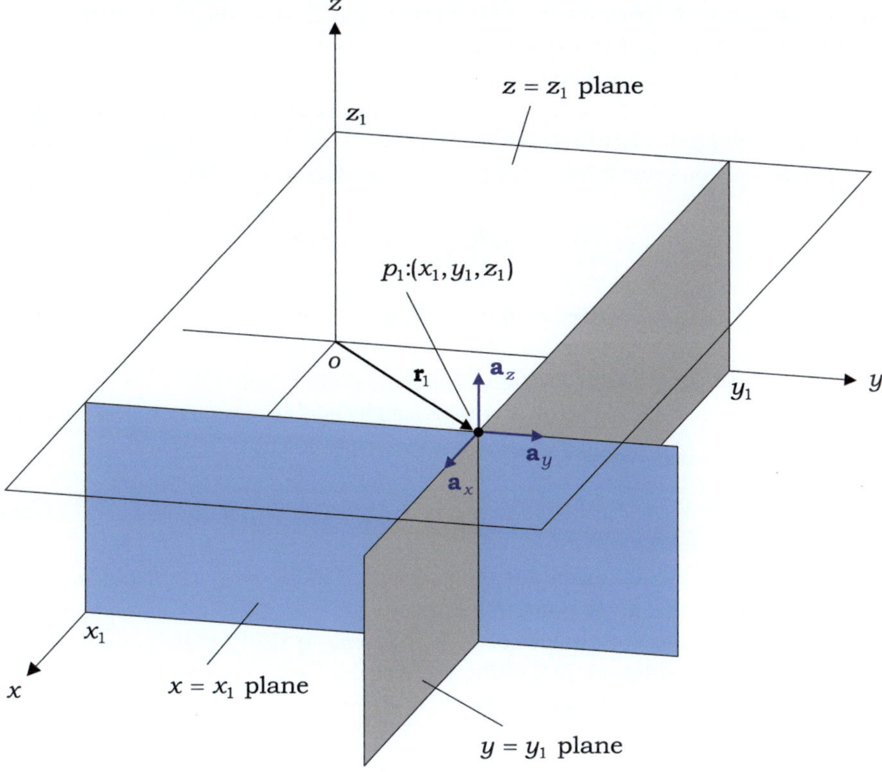

Fig. 1.16 Cartesian coordinate system

Fig. 1.17 Vector **A** is defined at point $p_1:(x_1, y_1, z_1)$ in the Cartesian coordinates

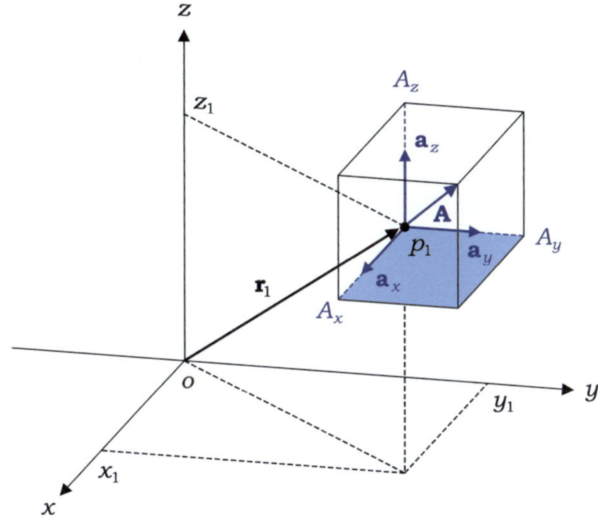

vectors are constant. However, the base vectors may also be functions of the space coordinates in cylindrical and spherical coordinate systems.

Now that we have learned how to express vector fields in component form, we are now ready to formulate the rules of vector algebra in terms of vector components. When two vector fields $\mathbf{A}(\mathbf{r})$ and $\mathbf{B}(\mathbf{r})$ coexist in a three-dimensional space, they can be expressed in component form in Cartesian coordinates as follows:

$$\mathbf{A}(\mathbf{r}) = A_x \, \mathbf{a}_x + A_y \, \mathbf{a}_y + A_z \, \mathbf{a}_z \tag{1.27a}$$

$$\mathbf{B}(\mathbf{r}) = B_x \, \mathbf{a}_x + B_y \, \mathbf{a}_y + B_z \, \mathbf{a}_z \tag{1.27b}$$

Again, the scalar components are functions of position in general.

The addition of $\mathbf{A}$ and $\mathbf{B}$ can be accomplished by adding similar components such that

$$\boxed{\mathbf{A} + \mathbf{B} = (A_x + B_x)\,\mathbf{a}_x + (A_y + B_y)\,\mathbf{a}_y + (A_z + B_z)\,\mathbf{a}_z} \tag{1.28}$$

This may serve as the definition for the addition of either two vector fields or two constituent vectors at a given point in the space.

Let us consider the dot product of $\mathbf{A} + \mathbf{B}$ and $\mathbf{a}_x$ to verify the x-component on the right-hand side of Eq. (1.28). Using the distributive law in Eq. (1.13b) and the orthonormality in Eq. (1.23), we obtain

$$(\mathbf{A} + \mathbf{B}) \cdot \mathbf{a}_x = \mathbf{A} \cdot \mathbf{a}_x + \mathbf{B} \cdot \mathbf{a}_x = A_x + B_x \tag{1.29}$$

Similarly, the y- and z-components can be verified. It can also be shown that the vector addition in Eq. (1.28) is consistent with the head-to-tail rule.

The dot product of $\mathbf{A}$ and $\mathbf{B}$ can be written in a literal sense as

$$\mathbf{A} \cdot \mathbf{B} = \left(A_x \, \mathbf{a}_x + A_y \, \mathbf{a}_y + A_z \, \mathbf{a}_z \right) \cdot \left(B_x \, \mathbf{a}_x + B_y \, \mathbf{a}_y + B_z \, \mathbf{a}_z \right) \tag{1.30}$$

This expression can be reduced to a much simpler form using the distributive law and orthonormality such that

$$\boxed{\mathbf{A} \cdot \mathbf{B} = A_x B_x + A_y B_y + A_z B_z} \tag{1.31}$$

The dot product of Eq. (1.31) must agree with Eq. (1.9) (see Example 1.6).

The cross product of $\mathbf{A}$ and $\mathbf{B}$ can be written in its literal sense as

$$\mathbf{A} \times \mathbf{B} = \left(A_x \mathbf{a}_x + A_y \mathbf{a}_y + A_z \mathbf{a}_z \right) \times \left(B_x \mathbf{a}_x + B_y \mathbf{a}_y + B_z \mathbf{a}_z \right) \tag{1.32}$$

Using the distributive law in Eq. (1.18c) and the cyclic relations in Eq. (1.24), the expression in Eq. (1.32) can be rearranged as

$$\boxed{\mathbf{A} \times \mathbf{B} = (A_y B_z - A_z B_y)\mathbf{a}_x + (A_z B_x - A_x B_z)\mathbf{a}_y + (A_x B_y - A_y B_x)\mathbf{a}_z} \quad (1.33)$$

Alternatively, $\mathbf{A} \times \mathbf{B}$ can be expressed as a determinant for easy recall.

$$\boxed{\mathbf{A} \times \mathbf{B} = \begin{vmatrix} \mathbf{a}_x & \mathbf{a}_y & \mathbf{a}_z \\ A_x & A_y & A_z \\ B_x & B_y & B_z \end{vmatrix}} \quad (1.34)$$

It is noteworthy that the right-hand side of Eq. (1.34) is unrelated to the determinant of a matrix but is calculated according to the determinant rule. For example, the x-component is equal to the multiplication of $\mathbf{a}_x$ and its **minor**, that is,

$$\mathbf{a}_x \cdot \mathbf{A} \times \mathbf{B} = \mathbf{a}_x \begin{vmatrix} A_y & A_z \\ B_y & B_z \end{vmatrix} \quad (1.35a)$$

Similarly,

$$\mathbf{a}_y \cdot \mathbf{A} \times \mathbf{B} = -\mathbf{a}_y \begin{vmatrix} A_x & A_z \\ B_x & B_z \end{vmatrix} \quad (1.35b)$$

$$\mathbf{a}_z \cdot \mathbf{A} \times \mathbf{B} = \mathbf{a}_z \begin{vmatrix} A_x & A_y \\ B_x & B_y \end{vmatrix} \quad (1.35c)$$

In several situations, the second point is positioned in the immediate vicinity of point $p_1:(x_1, y_1, z_1)$. Of course, it can be uniquely specified by its own coordinates such as $p_2:(x_2, y_2, z_2)$. However, it is often more convenient to express a neighboring point with respect to a given point using differential coordinates dx, dy, and dz, such as $p_2:(x_1 + dx, y_1 + dy, z_1 + dz)$. In this case, the three differentials constitute the **differential length vector** $d\mathbf{l}$, which is an infinitesimal vector drawn from p_1 to the neighboring point p_2. In Cartesian coordinates,

$$\boxed{d\mathbf{l} = dx\,\mathbf{a}_x + dy\,\mathbf{a}_y + dz\,\mathbf{a}_z} \quad \text{[m]} \quad (1.36)$$

The unit of $d\mathbf{l}$ is meter. In many cases, $d\mathbf{l}$ can be reduced to a simpler form. For example, dz vanishes in the constant z-plane.

The differential length vector is distinct from the distance vector in that the initial point of $d\mathbf{l}$ is the point of interest, whereas the terminal point of $\mathcal{R}$ is one (see Example 1.10).

The two neighboring points are associated with six planes of constant coordinate: the $x = x_1$, $y = y_1$, $z = z_1$, $x = x_1 + dx$, $y = y_1 + dy$, and $z = z_1 + dz$ planes. The six planes define a rectangular parallelepiped of sides dx, dy, and dz, and volume $dxdydz$ (see Fig. 1.18). In the Cartesian coordinate system, the **differential volume** is defined as

$$\boxed{dv = dx\,dy\,dz} \qquad [\text{m}^3] \tag{1.37}$$

The unit of dv is cubic meter. One of the three Cartesian coordinates is constant at each boundary surface of dv. Therefore, dv can provide the most consistent way of dividing a given volume into many volume elements in the Cartesian coordinate system, and vice versa.

The differential volume has six boundary surfaces with infinitesimal areas: $dx\,dy$, $dy\,dz$, and $dz\,dx$ (Fig. 1.18). We find it convenient to specify each surface area using the **differential area vector** $d\mathbf{s}$, whose magnitude represents the area, and whose

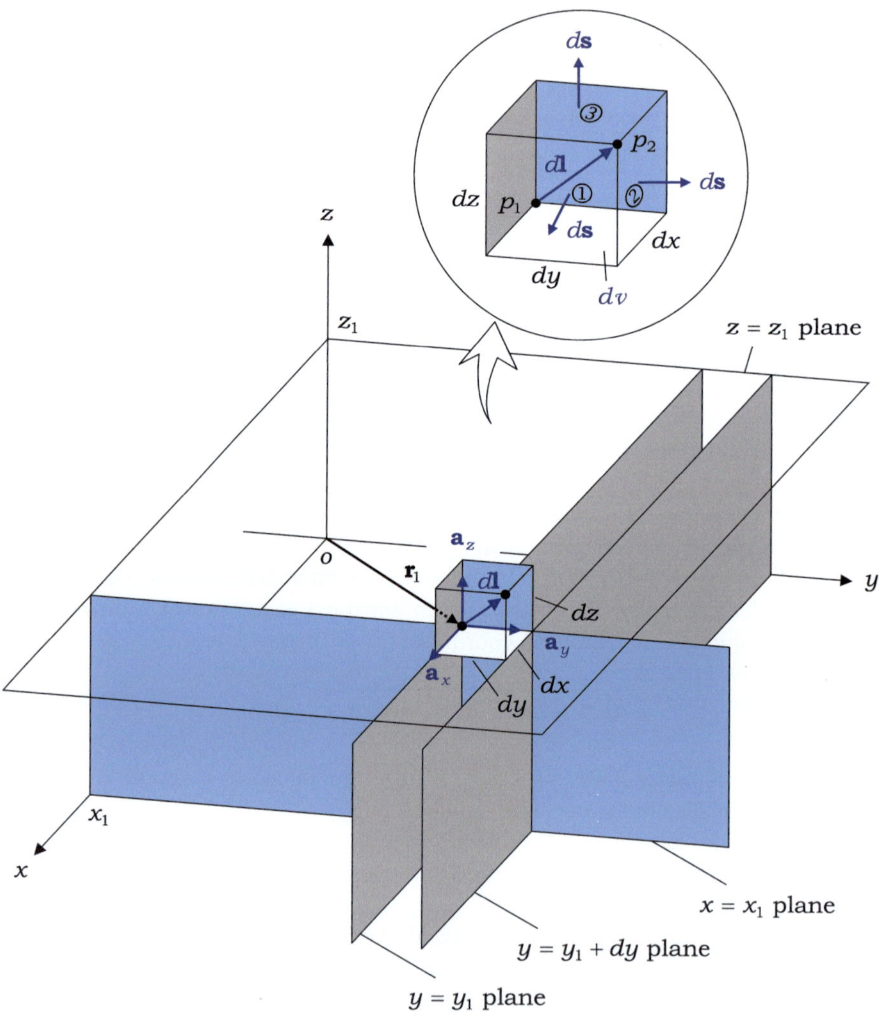

Fig. 1.18 Differential length vector $d\mathbf{l}$, differential area vector $d\mathbf{s}$, and differential volume dv in Cartesian coordinate system

unit vector points outward away from the volume perpendicular to the surface. The differential area vectors for the six faces of dv are

$$d\mathbf{s} = dy\,dz\,\mathbf{a}_x \quad (\text{face } ①) \tag{1.38a}$$

$$d\mathbf{s} = dx\,dz\,\mathbf{a}_y \quad (\text{face } ②) \tag{1.38b}$$

$$d\mathbf{s} = dx\,dy\,\mathbf{a}_z \quad (\text{face } ③) \tag{1.38c}$$

$$d\mathbf{s} = -dy\,dz\,\mathbf{a}_x \tag{1.38d}$$

$$d\mathbf{s} = -dx\,dz\,\mathbf{a}_y \tag{1.38e}$$

$$d\mathbf{s} = -dx\,dy\,\mathbf{a}_z \tag{1.38f}$$

where the last three represent those hidden from the sight behind the front face. The unit of $d\mathbf{s}$ is square meter. A differential area vector is particularly useful for dividing a plane with constant coordinates into several elements. However, if the orientation of a plane is arbitrary, a more general procedure should be followed to split the plane into pieces, as discussed in Chap. 2.

Example 1.5 Given point $p:(2, -3, \sqrt{3})$ in Cartesian coordinates, find (a) its position vector, and (b) distance from the origin.

Solution

(a) The scalar components of $\mathbf{r}$ are equal to the Cartesian coordinates of p. Thus,

$$\mathbf{r} = 2\mathbf{a}_x - 3\mathbf{a}_y + \sqrt{3}\,\mathbf{a}_z$$

(b) Taking the dot product of $\mathbf{r}$ with itself yields

$$\mathbf{r} \cdot \mathbf{r} = (2\mathbf{a}_x - 3\mathbf{a}_y + \sqrt{3}\,\mathbf{a}_z) \cdot (2\mathbf{a}_x - 3\mathbf{a}_y + \sqrt{3}\,\mathbf{a}_z)$$

Applying the distributive law and orthonormality, we obtain

$$\mathbf{r} \cdot \mathbf{r} = (2)(2) + (-3)(-3) + (\sqrt{3})(\sqrt{3}) = 16$$

The distance from the origin to p is represented by the magnitude of $\mathbf{r}$:

$$|\mathbf{r}| = \sqrt{\mathbf{r} \cdot \mathbf{r}} = \sqrt{16} = 4$$

Example 1.6 For two vectors, $\mathbf{A}$ and $\mathbf{B}$, meeting at an angle θ in a three-dimensional space, as shown in Fig. 1.19, show that $\mathbf{A} \cdot \mathbf{B}$ in Eq. (1.31) agrees with Eq. (1.9).

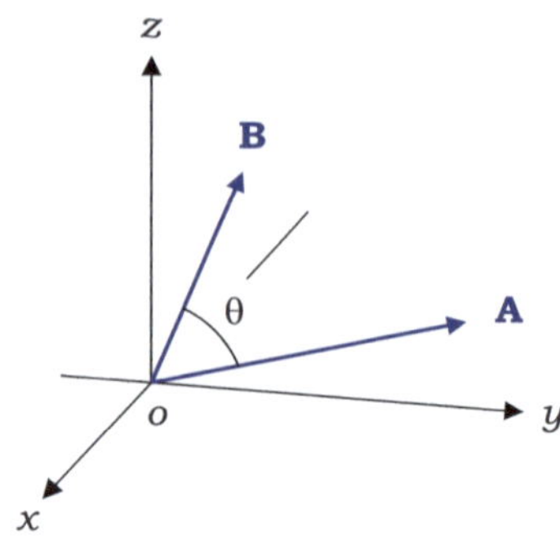

Fig. 1.19 Two vectors in three-dimensional space

Solution

Let $\mathbf{C} \equiv \mathbf{A} - \mathbf{B}$. Then, $\mathbf{C} \cdot \mathbf{C}$ yields

$$C^2 = (A_x - B_x)^2 + (A_y - B_y)^2 + (A_z - B_z)^2$$
$$= (A_x^2 + B_x^2 - 2A_x B_x) + (A_y^2 + B_y^2 - 2A_y B_y) + (A_z^2 + B_z^2 - 2A_z B_z)$$
$$(1.39a)$$

Applying the law of cosines to the triangle defined by $\mathbf{A}$, $\mathbf{B}$, and $\mathbf{A} - \mathbf{B}$ yields

$$C^2 = A^2 + B^2 - 2AB\cos\theta \tag{1.39b}$$

Equating Eqs. (1.39a) and (1.39b), with the aid of $A^2 = A_x^2 + A_y^2 + A_z^2$ and $B^2 = B_x^2 + B_y^2 + B_z^2$, we obtain

$$A_x B_x + A_y B_y + A_z B_z = AB\cos\theta \tag{1.39c}$$

Therefore, the two definitions of dot product are the same.

Example 1.7 Two vectors, $\mathbf{A} = 4\mathbf{a}_x + 3\mathbf{a}_y$ and $\mathbf{B} = \mathbf{a}_x + 2\mathbf{a}_y + 2\mathbf{a}_z$, are given at a point with position vector $\mathbf{r} = -\mathbf{a}_x + 2\mathbf{a}_y + 4\mathbf{a}_z$ in Cartesian coordinates. Find

(a) $\mathbf{A} \cdot \mathbf{B}$,
(b) $\mathbf{A} \times \mathbf{B}$, and
(c) θ_{AB} between $\mathbf{A}$ and $\mathbf{B}$.

Solution

(a) From Eq. (1.31),

$$\mathbf{A} \cdot \mathbf{B} = (4)(1) + (3)(2) + (0)(2) = 10$$

(b) From Eq. (1.34),

$$
\mathbf{A} \times \mathbf{B} = \begin{vmatrix} \mathbf{a}_x & \mathbf{a}_y & \mathbf{a}_z \\ 4 & 3 & 0 \\ 1 & 2 & 2 \end{vmatrix}
$$

$$
= \mathbf{a}_x(3 \times 2 - 0 \times 2) - \mathbf{a}_y(4 \times 2 - 0 \times 1) + \mathbf{a}_z(4 \times 2 - 3 \times 1)
$$

$$
= 6\mathbf{a}_x - 8\mathbf{a}_y + 5\mathbf{a}_z
$$

(c) The magnitudes of $\mathbf{A}$ and $\mathbf{B}$ are

$$
A = |\mathbf{A}| = \sqrt{(4)^2 + (3)^2} = 5
$$

$$
B = |\mathbf{B}| = \sqrt{(1)^2 + (2)^2 + (2)^2} = 3
$$

From Eq. (1.9),

$$
\cos\theta_{AB} = \frac{\mathbf{A} \cdot \mathbf{B}}{AB} = \left(\frac{10}{5 \times 3}\right), \text{ and thus, } \theta_{AB} = \cos^{-1}\left(\frac{2}{3}\right) = 48.2°
$$

Note that the position vector has no effect on $\mathbf{A} \cdot \mathbf{B}$, $\mathbf{A} \times \mathbf{B}$, or θ_{AB} in this case.

Example 1.8 Given vector field $\mathbf{F}(\mathbf{r}) = yz\,\mathbf{a}_x - x^2\mathbf{a}_y + y\,\mathbf{a}_z$ in Cartesian coordinates, find

(a) the constituent vector $\mathbf{A}$ at position $\mathbf{r} = \mathbf{a}_x + 2\mathbf{a}_y + 2\mathbf{a}_z$,
(b) the vector component of $\mathbf{A}$ in the direction parallel to $\mathbf{r}$, and
(c) the vector component of $\mathbf{A}$ in the direction perpendicular to $\mathbf{r}$.

Solution

(a) Inserting coordinates $(1, 2, 2)$ into $\mathbf{F}(\mathbf{r})$ yields

$$
\mathbf{A} = (2)(2)\,\mathbf{a}_x - (1)^2\mathbf{a}_y + (2)\,\mathbf{a}_z
$$

$$
= 4\mathbf{a}_x - \mathbf{a}_y + 2\mathbf{a}_z \tag{1.40a}
$$

(b) The magnitude of $\mathbf{r}$ is

$$
r = \sqrt{(1)^2 + (2)^2 + (2)^2} = 3
$$

A unit vector along $\mathbf{r}$ is

$$
\mathbf{a}_r = \frac{\mathbf{r}}{r} = \frac{1}{3}(\mathbf{a}_x + 2\mathbf{a}_y + 2\mathbf{a}_z)
$$

The scalar component of $\mathbf{A}$ along $\mathbf{r}$ is computed as

$$\mathbf{A} \cdot \mathbf{a}_r = (4\mathbf{a}_x - \mathbf{a}_y + 2\mathbf{a}_z) \cdot \frac{1}{3}(\mathbf{a}_x + 2\mathbf{a}_y + 2\mathbf{a}_z) = 2$$

The vector component of $\mathbf{A}$ parallel to $\mathbf{r}$ is, therefore,

$$\mathbf{A}_\parallel = \frac{2}{3}(\mathbf{a}_x + 2\mathbf{a}_y + 2\mathbf{a}_z) \tag{1.40b}$$

(c)   It is evident from Fig. 1.20 that

$$\mathbf{A}_\perp = \mathbf{A} - \mathbf{A}_\parallel \tag{1.40c}$$

Inserting Eqs. (1.40a, b) into Eq. (1.40c) yields

$$\mathbf{A}_\perp = \frac{1}{3}(10\mathbf{a}_x - 7\mathbf{a}_y + 2\mathbf{a}_z) \tag{1.40d}$$

Alternatively, we can express $\mathbf{A}_\perp$ as

$$\mathbf{A}_\perp = \mathbf{a}_r \times (\mathbf{A} \times \mathbf{a}_r) \tag{1.40e}$$

The cross product in the parentheses in Eq. (1.40e) is computed as follows:

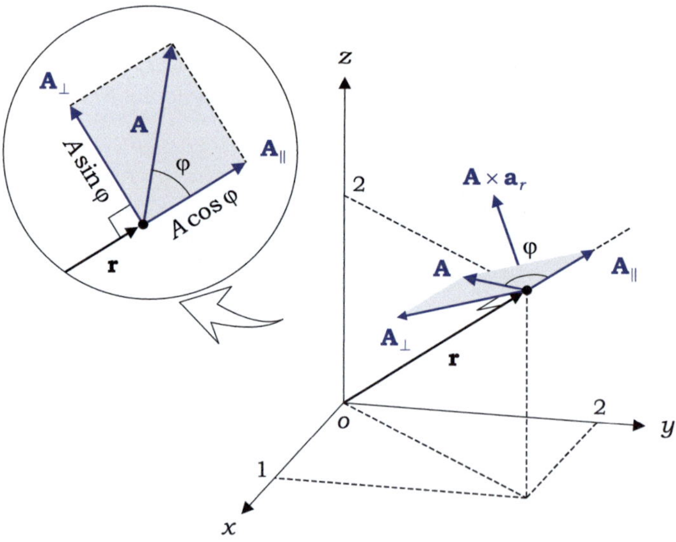

Fig. 1.20 Vector components of $\mathbf{A}$

$$\mathbf{A} \times \mathbf{a}_r = \frac{1}{3} \begin{vmatrix} \mathbf{a}_x & \mathbf{a}_y & \mathbf{a}_z \\ 4 & -1 & 2 \\ 1 & 2 & 2 \end{vmatrix}$$

$$= \frac{1}{3} \left[\mathbf{a}_x \{ (-1) \times 2 - 2 \times 2 \} - \mathbf{a}_y (4 \times 2 - 2 \times 1) + \mathbf{a}_z \{ 4 \times 2 - (-1) \times 1 \} \right]$$

$$= -2\mathbf{a}_x - 2\mathbf{a}_y + 3\mathbf{a}_z$$

Thus, the vector component of $\mathbf{A}$ perpendicular to $\mathbf{r}$ is

$$\mathbf{A}_\perp = \mathbf{a}_r \times (\mathbf{A} \times \mathbf{a}_r) = \frac{1}{3} \begin{vmatrix} \mathbf{a}_x & \mathbf{a}_y & \mathbf{a}_z \\ 1 & 2 & 2 \\ -2 & -2 & 3 \end{vmatrix}$$

$$= \frac{1}{3} (10\mathbf{a}_x - 7\mathbf{a}_y + 2\mathbf{a}_z) \tag{1.40f}$$

The results in Eqs. (1.40d) and (1.40f) are equal.

Example 1.9 A slanting plane contains two differential length vectors $d\mathbf{l}^1$ and $d\mathbf{l}^2$ that are parallel to and perpendicular to the xz-plane, respectively, as shown in Fig. 1.21. Express the following in terms of the differentials dx and dy: (a) $d\mathbf{l}^1$, (b) $d\mathbf{l}^2$, and (c) $d\mathbf{s}$.

Solution

(a) The line segment $\overline{ab}$ is expressed as $bx + az - ab = 0$. Differentiating both
sides of this equation with respect to x yields

$$bdx + adz = 0 \tag{1.41a}$$

Fig. 1.21 A slanting plane

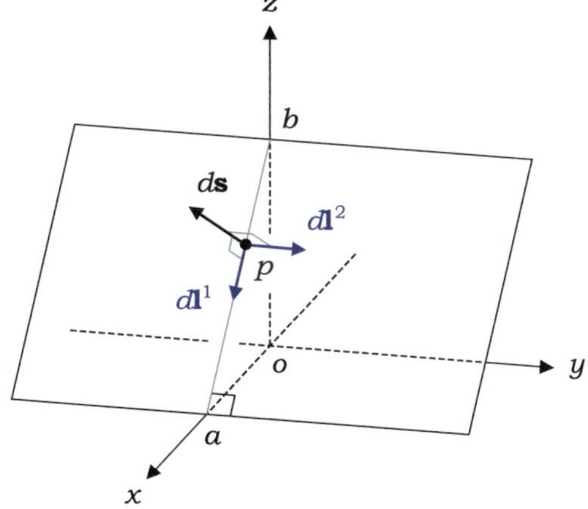

Inserting Eq. (1.41a) into Eq. (1.36) and using the fact that $d\mathbf{l}^1$ is perpendicular to $\mathbf{a}_y$, we obtain

$$d\mathbf{l}^1 = dx\,\mathbf{a}_x - (b/a)dx\,\mathbf{a}_z \qquad (1.41\text{b})$$

(b) $d\mathbf{l}^2$ is parallel to the y-axis. Setting $dx = dz = 0$ in Eq. (1.36) yields

$$d\mathbf{l}^2 = dy\,\mathbf{a}_y \qquad (1.41\text{c})$$

(c)   The differential area vector can be expressed as

$$d\mathbf{s} = d\mathbf{l}^1 \times d\mathbf{l}^2 = (b/a)dx\,dy\,\mathbf{a}_x + dx\,dy\,\mathbf{a}_z \qquad (1.41\text{d})$$

Exercise 1.14
For $\mathbf{A} = 2\mathbf{a}_x + \mathbf{a}_y + 3\mathbf{a}_z$, determine (a) $|\mathbf{A}|$, (b) $\mathbf{a}_A$, (c) the angles between $\mathbf{A}$ and the x-, y-, and z-axes, and (d) the necessary conditions for $\mathbf{A}$ to be a position vector.

Ans. (a) 3.742, (b) $\mathbf{a}_A = 0.534\mathbf{a}_x + 0.267\mathbf{a}_y + 0.802\mathbf{a}_z$, (c) 57.7°, 74.5°, and 36.7°, (d) The initial point is at the origin and the unit is meter.

Exercise 1.15
What is the projection of $\mathbf{A} = 4\mathbf{a}_x + 3\mathbf{a}_y + 8\mathbf{a}_z$ onto the xy-plane?

Ans. 5.

Exercise 1.16
Which boundary surfaces of dv in Fig. 1.18 are referred to in Eqs. (1.38d, e, f)?

Ans. Rear, left, and bottom surfaces.

Exercise 1.17
Specify the spatial point to which the following vector quantity belongs: (a) the constituent vector of a vector field, (b) $\mathbf{r}$, (c) $\boldsymbol{\mathcal{R}}$, (d) $d\mathbf{l}$, and (f) $d\mathbf{s}$.

Ans. (a) Initial, (b) Terminal, (c) Terminal, (d) Initial, and (f) Initial point.

1.3.2   Cylindrical Coordinate System

The cylindrical coordinate system uses three numbers, ρ, ϕ, and z, to define a point in a three-dimensional space. Point $p_1{:}(\rho_1, \phi_1, z_1)$ is located at the intersection of three surfaces: the cylindrical surface of radius ρ_1 centered on the z-axis, the half-plane hinged along the z-axis and rotated through angle ϕ_1, and the $z = z_1$ plane. Accordingly, ρ_1 is the *radial distance* from the z-axis, and ϕ_1 is the *azimuth* measured from the $+x$-axis in the xy-plane. The ranges of ρ, ϕ, and z are $0 \le \rho < \infty$ [m],

$0 \leq \phi \leq 2\pi$ [rad], and $-\infty < z < \infty$ [m], respectively. At point p_1, base vectors $\mathbf{a}_\rho$, $\mathbf{a}_\phi$, and $\mathbf{a}_z$ are defined such that they are perpendicular to the surfaces of constant ρ, ϕ, and z, and point in the directions of increase of ρ, ϕ, and z, respectively. Note that $\mathbf{a}_\rho$ and $\mathbf{a}_\phi$ vary with ϕ, whereas $\mathbf{a}_z$ is constant.

The base vectors in cylindrical coordinates obey orthonormality:

$$\mathbf{a}_\rho \cdot \mathbf{a}_\phi = \mathbf{a}_\phi \cdot \mathbf{a}_z = \mathbf{a}_z \cdot \mathbf{a}_\rho = 0 \tag{1.42a}$$

$$\mathbf{a}_\rho \cdot \mathbf{a}_\rho = \mathbf{a}_\phi \cdot \mathbf{a}_\phi = \mathbf{a}_z \cdot \mathbf{a}_z = 1 \tag{1.42b}$$

Fundamentally, the base vectors are three unit-vectors that are mutually perpendicular at a given point in space despite the fact that the orientations of $\mathbf{a}_\rho$ and $\mathbf{a}_\phi$ vary with ϕ.

In the right-handed cylindrical coordinate system, the base vectors follow cyclic relations:

$$\mathbf{a}_\rho \times \mathbf{a}_\phi = \mathbf{a}_z \tag{1.43a}$$

$$\mathbf{a}_\phi \times \mathbf{a}_z = \mathbf{a}_\rho \tag{1.43b}$$

$$\mathbf{a}_z \times \mathbf{a}_\rho = \mathbf{a}_\phi \tag{1.43c}$$

These relationships obey the right-hand rule: when the right fingers rotate from the first to the second vector, the thumb points in the direction of the third vector.

Following the same procedure used in the Cartesian coordinate system, the position vector of a point can be expanded in terms of the base vectors given at the point. It is evident from Fig. 1.22 that position vector $\mathbf{r}_1$ is perpendicular to $\mathbf{a}_\phi$, and thus has no ϕ-component. By omitting subscript 1 for generalization, the position vector in cylindrical coordinates can be expressed as

$$\boxed{\mathbf{r} = \rho\,\mathbf{a}_\rho + z\,\mathbf{a}_z} \tag{1.44}$$

Because the position vector in Eq. (1.44) is seemingly independent of ϕ, the question arises as to whether it can specify each point in a three-dimensional space. In fact, $\mathbf{a}_\rho$ is a function of ϕ, and thus $\rho\,\mathbf{a}_\rho$ can uniquely define all the points in a constant z-plane. Consequently, the position vector covers all points in a three-dimensional space. However, this is not the correct method to use a cylindrical coordinate system. If the geometry of a given configuration has **cylindrical symmetry**, it would look the same as we move around it, varying ϕ but keeping ρ and z constant. In other words, the entire system remains the same even if it is rotated about the z-axis, because the points on a circular cylinder around the z-axis cannot be distinguished. This feature is expressed as $\mathbf{r}$ in Eq. (1.44) that does not concern itself with an explicit ϕ-dependence, treating $\mathbf{a}_\rho$ as indistinguishable. In other words, in the presence of cylindrical symmetry, $\mathbf{a}_\rho$ can be considered a constant vector.

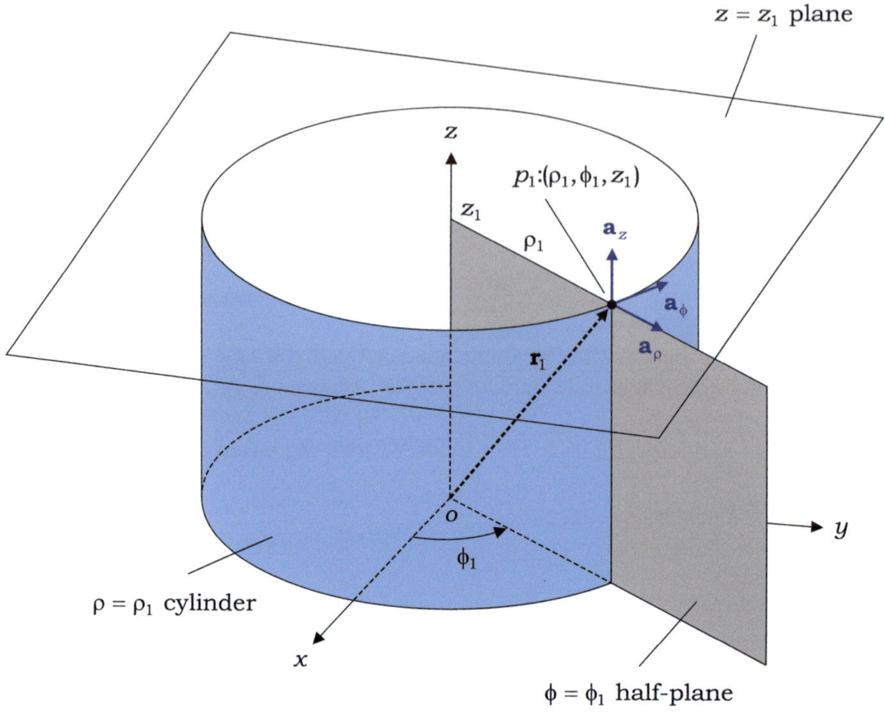

Fig. 1.22 Cylindrical coordinate system

For example, consider a rotating cylinder of radius a, as shown in Fig. 1.13. The tangential velocity of a point on the cylinder can be expressed as $\boldsymbol{v} = \omega\,\mathbf{a}_z \times \mathbf{r}$ [m/s], where $\mathbf{r}$ is the position vector. If the position vector is expressed in Cartesian coordinates and the relationship $x^2 + y^2 = a^2$ is used, $\boldsymbol{v}$ can be expressed as

$$
\boldsymbol{v} = \omega\,\mathbf{a}_z \times (x\,\mathbf{a}_x + y\,\mathbf{a}_y + z\,\mathbf{a}_z)
$$

$$
= \pm\omega\sqrt{a^2 - x^2}\,\mathbf{a}_x + \omega x\,\mathbf{a}_y \quad (+,\, y < 0;\; -,\, y > 0) \tag{1.45a}
$$

Next, taking into account the cylindrical symmetry of the rotating cylinder, the position vector in cylindrical coordinates is used to obtain

$$
\boldsymbol{v} = \omega\,\mathbf{a}_z \times (\rho\,\mathbf{a}_\rho + z\,\mathbf{a}_z)
$$

$$
= \omega a\,\mathbf{a}_\phi \tag{1.45b}
$$

It should be noted that the result of Eq. (1.45a) concerns a particular point on the cylinder, whereas that of Eq. (1.45b) applies to a general point on the cylinder.

When a vector field defines constituent vector $\mathbf{A}$ at point p_1, as shown in Fig. 1.23, it can be expanded in terms of the base vectors given at the point such that

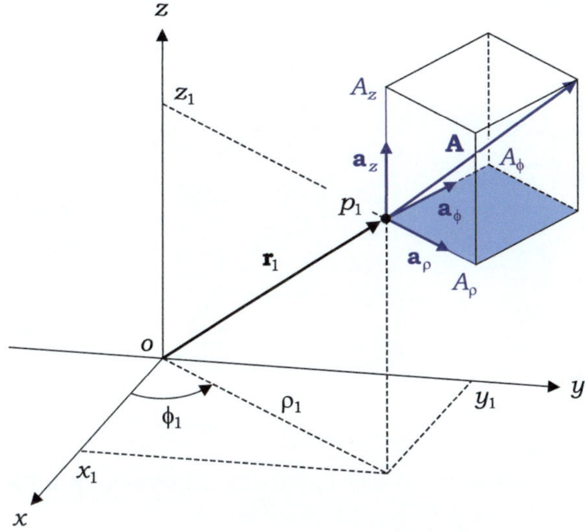

Fig. 1.23 Constituent vector **A** is defined at point $p_1{:}(\rho_1, \phi_1, z_1)$

$$\boxed{\mathbf{A} = A_\rho\,\mathbf{a}_\rho + A_\phi\,\mathbf{a}_\phi + A_z\,\mathbf{a}_z}\tag{1.46}$$

where the scalar components A_ρ, A_ϕ, and A_z are constants. Extending this to all points in the space leads to a general expression for the vector field, that is,

$$\boxed{\mathbf{A(r)} = A_\rho(\rho, \phi, z)\,\mathbf{a}_\rho + A_\phi(\rho, \phi, z)\,\mathbf{a}_\phi + A_z(\rho, \phi, z)\,\mathbf{a}_z}\tag{1.47}$$

where **r** is the position vector, and the unit vectors $\mathbf{a}_\rho$ and $\mathbf{a}_\phi$ are functions of ϕ.

Here, we briefly describe and discuss cylindrical symmetry. An object is said to possess cylindrical symmetry if *it looks the same as it is rotated about the z-axis or as we move around it, varying ϕ but keeping ρ and z constant.* Consider the vector fields $\mathbf{U(r)} = \mathbf{a}_\rho/\rho$, $\mathbf{V(r)} = \mathbf{a}_\phi/\rho$, and $\mathbf{W(r)} = \mathbf{a}_z/\rho$, which are defined in region $\rho > 0$ (see Fig. 1.24). All the examples exhibit cylindrical symmetry. Each vector field is indistinguishable on a surface of constant ρ in the sense that its constituent vector has the same magnitude and orientation with respect to the surface.

Symmetry considerations make it possible to determine the general form of the solution in advance, and significantly facilitate the final solution to the problem. Furthermore, to a certain degree, Gauss's and Ampere's laws require knowledge of the final solution before they can be applied. Let us consider the case where the source of a vector field is distributed in space to exhibit cylindrical symmetry. According to causality, the resultant vector field also has cylindrical symmetry such that it is independent of ϕ. Thus, the general form is expressed as $\mathbf{E(r)} = E_\rho(\rho, z)\,\mathbf{a}_\rho + E_\phi(\rho, z)\,\mathbf{a}_\phi + E_z(\rho, z)\,\mathbf{a}_z$. In many practical situations, cylindrical symmetry is accompanied by other symmetries such as translational symmetry along the z-axis and two-fold rotational symmetry about another coordinate axis,

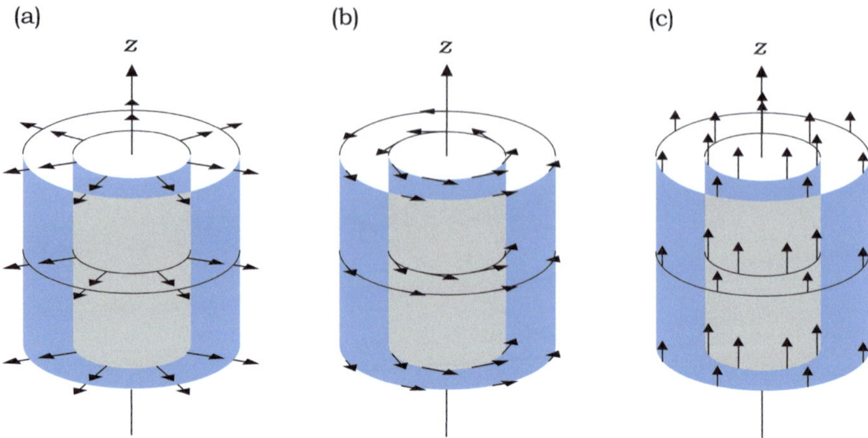

Fig. 1.24 Three vector fields with cylindrical symmetry: **a** $\mathbf{U}(\mathbf{r}) = \mathbf{a}_\rho/\rho$, **b** $\mathbf{V}(\mathbf{r}) = \mathbf{a}_\phi/\rho$, and **c** $\mathbf{W}(\mathbf{r}) = \mathbf{a}_z/\rho$

as in an infinitely long line charge. The former tells us that the source will look the same even if it is displaced in the z-direction, whereas the latter tells us that the source will look the same even if it is rotated about the x-axis by $180°$ or turned upside down. The translational symmetry further reduces the general form to $\mathbf{E}(\mathbf{r}) = E_\rho(\rho)\,\mathbf{a}_\rho + E_\phi(\rho)\,\mathbf{a}_\phi + E_z(\rho)\,\mathbf{a}_z$, and the two-fold rotational symmetry requires the ϕ- and z-components to vanish from the final form because they reverse their directions when turned upside down (Fig. 1.24). Consequently, in the presence of cylindrical, translational, and two-fold rotational symmetries, the final vector field must have the following form:

$$\boxed{\mathbf{E}(\mathbf{r}) = E_\rho(\rho)\,\mathbf{a}_\rho} \tag{1.48}$$

For example, $\mathbf{U}(\mathbf{r})$ in Fig. 1.24 has cylindrical, translational, and two-fold rotational symmetries, whereas $\mathbf{V}(\mathbf{r})$ and $\mathbf{W}(\mathbf{r})$ have only cylindrical and translational symmetries.

When two separate vector functions provide two vectors, $\mathbf{A}$ and $\mathbf{B}$, at a point in space, they can be expressed in component form, such that

$$\mathbf{A} = A_\rho\,\mathbf{a}_\rho + A_\phi\,\mathbf{a}_\phi + A_z\,\mathbf{a}_z \tag{1.49a}$$

$$\mathbf{B} = B_\rho\,\mathbf{a}_\rho + B_\phi\,\mathbf{a}_\phi + B_z\,\mathbf{a}_z \tag{1.49b}$$

where $\mathbf{a}_\rho$, $\mathbf{a}_\phi$, and $\mathbf{a}_z$ are the base vectors defined at the same spatial point where $\mathbf{A}$ and $\mathbf{B}$ are positioned.

The addition of $\mathbf{A}$ and $\mathbf{B}$ in cylindrical coordinates is given by

$$\boxed{\mathbf{A} + \mathbf{B} = \left(A_\rho + B_\rho\right)\mathbf{a}_\rho + \left(A_\phi + B_\phi\right)\mathbf{a}_\phi + (A_z + B_z)\mathbf{a}_z} \qquad (1.50)$$

By applying the distributive law to $(\mathbf{A} + \mathbf{B}) \cdot \mathbf{a}_\rho$, we can verify the ρ-component on the right-hand side of Eq. (1.50), among others.

The dot product of $\mathbf{A}$ and $\mathbf{B}$ in cylindrical coordinates is given by

$$\boxed{\mathbf{A} \cdot \mathbf{B} = A_\rho B_\rho + A_\phi B_\phi + A_z B_z} \qquad (1.51)$$

where we have used the distributive law and orthonormality.

The cross product of $\mathbf{A}$ and $\mathbf{B}$ in cylindrical coordinates can be expressed in the determinant form as follows:

$$\boxed{\mathbf{A} \times \mathbf{B} = \begin{vmatrix} \mathbf{a}_\rho & \mathbf{a}_\phi & \mathbf{a}_z \\ A_\rho & A_\phi & A_z \\ B_\rho & B_\phi & B_z \end{vmatrix}} \qquad (1.52)$$

This can be verified by direct substitution along with the distributive law of cross product and cyclic relations.

The vector operations in Eqs. (1.50)–(1.52) are defined for two vectors given at the same point in space. For example, if $\mathbf{A} = 2\mathbf{a}_\rho$ is given at p_1:(4, 0, 0), and $\mathbf{B} = 3\mathbf{a}_\phi$ is given at p_2:(4, $\pi/2$, 0) in cylindrical coordinates, one may first attempt to compute $\mathbf{A} \times \mathbf{B} = 6\mathbf{a}_z$. However, this is incorrect because $\mathbf{A}$ and $\mathbf{B}$ point in opposite directions at their respective points on the x- and y-axes. If $\mathbf{B}$ is expressed in terms of the base vectors at p_1, then it should be given by $\mathbf{B} = -3\mathbf{a}_\rho$. Thus, $\mathbf{A} \times \mathbf{B} = 0$, if performed at point p_1.

As in the Cartesian coordinate system, the differential length vector in cylindrical coordinates is an infinitesimal vector measured in meters. When it starts at point p_1:(ρ_1, ϕ_1, z_1) and ends at nearby point p_2:$(\rho_1 + d\rho, \phi_1 + d\phi, z_1 + dz)$, its projections onto the base vectors at point p_1 yield the differential lengths $d\rho$, $\rho_1 d\phi$, and dz. In this case, the differential angle $d\phi$ is converted into a differential length $\rho_1 d\phi$. Omitting 1 for generalization, the differential length vector in cylindrical coordinates is expressed as

$$\boxed{d\mathbf{l} = d\rho\,\mathbf{a}_\rho + \rho d\phi\,\mathbf{a}_\phi + dz\,\mathbf{a}_z} \quad \text{[m]} \qquad (1.53)$$

In many cases, $d\mathbf{l}$ can be reduced to a simpler form. For example, $d\rho$ vanishes on a cylindrical surface around the z-axis.

The two endpoints of $d\mathbf{l}$ are associated with six surfaces of constant coordinate (Fig. 1.25). These surfaces define a differential volume dv. Because $|d\mathbf{l}|$ is infinitely small, dv may be considered a rectangular parallelepiped with sides $d\rho$, $\rho_1 d\phi$, and dz and volume $\rho_1 d\rho d\phi dz$. In the cylindrical coordinate system, the differential volume is defined as

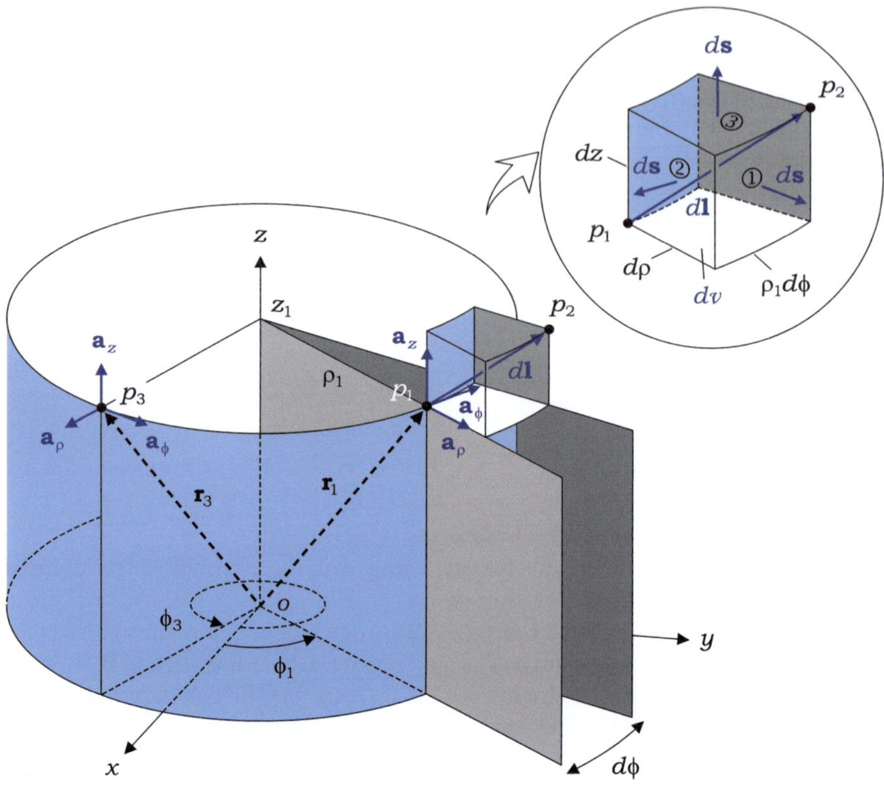

Fig. 1.25 Differential length vector $d\mathbf{l}$, differential area vector $d\mathbf{s}$, and differential volume dv in cylindrical coordinate system

$$\boxed{dv = \rho \, d\rho d\phi dz} \quad [\text{m}^3] \tag{1.54}$$

The differential volume is very useful for dividing a given volume into many volume elements in cylindrical coordinates, and vice versa.

Each of the six boundary surfaces of dv can be conveniently represented by the differential area vector $d\mathbf{s}$, whose magnitude gives the surface area, and whose unit vector is normal to the surface directed outward from dv. Referring to Fig. 1.25, the differential area vectors are expressed as

$$d\mathbf{s} = \rho_1 d\phi dz \, \mathbf{a}_\rho \quad (\text{face } ①) \tag{1.55a}$$

$$d\mathbf{s} = -d\rho dz \, \mathbf{a}_\phi \quad (\text{face } ②) \tag{1.55b}$$

$$d\mathbf{s} = \rho_1 d\phi d\rho \, \mathbf{a}_z \quad (\text{face } ③) \tag{1.55c}$$

$$ds = -\rho_1 d\phi dz\, \mathbf{a}_\rho \tag{1.55d}$$

$$ds = d\rho dz\, \mathbf{a}_\phi \tag{1.55e}$$

$$ds = -\rho_1 d\phi d\rho\, \mathbf{a}_z \tag{1.55f}$$

The three expressions in Eqs. (1.55d, e, f) are for those hidden from the sight behind the front face. The differential area vector is particularly useful for dividing a surface with constant coordinates into many elements of the area, and vice versa. However, for example, Eqs. (1.55a)–(1.55f) may not be applicable to the $x = 1$ plane or to a cylindrical surface lying along the x-axis, because these are not surfaces of constant coordinate in cylindrical coordinates.

Example 1.10 Given three points p_1, p_2, and p_3, as shown in Fig. 1.25, find the expressions for their position vectors $\mathbf{r}_1$, $\mathbf{r}_2$, and $\mathbf{r}_3$ in cylindrical coordinates.

Solution

Projections of $\mathbf{r}_1$ onto $\mathbf{a}_\rho$, $\mathbf{a}_\phi$, and $\mathbf{a}_z$ at p_1 lead to

$$\mathbf{r}_1 = \rho_1\, \mathbf{a}_\rho + z_1\, \mathbf{a}_z \tag{1.56a}$$

Following a similar procedure yields

$$\mathbf{r}_3 = \rho_1\, \mathbf{a}_\rho + z_1\, \mathbf{a}_z \tag{1.56b}$$

This has exactly the same form as $\mathbf{r}_1$, implying that possible cylindrical symmetry has already been incorporated into the expression.

Next, using the coordinates of p_1, point p_2 can be specified such as

$$p_2{:}(\rho_1 + d\rho,\ \phi_1 + d\phi,\ z_1 + dz) \tag{1.56c}$$

Subsequently, the position vector is expressed as

$$\mathbf{r}_2 = (\rho_1 + d\rho)\, \mathbf{a}_\rho + (z_1 + dz)\, \mathbf{a}_z \tag{1.56d}$$

With $\mathbf{r}_1 = \rho_1\mathbf{a}_\rho + z_1\mathbf{a}_z$ and $d\mathbf{l} = d\rho\, \mathbf{a}_\rho + \rho_1 d\phi\, \mathbf{a}_\phi + dz\, \mathbf{a}_z$, one might be tempted to write

$$\mathbf{r}_2 = \mathbf{r}_1 + d\mathbf{l} = (\rho_1 + d\rho)\, \mathbf{a}_\rho + \rho_1 d\phi\, \mathbf{a}_\phi + (z_1 + dz)\, \mathbf{a}_z\ (?) \tag{1.56e}$$

Although $\mathbf{r}_2 = \mathbf{r}_1 + d\mathbf{l}$ is always true in its abstract form, *the position vector must be expanded in terms of the base vectors provided at its terminal point.* The expression

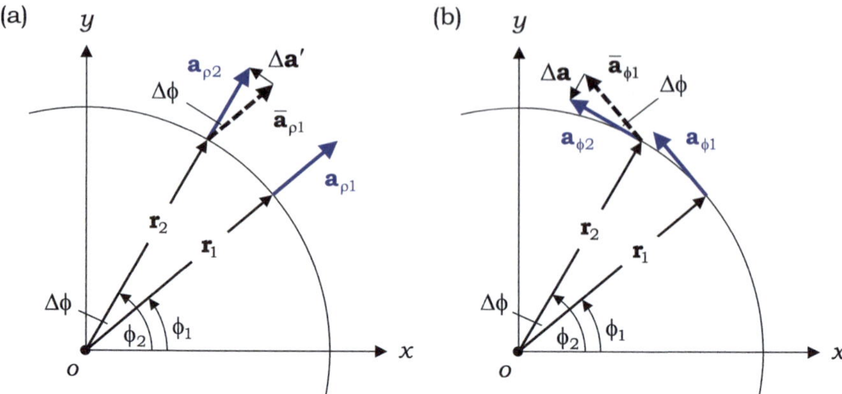

Fig. 1.26 **a** $\partial\mathbf{a}_\rho/\partial\phi$, and **b** $\partial\mathbf{a}_\phi/\partial\phi$

in Eq. (1.56e) is incorrect because the base vectors are those at point p_1, and not at p_2.

Example 1.11 Both $\mathbf{a}_\rho$ and $\mathbf{a}_\phi$ are differentiable because they are smooth functions of position. Verify the following partial derivatives:

(a) $\dfrac{\partial\mathbf{a}_\rho}{\partial\phi} = \mathbf{a}_\phi$, and (b) $\dfrac{\partial\mathbf{a}_\phi}{\partial\phi} = -\mathbf{a}_\rho$.

Solution

(a)   Refer to Fig. 1.26a. The rate of change of $\mathbf{a}_\rho$ with respect to ϕ at $\phi = \phi_1$ is, by definition,

$$\left.\frac{\partial\mathbf{a}_\rho}{\partial\phi}\right|_{\phi=\phi_1} = \lim_{\Delta\phi\to0}\frac{\mathbf{a}_{\rho2} - \mathbf{a}_{\rho1}}{\phi_2 - \phi_1} \equiv \lim_{\Delta\phi\to0}\frac{\Delta\mathbf{a}'}{\Delta\phi} \tag{1.57a}$$

where $\mathbf{a}_{\rho1}$ and $\mathbf{a}_{\rho2}$ are the unit vectors at $\phi = \phi_1$ and $\phi = \phi_2$, respectively. Let $\bar{\mathbf{a}}_{\rho1}$ be a displaced version of $\mathbf{a}_{\rho1}$. From the triangle formed by $\bar{\mathbf{a}}_{\rho1}$, $\mathbf{a}_{\rho2}$, and $\Delta\mathbf{a}'$, along with $|\bar{\mathbf{a}}_{\rho1}| = |\mathbf{a}_{\rho2}| = 1$, we can write

$$\left|\Delta\mathbf{a}'\right| = \left|\mathbf{a}_{\rho2}\right|\Delta\phi = \Delta\phi \tag{1.57b}$$

As $\Delta\phi \to 0$, it follows that $\mathbf{r}_2 \to \mathbf{r}_1$, $\mathbf{a}_{\rho2} \to \bar{\mathbf{a}}_{\rho1}$, and $\Delta\mathbf{a}'$ becomes perpendicular to $\mathbf{a}_{\rho1}$, pointing in the $\mathbf{a}_\phi$ direction. Thus, it follows that

$$\Delta\mathbf{a}' = \Delta\phi\,\mathbf{a}_\phi \tag{1.57c}$$

Substituting Eq. (1.57c) into Eq. (1.57a) gives

$$\boxed{\frac{\partial \mathbf{a}_\rho}{\partial \phi} = \mathbf{a}_\phi} \tag{1.57d}$$

(b) Refer to Fig. 1.26b. The rate of change of $\mathbf{a}_\phi$ with respect to ϕ at $\phi = \phi_1$ is expressed as

$$\left.\frac{\partial \mathbf{a}_\phi}{\partial \phi}\right|_{\phi=\phi_1} = \lim_{\Delta\phi \to 0} \frac{\mathbf{a}_{\phi 2} - \mathbf{a}_{\phi 1}}{\phi_2 - \phi_1} \equiv \lim_{\Delta\phi \to 0} \frac{\Delta \mathbf{a}}{\Delta\phi} \tag{1.57e}$$

where $\mathbf{a}_{\phi 1}$ and $\mathbf{a}_{\phi 2}$ are the unit vectors at $\phi = \phi_1$ and $\phi = \phi_2$, respectively. Let $\bar{\mathbf{a}}_{\phi 1}$ be a displaced version of $\mathbf{a}_{\phi 1}$. From the triangle formed by $\bar{\mathbf{a}}_{\phi 1}$, $\mathbf{a}_{\phi 2}$, and $\Delta \mathbf{a}$, along with $\left|\bar{\mathbf{a}}_{\phi 1}\right| = \left|\mathbf{a}_{\phi 2}\right| = 1$, we can write

$$|\Delta \mathbf{a}| = \left|\mathbf{a}_{\phi 2}\right| \Delta \phi = \Delta \phi \tag{1.57f}$$

As $\Delta \phi \to 0$, it follows that $\mathbf{r}_2 \to \mathbf{r}_1$, $\mathbf{a}_{\phi 2} \to \bar{\mathbf{a}}_{\phi 1}$, and $\Delta \mathbf{a}$ becomes perpendicular to $\mathbf{a}_{\phi 1}$, pointing in the direction of $-\mathbf{a}_\rho$. Thus,

$$\Delta \mathbf{a} = -\Delta \phi \, \mathbf{a}_\rho \tag{1.57g}$$

Substituting Eq. (1.57g) into Eq. (1.57e) gives

$$\boxed{\frac{\partial \mathbf{a}_\phi}{\partial \phi} = -\mathbf{a}_\rho} \tag{1.57h}$$

The results in Eqs. (1.57d) and (1.57h) can be verified by using the coordinate transformations discussed in the following section.

Example 1.12 A circular cylinder of radius a was sliced at $45°$, as shown in Fig. 1.27. Find an expression for $d\mathbf{l}$ along the perimeter of the elliptical cross section in (a) Cartesian, and (b) cylindrical coordinates.

Solution

(a) The cross-sectional ellipse is defined by

$$x^2 + y^2 = a^2 \tag{1.58a}$$

$$z = -x \tag{1.58b}$$

Differentiating Eqs. (1.58a) and (1.58b) with respect to x results in

$$x\,dx + y\,dy = 0 \tag{1.58c}$$

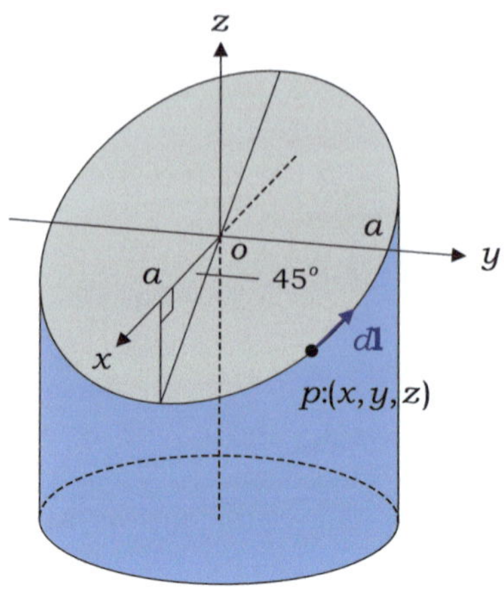

Fig. 1.27 Diagonally cut circular cylinder

$$dz = -dx \qquad\qquad (1.58\text{d})$$

Using these in the general expression for $d\mathbf{l}$ in Eq. (1.36) yields

$$d\mathbf{l} = dx\, \mathbf{a}_x - (x/y)dx\, \mathbf{a}_y - dx\, \mathbf{a}_z \qquad\qquad (1.58\text{e})$$

(b) The x-coordinate of a point on the elliptical perimeter is given by $x = a \cos \phi$, where ϕ is the azimuth angle. Differentiating this with respect to ϕ yields

$$dx = -a \sin \phi\, d\phi \qquad\qquad (1.58\text{f})$$

Inserting Eqs. (1.58d) and (1.58f) into the general expression for $d\mathbf{l}$ in Eq. (1.53) and setting $\rho = a$ and $d\rho = 0$, we obtain

$$d\mathbf{l} = ad\phi\, \mathbf{a}_\phi + ad\phi \sin \phi\, \mathbf{a}_z \qquad\qquad (1.58\text{g})$$

Example 1.13 It is convenient to use a mixed-coordinate system if the problem involves a distance vector. As shown in Fig. 1.28, the distance vector $\mathcal{R}$ is directed from point q in cylindrical coordinates to point p in Cartesian coordinates. Find an expression for $d\mathbf{l}' \times \mathcal{R}$, where $d\mathbf{l}'$ is the differential length vector along a circle of radius a.

Solution

The position vectors for points p and q are

$$\mathbf{r} = b\, \mathbf{a}_z \qquad\qquad (1.59\text{a})$$

$$\mathbf{r}' = a\, \mathbf{a}_{\rho'} \qquad\qquad (1.59\text{b})$$

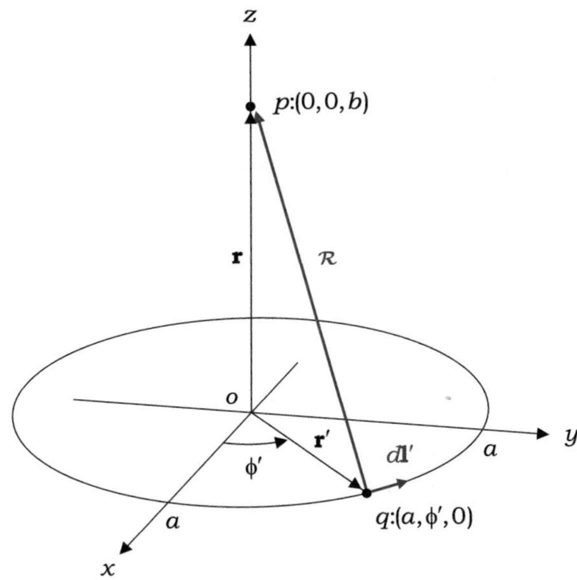

Fig. 1.28 Mixed-coordinate system. The prime symbol denotes the initial point of the distance vector

Thus, the distance vector is expressed as

$$\mathcal{R} = \mathbf{r} - \mathbf{r}' = b\,\mathbf{a}_z - a\,\mathbf{a}_{\rho'} \tag{1.59c}$$

The differential length vector along the circle is given by

$$d\mathbf{l}' = a\,d\phi'\,\mathbf{a}_{\phi'} \tag{1.59d}$$

With Eqs. (1.59c) and (1.59d), we can compute

$$d\mathbf{l}' \times \mathcal{R} = a\,d\phi'\,\mathbf{a}_{\phi'} \times (b\,\mathbf{a}_z - a\,\mathbf{a}_{\rho'})$$
$$= ab\,d\phi'\,\mathbf{a}_{\rho'} + a^2\,d\phi'\,\mathbf{a}_{z'} \tag{1.59e}$$

As the terminal point of $\mathcal{R}$ is the point of interest, Eq. (1.59e) should have been transformed into the unprimed or Cartesian coordinates. However, in most cases, it is more convenient to perform the conversion after all the other calculations are performed. The differential $d\phi'$ implies that the integration with respect to ϕ' follows.

Exercise 1.18
For point p_1:(1, $\sqrt{3}$, 5) in Cartesian coordinates, find (a) its coordinates, and (b) its position vector in cylindrical coordinates.

Ans. (a) p_1:(2, 60°, 5), (b) $\mathbf{r}_1 = 2\mathbf{a}_\rho + 5\mathbf{a}_z$.

Exercise 1.19
Which of the following vector fields does not have cylindrical symmetry?

(a) $\rho\,\mathbf{a}_z$, (b) $(\sin\phi)\,\mathbf{a}_\rho$, (c) $3\mathbf{a}_\phi$, (d) $(z-1)\,\mathbf{a}_\rho$, (e) $\mathbf{V}(\mathbf{r}) = \mathbf{a}_z \times \mathbf{r}$.

Ans. (b).

Exercise 1.20
What do the points referred to by the position vector $\mathbf{r} = 3\mathbf{a}_\rho + 4\mathbf{a}_z$ form in space?

Ans. A circle of radius 3 in the $z = 4$ plane.

Exercise 1.21
Given points p_1:$(2\sqrt{3}, 60°, 4)$ and p_2:$(2, 90°, 1)$ in cylindrical coordinates, express the distance vector from p_2 to p_1 in Cartesian coordinates.

Ans. $\mathcal{R} = \sqrt{3}\,\mathbf{a}_x + \mathbf{a}_y + 3\mathbf{a}_z$.

Exercise 1.22
Evaluate the integral $\int_{\phi=0}^{\pi} \mathbf{a}_\rho\,d\phi$.

Ans. $2\mathbf{a}_y$.

1.3.3   Spherical Coordinate System

The spherical coordinate system uses three numbers R, θ, and ϕ to define a point in a three-dimensional space. Point p_1:(R_1, θ_1, ϕ_1) is located at the intersection of three surfaces: the spherical surface of radius R_1 centered at the origin, the conical surface of half-angle θ_1 with the apex at the origin, and the half-plane hinged along the z-axis and rotated by angle ϕ_1. Accordingly, R_1 is the **radial distance** measured from the origin, θ_1 is the **polar angle** measured from the $+z$-axis, and ϕ_1 is the **azimuth angle** measured from the $+x$-axis in the xy-plane. The ranges of R, θ, and ϕ are $0 \le R < \infty$ [m], $0 \le \theta \le \pi$ [rad], and $0 \le \phi \le 2\pi$ [rad], respectively. At p_1, base vectors $\mathbf{a}_R$, $\mathbf{a}_\theta$, and $\mathbf{a}_\phi$ are defined in such a way that they are perpendicular to the surfaces of constant R, θ, and ϕ and point in the directions of increase of R, θ, and ϕ, respectively. The base vectors in spherical coordinates generally vary as functions of position.

The base vectors in spherical coordinates obey orthonormality:

$$\mathbf{a}_R \cdot \mathbf{a}_\theta = \mathbf{a}_\theta \cdot \mathbf{a}_\phi = \mathbf{a}_\phi \cdot \mathbf{a}_R = 0 \tag{1.60a}$$

$$\mathbf{a}_R \cdot \mathbf{a}_R = \mathbf{a}_\theta \cdot \mathbf{a}_\theta = \mathbf{a}_\phi \cdot \mathbf{a}_\phi = 1 \tag{1.60b}$$

Although the three base vectors may vary with the position, they are mutually perpendicular at every point in space.

In the right-handed spherical coordinate system, the base vectors follow cyclic relations:

$$\mathbf{a}_R \times \mathbf{a}_\theta = \mathbf{a}_\phi \tag{1.61a}$$

$$\mathbf{a}_\theta \times \mathbf{a}_\phi = \mathbf{a}_R \qquad (1.61\text{b})$$

$$\mathbf{a}_\phi \times \mathbf{a}_R = \mathbf{a}_\theta \qquad (1.61\text{c})$$

The three vectors in a cyclic permutation obey the right-hand rule: when the right fingers rotate from the first to the second vector, the thumb points in the direction of the third vector.

The position vector $\mathbf{r}_1$ is used to specify point $p_1{:}(R_1, \theta_1, \phi_1)$, as shown in Fig. 1.29. As the terminal point of $\mathbf{r}_1$ is the point of interest, $\mathbf{r}_1$ should be expanded in terms of the base vectors given at p_1. It is evident from Fig. 1.29 that $\mathbf{r}_1$ is perpendicular to both $\mathbf{a}_\theta$ and $\mathbf{a}_\phi$. Therefore, the position vector in spherical coordinates is expressed as

$$\boxed{\mathbf{r} = R\,\mathbf{a}_R} \qquad (1.62)$$

The position vector in Eq. (1.62) is seemingly independent of θ and ϕ. This is because the *spherical symmetry* has already been incorporated into the expression. An object is said to have spherical symmetry if it looks the same as we move around it in the $\mathbf{a}_\theta$ or $\mathbf{a}_\phi$ direction or rotate it about any axis passing through the center. In the presence of spherical symmetry, the points on a sphere centered at the origin are indistinguishable, as is well represented by $\mathbf{r}$ in Eq. (1.62). For example, the vector field $\mathbf{U}(\mathbf{r}) = \mathbf{a}_R$ shown in Fig. 1.30 has spherical symmetry; it appears the same as we move around it in any direction. In contrast, the vector fields $\mathbf{V}(\mathbf{r}) = \mathbf{a}_\theta$ and $\mathbf{W}(\mathbf{r}) = \mathbf{a}_\phi$ have no spherical symmetry because the direction of the vectors is reversed if the entire vector field is rotated about the x-axis by $180°$.

Symmetry considerations can provide considerable information regarding the solution to the problem. If the source is distributed in space to exhibit spherical symmetry, then the resultant vector field also exhibits spherical symmetry and is independent of θ and ϕ such that $\mathbf{D}(\mathbf{r}) = D_R(R)\,\mathbf{a}_R + D_\theta(R)\,\mathbf{a}_\theta + D_\phi(R)\,\mathbf{a}_\phi$. However, the terms $D_\theta(R)\,\mathbf{a}_\theta$ and $D_\phi(R)\,\mathbf{a}_\phi$ should vanish from the final expression because the direction is reversed when they are rotated about the x-axis by $180°$. Consequently, in the presence of spherical symmetry, the resultant vector field is expected to be of the form

$$\boxed{\mathbf{D}(\mathbf{r}) = D_R(R)\,\mathbf{a}_R} \qquad (1.63)$$

When vector $\mathbf{A}$ is defined at point p_1 in spherical coordinates as shown in Fig. 1.31, it can be expanded in terms of base vectors at the point such that

$$\boxed{\mathbf{A} = A_R\,\mathbf{a}_R + A_\theta\,\mathbf{a}_\theta + A_\phi\,\mathbf{a}_\phi} \qquad (1.64)$$

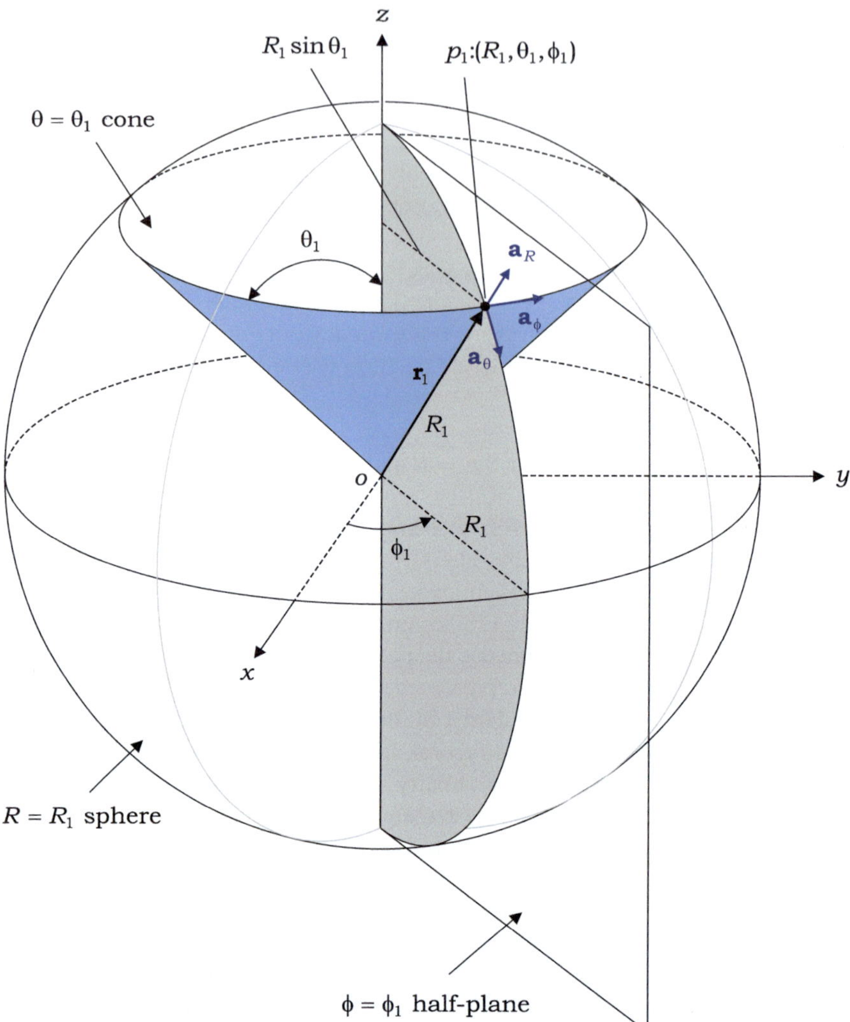

Fig. 1.29 Spherical coordinate system

When Eq. (1.64) is used for a vector field, the scalar components A_R, A_θ, and A_ϕ are generally functions of R, θ, and ϕ. Moreover, the base vectors $\mathbf{a}_R$, $\mathbf{a}_\theta$, and $\mathbf{a}_\phi$ vary with θ and ϕ.

When two vectors $\mathbf{A}$ and $\mathbf{B}$ are defined at the same point in space, they can be expressed in component form as follows:

$$\mathbf{A} = A_R\,\mathbf{a}_R + A_\theta\,\mathbf{a}_\theta + A_\phi\,\mathbf{a}_\phi \tag{1.65a}$$

Fig. 1.30 Vector field
$\mathbf{U}(\mathbf{r}) = \mathbf{a}_R$ with spherical
symmetry

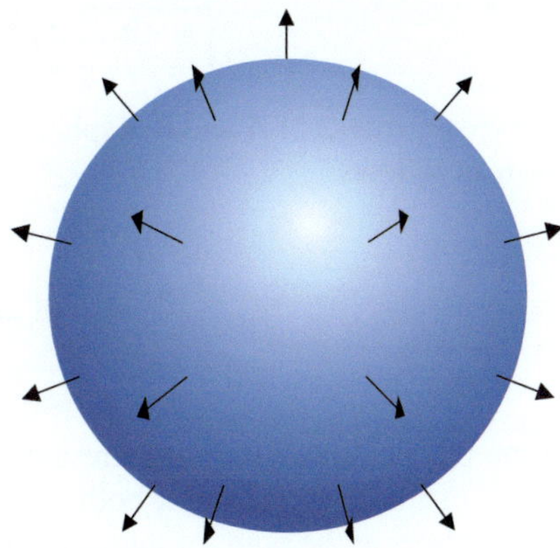

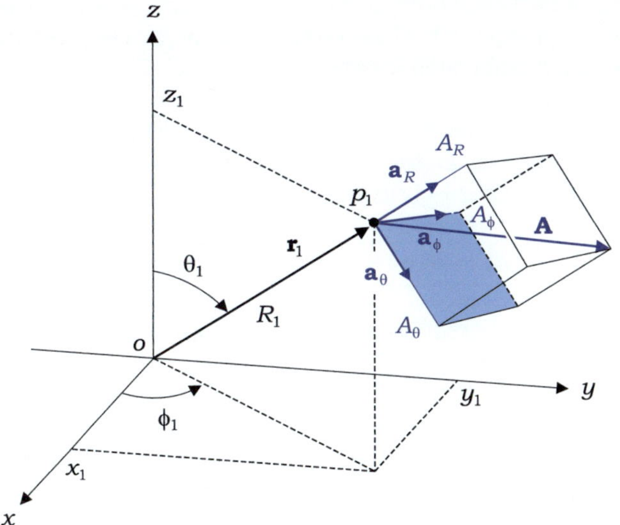

Fig. 1.31 Vector $\mathbf{A}$ is defined at p_1:(R_1, θ_1, ϕ_1) in spherical coordinates

$$\mathbf{B} = B_R\,\mathbf{a}_R + B_\theta\,\mathbf{a}_\theta + B_\phi\,\mathbf{a}_\phi \tag{1.65b}$$

where $\mathbf{a}_R$, $\mathbf{a}_\theta$, and $\mathbf{a}_\phi$ are the base vectors at the given point.

We are now ready to express the rules of the vector algebra in spherical coordinates. The addition of $\mathbf{A}$ and $\mathbf{B}$ in spherical coordinates is defined by

$$\boxed{\mathbf{A} + \mathbf{B} = (A_R + B_R)\mathbf{a}_R + (A_\theta + B_\theta)\mathbf{a}_\theta + \left(A_\phi + B_\phi\right)\mathbf{a}_\phi} \tag{1.66}$$

The dot product of $\mathbf{A}$ and $\mathbf{B}$ in spherical coordinates is defined by

$$\boxed{\mathbf{A} \cdot \mathbf{B} = A_R B_R + A_\theta B_\theta + A_\phi B_\phi}$$ (1.67)

The cross product of $\mathbf{A}$ and $\mathbf{B}$ in spherical coordinates can be expressed in determinant form as follows:

$$\boxed{\mathbf{A} \times \mathbf{B} = \begin{vmatrix} \mathbf{a}_R & \mathbf{a}_\theta & \mathbf{a}_\phi \\ A_R & A_\theta & A_\phi \\ B_R & B_\theta & B_\phi \end{vmatrix}}$$ (1.68)

Some physical laws may require a cross product of two vectors given at different points in space. In this case, both vectors should be expanded in terms of the same base vectors at the point of interest before the cross product is performed.

By definition, the differential length vector $d\mathbf{l}$ is an infinitesimal vector from point p_1:(R_1, θ_1, ϕ_1) to the neighboring point p_2:$(R_1 + dR, \theta_1 + d\theta, \phi_1 + d\phi)$. Again, $d\mathbf{l}$ must be expanded in terms of the base vectors given at p_1. The projections of $d\mathbf{l}$ onto $\mathbf{a}_R$, $\mathbf{a}_\theta$, and $\mathbf{a}_\phi$ yield differential lengths dR, $R_1 d\theta$, and $R_1 \sin\theta_1 d\phi$, respectively, as shown in Fig. 1.32. Evidently, the differential coordinates $d\theta$ and $d\phi$ are converted into differential lengths $R_1 d\theta$ and $R_1 \sin\theta_1 d\phi$, respectively. The multiplicative factors R_1 and $R_1 \sin\theta_1$ are called the **metric coefficients**. Accordingly, the differential length vector in spherical coordinates is expressed as

$$\boxed{d\mathbf{l} = dR\, \mathbf{a}_R + Rd\theta\, \mathbf{a}_\theta + R\sin\theta\, d\phi\, \mathbf{a}_\phi}$$ (1.69)

For example, on a spherical surface centered at the origin, dR vanishes and $d\mathbf{l}$ can be reduced to a simpler form.

The two endpoints of $d\mathbf{l}$ are associated with six surfaces of constant coordinate (Fig. 1.32). These surfaces define a differential volume dv, which can be conveniently used to divide a volume in spherical coordinates into many elements of volume in the form of dv, and vice versa. In view of the fact that $|d\mathbf{l}|$ is infinitesimally small, dv may be considered as a rectangular parallelepiped with sides dR, $R_1 d\theta$, and $R_1 \sin\theta_1 d\phi$ and volume $R_1^2 \sin\theta_1 dR d\theta d\phi$. In spherical coordinates, the differential volume is defined as

$$\boxed{dv = R^2 \sin\theta\, dR d\theta d\phi} \quad [\mathrm{m}^3]$$ (1.70)

It is important to note that the size of dv varies with R and θ.

The differential area vector $d\mathbf{s}$ can provide an efficient method for specifying the boundary surfaces of dv; its magnitude is equal to the surface area, and its unit vector is normal to the surface directed outward from dv. As dv is infinitesimal, each face of dv is considered to be rectangular. The differential area vectors for the six faces of dv are presented in Fig. 1.32, that is,

$$d\mathbf{s} = R_1^2 \sin\theta_1\, d\theta d\phi\, \mathbf{a}_R \quad (\text{face ①})$$ (1.71a)

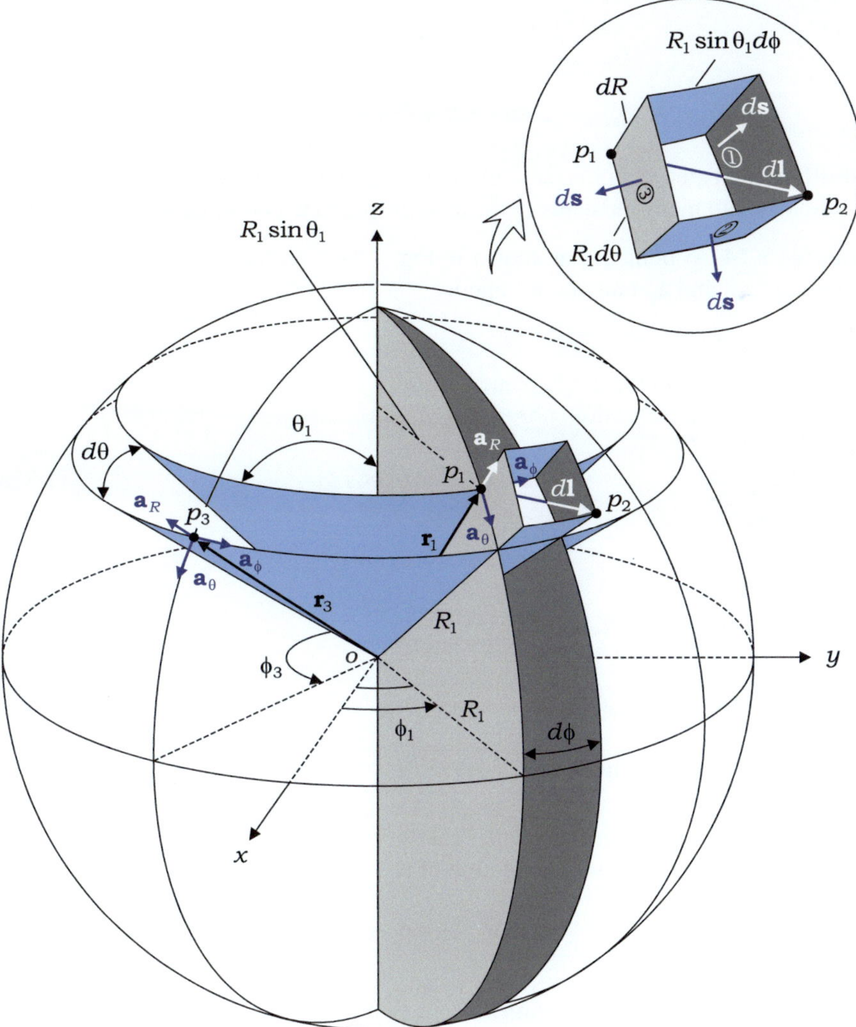

Fig. 1.32 Differential length vector $d\mathbf{l}$, differential area vector $d\mathbf{s}$, and differential volume dv in spherical coordinate system

$$d\mathbf{s} = R_1 \sin\theta_1 \, dR d\phi \, \mathbf{a}_\theta \quad (\text{face ②}) \tag{1.71b}$$

$$d\mathbf{s} = -R_1 \, dR d\theta \, \mathbf{a}_\phi \quad (\text{face ③}) \tag{1.71c}$$

$$d\mathbf{s} = -R_1^2 \sin\theta_1 \, d\theta d\phi \, \mathbf{a}_R \tag{1.71d}$$

$$d\mathbf{s} = -R_1 \sin\theta_1 \, dR d\phi \, \mathbf{a}_\theta \qquad (1.71e)$$

$$d\mathbf{s} = R_1 \, dR d\theta \, \mathbf{a}_\phi \qquad (1.71f)$$

The differential area vector is particularly useful for dividing a surface with constant coordinates into many elements of the area in spherical coordinates, and vice versa.

Example 1.14 At point p_1, as shown in Fig. 1.31, find the projections of the base vectors $\mathbf{a}_R$, $\mathbf{a}_\theta$, and $\mathbf{a}_\phi$ onto $\mathbf{a}_x$, $\mathbf{a}_y$, and $\mathbf{a}_z$.

Solution

Projection of $\mathbf{a}_R$ onto the $z = 0$ plane : $\sin\theta_1$.
From the two-step projection, projection of $\mathbf{a}_R$

$$\text{onto } \mathbf{a}_x \,:\, \sin\theta_1 \cos\phi_1 \qquad (1.72a)$$

$$\text{onto } \mathbf{a}_y \,:\, \sin\theta_1 \sin\phi_1 \qquad (1.72b)$$

$$\text{onto } \mathbf{a}_z \,:\, \cos\theta_1 \qquad (1.72c)$$

Combining these results leads to

$$\boxed{\mathbf{a}_R = \sin\theta_1 \cos\phi_1 \, \mathbf{a}_x + \sin\theta_1 \sin\phi_1 \, \mathbf{a}_y + \cos\theta_1 \, \mathbf{a}_z} \qquad (1.72d)$$

Projection of $\mathbf{a}_\theta$ onto the $z = 0$ plane : $\cos\theta_1$.
From the two-step projection, projection of $\mathbf{a}_\theta$

$$\text{onto } \mathbf{a}_x \,:\, \cos\theta_1 \cos\phi_1 \qquad (1.72e)$$

$$\text{onto } \mathbf{a}_y \,:\, \cos\theta_1 \sin\phi_1 \qquad (1.72f)$$

$$\text{onto } \mathbf{a}_z \,:\, -\sin\theta_1 \qquad (1.72g)$$

Combining these results leads to

$$\boxed{\mathbf{a}_\theta = \cos\theta_1 \cos\phi_1 \, \mathbf{a}_x + \cos\theta_1 \sin\phi_1 \, \mathbf{a}_y - \sin\theta_1 \, \mathbf{a}_z} \qquad (1.72h)$$

Without resorting to a two-step projection, we can obtain the projection of $\mathbf{a}_\phi$

$$\text{onto } \mathbf{a}_x \,:\, -\sin\phi_1 \qquad (1.72i)$$

$$\text{onto } \mathbf{a}_y : \cos \phi_1 \tag{1.72j}$$

$$\text{onto } \mathbf{a}_z : 0 \tag{1.72k}$$

Combining these results leads to

$$\boxed{\mathbf{a}_\phi = -\sin\phi_1 \mathbf{a}_x + \cos\phi_1 \mathbf{a}_y} \tag{1.72l}$$

The results in Eqs. (1.72d, h, l) are very useful for the coordinate transformation discussed in the following section.

Example 1.15 Given that $\mathbf{A}(\mathbf{r}) = \cos\phi\,\mathbf{a}_z$ in cylindrical coordinates and $\mathbf{B}(\mathbf{r}) = R^2\mathbf{a}_R$ in spherical coordinates, perform the following vector operations on a sphere of radius $R = 2$ centered at the origin: (a) $\mathbf{A} \cdot \mathbf{B}$, and (b) $\mathbf{A} \times \mathbf{B}$.

Solution

On the given sphere, the two vector fields are expressed as

$$\mathbf{A}(2, \theta, \phi) = \cos\phi\,\mathbf{a}_z$$

$$\mathbf{B}(2, \theta, \phi) = 4\mathbf{a}_R$$

With the help of Eq. (1.72d),

(a) $\mathbf{A} \cdot \mathbf{B} = 4\cos\phi(\mathbf{a}_z \cdot \mathbf{a}_R) = 4\cos\phi\cos\theta$

(b) $\mathbf{A} \times \mathbf{B} = 4\cos\phi(\mathbf{a}_z \times \mathbf{a}_R) = 4\cos\phi\sin\theta\,\mathbf{a}_\phi$

Example 1.16 Given a hemispherical shell with radii 1 and 3 (Fig. 1.33), express ds in spherical coordinates at the following points: (a) p_1:(2, $\pi/2$, 0), (b) p_2:(1, $3\pi/4$, $3\pi/2$), (c) p_3:(2, $\pi/4$, π), and (d) p_4:(3, $\pi/4$, $3\pi/2$).

Solution

(a) $d\mathbf{s}^1 = R\,dR\,d\theta\,\mathbf{a}_\phi = 2\,dR\,d\theta\,\mathbf{a}_\phi$

(b) $d\mathbf{s}^2 = -R^2\sin\theta\,d\theta\,d\phi\,\mathbf{a}_R = -\dfrac{1}{\sqrt{2}}d\theta\,d\phi\,\mathbf{a}_R$

(c) $d\mathbf{s}^3 = -R\,dR\,d\theta\,\mathbf{a}_\phi = -2\,dR\,d\theta\,\mathbf{a}_\phi$

(d) $d\mathbf{s}^4 = R^2\sin\theta\,d\theta\,d\phi\,\mathbf{a}_R = \dfrac{9}{\sqrt{2}}d\theta\,d\phi\,\mathbf{a}_R$

Exercise 1.23

Given p_1:(−4, 4, 7) in Cartesian coordinates, find (a) its spherical coordinates, and (b) its position vector in spherical coordinates.

Ans. (a) p_1:(9, 38.9°, 135°), (b) $\mathbf{r} = 9\mathbf{a}_R$.

Fig. 1.33 Hemispherical
shell

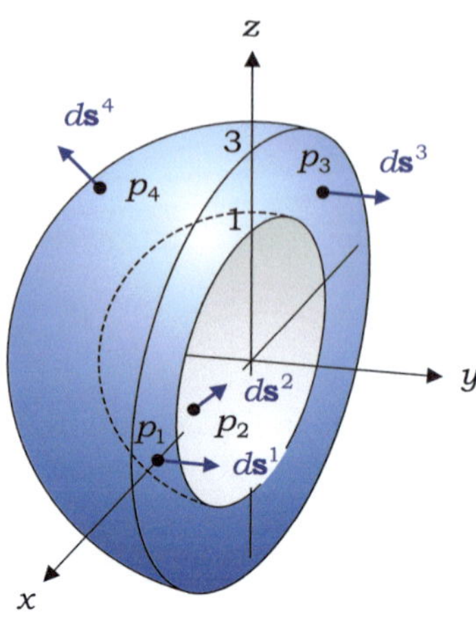

Exercise 1.24

On a sphere of radius $R = 3$ centered at the origin, find general expressions for (a)
$d\mathbf{s}$, and (b) $d\mathbf{l}$.

Ans. (a) $d\mathbf{s} = (9 \sin \theta) \, d\theta \, d\phi \, \mathbf{a}_R$, (b) $d\mathbf{l} = 3 d\theta \, \mathbf{a}_\theta + 3 \sin \theta \, d\phi \, \mathbf{a}_\phi$.

Exercise 1.25

Repeat Exercise 1.24(b) in Cartesian coordinates.

Ans. $d\mathbf{l} = dx \, \mathbf{a}_x + dy \, \mathbf{a}_y + dz \, \mathbf{a}_z$ with $x^2 + y^2 + z^2 = 9$ and $x \, dx + y \, dy + z \, dz = 0$.

Exercise 1.26

Which of the following vector fields has a spherical symmetry?
(a) $(1/R^2) \, \mathbf{a}_R$, (b) $\mathbf{a}_\phi \times \mathbf{a}_\theta$, and (c) $(\mathbf{r} \cdot \mathbf{a}_z) \, \mathbf{a}_R$.

Ans. (a), (b).

Review Questions

RQ 1.13	What is the orthogonal coordinate system?	[(1.23a,b)]
RQ 1.14	What is the right-handed coordinate system?	[(1.24a,b,c)]
RQ 1.15	What are the base vectors?	[Fig. 1.16]
RQ 1.16	Does operation $\mathbf{A} \cdot \mathbf{B}$ require $\mathbf{A}$ and $\mathbf{B}$ to be at the same spatial point?	[(1.42a,b)(1.51)]
RQ 1.17	What are the general expressions for the position vector?	[(1.25)(1.44)(1.62)]
RQ 1.18	What are the general expressions of $d\mathbf{l}$?	[(1.36)(1.53)(1.69)]
RQ 1.19	What are the general expressions of dv?	[(1.37)(1.54)(1.70)]
RQ 1.20	What are the differences between $d\mathbf{l}$ and $\mathcal{R}_{1-2}$?	[Exercise 1.17]
RQ 1.21	In what direction is $d\mathbf{s}$ directed on a closed surface?	[Fig. 1.18]

1.4 Coordinate Transformation

Thus far, we have discussed the three orthogonal coordinate systems in detail. Although different coordinate systems offer different methods for specifying a point and vector in space, one coordinate system may have a definite advantage over others in solving problems with symmetries. Because the points and vectors in space are independent of the coordinate system, we can select the coordinate system that best fits the geometry of the given problem. Moreover, a mixed-coordinate system is convenient for solving problems in which the source and field points must be independently specified. Subsequently, we may need to transform the coordinates of a point from one coordinate system to another, called *coordinate transformation*, and transform the components of a vector from one coordinate system to another, called ***coordinate transformation of the components of a vector***. The two topics are discussed in this section.

1.4.1 Cartesian-Cartesian Transformation

First, we examine the coordinate transformation from one Cartesian system to another. The former is called "unprimed" system and the latter is called "primed" system. The primed system is rotated about the z-axis by an angle φ with respect to the unprimed system, as shown in Fig. 1.34. Let us focus on a point specified as $p_1{:}(x_1, y_1, z_1)$ in unprimed coordinates, or $p_1{:}(x_1', y_1', z_1')$ in primed coordinates. The base vectors are denoted by $\mathbf{a}_x$, $\mathbf{a}_y$, and $\mathbf{a}_z$ in unprimed coordinates, and $\mathbf{a}_{x'}$, $\mathbf{a}_{y'}$, and $\mathbf{a}_{z'}$ in primed coordinates. If $\mathbf{A}$ is a vector at point p_1, it can be generally expressed either as

$$\mathbf{A} = A_x\,\mathbf{a}_x + A_y\,\mathbf{a}_y + A_z\,\mathbf{a}_z \tag{1.73a}$$

$$\mathbf{A} = A_{x'}\,\mathbf{a}_{x'} + A_{y'}\,\mathbf{a}_{y'} + A_{z'}\,\mathbf{a}_{z'} \equiv \mathbf{A}' \tag{1.73b}$$

The two expressions in Eqs. (1.73a) and (1.73b) refer to the same vector; they are expanded differently. To avoid confusion, the vector in Eq. (1.73b) is referred to as $\mathbf{A}'$.

The transformation of the components of $\mathbf{A}$ into the primed system can be accomplished simply by projecting $\mathbf{A}$ onto the base vectors in the primed coordinates. The scalar component $A_{x'}$ can be obtained from the projection of $\mathbf{A}$ onto $\mathbf{a}_{x'}$ as follows:

$$\mathbf{A} \cdot \mathbf{a}_{x'} = A_{x'}$$
$$= \left(A_x\,\mathbf{a}_x + A_y\,\mathbf{a}_y + A_z\,\mathbf{a}_z\right) \cdot \mathbf{a}_{x'} = A_x \cos\varphi + A_y \sin\varphi \tag{1.74}$$

where $\mathbf{a}_x \cdot \mathbf{a}_{x'} = \cos\varphi$, $\mathbf{a}_y \cdot \mathbf{a}_{x'} = \sin\varphi$, and $\mathbf{a}_z \cdot \mathbf{a}_{x'} = 0$, as can be seen from Fig. 1.34. Similarly, $A_{y'}$ and $A_{z'}$ are obtained from $\mathbf{A} \cdot \mathbf{a}_{y'}$ and $\mathbf{A} \cdot \mathbf{a}_{z'}$, respectively. Consequently, the transformation of the components of $\mathbf{A}$ from the unprimed to the

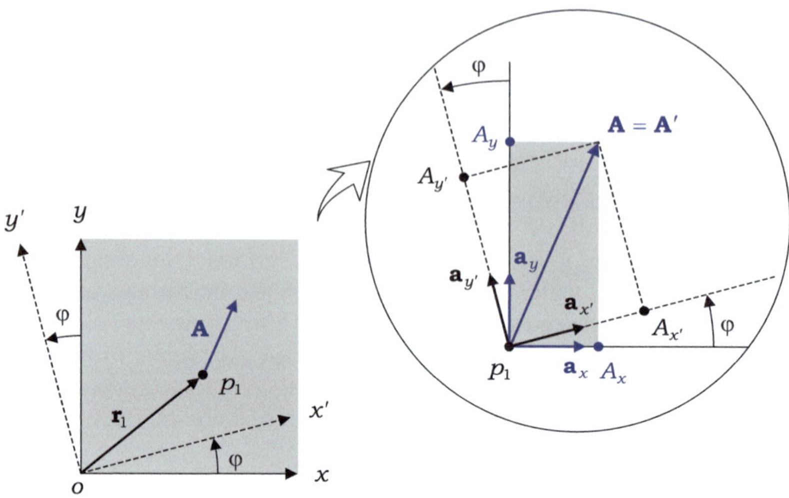

Fig. 1.34 Primed system is rotated about the z-axis relative to the unprimed system

primed system can be expressed in matrix form, as follows:

$$
\begin{bmatrix} A_{x'} \\ A_{y'} \\ A_{z'} \end{bmatrix} = \begin{bmatrix} \cos\varphi & \sin\varphi & 0 \\ -\sin\varphi & \cos\varphi & 0 \\ 0 & 0 & 1 \end{bmatrix} \begin{bmatrix} A_x \\ A_y \\ A_z \end{bmatrix}
\tag{1.75}
$$

where φ is the rotation angle of the primed system with respect to the unprimed system, and counterclockwise rotation corresponds to a positive value of φ. The 3×3 matrix in Eq. (1.75) is called the **transformation matrix**, denoted by $\mathbf{T}$. Then, the transformation can be written as $\mathbf{A}' = \mathbf{TA}$, a simple matrix multiplication.

It should be noted that the transformation matrix $\mathbf{T}$ in Eq. (1.75) is independent of the position of $\mathbf{A}$. This implies that $\mathbf{T}$ can be used for any vector including the position vector, which is expressed either as $\mathbf{r}_1 = x_1\mathbf{a}_x + y_1\mathbf{a}_y + z_1\mathbf{a}_z$ or as $\mathbf{r}'_1 = x'_1\mathbf{a}_{x'} + y'_1\mathbf{a}_{y'} + z'_1\mathbf{a}_{z'}$. By substituting $\mathbf{r}_1$ for $\mathbf{A}$ in Eq. (1.75), the coordinates of point p_1 can be transformed from the unprimed to the primed system such as

$$
\begin{bmatrix} x' \\ y' \\ z' \end{bmatrix} = \begin{bmatrix} \cos\varphi & \sin\varphi & 0 \\ -\sin\varphi & \cos\varphi & 0 \\ 0 & 0 & 1 \end{bmatrix} \begin{bmatrix} x \\ y \\ z \end{bmatrix}
\tag{1.76}
$$

where subscript 1 is omitted for generalization. The coordinate transformation in Eq. (1.75) or in Eq. (1.76) can be reversed by changing the sign of φ.

Example 1.17 Transform vector field $\mathbf{A} = 2x\,\mathbf{a}_x$ into the primed system, which is rotated about the z-axis by an angle φ.

Solution

Substituting $\mathbf{A} = 2x\,\mathbf{a}_x$ into Eq. (1.75), we have

$$
\begin{bmatrix} A_{x'} \\ A_{y'} \\ A_{z'} \end{bmatrix} = \begin{bmatrix} \cos\varphi & \sin\varphi & 0 \\ -\sin\varphi & \cos\varphi & 0 \\ 0 & 0 & 1 \end{bmatrix} \begin{bmatrix} 2x \\ 0 \\ 0 \end{bmatrix} \tag{1.77a}
$$

Thus,

$$
\begin{aligned}
\mathbf{A}' &= A_{x'}\,\mathbf{a}_{x'} + A_{y'}\,\mathbf{a}_{y'} + A_{z'}\,\mathbf{a}_{z'} \\
&= 2x\cos\varphi\,\mathbf{a}_{x'} - 2x\sin\varphi\,\mathbf{a}_{y'}
\end{aligned} \tag{1.77b}
$$

Substituting $-\varphi$ for φ and interchanging the roles of (x, y, z) and (x', y', z') in Eq. (1.76), we have

$$
x = x'\cos\varphi - y'\sin\varphi \tag{1.77c}
$$

Inserting Eq. (1.77c) into Eq. (1.77b) yields

$$
\mathbf{A} = 2(x'\cos\varphi - y'\sin\varphi)\cos\varphi\,\mathbf{a}_{x'} - 2(x'\cos\varphi - y'\sin\varphi)\sin\varphi\,\mathbf{a}_{y'} \tag{1.77d}
$$

The components of $\mathbf{A}$ are fully expressed in terms of the primed coordinates.

Exercise 1.27

Given $\mathbf{A} = 4\mathbf{a}_x + 2\mathbf{a}_y + 3\mathbf{a}_z$ defined at p:(2, 1, 3), transform its components into a primed system rotated by angle $\varphi = 30°$.

Ans. $\mathbf{A} = (2\sqrt{3} + 1)\,\mathbf{a}_{x'} + (-2 + \sqrt{3})\,\mathbf{a}_{y'} + 3\mathbf{a}_{z'}$.

Exercise 1.28

Derive the $\mathbf{T}$ matrix for the coordinate transformation from (x, y, z) to $(\overline{x}, \overline{y}, \overline{z})$ system, as shown below:

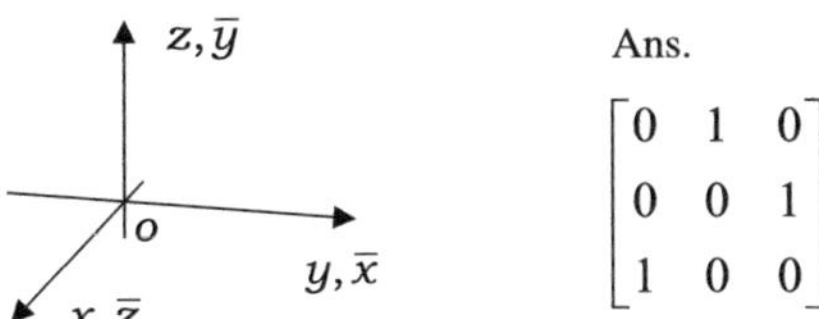

1.4.2 *Cylindrical-Cartesian Transformation*

Let us start by considering the vector $\mathbf{A}$ given at point p_1:(ρ_1, ϕ_1, z_1) in cylindrical coordinates, as shown in Fig. 1.23. It is apparent that the base vectors $\mathbf{a}_\rho$ and $\mathbf{a}_\phi$ are rotated about $\mathbf{a}_z$ by an angle ϕ_1 with respect to $\mathbf{a}_x$ and $\mathbf{a}_y$, respectively. A comparison of the coordinate axes in Fig. 1.23 with those in Fig. 1.34 shows that the components of $\mathbf{A}$ can be transformed from Cartesian to cylindrical system with the same transformation matrix given in Eq. (1.75) if φ is set to ϕ_1. Accordingly, the coordinate transformation of the components of a vector from Cartesian to cylindrical system can be expressed in matrix form as

$$\begin{bmatrix} A_\rho \\ A_\phi \\ A_z \end{bmatrix} = \begin{bmatrix} \cos(\phi_1) & \sin(\phi_1) & 0 \\ -\sin(\phi_1) & \cos(\phi_1) & 0 \\ 0 & 0 & 1 \end{bmatrix} \begin{bmatrix} A_x \\ A_y \\ A_z \end{bmatrix} \tag{1.78}$$

It is important to note that the angle ϕ_1 in Eq. (1.78) is *the ϕ-coordinate of the position of* $\mathbf{A}$. Also note that this coordinate transformation may be reversed simply by changing the sign of ϕ_1. Therefore, the transformation of the components of a vector from cylindrical to Cartesian system can be conducted as follows:

$$\begin{bmatrix} A_x \\ A_y \\ A_z \end{bmatrix} = \begin{bmatrix} \cos(-\phi_1) & \sin(-\phi_1) & 0 \\ -\sin(-\phi_1) & \cos(-\phi_1) & 0 \\ 0 & 0 & 1 \end{bmatrix} \begin{bmatrix} A_\rho \\ A_\phi \\ A_z \end{bmatrix} \tag{1.79}$$

Again, angle ϕ_1 is the ϕ-coordinate of the point where $\mathbf{A}$ is defined.

Although we can directly apply Eq. (1.79) to the coordinate transformation of a point from the cylindrical to the Cartesian system, the converse transformation expressed by Eq. (1.78) is always accompanied by an additional transformation performed using the Pythagorean theorem and trigonometry. To avoid mathematical complications, we simply use the relationship between the cylindrical and Cartesian coordinates of a point obtained from the Pythagorean theorem and trigonometry, that is,

$$\begin{aligned} x &= \rho \cos \phi \\ y &= \rho \sin \phi \\ z &= z \end{aligned} \tag{1.80}$$

and

$$\begin{aligned} \rho &= \sqrt{x^2 + y^2} \\ \phi &= \tan^{-1}(y/x) \\ z &= z \end{aligned} \tag{1.81}$$

where subscript 1 is omitted for generalization.

When the arctangent in Eq. (1.81) is used, care should be taken to ensure that ϕ is in the correct quadrant. Because the range of the principal value of the arctangent is usually $-\pi/2 < \phi \le \pi/2$, most electronic calculators give $-45°$ for a value of $\tan^{-1}(1/(-1))$ instead of $135°$.

Example 1.18 Express the following vectors in Cartesian coordinates:

(a) $\mathbf{A} = 2\mathbf{a}_\rho + 3\mathbf{a}_\phi + \sqrt{3}\,\mathbf{a}_z$ given at $(\rho, \phi, z) = (4, 60°, 5)$,
(b) $\mathbf{B} = 2\mathbf{a}_\rho + 3\mathbf{a}_\phi + \sqrt{3}\,\mathbf{a}_z$ given at $(x, y, z) = (\sqrt{2}, 2, 1)$, and
(c) $\mathbf{C}$ drawn from the origin to point $(\rho, \phi, z) = (3, 45°, 4)$.

Solution

(a)   From Eq. (1.79),

$$\begin{bmatrix} A_x \\ A_y \\ A_z \end{bmatrix} = \begin{bmatrix} \cos(-60°) & \sin(-60°) & 0 \\ -\sin(-60°) & \cos(-60°) & 0 \\ 0 & 0 & 1 \end{bmatrix}\begin{bmatrix} 2 \\ 3 \\ \sqrt{3} \end{bmatrix} = \begin{bmatrix} 1/2 & -\sqrt{3}/2 & 0 \\ \sqrt{3}/2 & 1/2 & 0 \\ 0 & 0 & 1 \end{bmatrix}\begin{bmatrix} 2 \\ 3 \\ \sqrt{3} \end{bmatrix}$$

Thus,

$$\mathbf{A} = \left(1 - \frac{3}{2}\sqrt{3}\right)\mathbf{a}_x + \left(\sqrt{3} + \frac{3}{2}\right)\mathbf{a}_y + \sqrt{3}\,\mathbf{a}_z$$

(b)   Letting $(x, y, z) = (\sqrt{2}, 2, 1)$ in Eq. (1.81) yields

$$\phi_1 = \tan^{-1}(2/\sqrt{2})$$

Then, it follows that

$$\sin\phi_1 = 2/\sqrt{6} \text{ and } \cos\phi_1 = 1/\sqrt{3} \qquad\qquad (1.82)$$

Inserting Eq. (1.82) into Eq. (1.79) yields

$$\begin{bmatrix} B_x \\ B_y \\ B_z \end{bmatrix} = \begin{bmatrix} \cos(-\phi_1) & \sin(-\phi_1) & 0 \\ -\sin(-\phi_1) & \cos(-\phi_1) & 0 \\ 0 & 0 & 1 \end{bmatrix}\begin{bmatrix} 2 \\ 3 \\ \sqrt{3} \end{bmatrix} = \begin{bmatrix} 1/\sqrt{3} & -2/\sqrt{6} & 0 \\ 2/\sqrt{6} & 1/\sqrt{3} & 0 \\ 0 & 0 & 1 \end{bmatrix}\begin{bmatrix} 2 \\ 3 \\ \sqrt{3} \end{bmatrix}$$

Thus,

$$\mathbf{B} = \left(\frac{2}{\sqrt{3}} - \sqrt{6}\right)\mathbf{a}_x + \left(\frac{4}{\sqrt{6}} + \sqrt{3}\right)\mathbf{a}_y + \sqrt{3}\,\mathbf{a}_z$$

(c) **C** is a position vector. Letting $(\rho, \phi, z) = (3, 45°, 4)$ in Eq. (1.80) yields

$$x = 3\cos 45° = 3/\sqrt{2}$$

$$y = 3\sin 45° = 3/\sqrt{2}$$

$$z = 4$$

Thus,

$$\mathbf{C} = \frac{3}{\sqrt{2}}\,\mathbf{a}_x + \frac{3}{\sqrt{2}}\,\mathbf{a}_y + 4\mathbf{a}_z$$

Exercise 1.29

Transform $\mathbf{a}_\phi$ into the Cartesian system using Eqs. (1.79) and (1.81), and then evaluate it at point $(x, y, z) = (1, 1, 1)$.

Ans. $(-1/\sqrt{2})\,\mathbf{a}_x + (1/\sqrt{2})\,\mathbf{a}_y$.

1.4.3 Spherical-Cartesian Transformation

When vector **A** is defined at point $p_1:(R_1, \theta_1, \phi_1)$ in spherical coordinates as shown in Fig. 1.31, it can be expanded in component form in either Cartesian or spherical coordinates as follows:

$$\mathbf{A} = A_x\,\mathbf{a}_x + A_y\,\mathbf{a}_y + A_z\,\mathbf{a}_z \tag{1.83a}$$

$$\mathbf{A} = A_R\,\mathbf{a}_R + A_\theta\,\mathbf{a}_\theta + A_\phi\,\mathbf{a}_\phi \equiv \mathbf{A}' \tag{1.83b}$$

The transformation of the components of $\mathbf{A}'$ into the Cartesian system can be accomplished by projecting $\mathbf{A}'$ onto $\mathbf{a}_x$, $\mathbf{a}_y$, and $\mathbf{a}_z$. For example, the dot product of $\mathbf{A}'$ and $\mathbf{a}_x$ results in the scalar component A_x, that is,

$$\begin{aligned}
A_x &= \mathbf{A}' \cdot \mathbf{a}_x \\
&= \left(A_R\,\mathbf{a}_R + A_\theta\,\mathbf{a}_\theta + A_\phi\,\mathbf{a}_\phi\right) \cdot \mathbf{a}_x \\
&= A_R \sin\theta_1 \cos\phi_1 + A_\theta \cos\theta_1 \cos\phi_1 - A_\phi \sin\phi_1
\end{aligned}$$

where $\mathbf{a}_R \cdot \mathbf{a}_x = \sin\theta_1 \cos\phi_1$, $\mathbf{a}_\theta \cdot \mathbf{a}_x = \cos\theta_1 \cos\phi_1$, and $\mathbf{a}_\phi \cdot \mathbf{a}_x = -\sin\phi_1$, as expressed in Eq. (1.72). By following the same procedure as that used for A_x, we obtain A_y and A_z from $\mathbf{A}' \cdot \mathbf{a}_y$ and $\mathbf{A}' \cdot \mathbf{a}_z$, respectively. Accordingly, the coordinate transformation of the components of **A** from spherical to Cartesian system can be expressed in matrix form as

$$
\begin{bmatrix} A_x \\ A_y \\ A_z \end{bmatrix} = \begin{bmatrix} \sin\theta_1\cos\phi_1 & \cos\theta_1\cos\phi_1 & -\sin\phi_1 \\ \sin\theta_1\sin\phi_1 & \cos\theta_1\sin\phi_1 & \cos\phi_1 \\ \cos\theta_1 & -\sin\theta_1 & 0 \end{bmatrix} \begin{bmatrix} A_R \\ A_\theta \\ A_\phi \end{bmatrix} \tag{1.84}
$$

The transformation matrix depends on θ_1 and ϕ_1, which are the spherical coordinates of the position of **A**.

Similarly, the transformation of the components of **A** from the Cartesian to spherical system is performed as follows:

$$
\begin{bmatrix} A_R \\ A_\theta \\ A_\phi \end{bmatrix} = \begin{bmatrix} \sin\theta_1\cos\phi_1 & \sin\theta_1\sin\phi_1 & \cos\theta_1 \\ \cos\theta_1\cos\phi_1 & \cos\theta_1\sin\phi_1 & -\sin\theta_1 \\ -\sin\phi_1 & \cos\phi_1 & 0 \end{bmatrix} \begin{bmatrix} A_x \\ A_y \\ A_z \end{bmatrix} \tag{1.85}
$$

The transformation matrix is a unitary matrix, that is, $\mathbf{T}^T = \mathbf{T}^{-1}$, where the superscript T represents the transpose, and -1 denotes the inverse. The transformation matrix in Eq. (1.85) is inversely related to that in Eq. (1.84), and is thus the transpose of the transformation matrix in Eq. (1.84).

The coordinate transformation of a point can be performed using the relationship between spherical and Cartesian coordinates, that is,

$$
\begin{aligned}
x &= R\sin\theta\cos\phi \\
y &= R\sin\theta\sin\phi \\
z &= R\cos\theta
\end{aligned} \tag{1.86}
$$

and

$$
\begin{aligned}
R &= \sqrt{x^2 + y^2 + z^2} \\
\theta &= \tan^{-1}\left(\sqrt{x^2 + y^2}/z\right) \\
\phi &= \tan^{-1}(y/x)
\end{aligned} \tag{1.87}
$$

where subscript 1 is omitted for generalization.

Example 1.19 After expanding $\mathbf{a}_R$ in terms of $\mathbf{a}_x$, $\mathbf{a}_y$, and $\mathbf{a}_z$, verify the following:

(a) $\dfrac{\partial \mathbf{a}_R}{\partial \theta} = \mathbf{a}_\theta$, and (b) $\dfrac{\partial \mathbf{a}_R}{\partial \phi} = \sin\theta\,\mathbf{a}_\phi$.

Solution

The coordinate transformation of $\mathbf{a}_R$ into the Cartesian system is expressed as follows:

$$\begin{bmatrix} A_x \\ A_y \\ A_z \end{bmatrix} = \begin{bmatrix} \sin\theta\cos\phi & \cos\theta\cos\phi & -\sin\phi \\ \sin\theta\sin\phi & \cos\theta\sin\phi & \cos\phi \\ \cos\theta & -\sin\theta & 0 \end{bmatrix}\begin{bmatrix} 1 \\ 0 \\ 0 \end{bmatrix}$$

Thus,

$$\begin{aligned} \mathbf{a}_R &= A_x\,\mathbf{a}_x + A_y\,\mathbf{a}_y + A_z\,\mathbf{a}_z \\ &= \sin\theta\cos\phi\,\mathbf{a}_x + \sin\theta\sin\phi\,\mathbf{a}_y + \cos\theta\,\mathbf{a}_z \end{aligned} \tag{1.88a}$$

Taking the derivatives of Eq. (1.88a) with respect to θ, we have

$$\frac{\partial \mathbf{a}_R}{\partial \theta} = \cos\theta\cos\phi\,\mathbf{a}_x + \cos\theta\sin\phi\,\mathbf{a}_y - \sin\theta\,\mathbf{a}_z \tag{1.88b}$$

Taking the derivatives of Eq. (1.88a) with respect to ϕ, we have

$$\frac{\partial \mathbf{a}_R}{\partial \phi} = -\sin\theta\sin\phi\,\mathbf{a}_x + \sin\theta\cos\phi\,\mathbf{a}_y \tag{1.88c}$$

(a) The coordinate transformation of the vector in Eq. (1.88b) back to the spherical system is performed as follows:

$$\begin{bmatrix} A_R \\ A_\theta \\ A_\phi \end{bmatrix} = \begin{bmatrix} \sin\theta\cos\phi & \sin\theta\sin\phi & \cos\theta \\ \cos\theta\cos\phi & \cos\theta\sin\phi & -\sin\theta \\ -\sin\phi & \cos\phi & 0 \end{bmatrix}\begin{bmatrix} \cos\theta\cos\phi \\ \cos\theta\sin\phi \\ -\sin\theta \end{bmatrix}$$

This gives

$$A_R = 0, \; A_\theta = 1, \; \text{and } A_\phi = 0.$$

Thus,

$$\begin{aligned} \frac{\partial \mathbf{a}_R}{\partial \theta} &= A_R\,\mathbf{a}_R + A_\theta\,\mathbf{a}_\theta + A_\phi\,\mathbf{a}_\phi \\ &= \mathbf{a}_\theta \end{aligned} \tag{1.88d}$$

(b) The coordinate transformation of the vector in Eq. (1.88c) back to the spherical system is performed as follows:

$$\begin{bmatrix} A_R \\ A_\theta \\ A_\phi \end{bmatrix} = \begin{bmatrix} \sin\theta\cos\phi & \sin\theta\sin\phi & \cos\theta \\ \cos\theta\cos\phi & \cos\theta\sin\phi & -\sin\theta \\ -\sin\phi & \cos\phi & 0 \end{bmatrix}\begin{bmatrix} -\sin\theta\sin\phi \\ \sin\theta\cos\phi \\ 0 \end{bmatrix}$$

This gives

$$A_R = 0, \ A_\theta = 0, \text{ and } A_\phi = \sin\theta \sin^2\phi + \sin\theta \cos^2\phi = \sin\theta$$

Thus,

$$\frac{\partial \mathbf{a}_R}{\partial \phi} = A_R\,\mathbf{a}_R + A_\theta\,\mathbf{a}_\theta + A_\phi\,\mathbf{a}_\phi$$

$$= \sin\theta\,\mathbf{a}_\phi \tag{1.88e}$$

Exercise 1.30

Transform $\mathbf{a}_R$ to the Cartesian system using Eqs. (1.84) and (1.87), and then evaluate it at point $(x, y, z) = (\sqrt{3}, 1, 2\sqrt{3})$.

Ans. $(\sqrt{3}/4)\,\mathbf{a}_x + (1/4)\,\mathbf{a}_y + (\sqrt{3}/2)\,\mathbf{a}_z$.

Review Questions

RQ 1.22	What is meant by φ in Eq. (1.75)?	[Fig. 1.34]
RQ 1.23	Does **T** in Eq. (1.75) also work on the position vector?	[(1.76)]
RQ 1.24	What is referred to by ϕ_1 in Eq. (1.79)?	[Fig. 1.23]
RQ 1.25	Relate Cartesian and cylindrical coordinates.	[(1.80)(1.81)]
RQ 1.26	What are referred to by θ_1 and ϕ_1 in Eq. (1.84)?	[Fig. 1.31]
RQ 1.27	Relate Cartesian and spherical coordinates.	[(1.86)(1.87)]

1.5 Problems

Rules of vector algebra in abstract form

1.1 Verify the associative law of vector addition expressed by Eq. (1.5a) for the three vectors, as shown in Fig. 1.35.

1.2 For a parallelogram formed by **A** and **B**,

  (a) express the two diagonals in terms of **A** and **B**, and
  (b) determine the condition under which the diagonals are mutually perpendicular.

Fig. 1.35 Three vectors
(Problem 1.1)

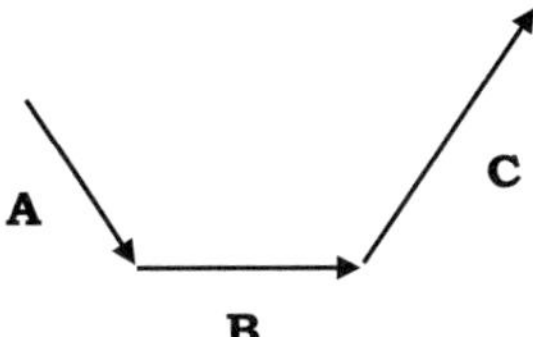

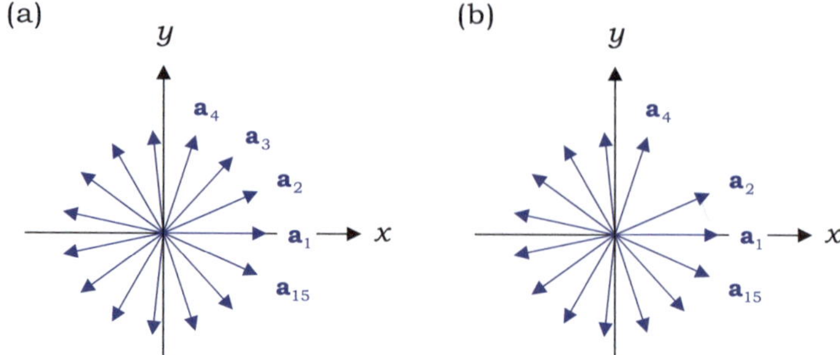

Fig. 1.36 **a** A total of 15 unit-vectors, **b** Vector $\mathbf{a}_3$ is missing (Problem 1.3)

Fig. 1.37 Distributive law
of dot product (Problem 1.4)

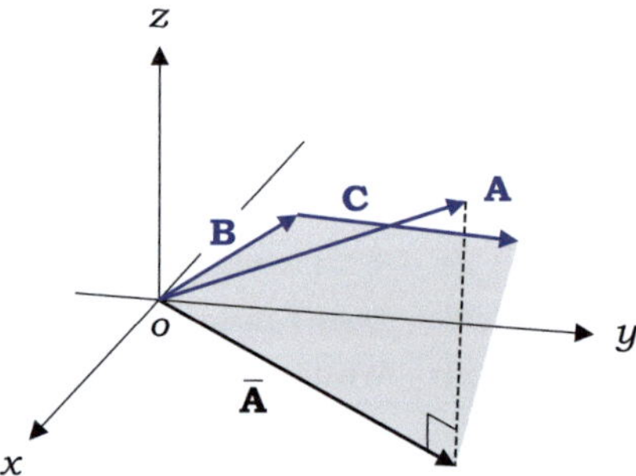

1.3   A total of 15 unit-vectors are spaced at 24° intervals so that no two vectors can
      cancel each other out (Fig. 1.36a). However, the $\mathbf{a}_3$ is not shown in Fig. 1.36b.
      Find the sum of the vectors for each case.
1.4   The distributive law of the dot product was proven for three coplanar vectors
      in Example 1.2 (see Fig. 1.9). Prove the law on three arbitrary vectors **A**, **B**,
      and **C**, as shown in Fig. 1.37. [Hint: Two-step projection.]
1.5   Verify the distributive law of the cross product graphically on the three coplanar
      vectors shown in Fig. 1.9.
1.6   A parallelogram formed by **A** and **B** is projected onto the $z = 0$ plane, as shown
      in Fig. 1.38. Explain that it is the same as that formed by the projections of **A**
      and **B** onto the $z = 0$ plane.
1.7   For a triangle formed by **A**, **B**, and $\mathbf{C} = \mathbf{A} - \mathbf{B}$, (a) show that the area of the
      parallelogram formed by any two vectors is the same, and (b) verify the law
      of sines when sides A, B, and C subtend angles θ_A, θ_B, and θ_C, respectively:

$$\frac{A}{\sin \theta_A} = \frac{B}{\sin \theta_B} = \frac{C}{\sin \theta_C}$$

1.8   Neither $\mathbf{A} \cdot \mathbf{B} = \mathbf{A} \cdot \mathbf{C}$ nor $\mathbf{A} \times \mathbf{B} = \mathbf{A} \times \mathbf{C}$ guarantee that $\mathbf{B} = \mathbf{C}$. Show that
      a combination of these two equations ensures that $\mathbf{B} = \mathbf{C}$.

Fig. 1.38 Parallelogram and its projection onto the $z = 0$ plane (Problems 1.6, 1.10, and 1.34)

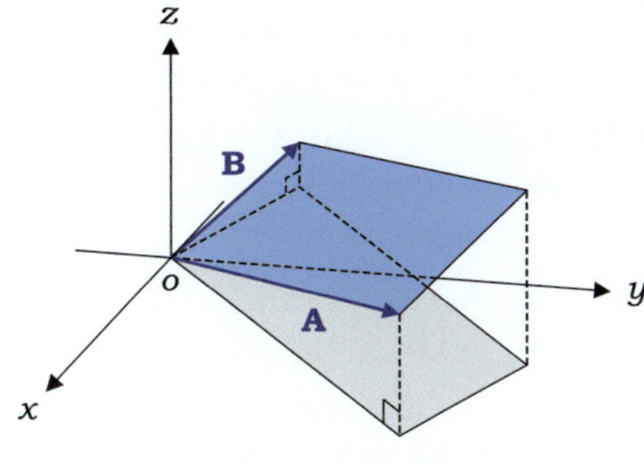

Fig. 1.39 A parallelepiped (Problem 1.9)

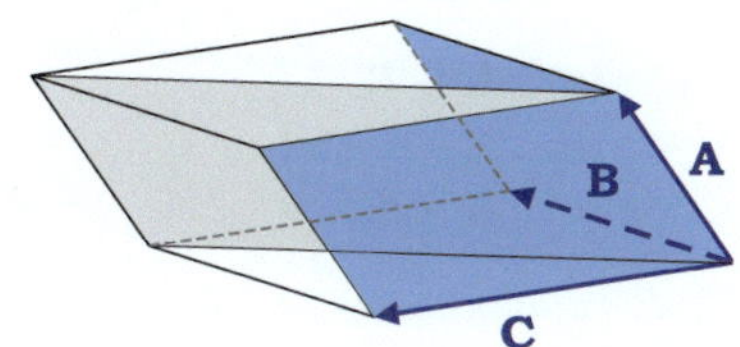

1.9   For a parallelepiped formed by **A**, **B**, and **C**, as shown in Fig. 1.39, show that the sum of the areas of the two adjoining side parallelograms is equal to that of the diagonal parallelogram.

1.10 Show that the parallelogram formed in the $z = 0$ plane, as shown in Fig. 1.38, has an area equal to the projection of $\mathbf{A} \times \mathbf{B}$ onto the z-axis. [Hint: Scalar triple product.]

Rules of vector algebra

1.11 Given $\mathbf{A} = 3\mathbf{a}_x + \mathbf{a}_y + 2\mathbf{a}_z$ and $\mathbf{B} = 2\mathbf{a}_x - 4\mathbf{a}_y$, compute $(\mathbf{A} \times \mathbf{B})(\mathbf{A} \cdot \mathbf{B})$.

1.12 Given $\mathbf{A} = 2\mathbf{a}_x + \mathbf{a}_y + 3\mathbf{a}_z$ and $\mathbf{B} = -\mathbf{a}_x + 2\mathbf{a}_y + \mathbf{a}_z$, determine the following:

   (a) $\mathbf{a}_A$,
   (b) $\mathbf{B} - \mathbf{A}$,
   (c) $\mathbf{A} \cdot \mathbf{B}$,
   (d) $\mathbf{A} \times \mathbf{B}$,
   (e) θ_{AB},
   (f) the vector component of **B** parallel to **A**, and
   (g) the vector component of **B** perpendicular to **A**.

1.13 Two vectors, $\mathbf{A} = 2\mathbf{a}_R + \mathbf{a}_\theta + \mathbf{a}_\phi$ and $\mathbf{B} = 3\mathbf{a}_R + 2\mathbf{a}_\theta - 4\mathbf{a}_\phi$, are given at point $p{:}(5, 30°, 45°)$ in spherical coordinates. Determine

   (a) $\mathbf{a}_A$,
   (b) $\mathbf{B} - \mathbf{A}$,
   (c) $\mathbf{A} \cdot \mathbf{B}$,
   (d) $\mathbf{A} \times \mathbf{B}$,
   (e) θ_{AB},

   (f)   the vector component of $\mathbf{B}$ parallel to $\mathbf{A}$, and

   (g)   the vector component of $\mathbf{B}$ perpendicular to $\mathbf{A}$.

1.14   Two vector fields, $\mathbf{A} = 3\mathbf{a}_\rho + (4\sin\phi)\,\mathbf{a}_\phi$ and $\mathbf{B} = \rho^2\mathbf{a}_\rho + (\rho + z + 2)\,\mathbf{a}_z$, coexist in three-dimensional space. Calculate the following at point $p_1{:}(2, 30°, 1)$ in cylindrical coordinates:

   (a) $\mathbf{A} + \mathbf{B}$,
   (b) $\mathbf{A} \cdot \mathbf{B}$, and
   (c) $\mathbf{A} \times \mathbf{B}$.

1.15   Given two vector fields $\mathbf{A} = 36R^{-2}\cos^2\phi\,\mathbf{a}_R$ and $\mathbf{B} = 27R^{-2}\mathbf{a}_R + 8\sin\theta\,\mathbf{a}_\phi$, compute the following at point $p{:}(3, \pi/4, \pi/6)$ in spherical coordinates:

   (a) $\mathbf{A} + \mathbf{B}$,
   (b) $\mathbf{A} \cdot \mathbf{B}$, and
   (c) $\mathbf{A} \times \mathbf{B}$.

Vector components

1.16   For a vector from the origin to a point with $(x, y, z) = (4, 3, 12)$, write its component form in (a) Cartesian, (b) cylindrical, and (c) spherical coordinates.

1.17   Given vector $\mathbf{A} = 5\mathbf{a}_x + 5\mathbf{a}_y + 5\mathbf{a}_z$, determine the vector components (a) parallel, and (b) perpendicular to a line defined by $x + 2y = 3$ and $z = 4$.

1.18   Given $\mathbf{A} = -2\mathbf{a}_x + 3\mathbf{a}_y + \sqrt{3}\,\mathbf{a}_z$, find a unit vector (a) along $\mathbf{A}$, and (b) perpendicular to both $\mathbf{A}$ and $\mathbf{a}_x$.

1.19   Vector $\mathbf{A} = \mathbf{a}_x + 2\mathbf{a}_y + 2\mathbf{a}_z$ is defined at $\mathbf{r} = 2\mathbf{a}_x + 3\mathbf{a}_y + \sqrt{3}\,\mathbf{a}_z$. What are the angles that (a) $\mathbf{A}$, and (b) $\mathbf{r}$ make with the three Cartesian axes?

1.20   A vector field is defined as $\mathbf{A} = (\rho/2)\,\mathbf{a}_\rho + (2\sin\phi)\,\mathbf{a}_\phi + z\,\mathbf{a}_z$ in cylindrical coordinates. Find a unit vector perpendicular to $\mathbf{A}$ and parallel to the xy-plane at a point with $\rho = 4$, $\phi = 30°$, and $z = 6$.

1.21   In the presence of a vector field $\mathbf{H}(\mathbf{r}) = (16/\rho^2)\,\mathbf{a}_\rho$, $\mathbf{A}$ is a constituent vector given at point $p_1{:}(2, 30°, 5)$ in cylindrical coordinates. Determine the vector component of $\mathbf{A}$ directed towards point $p_2{:}(2, 90°, 5)$.

1.22   Given vector field $\mathbf{F}(\mathbf{r}) = R\sin\phi\,\mathbf{a}_R + 3\cos\theta\,\mathbf{a}_\theta - 2R^2\mathbf{a}_\phi$ in spherical coordinates, at point $p_1{:}(1, 45°, 30°)$ on a spherical surface of unity radius, centered at the origin, determine the vector components (a) normal, and (b) tangential to the surface.

1.23   Repeat Problem 1.22 for the same point, but positioned on a conical surface with $\theta = 45°$.

Position and distance vectors

1.24   A straight line runs parallel to the z-axis and passes through point $p{:}(3, 4, 2)$ in Cartesian coordinates. Express the position vector of a point on the line.

1.25   Find an expression for the distance vector from an arbitrary point on the $z = 1$ plane to point $p{:}(0, 0, 5)$.

1.26   A straight line passes through points $p_1{:}(1, 1, 0)$ and $p_2{:}(3, 2, 2)$. Find

Fig. 1.40 Distance vector in cylindrical system (Problem 1.28)

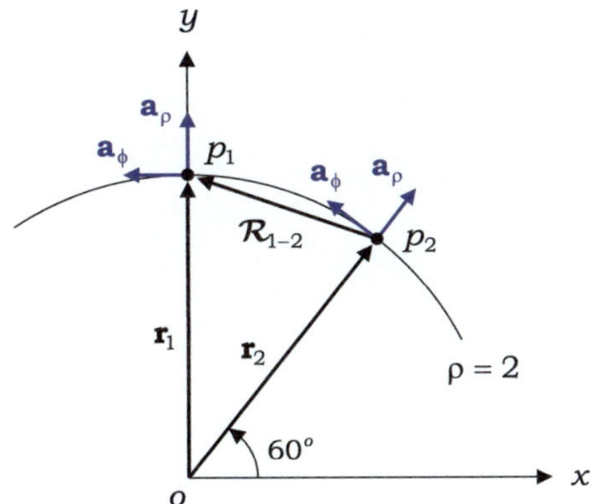

(a) the unit vector directed from p_1 to p_2, and

(b) the point on the line, being at a distance d from p_1.

1.27 Two straight lines intersect at right angles at point p_o. If the first passes through p_1:(1, -1, 3) and p_2:(-1, 1, 2), whereas the second passes through p_3:(2, 1, 1), find the distances of p_o (a) from p_1, and (b) from p_3.

1.28 Two points, p_1:(2, 90°, 0) and p_2:(2, 60°, 0), are positioned on a circle of radius $\rho = 2$ in cylindrical coordinates, as shown in Fig. 1.40. Express the distance vector $\mathcal{R}_{1-2}$ in terms of the base vectors at p_1.

Dot and cross products

1.29 Verify $\cos(\alpha - \beta) = \cos\alpha\cos\beta + \sin\alpha\sin\beta$ using two vectors residing in the first quadrant of the xy-plane.

1.30 For $\mathbf{E} = p\,(2\cos\theta\,\mathbf{a}_R + \sin\theta\,\mathbf{a}_\theta)$ with a constant p, show that this expression can be rewritten as $\mathbf{E} = 3(\mathbf{p}\cdot\mathbf{a}_R)\,\mathbf{a}_R - \mathbf{p}$ using the definition of $\mathbf{p} \equiv p\,\mathbf{a}_z$.

1.31 The sides of a triangle correspond to the three vectors $\mathbf{A} = 2\mathbf{a}_x + \mathbf{a}_y - 3\mathbf{a}_z$, $\mathbf{B} = -\mathbf{a}_x + 2\mathbf{a}_y + 5\mathbf{a}_z$, and $\mathbf{C} = \mathbf{a}_x + 3\mathbf{a}_y + 2\mathbf{a}_z$. Determine the interior angles.

1.32 If the triangle is defined by the three points p_1:(2, 0, 1), p_2:(-1, 1, 3), and p_3:(4, 2, -1) in Cartesian coordinates, determine its area.

1.33 A rectangular parallelepiped of height c has a square base on side a, as shown in Fig. 1.41. If the angle between the two face diagonals is $\gamma = 45°$, what is the ratio of c to a?

1.34 Repeat Problem 1.10 using the component form of $\mathbf{A}$ and $\mathbf{B}$.

1.35 For the four points p_1:(2, 1, 3), p_2:(4, 2, 1), p_3:(3, 4, 3), and p_4:(1, 3, 2) in Cartesian coordinates, as shown in Fig. 1.42, determine the angle made by the two triangles $\triangle p_1 p_2 p_4$ and $\triangle p_2 p_3 p_4$.

1.36 A laser beam propagated along $\mathbf{k}_i = -3\mathbf{a}_x + 4\mathbf{a}_y - 5\mathbf{a}_z$ and impinged on the $z = 0$ plane. After the reflection, the beam follows $\mathbf{k}_r$, which is in the plane formed by $\mathbf{k}_i$ and the normal to the $z = 0$ plane, as shown in Fig. 1.43. Using $\theta_i = \theta_r$ and $|\mathbf{k}_i| = |\mathbf{k}_r|$, express $\mathbf{k}_r$.

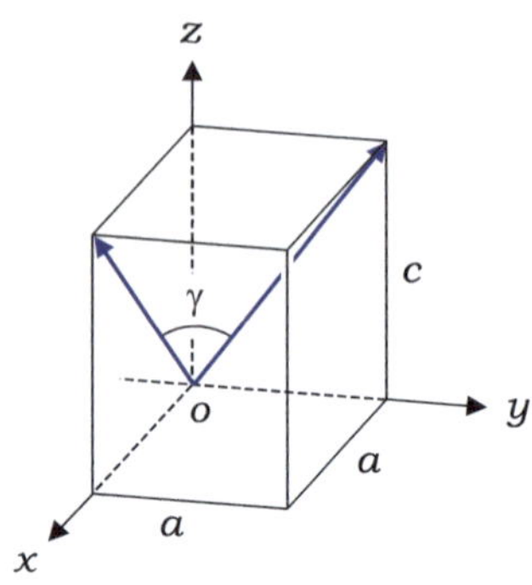

Fig. 1.41 Rectangular parallelepiped with square base (Problem 1.33)

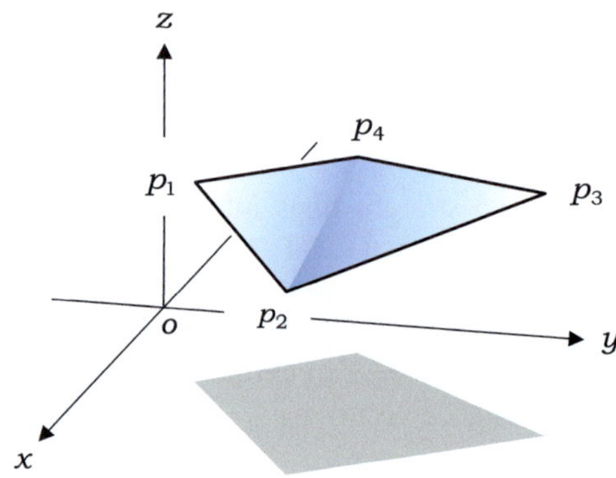

Fig. 1.42 Four points in three-dimensional space (Problem 1.35)

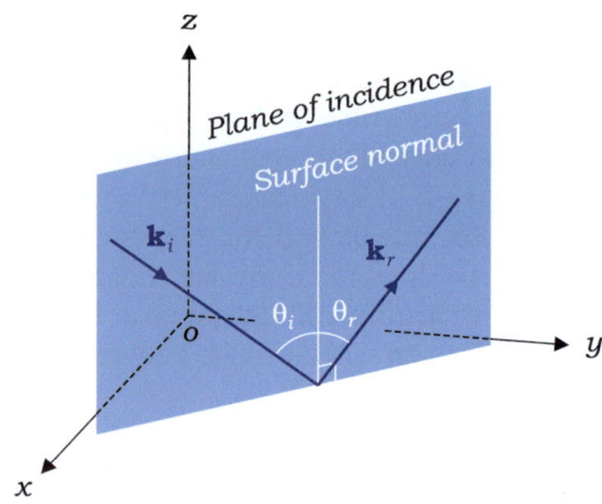

FIG. 1.43 Reflection of the laser beam (Problem 1.36)

Lines and planes

1.37　Find a unit vector along a straight line, defined by $3x - 4y = 12$.

1.38　A straight line may be defined either by $ax + by = c$ or $\mathbf{r} \cdot \mathbf{a}_n = p$ in the $z = 0$ plane, where $\mathbf{r} = x\,\mathbf{a}_x + y\,\mathbf{a}_y$ is the position vector of a point on the

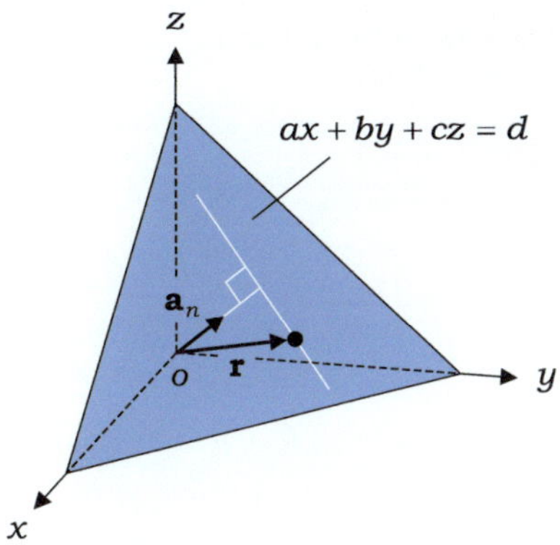

Fig. 1.44 A plane in three-dimensional space (Problem 1.40)

line, and $\mathbf{a}_n$ is a unit vector. (a) Express p in terms of a, b, and c. (b) What are the geometrical significances of $\mathbf{a}_n$ and p?

1.39   Two straight lines, $3x + 2y = 1$ and $-x + y = 4$, intersect in the $z = 0$ plane. Determine the smallest angle between them.

1.40   A plane may be expressed either by $ax + by + cz = d$ or $\mathbf{r} \cdot \mathbf{a}_n = p$, where $\mathbf{r} = x\,\mathbf{a}_x + y\,\mathbf{a}_y + z\,\mathbf{a}_z$ is the position vector of a point on the plane (see Fig. 1.44). (a) Express p in terms of a, b, c and d. (b) What are the geometrical significances of $\mathbf{a}_n$ and p?

1.41   Given a plane defined by $x + 2y + 2z = 8$ and line $3x + 2y = 5$, determine the angle that the line makes with a normal to the plane.

1.42   A plane is defined by three points $p_1 : (2, 0, 0)$, $p_2 : (0, 1, 0)$, and $p_3 : (0, 0, 1)$ in Cartesian coordinates. (a) Express it in the form $ax + by + cz = 1$. (b) Determine the perpendicular distance from the origin to the plane.

1.43   When a plane is normal to vector $\mathbf{k} = \mathbf{a}_x + 2\mathbf{a}_y + 4\mathbf{a}_z$ passing through point $p_1 : (3, 1, 2)$ in Cartesian coordinates, express the plane.

Differential area vector

1.44   A half-cylinder is defined by the $\rho = 4$, $\phi = 30°$, $\phi = 210°$, $z = 0$, and $z = 5$ surfaces, as shown in Fig. 1.45. Find the differential area vectors at points (a) p_1, (b) p_2 with $\rho = 2$, (c) p_3, and (d) p_4.

1.45   The object shown in Fig. 1.46 is bounded by the surfaces of constant coordinate: the $R = 6$, $\theta = 30°$, $\phi = 45°$, and $\phi = 225°$ surfaces. Find $d\mathbf{s}$ at points (a) p_1 with $\theta = 20°$, (b) p_2 with $R = 4$, (c) p_3 with $R = 4$, and (d) p_4 with $R = 5$.

1.46   Find an expression for $d\mathbf{s}$ for each of the following surfaces:

   (a) $x = 0$, $3 \le y \le 5$, and $0 \le z \le 2$,
   (b) $0 \le x \le 1$, $0 \le y \le (1 - x)$, and $z = 2$,
   (c) $1 \le \rho \le 2$, $0 \le \phi \le 30°$, and $z = 1$,
   (d) $\rho = 2$, $30° \le \phi \le 60°$, and $1 \le z \le 2$,
   (e) $2 \le \rho \le 4$, $\phi = 30°$, and $1 \le z \le 2$,
   (f) $R = 3$, $20° \le \theta \le 50°$, and $0 \le \phi \le 30°$,

Fig. 1.45 A half-cylinder
(Problem 1.44)

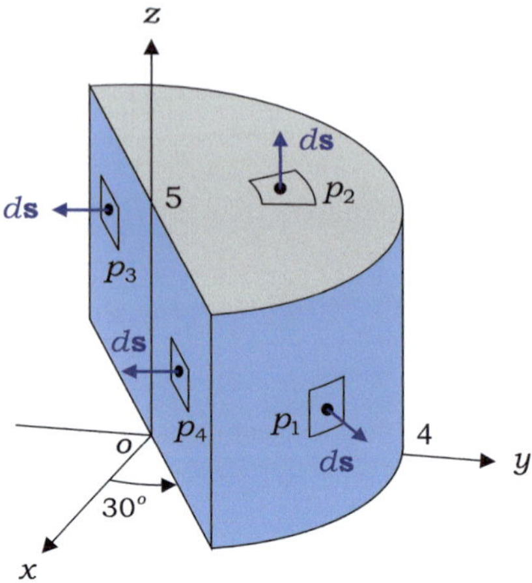

Fig. 1.46 A half-cone
(Problem 1.45)

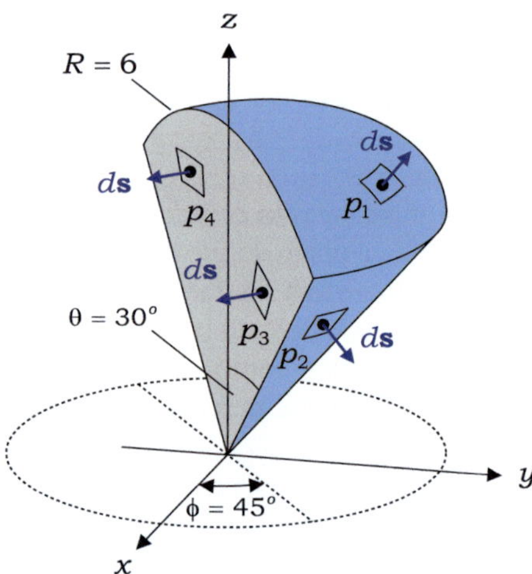

    (g) $2 \leq R \leq 5$, $30° \leq \theta \leq 60°$, and $\phi = 45°$, and

    (h) $1 \leq R \leq 2$, $\theta = 30°$, and $10° \leq \phi \leq 20°$.

Coordinate transformation

1.47 A line source with strength I lying along the z-axis produces $\mathbf{H} = (I/\rho)\,\mathbf{a}_\phi$ in cylindrical coordinates. If the source is linearly displaced to pass through point $p{:}(x_1, y_1, 0)$, express $\mathbf{H}$ in Cartesian coordinates.

1.48 A point source of strength q placed at the origin produces $\mathbf{E} = (q/R^2)\,\mathbf{a}_R$ in spherical coordinates. If the source is displaced to point $p{:}(x_1, y_1, z_1)$, express $\mathbf{E}$ in Cartesian coordinates.

1.49 Transform vector field $\mathbf{A} = 4(x + y)\,\mathbf{a}_x$ to a primed Cartesian system, which is rotated about the z-axis by $\varphi = 30°$ relative to the unprimed system.

1.50 Given point p:$(-2, 4, 3)$ in Cartesian coordinates, find its (a) cylindrical coordinates, and (b) spherical coordinates.

1.51 Transform $\mathbf{a}_x$ to the spherical system using $\mathbf{T}$ matrix, if it is given at point

  (a) $(x, y, z) = (1, -2, 3)$,
  (b) $(\rho, \phi, z) = (2, 30°, \sqrt{2})$, and
  (c) $(R, \theta, \phi) = (2, 45°, 60°)$.

1.52 For a vector $\mathbf{A} = \mathbf{a}_x + \mathbf{a}_y + 0.5\,\mathbf{a}_z$ given at point $(2,2,1)$, transform it into (a) cylindrical, and (b) spherical systems using the $\mathbf{T}$ matrix.

1.53 Transform vector field $\mathbf{H} = 2\mathbf{a}_\rho/\rho$ into the Cartesian system using a transformation matrix.

1.54 Transform $\mathbf{H} = \mathbf{a}_\theta/R^2$ into the Cartesian system using the $\mathbf{T}$ matrix.

1.55 Given two points p_1:$(1, 1, 3)$ and p_2:$(-1, 2, 1)$ in Cartesian coordinates, transform a unit vector directed from p_1 to p_2 into the cylindrical system.

1.56 Transform the differential length vector $d\mathbf{l} = dx\,\mathbf{a}_x + dy\,\mathbf{a}_y + dz\,\mathbf{a}_z$ into cylindrical coordinates using the transformation matrix.

1.57 Show that the position vector $\mathbf{r} = x\,\mathbf{a}_x + y\,\mathbf{a}_y + z\,\mathbf{a}_z$ may be transformed such that $\mathbf{r} = \rho\,\mathbf{a}_\rho + z\,\mathbf{a}_z$ using the transformation matrix. Explain why the use of $\mathbf{T}$ may not be of great interest.

1.58 As can be seen from Fig. 1.31, $\mathbf{a}_R$, $\mathbf{a}_\theta$, and $\mathbf{a}_\phi$ can be obtained by rotating $\mathbf{a}_x$, $\mathbf{a}_y$, and $\mathbf{a}_z$ about the z-axis by an angle ϕ, followed by the rotation of the primed system about the y'-axis by an angle θ. Show that the $\mathbf{T}$ matrix given in Eq. (1.85) results from this procedure.

1.59 Using the coordinate transformation of a vector from the unprimed to primed system, $\overline{\mathbf{A}}' = \mathbf{T}\overline{\mathbf{A}}$, verify that the dot product of two vectors $\mathbf{A}$ and $\mathbf{B}$ is independent of the coordinate system such that $\mathbf{A} \cdot \mathbf{B} = \mathbf{A}' \cdot \mathbf{B}'$. [Hint: $\mathbf{A} \cdot \mathbf{B} = \overline{\mathbf{A}}^T\overline{\mathbf{B}}$, where T denotes the transpose, and $\overline{\mathbf{A}}$ and $\overline{\mathbf{B}}$ are column matrices for the vectors.]

Chapter 2
Vector Calculus

Electromagnetics mostly deals with continuous distributions of vector quantities, as well as discrete vectors scattered over a region of space. Such a continuous distribution of vector quantities is called a vector field, and can be described well by a vector function. In most cases, the constituent of a vector field is related to a physical quantity directly observed at a point in space, or one observed in the immediate surroundings at the point. Because physical quantities should exhibit no abrupt change in magnitude or direction as a function of position, except at the boundary, vector functions in electromagnetics are always smooth and thus differentiable in a given region of space. Furthermore, the space rate of change of a vector quantity may have particular physical significance. In fact, several differential operators have been designed to relate the partial derivatives of a vector quantity at a point to the source present at that point.

In this chapter, we first discuss the integration of a vector field along a line or across a surface in three-dimensional space. Next, we define differential operators such as gradient, divergence, and curl, which are essential tools for describing electromagnetic phenomena. Although divergence and curl are based on a closed surface and closed-line integral of the vector field around a point in space, respectively, they are formulated in terms of the partial derivatives of the vector field at that point. In general, physical laws expressed in terms of derivatives, called the ***differential form*** or ***point form***, are particularly useful for describing local effects, for which the effect observed at a point is directly related to the cause found at the point. In contrast, it is more convenient to work with physical laws expressed in ***integral form*** to explain nonlocal effects, for which a localized source is connected to the sum of the effects observed in the region around the point. The definitions of divergence and curl directly lead to the divergence and Stokes's theorems, which are very useful for converting electromagnetic laws in differential form to integral form, and vice versa. Finally, we discuss Helmholtz's theorem, which ensures that the first estimate for the solution is only one if certain conditions are satisfied.

© The Author(s), under exclusive license to Springer Nature Switzerland AG 2024   67
Y. H. Lee, *Introduction to Engineering Electromagnetics*,
https://doi.org/10.1007/978-3-031-28659-9_2

2.1 Line and Surface Integrals

In the presence of the vector field $\mathbf{A}(\mathbf{r})$, the *line integral* of $\mathbf{A}$ along a path is defined as the sum of the products of the *tangential component* of $\mathbf{A}$ and the differential path as the differential path tends to zero. Line integral is the standard name. If the initial and terminal points of the *path of integration* coincide, forming a closed curve C, the integration is called the *closed-line integral* of $\mathbf{A}$ around C, or the *circulation* of $\mathbf{A}$ around C. The *surface integral* of $\mathbf{A}$ across a surface is defined as the sum of the products of the *normal component* of $\mathbf{A}$ and the differential surface area as the differential area tends to zero. If the surface has no opening, forming a closed surface, the integration is called the *closed-surface integral* of $\mathbf{A}$.

2.1.1 Curves

A curve may represent the path of an electric charge that moves spatially. In Cartesian coordinates, curve C can be defined using the position vector of point along the curve and parameter t as follows:

$$\mathbf{r}(t) = x(t)\,\mathbf{a}_x + y(t)\,\mathbf{a}_y + z(t)\,\mathbf{a}_z \tag{2.1}$$

where t varies from initial point $t = t_i$ to terminal point $t = t_t$. This is known as a *parametric representation* of a curve. The sense of increase in t determines the *positive direction* of C or the *direction of travel* on C.

A curve is considered *smooth* if the tangent lines show no abrupt change in direction along the curve. Therefore, $\mathbf{r}(t)$ is differentiable. Subsequently, the rate of change of $\mathbf{r}(t)$ with t at $t = t_1$ is defined as

$$\mathbf{r}'(t)\big|_{t=t_1} = \frac{d\mathbf{r}}{dt}\bigg|_{t=t_1} = \lim_{\Delta t \to 0} \frac{\mathbf{r}(t_1 + \Delta t) - \mathbf{r}(t_1)}{\Delta t} \tag{2.2}$$

The denominator on the right side of Eq. (2.2) is simply an increment of t, whereas the numerator is a vector subtraction of two position vectors: one for a point with $t = t_1$, and another for an advanced point with $t = t_1 + \Delta t$. Note that the numerator is equal to the differential length vector $d\mathbf{l}$ along C.

As shown in Fig. 2.1, the position vectors $\mathbf{r}(t_1)$ and $\mathbf{r}(t_1 + \Delta t)$ define the points p_1 and p_2 on C, respectively. Meanwhile, straight line L passes through two points. It is evident that L becomes tangent to C at point p_1 as $\Delta t \to 0$. Consequently, $\mathbf{r}'(t_1)$ *corresponds to the tangent vector at point p_1 on C.* Because $d\mathbf{l}$ along C is exactly equal to $d\mathbf{r}$, we have

$$d\mathbf{l} = d\mathbf{r} = \mathbf{r}'(t)\,dt \tag{2.3}$$

Fig. 2.1 Curve in
three-dimensional space

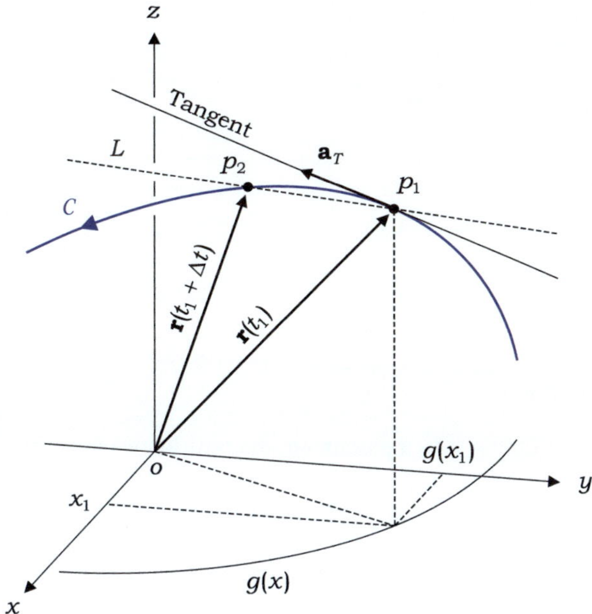

where $\mathbf{r}'(t) = d\mathbf{r}/dt$. It follows that $\mathcal{C}$ is smooth in space if $\mathbf{r}'(t)$ is a continuous function of t.

Therefore, a unit vector directed along the tangent to a curve can be expressed as

$$\mathbf{a}_T = \frac{d\mathbf{l}}{|d\mathbf{l}|} = \frac{\mathbf{r}'}{|\mathbf{r}'|} \tag{2.4}$$

The unit vector indicates the positive direction of the curve.

Alternatively, we may express a curve with the position vector by taking the x-coordinate as a parameter, that is,

$$\mathbf{r} = x\,\mathbf{a}_x + g(x)\,\mathbf{a}_y + h(x)\,\mathbf{a}_z \tag{2.5}$$

where $g(x)$ and $h(x)$ correspond to the projections of $\mathcal{C}$ onto the xy-plane and the xz-plane, respectively.

Example 2.1 When a circle is defined as $x^2+y^2 = 16$ (Fig. 2.2), write the parametric representation of the circle using (a) φ and (b) x as parameters. Then, find a unit vector tangent to the circle at p_1:$(2,\ 2\sqrt{3},\ 0)$ using the results of (a) and (b).

Fig. 2.2 Circular path with
a radius of 4

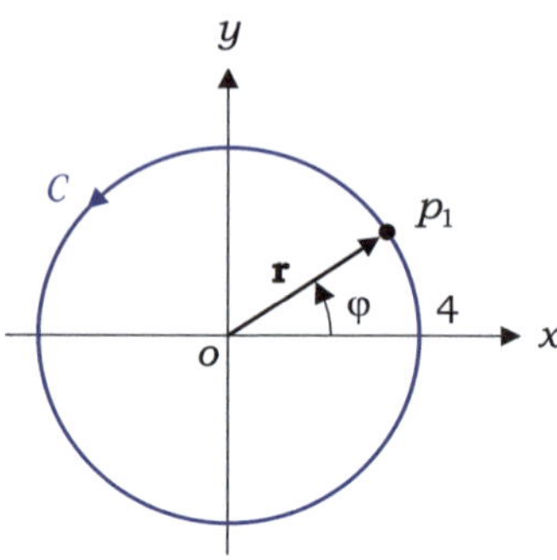

Solution

(a) Taking φ as a parameter, the position vector is expressed as

$$\mathbf{r}(\varphi) = x\,\mathbf{a}_x + y\,\mathbf{a}_y = 4\cos\varphi\,\mathbf{a}_x + 4\sin\varphi\,\mathbf{a}_y \qquad (2.6a)$$

(b) Taking x as a parameter, the position vector is expressed as

$$\mathbf{r}(x) = x\,\mathbf{a}_x + y\,\mathbf{a}_y = x\,\mathbf{a}_x \pm \sqrt{16 - x^2}\,\mathbf{a}_y \quad (+,\ y > 0;\ -,\ y < 0) \qquad (2.6b)$$

From Eq. (2.6a), we obtain

$$\mathbf{a}_T = \frac{\mathbf{r}'(\varphi)}{|\mathbf{r}'(\varphi)|} = \frac{-4\sin\varphi\,\mathbf{a}_x + 4\cos\varphi\,\mathbf{a}_y}{\sqrt{(4\sin\varphi)^2 + (4\cos\varphi)^2}} \qquad (2.6c)$$

At point p_1, letting $\varphi = 60^o$ in Eq. (2.6c), we obtain a unit tangent vector as

$$\mathbf{a}_T = -(\sqrt{3}/2)\,\mathbf{a}_x + (1/2)\,\mathbf{a}_y \qquad (2.6d)$$

Taking the derivative of Eq. (2.6b) with respect to x yields

$$\mathbf{r}'(x) = \mathbf{a}_x \mp \frac{x}{\sqrt{16 - x^2}}\,\mathbf{a}_y \qquad (2.6e)$$

At point p_1 with $y > 0$, letting $x = 2$ in Eq. (2.6e), we obtain

$$\mathbf{a}_T = \frac{\mathbf{r}'}{|\mathbf{r}'|} = (\sqrt{3}/2)\,\mathbf{a}_x - (1/2)\,\mathbf{a}_y \qquad (2.6f)$$

As φ increases in Eq. (2.6a), p_1 moves counterclockwise along the circle. In contrast, as x increases in Eq. (2.6b), p_1 moves clockwise. Accordingly, the unit vectors in Eqs. (2.6d) and (2.6f) exhibit the opposite signs.

Exercise 2.1

Find a parametric representation of the straight line parallel to the z-axis passing through point $(x, y, z) = (2, 1, 3)$.

Ans. $\mathbf{r}(t) = 2\mathbf{a}_x + \mathbf{a}_y + t\mathbf{a}_z \ (-\infty < t < \infty)$.

Exercise 2.2

Evaluate $|d\mathbf{r}/dx|$ if $\mathbf{r}(x)$ represents a straight line defined by $y = x$.

Ans. $\sqrt{2}$.

2.1.2 Line Integral

The line integral is an extension of the definite integral in the following three dimensions. The line integral of vector field $\mathbf{E}(\mathbf{r})$ along curve $\mathcal{C}$ is defined as

$$\boxed{\int_{\mathcal{C}} \mathbf{E} \cdot d\mathbf{l} = \int_{\mathcal{C}} E_t(x, y, z) \, dl}\tag{2.7}$$

where $d\mathbf{l}$ is the differential length vector along $\mathcal{C}$, and E_t is the component of $\mathbf{E}$ tangential to $\mathcal{C}$.

The line integral in Eq. (2.7) can be converted to a definite integral using $d\mathbf{l}$ expressed in parametric form. Referring to Fig. 2.3, a given curve is first divided into N line segments, represented by incremental length vectors $\Delta\mathbf{l}_i$. Then the line integral of $\mathbf{E}$ along $\mathcal{C}$ is expressed as

$$\int_{\mathcal{C}} \mathbf{E} \cdot d\mathbf{l} = \lim_{\substack{N \to \infty \\ |\Delta\mathbf{l}| \to 0}} \sum_{i=1}^{i=N} \mathbf{E}(\mathbf{r}_i) \cdot \Delta\mathbf{l}_i \tag{2.8}$$

In the limit as $N \to \infty$, $\Delta\mathbf{l}$ becomes a differential length vector such that

$$d\mathbf{l} = d\mathbf{r} = \mathbf{r}' \, dt \tag{2.9}$$

Upon inserting Eq. (2.9) into Eq. (2.8), the line integral can be expressed as

$$\boxed{\int_{\mathcal{C}} \mathbf{E} \cdot d\mathbf{l} = \int_{A}^{B} \mathbf{E} \cdot \mathbf{r}' \, dt}\tag{2.10}$$

where A and B are the initial and terminal points of $\mathcal{C}$, respectively, and t is the parameter. Notably, the right-hand side of Eq. (2.10) is a definite integral.

Fig. 2.3 Path of integration
in the vector field $\mathbf{E}(\mathbf{r})$

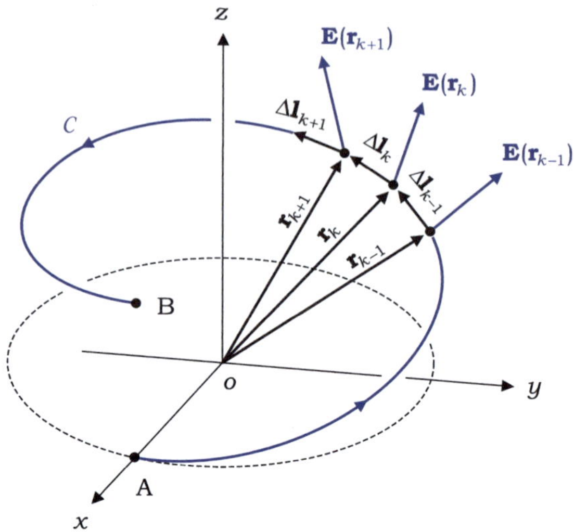

In the three coordinate systems, the line integral of $\mathbf{E}$ along C is expressed as
follows:

In Cartesian coordinates,

$$
\begin{aligned}
\int_{C} \mathbf{E} \cdot d\mathbf{l} &= \int_{C} \left(E_x\,\mathbf{a}_x + E_y\,\mathbf{a}_y + E_z\,\mathbf{a}_z\right) \cdot \left(dx\,\mathbf{a}_x + dy\,\mathbf{a}_y + dz\,\mathbf{a}_z\right) \\
&= \int_{x_1}^{x_2} E_x dx + \int_{y_1}^{y_2} E_y dy + \int_{z_1}^{z_2} E_z dz
\end{aligned}
$$

(2.11a)

In cylindrical coordinates,

$$
\begin{aligned}
\int_{C} \mathbf{E} \cdot d\mathbf{l} &= \int_{C} \left(E_\rho\,\mathbf{a}_\rho + E_\phi\,\mathbf{a}_\phi + E_z\,\mathbf{a}_z\right) \cdot \left(d\rho\,\mathbf{a}_\rho + \rho d\phi\,\mathbf{a}_\phi + dz\,\mathbf{a}_z\right) \\
&= \int_{\rho_1}^{\rho_2} E_\rho d\rho + \int_{\phi_1}^{\phi_2} E_\phi \rho d\phi + \int_{z_1}^{z_2} E_z dz
\end{aligned}
$$

(2.11b)

In spherical coordinates,

$$
\begin{aligned}
\int_{C} \mathbf{E} \cdot d\mathbf{l} &= \int_{C} \left(E_R\,\mathbf{a}_R + E_\theta\,\mathbf{a}_\theta + E_\phi\,\mathbf{a}_\phi\right) \cdot \left(dR\,\mathbf{a}_R + Rd\theta\,\mathbf{a}_\theta + R\sin\theta\,d\phi\,\mathbf{a}_\phi\right) \\
&= \int_{R_1}^{R_2} E_R\,dR + \int_{\theta_1}^{\theta_2} E_\theta\,Rd\theta + \int_{\phi_1}^{\phi_2} E_\phi R\sin\theta\,d\phi
\end{aligned}
$$

(2.11c)

where subscripts 1 and 2 denote the initial and terminal points of $\mathcal{C}$, respectively.

For example, if the path of integration lies in the $z = z_o$ plane and is defined by $y = f(x)$ in Cartesian coordinates, then $d\mathbf{l}$ is reduced to $d\mathbf{l} = dx\,\mathbf{a}_x + dy\,\mathbf{a}_y$. In this case, the line integral of $\mathbf{E}$ along the path is expressed as

$$\int_{\mathcal{C}} \mathbf{E} \cdot d\mathbf{l} = \int_{\mathcal{C}} \left(E_x \mathbf{a}_x + E_y \mathbf{a}_y + E_z \mathbf{a}_z\right) \cdot \left(dx\,\mathbf{a}_x + dy\,\mathbf{a}_y\right)$$

$$= \int_{x_1}^{x_2} E_x(x, f(x), z_o)\,dx + \int_{y_1}^{y_2} E_y(f^{-1}(y), y, z_o)\,dy \qquad (2.12)$$

where it is assumed that the initial point is at (x_1, y_1, z_o), and the terminal point is at (x_2, y_2, z_o). Moreover, if the path of integration is a straight line parallel to the x-axis passing through point $(0, y_o, z_o)$, the line integral of $\mathbf{E}$ is further reduced to

$$\int_{\mathcal{C}} \mathbf{E} \cdot d\mathbf{l} = \int_{x_1}^{x_2} E_x(x, y_o, z_o)\,dx \qquad (2.13)$$

where the initial point is at (x_1, y_o, z_o), and the terminal point is at (x_2, y_o, z_o).

If the line integral in Eq. (2.13) is conducted in the $-x$-direction, it is necessary to set $d\mathbf{l} = dx\,\mathbf{a}_x$, not $d\mathbf{l} = -dx\,\mathbf{a}_x$, and switch the roles of x_1 and x_2 such that the integration is conducted from x_2 to x_1. Similarly, $d\mathbf{l}$ in Eqs. (2.11a)–(2.11c) should remain unchanged even if the direction of travel on $\mathcal{C}$ is reversed. The reverse direction should be addressed using integration limits. Note that the upper limit of integration in the line integral may be less than the lower limit, but the upper limit is always greater than the lower limit in the definite integral, which is graphically defined as the net area under the curve of a function.

When the line integral of $\mathbf{E}$ is applied around a closed path, a small circle is set on the integral sign to denote it.

$$\boxed{\oint_{\mathcal{C}} \mathbf{E} \cdot d\mathbf{l} = \oint_{\mathcal{C}} E_t\,dl} \qquad (2.14)$$

The closed-line integral is also referred to as the circulation of $\mathbf{E}$ around $\mathcal{C}$. The direction of travel on $\mathcal{C}$ is called the *positive direction* of $\mathcal{C}$.

In most cases, a closed-line integral is applied along the closed contour of a surface that is not necessarily flat. Ordinarily, the direction of $d\mathbf{l}$ along a contour may be chosen for the convenience of calculation. If the direction of $d\mathbf{s}$ on a given surface is set, the direction of $d\mathbf{l}$ along the contour is determined by the **right-hand rule**, and vice versa. According to the right-hand rule, the right thumb points in the direction of $d\mathbf{s}$ when the four fingers advance in the direction of $d\mathbf{l}$ (Fig. 2.4). If the closed-line integral is to be conducted in the reverse direction, the roles of the upper and lower limits of integration are interchanged with the expression of $d\mathbf{l}$ unchanged.

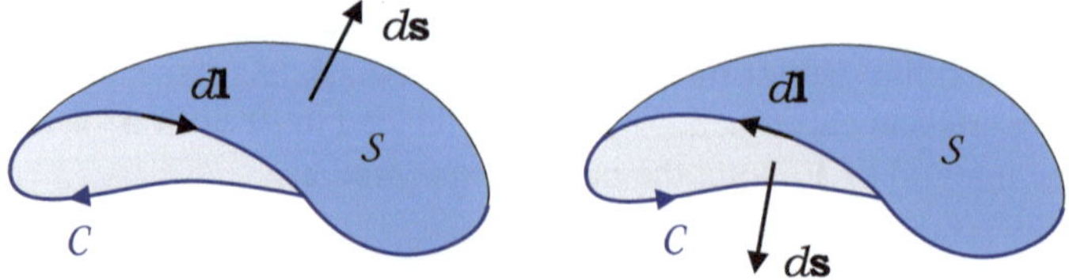

Fig. 2.4 Right-hand rule for the directions of $d\mathbf{l}$ and $d\mathbf{s}$

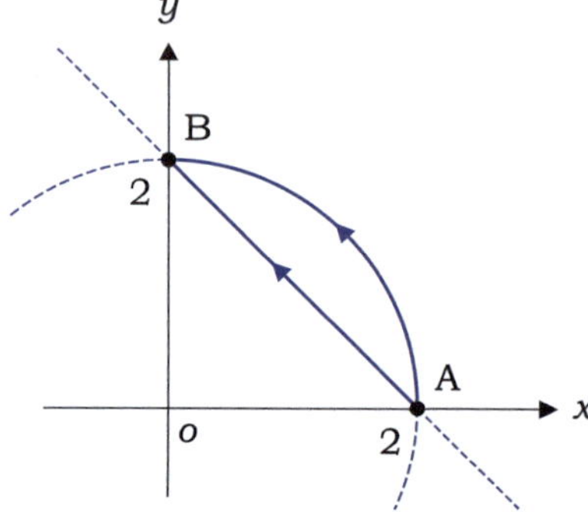

Fig. 2.5 Two paths of integration taken between points A and B

Example 2.2 Given the vector field $\mathbf{E} = y^2\mathbf{a}_x + (2xy + 4y)\,\mathbf{a}_y$, compute the line integral of $\mathbf{E}$ from point A:(2, 0, 0) to point B:(0, 2, 0) along (a) straight line $y = -x + 2$, and (b) an arc of circle $x^2 + y^2 = 4$, as shown in Fig. 2.5.

Solution

The line integral is expressed as

$$\int_C \mathbf{E} \cdot d\mathbf{l} = \int_A^B \left[y^2\mathbf{a}_x + (2xy + 4y)\,\mathbf{a}_y \right] \cdot \left(dx\,\mathbf{a}_x + dy\,\mathbf{a}_y + dz\,\mathbf{a}_z \right)$$

$$= \int_2^0 y^2\,dx + \int_0^2 (2xy + 4y)\,dy \tag{2.15}$$

(a) Using $y = -x + 2$ in Eq. (2.15), we obtain

$$\int_C \mathbf{E} \cdot d\mathbf{l} = \int_2^0 (-x + 2)^2\,dx + \int_0^2 [2y(2 - y) + 4y]\,dy = 8$$

(b) Using $x^2 + y^2 = 4$ in Eq. (2.15), we obtain

$$\int_C \mathbf{E} \cdot d\mathbf{l} = \int_2^0 (4 - x^2)\,dx + \int_0^2 \left[+2y\sqrt{4 - y^2} + 4y \right] dy = 8$$

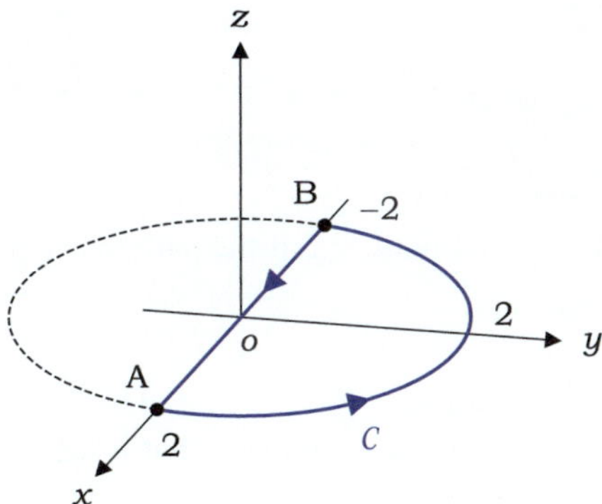

Fig. 2.6 Closed path of integration

These results are equal, implying that the line integral of **E** is independent of integration path. This field is called the ***conservative field***.

Example 2.3 Determine the circulation of $\mathbf{E} = \rho^2 \mathbf{a}_\rho + 3 \sin \phi \, \mathbf{a}_\phi + 5z \, \mathbf{a}_z$ around the closed path, as shown in Fig. 2.6.

Solution

The closed-line integral is divided into two parts:

$$\oint_C \mathbf{E} \cdot d\mathbf{l} = \int_A^B \mathbf{E} \cdot d\mathbf{l} + \int_B^A \mathbf{E} \cdot d\mathbf{l} \tag{2.16a}$$

Along the semicircle, $d\mathbf{l} = 2d\phi \, \mathbf{a}_\phi$. Thus, the first part on the right-hand side of Eq. (2.16a) becomes

$$\begin{aligned}
\int_A^B \mathbf{E} \cdot d\mathbf{l} &= \int_A^B (\rho^2 \mathbf{a}_\rho + 3 \sin \phi \, \mathbf{a}_\phi + 5z \, \mathbf{a}_z) \cdot (2d\phi \, \mathbf{a}_\phi) \\
&= \int_{\phi=0}^{\phi=\pi} 6(\sin \phi) \, d\phi = 12
\end{aligned} \tag{2.16b}$$

Along the straight line, $d\mathbf{l} = dx \, \mathbf{a}_x$, $\rho^2 = x^2$, and $\mathbf{a}_\phi \cdot \mathbf{a}_x = \mathbf{a}_z \cdot \mathbf{a}_x = 0$. Careful examination reveals that $\mathbf{a}_\rho \cdot \mathbf{a}_x = -1$ along $\overline{Bo}$, but $\mathbf{a}_\rho \cdot \mathbf{a}_x = 1$ along $\overline{oA}$. The second part on the right-hand side of Eq. (2.16a) then becomes

$$\begin{aligned}
\int_B^A \mathbf{E} \cdot d\mathbf{l} &= \int_B^A \left(\rho^2 \mathbf{a}_\rho + 3 \sin \phi \, \mathbf{a}_\phi + 5z \, \mathbf{a}_z\right) \cdot (dx \, \mathbf{a}_x) \\
&= \int_{x=-2}^{x=0} -x^2 dx + \int_{x=0}^{x=2} x^2 dx = 0
\end{aligned} \tag{2.16c}$$

Thus, the closed-line integral of **E** is given by

$$\oint_C \mathbf{E} \cdot d\mathbf{l} = 12$$

Exercise 2.3

Repeat Example 2.3 if the path of integration is in the $z = 4$ plane.

Ans. 12.

Exercise 2.4

If $\mathbf{E}$ is in volts per meter, what is the unit of line integral of $\mathbf{E}$?

Ans. Volt.

2.1.3 Surface Integral

An open surface S is generally expressed as $z = f(x, y)$ or $g(x, y, z) = k$ with a constant k in Cartesian coordinates. The position vector of point on the surface can then be expressed as

$$\boxed{\mathbf{r} = x\,\mathbf{a}_x + y\,\mathbf{a}_y + f(x, y)\,\mathbf{a}_z} \tag{2.17}$$

This is referred to as a ***parametric representation*** of a surface.

The parametric representation of Eq. (2.17) can be used to obtain a general expression for the differential area vector of an arbitrary surface. The slanting plane S shown in Fig. 2.7 is considered to be an arbitrary surface, to which $d\mathbf{s}$ of Eq. (1.38) could not be applied. It is evident from Fig. 2.7 that point p_1 on S can be specified by the intersection of three planes: the S, $x = x_1$, and $y = y_1$ planes, where x_1 and y_1 are the Cartesian coordinates of p_1. Furthermore, lines $\mathcal{L}_1$ and $\mathcal{L}_2$ are the intersections of S with the $y = y_1$ and $x = x_1$ planes, respectively. With x and y taken as parameters, $\mathcal{L}_1$ and $\mathcal{L}_2$ can be expressed in parametric form as

$$\mathbf{r}(x) = x\,\mathbf{a}_x + y_1\,\mathbf{a}_y + f(x, y_1)\,\mathbf{a}_z \quad (\text{for } \mathcal{L}_1) \tag{2.18a}$$

$$\mathbf{r}(y) = x_1\,\mathbf{a}_x + y\,\mathbf{a}_y + f(x_1, y)\,\mathbf{a}_z \quad (\text{for } \mathcal{L}_2) \tag{2.18b}$$

Next, using the partial derivatives of Eqs. (2.18a) and (2.18b) with respect to x and y, the differential length vectors along the positive directions of $\mathcal{L}_1$ and $\mathcal{L}_2$ can be expressed as

$$d\mathbf{l}^1 = \left.\frac{\partial \mathbf{r}}{\partial x}\right|_{y=y_1} dx \equiv \mathbf{r}'_x\, dx \tag{2.19a}$$

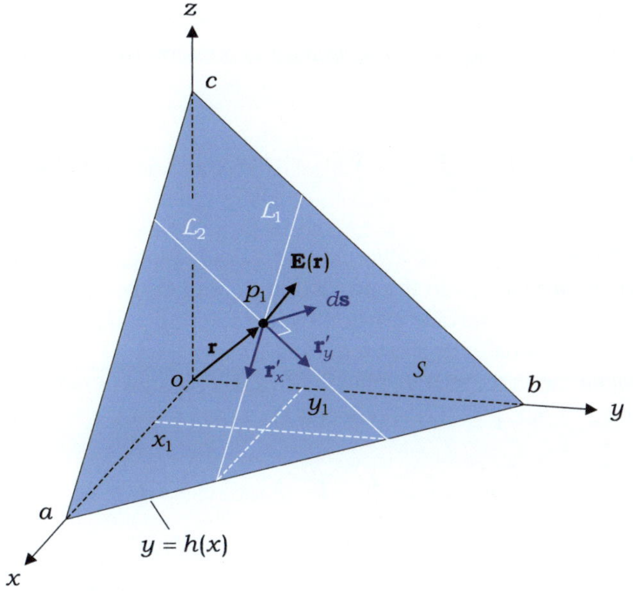

Fig. 2.7 Differential area vector, $d\mathbf{s} = (\mathbf{r}'_x \times \mathbf{r}'_y)\, dx\, dy$

$$d\mathbf{l}^2 = \left.\frac{\partial \mathbf{r}}{\partial y}\right|_{x=x_1} dy \equiv \mathbf{r}'_y\, dy \tag{2.19b}$$

where superscripts 1 and 2 denote $\mathcal{L}_1$ and $\mathcal{L}_2$, respectively, and $\mathbf{r}'_x$ and $\mathbf{r}'_y$ denote the partial derivatives of $\mathbf{r}$ with respect to x and y, respectively. In view of the fact that $d\mathbf{l}^1$ and $d\mathbf{l}^2$ form a parallelogram with area $|d\mathbf{l}^1 \times d\mathbf{l}^2|$, the differential area vector on S can be expressed as

$$\boxed{d\mathbf{s} = d\mathbf{l}^1 \times d\mathbf{l}^2 = (\mathbf{r}'_x \times \mathbf{r}'_y)\, dx\, dy} \tag{2.20}$$

This can be applied to any surface regardless of the surface curvature. Note that the differential area vectors discussed in Chap. 1 are special cases of Eq. (2.20). For example, Eq. (2.20) is reduced to $d\mathbf{s} = (\mathbf{a}_x \times \mathbf{a}_y)\, dx\, dy = dx\, dy\, \mathbf{a}_z$ in the $z = 1$ plane.

The surface integral of $\mathbf{E}$ over surface S is defined as

$$\int_S \mathbf{E(r)} \cdot d\mathbf{s} = \int_S E_n\, ds \tag{2.21}$$

where $d\mathbf{s}$ is the differential area vector on S, and E_n is the component of $\mathbf{E}$ normal to S. *The surface integral of $\mathbf{E}$ over S is the sum of the products of the normal* *component of* $\mathbf{E}$ *and the differential area as the differential area tends to zero.* By inserting Eq. (2.20) into Eq. (2.21), a general expression for the surface integral is obtained as

$$\int_S \mathbf{E}(\mathbf{r}) \cdot d\mathbf{s} = \int_{x=0}^{x=a} \int_{y=0}^{y=h(x)} \mathbf{E}(x, y, f(x, y)) \cdot \left(\mathbf{r}'_x \times \mathbf{r}'_y\right) dy\,dx \qquad (2.22)$$

Notably, the surface integral is converted into a double integral over a region in the xy-plane, which corresponds to the projection of S onto the xy-plane. Specifically, Eq. (2.22) represents the surface integral over a slanting plane in the first quadrant of the Cartesian coordinates as shown in Fig. 2.7.

If S is a rectangle in the $z = z_o$ plane with sides parallel to the x- or y-axis, then the surface integral in Eq. (2.22) is reduced to

$$\int_S \mathbf{E} \cdot d\mathbf{s} = \int_{x_1}^{x_2} \int_{y_1}^{y_2} E_z(x, y, z_o)\, dy\,dx \qquad (2.23)$$

where $d\mathbf{s} = dx\,dy\,\mathbf{a}_z$, and x and y are independent space coordinates.

Example 2.4 Find the area of the surface S, as shown in Fig. 2.7, if $a = c = 1$ and $b = 1/2$.

Solution

The surface S is expressed as

$$x + 2y + z = 1 \qquad (2.24a)$$

The parametric representation of S is

$$\mathbf{r} = x\,\mathbf{a}_x + y\,\mathbf{a}_y + (1 - x - 2y)\,\mathbf{a}_z$$

The partial derivatives of $\mathbf{r}$ with respect to x and y are

$$\mathbf{r}'_x = \mathbf{a}_x - \mathbf{a}_z \qquad (2.24b)$$

$$\mathbf{r}'_y = \mathbf{a}_y - 2\mathbf{a}_z \qquad (2.24c)$$

Inserting Eqs. (2.24b, c) into Eq. (2.20) yields

$$d\mathbf{s} = (\mathbf{a}_x - \mathbf{a}_z) \times (\mathbf{a}_y - 2\mathbf{a}_z)\, dx\,dy = (\mathbf{a}_x + 2\mathbf{a}_y + \mathbf{a}_z)\, dx\,dy \qquad (2.24d)$$

The projection of S onto the xy-plane is a right triangle with hypotenuse defined by

$$x + 2y = 1 \qquad (2.24e)$$

Fig. 2.8 Hemispherical surface

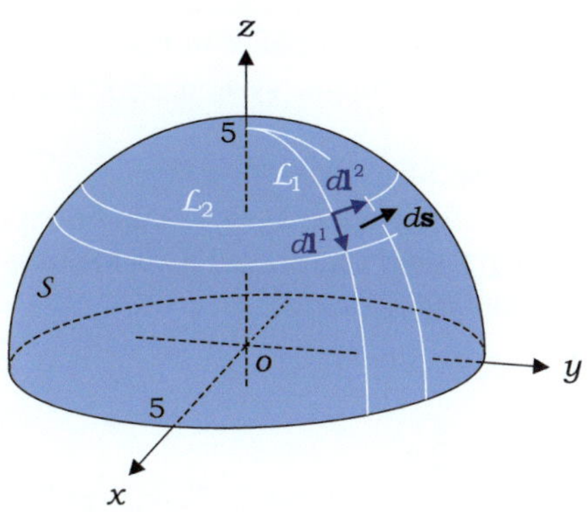

Using Eqs. (2.24d) and (2.24e), the area of $\mathcal{S}$ is calculated as

$$A = \int_{\mathcal{S}} |ds| = \int_{x=0}^{x=1} \int_{y=0}^{y=(1-x)/2} \sqrt{6}\, dy\, dx = \frac{\sqrt{6}}{4}$$

Example 2.5 Compute the surface integral of $\mathbf{A} = 2\mathbf{a}_z$ over a hemisphere of radius 5, as shown in Fig. 2.8, first expressing $d\mathbf{s}$ in terms of $d\mathbf{l}^1$ and $d\mathbf{l}^2$.

Solution

The differential length vectors $d\mathbf{l}^1$ and $d\mathbf{l}^2$ in spherical coordinates are

$$d\mathbf{l}^1 = R\, d\theta\, \mathbf{a}_\theta = 5\, d\theta\, \mathbf{a}_\theta \tag{2.25a}$$

$$d\mathbf{l}^2 = R \sin\theta\, d\phi\, \mathbf{a}_\phi = 5 \sin\theta\, d\phi\, \mathbf{a}_\phi \tag{2.25b}$$

Thus, the differential area vector on the hemisphere is

$$d\mathbf{s} = d\mathbf{l}^1 \times d\mathbf{l}^2 = 25 \sin\theta\, d\theta\, d\phi\, \mathbf{a}_R \tag{2.25c}$$

Using $\mathbf{a}_z \cdot \mathbf{a}_R = \cos\theta$, the surface integral of $\mathbf{A}$ is obtained as

$$\int_{\mathcal{S}} \mathbf{A} \cdot d\mathbf{s} = \int_{\mathcal{S}} (2\,\mathbf{a}_z) \cdot (25 \sin\theta\, d\theta\, d\phi\, \mathbf{a}_R) = 50 \int_{\phi=0}^{2\pi} \int_{\theta=0}^{\pi/2} \cos\theta \sin\theta\, d\theta\, d\phi$$

$$= 50\pi$$

Because the hemisphere in the problem is a surface of constant coordinate in the spherical coordinate system, $d\mathbf{s}$ in Eq. (2.25c) is the same as that in Eq. (1.71a).

Exercise 2.5

A plane $6x + 3y + 2z = 12$ can be expressed in vector form as $\mathbf{a}_n \cdot \mathbf{r} = c$, where $\mathbf{r}$ is the position vector, and $\mathbf{a}_n$ is the unit vector. What is the physical significance of constant c?

Ans. Perpendicular distance from origin to plane.

Exercise 2.6

Given a sphere of radius 5 centered at the origin, determine the surface area in the region $z \geq 4$.

Ans. 10π.

Exercise 2.7

If $\mathbf{D}$ is in coulombs per square meter, what is the unit of surface integral of $\mathbf{D}$?

Ans. Coulomb.

Review Questions

RQ 2.1	Write a general expression for the parametric representation of a curve.	[(2.1)]
RQ 2.2	Express a unit tangent vector in the positive direction of a curve.	[(2.4)]
RQ 2.3	Which component of $\mathbf{A}$ is used in the line integral of $\mathbf{A}$?	[(2.7)]
RQ 2.4	What causes the line integral to reduce to a definite integral?	[(2.9)]
RQ 2.5	State the right-hand rule for $d\mathbf{l}$ and $d\mathbf{s}$ directions.	[Fig. 2.4]
RQ 2.6	Write a general expression for the parametric representation of a surface.	[(2.17)]
RQ 2.7	Which component of $\mathbf{A}$ is used in its surface integral?	[(2.21)]
RQ 2.8	What causes the surface integral to reduce to a double integral?	[(2.20)]
RQ 2.9	What causes the general expression in Eq. (2.20) to reduce to $d\mathbf{s}$ in spherical coordinates?	[Fig. 2.8]

2.2   Directional Derivative and Gradient

When a scalar field exists in a three-dimensional space, it is certain that the field is a continuous function of position, as the quantities observed at points in the space are real physical quantities. In many cases, the space rate of change of a scalar field may have its own physical significance in addition to that of the field itself. In this section, we are concerned with the space rate of change of a scalar field, which generally varies with the direction along which the derivative is taken. The ***directional derivative*** of the scalar field $V(\mathbf{r})$ in the $\mathbf{a}_l$-direction at point p is defined as the space rate of change of V in the direction of $\mathbf{a}_l$ at point p. The maximum directional derivative of V at point p and the direction along which it occurs are combined into a vector quantity, called the **gradient** of V at p.

A directional derivative is denoted by dV/dl, and represents the space rate of change of V in the direction of increase of the differential length dl. At a point with position vector $\mathbf{r}_1$, the directional derivative of V in $\mathbf{a}_l$-direction is defined as

$$\left.\frac{dV}{dl}\right|_{\mathbf{r}_1, \mathbf{a}_l} = \lim_{dl \to 0} \frac{V(\mathbf{r}_1 + dl\,\mathbf{a}_l) - V(\mathbf{r}_1)}{dl} \tag{2.26}$$

where dl is the differential length in the direction of $\mathbf{a}_l$. Note that the directional derivative depends not only on the position, but also on the direction.

The gradient of V at point p is a vector quantity whose magnitude is the greatest possible space rate of charge of V at p, and its unit vector points in the direction of the maximum space rate of change. The gradient of V is expressed as

$$\boxed{grad\,V = \frac{dV}{dn}\mathbf{a}_n} \tag{2.27}$$

where $\mathbf{a}_n$ is the unit vector, and dn is the differential length in the $\mathbf{a}_n$-direction.

If a scalar field $V(\mathbf{r})$ is induced by a localized source, it must be a continuous function of position in the region surrounding the source. This results from the fact that the spatial variation of V is entirely governed by the distance between a given point and the source, which is a continuous quantity in space. Therefore, points with constant V (i.e., V_1) should form a smooth surface around the source. Similarly, the surface of $V = V_1 + dV$ is formed in the immediate vicinity of the surface of $V = V_1$, as shown in Fig. 2.9. Because V is a single-valued function, the two surfaces should never cross for nonzero dV.

Suppose we make a short expedition from point p_1 on the surface of V_1 to point p_2 on the surface of $V_1 + dV$. The difference in V observed during this short trip is dV. This is also true for points p_3 and p_4, which are located on the same surface as point p_2, although the travel distance varies with destination. As line segment $\overline{p_1 p_2}$ is perpendicular to both surfaces, it corresponds to the shortest distance between the two surfaces measured at p_1. Evidently, the directional derivative of V at p_1 is maximal in the direction along $\overline{p_1 p_2}$. *The gradient of V at a point in space is normal to the surface of constant V that passes through the point.*

The gradient of V provides a convenient way to determine the directional derivative of V in an arbitrary direction without actually computing the space rate of change of V at that point. With the aid of the chain rule in calculus, the directional derivative of V can be expressed as

$$\frac{dV}{dl} = \frac{dV}{dn}\frac{dn}{dl} = \frac{dV}{dn}\cos\gamma = \frac{dV}{dn}\mathbf{a}_n \cdot \mathbf{a}_l \tag{2.28}$$

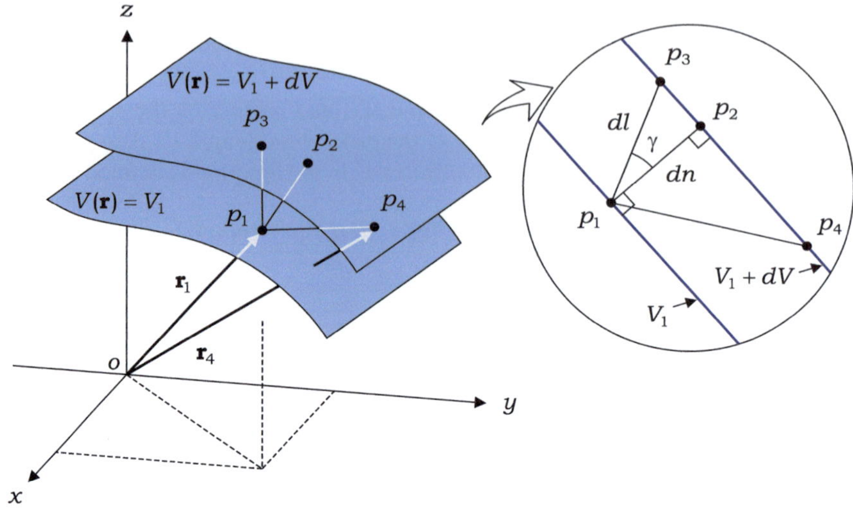

Fig. 2.9 Surfaces of constant V

where $dn/dl = \cos\gamma$ and $\mathbf{a}_n \cdot \mathbf{a}_l = \cos\gamma$, as can be seen from Fig. 2.9. Note that $\mathbf{a}_n$ is normal to the surface of constant V, and $\mathbf{a}_l$ is along the direction of increase in dl.

Combining Eqs. (2.27) and (2.28), the directional derivative of V in the direction of $\mathbf{a}_l$ is given by

$$\boxed{\frac{dV}{dl} = grad\, V \cdot \mathbf{a}_l \equiv \nabla V \cdot \mathbf{a}_l} \tag{2.29}$$

The gradient of V is denoted by either *grad V* or ∇V (read "del" V). *The directional derivative of V in the direction of* $\mathbf{a}_l$ *is equal to the dot product of* ∇V *and* $\mathbf{a}_l$. From Eq. (2.29), it follows that

$$\boxed{dV = \nabla V \cdot dl\, \mathbf{a}_l = \nabla V \cdot d\mathbf{l}} \tag{2.30}$$

where $d\mathbf{l}$ is the differential length vector in the $\mathbf{a}_l$-direction. *The differential of V in the* $\mathbf{a}_l$-*direction is equal to the dot product of* ∇V *and* $d\mathbf{l}$. Note that Eqs. (2.29) and (2.30) constitute two cases, in which we can take complete advantage of the gradient.

From calculus, the total differential of V in Cartesian coordinates is given by

$$dV = \frac{\partial V}{\partial x}dx + \frac{\partial V}{\partial y}dy + \frac{\partial V}{\partial z}dz$$

$$= \left(\frac{\partial V}{\partial x}\mathbf{a}_x + \frac{\partial V}{\partial y}\mathbf{a}_y + \frac{\partial V}{\partial z}\mathbf{a}_z\right) \cdot (dx\,\mathbf{a}_x + dy\,\mathbf{a}_y + dz\,\mathbf{a}_z)$$

$$= \left[\frac{\partial V}{\partial x}\mathbf{a}_x + \frac{\partial V}{\partial y}\mathbf{a}_y + \frac{\partial V}{\partial z}\mathbf{a}_z \right] \cdot d\mathbf{l} \tag{2.31}$$

where the dot product is used to separate the term on the right-hand side into two parts (No physics is involved in this process).

Comparison of Eq. (2.30) with Eq. (2.31) leads to the following expression for ∇V in Cartesian coordinates:

$$\nabla V = \frac{\partial V}{\partial x}\mathbf{a}_x + \frac{\partial V}{\partial y}\mathbf{a}_y + \frac{\partial V}{\partial z}\mathbf{a}_z = \left(\mathbf{a}_x \frac{\partial}{\partial x} + \mathbf{a}_y \frac{\partial}{\partial y} + \mathbf{a}_z \frac{\partial}{\partial z} \right) V \tag{2.32}$$

In view of Eq. (2.32), the vector operator ∇ is defined as

$$\boxed{\nabla \equiv \mathbf{a}_x \frac{\partial}{\partial x} + \mathbf{a}_y \frac{\partial}{\partial y} + \mathbf{a}_z \frac{\partial}{\partial z}} \tag{2.33}$$

Although this has the form of a vector, ∇ is not a vector in the sense that it has neither magnitude nor direction, and never conforms with the dot or the cross product defined in Eqs. (1.9) and (1.17). However, the del operator follows the rules of the vector algebra in Eqs. (1.31) and (1.34) and the rule in Eq. (2.32), if instructions on how to differentiate the operand are provided.

In cylindrical coordinates, the total differential of V at $p_1:(\rho_1, \phi_1, z_1)$ is

$$\begin{aligned} dV &= \frac{\partial V}{\partial \rho}d\rho + \frac{\partial V}{\partial \phi}d\phi + \frac{\partial V}{\partial z}dz \\ &= \left(\frac{\partial V}{\partial \rho}\mathbf{a}_\rho + \frac{\partial V}{\partial \phi}\mathbf{a}_\phi + \frac{\partial V}{\partial z}\mathbf{a}_z \right) \cdot (d\rho\, \mathbf{a}_\rho + d\phi\, \mathbf{a}_\phi + dz\, \mathbf{a}_z) \\ &= \left[\frac{\partial V}{\partial \rho}\mathbf{a}_\rho + \frac{1}{\rho_1}\frac{\partial V}{\partial \phi}\mathbf{a}_\phi + \frac{\partial V}{\partial z}\mathbf{a}_z \right] \cdot d\mathbf{l} \end{aligned} \tag{2.34}$$

By following the procedure described in Eq. (2.31), the right-hand side of Eq. (2.34) was separated into two parts such that the second part is equal to the differential length vector in cylindrical coordinates. Comparison of Eq. (2.30) with Eq. (2.34) leads to ∇V in cylindrical coordinates, as indicated by brackets.

Similarly, the total differential of V in spherical coordinates is

$$\begin{aligned} dV &= \frac{\partial V}{\partial R}dR + \frac{\partial V}{\partial \theta}d\theta + \frac{\partial V}{\partial \phi}d\phi \\ &= \left(\frac{\partial V}{\partial R}\mathbf{a}_R + \frac{\partial V}{\partial \theta}\mathbf{a}_\theta + \frac{\partial V}{\partial \phi}\mathbf{a}_\phi \right) \cdot (dR\, \mathbf{a}_R + d\theta\, \mathbf{a}_\theta + d\phi\, \mathbf{a}_\phi) \\ &= \left[\frac{\partial V}{\partial R}\mathbf{a}_R + \frac{1}{R_1}\frac{\partial V}{\partial \theta}\mathbf{a}_\theta + \frac{1}{R_1 \sin \theta_1}\frac{\partial V}{\partial \phi}\mathbf{a}_\phi \right] \cdot d\mathbf{l} \end{aligned} \tag{2.35}$$

The term in brackets is ∇V in spherical coordinates.

Even if there is no specific form of ∇ in the cylindrical or spherical system, ∇V is used to represent the gradient of V in these coordinates.

The gradient of V is expressed in Cartesian, cylindrical, and spherical coordinates as follows:

$$\nabla V = \frac{\partial V}{\partial x}\mathbf{a}_x + \frac{\partial V}{\partial y}\mathbf{a}_y + \frac{\partial V}{\partial z}\mathbf{a}_z \quad \text{(Cartesian)} \tag{2.36a}$$

$$\nabla V = \frac{\partial V}{\partial \rho}\mathbf{a}_\rho + \frac{1}{\rho}\frac{\partial V}{\partial \phi}\mathbf{a}_\phi + \frac{\partial V}{\partial z}\mathbf{a}_z \quad \text{(cylindrical)} \tag{2.36b}$$

$$\nabla V = \frac{\partial V}{\partial R}\mathbf{a}_R + \frac{1}{R}\frac{\partial V}{\partial \theta}\mathbf{a}_\theta + \frac{1}{R \sin\theta}\frac{\partial V}{\partial \phi}\mathbf{a}_\phi \quad \text{(spherical)} \tag{2.36c}$$

where subscript 1 is omitted for generalization.

The gradient of V can be expressed in general orthogonal curvilinear coordinates (u, v, w) as follows:

$$\nabla V = \left(\mathbf{a}_u \frac{1}{h_1}\frac{\partial}{\partial u} + \mathbf{a}_v \frac{1}{h_2}\frac{\partial}{\partial v} + \mathbf{a}_w \frac{1}{h_3}\frac{\partial}{\partial w} \right) V \tag{2.37}$$

where the metric coefficients are given by

$$h_1 = 1, \ h_2 = 1, \ h_3 = 1 \qquad (u, v, w) = (x, y, z) \tag{2.38a}$$

$$h_1 = 1, \ h_2 = \rho, \ h_3 = 1 \qquad (u, v, w) = (\rho, \phi, z) \tag{2.38b}$$

$$h_1 = 1, \ h_2 = R, \ h_3 = R \sin\theta \qquad (u, v, w) = (R, \theta, \phi) \tag{2.38c}$$

The total differential of V involves differential coordinates, whereas the gradient of V is defined in terms of differential lengths. Therefore, metric coefficients are required to convert differential angles into differential lengths.

Example 2.6 Given the scalar field $V(\mathbf{r}) = x^2 + 4yz$, find the following at point $p{:}(4, -1, 3)$:

(a) The gradient of V, and
(b) The directional derivative of V in the direction of $\mathbf{A} = 3\mathbf{a}_x - 2\mathbf{a}_y - \sqrt{3}\,\mathbf{a}_z$.

Solution

(a) From Eq. (2.36a),

$$\nabla V = \frac{\partial V}{\partial x}\mathbf{a}_x + \frac{\partial V}{\partial y}\mathbf{a}_y + \frac{\partial V}{\partial z}\mathbf{a}_z = 2x\,\mathbf{a}_x + 4z\,\mathbf{a}_y + 4y\,\mathbf{a}_z \qquad (2.39)$$

Inserting the coordinates of p into Eq. (2.39) yields

$$\nabla V = 8\mathbf{a}_x + 12\mathbf{a}_y - 4\mathbf{a}_z$$

(b) The unit vector in the direction of $\mathbf{A}$ is

$$\mathbf{a}_A = \frac{3}{4}\mathbf{a}_x - \frac{1}{2}\mathbf{a}_y - \frac{\sqrt{3}}{4}\mathbf{a}_z$$

The directional derivative of V in the direction of $\mathbf{a}_A$ is

$$\frac{dV}{dl} = \nabla V \cdot \mathbf{a}_A = (8\mathbf{a}_x + 12\mathbf{a}_y - 4\mathbf{a}_z) \cdot \left(\frac{3}{4}\mathbf{a}_x - \frac{1}{2}\mathbf{a}_y - \frac{\sqrt{3}}{4}\mathbf{a}_z\right) = \sqrt{3}$$

Example 2.7 Express $\nabla \frac{1}{\mathcal{R}}$ in terms of distance vector $\mathcal{R}$ and its magnitude given by $\mathcal{R} = |\mathbf{r} - \mathbf{r}'|$.

Solution

The position vectors in the unprimed and primed coordinates are

$$\mathbf{r} = x\,\mathbf{a}_x + y\,\mathbf{a}_y + z\,\mathbf{a}_z$$

$$\mathbf{r}' = x'\,\mathbf{a}_x + y'\,\mathbf{a}_y + z'\,\mathbf{a}_z$$

Thus, the distance vector is given by

$$\mathcal{R} = \mathbf{r} - \mathbf{r}' = (x - x')\,\mathbf{a}_x + (y - y')\,\mathbf{a}_y + (z - z')\,\mathbf{a}_z$$

The reciprocal of $|\mathbf{r} - \mathbf{r}'|$ is

$$\frac{1}{|\mathbf{r} - \mathbf{r}'|} = \left[(x - x')^2 + (y - y')^2 + (z - z')^2\right]^{-1/2}$$

The partial derivatives of $1/|\mathbf{r} - \mathbf{r}'|$ with respect to x, y, and z are

$$\frac{\partial}{\partial x}\frac{1}{|\mathbf{r} - \mathbf{r}'|} = -(x - x')\left[(x - x')^2 + (y - y')^2 + (z - z')^2\right]^{-3/2} \qquad (2.40a)$$

$$\frac{\partial}{\partial y}\frac{1}{|\mathbf{r}-\mathbf{r'}|} = -(y-y')\left[(x-x')^2 + (y-y')^2 + (z-z')^2\right]^{-3/2} \tag{2.40b}$$

$$\frac{\partial}{\partial z}\frac{1}{|\mathbf{r}-\mathbf{r'}|} = -(z-z')\left[(x-x')^2 + (y-y')^2 + (z-z')^2\right]^{-3/2} \tag{2.40c}$$

Inserting Eqs. (2.40a)–(2.40c) into Eq. (2.36a) yields

$$\nabla\frac{1}{|\mathbf{r}-\mathbf{r'}|} = -\frac{(x-x')\,\mathbf{a}_x + (y-y')\,\mathbf{a}_y + (z-z')\,\mathbf{a}_z}{\left[(x-x')^2 + (y-y')^2 + (z-z')^2\right]^{3/2}}$$

Thus,

$$\boxed{\nabla\frac{1}{\mathcal{R}} = -\frac{\mathcal{R}}{\mathcal{R}^3}} \tag{2.41}$$

The del operator in Eq. (2.41) acts on x, y, and z, whereas x', y', and z' are treated as constants.

Example 2.8 Find the outward unit normal to sphere $x^2 + y^2 + z^2 = 9$ at point $p{:}(2, 2, 1)$.

Solution

Consider a family of spheres, defined as $f(x, y, z) = x^2 + y^2 + z^2$. The sphere in the problem is a member as specified by $f(x, y, z) = 9$.

Because ∇f is normal to the surface of constant f, the unit normal vector can be obtained as follows:

$$\frac{\nabla f}{|\nabla f|} = \frac{x\,\mathbf{a}_x + y\,\mathbf{a}_y + z\,\mathbf{a}_z}{\sqrt{x^2 + y^2 + z^2}} \tag{2.42a}$$

Setting $(x, y, z) = (2, 2, 1)$ in Eq. (2.42a), the unit normal is given by

$$\frac{1}{3}\left(2\mathbf{a}_x + 2\mathbf{a}_y + \mathbf{a}_z\right) \tag{2.42b}$$

This is directed outward from the sphere at point p.

Alternatively, taking the spherical symmetry into account, the family of spheres is expressed as $g(R, \theta, \phi) = R^2$ in spherical coordinates, and compute

$$\nabla g = \frac{\partial g}{\partial R}\mathbf{a}_R + \frac{1}{R}\frac{\partial g}{\partial \theta}\mathbf{a}_\theta + \frac{1}{R\sin\theta}\frac{\partial g}{\partial \phi}\mathbf{a}_\phi = 2R\,\mathbf{a}_R \tag{2.42c}$$

Fig. 2.10 A quarter circle

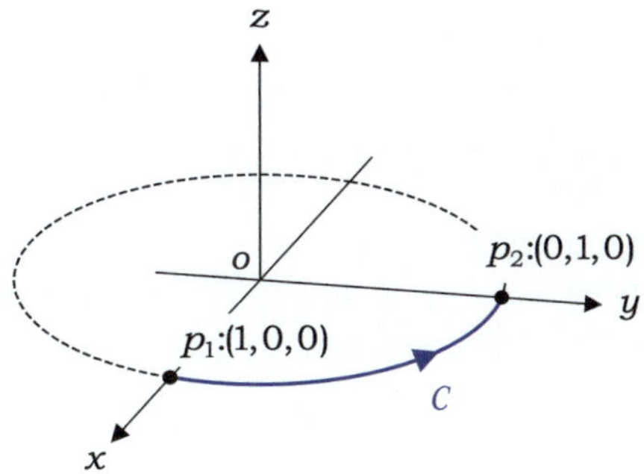

The outward unit normal is simply $\mathbf{a}_R$.

Example 2.9 Given a scalar field $V = (x^2 + 3)\, y^2$, using the quarter shown in Fig. 2.10,

(a) evaluate $\int_C \nabla V \cdot d\mathbf{l}$ from p_1:$(1, 0, 0)$ to p_2:$(0, 1, 0)$, and
(b) show that the line integral of ∇V is independent of the path of integration, irrespective of V.

Solution

(a) In Cartesian coordinates,

$$\nabla V = 2xy^2 \mathbf{a}_x + 2y(x^2 + 3)\, \mathbf{a}_y$$

Using the path of integration expressed by $x^2 + y^2 = 1$, the line integral is performed as

$$\int_{p_1}^{p_2} \nabla V \cdot d\mathbf{l} = \int_{(1,0,0)}^{(0,1,0)} \left[2xy^2 \mathbf{a}_x + 2y(x^2 + 3)\, \mathbf{a}_y \right] \cdot (dx\, \mathbf{a}_x + dy\, \mathbf{a}_y)$$

$$= \int_{x=1}^{x=0} 2x(1 - x^2)\, dx + \int_{y=0}^{y=1} 2y(1 - y^2 + 3)\, dy = 3$$

(b) In general, the line integral of the gradient of a scalar field is expressed as

$$\int_{p_1}^{p_2} \nabla V \cdot d\mathbf{l} = \int_{p_1}^{p_2} \left(\frac{\partial V}{\partial x} \mathbf{a}_x + \frac{\partial V}{\partial y} \mathbf{a}_y + \frac{\partial V}{\partial z} \mathbf{a}_z \right) \cdot (dx\, \mathbf{a}_x + dy\, \mathbf{a}_y + dz\, \mathbf{a}_z)$$

$$= \int_{p_1}^{p_2} \frac{\partial V}{\partial x} dx + \frac{\partial V}{\partial y} dy + \frac{\partial V}{\partial z} dz = \int_{p_1}^{p_2} dV$$

$$= V(p_2) - V(p_1)$$

The line integral of the gradient of a scalar field is independent of the integration path and depends only on the initial and terminal points of the integration path.

Exercise 2.8
If the scalar field V is in volts, what is the unit of ∇V?

Ans. Volt per meter.

Exercise 2.9
If $\nabla V = -3\mathbf{a}_x$ at a point in space, determine (a) the maximum space rate of change of V at the point, and (b) the direction in which it occurs.

Ans. (a) 3, (b) $-x$-direction.

Exercise 2.10
Given $\nabla V = 10\mathbf{a}_\rho + 2\mathbf{a}_\phi$ at p:(4, 30°, 5) in cylindrical coordinates, determine (a) $dV/d\rho$, and (b) $dV/d\phi$ at p.

Ans. (a) 10, (b) 8.

Exercise 2.11
Keeping in mind that $\mathbf{A} \cdot \mathbf{C} = \mathbf{B} \cdot \mathbf{C}$ does not ensure that $\mathbf{A} = \mathbf{B}$, explain why an expression for ∇V can be obtained from $\nabla V \cdot d\mathbf{l} = [..] \cdot d\mathbf{l}$, as shown in Eqs. (2.30) and (2.31)?

Ans. $d\mathbf{l}$ is arbitrary.

Review Questions

RQ 2.10	What are the magnitude and direction of ∇V?	[(2.27)]
RQ 2.11	Does $V = 0$ at a point in space imply $\nabla V = 0$ at that point?	[(2.27)]
RQ 2.12	How is the directional derivative of V related to ∇V?	[(2.29)]
RQ 2.13	How is the differential dV related to ∇V?	[(2.30)]
RQ 2.14	Define the del operator in Cartesian coordinates.	[(2.33)]
RQ 2.15	Express ∇V in the three coordinate systems.	[(2.36a,b,c)]
RQ 2.16	Write the metric coefficients for the three coordinate systems.	[(2.38a,b,c)]

2.3 Flux and Flux Density

If a vector field is induced by a localized source, its constituent vectors are specified by using the magnitude and direction of the distance vector drawn from the source to a given point. As the distance vector is itself a smooth function of position, so is the vector field. This means that the constituent vectors should exhibit no abrupt changes in magnitude or direction as a function of position. Subsequently, we can draw smooth lines connecting the constituent vectors in three-dimensional space, called ***field lines*** or ***flux lines***. Field lines never pass across each other if the given vector field is a single-valued function of position. Graphically, the tangent to the field line indicates the direction of the constituent vector at a given point and the density of the field lines in a small region surrounding the point represents the magnitude of the constituent vector.

Flux density generally refers to the amount of certain quantities passing through a unit area per unit time. The **total flux**, or simply **flux**, is the total scalar quantity passing through a given surface per unit time. The flux density is a vector quantity, whereas the flux is a scalar quantity. Electromagnetics relies heavily on electric flux density and magnetic flux density. In this case, flux density refers to the number of field lines through a unit area, and total flux refers to the total number of field lines through a given surface. It is important to note that the unit area for the flux density should be taken in a plane perpendicular to the direction of flow.

As an example of flux density, we consider an electron cloud with density n_e [m^{-3}] flowing at a constant velocity $\boldsymbol{v}$ [m/s]. The flux density of the flow of electrons is defined as the product of number density and velocity, that is,

$$\mathbf{N} = n_e \boldsymbol{v} \quad [\text{m}^{-2} \cdot \text{s}^{-1}] \tag{2.43}$$

When the electron cloud passes through surface $\mathcal{S}$ (see Fig. 2.11), differential area $|d\mathbf{s}|$ in $\mathcal{S}$ is equivalent to differential area $|d\mathbf{s}| \cos\gamma$ in the cross section. Thus, the differential flux passing through $|d\mathbf{s}|$ is given by

$$d\Psi = |\mathbf{N}||d\mathbf{s}| \cos\gamma = |\mathbf{N}||d\mathbf{s}|\, \mathbf{a}_N \cdot \mathbf{a}_s = \mathbf{N} \cdot d\mathbf{s}$$

where $\cos\gamma$ is replaced by the dot product of $\mathbf{a}_N$ and $\mathbf{a}_s$, which are the unit vectors in the directions of $\mathbf{N}$ and $d\mathbf{s}$, respectively. Thus, the total flux through $\mathcal{S}$ is expressed as

$$\Psi = \int_{\mathcal{S}} d\Psi = \int_{\mathcal{S}} \mathbf{N} \cdot d\mathbf{s} \tag{2.44}$$

Fig. 2.11 Differential flux passing through differential area $|d\mathbf{s}|$

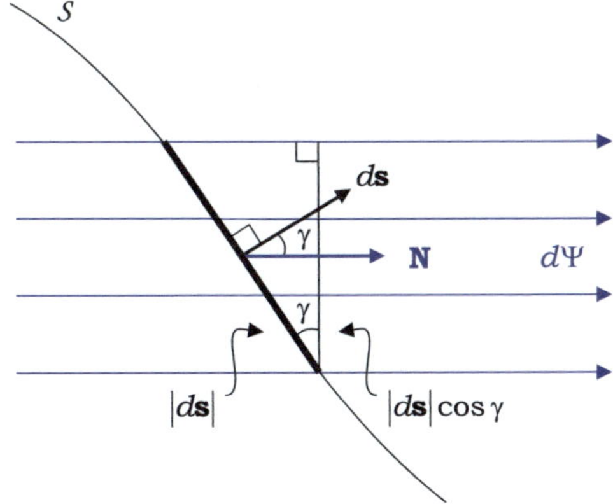

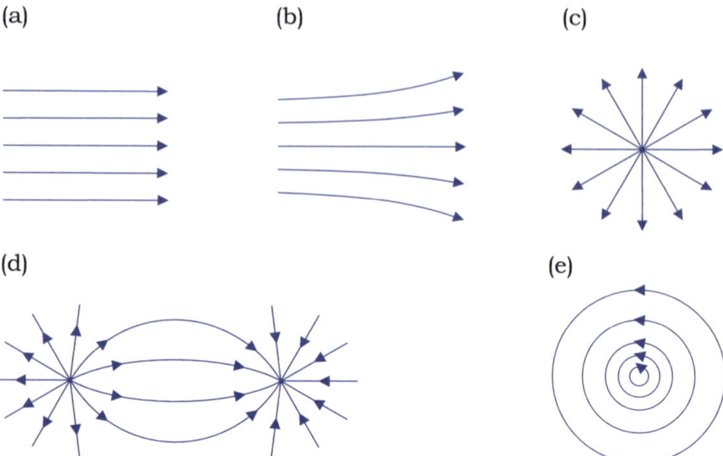

Fig. 2.12 Field lines

The total flux through the surface S is equal to the surface integral of the flux density over S. The dot product indicates that the flux density is defined as the flux per unit area in the cross section.

Several field lines are shown in Fig. 2.12: (a) uniform field lines, (b) slightly diverging field lines, (c) radial field lines due to a divergence source, (d) field lines starting at a source and ending at a sink, and (e) concentric field lines due to a circulation source.

Exercise 2.12
Refer to Fig. 2.12c. Express the flux density in spherical coordinates when the total flux is A_o.

Ans. $\mathbf{A} = A_o \mathbf{a}_R / 4\pi R^2$.

Exercise 2.13
Refer to Fig. 2.12e. Express the vector field $\mathbf{B}$ in cylindrical coordinates when the circulation of $\mathbf{B}$ is B_o.

Ans. $\mathbf{B} = B_o \mathbf{a}_\phi / 2\pi\rho$.

Review Questions

RQ 2.17	What, in general, is the unit of flux density?	[(2.43)]
RQ 2.18	Distinguish between flux and flux density.	[(2.44)]
RQ 2.19	What is the significance of the dot product in Eq. (2.44)?	[Fig. 2.11]

2.4 Divergence and Divergence Theorem

Divergence is a differential operator acting on flux density, which is a vector field that defines a scalar quantity through a unit cross-sectional area. The divergence is devised to relate the net outward flux through a closed surface to the source enclosed by that surface, based on the premise that the total flux is conserved in a three-dimensional space, and the flux density obeys the principle of superposition. Accordingly, the net outward flux is independent of the closed surface used for computation only if the surface encloses the same source. The net outward flux is zero if the closed surface excludes this source. In this case, the outward flux through the surface cancels the inward flux. From the definition of divergence, we can derive the ***divergence theorem***, which is useful for converting a closed-surface integral into a volume integral, and vice versa.

2.4.1 Divergence of the Flux Density

The divergence of the flux density $\mathbf{D}$, denoted by $div\,\mathbf{D}$, results in a scalar field in a given region. The divergence involves an imaginary surface surrounding a given point in space such that *the divergence of* $\mathbf{D}$ *at point* p_1 *is the ratio of the net outward flux through a closed surface centered at* p_1 *to the enclosed volume as the volume shrinks to a point at* p_1. The divergence of $\mathbf{D}$ at point p_1 is mathematically expressed as

$$div\,\mathbf{D} \equiv \lim_{\Delta v \to 0} \frac{\oint_{\mathcal{S}} \mathbf{D} \cdot d\mathbf{s}}{\Delta v} \tag{2.45}$$

where a small circle on the integration sign signifies that $\mathcal{S}$ is a closed surface surrounding a given point, Δv is the volume bounded by $\mathcal{S}$, and $d\mathbf{s}$ is the differential area vector of $\mathcal{S}$ pointing out of the enclosed volume. In short, divergence is *the net outward flux per unit volume at a given point in the space.*

We consider a differential volume $\Delta v = \Delta x \Delta y \Delta z$ enclosed by a rectangular parallelepiped centered at p_1, as shown in Fig. 2.13. The closed-surface integral in Eq. (2.45) is first broken into six parts:

$$\lim_{\Delta v \to 0} \oint_{\mathcal{S}} \mathbf{D} \cdot d\mathbf{s}$$

$$= \lim_{\Delta v \to 0} \left[\int_{\mathcal{S}^1} \mathbf{D} \cdot d\mathbf{s} + \int_{\mathcal{S}^2} \mathbf{D} \cdot d\mathbf{s} + \int_{\mathcal{S}^3} \mathbf{D} \cdot d\mathbf{s} + \int_{\mathcal{S}^4} \mathbf{D} \cdot d\mathbf{s} + \int_{\mathcal{S}^5} \mathbf{D} \cdot d\mathbf{s} + \int_{\mathcal{S}^6} \mathbf{D} \cdot d\mathbf{s} \right]$$

$$= \lim_{\Delta v \to 0} \left[\mathbf{D}^1 \cdot d\mathbf{s}^1 + \mathbf{D}^2 \cdot d\mathbf{s}^2 + \mathbf{D}^3 \cdot d\mathbf{s}^3 + \mathbf{D}^4 \cdot d\mathbf{s}^4 + \mathbf{D}^5 \cdot d\mathbf{s}^5 + \mathbf{D}^6 \cdot d\mathbf{s}^6 \right] \tag{2.46}$$

where $\mathbf{D}^i$ denotes the value of $\mathbf{D}$ at the center of the i-th face, and $d\mathbf{s}^i$ is the differential area vector for the face. In the limit as $\Delta v \to 0$, $\mathbf{D}$ is essentially constant over surface

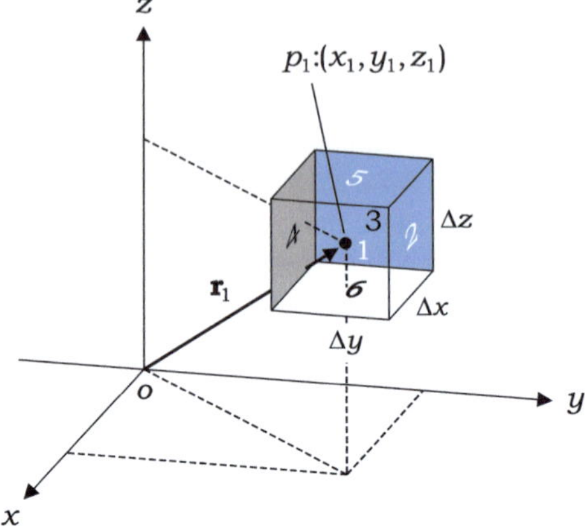

Fig. 2.13 Rectangular parallelepiped with volume $\Delta v = \Delta x \, \Delta y \, \Delta z$

$\mathcal{S}^i$; thus, the integral of $\mathbf{D}$ over $\mathcal{S}^i$ is simply given by $\mathbf{D}^i \cdot d\mathbf{s}^i$, representing the outward flux through the i-th face.

As Δv eventually shrinks to a point at p_1, it would be more convenient to expand $\mathbf{D}^i$ in the Taylor series centered at p_1 such as

$$\mathbf{D}^1 \cdot d\mathbf{s}^1 = \left[\mathbf{D}_o + \frac{\Delta x}{2} \frac{\partial \mathbf{D}}{\partial x}\bigg|_{\text{at } p_1} \right] \cdot d\mathbf{s}^1, \qquad d\mathbf{s}^1 = \Delta y \Delta z \, \mathbf{a}_x \tag{2.47a}$$

$$\mathbf{D}^2 \cdot d\mathbf{s}^2 = \left[\mathbf{D}_o + \frac{\Delta y}{2} \frac{\partial \mathbf{D}}{\partial y}\bigg|_{\text{at } p_1} \right] \cdot d\mathbf{s}^2, \qquad d\mathbf{s}^2 = \Delta x \Delta z \, \mathbf{a}_y \tag{2.47b}$$

$$\mathbf{D}^3 \cdot d\mathbf{s}^3 = \left[\mathbf{D}_o - \frac{\Delta x}{2} \frac{\partial \mathbf{D}}{\partial x}\bigg|_{\text{at } p_1} \right] \cdot d\mathbf{s}^3, \qquad d\mathbf{s}^3 = -\Delta y \Delta z \, \mathbf{a}_x \tag{2.47c}$$

$$\mathbf{D}^4 \cdot d\mathbf{s}^4 = \left[\mathbf{D}_o - \frac{\Delta y}{2} \frac{\partial \mathbf{D}}{\partial y}\bigg|_{\text{at } p_1} \right] \cdot d\mathbf{s}^4, \qquad d\mathbf{s}^4 = -\Delta x \Delta z \, \mathbf{a}_y \tag{2.47d}$$

$$\mathbf{D}^5 \cdot d\mathbf{s}^5 = \left[\mathbf{D}_o + \frac{\Delta z}{2} \frac{\partial \mathbf{D}}{\partial z}\bigg|_{\text{at } p_1} \right] \cdot d\mathbf{s}^5, \qquad d\mathbf{s}^5 = \Delta x \Delta y \, \mathbf{a}_z \tag{2.47e}$$

$$\mathbf{D}^6 \cdot d\mathbf{s}^6 = \left[\mathbf{D}_o - \frac{\Delta z}{2} \frac{\partial \mathbf{D}}{\partial z}\bigg|_{\text{at } p_1} \right] \cdot d\mathbf{s}^6, \qquad d\mathbf{s}^6 = -\Delta x \Delta y \, \mathbf{a}_z \tag{2.47f}$$

where higher-order terms containing $(\Delta x)^2$, $(\Delta y)^2$, and $(\Delta z)^2$ and those with higher power are ignored. The negative sign inside brackets indicates that the given face is retreated from p_1 along the coordinate axis, whereas the negative sign in the expression for $d\mathbf{s}$ indicates that $d\mathbf{s}$ points out of the volume. It should be noted that although $\mathbf{D}_o$ and the partial derivatives of $\mathbf{D}$ have values specific to the center of Δv, the quantity in brackets is the value of $\mathbf{D}$ at the center of a face of Δv.

By combining the six expressions in Eqs. (2.47a)–(2.47f), we obtain

$$
\mathbf{D}^1 \cdot d\mathbf{s}^1 + \mathbf{D}^2 \cdot d\mathbf{s}^2 + \mathbf{D}^3 \cdot d\mathbf{s}^3 + \mathbf{D}^4 \cdot d\mathbf{s}^4 + \mathbf{D}^5 \cdot d\mathbf{s}^5 + \mathbf{D}^6 \cdot d\mathbf{s}^6
$$

$$
= \Delta x \Delta y \Delta z \left[\left(\left. \frac{\partial \mathbf{D}}{\partial x} \right|_{\text{at } p_1} \right) \cdot \mathbf{a}_x + \left(\left. \frac{\partial \mathbf{D}}{\partial y} \right|_{\text{at } p_1} \right) \cdot \mathbf{a}_y + \left(\left. \frac{\partial \mathbf{D}}{\partial z} \right|_{\text{at } p_1} \right) \cdot \mathbf{a}_z \right]
$$

$$
= \Delta x \Delta y \Delta z \left[\left. \frac{\partial D_x}{\partial x} \right|_{\text{at } p_1} + \left. \frac{\partial D_y}{\partial y} \right|_{\text{at } p_1} + \left. \frac{\partial D_z}{\partial z} \right|_{\text{at } p_1} \right] \tag{2.48}
$$

where $\partial \mathbf{D}/\partial x = (\partial D_x/\partial x)\,\mathbf{a}_x + (\partial D_y/\partial x)\,\mathbf{a}_y + (\partial D_z/\partial x)\,\mathbf{a}_z$ among others.

Substituting Eqs. (2.46) and (2.48) into Eq. (2.45) and using $\Delta v = \Delta x \Delta y \Delta z$, the divergence of $\mathbf{D}$ in Cartesian coordinates is expressed as

$$
\boxed{\; div\,\mathbf{D} = \frac{\partial D_x}{\partial x} + \frac{\partial D_y}{\partial y} + \frac{\partial D_z}{\partial z} \;} \tag{2.49}
$$

Note that $div\,\mathbf{D}$ constitutes a scalar field in three-dimensional space. Although $div\,\mathbf{D}$ is expressed in terms of partial derivatives of $\mathbf{D}$, it is inherently a closed-surface integral of $\mathbf{D}$ per unit volume, representing the net outward flux per unit volume. Accordingly, a nonzero divergence at a point in space can indicate the existence of a source or sink at that point. If $div\,\mathbf{D}$ is zero, then $\mathbf{D}$ is said to be **divergenceless**.

The del operator enables us to write the divergence of $\mathbf{D}$ in a more compact form as follows:

$$
\boxed{\; div\,\mathbf{D} = \nabla \cdot \mathbf{D} \;} \tag{2.50}
$$

A similar procedure can be followed to obtain the expressions for $div\,\mathbf{D}$ in other coordinate systems. The divergence of $\mathbf{D}$ can be expressed in terms of the general orthogonal curvilinear coordinates (u, v, w) as

$$
\boxed{\; \nabla \cdot \mathbf{D} = \frac{1}{h_1 h_2 h_3} \left[\frac{\partial}{\partial u}(h_2 h_3 D_u) + \frac{\partial}{\partial v}(h_1 h_3 D_v) + \frac{\partial}{\partial w}(h_1 h_2 D_w) \right] \;} \tag{2.51}
$$

where h_1, h_2, and h_3 are the metric coefficients given in Eq. (2.38a, b, c). In the three coordinate systems,

$$
\boxed{\; \nabla \cdot \mathbf{D} = \frac{\partial D_x}{\partial x} + \frac{\partial D_y}{\partial y} + \frac{\partial D_z}{\partial z} \;} \quad \text{(Cartesian)} \tag{2.52a}
$$

$$\boxed{\nabla \cdot \mathbf{D} = \frac{1}{\rho}\frac{\partial}{\partial \rho}(\rho D_\rho) + \frac{1}{\rho}\frac{\partial D_\phi}{\partial \phi} + \frac{\partial D_z}{\partial z}} \quad \text{(cylindrical)} \tag{2.52b}$$

$$\boxed{\nabla \cdot \mathbf{D} = \frac{1}{R^2}\frac{\partial}{\partial R}(R^2 D_R) + \frac{1}{R \sin\theta}\frac{\partial}{\partial \theta}(D_\theta \sin\theta) + \frac{1}{R \sin\theta}\frac{\partial D_\phi}{\partial \phi}} \quad \text{(spherical)}$$

$$\tag{2.52c}$$

Although the del operator is defined in the Cartesian coordinates, the notation $\nabla \cdot \mathbf{D}$ is used to denote the divergence of $\mathbf{D}$ in the cylindrical and spherical coordinates as well, without suggesting dot product of ∇ and $\mathbf{D}$.

Example 2.10 Find the divergence of the following vector fields in region $R > 0$:

(a) $\mathbf{D} = \dfrac{1}{R^2}\mathbf{a}_R$, and $\tag{2.53a}$

(b) $\mathbf{A} = \dfrac{1}{R}\mathbf{a}_R.$ $\tag{2.53b}$

Solution

(a)   From (2.52c),

$$\nabla \cdot \mathbf{D} = \frac{1}{R^2}\frac{\partial}{\partial R}\left(R^2 \frac{1}{R^2}\right) = 0 \tag{2.53c}$$

A vector field with zero divergence is called a ***solenoidal field***.

(b)   Similarly,

$$\nabla \cdot \mathbf{A} = \frac{1}{R^2}\frac{\partial}{\partial R}\left(R^2 \frac{1}{R}\right) = \frac{1}{R^2} \tag{2.53d}$$

The divergence of $\mathbf{D}$ and divergence of $\mathbf{A}$ are certainly not the same, even if the flux lines of $\mathbf{D}$ and $\mathbf{A}$ appear to be the same when drawn on the paper. In fact, they appear as shown in Fig. 2.12c. A zero divergence of $\mathbf{D}$ ensures that there is no source or sink in region $R > 0$, whereas a positive divergence of $\mathbf{A}$ indicates that the source of $\mathbf{A}$ is distributed throughout the given region. The question then arises as to the source of $\mathbf{D}$. The surface integral of $\mathbf{D}$ over any sphere centered at the origin yields the same value, implying that a point source is located at the origin (see Exercise 2.17).

The simplest possible ***divergence source*** takes the form of a point in three-dimensional space. This produces a vector field that is directed away from the source and has a magnitude that is inversely proportional to the squared radial distance based on the conservation of the total flux, as shown in Eq. (2.53a). This vector field has a ***singularity*** at the source point where no derivatives of the field can be taken.

However, the point of observation can never coincide with the source point, because there is no way to bring a field detector to the point where it is already occupied by the source. If a large number of point sources exist in a given region, it is more convenient to treat them as a continuous distribution of sources, called a ***distributed source***, which is specified by the number of sources per unit volume. Accordingly, the divergence of the flux density is directly related to the value of the distributed source at a given point, under the condition that a unit source produces a unity flux.

Example 2.11 Verify vector identity $\nabla \cdot (V\mathbf{A}) = (\nabla V) \cdot \mathbf{A} + V(\nabla \cdot \mathbf{A})$ using direct substitution.

Solution

Using a general expression $\mathbf{A} = A_x\,\mathbf{a}_x + A_y\,\mathbf{a}_y + A_z\,\mathbf{a}_z$, we write

$$
\begin{aligned}
\nabla \cdot (V\mathbf{A}) &= \left[\frac{\partial}{\partial x}(V A_x) + \frac{\partial}{\partial y}(V A_y) + \frac{\partial}{\partial z}(V A_z) \right] \\
&= \left[\frac{\partial V}{\partial x} A_x + \frac{\partial V}{\partial y} A_y + \frac{\partial V}{\partial z} A_z \right] + \left[V \frac{\partial A_x}{\partial x} + V \frac{\partial A_y}{\partial y} + V \frac{\partial A_z}{\partial z} \right]
\end{aligned}
$$

In vector notation,

$$
\begin{aligned}
\nabla \cdot (V\mathbf{A}) &= \left(\frac{\partial V}{\partial x}\mathbf{a}_x + \frac{\partial V}{\partial y}\mathbf{a}_y + \frac{\partial V}{\partial z}\mathbf{a}_z \right) \cdot \left(A_x\,\mathbf{a}_x + A_y\,\mathbf{a}_y + A_z\,\mathbf{a}_z \right) \\
&\quad + V\left(\frac{\partial A_x}{\partial x} + \frac{\partial A_y}{\partial y} + \frac{\partial A_z}{\partial z} \right)
\end{aligned}
$$

Thus,

$$
\boxed{\nabla \cdot (V\mathbf{A}) = (\nabla V) \cdot \mathbf{A} + V(\nabla \cdot \mathbf{A})}
\tag{2.54}
$$

Exercise 2.14
Find the divergence of the following fields at point $(x, y, z) = (1, 1, \sqrt{6})$:
(a) $\mathbf{A} = x^2\,\mathbf{a}_x + 3z\,\mathbf{a}_y + yz\,\mathbf{a}_z$, and (b) $\mathbf{B} = (3\cos\phi / R^2)\,\mathbf{a}_R + (4\sin\theta / R)\,\mathbf{a}_\theta - 2\mathbf{a}_\phi$.

Ans. (a) 3, (b) $\sqrt{3}/2$.

Exercise 2.15
If $\mathbf{D}$ is in coulombs per square meter, what is the unit of $\nabla \cdot \mathbf{D}$?

Ans. Coulomb per cubic meter.

Exercise 2.16
Given a vector field $\mathbf{D}(\mathbf{r}) = \mathbf{A}\cos(\mathbf{k} \cdot \mathbf{r})$ with position vector $\mathbf{r}$ and constant vectors $\mathbf{A}$ and $\mathbf{k}$, under what conditions does $\mathbf{D}$ become solenoidal?

Ans. $\mathbf{A} \perp \mathbf{k}$.

Exercise 2.17
For $\mathbf{D}$ and $\mathbf{A}$, as given in Eqs. (2.53a) and (2.53b), find the magnitude of the source located at the origin by assuming that a unit source produces a unity flux.

Ans. (a) $\oint_S \mathbf{D} \cdot d\mathbf{s} = 4\pi$, (b) $\oint_S \mathbf{A} \cdot d\mathbf{s} = 4\pi R$, $\lim_{\Delta v \to 0} \oint_S \mathbf{A} \cdot d\mathbf{s} \Rightarrow 0$.

2.4.2 Divergence Theorem

The ***divergence theorem*** follows directly from the definition of the divergence. *The integral of $\nabla \cdot \mathbf{D}$ over a volume is equal to the closed-surface integral of $\mathbf{D}$ over the volume boundary.* The divergence theorem is expressed as

$$\int_V \nabla \cdot \mathbf{D}\, dv = \oint_S \mathbf{D} \cdot d\mathbf{s} \tag{2.55}$$

where S is the surface bounding volume V, and $d\mathbf{s}$ is the differential area vector on S pointing outward from the volume.

To verify the divergence theorem, we begin with a finite volume V bounded by the surface S, as shown in Fig. 2.14. First, the volume is subdivided into a large number of infinitesimal elements. The three volume elements Δv_{k-1}, Δv_k, and Δv_{k+1} are positioned next to each other, with their top sides coincident with the boundary surface. Moreover, the sides of Δv_k are shared by five adjoining elements, except for the top side. On the shared side, the two unit surface normals are directed out of their respective volume elements, pointing in opposite directions.

Let us construct a divergence of $\mathbf{D}$ over Δv_k such that

$$(\nabla \cdot \mathbf{D})_{\Delta v_k} = \lim_{\Delta v_k \to 0} \frac{\oint_{\Delta s_k} \mathbf{D} \cdot d\mathbf{s}}{\Delta v_k} \tag{2.56}$$

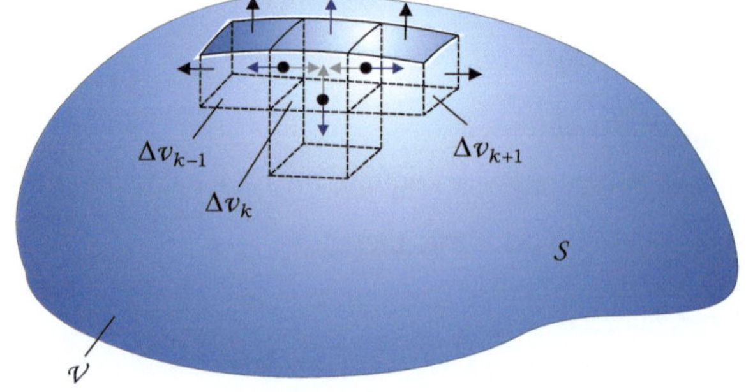

Fig. 2.14 Subdivision of volume V bounded by surface S

where Δs_k is the surface bounding differential volume Δv_k. Multiplying both sides of Eq. (2.56) with Δv_k gives

$$\lim_{\Delta v_k \to 0}\left[(\nabla \cdot \mathbf{D})_{\Delta v_k} \Delta v_k\right] = \lim_{\Delta s_k \to 0}\left[\oint_{\Delta s_k} \mathbf{D} \cdot d\mathbf{s}\right] \tag{2.57}$$

The brackets on the right-hand side of Eq. (2.57) represent the flux leaving Δv_k. Because the flux is conserved in space, it eventually reaches boundary surface $\mathcal{S}$ and crosses it without loss. By extending this argument to all elements of the volume, we can conclude that the total flux passing through $\mathcal{S}$ is equal to the sum of all fluxes leaving the individual volume elements. That is,

$$\oint_{\mathcal{S}} \mathbf{D} \cdot d\mathbf{s} = \sum_{k=1}^{k=N}\left[\oint_{\Delta s_k} \mathbf{D} \cdot d\mathbf{s}\right] \tag{2.58}$$

This relation can also be verified mathematically by noting that its right-hand side is equal to the sum of the surface integrals of $\mathbf{D}$ conducted over all sides of all the volume elements. If a side is shared by two adjoining elements, it does not contribute to the summation because the surface integral is conducted over the same side twice but uses two differential area vectors pointing in opposite directions. Therefore, the right-hand side of Eq. (2.58) is reduced to the surface integral of $\mathbf{D}$ over $\mathcal{S}$, as shown on the left-hand side of Eq. (2.58).

Next, extending Eq. (2.57) to all the volume elements and combining it with Eq. (2.58), we have

$$\lim_{N \to \infty} \sum_{k=1}^{k=N}\left[(\nabla \cdot \mathbf{D})_{\Delta v_k} \Delta v_k\right] = \oint_{\mathcal{S}} \mathbf{D} \cdot d\mathbf{s} \tag{2.59}$$

In the limit as $N \to \infty$ and $\Delta v_k \to 0$, the left-hand side of Eq. (2.59), by definition, is the volume integral of $\nabla \cdot \mathbf{D}$ over the volume $\mathcal{V}$. Thus, the divergence theorem is verified.

Example 2.12 For flux density $\mathbf{A}(\mathbf{r}) = x^2 y\,\mathbf{a}_x - x^2 y\,\mathbf{a}_y + z\,\mathbf{a}_z$, verify the divergence theorem over a cube of two units on each side centered at the origin, as shown in Fig. 2.15a.

Solution

The surface integrals of $\mathbf{A}$ over the six sides of the cube are

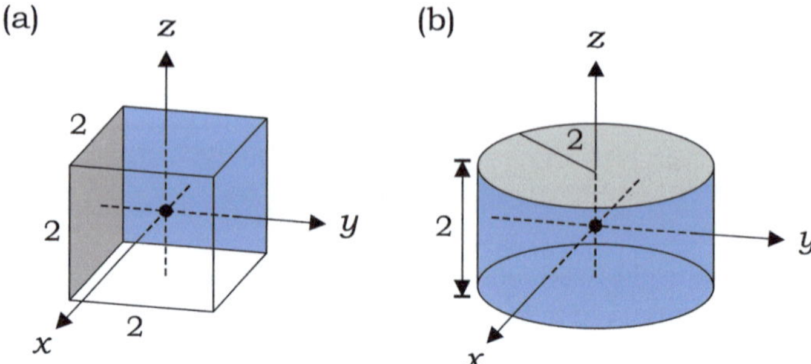

Fig. 2.15 Cube and cylinder

$$\int_{S_1} \mathbf{A} \cdot d\mathbf{s} = \int_{z=-1}^{z=1} \int_{y=-1}^{y=1} \left(x^2 y \mathbf{a}_x - x^2 y \mathbf{a}_y + z \mathbf{a}_z\right) \cdot (dy\,dz\,\mathbf{a}_x)$$

$$= (1)^2 \int_{y=-1}^{y=1} y\,dy \int_{z=-1}^{z=1} dz = 0 \quad (\text{side at } x = 1)$$

$$\int_{S_2} \mathbf{A} \cdot d\mathbf{s} = \int_{z=-1}^{z=1} \int_{y=-1}^{y=1} \mathbf{A} \cdot (-dy\,dz\,\mathbf{a}_x)$$

$$= -(-1)^2 \int_{y=-1}^{y=1} y\,dy \int_{z=-1}^{z=1} dz = 0 \quad (x = -1)$$

$$\int_{S_3} \mathbf{A} \cdot d\mathbf{s} = \int_{z=-1}^{z=1} \int_{x=-1}^{x=1} \mathbf{A} \cdot (dx\,dz\,\mathbf{a}_y)$$

$$= -(1) \int_{x=-1}^{x=1} x^2\,dx \int_{z=-1}^{z=1} dz = -\frac{4}{3} \quad (y = 1)$$

$$\int_{S_4} \mathbf{A} \cdot d\mathbf{s} = \int_{z=-1}^{z=1} \int_{x=-1}^{x=1} \mathbf{A} \cdot (-dx\,dz\,\mathbf{a}_y)$$

$$= (-1) \int_{x=-1}^{x=1} x^2\,dx \int_{z=-1}^{z=1} dz = -\frac{4}{3} \quad (y = -1)$$

$$\int_{S_5} \mathbf{A} \cdot d\mathbf{s} = \int_{y=-1}^{y=1} \int_{x=-1}^{x=1} \mathbf{A} \cdot (dx\,dy\,\mathbf{a}_z)$$

$$= (1) \int_{x=-1}^{x=1} dx \int_{y=-1}^{y=1} dy = 4 \quad (z = 1)$$

$$\int_{S_6} \mathbf{A} \cdot d\mathbf{s} = \int_{y=-1}^{y=1} \int_{x=-1}^{x=1} \mathbf{A} \cdot (-dx\,dy\,\mathbf{a}_z)$$

$$= -(-1) \int_{x=-1}^{x=1} dx \int_{y=-1}^{y=1} dy = 4 \quad (z = -1)$$

Combining the above results, the closed-surface integral of $\mathbf{A}$ is obtained as

$$\oint_S \mathbf{A} \cdot d\mathbf{s} = \frac{16}{3} \tag{2.60a}$$

Next, in Cartesian coordinates, the divergence of $\mathbf{A}$ is given by

$$\nabla \cdot \mathbf{A} = \frac{\partial}{\partial x}(x^2 y) + \frac{\partial}{\partial y}(-x^2 y) + \frac{\partial}{\partial z}(z) = 2xy - x^2 + 1$$

The volume integral of $\nabla \cdot \mathbf{A}$ over the cube is calculated as

$$\int_V \nabla \cdot \mathbf{A} \, dv = \int_{z=-1}^{z=1} \int_{y=-1}^{y=1} \int_{x=-1}^{x=1} (2xy - x^2 + 1) \, dx\,dy\,dz$$

$$= 2 \int_{x=-1}^{x=1} x\,dx \int_{y=-1}^{y=1} y\,dy \int_{z=-1}^{z=1} dz - \int_{x=-1}^{x=1} x^2\,dx \int_{y=-1}^{y=1} dy \int_{z=-1}^{z=1} dz$$

$$+ \int_{x=-1}^{x=1} dx \int_{y=-1}^{y=1} dy \int_{z=-1}^{z=1} dz = \frac{16}{3} \tag{2.60b}$$

The results in Eqs. (2.60a) and (2.60b) are equal. Thus, the divergence theorem is verified.

Example 2.13 For flux density $\mathbf{B}(\mathbf{r}) = \sin\phi\,\mathbf{a}_\phi$, verify the divergence theorem over a cylinder of radius 2 and height 2, centered at the origin, as shown in Fig. 2.15b.

Solution

The closed-surface integral of $\mathbf{B}$ is separated into three parts:

$$\int_{cylin} \mathbf{B} \cdot d\mathbf{s} = \int_{z=-1}^{z=1} \int_{\phi=0}^{\phi=2\pi} \sin\phi\,\mathbf{a}_\phi \cdot (2d\phi\,dz\,\mathbf{a}_\rho) = 0 \quad \text{(cylindrical surface)}$$

$$\int_{top} \mathbf{B} \cdot d\mathbf{s} = \int_{\phi=0}^{\phi=2\pi} \int_{\rho=0}^{\rho=2} \sin\phi\,\mathbf{a}_\phi \cdot (\rho\,d\rho\,d\phi\,\mathbf{a}_z) = 0 \quad \text{(top plate)}$$

$$\int_{bottom} \mathbf{B} \cdot d\mathbf{s} = \int_{\phi=0}^{\phi=2\pi} \int_{\rho=0}^{\rho=2} \sin\phi \, \mathbf{a}_\phi \cdot (-\rho d\rho d\phi \, \mathbf{a}_z) = 0 \quad \text{(bottom plate)}$$

Thus,

$$\oint_S \mathbf{B} \cdot d\mathbf{s} = \int_{cylind} \mathbf{B} \cdot d\mathbf{s} + \int_{top} \mathbf{B} \cdot d\mathbf{s} + \int_{bottom} \mathbf{B} \cdot d\mathbf{s} = 0 \qquad (2.61a)$$

In cylindrical coordinates, the divergence of $\mathbf{B}$ is

$$\nabla \cdot \mathbf{B} = \frac{1}{\rho}\frac{\partial}{\partial\rho}(\rho B_\rho) + \frac{1}{\rho}\frac{\partial B_\phi}{\partial\phi} + \frac{\partial B_z}{\partial z} = \frac{1}{\rho}\cos\phi \qquad (2.61b)$$

Thus, the volume integral of $\nabla \cdot \mathbf{B}$ over the cylinder is calculated as

$$\int_V \nabla \cdot \mathbf{B} \, dv = \int_{z=-1}^{z=1}\int_{\phi=0}^{\phi=2\pi}\int_{\rho=0}^{\rho=2} \frac{1}{\rho}(\cos\phi)\rho d\rho d\phi dz = 4\sin\phi\Big|_{\phi=0}^{\phi=2\pi} = 0$$

$$(2.61c)$$

Two results in Eqs. (2.61a) and (2.61c) are equal. Thus, the divergence theorem is verified.

Although the volume integral in Eq. (2.61c) vanishes, its integrand is not zero inside the cylinder, as shown in Eq. (2.61b). This implies that fluxes emanate from the distributed source in region $-\pi/2 < \phi < \pi/2$ ($\nabla \cdot \mathbf{B} > 0$) and terminate at the distributed sink in region $\pi/2 < \phi < 3\pi/2$ ($\nabla \cdot \mathbf{B} < 0$). Divergence provides a simple method for checking for the existence of a distributed source or sink in space.

Example 2.14 Given $\mathbf{A}(\mathbf{r}) = (1/R^2)\,\mathbf{a}_R$, verify the divergence theorem over the region between two concentric spheres of radii R_1 and R_2, as shown in Fig. 2.16.

Solution

The differential area vector always points outward such that $d\mathbf{s}$ is along $\mathbf{a}_R$ on the outer sphere, but along $-\mathbf{a}_R$ on the inner sphere.

Fig. 2.16 Hollow sphere

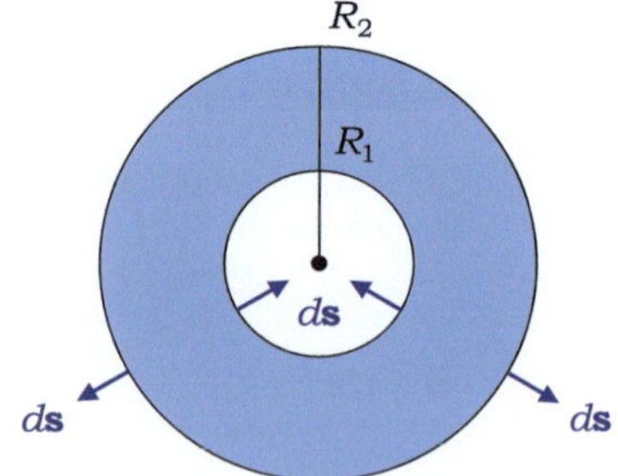

The surface integrals of $\mathbf{A}$ over the spheres are given by

$$\int_{outer} \mathbf{A} \cdot d\mathbf{s} = \int_{\phi=0}^{\phi=2\pi} \int_{\theta=0}^{\theta=\pi} \frac{1}{R_2^2} \mathbf{a}_R \cdot (R_2^2 \sin\theta \, d\theta \, d\phi \, \mathbf{a}_R) = 4\pi \qquad (2.62a)$$

$$\int_{inner} \mathbf{A} \cdot d\mathbf{s} = \int_{\phi=0}^{\phi=2\pi} \int_{\theta=0}^{\theta=\pi} \frac{1}{R_1^2} \mathbf{a}_R \cdot (-R_1^2 \sin\theta \, d\theta \, d\phi \, \mathbf{a}_R) = -4\pi \qquad (2.62b)$$

Thus, the closed-surface integral of $\mathbf{A}$ is

$$\oint_S \mathbf{A} \cdot d\mathbf{s} = 4\pi - 4\pi = 0 \qquad (2.62c)$$

Next, the divergence of $\mathbf{A}$ in spherical coordinates is

$$\nabla \cdot \mathbf{A} = \frac{1}{R^2} \frac{\partial}{\partial R}\left(R^2 \frac{1}{R^2} \right) = 0 \qquad (2.62d)$$

Thus, the divergence theorem is verified.

We see from Eqs. (2.62a) and (2.62b), a net flux of 4π enters the given region through the inner sphere, as indicated by the negative value, whereas a net flux of 4π leaves the region through the outer sphere, as indicated by the positive value. Furthermore, $\nabla \cdot \mathbf{A} = 0$ in the given region, implying that no fluxes are generated or annihilated within the region.

Exercise 2.18

Under what conditions does the closed-surface integral of $\mathbf{A}$ vanish even if $\mathbf{A}$ is nonzero in the enclosed volume $\mathcal{V}$?

Ans. $\mathcal{V}$ excludes the source and sink or all fluxes from the source in $\mathcal{V}$ terminate at the sink in $\mathcal{V}$.

Review Questions

RQ 2.20	Define $div\,\mathbf{D}$.	[(2.45)]
RQ 2.21	Does $\nabla \cdot \mathbf{A} = 0$ at a point directly lead to $\mathbf{A} = 0$ at this point?	[(2.45)]
RQ 2.22	Express $div\,\mathbf{D}$ in the three coordinates systems.	[(2.52a,b,c)]
RQ 2.23	What is the solenoidal field?	[(Example 2.10)]
RQ 2.24	State the divergence theorem in words.	[(2.55)]
RQ 2.25	Is $\nabla \cdot \mathbf{D}$ directly related to the distributed source of $\mathbf{D}$?	[(2.57)]
RQ 2.26	What is the unit of the distributed source of the flux density?	[Exercise 2.15]
RQ 2.27	How can a point source of the flux density be detected?	[Exercise 2.17]
RQ 2.28	What are the practical applications of divergence theorem?	[(2.55)]

2.5 Curl and Stokes's Theorem

As stated previously, the *circulation of* **H** *around* C *at* point p_1 is the closed-line integral of the vector field around loop C, centered at p_1. The simplest possible **circulation source** takes the form of a straight line that produces field lines in the form of concentric circles around it. The discussion in this section is based on the premise that the circulation of vector field is conserved in space only if the integration path encloses the same circulation source. **Curl** is a differential operator acting on a vector field, which is typically measured in units of certain quantities per unit length. This is useful for relating the circulation of the vector field to the source enclosed by the integration path. Stokes's theorem follows directly from the definition of the curl and is particularly useful for converting a closed-line integral into a surface integral, and vice versa.

2.5.1 Curl of the Vector Field

The curl of the vector field **H** is denoted as *curl* **H** and yields a vector at a given point in space. The definition of curl involves an imaginary loop centered at a given point, and states that *the component of curl* **H** *in the* $\mathbf{a}_k$-*direction at point* p_1 *is the ratio of the circulation of* **H** *around a loop, which is centered at* p_1 *and oriented so as to make its surface perpendicular to* $\mathbf{a}_k$, *to the loop area as the loop shrinks to a point at* p_1. Expressed mathematically, the k-component of *curl* **H** is defined as

$$(curl\ \mathbf{H}) \cdot \mathbf{a}_k = \lim_{\Delta s \to 0} \frac{\oint_C \mathbf{H} \cdot d\mathbf{l}}{\Delta s} \tag{2.63}$$

where C is the closed loop centered at p_1 with its surface perpendicular to $\mathbf{a}_k$, and Δs is the area of the loop. The connection between the differential area vector $d\mathbf{s}$ on Δs and the differential length vector $d\mathbf{l}$ on C is governed by the right-hand rule; the right thumb points in the direction of $d\mathbf{s}$ when the fingers advance towards $d\mathbf{l}$.

Combining the x-, y-, and z-components obtained using Eq. (2.63) leads to the curl of **H** in Cartesian coordinates. Alternatively, the curl of **H** can be defined as follows: *the curl of* **H** *at point* p_1 *is a vector whose magnitude is the maximum possible circulation of* **H** *per unit area observed at* p_1, *and whose unit vector is normal to the loop surface oriented so as to yield the maximum.* Expressed mathematically,

$$curl\ \mathbf{H} = \lim_{\Delta s \to 0} \frac{\oint_C \mathbf{H} \cdot d\mathbf{l}}{\Delta s} \mathbf{a}_n \tag{2.64}$$

where $\mathcal{C}$ is the loop centered at p_1, oriented to yield the maximum circulation per unit area, Δs is the area of the loop, and $\mathbf{a}_n$ is a unit normal to the loop surface. Again, $d\mathbf{s}$ on Δs and $d\mathbf{l}$ on $\mathcal{C}$ are related by the right-hand rule.

In Eq. (2.63), when a unit vector $\mathbf{a}_k$ is given, the loop surface is oriented perpendicular to $\mathbf{a}_k$ for the component of *curl* $\mathbf{H}$ in the $\mathbf{a}_k$-direction. Meanwhile, in Eq. (2.64), the loop is first oriented to yield the maximum circulation per unit area. Subsequently, the unit normal to the loop surface, $\mathbf{a}_n$, is identified as the direction of *curl* $\mathbf{H}$. At this point, the question arises as to whether the *curl* $\mathbf{H}$ obtained using Eq. (2.63) corresponds to the maximum possible circulation of $\mathbf{H}$ per unit area at a given point, as expressed in Eq. (2.64). This issue will be addressed in the following sections.

First, the component of *curl* $\mathbf{H}$ in an arbitrary direction $\mathbf{a}_k$ at point p_1 is obtained by taking a rectangular loop as the integration path, which is centered at p_1 and oriented perpendicular to $\mathbf{a}_k$, as shown in Fig. 2.17. For simplicity, we assume that $\mathbf{a}_k = \sin\gamma\,\mathbf{a}_x + \cos\gamma\,\mathbf{a}_z$. As the rectangle shrinks to a point at p_1, the circulation of $\mathbf{H}$ around the rectangle can be expressed as

$$\lim_{\Delta s \to 0} \oint_{\mathcal{C}} \mathbf{H} \cdot d\mathbf{l} = \lim_{\Delta s \to 0}\left[\mathbf{H}^1 \cdot d\mathbf{l}^1 + \mathbf{H}^2 \cdot d\mathbf{l}^2 + \mathbf{H}^3 \cdot d\mathbf{l}^3 + \mathbf{H}^4 \cdot d\mathbf{l}^4\right] \qquad (2.65)$$

where superscript 1 denotes side ① along which $\mathbf{H}$ is assumed to be constant, $\mathbf{H}^1$, and so on. Again, as the loop eventually shrinks to a point, it is more convenient to expand $\mathbf{H}$ in the Taylor series centered at p_1. That is,

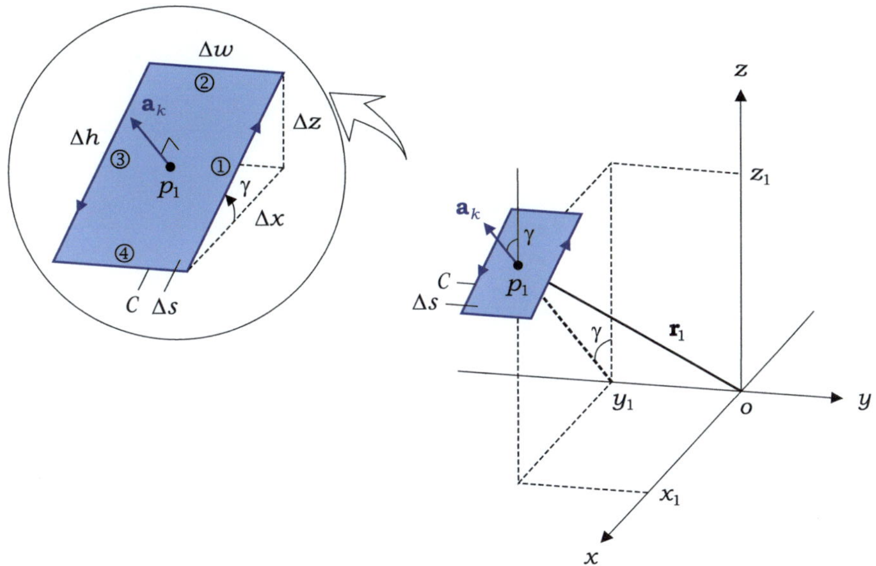

Fig. 2.17 Rectangular loop used for the calculation of $(curl\,\mathbf{H}) \cdot \mathbf{a}_k$

$$\mathbf{H}^1 \cdot d\mathbf{l}^1 = \left[\mathbf{H}_o + \frac{\Delta w}{2} \left. \frac{\partial \mathbf{H}}{\partial y} \right|_{\text{at } p_1} \right] \cdot d\mathbf{l}^1 \quad (d\mathbf{l}^1 = -\Delta x \, \mathbf{a}_x + \Delta z \, \mathbf{a}_z) \qquad (2.66a)$$

$$\mathbf{H}^2 \cdot d\mathbf{l}^2 = \left[\mathbf{H}_o - \frac{\Delta x}{2} \left. \frac{\partial \mathbf{H}}{\partial x} \right|_{\text{at } p_1} + \frac{\Delta z}{2} \left. \frac{\partial \mathbf{H}}{\partial z} \right|_{\text{at } p_1} \right] \cdot d\mathbf{l}^2 \quad (d\mathbf{l}^2 = -\Delta w \, \mathbf{a}_y) \quad (2.66b)$$

$$\mathbf{H}^3 \cdot d\mathbf{l}^3 = \left[\mathbf{H}_o - \frac{\Delta w}{2} \left. \frac{\partial \mathbf{H}}{\partial y} \right|_{\text{at } p_1} \right] \cdot d\mathbf{l}^3 \quad (d\mathbf{l}^3 = \Delta x \, \mathbf{a}_x - \Delta z \, \mathbf{a}_z) \qquad (2.66c)$$

$$\mathbf{H}^4 \cdot d\mathbf{l}^4 = \left[\mathbf{H}_o + \frac{\Delta x}{2} \left. \frac{\partial \mathbf{H}}{\partial x} \right|_{\text{at } p_1} - \frac{\Delta z}{2} \left. \frac{\partial \mathbf{H}}{\partial z} \right|_{\text{at } p_1} \right] \cdot d\mathbf{l}^4 \quad (d\mathbf{l}^4 = \Delta w \, \mathbf{a}_y) \quad (2.66d)$$

where higher-order terms in the Taylor series expansion are ignored. Although $\mathbf{H}_o$ and the partial derivatives of $\mathbf{H}$ have values specific to point p_1, the quantity in brackets is the value of $\mathbf{H}$ on the side of the rectangle. Combining the expressions in Eqs. (2.66a)–(2.66d) gives

$$(\mathbf{H}^1 \cdot d\mathbf{l}^1 + \mathbf{H}^3 \cdot d\mathbf{l}^3) + (\mathbf{H}^2 \cdot d\mathbf{l}^2 + \mathbf{H}^4 \cdot d\mathbf{l}^4)$$

$$= -\Delta w \Delta x \left(\mathbf{a}_x \cdot \left. \frac{\partial \mathbf{H}}{\partial y} \right|_{p_1} \right) + \Delta w \Delta z \left(\mathbf{a}_z \cdot \left. \frac{\partial \mathbf{H}}{\partial y} \right|_{p_1} \right)$$

$$+ \Delta w \Delta x \left(\mathbf{a}_y \cdot \left. \frac{\partial \mathbf{H}}{\partial x} \right|_{p_1} \right) - \Delta w \Delta z \left(\mathbf{a}_y \cdot \left. \frac{\partial \mathbf{H}}{\partial z} \right|_{p_1} \right)$$

$$= \Delta w \Delta z \left(\left. \frac{\partial H_z}{\partial y} \right|_{p_1} - \left. \frac{\partial H_y}{\partial z} \right|_{p_1} \right) + \Delta w \Delta x \left(\left. \frac{\partial H_y}{\partial x} \right|_{p_1} - \left. \frac{\partial H_x}{\partial y} \right|_{p_1} \right) \qquad (2.67)$$

where $\partial \mathbf{H}/\partial x = (\partial H_x/\partial x) \, \mathbf{a}_x + (\partial H_y/\partial x) \, \mathbf{a}_y + (\partial H_z/\partial x) \, \mathbf{a}_z$, etc.

Substituting Eqs. (2.65) and (2.67) into Eq. (2.63) and using $\Delta s = \Delta w \Delta h$, we obtain

$$(curl \, \mathbf{H})_k = \lim_{\Delta s \to 0} \left\{ \frac{\Delta z}{\Delta h} \left(\left. \frac{\partial H_z}{\partial y} \right|_{p_1} - \left. \frac{\partial H_y}{\partial z} \right|_{p_1} \right) + \frac{\Delta x}{\Delta h} \left(\left. \frac{\partial H_y}{\partial x} \right|_{p_1} - \left. \frac{\partial H_x}{\partial y} \right|_{p_1} \right) \right\}$$

$$= \sin \gamma \left(\frac{\partial H_z}{\partial y} - \frac{\partial H_y}{\partial z} \right) + \cos \gamma \left(\frac{\partial H_y}{\partial x} - \frac{\partial H_x}{\partial y} \right) \qquad (2.68)$$

where $\Delta z/\Delta h = \sin \gamma$ and $\Delta x/\Delta h = \cos \gamma$, as can be seen from Fig. 2.17.

Setting $\gamma = 90^o$ in Eq. (2.68), the unit vector $\mathbf{a}_k$ is directed along $\mathbf{a}_x$ (Fig. 2.17). Thus, the x-component of the curl of $\mathbf{H}$ is obtained as

$$(curl\,\mathbf{H})_x = \frac{\partial H_z}{\partial y} - \frac{\partial H_y}{\partial z} \qquad (2.69a)$$

Similarly, setting $\gamma = 0^o$ in Eq. (2.68) yields the z-component, that is,

$$(curl\,\mathbf{H})_z = \frac{\partial H_y}{\partial x} - \frac{\partial H_x}{\partial y} \qquad (2.69b)$$

A similar procedure can be followed to obtain the y-component such that

$$(curl\,\mathbf{H})_y = \frac{\partial H_x}{\partial z} - \frac{\partial H_z}{\partial x} \qquad (2.69c)$$

Next, inserting Eqs. (2.69a, b) into Eq. (2.68), along with the relations $\mathbf{a}_x \cdot \mathbf{a}_k = \sin\gamma$ and $\mathbf{a}_z \cdot \mathbf{a}_k = \cos\gamma$, and extending Eq. (2.68) to an arbitrary direction $\mathbf{a}_t$ in three-dimensional space to include the y-component, we have

$$\begin{aligned}
(curl\,\mathbf{H})_t &= (curl\,\mathbf{H})_x\,\mathbf{a}_x \cdot \mathbf{a}_t + (curl\,\mathbf{H})_y\,\mathbf{a}_y \cdot \mathbf{a}_t + (curl\,\mathbf{H})_z\,\mathbf{a}_z \cdot \mathbf{a}_t \\
&= \left[(curl\,\mathbf{H})_x\,\mathbf{a}_x + (curl\,\mathbf{H})_y\,\mathbf{a}_y + (curl\,\mathbf{H})_z\,\mathbf{a}_z \right] \cdot \mathbf{a}_t \qquad (2.70)
\end{aligned}$$

Here, let us remind ourselves that $curl\,\mathbf{H}$ in the $\mathbf{a}_t$-direction, as expressed in Eq. (2.70), was obtained using Eq. (2.63), and note that the expression in brackets on the right-hand side of Eq. (2.70) is $curl\,\mathbf{H}$ in the component form. If the unit vector $\mathbf{a}_t$ is taken along the direction of $curl\,\mathbf{H}$, the right-hand side of Eq. (2.70) is the greatest, and the circulation of $\mathbf{H}$ per unit area on the left-hand side of Eq. (2.70) is the maximum. Consequently, $curl\,\mathbf{H}$ in component form corresponds to the maximum possible circulation of $\mathbf{H}$ per unit area at a given point.

Combining the three components in Eqs. (2.69a, b, c) leads to $curl\,\mathbf{H}$ in Cartesian coordinates:

$$curl\,\mathbf{H} = \left(\frac{\partial H_z}{\partial y} - \frac{\partial H_y}{\partial z} \right)\mathbf{a}_x + \left(\frac{\partial H_x}{\partial z} - \frac{\partial H_z}{\partial x} \right)\mathbf{a}_y + \left(\frac{\partial H_y}{\partial x} - \frac{\partial H_x}{\partial y} \right)\mathbf{a}_z \qquad (2.71)$$

In the component form, $curl\,\mathbf{H}$ is expressed as

$$curl\,\mathbf{H} = \nabla \times \mathbf{H} = \begin{vmatrix} \mathbf{a}_x & \mathbf{a}_y & \mathbf{a}_z \\ \dfrac{\partial}{\partial x} & \dfrac{\partial}{\partial y} & \dfrac{\partial}{\partial z} \\ H_x & H_y & H_z \end{vmatrix}$$

In general, for the orthogonal curvilinear coordinates (u, v, w), the curl of $\mathbf{H}$ is expressed as

$$\nabla \times \mathbf{H} = \frac{1}{h_1 h_2 h_3} \begin{vmatrix} h_1\,\mathbf{a}_u & h_2\,\mathbf{a}_v & h_3\,\mathbf{a}_w \\ \dfrac{\partial}{\partial u} & \dfrac{\partial}{\partial v} & \dfrac{\partial}{\partial w} \\ h_1 H_u & h_2 H_v & h_3 H_w \end{vmatrix} \tag{2.72}$$

where h_1, h_2, and h_3 are the metric coefficients given in Eq. (2.38).

In the three coordinate systems, the curl of $\mathbf{H}$ is given by

$$\nabla \times \mathbf{H} = \begin{vmatrix} \mathbf{a}_x & \mathbf{a}_y & \mathbf{a}_z \\ \dfrac{\partial}{\partial x} & \dfrac{\partial}{\partial y} & \dfrac{\partial}{\partial z} \\ H_x & H_y & H_z \end{vmatrix} \quad \text{(Cartesian)} \tag{2.73a}$$

$$\nabla \times \mathbf{H} = \frac{1}{\rho} \begin{vmatrix} \mathbf{a}_\rho & \rho\,\mathbf{a}_\phi & \mathbf{a}_z \\ \dfrac{\partial}{\partial \rho} & \dfrac{\partial}{\partial \phi} & \dfrac{\partial}{\partial z} \\ H_\rho & \rho H_\phi & H_z \end{vmatrix} \quad \text{(cylindrical)} \tag{2.73b}$$

$$\nabla \times \mathbf{H} = \frac{1}{R^2 \sin\theta} \begin{vmatrix} \mathbf{a}_R & R\,\mathbf{a}_\theta & R\sin\theta\,\mathbf{a}_\phi \\ \dfrac{\partial}{\partial R} & \dfrac{\partial}{\partial \theta} & \dfrac{\partial}{\partial \phi} \\ H_R & R H_\theta & R\sin\theta\,H_\phi \end{vmatrix} \quad \text{(spherical)} \tag{2.73c}$$

Note that the curl of $\mathbf{H}$ is also denoted by $\nabla \times \mathbf{H}$ in cylindrical and spherical coordinates, without suggesting a cross product of ∇ and $\mathbf{H}$.

Example 2.15 Find the curl of the fields in region $\rho > 0$ in cylindrical coordinates:
(a) $\mathbf{F}_1 = \rho^{-2}\mathbf{a}_\phi$, (b) $\mathbf{F}_2 = \rho^{-1}\mathbf{a}_\phi$, (c) $\mathbf{F}_3 = \mathbf{a}_\phi$, and (d) $\mathbf{F}_4 = \rho^2\mathbf{a}_\phi$.

Solution

From Eq. (2.73b),

$$\text{(a)} \quad \nabla \times \mathbf{F}_1 = \frac{1}{\rho}\begin{vmatrix} \mathbf{a}_\rho & \rho\,\mathbf{a}_\phi & \mathbf{a}_z \\ \dfrac{\partial}{\partial\rho} & \dfrac{\partial}{\partial\phi} & \dfrac{\partial}{\partial z} \\ 0 & 1/\rho & 0 \end{vmatrix} = -\frac{1}{\rho^3}\mathbf{a}_z$$

$$\text{(b)} \quad \nabla \times \mathbf{F}_2 = \frac{1}{\rho}\begin{vmatrix} \mathbf{a}_\rho & \rho\,\mathbf{a}_\phi & \mathbf{a}_z \\ \dfrac{\partial}{\partial\rho} & \dfrac{\partial}{\partial\phi} & \dfrac{\partial}{\partial z} \\ 0 & 1 & 0 \end{vmatrix} = 0$$

$$\text{(c)} \quad \nabla \times \mathbf{F}_3 = \frac{1}{\rho}\begin{vmatrix} \mathbf{a}_\rho & \rho\,\mathbf{a}_\phi & \mathbf{a}_z \\ \dfrac{\partial}{\partial\rho} & \dfrac{\partial}{\partial\phi} & \dfrac{\partial}{\partial z} \\ 0 & \rho & 0 \end{vmatrix} = \frac{1}{\rho}\mathbf{a}_z$$

$$\text{(d)} \quad \nabla \times \mathbf{F}_4 = \frac{1}{\rho}\begin{vmatrix} \mathbf{a}_\rho & \rho\,\mathbf{a}_\phi & \mathbf{a}_z \\ \dfrac{\partial}{\partial\rho} & \dfrac{\partial}{\partial\phi} & \dfrac{\partial}{\partial z} \\ 0 & \rho^3 & 0 \end{vmatrix} = 3\rho\,\mathbf{a}_z$$

The field lines of these vector fields are shown in Fig. 2.18. The curl of $\mathbf{F}_2$ is zero within the given region. Such a field is called an ***irrotational*** or ***conservative field***.

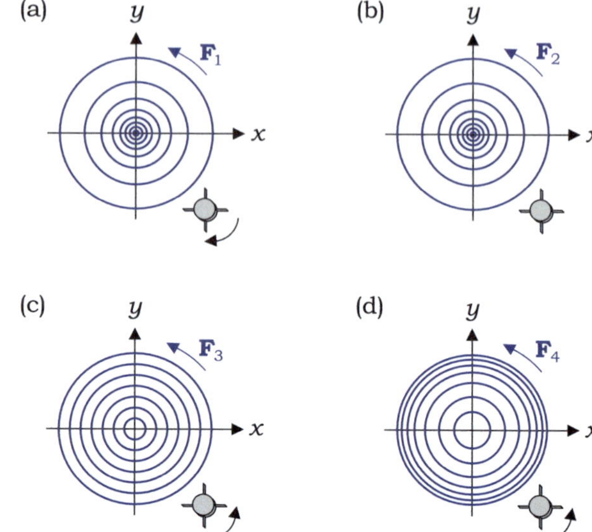

Fig. 2.18 Field lines of the four vector fields in Example 2.15. A paddle wheel detects the presence of a circulation source at a given point

However, it can be shown that the circulation of $\mathbf{F}_2$ around any loop enclosing the z-axis is nonzero, which suggests that $\mathbf{F}_2$ is produced by a circulation source residing along the z-axis.

The nonzero curls of $\mathbf{F}_1$, $\mathbf{F}_3$, and $\mathbf{F}_4$ indicate that these fields are produced by circulation sources distributed over the given region, regardless of whether an additional source exists along the z-axis. Taking the circulation of $\mathbf{F}_3$ around a small circle of radius δ centered on the z-axis, we obtain

$$\oint_C \mathbf{F}_3 \cdot d\mathbf{l} = 2\pi\delta$$

This becomes zero as $\delta \to 0$; thus, there is no extra source of $\mathbf{F}_3$ along the z-axis.

A paddle wheel may be used to detect the presence of a circulation source at a point in space; nonzero circulation causes the wheel to rotate. The direction of the curl and the direction of rotation of the wheel follow the right-hand rule: the right thumb points in the direction of the curl when the fingers follow the rotation of the wheel.

Example 2.16 Verify vector identity $\nabla \times (V\mathbf{A}) = (\nabla V) \times \mathbf{A} + V(\nabla \times \mathbf{A})$ by direct substitution.

Solution

Using $\mathbf{A} = A_x\,\mathbf{a}_x + A_y\,\mathbf{a}_y + A_z\,\mathbf{a}_z$ on the left-hand side of the identity, we obtain

$$\nabla \times (V\mathbf{A}) = \begin{vmatrix} \mathbf{a}_x & \mathbf{a}_y & \mathbf{a}_z \\ \dfrac{\partial}{\partial x} & \dfrac{\partial}{\partial y} & \dfrac{\partial}{\partial z} \\ VA_x & VA_y & VA_z \end{vmatrix}$$

$$= \mathbf{a}_x\left(\frac{\partial}{\partial y}VA_z - \frac{\partial}{\partial z}VA_y\right) + \mathbf{a}_y\left(\frac{\partial}{\partial z}VA_x - \frac{\partial}{\partial x}VA_z\right) + \mathbf{a}_z\left(\frac{\partial}{\partial x}VA_y - \frac{\partial}{\partial y}VA_x\right)$$

$$= \mathbf{a}_x\left(\frac{\partial V}{\partial y}A_z - \frac{\partial V}{\partial z}A_y\right) + \mathbf{a}_y\left(\frac{\partial V}{\partial z}A_x - \frac{\partial V}{\partial x}A_z\right) + \mathbf{a}_z\left(\frac{\partial V}{\partial x}A_y - \frac{\partial V}{\partial y}A_x\right)$$

$$+ \mathbf{a}_x V\left(\frac{\partial A_z}{\partial y} - \frac{\partial A_y}{\partial z}\right) + \mathbf{a}_y V\left(\frac{\partial A_x}{\partial z} - \frac{\partial A_z}{\partial x}\right) + \mathbf{a}_z V\left(\frac{\partial A_y}{\partial x} - \frac{\partial A_x}{\partial y}\right)$$

Rearranging this,

$$\nabla \times (V\mathbf{A}) = \begin{vmatrix} \mathbf{a}_x & \mathbf{a}_y & \mathbf{a}_z \\ \dfrac{\partial V}{\partial x} & \dfrac{\partial V}{\partial y} & \dfrac{\partial V}{\partial z} \\ A_x & A_y & A_z \end{vmatrix} + V\begin{vmatrix} \mathbf{a}_x & \mathbf{a}_y & \mathbf{a}_z \\ \dfrac{\partial}{\partial x} & \dfrac{\partial}{\partial y} & \dfrac{\partial}{\partial z} \\ A_x & A_y & A_z \end{vmatrix} = (\nabla V) \times \mathbf{A} + V(\nabla \times \mathbf{A})$$

Thus, the vector identity is verified. For future reference,

$$\boxed{\nabla \times (V\mathbf{A}) = (\nabla V) \times \mathbf{A} + V(\nabla \times \mathbf{A})} \tag{2.74}$$

Exercise 2.19
Find the curl of vector field $\mathbf{H}(\mathbf{r}) = \mathbf{A}\sin(\mathbf{k}\cdot\mathbf{r})$ with position vector $\mathbf{r}$ and constant vectors $\mathbf{A}$ and $\mathbf{k}$.

Ans. $\nabla \times \mathbf{H} = (\mathbf{k} \times \mathbf{A})\cos(\mathbf{k}\cdot\mathbf{r})$.

Exercise 2.20
If $\mathbf{H}$ is in amperes per meter, what is the unit of $\nabla \times \mathbf{H}$?

Ans. Ampere per square meter.

Exercise 2.21

Assuming a unit source produces a unity circulation, determine the magnitude of the circulation source for $\mathbf{F}_2$ in Example 2.15, which would be along the z-axis.

Ans. $\displaystyle\lim_{\delta\to 0}\oint_C \mathbf{F}_2 \cdot d\mathbf{l} = \lim_{\delta\to 0}\frac{1}{\delta}2\pi\delta = 2\pi.$

2.5.2   Stokes's Theorem

Stokes's theorem follows directly from the definition of curl, stating that *the integration of $\nabla \times \mathbf{H}$ over an open surface is equal to the line integral of $\mathbf{H}$ around the closed contour of the surface.* Namely,

$$\int_S (\nabla \times \mathbf{H}) \cdot d\mathbf{s} = \oint_C \mathbf{H} \cdot d\mathbf{l} \tag{2.75}$$

where the surface S is bounded by loop C. The directions of $d\mathbf{l}$ along C and $d\mathbf{s}$ on S are governed by the right-hand rule.

To verify Stokes's theorem, a given surface S is subdivided into many elements of the area, as illustrated in Fig. 2.19. The three elements Δs_{k-1}, Δs_k, and Δs_{k+1} are positioned next to each other, with their top sides coincident with the contour of S or loop C. Therefore, the direction of travel on Δc_k should be counterclockwise in accordance with the positive direction of C, as shown in Fig. 2.19. Subsequently, the unit normal to Δs_k or $\mathbf{a}_k$ is determined by the right-hand rule, and is directed along the unit normal to S or $\mathbf{a}_s$. As the circumnavigations along Δc_k and Δc_{k+1} are both conducted in the counterclockwise direction, part of the contour shared by Δs_k and Δs_{k+1} is traversed twice in opposite directions.

Let us construct the curl of $\mathbf{H}$ over the surface element Δs_k following the definition in Eq. (2.63), that is,

$$\lim_{\Delta s_k\to 0}[(\nabla \times \mathbf{H}) \cdot \mathbf{a}_k \Delta s_k] = \lim_{\Delta c_k\to 0}\left[\oint_{\Delta c_k} \mathbf{H} \cdot d\mathbf{l}\right] \tag{2.76}$$

Fig. 2.19 Open surface S
with closed contour C

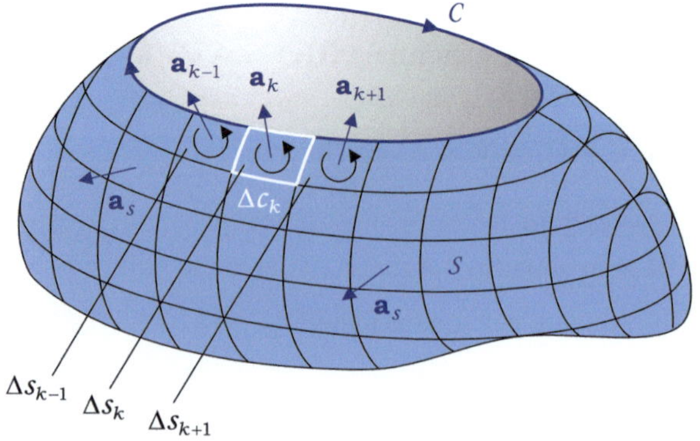

where Δc_k is the contour of Δs_k, and $\mathbf{a}_k$ is the unit normal to Δs_k. Because $\mathbf{a}_k \Delta s_k$ on the left-hand side of Eq. (2.76) is a differential area vector, the quantity in brackets corresponds to the flux passing through Δs_k. The total flux through the entire surface S is then given by the sum of the fluxes through all surface elements, and is expressed as

$$\lim_{\substack{N \to \infty \\ \Delta s_k \to 0}} \left[\sum_{k=1}^{N} (\nabla \times \mathbf{H}) \cdot (\mathbf{a}_k \Delta s_k) \right] = \lim_{\substack{N \to \infty \\ \Delta c_k \to 0}} \left[\sum_{k=1}^{N} \oint_{\Delta c_k} \mathbf{H} \cdot d\mathbf{l} \right] \tag{2.77}$$

The left-hand side of Eq. (2.77), by definition, is the surface integral of $\nabla \times \mathbf{H}$ over S. Meanwhile, the right-hand side is the sum of the line integrals conducted around all contours of all the surface elements. Because there is no contribution to the summation from the contour shared by the two adjoining elements, the right-hand side is reduced to the closed-line integral of $\mathbf{H}$ around loop C. Thus, Stokes's theorem is verified.

Leaving aside physical significance, if any, Stokes's theorem is used to convert a surface integral into a closed-line integral, and vice versa.

Example 2.17 For the vector field $\mathbf{A} = \rho \cos \phi \, \mathbf{a}_\phi$ in cylindrical coordinates, verify Stokes's theorem over a quarter-disk of radius a, as shown in Fig. 2.20.

Solution

In cylindrical coordinates,

$$\nabla \times \mathbf{A} = \frac{1}{\rho} \begin{vmatrix} \mathbf{a}_\rho & \rho \, \mathbf{a}_\phi & \mathbf{a}_z \\ \dfrac{\partial}{\partial \rho} & \dfrac{\partial}{\partial \phi} & \dfrac{\partial}{\partial z} \\ 0 & \rho^2 \cos \phi & 0 \end{vmatrix} = 2 \cos \phi \, \mathbf{a}_z$$

Integrating $\nabla \times \mathbf{A}$ over the given surface yields

Fig. 2.20 A quarter of disk

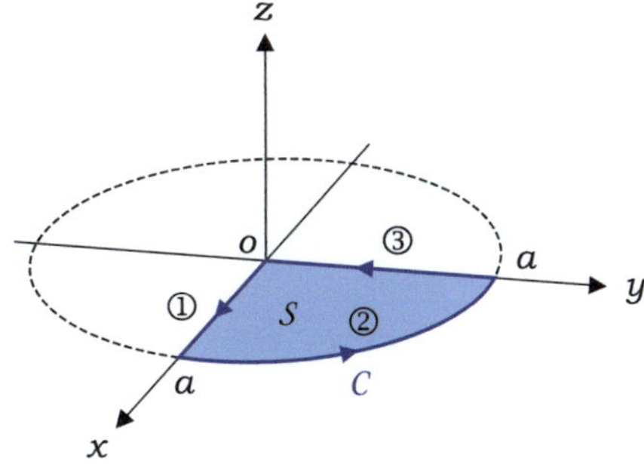

$$\int_S (\nabla \times \mathbf{A}) \cdot d\mathbf{s} = \int_{\phi=0}^{\phi=\pi/2} \int_{\rho=0}^{\rho=a} (2\cos\phi\, \mathbf{a}_z) \cdot (\rho d\rho d\phi\, \mathbf{a}_z) = a^2 \qquad (2.78a)$$

The circulation of $\mathbf{A}$ around C is separated into three parts:

$$\oint_C \mathbf{A} \cdot d\mathbf{l} = \int_{①} \mathbf{A} \cdot d\mathbf{l} + \int_{②} \mathbf{A} \cdot d\mathbf{l} + \int_{③} \mathbf{A} \cdot d\mathbf{l}$$

where

$$\int_{①} \mathbf{A} \cdot d\mathbf{l} = \int_{\rho=0}^{\rho=a} (\rho \cos 0^o\, \mathbf{a}_\phi) \cdot (d\rho\, \mathbf{a}_\rho) = 0$$

$$\int_{②} \mathbf{A} \cdot d\mathbf{l} = \int_{\phi=0}^{\phi=\pi/2} (a \cos\phi\, \mathbf{a}_\phi) \cdot (ad\phi\, \mathbf{a}_\phi) = a^2$$

$$\int_{③} \mathbf{A} \cdot d\mathbf{l} = \int_{\rho=a}^{\rho=0} (\rho \cos 90^o\, \mathbf{a}_\phi) \cdot (d\rho\, \mathbf{a}_\rho) = 0$$

Thus, the closed-line integral of $\mathbf{A}$ is

$$\oint_C \mathbf{A} \cdot d\mathbf{l} = a^2 \qquad (2.78b)$$

Two results in Eqs. (2.78a) and (2.78b) are equal. Thus, Stokes's theorem is verified.

Exercise 2.22
Verify the distributive law $\nabla \times (\mathbf{A} + \mathbf{B}) = \nabla \times \mathbf{A} + \nabla \times \mathbf{B}$ using Stokes's theorem.

Ans.

$$\int_S [\nabla \times (\mathbf{A} + \mathbf{B})] \cdot d\mathbf{s} = \oint_C \mathbf{A} \cdot d\mathbf{l} + \oint_C \mathbf{B} \cdot d\mathbf{l}$$

$$= \int_S [(\nabla \times \mathbf{A}) + (\nabla \times \mathbf{B})] \cdot d\mathbf{s}.$$

Review Questions

RQ 2.29 In what manner is the circulation of $\mathbf{A}$ incorporated into *curl* $\mathbf{A}$? [(2.64)]

RQ 2.30 If $\nabla \times \mathbf{A} = 0$ at a point, does it mean $\mathbf{A} = 0$ at that point? [(2.64)]

RQ 2.31 Define a conservative field. [Fig. 2.18]

RQ 2.32 Does *curl* $\mathbf{A}$ in component form correspond to the maximum circulation of $\mathbf{A}$ per unit area at a given point? [(2.70)]

RQ 2.33 State Stokes's theorem in words. [(2.75)]

RQ 2.34   If the unit of circulation of **H** is Ampere, then what is the   [(2.76)]
          unit of $\nabla \times \mathbf{H}$?

RQ 2.35   Explain how to detect a discrete circulation source.   [Exercise 2.21]

2.6 Dual Operation of ∇

The gradient, divergence, and curl in the previous sections incorporated special arrangements of partial derivatives in such a way that they each have particular significance. At a point in the space, the gradient represents the maximum directional derivative of the scalar field. In the presence of a vector field, divergence represents the net outward flux per unit volume and curl represents the maximum circulation per unit area. We also observed that it is convenient to use the del operator to represent these operations.

We may let a second del operator act on the gradient, divergence, or curl, if allowed to do so. Although the del operator is defined in Cartesian coordinates, the second operation denotes an additional gradient, divergence, or curl, in cylindrical and spherical coordinates. The dual operation of ∇ involves the second partial derivative of the operand. Even if a dual operation does not convey any physical significance, it is useful for expressing mathematical identity. Frequently encountered dual operations are as follows:

(a) The divergence of the gradient of a scalar field results in a scalar field. In Cartesian coordinates,

$$\nabla \cdot \nabla V = \nabla \cdot \left(\frac{\partial V}{\partial x}\mathbf{a}_x + \frac{\partial V}{\partial y}\mathbf{a}_y + \frac{\partial V}{\partial z}\mathbf{a}_z \right) = \frac{\partial^2 V}{\partial x^2} + \frac{\partial^2 V}{\partial y^2} + \frac{\partial^2 V}{\partial z^2} \quad (2.79)$$

This is also referred to as the **_Laplacian_** of V.

The **_Laplacian operator_** is defined in Cartesian coordinates as

$$\nabla^2 = \frac{\partial^2}{\partial x^2} + \frac{\partial^2}{\partial y^2} + \frac{\partial^2}{\partial z^2}$$

In general, for the orthogonal curvilinear coordinates (u,v,w), the Laplacian operator is given by

$$\nabla^2 = \nabla \cdot \nabla = \frac{1}{h_1 h_2 h_3}\left[\mathbf{a}_u \frac{\partial}{\partial u}(h_2 h_3) + \mathbf{a}_v \frac{\partial}{\partial v}(h_1 h_3) + \mathbf{a}_w \frac{\partial}{\partial w}(h_1 h_2) \right]$$
$$\cdot \left(\mathbf{a}_u \frac{1}{h_1}\frac{\partial}{\partial u} + \mathbf{a}_v \frac{1}{h_2}\frac{\partial}{\partial v} + \mathbf{a}_w \frac{1}{h_3}\frac{\partial}{\partial w} \right) \quad (2.80)$$

where the metric coefficients are

$$h_1 = 1, \; h_2 = 1, \;\; h_3 = 1 \qquad (u, v, w) = (x, y, z) \qquad \text{(Cartesian)}$$

$$h_1 = 1, \; h_2 = \rho, \;\; h_3 = 1 \qquad (u, v, w) = (\rho, \phi, z) \quad \text{(cylindrical)}$$

$$h_1 = 1, \; h_2 = R, \; h_3 = R \sin\theta \;\; (u, v, w) = (R, \theta, \phi) \quad \text{(spherical)}$$

When applying the Laplacian operator, the dot product precedes any differentiation. For example,

$$\mathbf{a}_u \frac{\partial}{\partial u} \; \cdot \; \mathbf{a}_v \frac{\partial}{\partial v} = \mathbf{a}_u \cdot \mathbf{a}_v \frac{\partial}{\partial u} \frac{\partial}{\partial v} \neq \mathbf{a}_u \cdot \frac{\partial}{\partial u}\left(\mathbf{a}_v \frac{\partial}{\partial v}\right)$$

The Laplacian of V is expressed in the three coordinate systems as follows:

$$\boxed{\nabla^2 V = \frac{\partial^2 V}{\partial x^2} + \frac{\partial^2 V}{\partial y^2} + \frac{\partial^2 V}{\partial z^2}} \quad \text{(Cartesian)} \qquad (2.81a)$$

$$\boxed{\nabla^2 V = \frac{1}{\rho}\frac{\partial}{\partial \rho}\left(\rho \frac{\partial V}{\partial \rho}\right) + \frac{1}{\rho^2}\frac{\partial^2 V}{\partial \phi^2} + \frac{\partial^2 V}{\partial z^2}} \quad \text{(cylindrical)} \qquad (2.81b)$$

$$\boxed{\nabla^2 V = \frac{1}{R^2}\frac{\partial}{\partial R}\left(R^2 \frac{\partial V}{\partial R}\right) + \frac{1}{R^2 \sin\theta}\frac{\partial}{\partial \theta}\left(\sin\theta \frac{\partial V}{\partial \theta}\right) + \frac{1}{R^2 \sin^2\theta}\frac{\partial^2 V}{\partial \phi^2}}$$

$$\text{(spherical)} \qquad (2.81c)$$

(b)   The curl of the gradient of any scalar field is identically zero:

$$\boxed{\nabla \times (\nabla V) = 0} \qquad (2.82)$$

This can be readily verified by direct substitution of the gradient followed by that of the curl in Cartesian coordinates.

Alternatively, by applying Stokes's theorem to the left-hand side of Eq. (2.82) and using Eq. (2.30), we have

$$\int_S [\nabla \times (\nabla V)] \cdot d\mathbf{s} = \oint_C \nabla V \cdot d\mathbf{l} = \oint_C dV \qquad (2.83)$$

The closed-line integral on the right side of Eq. (2.83) is zero, because the initial and terminal points are the same on a closed path. Thus, the surface integral on the left-hand side should vanish regardless of S, and the term in brackets should be zero at every point in a given region. The vector identity is thus verified.

(c) The divergence of the curl of any vector field is zero:

$$\boxed{\nabla \cdot (\nabla \times \mathbf{A}) = 0} \tag{2.84}$$

This can be verified by direct substitution. Alternatively, by applying the divergence and Stokes's theorems to the left-hand side of Eq. (2.84), we have

$$\int_V [\nabla \cdot (\nabla \times \mathbf{A})]\, dv = \oint_S (\nabla \times \mathbf{A}) \cdot d\mathbf{s} = \oint_C \mathbf{A} \cdot d\mathbf{l} \tag{2.85}$$

Here, volume V is bounded by closed surface S with no openings or contours ($C = 0$). Thus, the closed-line integral on the right side vanishes. The volume integral on the left side should also vanish, regardless of V; thus, the term in brackets should be zero at every point in a given region. The vector identity is thus verified.

(d) The gradient of the divergence of any vector field results in a vector field. In Cartesian coordinates,

$$\begin{aligned}
\nabla\nabla \cdot \mathbf{E} &= \left(\mathbf{a}_x \frac{\partial}{\partial x} + \mathbf{a}_y \frac{\partial}{\partial y} + \mathbf{a}_z \frac{\partial}{\partial z}\right)\left(\frac{\partial E_x}{\partial x} + \frac{\partial E_y}{\partial y} + \frac{\partial E_z}{\partial z}\right) \\
&= \left(\frac{\partial^2 E_x}{\partial x^2} + \frac{\partial^2 E_y}{\partial x \partial y} + \frac{\partial^2 E_z}{\partial x \partial z}\right)\mathbf{a}_x + \left(\frac{\partial^2 E_x}{\partial y \partial x} + \frac{\partial^2 E_y}{\partial y^2} + \frac{\partial^2 E_z}{\partial y \partial z}\right)\mathbf{a}_y \\
&\quad + \left(\frac{\partial^2 E_x}{\partial z \partial x} + \frac{\partial^2 E_y}{\partial z \partial y} + \frac{\partial^2 E_z}{\partial z^2}\right)\mathbf{a}_z
\end{aligned} \tag{2.86}$$

In most cases, taking the gradient of $\nabla \cdot \mathbf{E}$ is much easier than directly evaluating the right-hand side of Eq. (2.86).

(e) The curl of the curl of $\mathbf{E}$ is equal to the gradient of the divergence of $\mathbf{E}$ minus the Laplacian of $\mathbf{E}$. That is,

$$\nabla \times \nabla \times \mathbf{E} = \nabla\nabla \cdot \mathbf{E} - \nabla^2 \mathbf{E} \tag{2.87}$$

The Laplacian of $\mathbf{E}$ or $\nabla^2 \mathbf{E}$ represents the divergence of the gradient of the scalar components of $\mathbf{E}$. In Cartesian coordinates,

$$\nabla^2 \mathbf{E} = (\nabla^2 E_x)\,\mathbf{a}_x + (\nabla^2 E_y)\,\mathbf{a}_y + (\nabla^2 E_z)\,\mathbf{a}_z \tag{2.88}$$

In cylindrical and spherical coordinates, $\nabla^2 \mathbf{E}$ can be obtained using the following relation:

$$\nabla^2 \mathbf{E} = \nabla\nabla \cdot \mathbf{E} - \nabla \times \nabla \times \mathbf{E}$$

Exercise 2.23

Find the Laplacian of $V = \sin x \cos y \sin z$.

Ans. $-3 \sin x \cos y \sin z$.

Exercise 2.24

Find $\nabla^2 \mathbf{A}$ if $\mathbf{A} = (-2\mathbf{a}_x + \mathbf{a}_y + 3\mathbf{a}_z) \cos(x + 2y)$.

Ans. $\mathbf{A} = (10\mathbf{a}_x - 5\mathbf{a}_y - 15\mathbf{a}_z) \cos(x + 2y)$.

Exercise 2.25

Does the following make sense? (a) $\nabla \nabla V$, (b) $\nabla \times \nabla \cdot \mathbf{D}$, and (c) $\nabla \cdot \nabla \times \mathbf{A}$.

Ans. (a) No, (b) No, (c) Yes.

Exercise 2.26

Verify the identities in Eqs. (2.82) and (2.84) through direct substitution.

2.7   Helmholtz's Theorem

A vector field is uniquely defined in a region of space by specifying its divergence and curl within the region and its normal component over the boundary. If the boundary is at infinity, where the vector field diminishes to zero, then it can be uniquely determined by its divergence and curl.

Vector fields can be classified into four groups according to their vanishing or non-vanishing divergence and curl as follows:

$$(1) \quad \nabla \cdot \mathbf{A} = 0 \text{ and } \nabla \times \mathbf{A} = 0 \tag{2.89a}$$

$$(2) \quad \nabla \cdot \mathbf{A} = 0 \text{ and } \nabla \times \mathbf{A} \neq 0 \tag{2.89b}$$

$$(3) \quad \nabla \cdot \mathbf{A} \neq 0 \text{ and } \nabla \times \mathbf{A} = 0 \tag{2.89c}$$

$$(4) \quad \nabla \cdot \mathbf{A} \neq 0 \text{ and } \nabla \times \mathbf{A} \neq 0 \tag{2.89d}$$

A vector field is said to be ***solenoidal*** if its divergence is zero, and ***irrotational*** if its curl is zero.

Example 2.18 Show that vector field $\mathbf{A}$ must be a multi-valued function of position if its curl is given by $\nabla \times \mathbf{A} = \mathbf{B}$, but its divergence is not specified.

Solution

Using Eq. (2.82), we can write

$$\nabla \times \mathbf{A} + \nabla \times (\nabla V) = \mathbf{B} \quad \rightarrow \quad \nabla \times (\mathbf{A} + \nabla V) = \mathbf{B}$$

Because $\mathbf{A} + \nabla V$ satisfies the curl equation for any scalar field V, the solution to the curl equation is a multi-valued function.

Example 2.19 Show that vector field $\mathbf{B}$ must be a multi-valued function of position if its divergence is given by $\nabla \cdot \mathbf{B} = 0$, but its curl is not specified.

Solution

Using Eq. (2.84), we can write

$$\nabla \cdot \mathbf{B} + \nabla \cdot (\nabla \times \mathbf{K}) = 0 \quad \rightarrow \quad \nabla \cdot (\mathbf{B} + \nabla \times \mathbf{K}) = 0$$

Because $\mathbf{B} + \nabla \times \mathbf{K}$ is solenoidal for any vector field $\mathbf{K}$, the solution to the divergence equation is a multi-valued function.

Exercise 2.27
Determine if the following vector fields are solenoidal, irrotational, or both:
(a) $\mathbf{A} = -2xy\,\mathbf{a}_x + y^2\,\mathbf{a}_y + x^2\,\mathbf{a}_z$, (b) $\mathbf{B} = \cos\phi\,\mathbf{a}_\rho - \sin\phi\,\mathbf{a}_\phi$, and
(c) $\mathbf{C} = R^2 \sin^3\theta\,\mathbf{a}_R + R^2 \sin^2\theta \cos\theta\,\mathbf{a}_\theta$.

Ans.(a) Solenoidal, (b) Solenoidal and irrotational, (c) Irrotational.

Review Questions

RQ 2.36 State Helmholtz's theorem in words.

RQ 2.37 What is the solenoidal field? [(2.89a,b)]

RQ 2.38 What is the irrotational field? [(2.89c)]

2.8 Problems

Curves and surfaces

2.1 For an ellipse expressed as $x^2/16 + y^2/4 = 1$, as shown in Fig. 2.21, write a parametric representation by taking ϕ as a parameter, and determine the unit tangent vector at point p_1:$(2, \sqrt{3}, 0)$.

2.2 Repeat Problem 2.1 by taking x as a parameter. Determine the direction of travel on the ellipse in this case.

2.3 A circular loop of radius 0.5, C, is centered at p:$(2, 1, 0)$ in the xy-plane. Find

(a) a parametric representation of C, and
(b) expression for $d\mathbf{l}$ on C using φ as a parameter (Fig. 2.22).

Fig. 2.21 An ellipse
(Problems 2.1 and 2.2)

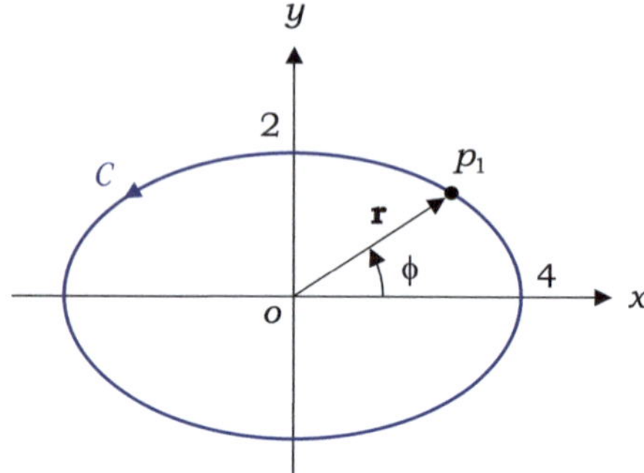

Fig. 2.22 Circular loop in
the xy-plane (Problem 2.3)

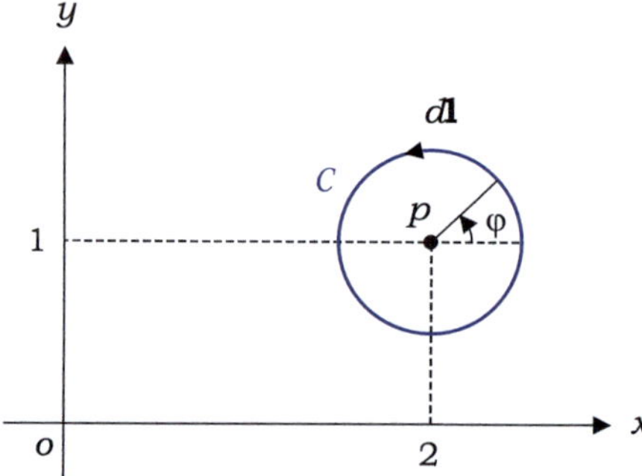

Fig. 2.23 A hemisphere
(Problem 2.4)

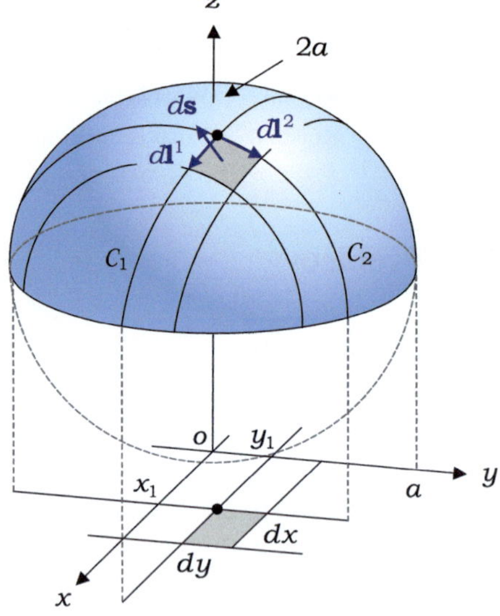

2.4 On a hemisphere of radius a, with its vertex at $z = 2a$, the $y = y_1$ and $x = x_1$ planes trace curves C_1 and C_2, respectively, as shown in Fig. 2.23. Express the following in parametric form: (a) C_1 and C_2, (b) $d\mathbf{l}^1$ and $d\mathbf{l}^2$ along C_1 and C_2, and (c) $d\mathbf{s}$.

2.5 For a long cylinder of radius a lying along the x-axis, (a) find a parametric representation of the surface for $z \geq 0$, and (b) an expression for $d\mathbf{s}$.

2.6 Given an ellipsoid $f = x^2 + y^2 + 4z^2 = 16$, as shown in Fig. 2.24, find (a) a parametric representation of the surface for $z \geq 0$ using ρ and ϕ as parameters, and (b) an expression for $d\mathbf{s}$ in cylindrical coordinates.

Line integral

2.7 Given vector field $\mathbf{E} = (x + y)^2(\mathbf{a}_x + \mathbf{a}_y)$, determine the line integral of $\mathbf{E}$ from $A:(-1, 0, 0)$ to $B:(1, 2, 0)$ along paths (a) C_1, and (b) C_2, as shown in Fig. 2.25.

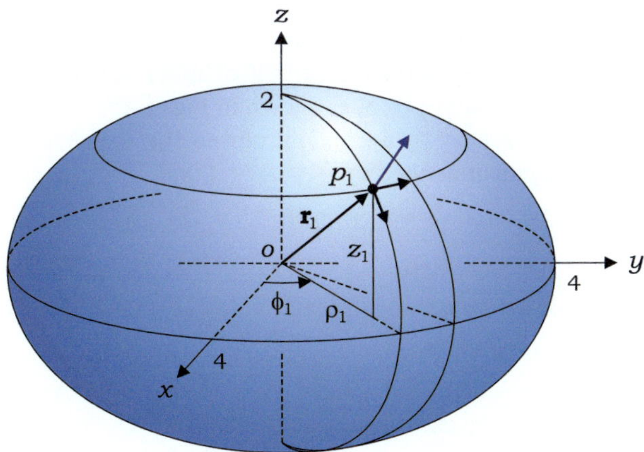

Fig. 2.24 An ellipsoidal surface (Problem 2.6)

Fig. 2.25 Two paths of integration (Problem 2.7)

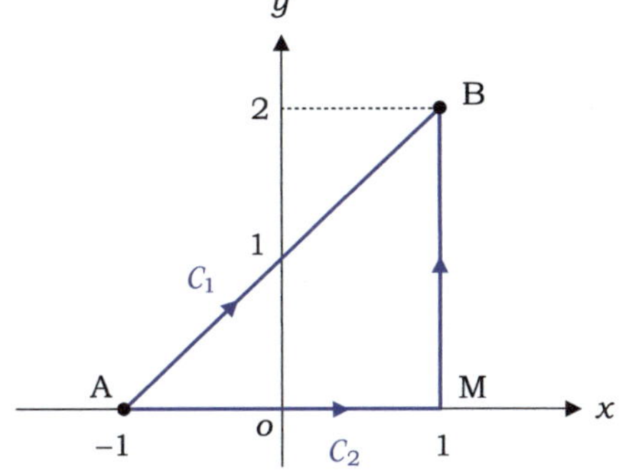

2.8   Find the line integral of $\mathbf{E} = y\,\mathbf{a}_x + z\,\mathbf{a}_y + x\,\mathbf{a}_z$ from point A:$(1, 0, 0)$ to point B:$(0, 1, \pi/2)$ along a spiral defined by $\mathbf{r}(t) = \cos t\,\mathbf{a}_x + \sin t\,\mathbf{a}_y + t\,\mathbf{a}_z$. [Hint: $\int t \cos t\, dt = \cos t + t \sin t$.]

2.9   For an ellipse $x^2/4 + y^2 = 1$, lying in the $z = 0$ plane,

(a)   express it in parametric form, and
(b)   find the line integral of $\mathbf{E} = (1/\rho)\,\mathbf{a}_\rho$ from A:$(2, 0, 0)$ to B:$(0, 1, 0)$ along the ellipse in the counterclockwise direction.

2.10   Given $\mathbf{A} = R \sin\theta \cos\phi\,\mathbf{a}_R + \cos\theta \cos\phi\,\mathbf{a}_\theta - \sin\phi\,\mathbf{a}_\phi$, determine the closed-line integral of $\mathbf{A}$ around loop $\mathcal{C}$, as shown in Fig. 2.26.

2.11   Determine the line integral of $\mathbf{A} = \mathbf{a}_x$ along an arc defined by $R = 4, \theta = 30^o$, and $0 \le \phi \le \pi/2$ in spherical coordinates.

Surface integral

2.12   Find the area of the circular strip formed on a sphere of radius 5, as shown in Fig. 2.27.

2.13   Given $\mathbf{A} = \rho\,\mathbf{a}_\rho - 3\rho \cos\phi\,\mathbf{a}_\phi - 2z\,\mathbf{a}_z$, evaluate the closed-surface integral of $\mathbf{A}$ over the bounding surface of a half-cylinder defined by $0 \le \rho \le 4$, $30^o \le \phi \le 210^o$, and $0 \le z \le 5$, as shown in Fig. 1.45.

2.14   Given $\mathbf{D} = 3R^{-2}\,\mathbf{a}_R + \cos^2\phi\,\mathbf{a}_\phi$, find the closed-surface integral of $\mathbf{D}$ over the boundary of a segment defined by $0 \le R \le 4$ and $30^o \le \phi \le 60^o$, as shown in Fig. 2.28.

Gradient

2.15   Find the gradient of the following scalar fields:

(a) $U = 3z e^{2x+y} + 10$,
(b) $V = 5\rho \sin\phi - \ln(z^2 + 1)$, and
(c) $W = R^{-2} \sin\theta \cos\phi$.

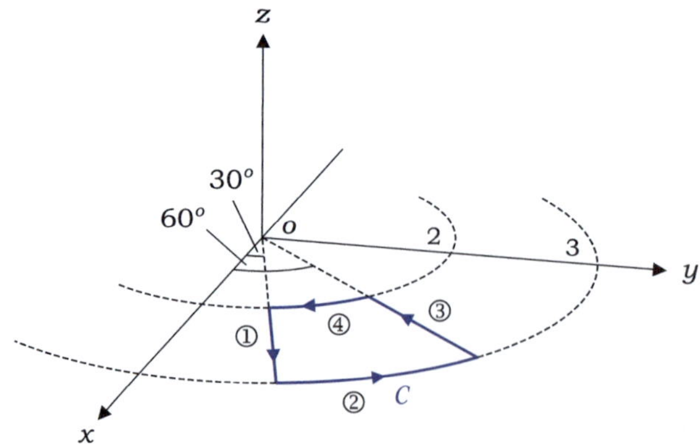

Fig. 2.26   Closed path of integration (Problem 2.10)

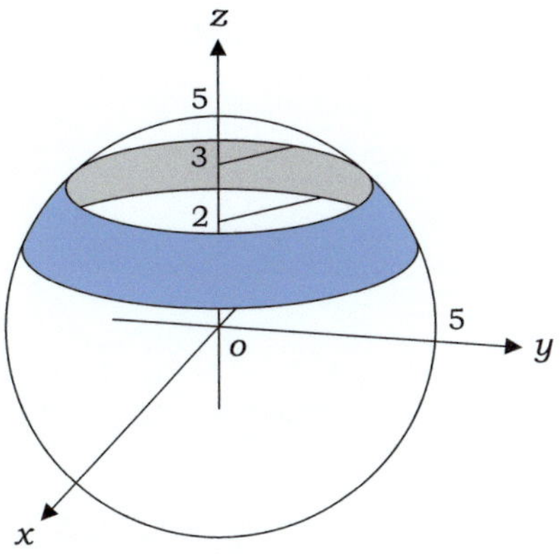

Fig. 2.27 Circular strip on a sphere (Problem 2.12)

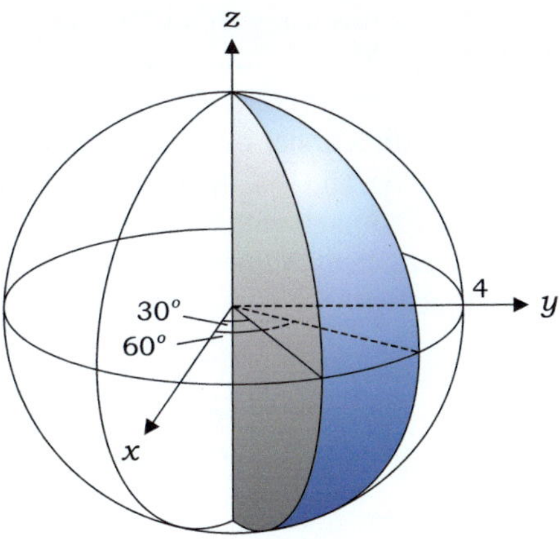

Fig. 2.28 Segment of a sphere (Problem 2.14)

2.16 For U, V, and W in Problem 2.15, evaluate the line integral of the gradient of each field from p_1:(2, 2, 0) to p_2:(0, 0, 2) in Cartesian coordinates along the trace on ellipsoid $x^2 + y^2 + 2z^2 = 8$ caused by the $\phi = 45^o$ plane:

(a) $\int_{p_1}^{p_2} \nabla U \cdot d\mathbf{l}$ in Cartesian coordinates,

(b) $\int_{p_1}^{p_2} \nabla V \cdot d\mathbf{l}$ in cylindrical coordinates, and

(c) $\int_{p_1}^{p_2} \nabla W \cdot d\mathbf{l}$ in spherical coordinates.

2.17 Do the following expressions make sense?

(a) $\nabla(UV) = V \nabla U + U \nabla V$,

(b) $\nabla(U + V) = \nabla U + \nabla V$,

(c) $\nabla(U^4) = 4U^3(\nabla U)$, and

(d) $\nabla(1/U) = -(\nabla U)/U^2$.

2.18 A scalar function $f(x, y)$ can be expanded in a Taylor series about point p:$(x_o, y_o, 0)$ in the $z = 0$ plane. With higher-order terms ignored, we have

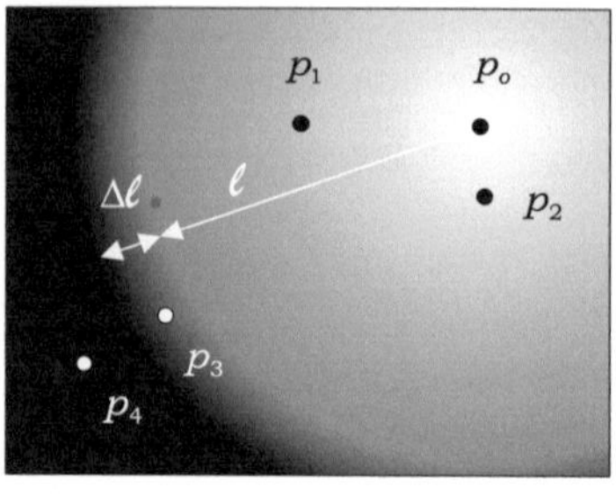

Fig. 2.29 An 8-bit grayscale image (Problem 2.21)

$$f(x, y) = f(x_o, y_o) + (x - x_o)\left.\frac{\partial f}{\partial x}\right|_{\substack{x=x_o\\y=y_o}} + (y - y_o)\left.\frac{\partial f}{\partial y}\right|_{\substack{x=x_o\\y=y_o}}$$

Verify this using $df = \nabla f \cdot d\mathbf{l}$.

2.19 Given $\nabla W = 5\mathbf{a}_R - 2\mathbf{a}_\theta + 3\mathbf{a}_\phi$ at $p{:}(4, 30^o, 45^o)$ in spherical coordinates, determine (a) dW/dR, (b) $dW/d\theta$, and (c) $dW/d\phi$ at p.

2.20 The directional derivatives of W at point p along the directions of $\mathbf{a}_x + \mathbf{a}_y$, $\mathbf{a}_x - \mathbf{a}_z$, and $-\mathbf{a}_y + \mathbf{a}_z$ are -4, 6, and 6, respectively. Determine ∇W at p.

2.21 For an image with 8-bit grayscale in Fig. 2.29, the gray level is $s = 255$ at p_o, and decreases linearly with the radial distance down to $s = 128$ for $R \le \ell$. Meanwhile, $s = 0$ for $R \ge \ell + \Delta\ell$. Find

    (a) the point with $\nabla s = 0$,
    (b) the point at which $|\nabla s|$ is the greatest, and
    (c) the direction of ∇s at points $p_0 \sim p_4$.

2.22 Take the gradient of $V(\mathbf{r}) = 2z$ in two different ways, as described below:

    (a) Find ∇V in Cartesian coordinates.
    (b) Transform V from Cartesian to spherical system and determine ∇V.
    (c) Transform ∇V in (b) back into the Cartesian system.

2.23 Transform ∇V expressed in the Cartesian coordinates into the (x', y', z')-system, which is rotated about the z-axis by an angle φ, to show

$$\nabla V = \frac{\partial V}{\partial x'}\mathbf{a}_{x'} + \frac{\partial V}{\partial y'}\mathbf{a}_{y'} + \frac{\partial V}{\partial z'}\mathbf{a}_{z'}$$

2.24 Given $f(\mathbf{r}) = \mathbf{k} \cdot \mathbf{r}$ in Cartesian coordinates, where $\mathbf{r}$ is the position vector and $\mathbf{k}$ is a constant vector, determine (a) ∇f, and (b) the geometric shape of $f(\mathbf{r}) = c_1$ (constant). (c) What is the significance of c_1?

2.25 For two families of circles given in the $z = 0$ plane, $f(x, y) = x^2 + y^2 = c_1$ and $g(x, y) = (x - 6)^2 + y^2 = c_2$, with constants c_1 and c_2, find a smaller angle between the two circles at point $p{:}(3, 4, 0)$.

2.26 For an ellipsoid $x^2 + y^2 + z^2/2 = 12$, find, at $p_1{:}(3, 1, 2)$, expressions for

(a) the outward unit normal to the ellipsoid, and
(b) tangent plane.

2.27 *The gradient can be used to determine the differential area vector $d\mathbf{s}$.* Consider a family of ellipsoids $f = \rho^2 + 4z^2$ (see Fig. 2.24). Starting with the observation that ∇f is parallel to $d\mathbf{s}$, whose z-component is given by $d\mathbf{s} \cdot \mathbf{a}_z = \rho\, d\rho\, d\phi$, find the complete expression for $d\mathbf{s}$ on an ellipsoid specified by $f = 16$.

2.28 A line crosses a family of ellipsoids $f = x^2 + y^2 + 4z^2$ at right angles and passes through point $p{:}(2, 6, 32)$. Express the line in parametric form using x as a parameter.

Flux and flux density

2.29 Given the flux density $\mathbf{D} = \mathbf{a}_R / R$, calculate the outward flux through a sphere of radius a centered at the origin.

2.30 If the divergence of flux density $\mathbf{A}$ is $\nabla \cdot \mathbf{A} = e^{-2R}/ R^2$, calculate the outward flux through a sphere of radius a centered at the origin.

Divergence and divergence theorem

2.31 Verify the following vector identities by direct substitution:

(a) $\nabla \cdot (f\mathbf{A}) = f(\nabla \cdot \mathbf{A}) + \mathbf{A} \cdot (\nabla f)$, and
(b) $\nabla \cdot (\mathbf{A} \times \mathbf{B}) = \mathbf{B} \cdot (\nabla \times \mathbf{A}) - \mathbf{A} \cdot (\nabla \times \mathbf{B})$.

2.32 Find the divergence of the following vector fields:

(a) $\mathbf{A} = \dfrac{1}{\sqrt{x^2+y^2+z^2}}(\mathbf{a}_x + \mathbf{a}_y + \mathbf{a}_z)$ (for $x^2 + y^2 + z^2 > 0$),
(b) $\mathbf{B} = (\ln \rho)\,\mathbf{a}_\rho + \rho \cos \phi\, \mathbf{a}_\phi + z^3\mathbf{a}_z$, and
(c) $\mathbf{C} = \frac{1}{R^3}\,\mathbf{a}_R + Re^{-R} \sin \theta\, \mathbf{a}_\theta + (R \sin \theta \cos \phi)^2\mathbf{a}_\phi$.

2.33 Take the divergence of the vector field $\mathbf{D} = x\,\mathbf{a}_x$ in two different ways:

(a) Compute $\nabla \cdot \mathbf{D}$ in Cartesian coordinates.
(b) Transform $\mathbf{D}$ into spherical coordinates to determine the divergence.

2.34 For $\mathbf{A} = x^2\,\mathbf{a}_x + y^2\,\mathbf{a}_y + z^2\,\mathbf{a}_z$, verify the divergence theorem over a unit cube $(0 \le x \le 1,\ 0 \le y \le 1,\ 0 \le z \le 1)$ by computing

(a) $\oint_S \mathbf{A} \cdot d\mathbf{s}$, and (b) $\int_V \nabla \cdot \mathbf{A}\, dv$.

2.35 For $\mathbf{A} = \rho \sin \phi\, (\sin \phi\, \mathbf{a}_\rho + \cos \phi\, \mathbf{a}_\phi)$, verify the divergence theorem over the volume enclosed by the surfaces with $\rho = 2$, $\rho = 4$, $z = 0$, and $z = 3$ by computing

(a) $\oint_S \mathbf{A} \cdot d\mathbf{s}$, and (b) $\int_V \nabla \cdot \mathbf{A}\, dv$.

2.36 For $\mathbf{D} = (\sin \phi/R)\,\mathbf{a}_R + \cos \phi\, \mathbf{a}_\phi$, verify the divergence theorem for the volume defined by $1 \le R \le 2$ and $\pi \le \phi \le 2\pi$ (Fig. 2.30), by computing

(a) $\oint_S \mathbf{D} \cdot d\mathbf{s}$, and (b) $\int_V \nabla \cdot \mathbf{D}\, dv$.

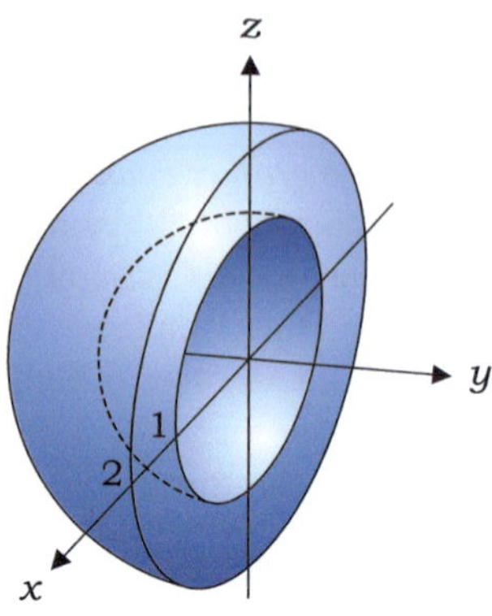
Fig. 2.30 Hemispherical shell (Problem 2.36)

2.37   For $\mathbf{A} = y\,\mathbf{a}_y$, verify the divergence theorem over the half-cylinder formed by surfaces $\rho = 2$, $y = 0$, $z = 0$, and $z = 3$ in the $y \geq 0$ region.

2.38   For $\mathbf{D} = 3z^2\mathbf{a}_z$, verify the divergence theorem over the volume bounded by the $z = 2$ plane and the cone of half-angle $\theta = 30^o$ in spherical coordinates by computing

(a) $\oint_S \mathbf{D} \cdot d\mathbf{s}$, and (b) $\int_V \nabla \cdot \mathbf{D}\, dv$.

Curl and Stokes's theorem

2.39   Take the curl of the vector field $\mathbf{A} = x^2\,\mathbf{a}_y - z\,\mathbf{a}_z$ in two different ways:

(a)   Compute $\nabla \times \mathbf{A}$ in Cartesian coordinates.
(b)   Transform $\mathbf{A}$ into cylindrical coordinates, and then compute $\nabla \times \mathbf{A}$.

2.40   Show, by Stokes's theorem, that if $\mathbf{H}$ is an irrotational field, (a) the closed-line integral of $\mathbf{H}$ must always vanish, and (b) the line integral of $\mathbf{H}$ is independent of the path of integration.

2.41   Given $\mathbf{H} = [-(y-2)\,\mathbf{a}_x + x\,\mathbf{a}_y]/[x^2 + (y-2)^2]$ in the xy-plane, show that the circulation of $\mathbf{H}$ is independent of the closed path of integration (see Fig. 2.31) only if the loop encloses the singular point at $p{:}(0, 2, 0)$ by following these steps:

(a)   Show $\nabla \times \mathbf{H} = 0$ in the region excluding p.
(b)   Show $\oint_{C_1} \mathbf{H} \cdot d\mathbf{l} = 0$ (C_1 is arbitrary, excluding p).
(c)   Show $\oint_{C_2} \mathbf{H} \cdot d\mathbf{l} = 2\pi$ (C_2 is arbitrary, enclosing p).

2.42   For the vector field $\mathbf{A} = (x - 2y^2)\,\mathbf{a}_x + (2xy + y^2)\,\mathbf{a}_y$, verify Stokes's theorem for the triangle in Fig. 2.32, by computing

(a) $\oint_C \mathbf{A} \cdot d\mathbf{l}$, and (b) $\int_S (\nabla \times \mathbf{A}) \cdot d\mathbf{s}$.

2.43   Given $\mathbf{H} = \rho^2\mathbf{a}_\rho + \rho \sin(\phi/2)\,\mathbf{a}_\phi$, verify Stokes's theorem for an annulus in the xy-plane, as shown in Fig. 2.33a, by computing

(a) $\int_S (\nabla \times \mathbf{H}) \cdot d\mathbf{s}$, and (b) $\oint_C \mathbf{H} \cdot d\mathbf{l}$.

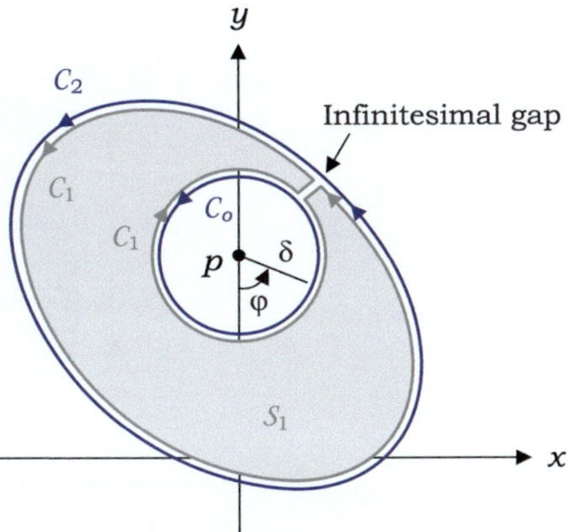

Fig. 2.31 Closed loop C_1 excludes point p and bounds surface S_1. As the gap shrinks to zero, C_1 splits into C_2 and C_o. The latter has a reverse direction of travel compared with that of C_1 (Problem 2.41)

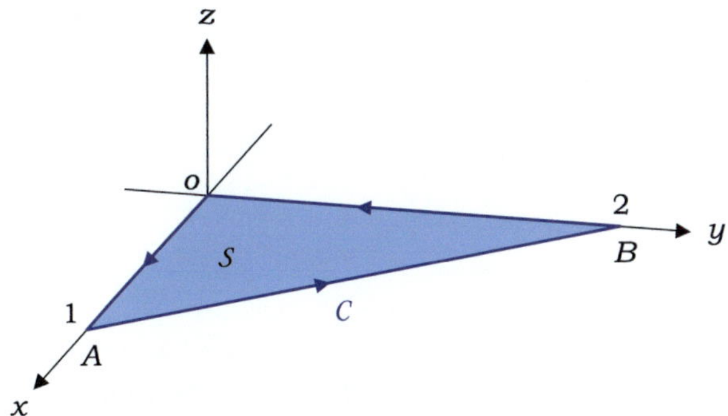

Fig. 2.32 Triangular region in the xy-plane (Problem 2.42)

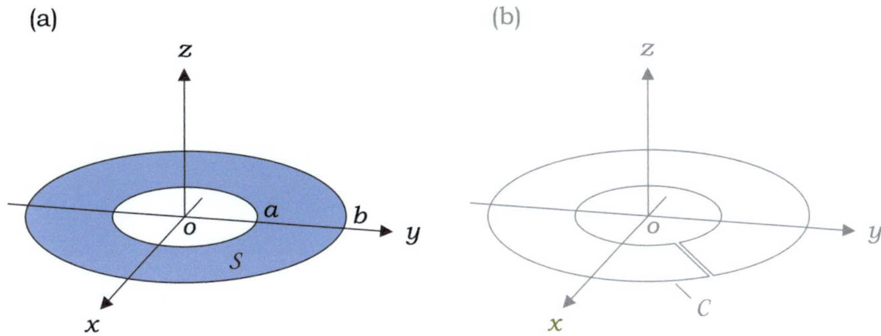

Fig. 2.33 Annulus (Problem 2.43)

Fig. 2.34 Can with an opening on top (Problem 2.44)

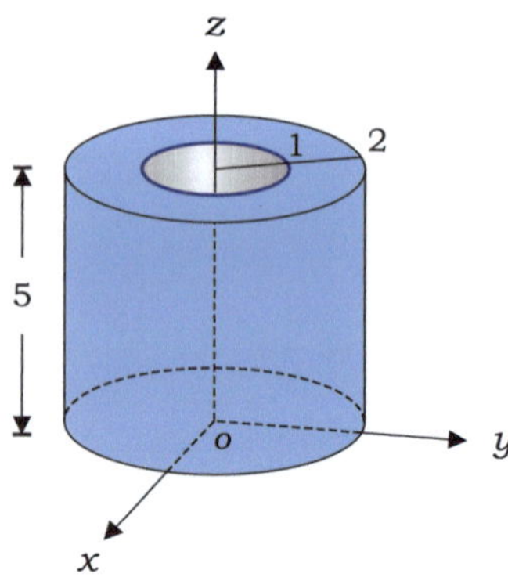

Fig. 2.35 Hemispherical surface (Problem 2.45)

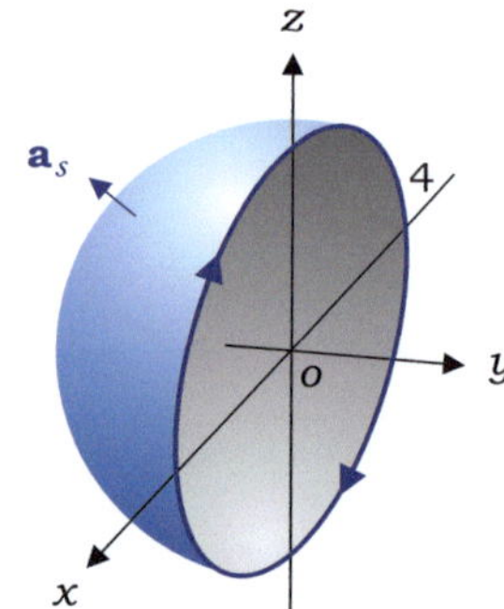

2.44 Given $\mathbf{B} = (3\rho)^{-1}\mathbf{a}_\rho + \rho\cos^2\phi\,\mathbf{a}_\phi + \rho(1+\sin\phi)\,\mathbf{a}_z$, verify Stokes's theorem for a can with a circular opening on the top, as shown in Fig. 2.34.

2.45 Given $\mathbf{H} = 2R^{-2}\cos\theta\,\mathbf{a}_R + R^{-2}\sin\theta\cos\phi\,\mathbf{a}_\theta$, verify Stokes's theorem for a hemispherical surface of radius 4, as shown in Fig. 2.35, by computing

   (a) $\oint_C \mathbf{H}\cdot d\mathbf{l}$, and (b) $\int_S (\nabla\times\mathbf{H})\cdot d\mathbf{s}$.

2.46 For $\mathbf{H} = R\,\mathbf{a}_R + 3R\sin\theta\,\mathbf{a}_\phi$, verify Stokes's theorem for a conical surface, as shown in Fig. 2.36, by computing

   (a) $\oint_C \mathbf{H}\cdot d\mathbf{l}$, and (b) $\int_S (\nabla\times\mathbf{H})\cdot d\mathbf{s}$.

Laplacian

2.47 Find the Laplacian of the following scalar fields:

   (a) $V = x^2 z + yz^2$,
   (b) $V = \rho^2 z\sin\phi$, and
   (c) $V = R^2\sin^2\theta\cos\phi$.

Fig. 2.36 Conical surface
(Problem 2.46)

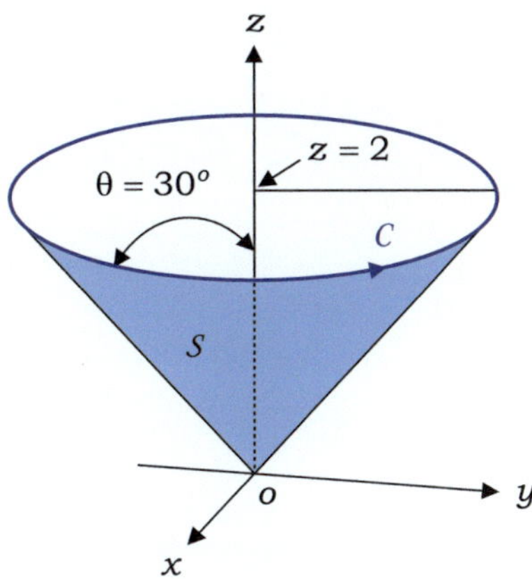

Helmholtz's theorem

2.48 Determine if the following vector fields are solenoidal, irrotational, both, or
neither:

    (a) $\mathbf{A} = x\,\mathbf{a}_x + y\,\mathbf{a}_y + z\,\mathbf{a}_z$,
    (b) $\mathbf{B} = (x\,\mathbf{a}_x + y\,\mathbf{a}_y)/(x^2 + y^2)$,
    (c) $\mathbf{C} = \cos\phi\,\mathbf{a}_\rho/\rho$,
    (d) $\mathbf{D} = 3\sin\phi\,\mathbf{a}_\phi$,
    (e) $\mathbf{E} = \sin\theta\,\mathbf{a}_\theta/R$, and
    (f) $\mathbf{F} = 4\,\mathbf{a}_R/R^2$.

2.49 Consider vector field $\mathbf{E}(x, y, z) = \mathbf{A}\cos(\mathbf{k} \cdot \mathbf{r})$, where $\mathbf{r}$ is the position vector
and $\mathbf{A}$ and $\mathbf{k}$ are constant vectors. Under what conditions is $\mathbf{E}$ (a) solenoidal,
and (b) irrotational?

2.50 In region $\rho \geq b$, in cylindrical coordinates, $\nabla \cdot \mathbf{H} = 0$ and $\nabla \times \mathbf{H} = 0$.
Furthermore, $\mathbf{H} = (A/b)\,\mathbf{a}_\phi$ with a constant A at the boundary at $\rho = b$.
Starting with a reasonable guess of the solution based on Helmholtz's theorem,
find $\mathbf{H}$ for $\rho \geq b$.

2.51 In region $R \geq a$, in spherical coordinates, $\nabla \cdot \mathbf{E} = 0$ and $\nabla \times \mathbf{E} = 0$.
Furthermore, $\mathbf{E} = (A/a^2)\,\mathbf{a}_R$ with a constant A at the boundary at $R = a$.
Starting with a reasonable guess of the solution based on Helmholtz's theorem,
find $\mathbf{E}$ in the region $R \geq a$.

Chapter 3
Electrostatics

Electrostatics is a branch of electromagnetics concerned with the electric phenomena caused by electric charges at rest, which are also known as static electric charges. Although the adjectives "static", "stationary," and "steady" refer to quantities that are independent of time, they are generally functions of position. The other two branches are magnetostatics and electrodynamics. The former involves static magnetic fields, whereas the latter involves time-varying electric and magnetic fields. In this chapter, we focus on electrostatics, which is not a simple case of electromagnetics. A complete mastery of electrostatics is essential not only for a thorough understanding of the general laws of electromagnetics but also for understanding and designing such devices as inkjet printers, liquid crystal displays, field-effect transistors, electroplating machines, electrocardiogram recorders, to name just a few.

Electrostatics begins with the electric force between two electric charges, which was experimentally observed and formulated by Charles Augustin de Coulomb (1785). This is now known as Coulomb's law, and is the fundamental law of electrostatics. The electric force due to the electric charge can be treated as a continuous vector quantity in space, according to the concept of an electric field. The electric field is differentiable and integrable. The physical significance of the derivatives and integrals of the electric field is one of the main topics discussed in this chapter. The principle of superposition allows us to extend the electric field of a point charge to multiple charges, and even to a continuous distribution of charges. Although the ***electric flux density*** may appear as a redundant replica of the electric field, which is the product of the electric field and a constant in a simple medium, it is directly related to the free charge of the material. Differential operators are used to formulate the conservation of energy and electric flux in terms of the curl of the electric field and divergence of the electric flux density, respectively. These are the two fundamental relations in electrostatics required for uniquely specifying the static electric field in a given region of space, in accordance with Helmholtz's theorem. The topics discussed in the second half of this chapter include the electric potential, electrostatic potential energy, boundary value problems, and capacitance.

© The Author(s), under exclusive license to Springer Nature Switzerland AG 2024

Y. H. Lee, *Introduction to Engineering Electromagnetics*,

https://doi.org/10.1007/978-3-031-28659-9_3

Most electrostatic relations can be derived from Coulomb's and Gauss's laws in conjunction with the principle of superposition. In this case, derivations can be facilitated by following a particular line of reasoning that may be either physical or mathematical in origin. Certain steps may correspond to simple applications of mathematical theorems in a casual manner. Students often try to grasp any possible physical significance that might be attached to no more than standard mathematical procedures. However, this makes it difficult to understand the problem. Students should study how a particular line of reasoning follows. At the same time, students should be able to apply mathematical rules and theorems in a casual manner if they are correct.

3.1 Coulomb's Law

Static electricity dates back to Ancient Greece. The ancients observed that a piece of amber rubbed on fur or silk attracted straw, lint, feathers, and so on. The word electron is a Greek word for amber. It took many centuries before magic revealed its connection to static electric charges. According to a simple atomic model, an atom consists of a positively charged nucleus and negatively charged electrons orbiting around it. Rubbing the amber against fur removes bound electrons from the fur and gives them to the amber, causing the fur to be positively charged and the amber to be negatively charged.

Coulomb's law is a fundamental law of electrostatics established through elaborate experiments on two charged objects separated in space. It states that *the electric force exerted on a charge due to another charge is proportional to the product of the two charges and inversely proportional to the square of the distance between them.* Coulomb's law can be mathematically expressed as

$$F = \frac{1}{4\pi\varepsilon_0} \frac{q_1 q_2}{\mathcal{R}^2} \quad [\text{N}] \tag{3.1}$$

where q_1 and q_2 are the charges measured in coulombs [C], and $\mathcal{R}$ is the distance between the charges measured in meters [m]. The proportionality constant ε_0 is the ***permittivity of free space*** (or vacuum), which is measured in farads per meter [F/m], and has the following value:

$$\varepsilon_0 = 8.854 \times 10^{-12} \quad [\text{F/m}] \tag{3.2}$$

It is acceptable to use an approximate number $\varepsilon_0 \cong (1/36\pi) \times 10^{-9}$ [F/m], ignoring small errors. In a material medium, ε_0 should be replaced by a material constant ε called ***permittivity***, which is discussed in later sections.

Two charges of the same polarity repel each other along the line joining them, whereas two charges of the opposite polarity attract each other. The ***Coulomb force*** is another term for the ***electric force*** between two electric charges.

Using vector notation, the electric force exerted on charge q_1 due to charge q_2 can be expressed as

$$\boxed{\mathbf{F}_1 = \frac{q_1 q_2}{4\pi\varepsilon_0 \mathcal{R}_{1-2}^2}\mathbf{a}_{1-2}} \qquad [\text{N}] \qquad\qquad (3.3)$$

where $\mathcal{R}_{1-2}$ and $\mathbf{a}_{1-2}$ are the magnitude and unit vector of the distance vector from the position of q_2 to that of q_1.

The distance vector is mathematically expressed as

$$\mathcal{R}_{1-2} = \mathbf{r}_1 - \mathbf{r}_2 \qquad\qquad (3.4)$$

where subscript $1-2$ on $\mathcal{R}$ denotes that $\mathcal{R}_{1-2}$ is a vector "from 2 to 1," while mimicking the subtraction of position vector $\mathbf{r}_2$ from position vector $\mathbf{r}_1$ on the right-hand side. Throughout the text, the subscript $a–b$ denotes something from b to a, whereas the subscript ab represents something from a to b. The unit vector of $\mathcal{R}_{1-2}$ can be expressed in terms of the position vectors $\mathbf{r}_1$ and $\mathbf{r}_2$ as follows:

$$\mathbf{a}_{1-2} = \frac{\mathcal{R}_{1-2}}{\mathcal{R}_{1-2}} = \frac{\mathbf{r}_1 - \mathbf{r}_2}{|\mathbf{r}_1 - \mathbf{r}_2|} \qquad\qquad (3.5)$$

Inserting Eq. (3.5) into Eq. (3.3) leads to another expression for the electric force on the point charge q_1 due to the point charge q_2. That is,

$$\boxed{\mathbf{F}_1 = \frac{q_1 q_2}{4\pi\varepsilon_0}\frac{\mathbf{r}_1 - \mathbf{r}_2}{|\mathbf{r}_1 - \mathbf{r}_2|^3}} \qquad [\text{N}] \qquad\qquad (3.6)$$

Again, $\mathbf{r}_1$ and $\mathbf{r}_2$ are the position vectors for points p_1 and p_2, respectively, where charges q_1 and q_2 are located (Fig. 3.1). In this case, p_1 is called the *field point* and p_2 is the *source point*. In our notation, the position vectors of the field and source points are denoted by $\mathbf{r}_1$ and $\mathbf{r}_2$, respectively, or $\mathbf{r}$ and $\mathbf{r}'$.

Because $\mathbf{F}_1$ in Eq. (3.6) is a vector specific to point p_1, it should be expressed in terms of the coordinates of p_1 and expanded in terms of the base vectors given at p_1. Accordingly, the distance vector $\mathbf{r}_1 - \mathbf{r}_2$ on the right side of Eq. (3.6) should also be expanded in terms of the base vectors at p_1, even if $\mathbf{r}_2$ is expressed in terms of the coordinates and base vectors given at p_2. Because the field and source points are independent of each other, it is most convenient to adopt a mixed-coordinate system, in which the field point is defined in terms of unprimed coordinates, whereas the source point is defined in terms of primed coordinates.

The electric force is a mutual force. The force acting on q_2 due to q_1 has the same magnitude as $\mathbf{F}_1$ but points in the direction opposite to $\mathbf{F}_1$, that is, $\mathbf{F}_2 = -\mathbf{F}_1$. Expressed mathematically,

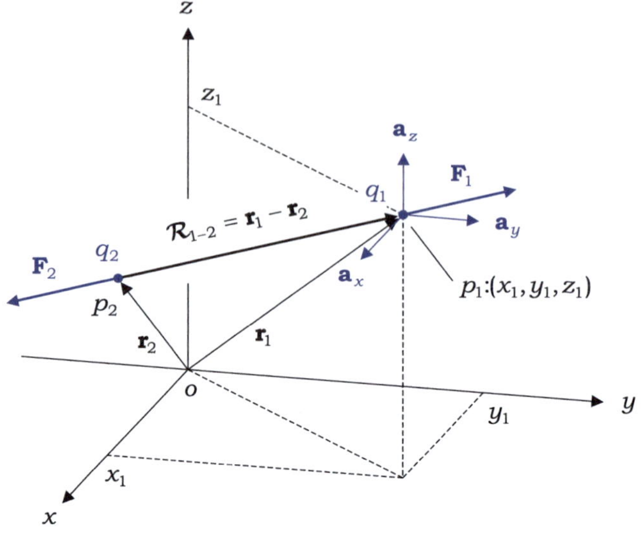

Fig. 3.1 Electric forces on point charges q_1 and q_2

$$\mathbf{F}_2 = \frac{q_2 q_1}{4\pi\varepsilon_0} \frac{\mathbf{r}_2 - \mathbf{r}_1}{|\mathbf{r}_2 - \mathbf{r}_1|^3} = -\mathbf{F}_1 \tag{3.7}$$

Point p_2 is now a field point. Thus, $\mathbf{F}_2$ should be expressed in terms of the base vectors at p_2, which are generally not the same as those at p_1.

Example 3.1 Determine the electric force acting on the point charge q_1 located at point $(2, 3, 5)$ caused by another point charge q_2 located at point $(1, 1, 3)$ in Cartesian coordinates. All distances are in meters.

Solution

The position vectors for q_1 and q_2 are

$$\mathbf{r}_1 = 2\mathbf{a}_x + 3\mathbf{a}_y + 5\mathbf{a}_z$$

$$\mathbf{r}_2 = \mathbf{a}_x + \mathbf{a}_y + 3\mathbf{a}_z$$

Thus, the distance vector is

$$\mathcal{R}_{1-2} = \mathbf{r}_1 - \mathbf{r}_2 = \mathbf{a}_x + 2\mathbf{a}_y + 2\mathbf{a}_z$$

Subsequently, the magnitude and unit vector of $\mathcal{R}_{1-2}$ are obtained as

$$\mathcal{R}_{1-2} = |\mathbf{r}_1 - \mathbf{r}_2| = \sqrt{1 + 2^2 + 2^2} = 3 \tag{3.8a}$$

$$\mathbf{a}_{1-2} = \frac{1}{3}\mathbf{a}_x + \frac{2}{3}\mathbf{a}_y + \frac{2}{3}\mathbf{a}_z \tag{3.8b}$$

Substituting Eq. (3.8) into Eq. (3.3) yields

$$\mathbf{F}_1 = \frac{q_1 q_2}{108\pi\varepsilon_0}(\mathbf{a}_x + 2\mathbf{a}_y + 2\mathbf{a}_z)$$

Exercise 3.1
A point charge q experiences an electric force of 2 [pN] at a distance of 10 [cm] from the other point charge. At what distance is the force on q reduced by 50%?

Ans. 14.14 [cm].

3.2 Electric Field Intensity

The electric force between the two point-charges may be viewed as if the effect of charge q' experienced by charge q is predetermined at every point in space and charge q only occurs at one of these points. Generalizing from this, for an electric charge positioned at a point in space, we can express its predetermined effect on unit charge as a continuous function of position. The **electric field intensity** of an electric charge is defined as *the electric force exerted on a unit of positive test charge due to the given charge*. Mathematically, for a point charge q located at a point with position vector $\mathbf{r}'$, the electric field intensity at position $\mathbf{r}$ is expressed as

$$\boxed{\mathbf{E}(\mathbf{r}) = \frac{q}{4\pi\varepsilon_0}\frac{\mathbf{r} - \mathbf{r}'}{|\mathbf{r} - \mathbf{r}'|^3}} \quad [\text{V/m}] \tag{3.9}$$

which is measured in volts per meter. Note that the electric field intensity is not defined at point $\mathbf{r} = \mathbf{r}'$, which is already occupied by the source charge. From a mathematical perspective, Eq. (3.9) is said to define the **electric field** of a point charge q positioned at $\mathbf{r}'$, and is illustrated in Fig. 3.2, where shade was used for visual effects.

Example 3.2 Express the electric field of a point charge q placed at the origin in terms of (a) Cartesian, and (b) spherical coordinates.

Solution

(a) The position vectors of field and source points are

$$\mathbf{r} = x\,\mathbf{a}_x + y\,\mathbf{a}_y + z\,\mathbf{a}_z$$

$$\mathbf{r}' = 0$$

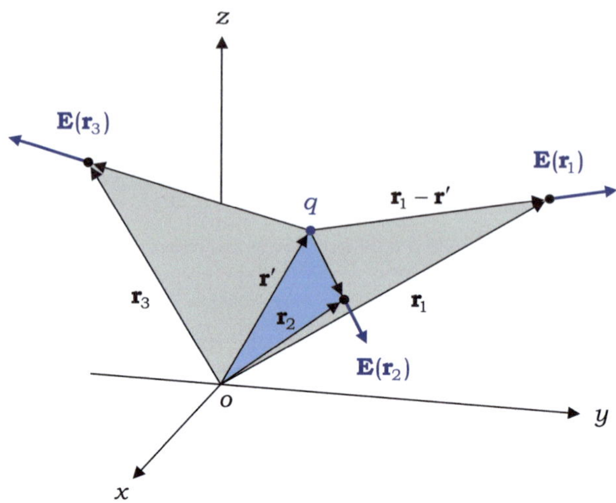

Fig. 3.2 Electric field due to point charge q positioned at $\mathbf{r}'$

Thus, the magnitude of the distance vector is

$$|\mathcal{R}| = |\mathbf{r} - \mathbf{r}'| = \sqrt{x^2 + y^2 + z^2}$$

In the Cartesian coordinate system, using Eq. (3.9), the electric field due to q located at the origin is expressed as

$$\mathbf{E}(\mathbf{r}) = \frac{q}{4\pi\varepsilon_0} \frac{x\,\mathbf{a}_x + y\,\mathbf{a}_y + z\,\mathbf{a}_z}{[x^2 + y^2 + z^2]^{3/2}} \qquad (3.10)$$

(b)   For a source charge placed at the origin in the spherical coordinate system, the distance vector from the source to field point is

$$\mathcal{R} = R\,\mathbf{a}_R$$

Thus, in the spherical coordinate system, the electric field of charge q placed at the origin is expressed as

$$\mathbf{E}(\mathbf{r}) = \frac{q}{4\pi\varepsilon_0 R^2}\,\mathbf{a}_R \qquad (3.11)$$

A point charge at the origin has spherical symmetry; it looks the same as we move around it, varying θ and ϕ while maintaining a constant R. As discussed in Chap. 1, the resulting vector field should be independent of θ and ϕ and have neither the θ- nor the ϕ-component. The result in Eq. (3.11) is entirely consistent with the outcome of symmetry considerations.

Exercise 3.2
Show that $\mathbf{E}$ due to a point charge at the origin is both irrotational and solenoidal in the region $R > 0$.

Ans. $\nabla \times \mathbf{E} = 0$ and $\nabla \cdot \mathbf{E} = 0$ for $\mathbf{E}$, as expressed in Eq. (3.11).

3.2.1 Electric Field Due to Multiple Charges

The electric field intensity follows the principle of superposition. That is, *the total electric field intensity due to multiple charges is equal to the vector sum of the electric field intensities of all the individual charges.* This is because the electric field intensity of a single charge remains the same and is unaffected by the presence of other charges. When there are N point charges, $q_1, q_2, \ldots$, and q_N, which are located at positions $\mathbf{r}_1, \mathbf{r}_2, \ldots$, and $\mathbf{r}_N$, the electric field intensity at position $\mathbf{r}$ is expressed as

$$\mathbf{E}(\mathbf{r}) = \sum_{i=1}^{N} \frac{q_i}{4\pi\varepsilon_0} \frac{\mathbf{r} - \mathbf{r}_i}{|\mathbf{r} - \mathbf{r}_i|^3} \tag{3.12}$$

For example, as illustrated in Fig. 3.3, the electric field intensity at position $\mathbf{r}$ is caused by two point-charges q_1 and q_2.

Example 3.3 Determine $\mathbf{E}$ due to the two point-charges $+q$ and $-q$ separated by a small distance d, as shown in Fig. 3.4, which is called the ***electric dipole***.

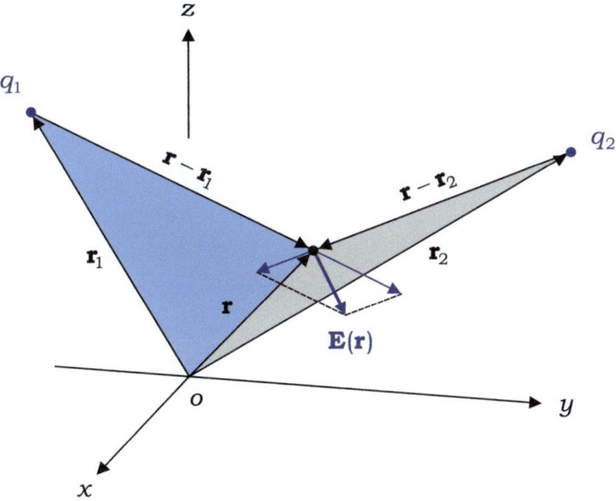

Fig. 3.3 Electric field intensity due to two point-charges. Shade is used for visual effects

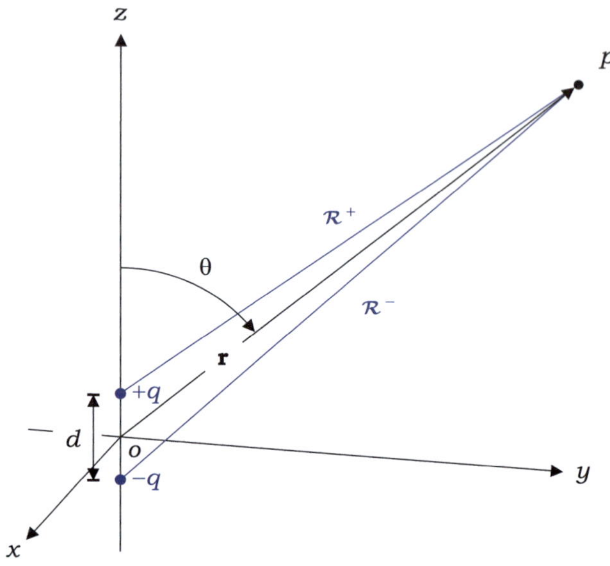

Fig. 3.4 An electric dipole consists of two identical charges of opposite polarity, separated by a small distance

Solution

The electric field intensity of the electric dipole can be expressed as

$$\mathbf{E}(\mathbf{r}) = \frac{q}{4\pi\varepsilon_0}\left[\frac{\mathcal{R}^+}{(\mathcal{R}^+)^3} - \frac{\mathcal{R}^-}{(\mathcal{R}^-)^3}\right] \tag{3.13a}$$

The two distance vectors are given by

$$\mathcal{R}^+ \equiv \mathbf{r} - \mathbf{r}^+ = x\,\mathbf{a}_x + y\,\mathbf{a}_y + (z - d/2)\,\mathbf{a}_z \tag{3.13b}$$

$$\mathcal{R}^- \equiv \mathbf{r} - \mathbf{r}^- = x\,\mathbf{a}_x + y\,\mathbf{a}_y + (z + d/2)\,\mathbf{a}_z \tag{3.13c}$$

where the field point is specified by $\mathbf{r} = x\,\mathbf{a}_x + y\,\mathbf{a}_y + z\,\mathbf{a}_z$, and the two source points are expressed as $\mathbf{r}^+ = (d/2)\,\mathbf{a}_z$ and $\mathbf{r}^- = (-d/2)\,\mathbf{a}_z$.

Applying the law of cosines to the two triangles $(+q)op$ and $(-q)op$ yields

$$\mathcal{R}^+ = \sqrt{r^2 + (d/2)^2 - 2r(d/2)\cos\theta} \tag{3.13d}$$

$$\mathcal{R}^- = \sqrt{r^2 + (d/2)^2 - 2r(d/2)\cos(\pi - \theta)} \tag{3.13e}$$

For $|\mathbf{r}| \gg d$, term $(d/2)^2$ in the radicands of Eqs. (3.13d) and (3.13e) is negligibly small compared with r^2. Therefore, by applying binomial expansion, that is, $(1 \pm a)^n \approx 1 \pm na$ for $a \ll 1$ and a real n, we obtain

$$\frac{1}{(\mathcal{R}^+)^3} \cong \frac{1}{[r^2 - rd\cos\theta]^{3/2}} \cong \frac{1}{r^3}\left[1 + \frac{3}{2}\frac{d}{r}\cos\theta\right] \tag{3.13f}$$

$$\frac{1}{(\mathcal{R}^-)^3} \cong \frac{1}{[r^2 + rd\cos\theta]^{3/2}} \cong \frac{1}{r^3}\left[1 - \frac{3}{2}\frac{d}{r}\cos\theta\right] \tag{3.13g}$$

Inserting Eqs. (3.13b, c, f, g) into Eq. (3.13a), along with $|\mathbf{r}| = R$, yields

$$\mathbf{E}(\mathbf{r}) = \frac{q}{4\pi\varepsilon_0 R^3}\left[-d\,\mathbf{a}_z + \left(x\,\mathbf{a}_x + y\,\mathbf{a}_y + z\,\mathbf{a}_z\right)3\frac{d}{R}\cos\theta\right] \tag{3.13h}$$

This can be transformed into spherical coordinates using the following relationship:

$$x\,\mathbf{a}_x + y\,\mathbf{a}_y + z\,\mathbf{a}_z = R\,\mathbf{a}_R \tag{3.13i}$$

$$-d\,\mathbf{a}_z = -d\cos\theta\,\mathbf{a}_R + d\sin\theta\,\mathbf{a}_\theta \tag{3.13j}$$

Thus, the electric field intensity of an electric dipole placed at the origin is expressed as

$$\mathbf{E}(\mathbf{r}) = \frac{qd}{4\pi\varepsilon_0 R^3}[2\cos\theta\,\mathbf{a}_R + \sin\theta\,\mathbf{a}_\theta] \tag{3.14}$$

An electric dipole is the simplest atomic model of electricity. It is very useful for examining the interaction between the material medium and the external electric field, and for specifying the electrical properties of a material.

Exercise 3.3
Two identical point charges are positioned on the z-axis: one at $z = -1$ and the other at $z = 3$. Locate the point at which $\mathbf{E}$ is zero.

Ans. $z = 1$.

Exercise 3.4
If the z-component of the electric field intensity at point $(2, 2, 1)$ is $E_z = 1\,[\text{V/m}]$ in the presence of a point charge q at the origin, what is the value of q?

Ans. 3 [nC].

Exercise 3.5
What are you basing the following statement on: "Because $\mathbf{E}$ of a single charge is a continuous function of position, so is the total $\mathbf{E}$ due to multiple charges".

Ans. Principle of superposition.

3.2.2 Electric Field Due to Continuous Charge Distribution

When a very large number of point charges are gathered in a small region of space such that the spacing between adjacent charges is much smaller than the distance to the field point, the discrete nature of the electric charges is ignored, and the source charge is considered as a continuous charge distribution. To find the electric field at the field point, the charged region is first subdivided into many differential elements of volume in such a way that the differential volume is large enough to contain many point charges, but small enough to be regarded as a point, as viewed from the field point. Then, the electric field due to the charge within each differential volume is computed by invoking Coulomb's law. The total electric field is obtained by adding all contributions from all volume elements, in accordance with the superposition principle. In the limit as the differential volume shrinks to zero, the summation becomes a volume integral.

For charges filling a volume, we define the **volume charge density** ρ_v as the charge per unit volume $[C/m^3]$. When the source charges are spread over a surface, it is convenient to work with the **surface charge density** ρ_s, which is defined as the charge per unit area $[C/m^2]$. If charges form a line, it is more appropriate to use the **line charge density**, ρ_ℓ, which is defined as the charge per unit length $[C/m]$. Note that these charge densities are generally continuous functions of the coordinates. Accordingly, the electric fields of the charge densities ρ_v, ρ_s, and ρ_ℓ are given by volume, surface, and line integrals, respectively.

The volume charge density at position $\mathbf{r}$ is defined as

$$\rho_v(\mathbf{r}) = \lim_{\Delta v \to 0} \frac{\Delta q}{\Delta v} \quad [C/m^3] \tag{3.15}$$

where Δv is the differential volume centered at point $\mathbf{r}$ and carrying charge Δq. The volume charge density represents the smoothed-out average density. Therefore, it must be a continuous function of position and can be conveniently used to express electromagnetic laws in a **point form** or **differential form**.

For a volume charge of density ρ_v $[C/m^3]$, the charge within the differential volume centered at point $\mathbf{r}'_i$ is given by $q_i = \rho_v(\mathbf{r}'_i)\,\Delta v(\mathbf{r}'_i)$. Because it is treated as a point charge, as viewed from the field point, the electric field intensity at point $\mathbf{r}$ can be expressed as

$$\mathbf{E}(\mathbf{r}) = \lim_{\substack{N \to \infty \\ \Delta v \to 0}} \sum_{i=1}^{N} \frac{\rho_v(\mathbf{r}'_i)\,\Delta v(\mathbf{r}'_i)}{4\pi\varepsilon_0}\,\frac{\mathbf{a}_{\mathcal{R}}}{\left|\mathbf{r} - \mathbf{r}'_i\right|^2} \tag{3.16}$$

where $\mathbf{a}_{\mathcal{R}}$ is a unit vector in the direction of distance vector $\mathcal{R} = \mathbf{r} - \mathbf{r}'_i$ (Fig. 3.5). The right side of Eq. (3.16) is by definition a volume integral. Therefore, the electric field of the volume charge density ρ_v is given by

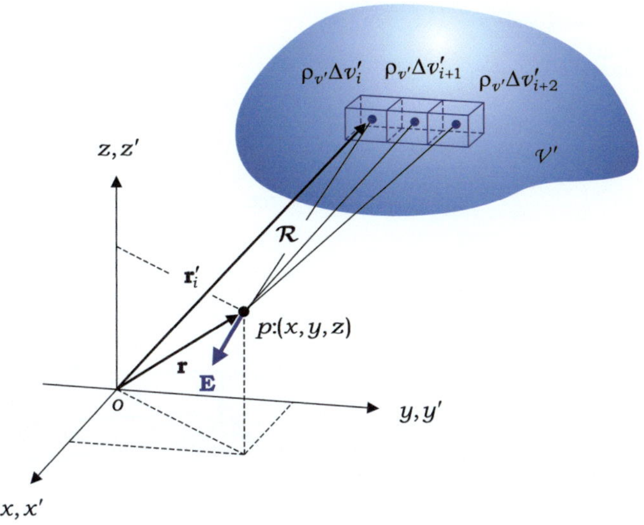

Fig. 3.5 Electric field due to volume charge density

$$\boxed{\mathbf{E}(\mathbf{r}) = \frac{1}{4\pi\varepsilon_0} \int_{\mathcal{V}'} \frac{\rho_{v'}}{\mathcal{R}^2}\mathbf{a}_\mathcal{R}\, dv'} \quad [\text{V/m}] \tag{3.17}$$

Here, $\mathcal{R}$ and $\mathbf{a}_\mathcal{R}$ depend on both $\mathbf{r}$ and $\mathbf{r}'$ through distance vector $\mathcal{R} = \mathbf{r} - \mathbf{r}'$, which is expressed in mixed coordinates. Although the coordinate axes of the primed and unprimed systems coincide, the field and source points are separate and independent. Therefore, the volume integral in Eq. (3.17) is conducted in primed coordinates with the field point held fixed. For the same reason, the unit vector $\mathbf{a}_\mathcal{R}$ cannot be taken outside the integral sign.

If the source charge is distributed over a surface with density ρ_s, the electric field intensity at the field point can be obtained by summing the contributions of all differential charges in all differential areas of the surface. The charge within the differential area is equal to $\rho_s'\, ds'$, and is considered a point charge. By invoking Coulomb's law and the superposition principle, the electric field of the surface charge density can be expressed in the form of a surface integral, that is,

$$\boxed{\mathbf{E}(\mathbf{r}) = \frac{1}{4\pi\varepsilon_0} \int_{\mathcal{S}'} \frac{\rho_{s'}}{\mathcal{R}^2}\mathbf{a}_\mathcal{R}\, ds'} \quad [\text{V/m}] \tag{3.18}$$

For a continuous charge formed into a line with density ρ_ℓ, which is not necessarily a straight line, the electric field intensity at the field point can be obtained by adding the contributions of all differential charges of all differential lengths along the line. The charge within the differential length is $\rho_{\ell'}\, dl'$, and is considered to be a point charge. By invoking Coulomb's law and superposition principle, the electric field of the line charge density can be expressed in the form of a line integral, that is,

Fig. 3.6 Infinitely long and
straight line charge

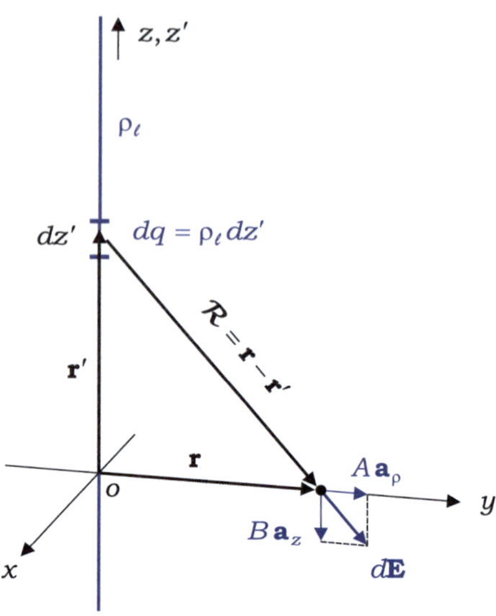

$$\boxed{\mathbf{E}(\mathbf{r}) = \frac{1}{4\pi\varepsilon_0} \int_{\mathcal{L}'} \frac{\rho_{\ell'}}{\mathcal{R}^2} \mathbf{a}_\mathcal{R} \, dl'} \quad [\text{V/m}] \tag{3.19}$$

where $\mathcal{R} = \mathcal{R}\,\mathbf{a}_\mathcal{R} = |\mathbf{r} - \mathbf{r}'|\,\mathbf{a}_\mathcal{R}$.

Example 3.4 Find the electric field of an infinitely long and straight line charge of uniform density ρ_ℓ coincident with the z-axis, as shown in Fig. 3.6.

Solution

In view of the cylindrical symmetry around the line charge, the field point can be a point on the y-axis without loss of generality. Then, the position vectors for the field and source points are

$$\mathbf{r} = \rho\,\mathbf{a}_\rho$$

$$\mathbf{r}' = z'\mathbf{a}_{z'}$$

The distance vector is expressed in terms of the base vectors $\mathbf{a}_\rho$, $\mathbf{a}_\phi$, and $\mathbf{a}_z$ defined at the field point such as

$$\mathcal{R} = \mathbf{r} - \mathbf{r}' = \rho\,\mathbf{a}_\rho - z'\,\mathbf{a}_z \tag{3.20a}$$

The differential charge at the source point is

$$dq = \rho_\ell\,dz'$$

By applying Coulomb's law, $d\mathbf{E}$ at $\mathbf{r}$ due to dq at $\mathbf{r}'$ can be obtained as follows:

$$d\mathbf{E} = \frac{dq}{4\pi\varepsilon_0}\frac{\mathcal{R}}{\mathcal{R}^3} = \frac{\rho_\ell\,dz'}{4\pi\varepsilon_0}\frac{\rho\,\mathbf{a}_\rho - z'\,\mathbf{a}_z}{(\rho^2 + z'^2)^{3/2}} \tag{3.20b}$$

From Fig. 3.6, the two differential charges positioned at $z' = a$ and $z' = -a$ jointly cause the z-component of $\mathbf{E}$ to vanish at the field point. Taking only the ρ-component of $\mathbf{E}$ into account, we write

$$\mathbf{E} = \int d\mathbf{E} = \frac{\rho_\ell\rho\,\mathbf{a}_\rho}{4\pi\varepsilon_0}\int_{-\infty}^{\infty}\frac{dz'}{(\rho^2 + z'^2)^{3/2}} = \frac{\rho_\ell\rho\,\mathbf{a}_\rho}{4\pi\varepsilon_0}\left[\frac{z'/\rho^2}{\sqrt{\rho^2 + z'^2}}\Bigg|_{z'=-\infty}^{z'=\infty}\right] \tag{3.20c}$$

Thus, in cylindrical coordinates, the electric field of the line charge density is expressed as

$$\boxed{\mathbf{E} = \frac{\rho_\ell}{2\pi\varepsilon_0\,\rho}\mathbf{a}_\rho}\ \ [\text{V/m}] \tag{3.21}$$

The line charge density ρ_ℓ should not be confused with radial distance ρ.

For future reference,

$$\boxed{\int\frac{dx}{(x^2 + a^2)^{3/2}} = \frac{x/a^2}{\sqrt{x^2 + a^2}}} \tag{3.22}$$

The line charge shown in Fig. 3.6 is further examined for symmetry. It appears to be the same even if it is rotated about the z-axis, linearly displaced in the z-direction, or rotated $180°$ about the x-axis. In view of this, the line charge is said to have cylindrical symmetry, translational symmetry in the z-direction, and two-fold rotational symmetry about the x-axis. In the presence of these symmetries, the resulting vector field or electric field should be independent of ϕ and z, and have neither the ϕ- nor z-components, such that $\mathbf{E}(\mathbf{r}) = E_\rho(\rho)\,\mathbf{a}_\rho$. In fact, the result of Eq. (3.21) is in this form.

Example 3.5 A line charge of uniform density $\rho_{\ell o}$ forms a circle of radius a in the $z = 0$ plane, with the center at the origin, as shown in Fig. 3.7.

(a) Find $\mathbf{E}$ at p_1:$(0,\ 0,\ b)$ along the z-axis.
(b) Show that the charge can be regarded as a point charge at the origin as $b \to \infty$.

Solution

(a) The position vectors for the field and source points are

$$\mathbf{r} = b\,\mathbf{a}_z$$

$$\mathbf{r}' = a\,\mathbf{a}_{\rho'} = a(\cos\phi_1\,\mathbf{a}_x + \sin\phi_1\,\mathbf{a}_y)$$

The distance vector is expressed in mixed coordinates as

$$\mathcal{R} = \mathbf{r} - \mathbf{r}' = b\,\mathbf{a}_z - a\,\mathbf{a}_{\rho'}$$

The differential charge at the source point is

$$dq = \rho_{\ell o}\,a\,d\phi'$$

Thus, the differential electric field at the field point is

$$d\mathbf{E} = \frac{dq}{4\pi\varepsilon_0}\frac{\mathcal{R}}{\mathcal{R}^3} = \frac{\rho_{\ell o}\,a\,d\phi'}{4\pi\varepsilon_0}\frac{(-a\,\mathbf{a}_{\rho'} + b\,\mathbf{a}_z)}{(a^2 + b^2)^{3/2}} \tag{3.23a}$$

Although $d\mathbf{E}$ due to dq at $\phi' = \phi_1$ has exactly the same form as that due to dq at $\phi' = \phi_1 + \pi$, the unit vector $\mathbf{a}_{\rho'}$ at $\phi' = \phi_1$ is opposite to $\mathbf{a}_{\rho'}$ at $\phi' = \phi_1 + \pi$. Therefore, the total $\mathbf{E}$ should have no ρ-component such that

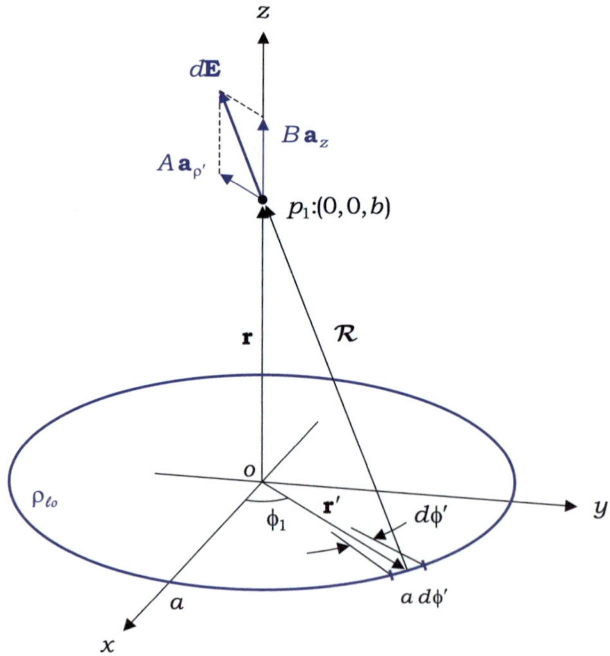

Fig. 3.7 Line charge in the form of circle

$$\mathbf{E} = \int d\mathbf{E} = \frac{\rho_{\ell o}}{4\pi\varepsilon_0} \frac{ab\,\mathbf{a}_z}{(a^2 + b^2)^{3/2}} \int_{\phi'=0}^{\phi'=2\pi} d\phi' = \frac{\rho_{\ell o}}{2\varepsilon_0} \frac{ab\,\mathbf{a}_z}{(a^2 + b^2)^{3/2}} \qquad (3.23b)$$

(b) The total charge on the circle is

$$Q = 2\pi a\,\rho_{\ell o} \qquad (3.23c)$$

Inserting Eq. (3.23c) into Eq. (3.23b) and using $(a^2 + b^2)^{3/2} \approx b^3$ for $b \to \infty$, we obtain

$$\mathbf{E} = \frac{Q}{4\pi\varepsilon_0 b^2}\mathbf{a}_z \qquad (3.23d)$$

This is the same as the electric field intensity of a point charge at the origin.

Exercise 3.6

An infinitely long line charge of uniform density ρ_ℓ is parallel to the z-axis and passes through point $(0, 2, 0)$ in Cartesian coordinates. Determine $\mathbf{E}$ at $(x_1, y_1, 0)$.

Ans. $\mathbf{E} = (\rho_\ell/2\pi\varepsilon_0)[x_1\,\mathbf{a}_x + (y_1 - 2)\,\mathbf{a}_y]/[x_1^2 + (y_1 - 2)^2]$.

Exercise 3.7

Is it true that the $\mathbf{E}$ of a line charge is solenoidal $(\nabla \cdot \mathbf{E} = 0)$ only because the $\mathbf{E}$ of a point charge is so.

Ans. Yes, $\mathbf{E}$ obeys the superposition principle and the divergence is distributive.

Exercise 3.8

Find the equivalent line charge density for an infinitely long cylinder with radius a carrying ρ_v [C/m^3], as viewed from a distant point.

Ans. $\rho_v \pi a^2$ [C/m].

Review Questions

RQ 3.1	State Coulomb's law.	[(3.1)]
RQ 3.2	Define the electric field intensity of the point charge.	[(3.9)]
RQ 3.3	What are the source and field points?	[Fig. 3.2]
RQ 3.4	State the principle of superposition of the electric field.	[(3.12)]
RQ 3.5	What is an electric dipole?	[Fig. 3.4]
RQ 3.6	What are the units of ρ_v, ρ_s, and ρ_ℓ?	[(3.17)(3.18)(3.19)]
RQ 3.7	In what ways do the electric fields of the point charge, electric dipole, and long line charge vary with distance?	[(3.11)(3.14)(3.21)]
RQ 3.8	What can be used to relate the source in primed coordinates to the electric field in unprimed coordinates?	[Fig. 3.5]

3.3 Electric Flux Density and Gauss's Law

The simplest source of an electric field is an electron, which, in most cases, is considered to be a point charge. A positive point charge produces an electric field radially outwards and its constituent vectors are continuous in space, except for the charge position. In other words, there is no abrupt change in magnitude or direction as the electric field varies with position. If such an electric field is depicted by a random sample of vectors drawn on paper, as shown in Fig. 3.8a, it may give the incorrect impression that the given region is randomly scattered with discrete vectors. Alternatively, we may draw differential directed lines along the electric field vectors such that they eventually form smooth lines, called *electric field lines* or *electric flux lines*, which are more appropriate for exhibiting the continuous nature of the electric field (see Fig. 3.8b). In general, electric field lines are smooth directed lines extending between the positive and negative charges. The tangent to the field line indicates the direction of the electric field at a given point and the density of the field lines in the small region surrounding the point represents the magnitude of the electric field. The electric field lines do not cross because the electric field is a single-valued function of position.

3.3.1 Electric Flux Density

The concept of the electric flux begins with an electrostatic induction experiment, as illustrated in Fig. 3.9. The space between the two perfectly conducting concentric spheres was filled with a *dielectric* or insulating material so that no free charges could be in motion in the gap. When a net charge $+Q$ is imparted to the inner conductor, the charge is uniformly distributed on the conductor surface such that no electric field exists inside the inner conductor. Otherwise, the electric field instantaneously accelerates the charge and generates an infinite conduction current, causing the charge to redistribute until there is no internal electric field. Immediately after the outer

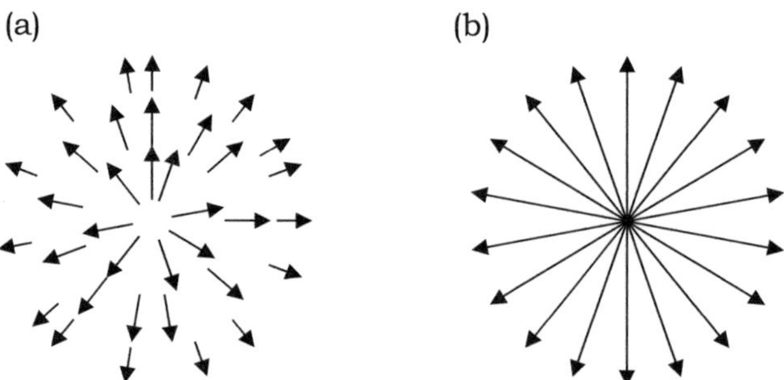

Fig. 3.8 **a** Electric field vectors, and **b** Electric field lines of point charge

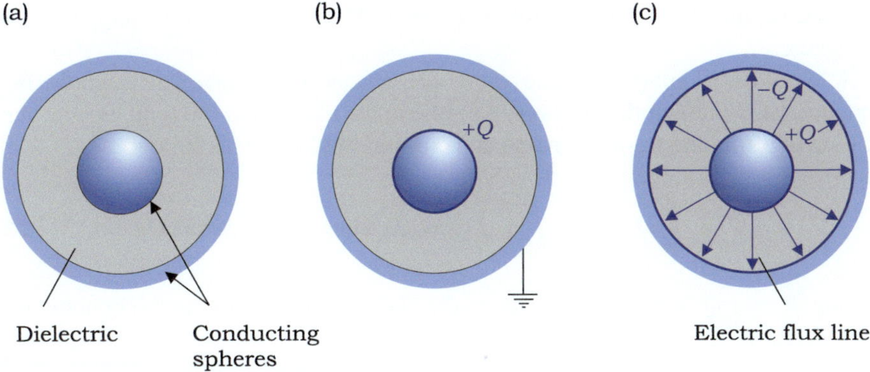

Fig. 3.9 Electrostatic induction. The electric flux is equal to the net charge $+Q$

conductor was temporarily connected to the ground for discharge, an induced charge $-Q$ was observed on the outer conductor independent of the dielectric in the gap. This electrostatic induction was interpreted by early pioneers as the result of some sort of ***displacement***, also called ***electric flux***, that starts at the inner conductor and ends at the outer conductor.

The total electric flux Ψ from the inner conductor equals the net charge of the conductor. That is,

$$\Psi = +Q \quad [\text{C}] \tag{3.24}$$

which is in coulombs. Here, we define the electric flux density $\mathbf{D}$ as the electric flux per unit area, which is given in coulombs per square meter. The electric flux density constitutes a vector field that is a continuous function of position in homogeneous media. It is important to note that the magnitude of $\mathbf{D}$ represents *the net electric flux crossing a unit area in the cross section* or in the plane perpendicular to the direction of $\mathbf{D}$.

We briefly digress and explore a perfectly conducting sphere with a uniform surface charge for symmetry. It has spherical symmetry; it looks the same as we move around it, varying the spherical coordinates θ and ϕ while keeping R constant. As discussed in Chap. 1, spherical symmetry ensures that the resultant vector field is independent of θ and ϕ and has neither the θ- nor the ϕ-component. Consequently, the electric flux density of a uniform spherical surface charge is expected to have the form $\mathbf{D} = G(R)\,\mathbf{a}_R$.

Next, we consider two concentric spheres, as shown in Fig. 3.9c, which certainly has a spherical symmetry. To satisfy the requirement that the electric field vanishes inside perfect conductors, the net and induced charges should at the very least distribute themselves in such a way that they conform to spherical symmetry. As the electric flux lines extend between two charges of opposite polarity, the charges reside on the surface of the inner conductor and the inner surface of the outer sphere.

In this case, the electric flux density is given by $\mathbf{D} = G(R)\,\mathbf{a}_R$ in the gap, and zero elsewhere.

A hypothetical sphere of radius R_o is constructed in the space between the two conductors. By integrating $\mathbf{D}$ over this sphere, we can obtain the total electric flux originating from the positive charge and crossing the sphere, as follows:

$$\Psi = \oint_{S} \mathbf{D} \cdot d\mathbf{s} = \int_{\phi=0}^{2\pi} \int_{\theta=0}^{\pi} D_R R_o^2 \sin\theta \, d\theta \, d\phi = 4\pi R_o^2 D_R \qquad (3.25)$$

Setting $\Psi = +Q$ in Eq. (3.25) leads to the electric flux density on the hypothetical sphere, that is,

$$\mathbf{D} = D_R\,\mathbf{a}_R = \frac{+Q}{4\pi R_o^2}\,\mathbf{a}_R \qquad (3.26)$$

There is no change in $\mathbf{D}$, as shown in Eq. (3.26), even if the gap is filled with a different dielectric or the inner sphere shrinks to a point at the origin, as the outer sphere is expanded to infinity while maintaining the amount of charge unchanged. Therefore, the electric flux density of point charge q at the origin is given by

$$\boxed{\mathbf{D} = \frac{q}{4\pi R^2}\,\mathbf{a}_R} \quad [\mathrm{C/m^2}] \qquad (3.27)$$

It is important to note that Eq. (3.27) always holds regardless of the dielectric surrounding the charge. Furthermore, $\mathbf{D}$ follows the ***inverse square law***. Therefore, the integral of $\mathbf{D}$ over any sphere around the charge is constant.

Borrowing from Sect. 3.2, the electric field due to point charge q at the origin in free space is given by

$$\mathbf{E} = \frac{q}{4\pi\varepsilon_0 R^2}\,\mathbf{a}_R \qquad (3.11)$$

Comparison of Eq. (3.27) with Eq. (3.11) leads to a ***constitutive relation*** in free space, that is,

$$\boxed{\mathbf{D} = \varepsilon_0 \mathbf{E}} \quad [\mathrm{C/m^2}] \qquad (3.28)$$

where ε_0 is the permittivity of free space.

Following the same procedure as that used for $\mathbf{E}$ of a continuous charge distribution, we can obtain the $\mathbf{D}$ of the volume, surface, and line charges in free space as follows:

$$\boxed{\mathbf{D}(\mathbf{r}) = \frac{1}{4\pi} \int_{\mathcal{V}'} \frac{\rho_{v'}}{\mathcal{R}^2}\,\mathbf{a}_{\mathcal{R}}\, dv'} \quad \text{(volume charge)} \qquad (3.29a)$$

$$\boxed{\mathbf{D}(\mathbf{r}) = \frac{1}{4\pi} \int_{\mathcal{S}'} \frac{\rho_{s'}}{\mathcal{R}^2} \mathbf{a}_{\mathcal{R}} \, ds'} \quad \text{(surface charge)} \qquad (3.29b)$$

$$\boxed{\mathbf{D}(\mathbf{r}) = \frac{1}{4\pi} \int_{\mathcal{L}'} \frac{\rho_{\ell'}}{\mathcal{R}^2} \mathbf{a}_{\mathcal{R}} \, d\ell'} \quad \text{(line charge)} \qquad (3.29c)$$

where $\mathcal{R}$ and $\mathbf{a}_{\mathcal{R}}$ are the magnitude and unit vector of the distance vector $\mathcal{R}$, respectively, and ρ_v, ρ_s, and ρ_ℓ are the volume, surface, and line charge densities, respectively.

In homogeneous media, the expressions for $\mathbf{D}$ are the same as those in Eqs. (3.29a, b, c). However, the constitutive relationship in Eq. (3.28) should be modified. In addition, in finite material media, the presence of the boundary may reshape $\mathbf{D}$ in the given region, even if the total electric flux remains the same. These topics will be discussed in the following sections.

Exercise 3.9
If a net charge of $+Q$ [C] is imparted to a perfectly conducting cube, what is the total electric flux leaving the cube?

Ans. $+Q$ [C].

Exercise 3.10
Find the divergence and curl of the electric flux density induced by a point charge placed at the origin in free space.

Ans. $\nabla \cdot \mathbf{D} = 0$ and $\nabla \times \mathbf{D} = 0$, except for the origin.

Exercise 3.11
Given the expressions for line $\mathbf{r}(t) = x(t)\,\mathbf{a}_x + y(t)\,\mathbf{a}_y$ and electric flux density $\mathbf{D} = D_x\,\mathbf{a}_x + D_y\,\mathbf{a}_y$ in the $z = 0$ plane, write a differential equation for the electric flux lines in this plane.

Ans. $\dfrac{dy/dt}{dx/dt} = \dfrac{dy}{dx} = \dfrac{D_y}{D_x}$.

3.3.2 Gauss's Law

The electric flux density of a point charge is solenoidal except at the charge position, and obeys the inverse square law. As the surface area of a sphere varies with the square of its radius, the integration of the electric flux density over a sphere centered at the charge remains the same, irrespective of the radius of the sphere. In other words, the total electric flux is conserved. This is also the case for a closed surface with arbitrary shape. To prove this, we consider a point charge q and a closed surface $\mathcal{S}_1$ that excludes the charge, as shown in Fig. 3.10a. Surface $\mathcal{S}_1$ consists of three parts: an outer surface of arbitrary shape, a small sphere centered at q, and a thin hollow pipe connecting the two surfaces. Because $\mathbf{D}$ is solenoidal, the integration

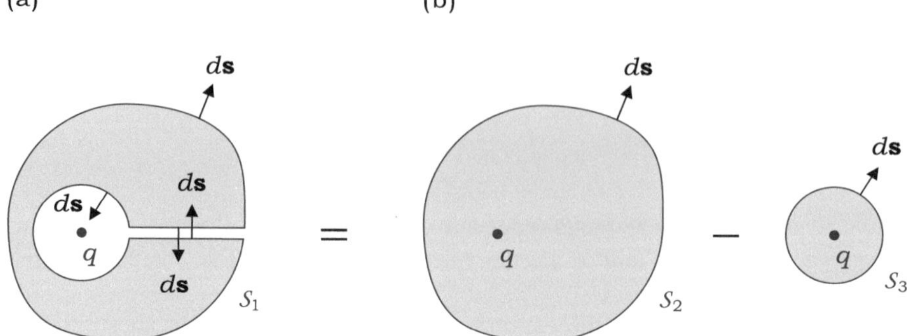

Fig. 3.10 a Closed surface S_1 excludes charge q. **b** As the connecting pipe collapses, S_1 splits into two closed surfaces. The reversal of $d\mathbf{s}$ on S_3 with respect to the corresponding $d\mathbf{s}$ on S_1 is compensated by subtraction

of $\nabla \cdot \mathbf{D}$ over the volume bounded by S_1 (the shaded region in Fig. 3.10a) is zero. Subsequently, the divergence theorem ensures that the integration of $\mathbf{D}$ over S_1 is zero. When the radius of the connecting pipe is very small, the integration over the pipe wall vanishes, because the two differential area vectors of the opposite walls point in opposite directions. Thus, the pipe walls do not contribute to the surface integral. Under this condition, the integration over S_1 can be obtained using surfaces S_2 and S_3, as shown in Fig. 3.10b. The integration over S_1 is zero, because S_1 encloses no charge. Meanwhile, the integration over the spherical surface S_3 is simply q, as stated previously. Consequently, the integration of $\mathbf{D}$ over an arbitrary surface S_2 around the charge is equal to q.

Extending this procedure to multiple charges and applying the principle of superposition, what it comes down to is Gauss's law, which states that *the net outward electric flux through any closed surface is equal to the charge enclosed within the surface.* Expressed mathematically, Gauss's law is expressed as

$$\boxed{\oint_S \mathbf{D} \cdot d\mathbf{s} = Q} \quad [\text{C}] \tag{3.30}$$

A small circle on the integral sign indicates that the integration is conducted over a closed surface. The dot product in the integrand converts the differential area $|d\mathbf{s}|$ in S into an equivalent area in the cross section of $\mathbf{D}$. Note that $d\mathbf{s}$ is always directed outward from the enclosed volume. A positive value of the closed-surface integral indicates that the total flux of Q [C] exits the enclosed volume. Gauss's law can be expressed in terms of volume charge density as

$$\boxed{\oint_S \mathbf{D} \cdot d\mathbf{s} = \int_v \rho_v \, dv} \quad [\text{C}] \tag{3.31}$$

where $\mathcal{V}$ is the volume bounded by the surface $\mathcal{S}$.

Let us remind ourselves that Coulomb's law is the fundamental law of electro-statics that can be applied anytime to determine $\mathbf{E}$ of a charge distribution. Gauss's law is particularly useful for determining $\mathbf{D}$ when a given charge distribution has symmetry. The application of Gauss's law is subject to the proper choice of a closed surface called a **Gaussian surface**. This hypothetical surface is constructed only for mathematical purposes. Using a Gaussian surface, the closed-surface integral in Eq. (3.31) can be drastically reduced to a simple algebraic equation. In short, the Gaussian surface $\mathcal{S}$ is chosen in such a way that $\mathbf{D}$ *is constant and normal to $\mathcal{S}$* ($\mathbf{D} \cdot d\mathbf{s}$ is constant over $\mathcal{S}$) *or tangential to $\mathcal{S}$ otherwise* ($\mathbf{D} \cdot d\mathbf{s}$ vanishes).

Gauss's law in **integral form**, as shown in Eq. (3.31), can be converted into a point or differential form. Consider an infinitesimal volume Δv bounded by closed surface Δs. After dividing both sides of Eq. (3.30) by Δv, taking the limit as $\Delta v \to 0$ gives

$$\lim_{\Delta v \to 0} \frac{1}{\Delta v} \oint_{\Delta s} \mathbf{D} \cdot d\mathbf{s} = \lim_{\Delta v \to 0} \frac{Q}{\Delta v} \tag{3.32}$$

The left-hand side of Eq. (3.32) is, by definition, the divergence of $\mathbf{D}$ and the right-hand side is the volume charge density. Therefore, Gauss's law can be expressed in point form, as

$$\boxed{\nabla \cdot \mathbf{D} = \rho_v} \quad [\mathrm{C/m^3}] \tag{3.33}$$

Alternatively, the point form of Gauss's law can be obtained by applying the divergence theorem to the left-hand side of Eq. (3.31), that is,

$$\int_{\mathcal{V}} \nabla \cdot \mathbf{D} \, dv = \int_{\mathcal{V}} \rho_v \, dv \tag{3.34}$$

Because this holds for any volume enclosing all charges under consideration, the integrands must be equal; that is, $\nabla \cdot \mathbf{D} = \rho_v$.

Example 3.6 For a uniform volume charge density ρ_{vo} in the form of a spherical shell with radii a and b, as shown in Fig. 3.11, determine the electric field everywhere.

Solution

The given volume charge has spherical symmetry, from which an electric flux density of the form $\mathbf{D} = D_R(R)\,\mathbf{a}_R$ is expected. A spherical surface that is concentric with charge is an appropriate Gaussian surface.

(a) In region $R \geq b$ (Fig. 3.11a), the net outward flux through a spherical Gaussian surface of radius R_1 is

$$\oint_{\mathcal{S}} \mathbf{D} \cdot d\mathbf{s} = \int_{\phi=0}^{2\pi} \int_{\theta=0}^{\pi} D_R R_1^2 \sin\theta \, d\theta \, d\phi = D_R 4\pi R_1^2 \tag{3.35a}$$

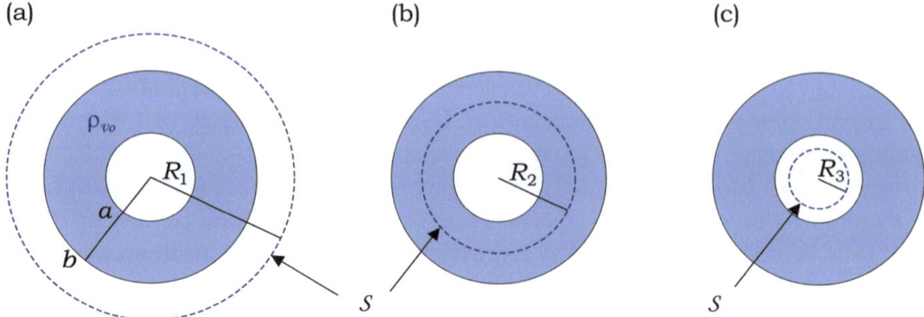

Fig. 3.11 Uniform volume charge in the form of hollow sphere

where $ds = R_1^2 \sin\theta \, d\theta \, d\phi \, \mathbf{a}_R$. The total charge within the Gaussian surface is

$$Q = \int_V \rho_v \, dv = \int_{\phi=0}^{2\pi} \int_{\theta=0}^{\pi} \int_{R=a}^{b} \rho_{vo} R^2 \sin\theta \, dR \, d\theta \, d\phi = \rho_{vo} \frac{4\pi}{3}(b^3 - a^3)$$

(3.35b)

where $dv = R^2 \sin\theta \, dR \, d\theta \, d\phi$. Equating Eqs. (3.35a) and (3.35b) yields

$$D_R = \frac{\rho_{vo}}{3R_1^2}(b^3 - a^3)$$

(3.35c)

Therefore, in region $R \geq b$,

$$\mathbf{D} = D_R \mathbf{a}_R = \frac{\rho_{vo}}{3R^2}(b^3 - a^3)\, \mathbf{a}_R$$

(3.35d)

$$\mathbf{E} = \frac{\mathbf{D}}{\varepsilon_0} = \frac{\rho_{vo}}{3\varepsilon_0 R^2}(b^3 - a^3)\, \mathbf{a}_R$$

(3.35e)

where subscript 1 is omitted for generalization.

(b) In region $a \leq R \leq b$ (Fig. 3.11b), using a Gaussian surface of radius R_2, we obtain

$$\oint_S \mathbf{D} \cdot d\mathbf{s} = \oint_S D_R \mathbf{a}_R \cdot d\mathbf{s} = D_R 4\pi R_2^2$$

(3.35f)

The enclosed charge is

$$Q = \int_V \rho_v \, dv = \int_{\phi=0}^{2\pi} \int_{\theta=0}^{\pi} \int_{R=a}^{R_2} \rho_{vo} R^2 \sin\theta \, dR \, d\theta \, d\phi$$

$$= \rho_{vo} \frac{4\pi}{3}(R_2^3 - a^3)$$

(3.35g)

Equating Eqs. (3.35f) and (3.35g) yields

$$D_R = \frac{\rho_{vo}}{3R_2^2}(R_2^3 - a^3) \tag{3.35h}$$

Therefore, in region $a \le R \le b$,

$$\mathbf{D} = \frac{\rho_{vo}}{3R^2}(R^3 - a^3)\,\mathbf{a}_R \tag{3.35i}$$

$$\mathbf{E} = \frac{\mathbf{D}}{\varepsilon_0} = \frac{\rho_{vo}}{3\varepsilon_0 R^2}(R^3 - a^3)\,\mathbf{a}_R \tag{3.35j}$$

where subscript 2 is omitted for generalization.

(c) In region $R \le a$ (Fig. 3.11c), the Gaussian surface encloses no charge. Therefore, the closed-surface integral of $\mathbf{D}$ disappears. Because $\mathbf{D}$ is initially constant over the Gaussian surface, it follows that, in the given region,

$$\mathbf{D} = 0 = \mathbf{E} \tag{3.35k}$$

However, in general, even if the closed-surface integral of $\mathbf{D}$ is zero, $\mathbf{D}$ is not necessarily zero at every point on the surface.

Next, rearranging Eq. (3.35e) gives

$$\mathbf{E} = \frac{1}{4\pi\varepsilon_0 R^2}\left[\rho_{vo}\frac{4\pi}{3}(b^3 - a^3)\right]\mathbf{a}_R = \frac{Q}{4\pi\varepsilon_0 R^2}\mathbf{a}_R \tag{3.35l}$$

where the term in brackets represents the total charge Q. From Eq. (3.35l), the electric field in region $R \ge b$ is the same as if all charges are concentrated at the center. This is because the given volume charge possesses the same spherical symmetry as the point charge at the center.

Example 3.7 When an infinitely long and straight line charge of uniform density ρ_ℓ lies along the z-axis in free space, as shown in Fig. 3.12, determine the electric field using Gauss's law.

Solution

The line charge has cylindrical, translational, and two-fold rotational symmetries, which ensures that $\mathbf{D}$ has the form $\mathbf{D} = D_\rho(\rho)\,\mathbf{a}_\rho$ (see Example 3.4). Therefore, the closed cylinder is an appropriate Gaussian surface.

On the side surface, using $d\mathbf{s} = \rho_1\,d\phi dz\,\mathbf{a}_\rho$, we have

$$\mathbf{D} \cdot d\mathbf{s} = D_\rho \rho_1\,d\phi dz$$

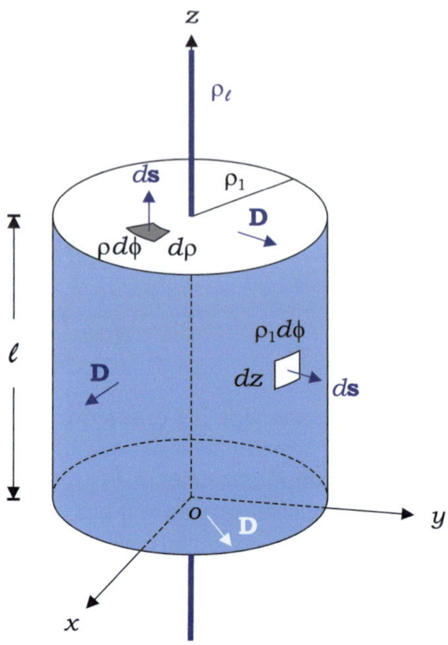

Fig. 3.12 Infinitely long line charge and Gaussian surface

On the top and bottom surfaces, **D** is normal to ds. Thus,

$$\mathbf{D} \cdot d\mathbf{s} = 0$$

The net outward flux through the Gaussian surface is

$$\oint_S \mathbf{D} \cdot d\mathbf{s} = \int_{side} \mathbf{D} \cdot d\mathbf{s} + \int_{top} \mathbf{D} \cdot d\mathbf{s} + \int_{bottom} \mathbf{D} \cdot d\mathbf{s}$$

$$= \int_{z=-\ell/2}^{\ell/2} \int_{\phi=0}^{2\pi} D_\rho \rho_1 \, d\phi dz = 2\pi D_\rho \rho_1 \ell \tag{3.36a}$$

The enclosed charge is given by

$$Q = \int_{z=-\ell/2}^{z=\ell/2} \rho_\ell \, dz = \rho_\ell \ell \tag{3.36b}$$

Equating Eqs. (3.36a) and (3.36b) yields

$$D_\rho = \frac{\rho_\ell}{2\pi \rho_1} \tag{3.36c}$$

By omitting subscript 1 for generalization, we have

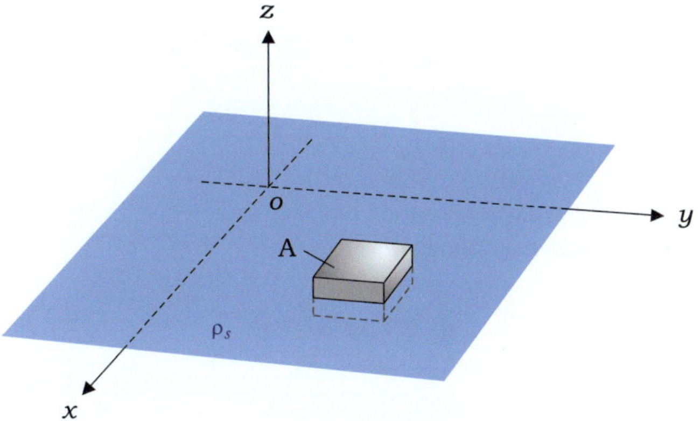

Fig. 3.13 Infinite sheet with uniform surface charge

$$\mathbf{D} = D_\rho\, \mathbf{a}_\rho = \frac{\rho_\ell}{2\pi\rho}\mathbf{a}_\rho \tag{3.36d}$$

$$\mathbf{E} = \frac{\mathbf{D}}{\varepsilon_0} = \frac{\rho_\ell}{2\pi\varepsilon_0\rho}\mathbf{a}_\rho \tag{3.36e}$$

The line charge density ρ_ℓ should not be confused with the radial distance ρ in the cylindrical coordinates. $\mathbf{E}$ in Eq. (3.36e) is identical to that in Eq. (3.21) that was obtained using Coulomb's law.

Example 3.8 An infinite sheet with uniform surface charge density ρ_s is coincident with the $z = 0$ plane, as shown in Fig. 3.13. Find its electric field using Gauss's law.

Solution

Let us explore the surface charge for symmetry. It exhibits translational symmetry in the x- and y-directions; thus, $\mathbf{D}$ is independent of x and y. It also has rotational symmetry about the z-axis; thus, $\mathbf{D}$ has neither the x- nor y-components. In addition, it has a two-fold rotational symmetry about the x-axis. Consequently, $\mathbf{D}$ is expected to have the following form:

$$\mathbf{D} = D_z(z)\,\mathbf{a}_z \text{ for } z > 0, \text{ and } \mathbf{D} = -D_z(z)\,\mathbf{a}_z \text{ for } z < 0 \tag{3.37a}$$

In view of Eq. (3.37a), we choose a rectangular parallelepiped as the Gaussian surface with its upper half in region $z > 0$ and lower half in region $z < 0$, as illustrated in Fig. 3.13.

The electric flux through each face of the parallelepiped is

$$\int_T \mathbf{D} \cdot d\mathbf{s} = \int_T D_z\,\mathbf{a}_z \cdot dx dy\, \mathbf{a}_z = A D_z$$

$$\int_B \mathbf{D} \cdot d\mathbf{s} = \int_B -D_z\,\mathbf{a}_z \cdot (-dx\,dy)\,\mathbf{a}_z = A D_z$$

$$\int_S \mathbf{D} \cdot d\mathbf{s} = 0$$

where T, B, and S denote the top, bottom, and side face, respectively. The net outward flux through the Gaussian surface is thus

$$\oint_S \mathbf{D} \cdot d\mathbf{s} = 2 A D_z \tag{3.37b}$$

The total charge within the Gaussian surface is

$$Q = \int_A \rho_s\, ds = \rho_s A \tag{3.37c}$$

Equating Eqs. (3.37b) and (3.37c) yields

$$D_z = \frac{\rho_s}{2} \tag{3.37d}$$

Inserting Eq. (3.37d) into Eq. (3.37a), in the region $z > 0$, we obtain

$$\mathbf{D} = \frac{\rho_s}{2}\mathbf{a}_z \quad [\text{C/m}^2] \tag{3.38a}$$

$$\boxed{\mathbf{E} = \frac{\rho_s}{2\varepsilon_0}\mathbf{a}_z} \quad [\text{V/m}] \tag{3.38b}$$

Similarly, in region $z < 0$, we obtain

$$\mathbf{D} = -\frac{\rho_s}{2}\mathbf{a}_z \quad [\text{C/m}^2] \tag{3.38c}$$

$$\boxed{\mathbf{E} = -\frac{\rho_s}{2\varepsilon_0}\mathbf{a}_z} \quad [\text{V/m}] \tag{3.38d}$$

If the sheet were finite in extent, it would have been impossible to construct a Gaussian surface and use Gauss's law because of the lack of symmetry. Nevertheless, the electric field can be obtained using Coulomb's law in numerical methods.

Exercise 3.12

If the $z = 3$ and $z = 0$ planes carry surface charges of uniform densities ρ_s and $-\rho_s$, respectively, in free space, determine $\mathbf{E}$ everywhere.

Ans. $\mathbf{E} = -(\rho_s/\varepsilon_0)\,\mathbf{a}_z$ for $0 < z < 3$, $\mathbf{E} = 0$ for $z > 3$ or $z < 0$.

Exercise 3.13
If the electric flux density is $\mathbf{D} = 2\rho\,\mathbf{a}_\rho$ in cylindrical coordinates, determine ρ_v in a given region.

Ans. $\rho_v = 4\,[\text{C/m}^3]$.

Exercise 3.14
If a sphere is uniformly filled with (a) identical point charges, and (b) volume charge of density ρ_{vo}, determine $\nabla \cdot \mathbf{D}$ inside the sphere.

Ans. (a) Zero, excluding the charge position, (b) ρ_{vo}.

Review Questions

RQ 3.9	Express $\mathbf{D}$ of the point charge in spherical coordinates.	[(3.27)]
RQ 3.10	Relate $\mathbf{D}$ to $\mathbf{E}$ in free space.	[(3.28)]
RQ 3.11	Express $\mathbf{D}$ for continuous charge distribution.	[(3.29a,b,c)]
RQ 3.12	Express Gauss's law in integral form.	[(3.30)(3.31)]
RQ 3.13	Express Gauss's law in point form.	[(3.33]
RQ 3.14	In what manner is a Gaussian surface constructed?	[Fig. 3.12]
RQ 3.15	Is the divergence of $\mathbf{D}$ of a point charge zero everywhere?	[(3.27)]
RQ 3.16	Are the electric field lines always smooth in free space?	[Fig. 3.8, (3.12)]

3.4 Electric Potential

The electric field of the charge distribution can be obtained using Coulomb's or Gauss's law. However, Coulomb's law involves a vectorial sum that may not be straightforward in many cases and the application of Gauss's law is limited to charge distributions with symmetries. The *electric potential* provides another method to determine the electric field. Although the electric potential owes its origin to a vector identity, it has the physical significance of the work done in moving a unit charge from infinity to a point. The electric potential is a scalar quantity that can be obtained from the charge distribution or from another electric potential. Thus, it is much easier to work with. Furthermore, it plays a major role in boundary-value problems, in which the electric potential in a given region is obtained from the charge distribution or electric potential at the boundary. The electric field in the given region is then obtained by taking the negative gradient of the electric potential. The electric potential can be computed using a line or a definite integral. Therefore, care should be taken to avoid confusion between these two integrals. The electric potential has the same meaning as the voltage at a node in an electric circuit.

3.4.1   *Work Done in Moving Charges*

In the presence of an electric field, the charge can be moved from one point to another only after an external force is applied to counteract the electric force. Consider the situation in which we move charge q in the direction of unit vector $\mathbf{a}_l$ in an electric field $\mathbf{E}$. First, the electric force acting on q is given by

$$\mathbf{F}_e = q\mathbf{E} \quad [\text{N}] \tag{3.39}$$

An external force on the charge $\mathbf{F}_{ext}$ is necessary to cancel $\mathbf{F}_e$. That is,

$$\mathbf{F}_e + \mathbf{F}_{ext} = 0 \tag{3.40}$$

or

$$\mathbf{F}_{ext} = -q\mathbf{E} \quad [\text{N}] \tag{3.41}$$

Next, when charge q is displaced by a distance dl along $\mathbf{a}_l$, the **work done** or the energy expended in moving the charge is

$$dW = \mathbf{F}_{ext} \cdot \mathbf{a}_l \, dl = -q(\mathbf{E} \cdot \mathbf{a}_l) \, dl \tag{3.42}$$

where $\mathbf{E}$ is assumed to be constant along dl. At this point, the question arises as to how the charge can be displaced when it is at rest with no net force on it. An additional external force $\Delta\mathbf{F}$ should be applied so that the charge is accelerated to gain speed and change its position. When the charge reaches the midpoint of dl, a counterforce $-\Delta\mathbf{F}$ is applied to decelerate the charge and stop it. This process does not require any additional work done, as shown in Eq. (3.42).

Substituting $d\mathbf{l} = dl\, \mathbf{a}_l$ into Eq. (3.42) leads to the work done in moving charge q by distance $d\mathbf{l}$ in the electric field $\mathbf{E}$, that is,

$$\boxed{dW = -q\mathbf{E} \cdot d\mathbf{l}} \quad [\text{J}] \tag{3.43}$$

The total work done to carry charge q from point B to point A in an electric field $\mathbf{E}$ is given by the following line integral:

$$\boxed{W = \int_B^A dW = -q \int_B^A \mathbf{E} \cdot d\mathbf{l}} \quad [\text{J}] \tag{3.44}$$

which is a scalar quantity given in joules. The sign of W indicates a sense of direction. A positive value indicates that the work has been done by an external force or by us, whereas a negative value indicates that the work has been done by an electric field.

Exercise 3.15

In the presence of $\mathbf{E} = E_o\mathbf{a}_z$, what is the work done in moving charge q in the directions of (a) $\mathbf{a}_x$, (b) $\mathbf{a}_z$, and (c) $3\mathbf{a}_y - 4\mathbf{a}_z$ by distance dl?

Ans. (a) 0, (b) $(-qE_o)\,dl$, (c) $(0.8qE_o)\,dl$.

Exercise 3.16

If a charge is displaced by a unit distance in the $\mathbf{a}_l$-direction in the presence of $\mathbf{E} = -2\mathbf{a}_x + \mathbf{a}_y$, determine $\mathbf{a}_l$, along which the work done is maximum.

Ans. $\mathbf{a}_l = (2/\sqrt{5})\,\mathbf{a}_x - (1/\sqrt{5})\,\mathbf{a}_y$.

3.4.2 Electric Potential and Potential Difference

The *potential difference* V_{A-B} is defined as the work done in carrying a unit positive charge from initial point B to terminal point A. In an electric field $\mathbf{E}$, the potential difference between A and B is defined as

$$\boxed{V_{A-B} = -\int_B^A \mathbf{E} \cdot d\mathbf{l}} \quad \text{[V]} \qquad (3.45)$$

which is measured in volts. As mentioned earlier, in our system of notations, the subscript $A-B$ stands for "something from B to A," and, in the present case, shows the direction of the line integral. V_{A-B} is referred to as the electric potential at point A with reference to point B, or simply the potential difference between A and B.

The ***absolute electric potential*** or ***electric potential***, is the potential difference between a point and a ***zero-reference point***, which is usually taken at infinity. Otherwise, the zero-reference point is specified as a point, line, or surface within the given region. The electric potential at point A is defined as

$$\boxed{V_A = -\int_\infty^A \mathbf{E} \cdot d\mathbf{l}} \quad \text{[V]} \qquad (3.46)$$

The electric potential forms a scalar field that varies as a function of position.

The potential difference can be expressed in terms of the electric potential such as

$$\boxed{V_{A-B} = V_A - V_B = -\int_\infty^A \mathbf{E} \cdot d\mathbf{l} + \int_\infty^B \mathbf{E} \cdot d\mathbf{l}} \quad \text{[V]} \qquad (3.47)$$

Note that the potential difference is independent of the zero-reference point.

Consider the simple case of a point charge q located at the origin. Substituting the electric field in Eq. (3.11) into Eq. (3.46) leads to an electric potential at point (R_1, θ_1, ϕ_1):

$$V = -\int_{\infty}^{A} \mathbf{E} \cdot d\mathbf{l} = -\int_{\infty}^{R_1} \frac{q}{4\pi\varepsilon_0 R^2} dR = \frac{q}{4\pi\varepsilon_0 R_1} \tag{3.48}$$

where $d\mathbf{l} = dR\, \mathbf{a}_R + Rd\theta\, \mathbf{a}_\theta + (R\sin\theta)\, d\phi\, \mathbf{a}_\phi$, and the zero-reference point is set to infinity. Thus, the electric potential due to point charge q located at the origin is

$$\boxed{V = \frac{q}{4\pi\varepsilon_0 R}} \quad [\text{V}] \tag{3.49}$$

where subscript 1 is omitted for generalization. Note that V in Eq. (3.49) depends only on R, and is independent of the integration path. In addition, the electric potential of a point charge varies as $1/R$, whereas the electric field varies as $1/R^2$.

Referring to Fig. 3.14, we demonstrate that the path independence of the potential difference is a direct consequence of the law of energy conservation. Consider two separate paths established between points A and B in electric field $\mathbf{E}$ of a point charge. The irrotational nature of $\mathbf{E}$ ensures that its circulation around a closed path, consisting of path ① and the reversal of path ②, is zero. Therefore, it follows that the line integral of $\mathbf{E}$ along path ① is equal to that along path ②. From a physics point of view, if the potential difference V_{A-B} depends on the path, one would carry a charge from point B to A along the path that would demand less work done and return to B along the other to gain net energy. This violates the law of conservation of energy. *The potential difference is independent of the integration path.* With the superposition principle, the path independence can be extended to the electric field of a continuous charge distribution.

Fig. 3.14 Potential difference V_{A-B} is independent of the integration path

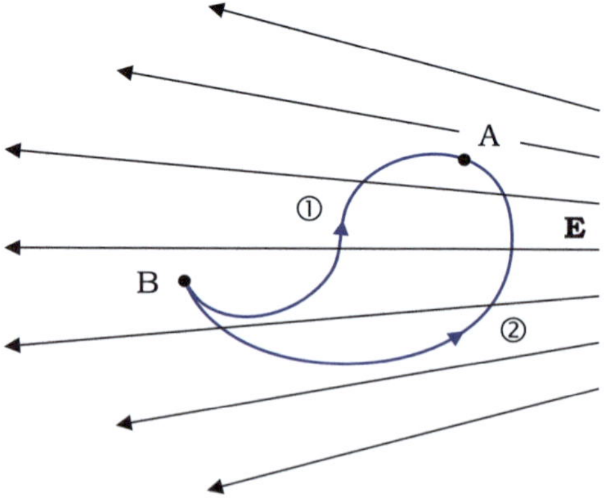

The electric potential follows the superposition principle. Accordingly, the total electric potential of N point charges, $q_1, q_2, \ldots$, and q_N, at position vectors $\mathbf{r}_1, \mathbf{r}_2, \ldots$, and $\mathbf{r}_N$, respectively, is given by

$$V = \frac{1}{4\pi\varepsilon_0} \sum_{i=1}^{N} \frac{q_i}{|\mathbf{r} - \mathbf{r}_i|} \tag{3.50}$$

For a continuous charge distribution, q_i in Eq. (3.50) is replaced with $\rho_{v'}\, dv'$, $\rho_{s'}\, ds'$, or $\rho_{\ell'}\, d\ell'$, depending on the type of charge distribution, and $\mathcal{R} = \mathbf{r} - \mathbf{r}'$ is substituted for $\mathbf{r} - \mathbf{r}_i$. Taking the limit as $N \to \infty$, the electric potential is expressed as

$$V = \frac{1}{4\pi\varepsilon_0} \int_{V'} \frac{\rho_{v'}}{\mathcal{R}}\, dv' \qquad \text{(volume charge)} \tag{3.51a}$$

$$V = \frac{1}{4\pi\varepsilon_0} \int_{S'} \frac{\rho_{s'}}{\mathcal{R}}\, ds' \qquad \text{(surface charge)} \tag{3.51b}$$

$$V = \frac{1}{4\pi\varepsilon_0} \int_{\mathcal{L}'} \frac{\rho_{\ell'}}{\mathcal{R}}\, d\ell' \qquad \text{(line charge)} \tag{3.51c}$$

where the zero-reference point is taken at infinity. It is important to note that a line integral is used for the electric potential expressed in Eq. (3.46). However, definite integrals are used in Eq. (3.51c). For a positive source charge, the definite integral always yields a positive electric potential, whereas the line integral may result in a negative value, depending on the choice of the zero-reference point.

Example 3.9 When an infinitely long line with uniform charge density ρ_ℓ is coincident with the z-axis in free space, as shown in Fig. 3.15, determine V_{A-B} between points $A:(\rho_A, \phi_A, 0)$ and $B:(\rho_B, \phi_B, 0)$ in cylindrical coordinates.

Solution

Two different methods are applied using Eqs. (3.45) and (3.51c).

(1) Inserting $\mathbf{E}$ of ρ_ℓ given in Eq. (3.36e) into Eq. (3.45) results in

$$V_{A-B} = -\int_B^A \mathbf{E} \cdot d\mathbf{l} = -\int_B^A \frac{\rho_\ell}{2\pi\varepsilon_0 \rho}\, \mathbf{a}_\rho \cdot \left(d\rho\, \mathbf{a}_\rho + \rho d\phi\, \mathbf{a}_\phi + dz\, \mathbf{a}_z \right)$$

$$= -\int_{\rho=\rho_B}^{\rho=\rho_A} \frac{\rho_\ell}{2\pi\varepsilon_0 \rho}\, d\rho$$

Thus, the potential difference is

$$V_{A-B} = \frac{\rho_\ell}{2\pi\varepsilon_0} \ln\left(\frac{\rho_B}{\rho_A}\right) \tag{3.52}$$

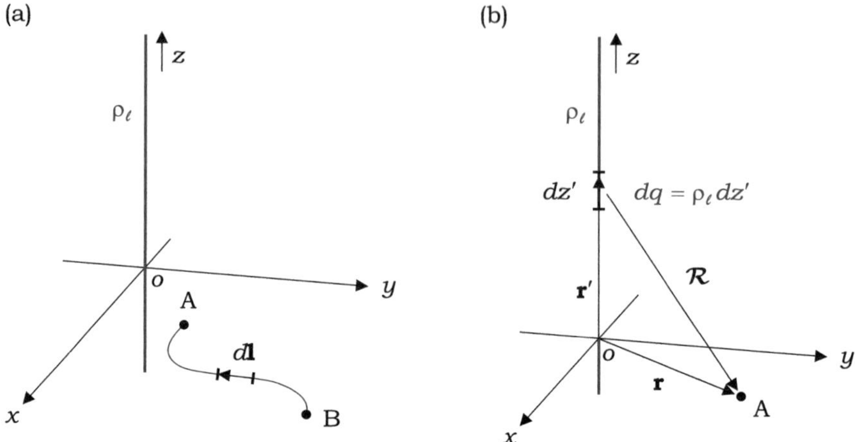

Fig. 3.15 Infinitely long line charge

where ρ_A and ρ_B are the radial distances in cylindrical coordinates, and ρ_ℓ is the line charge density.

(2) The position vectors for the field and source points are $\mathbf{r} = \rho_A \mathbf{a}_\rho$ and $\mathbf{r}' = z' \mathbf{a}_z$, respectively. Thus, the distance vector is

$$\mathcal{R} = \mathbf{r} - \mathbf{r}' = \rho_A \mathbf{a}_\rho - z' \mathbf{a}_z \quad \text{with} \ \mathcal{R} = \sqrt{\rho_A^2 + z'^2}$$

Letting $d\mathbf{l}' = dz'$ in Eq. (3.51c), the electric potential at point A is

$$V_A = \frac{1}{4\pi\varepsilon_0} \int_{z'=-\infty}^{z'=\infty} \frac{\rho_\ell dz'}{\sqrt{\rho_A^2 + z'^2}} = \frac{\rho_\ell}{4\pi\varepsilon_0} \ln\left[z' + \sqrt{\rho_A^2 + z'^2}\right]\Bigg|_{z'=-\infty}^{z'=\infty} = \infty$$

Similarly, the electric potential at point B is

$$V_B = \frac{1}{4\pi\varepsilon_0} \int_{z'=-\infty}^{z'=\infty} \frac{\rho_\ell dz'}{\sqrt{\rho_B^2 + z'^2}} = \infty$$

In this case, the potential difference cannot be obtained from the difference between two absolute electric potentials.

For future reference

$$\boxed{\int \frac{dx}{\sqrt{x^2 + a^2}} = \ln\left(x + \sqrt{x^2 + a^2}\right)} \tag{3.53}$$

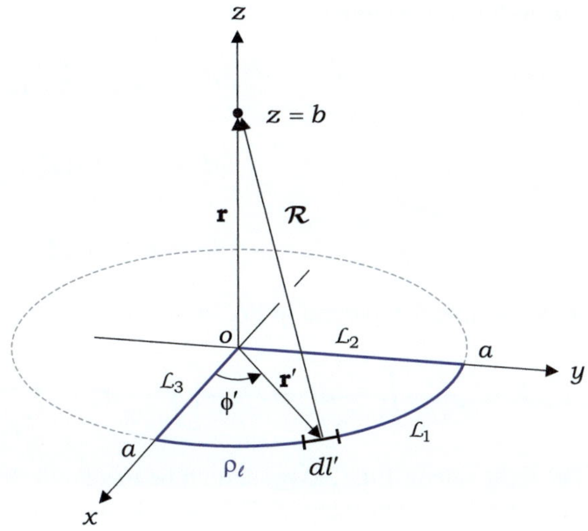

Fig. 3.16 Line charge around a quarter of disk

Example 3.10 A line charge of uniform density ρ_ℓ is formed around a quarter of a disk with radius a in the $z = 0$ plane, as shown in Fig. 3.16. Find V at a distance b above the origin.

Solution

The position vector for field point is

$$\mathbf{r} = b\,\mathbf{a}_z$$

In mixed coordinates, the position vector for source point is

$$\mathbf{r}' = a\,\mathbf{a}_{\rho'} \quad (\text{on } \mathcal{L}_1)$$

$$\mathbf{r}' = y'\,\mathbf{a}_{y'} \quad (\text{on } \mathcal{L}_2)$$

$$\mathbf{r}' = x'\,\mathbf{a}_{x'} \quad (\text{on } \mathcal{L}_3)$$

The magnitude of the distance vector is

$$\mathcal{R} = \left|\mathbf{r} - \mathbf{r}'\right| = \left|b\,\mathbf{a}_z - a\,\mathbf{a}_{\rho'}\right| = \sqrt{b^2 + a^2} \quad (\text{on } \mathcal{L}_1) \tag{3.54a}$$

$$\mathcal{R} = \left|\mathbf{r} - \mathbf{r}'\right| = \left|b\,\mathbf{a}_z - y'\,\mathbf{a}_{y'}\right| = \sqrt{b^2 + y'^2} \quad (\text{on } \mathcal{L}_2) \tag{3.54b}$$

$$\mathcal{R} = \left|\mathbf{r} - \mathbf{r}'\right| = \left|b\,\mathbf{a}_z - x'\,\mathbf{a}_{x'}\right| = \sqrt{b^2 + x'^2} \quad (\text{on } \mathcal{L}_3) \tag{3.54c}$$

The differential length is

$$dl' = a\,d\phi' \quad (\text{on } \mathcal{L}_1) \tag{3.54d}$$

$$dl' = dy' \quad (\text{on } \mathcal{L}_2) \tag{3.54e}$$

$$dl' = dx' \quad (\text{on } \mathcal{L}_3) \tag{3.54f}$$

By separating the integral in Eq. (3.51c) into three parts, we have

$$V = \frac{1}{4\pi\varepsilon_0} \int_{\mathcal{L}'} \frac{\rho_\ell}{R} dl' = \frac{\rho_\ell}{4\pi\varepsilon_0}\left[\int_{\mathcal{L}_1} \frac{1}{R} dl' + \int_{\mathcal{L}_2} \frac{1}{R} dl' + \int_{\mathcal{L}_3} \frac{1}{R} dl'\right] \tag{3.54g}$$

The right side of Eq. (3.54g) should be treated as three definite integrals, not three parts of a closed-line integral. This assertion is correct, because a positive charge must produce a positive electric potential. Accordingly, the integral in Eq. (3.54g) should be carried out along the line charge in the direction of increase of the coordinate such that

$$
\begin{aligned}
V &= \frac{\rho_\ell}{4\pi\varepsilon_0}\left[\int_{\phi'=0}^{\pi/2} \frac{1}{\sqrt{b^2+a^2}}\,a\,d\phi' + \int_{y'=0}^{a} \frac{1}{\sqrt{b^2+y'^2}}dy' + \int_{x'=0}^{a} \frac{1}{\sqrt{b^2+x'^2}}dx'\right]\\
&= \frac{\rho_\ell}{4\pi\varepsilon_0}\left[\frac{a\pi}{2\sqrt{b^2+a^2}} + \ln\left(y' + \sqrt{b^2+y'^2}\right)\Big|_{y'=0}^{a} + \ln\left(x' + \sqrt{b^2+x'^2}\right)\Big|_{x'=0}^{a}\right]
\end{aligned}
$$

Thus,

$$V = \frac{\rho_\ell}{4\pi\varepsilon_0}\left[\frac{a\pi}{2\sqrt{b^2+a^2}} + 2\ln\left(a + \sqrt{b^2+a^2}\right) - 2\ln b\right]$$

In short, the potential difference can be obtained from a given electric field or charge distribution. The former involves a line integral whereas the latter involves a definite integral.

Exercise 3.17
For a point charge at point (1, 1, 0), determine V at point (2, 0, 0) if $V = 10\,[\text{V}]$ at point (4, 0, 0).

Ans. $V = 22.4\,[\text{V}]$.

Exercise 3.18
Determine V at distance b above the center of a circular loop of radius a carrying line charge density ρ_ℓ (see Fig. 3.7).

Ans. $V = (\rho_\ell a/2\varepsilon_0)(a^2 + b^2)^{-1/2}$.

Exercise 3.19

If the electric potential depends on $R\cos\theta$ in spherical coordinates, what would be a possible source?

Ans. Infinite charged sheet in the $z = 0$ plane.

3.4.3 The Conservative Field

The path independence of the electric potential can be fully expressed in terms of a closed-line integral or curl of the electric field. Consider the case illustrated in Fig. 3.14, where a unit positive charge moved from B to A along path ① and returned to B along the reversal of path ② in the presence of an electric field $\mathbf{E}$. The work done during this round trip is expressed as

$$V^{①}_{A-B} + V^{②'}_{B-A} \Rightarrow -\int_{B}^{A\,①} \mathbf{E}\cdot d\mathbf{l} - \int_{A}^{B\,②'} \mathbf{E}\cdot d\mathbf{l} \tag{3.55}$$

where ②′ denotes the reversal of path ②. Mathematically, $V^{②'}_{B-A}$ is equal to $-V^{②}_{A-B}$, whereas $V^{①}_{A-B}$ is equal to $V^{②}_{A-B}$ because of the path independence. Therefore, the left-hand side of Eq. (3.55) vanishes. The right-hand side of Eq. (3.55) is by definition the negative of the circulation of $\mathbf{E}$ around a closed path formed by ① and ②′. That is,

$$\boxed{\oint_{C} \mathbf{E}\cdot d\mathbf{l} = 0} \tag{3.56}$$

As the path between A and B may be arbitrary, we conclude that the closed-line integral of $\mathbf{E}$ is always zero. It should be noted that Eq. (3.56) is one of the fundamental relations of static electric fields, because it is based on the law of energy conservation.

Next, by applying Stokes's theorem to Eq. (3.56), we obtain

$$\int_{S} (\nabla \times \mathbf{E}) \cdot d\mathbf{s} = 0$$

where S is any surface bounded by C. Because this relation holds true for any surface, the integrand itself should be zero, that is,

$$\boxed{\nabla \times \mathbf{E} = 0} \tag{3.57}$$

This indicates that the static electric fields are **irrotational** or **conservative**. It is important to note that Eqs. (3.56) and (3.57) are only true for static electric fields. Under time-varying conditions, the closed-line integral, and thus, the curl of the electric field, is generally nonzero.

Finally, the relationships in Eqs. (3.33) and (3.57) are *two fundamental relations for a static electric field*, as Helmholtz's theorem requires that both the divergence and curl of a vector field be specified for unique determination of the field.

Exercise 3.20
For $\mathbf{E}$ of an electric dipole given in Eq. (3.14), show that $\nabla \times \mathbf{E} = 0$.

Exercise 3.21
If term $\mathbf{a}_\theta$ is omitted from Eq. (3.14), can the rest be a static electric field from another source?

Ans. No, $\nabla \times \mathbf{E} \neq 0$.

Exercise 3.22
For $\mathbf{E}$ of the volume charge density ρ_v in Eq. (3.17), show $\nabla \times \mathbf{E} = 0$.

Ans. $\nabla \times (\mathcal{R}/\mathcal{R}^3) = 0$.

3.4.4 E as the Negative Gradient of V

The differential of the electric potential represents the work done in moving a unit charge by a differential distance in the electric field. Expressed mathematically,

$$dV = -\mathbf{E} \cdot d\mathbf{l} \tag{3.58}$$

As discussed in Sect. 2.2, the differential of V can be expressed in terms of its gradient such that

$$dV = \nabla V \cdot d\mathbf{l} \tag{3.59}$$

which is a mathematical theorem derived from the definition of gradient. Comparing Eq. (3.58) with Eq. (3.59) gives

$$\boxed{\mathbf{E} = -\nabla V} \quad [\text{V/m}] \tag{3.60}$$

The static electric field is equal to the negative gradient of the electric potential. This relationship enables us to obtain $\mathbf{E}$ first by calculating V from a given charge distribution and then taking the negative gradient of V. The relationship in Eq. (3.60) conforms to the irrotational nature of $\mathbf{E}$, as the curl of the gradient of a scalar function is always zero, that is, $\nabla \times \mathbf{E} = -\nabla \times \nabla V = 0$.

The electric potential must be a smooth function of position in homogenous media; otherwise, its derivative, or the electric field, would change abruptly from point to point. Subsequently, the points of constant V should form a smooth surface in a given region called an **equipotential surface**. Moreover, the electric field lines are always

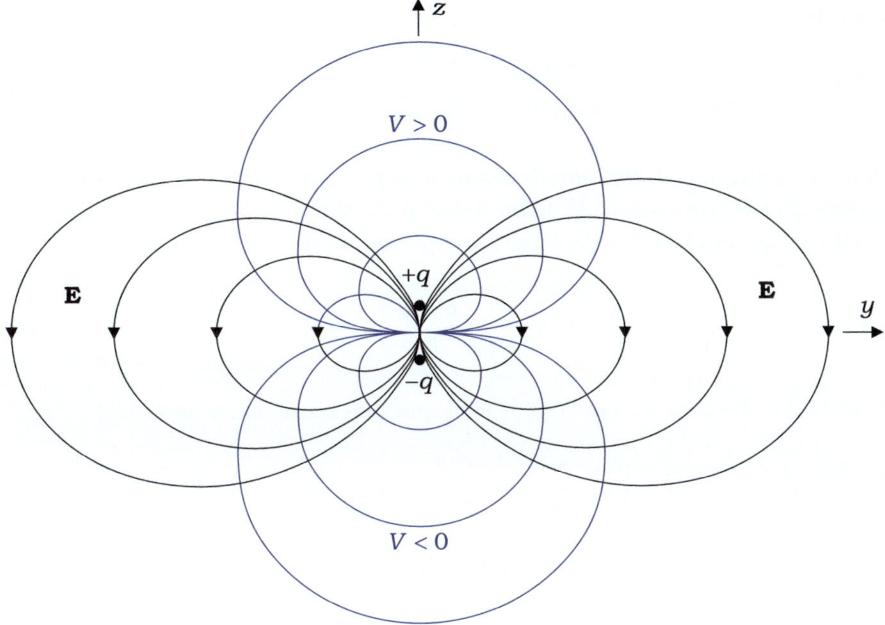

Fig. 3.17 Electric field lines (black lines) and equipotential surfaces (blue lines) of an electric dipole

perpendicular to the equipotential surface because no work is required to move a charge on an equipotential surface.

Example 3.11 For an electric dipole located at the origin in free space (Fig. 3.4), determine its electric potential and electric field.

Solution

Using $\mathcal{R}^+$ and $\mathcal{R}^-$, as shown in Fig. 3.4, which represent the distances from the charges to the field point, and applying the superposition principle, we write

$$V = \frac{q}{4\pi\varepsilon_0}\left(\frac{1}{\mathcal{R}^+} - \frac{1}{\mathcal{R}^-}\right) \tag{3.61a}$$

At a large distance from the dipole, $r \gg d$, applying binomial approximations to Eqs. (3.13d) and (3.13e) yields

$$\mathcal{R}^+ \cong r - \frac{1}{2}d\cos\theta \tag{3.61b}$$

$$\mathcal{R}^- \cong r + \frac{1}{2}d\cos\theta \tag{3.61c}$$

Inserting these relations into Eq. (3.61a), we obtain

$$V \cong \frac{q}{4\pi\varepsilon_0} \frac{d\cos\theta}{\left(r - \frac{1}{2}d\cos\theta\right)\left(r + \frac{1}{2}d\cos\theta\right)} \cong \frac{1}{4\pi\varepsilon_0} \frac{qd\cos\theta}{r^2} \qquad (3.61d)$$

Here, we define the **electric dipole moment** as $\mathbf{p} = q\,\mathbf{d}$, where $\mathbf{d}$ is the vector from negative to positive charges. With the aid of $\mathbf{p}$ and the position vector $\mathbf{r} = R\,\mathbf{a}_R$, the electric potential of an electric dipole is expressed in spherical coordinates as

$$\boxed{V = \frac{1}{4\pi\varepsilon_0} \frac{\mathbf{p}\cdot\mathbf{a}_R}{R^2}} \qquad [\text{V}] \qquad (3.62)$$

Next, taking the negative gradient of V in spherical coordinates, we obtain

$$\mathbf{E} = -\nabla V = -\left(\mathbf{a}_R \frac{\partial}{\partial R} + \mathbf{a}_\theta \frac{1}{R}\frac{\partial}{\partial\theta} + \mathbf{a}_\phi \frac{1}{R\sin\theta}\frac{\partial}{\partial\phi}\right)\left(\frac{qd\cos\theta}{4\pi\varepsilon_0 R^2}\right)$$

For $R \gg d$, the electric field of an electric dipole is expressed as

$$\boxed{\mathbf{E} = \frac{p}{4\pi\varepsilon_0 R^3}(2\cos\theta\,\mathbf{a}_R + \sin\theta\,\mathbf{a}_\theta)} \qquad [\text{V/m}] \qquad (3.63)$$

where p is the magnitude of the dipole moment, and R is the distance from the origin. Obviously, Eq. (3.63) is the same as that in Eq. (3.14). The $\mathbf{E}$ value of an electric dipole is inversely proportional to the cube of R, whereas the $\mathbf{E}$ value of a point charge is inversely proportional to the square of R.

Numerical methods were employed to compute the electric field lines and equipotential surfaces of the electric dipole using Eq. (3.63), and plotted to scale in the yz-plane, as shown in Fig. 3.17. As expected, the electric field lines intersect equipotential surfaces at right angles. It is worth noting that *the $z = 0$ plane is an equipotential surface with $V = 0$*, as evident from Eq. (3.62), because $\mathbf{p}\cdot\mathbf{a}_R = 0$ in the $z = 0$ plane.

Example 3.12 A uniform surface charge density ρ_s forms a disk of radius a centered at the origin in the $z = 0$ plane (Fig. 3.18). Determine V and $\mathbf{E}$ at point $(0, 0, b)$.

Solution

With $\mathbf{r} = b\,\mathbf{a}_z$ and $\mathbf{r}' = \rho'\,\mathbf{a}_{\rho'}$, the distance from the source to the field point is

$$\mathcal{R} = |\mathbf{r} - \mathbf{r}'| = |b\,\mathbf{a}_z - \rho'\,\mathbf{a}_{\rho'}| = \sqrt{b^2 + \rho'^2}$$

The differential area at the source point is

Fig. 3.18 Disk with uniform surface charge density ρ_s

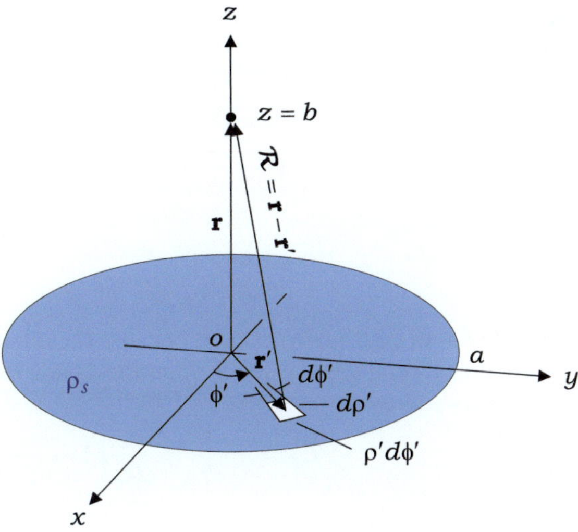

$$ds' = \rho' d\rho' d\phi'$$

From Eq. (3.51b),

$$V = \frac{1}{4\pi\varepsilon_0} \int_{S'} \frac{\rho_s}{\mathcal{R}} ds' = \frac{\rho_s}{4\pi\varepsilon_0} \int_{\phi'=0}^{2\pi} \int_{\rho'=0}^{a} \frac{\rho'}{\sqrt{b^2 + \rho'^2}} d\rho' d\phi'$$

$$= \frac{\rho_s}{2\varepsilon_0}\left[\sqrt{b^2 + a^2} - b\right] \tag{3.64a}$$

Next, we replace b in Eq. (3.64a) with z for generalization such as

$$V = \frac{\rho_s}{2\varepsilon_0}\left[\sqrt{z^2 + a^2} - z\right] \tag{3.64b}$$

Because this was obtained by setting $x = y = 0$, it has no directional derivative in the $\mathbf{a}_x$- or $\mathbf{a}_y$-direction. Nevertheless, $\mathbf{E}$ can be obtained by taking the negative

Fig. 3.19 Equipotential surfaces (blue) and electric field lines (black)

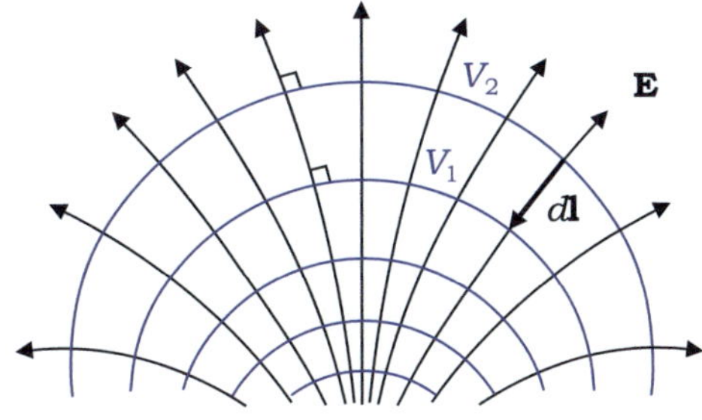

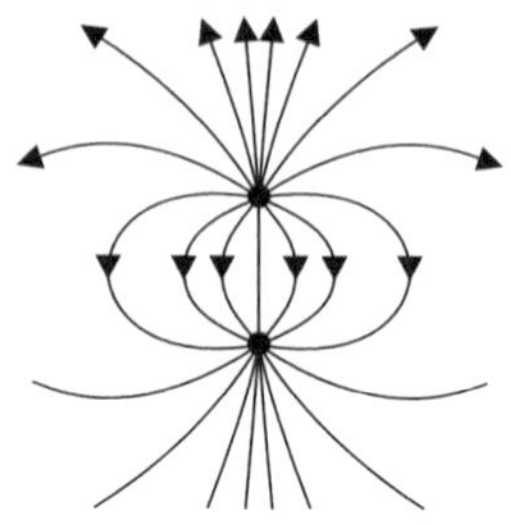

Fig. 3.20 Near field pattern
of the electric dipole

gradient of V in Eq. (3.64b). This is because the rotational symmetry about the z-axis ensures that $\mathbf{E}$ on the z-axis has neither the x- nor y-components. Therefore,

$$E_z = -\frac{dV}{dz} = \frac{\rho_s}{2\varepsilon_0}\left[1 - \frac{z}{\sqrt{z^2 + a^2}}\right]$$

At point $(0, 0, b)$, the electric field is thus

$$\mathbf{E} = \frac{\rho_s}{2\varepsilon_0}\left[1 - \frac{b}{\sqrt{b^2 + a^2}}\right]\mathbf{a}_z \qquad (3.64c)$$

Exercise 3.23

Given an electric potential $V = 2xy^2 + 3z^2$, determine (a) the electric field, and (b) the volume charge density in a given region.

Ans. (a) $\mathbf{E} = -2y^2\mathbf{a}_x - 4xy\,\mathbf{a}_y - 6z\,\mathbf{a}_z$ [V/m], (b) $\rho_v = -\varepsilon_0(4x + 6)$ [C/m^3].

Exercise 3.24

Is $V_1 > V_2$ in Fig. 3.19?

Ans. Yes, $V_1 - V_2 = -\mathbf{E} \cdot d\mathbf{l}$.

Exercise 3.25

The far-field pattern of the electric dipole is shown in Fig. 3.17. Next, sketch the near-field pattern of the electric dipole by referring to Eq. (3.61a).

Ans. Fig. 3.20.

Review Questions

RQ 3.17	What is meant by the negative value of the work done?	[(3.43)]
RQ 3.18	State potential difference and electric potential in words.	[(3.47)]
RQ 3.19	What is the zero-reference point?	[(3.46)]
RQ 3.20	Does $\mathbf{E} = 0$ at a point in space mean $V = 0$ at that point?	[(3.46)]
RQ 3.21	State two integral methods for computing V.	[(3.46)(3.51)]

RQ 3.22	What is the physical law underlying Eq. (3.46)?	[(3.43)]
RQ 3.23	What is the principle underlying Eq. (3.51a, b, c)?	[(3.50)]
RQ 3.24	In what manner does V vary with R for a point charge?	[(3.49)]
RQ 3.25	In what manner does V vary with R for an electric dipole?	[(3.62)]
RQ 3.26	Define the conservative field.	[(3.57)]
RQ 3.27	What is the physical law underlying $\nabla \times \mathbf{E} = 0$?	[Fig. 3.14]
RQ 3.28	Write two fundamental relationships for the electrostatic field.	[(3.33)(3.57)]
RQ 3.29	What is the significance of the negative sign in $\mathbf{E} = -\nabla V$?	[(3.43)]

3.5 Dielectric in Static Electric Field

To this point, our discussion has focused on static electric fields in free space. We now turn our attention to material media placed in an externally applied electric field. In a primitive atomic model of matter, a material is viewed as a three-dimensional array of atoms in free space. Moreover, according to the shell model of an atom, electrons occupy shells around the nucleus in an ordered manner. The electrons in the outermost shell are called **valence electrons**, and are responsible for the electrical properties of the material. Based on their electrical properties, materials can be classified into three categories: conductors, semiconductors, and insulators. In conductors, the binding force on the valence electrons is so weak that the electrons can easily detach from the atom and migrate from one atom to another. These electrons are called **free electrons** or **conduction electrons**. In the presence of an externally applied electric field, free electrons are accelerated over a short period of time before colliding with impurities or imperfections in the lattice, and are scattered in random directions. Owing to repeated acceleration and random scattering, free electrons may move at a constant speed and constitute a steady conduction current in the conductor. In insulators or dielectrics, valence electrons are tightly bound by atomic forces and may not be freed by an external electric field of moderate strength. Nevertheless, they are displaced with respect to the much heavier positively charged nucleus by an external electric field, resulting in charge separation. These separated charges are called the **bound charges**. However, they do not contribute to the conduction current. Semiconductors contain a relatively small number of free electrons and are positioned between the conductor and insulator from the conductivity standpoint.

Although the atoms of a dielectric may be electrically neutral under normal conditions, an externally applied electric field induces electric dipoles in the material regardless of whether the dipole moment is strong enough to be detected. At the macroscopic scale, where the discrete nature of the electric dipole is completely ignored, the sum of the electric fields produced by the individual electric dipoles constitutes a **polarization field** that is directed opposite to the external field under static conditions. The sum of the external and polarization fields is called the **internal** *electric field*, which is always smaller than the external field. The ratio of the external field to the internal field is defined as the ***relative permittivity*** of the material. It is a characteristic parameter of a material that provides a measure of its electrical properties.

The polar molecules of some dielectrics have permanent dipole moments even in the absence of an external electric field, which originate from the unequal sharing of the valence electrons within a bond. For example, water molecules are polar and exhibit permanent dipole moment. Each hydrogen–oxygen bond in the bent structure is polar and covalent. In the absence of an externally applied electric field, the polar molecules are randomly oriented and yield no net dipole moment. However, in the presence of an external field, molecules align with the field, resulting in a net dipole moment. Although nonpolar molecules return to their original neutral state when the external field is removed, polar molecules may or may not return to the initial random state, depending on the material.

3.5.1 *Electric Polarization*

An electric dipole consists of two identical charges of opposite polarity, $+q$ and $-q$, separated by a small distance d. The ***electric dipole moment*** is defined as

$$\mathbf{p} = q\mathbf{d} \quad [\text{C} \cdot \text{m}] \tag{3.65}$$

where $\mathbf{d}$ is the distance vector from $-q$ to $+q$.

On a macroscopic scale, in which the discrete nature of the electric dipole is completely disregarded, it is convenient to define the ***electric polarization***, $\mathbf{P}$, also called the ***dipole moment density***, as

$$\mathbf{P}(\mathbf{r}) = \lim_{\Delta v \to 0} \frac{1}{\Delta v} \sum_{i=1}^{n\Delta v} \mathbf{p}_i \quad [\text{C/m}^2] \tag{3.66}$$

where n is the number density of electric dipoles or the number of atoms per unit volume, and Δv is the differential volume centered at $\mathbf{r}$. Note that dipole moments $\mathbf{p}_i$ are discrete vectors in space, whereas electric polarization $\mathbf{P}(\mathbf{r})$ is a vector field given by a continuous function of position. If the material consists of one type of atom such that the atoms carry the same electric dipole moment $\mathbf{p} = q\mathbf{d}$, then the electric polarization is simply

$$\mathbf{P} = nq\mathbf{d} \tag{3.67}$$

Electric polarization induces an electric charge on the surface of the material. Suppose an external electric field is applied to a dielectric along the direction of $\mathbf{a}_p$, and the induced electric dipoles are all aligned along $\mathbf{a}_p$, as shown in Fig. 3.21. In

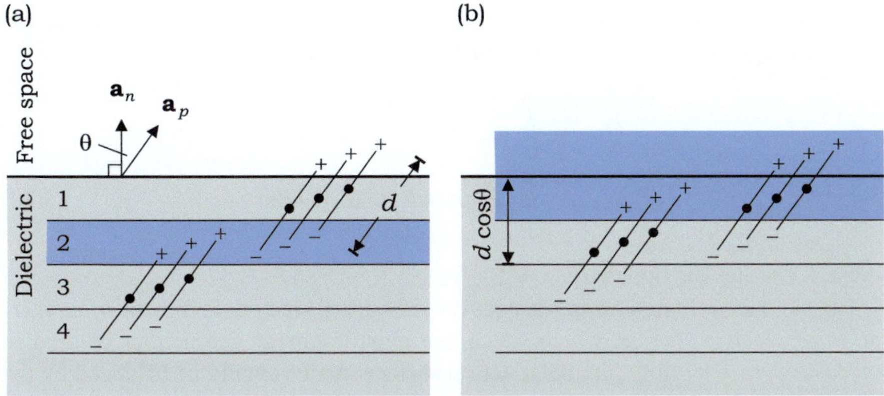

Fig. 3.21 Surface charges due to electric polarization. **a** No net charge in layer 2. **b** Net positive charges in layer 1 and in free space

this case, the material has a nonzero electric polarization and is said to be ***polarized***. It must be an ***isotropic*** material because otherwise, in an ***anisotropic*** material, the induced dipole moments are not parallel to the external field. In Fig. 3.21, the black dot denotes the center of an electric dipole consisting of two charges, denoted by $+$ and $-$, separated by distance d. For the sake of argument, the polarized material is sliced into hypothetical layers of thickness $(d/2)\cos\theta$, where θ is the angle between $\mathbf{a}_p$ and $\mathbf{a}_n$, which is a unit normal to the material surface. From Fig. 3.21a, we can see that layer 2 contains equal amounts of positive and negative charges stemming from the dipoles, with the center positioned in layers 1 or 3. In contrast, there are net positive charges in layer 1 and free space that stem from the dipoles, with the center positioned in layers 1 or 2. These net charges constitute the surface charges of the material.

For differential area Δs on the dielectric surface, the net induced charge within Δs is obtained as

$$\Delta Q = qn\,[d\,\Delta s\cos\theta]$$
$$= (qnd)(\Delta s)\,\mathbf{a}_p \cdot \mathbf{a}_n = \mathbf{P} \cdot \mathbf{a}_n(\Delta s) \tag{3.68}$$

where $\mathbf{P} = (qnd)\,\mathbf{a}_p$. The brackets in Eq. (3.68) represent the volume of a rectangular parallelepiped with base Δs and height $d\cos\theta$. By dividing both sides of Eq. (3.68) by Δs and taking the limit as $\Delta s \to 0$, the ***polarization surface charge density*** is obtained as

$$\boxed{\rho_{ps} = \mathbf{P} \cdot \mathbf{a}_n} \quad [\text{C/m}^2] \tag{3.69}$$

where $\mathbf{P}$ is the electric polarization on the material surface, and $\mathbf{a}_n$ is the outward unit normal to the material surface.

The ***polarization surface charge*** is obtained by integrating the polarization surface charge density over the material boundary such as

$$Q_{ps} = \oint_{\mathcal{S}} \rho_{ps}\, ds$$
$$= \oint_{\mathcal{S}} (\mathbf{P} \cdot \mathbf{a}_n)\, ds = \oint_{\mathcal{S}} \mathbf{P} \cdot d\mathbf{s} \qquad (3.70)$$

where $\mathcal{S}$ denotes the boundary.

The net charge in a dielectric is conserved in accordance with the ***law of conservation of electric charge***, which states that electric charge can neither be created nor destroyed. Therefore, the polarization surface charge should be balanced by the volume charge within the dielectric, which is called the ***polarization volume charge*** and is denoted by Q_{pv}. That is,

$$Q_{pv} = -Q_{ps} = -\oint_{\mathcal{S}} \mathbf{P} \cdot d\mathbf{s}$$
$$= -\int_{\mathcal{V}} \nabla \cdot \mathbf{P}\, dv \qquad (3.71)$$

where the divergence theorem is applied. In view of the volume integral in Eq. (3.71), the ***polarization volume charge density*** is defined as

$$\boxed{\rho_{pv} = -\nabla \cdot \mathbf{P}} \qquad [\mathrm{C/m^3}] \qquad (3.72)$$

If $\mathbf{P}$ is constant in the dielectric, ρ_{pv} is zero in the interior; thus, both Q_{pv} and Q_{ps} are zero. Nevertheless, the polarization surface charge density may not be zero at all the points on the boundary surface.

The polarization charge densities ρ_{ps} and ρ_{pv} jointly produce a ***polarization field*** such that

$$\mathbf{E}_p = \frac{1}{4\pi\varepsilon_0} \oint_{\mathcal{S}'} \frac{\rho_{ps'}}{\mathcal{R}^2} \mathbf{a}_\mathcal{R}\, ds' + \frac{1}{4\pi\varepsilon_0} \int_{\mathcal{V}'} \frac{\rho_{pv'}}{\mathcal{R}^2} \mathbf{a}_\mathcal{R}\, dv' \qquad (3.73)$$

where $\mathcal{R} = \mathbf{r} - \mathbf{r}' = \mathcal{R}\,\mathbf{a}_R$ is the distance vector. It is noteworthy that ε_0 in Eq. (3.73) indicates that the polarization charges are assumed to be placed in free space for the calculation of $\mathbf{E}_p$.

Example 3.13 A dielectric cylinder of radius a and height b lies along the z-axis, carrying a uniform electric polarization of $\mathbf{P} = P_o\,\mathbf{a}_z$. Determine

(a) ρ_{ps} and ρ_{pv}, and
(b) Q_{ps} and Q_{pv}.

Solution

(a) From Eq. (3.69),

$$\rho_{ps} = \mathbf{P} \cdot \mathbf{a}_n = (P_o\,\mathbf{a}_z) \cdot \mathbf{a}_\rho = 0 \quad \text{(side surface)}$$

$$\rho_{ps} = (P_o\,\mathbf{a}_z) \cdot (\pm \mathbf{a}_z) = \pm P_o \quad (\pm,\ \text{top and bottom surfaces})$$

From Eq. (3.72),

$$\rho_{pv} = -\nabla \cdot \mathbf{P} = -\nabla \cdot (P_o\,\mathbf{a}_z) = 0$$

(b) The polarization surface charge is

$$Q_{ps} = \oint_{S} \rho_{ps}\,ds = \int_{side} \rho_{ps}\,ds + \int_{top} \rho_{ps}\,ds + \int_{bottom} \rho_{ps}\,ds$$
$$= 0 + P_o \pi a^2 - P_o \pi a^2 = 0$$

The polarization volume charge is

$$Q_{pv} = \int_{\mathcal{V}} \rho_{pv}\,dv = 0$$

The electric polarization induces charges only on the top and bottom surfaces.

Exercise 3.26
Determine ρ_{ps} for a dielectric sphere with constant $\mathbf{P} = P_o\,\mathbf{a}_z$.

Ans. $\rho_{ps} = P_o \cos\theta$.

Exercise 3.27
If the electric dipoles in Exercise 3.26 have the same dipole moment $\mathbf{p} = q d\,\mathbf{a}_z$,
what is the volume charge density of the positive bound charges?

Ans. $\rho_v = P_o/d\,[\text{C/m}^3]$.

Exercise 3.28
What is the difference, if any, between the two $\mathbf{P}$'s in Eqs. (3.69) and (3.72)?

Ans. The former is for the surface, whereas the latter is for the interior.

3.5.2 *Dielectric Constant*

If a net volume charge of density $\rho_v\,[\text{C/m}^3]$ is injected into a dielectric, its electric
field behaves as an external field and induces electric polarization $\mathbf{P}$ and polarization

volume charge density ρ_{pv} in the material. The sum of the external field due to ρ_v and the polarization field due to ρ_{pv} constitutes the internal electric field. Gauss's law can be used to relate the internal field $\mathbf{E}$ to these charges such that

$$\nabla \cdot (\varepsilon_0 \mathbf{E}) = \rho_v + \rho_{pv}$$
$$= \rho_v - \nabla \cdot \mathbf{P} \tag{3.74}$$

where we have used Eq. (3.72). It is evident from ε_0 in Eq. (3.74) that both ρ_v and ρ_{pv} are treated as if they are placed in free space. Rewriting Eq. (3.74) gives

$$\nabla \cdot (\varepsilon_0 \mathbf{E} + \mathbf{P}) = \rho_v \tag{3.75}$$

At this point, we redefine the electric flux density in the dielectric as

$$\boxed{\mathbf{D} = \varepsilon_0 \mathbf{E} + \mathbf{P}} \quad [\text{C/m}^2] \tag{3.76}$$

The newly defined electric flux density enables us to express Gauss's law in the material as

$$\boxed{\nabla \cdot \mathbf{D} = \rho_v} \quad [\text{C/m}^3] \tag{3.77}$$

where ρ_v is the volume density of the ***free charges***, apart from the induced polarization charge. Gauss's law in Eq. (3.77) has the same form as that in free space. In other words, *Gauss's law is independent of the material*. By applying the divergence theorem to Eq. (3.77), Gauss's law is expressed in integral form as

$$\boxed{\oint_S \mathbf{D} \cdot d\mathbf{s} = Q} \tag{3.78}$$

Gauss's law states that the net outward electric flux through any closed surface is equal to *the free charge* enclosed by the surface irrespective of the material surrounding the charge.

In a homogenous, linear, and isotropic material, called a ***simple medium***, the electric polarization is directly proportional to the internal electric field such that

$$\boxed{\mathbf{P} = \varepsilon_0 \chi_e \mathbf{E}} \quad [\text{C/m}^2] \tag{3.79}$$

where χ_e is the ***electric susceptibility***. It is important to note that $\mathbf{E}$ in Eq. (3.79) is the internal field, not the external field. In general, χ_e may vary with position (in inhomogeneous materials), depend on $|\mathbf{E}|$, $|\mathbf{E}|^2$, etc. (in nonlinear materials), and be a second-rank tensor (in anisotropic materials) such that $\mathbf{P}$ may not be parallel to $\mathbf{E}$, for example, $P_x = \varepsilon_0 \chi_e' E_x$, $P_y = \varepsilon_0 \chi_e'' E_y$, and $P_z = \varepsilon_0 \chi_e''' E_z$ in the anisotropic material. In this text, we deal only with simple media, for which χ_e is a constant.

Inserting Eq. (3.79) into Eq. (3.76) gives

$$\mathbf{D} = \varepsilon_0 (1 + \chi_e)\, \mathbf{E} \equiv \varepsilon_0 \varepsilon_r\, \mathbf{E} \equiv \varepsilon \mathbf{E} \quad [\text{C/m}^2] \qquad (3.80)$$

Thus, the constitutive relation of the material is given by

$$\boxed{\mathbf{D} = \varepsilon \mathbf{E}} \qquad (3.81)$$

In Eqs. (3.80) and (3.81), ε is the permittivity of the material, ε_r is the relative permittivity (or *dielectric constant*), and ε_0 is the permittivity of free space. Dielectric constant is a measure of the electrical properties of a material. If ε_r is known for a material, its electric susceptibility can be obtained as $\chi_e = \varepsilon_r - 1$.

One might try to find the internal electric field by determining the electric polarization induced in the material. However, as shown in Eq. (3.79), the electric polarization can be determined only when the internal field in the material is already known. This regression can be avoided if the electric flux density due to free charge is first obtained using Gauss's law. The internal field can then be obtained using Eq. (3.81).

For example, consider the case illustrated in Fig. 3.22, in which a dielectric slab of finite extent is placed normal to a uniform external field. One might assume that electric polarization is uniform in the dielectric. However, even if it is uniform and induces uniform polarization surface charges on the top and bottom surfaces, the resulting polarization field is not uniform in the interior because the polarization surface charge of a finite extent inevitably involves fringing of its electric field at the edges. Consequently, the electric field is not uniform in the interior. In this case, the electric polarization and internal electric field can be obtained using only an iterative method.

If a strong external field is applied to the dielectric, electrons are completely detached from the host atoms and accelerated. This material is now highly conductive. The accelerated electrons collide with the lattice atoms in a violent manner and inflict permanent damage on the lattice structure. Such an event is called *dielectric breakdown*. The *dielectric strength* is the maximum electric field intensity that can be applied to a dielectric without causing a dielectric breakdown. The dielectric strength depends not only on the material composition but also on other factors, such as temperature and humidity. In air, the dielectric strength is approximately 3 [MV/m], beyond which sparks can occur.

Fig. 3.22 Finite dielectric slab is placed normal to a uniform electric field

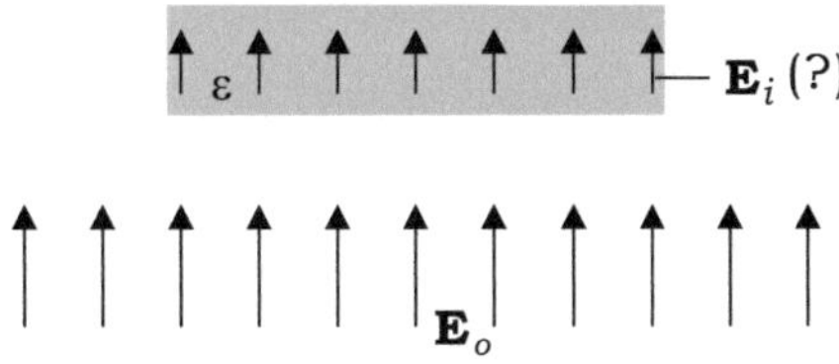

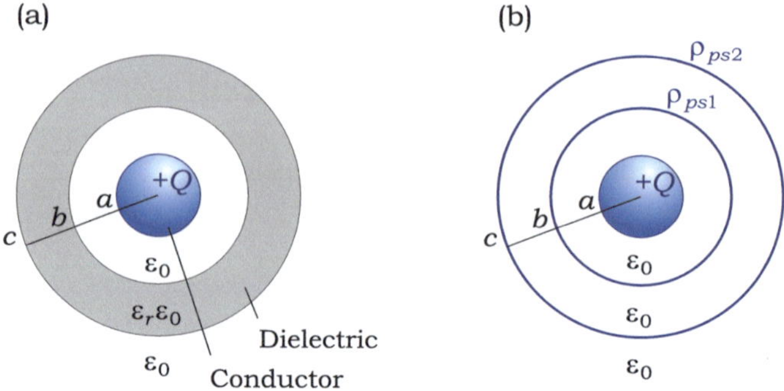

Fig. 3.23 Net spherical charge enclosed in dielectric shell

Example 3.14 A net charge of $+Q$ [C] is imparted to a perfectly conducting sphere of radius a, which is then enclosed by a concentric dielectric shell of radii b and c with dielectric constant ε_r (Fig. 3.23a). Determine ρ_{ps} induced on the dielectric.

Solution

(1) The charge is uniformly distributed on the sphere to avoid breaking the spherical symmetry of the system and cause the electric field to vanish inside the conductor. Thus, $\mathbf{D}$ is expected to be of the form $\mathbf{D} = D_R(R)\,\mathbf{a}_R$ everywhere.

In region $b \le R \le c$, from Gauss's law,

$$4\pi R^2 D_R = 4\pi R^2 (\varepsilon_0 \varepsilon_r E_R) = +Q$$

This gives

$$\mathbf{E} = E_R\,\mathbf{a}_R = \frac{Q}{4\pi\varepsilon_0\varepsilon_r R^2}\,\mathbf{a}_R \tag{3.82a}$$

Inserting Eq. (3.82a) into Eq. (3.79), along with $\chi_e = \varepsilon_r - 1$, we obtain

$$\mathbf{P} = \varepsilon_0 \chi_e \mathbf{E} = \frac{\varepsilon_r - 1}{\varepsilon_r}\frac{Q}{4\pi R^2}\,\mathbf{a}_R \tag{3.82b}$$

Inserting Eq. (3.82b) into Eq. (3.69) and letting $\mathbf{a}_n = -\mathbf{a}_R$ for $R = b$, we obtain

$$\rho_{ps1} = \mathbf{P} \cdot \mathbf{a}_n = -\frac{\varepsilon_r - 1}{\varepsilon_r}\frac{Q}{4\pi b^2} \quad (\text{at } R = b) \tag{3.82c}$$

Similarly, letting $\mathbf{a}_n = \mathbf{a}_R$ for $R = c$, we obtain

$$\rho_{ps2} = \mathbf{P} \cdot \mathbf{a}_n = \frac{\varepsilon_r - 1}{\varepsilon_r}\frac{Q}{4\pi c^2} \quad (\text{at } R = c) \tag{3.82d}$$

(2) With respect to the electric field, the dielectric shell may be replaced by polarization charges residing in the free space, as shown in Fig. 3.23b. Applying Gauss's law in the region $b < R < c$, we obtain

$$4\pi R^2 (\varepsilon_0 E_R) = Q + 4\pi b^2 \rho_{ps1} \qquad (3.83a)$$

Inserting Eq. (3.82c) into Eq. (3.83a) yields

$$\mathbf{E} = \frac{Q}{4\pi\varepsilon_0\varepsilon_r R^2}\, \mathbf{a}_R \quad (b < R < c) \qquad (3.83b)$$

This result is equal to that shown in Eq. (3.82a). In this region, $\mathbf{D}$ can be obtained simply by multiplying $\mathbf{E}$ by the permittivity $\varepsilon = \varepsilon_0\varepsilon_r$. However, $\mathbf{D}$ cannot be used instead of $\mathbf{E}$ in Eq. (3.83a).

Example 3.15 In general, the internal field of a dielectric does not follow the externally applied electric field. However, the internal field of a dielectric sphere placed under a uniform external field is also uniform, as illustrated in Fig. 3.24. Relate the internal field to a uniform external field.

Solution

First, we consider two overlapping spheres of radius a with uniform volume charge densities ρ_v and $-\rho_v$, as shown in Fig. 3.25a. We wish to determine $\mathbf{E}$ in the region of overlap and the surface charge density on a sphere of radius a as the separation d approaches zero (Fig. 3.25b).

Let us investigate the electric field within the lower sphere, denoted $\mathbf{E}_1$, produced only by charge $-\rho_v$ with the upper sphere removed. Setting $\rho_{vo} = -\rho_v$ and $a = 0$ in Eq. (3.35j), we obtain

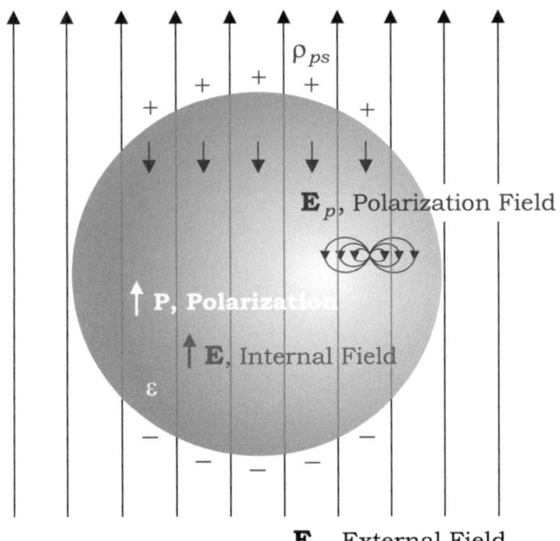

Fig. 3.24 The internal field is the sum of the external and polarization fields

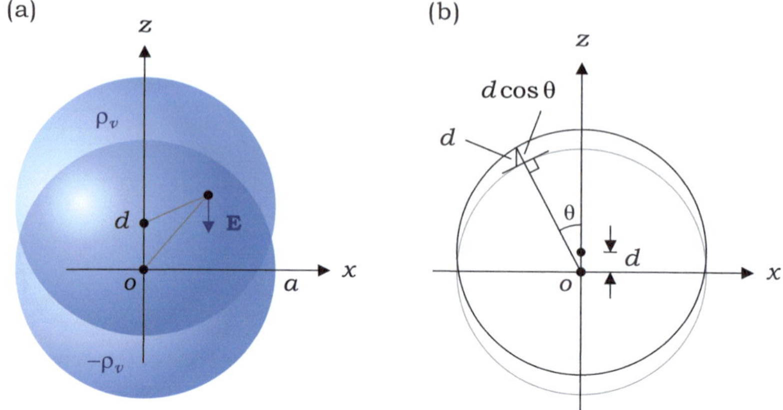

Fig. 3.25 Two overlapping volume charges

$$\mathbf{E}_1 = -\frac{\rho_v R}{3\varepsilon_0}\,\mathbf{a}_R = -\frac{\rho_v}{3\varepsilon_0}\left(\rho\,\mathbf{a}_\rho + z\,\mathbf{a}_z\right) \tag{3.84a}$$

where $R\,\mathbf{a}_R = \rho\,\mathbf{a}_\rho + z\,\mathbf{a}_z$ is used. Because the ρ-component of $\mathbf{E}_1$ is independent of z, it is canceled by the electric field of ρ_v within the overlap region. Therefore, the general expression of $\mathbf{E}$ in the overlapping region is $\mathbf{E} = E_z(z)\,\mathbf{a}_z$. As there is no charge in the overlapping region, $\nabla \cdot \mathbf{D} = 0$. Using $\mathbf{E}$ in this equation leads to the conclusion that E_z is constant in this region. The electric field intensity at point $(0,\ 0,\ d/2)$ represents $\mathbf{E}$ in the overlapping region. From Eq. (3.84a),

$$\mathbf{E} = \mathbf{E}_1(z = d/2) + \mathbf{E}_2(z = d/2) = \frac{-\rho_v d}{3\varepsilon_0}\,\mathbf{a}_z \tag{3.84b}$$

where $\mathbf{E}_2$ is caused by ρ_v. The thickness of the charged region on the sphere is $d\cos\theta$, as shown in Fig. 3.25b. Thus, the surface charge density is given by

$$\rho_s = \rho_v d\cos\theta \tag{3.84c}$$

Consequently, Eqs. (3.84b) and (3.84c) lead us to the conclusion that when a sphere has a surface charge density $\rho_s = \rho_o\cos\theta$ with constant ρ_o, it induces a uniform electric field within the sphere, that is,

$$\boxed{\mathbf{E} = -\frac{\rho_o}{3\varepsilon_0}\,\mathbf{a}_z} \tag{3.85}$$

We are now ready to solve this problem, which ensures that the internal field, $\mathbf{E}$, is uniform in the dielectric sphere (Fig. 3.24). It follows that the electric polarization and polarization surface charge density are given by

$$\mathbf{P} = (\varepsilon - \varepsilon_0)\mathbf{E} \tag{3.86a}$$

$$\rho_{ps} = \mathbf{P} \cdot \mathbf{a}_n = (\varepsilon - \varepsilon_0)E \cos\theta \tag{3.86b}$$

Next, comparing Eq. (3.86b) with Eq. (3.84c), along with Eq. (3.84b), indicates that the resulting polarization field is uniform in the sphere such that

$$\mathbf{E}_p = -\frac{(\varepsilon - \varepsilon_0)}{3\varepsilon_0} E\,\mathbf{a}_z \tag{3.86c}$$

Inserting Eq. (3.86c) into the relationship $\mathbf{E} = \mathbf{E}_o + \mathbf{E}_p$, the internal field of a dielectric sphere placed in the uniform external field $\mathbf{E}_o$ is obtained as

$$\boxed{\mathbf{E} = \frac{3}{\chi + 3}\mathbf{E}_o} \tag{3.87}$$

The initial assumption of a uniform internal field is justified by the results of Eq. (3.87). Although the electric field is uniform inside the sphere, the sum of the external and polarization fields is never uniform outside the sphere.

Exercise 3.29
Explain why ε_r of water decreases with temperature under static conditions.

Ans. Random orientation of water molecules due to thermal agitation.

Exercise 3.30
Explain why ρ_{ps} does not have to be included in Eq. (3.74).

Ans. $\nabla \cdot (\varepsilon_0\,\mathbf{E}_1) = 0$ for $\mathbf{E}_1$ of ρ_{ps}.

Exercise 3.31
Show that $\rho_{pv} = 0$ in a simple medium if there is no free charge inside.

Ans. $\nabla \cdot \mathbf{D} = \nabla \cdot (\varepsilon_r \mathbf{P}/\chi_e) = 0$, and thus, $\nabla \cdot \mathbf{P} = 0$.

Exercise 3.32
A small dielectric cylinder is placed in free space, carrying permanent electric polarization $\mathbf{P} = P_o\,\mathbf{a}_z$. Draw the electric field and flux lines.

Ans. Fig. 3.26.

3.5.3 Boundary Conditions at the Dielectric Interface

In the previous sections, we derived two fundamental relations for static electric fields based on Coulomb's law and the irrotational nature of the electric field. That is,

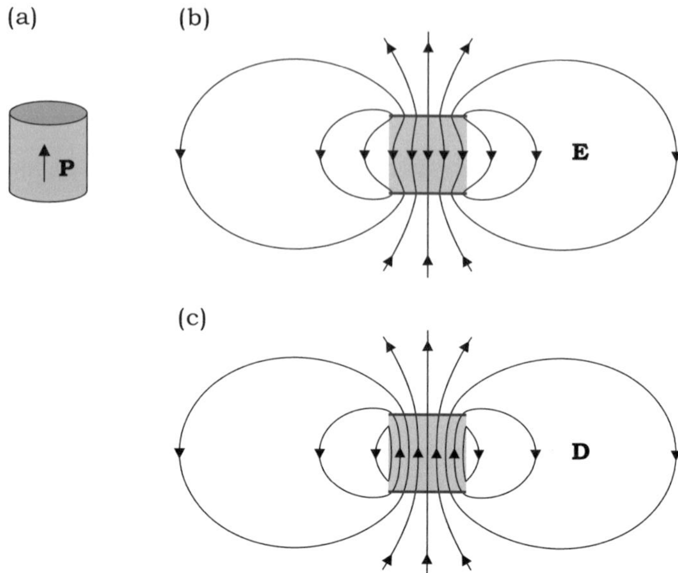

Fig. 3.26 **a** Dielectric cylinder with uniform **P**. **b** Electric field lines of ρ_{ps}. **c** Electric flux lines, $\mathbf{D} = \varepsilon_0\,\mathbf{E} + \mathbf{P}$. **E** and **D** are generally opposite inside the cylinder, but exhibit the same spatial variation outside

$$\oint_{S} \mathbf{D} \cdot d\mathbf{s} = Q \text{ and } \oint_{C} \mathbf{E} \cdot d\mathbf{l} = 0 \qquad (3.88)$$

These provide clear definitions of the divergence and curl of the electric field, which are required for the unique determination of the electric field in a given region in accordance with Helmholtz's theorem. It is important to remember that these relationships hold true in free space, material media, and even in regions consisting of two dissimilar materials. Although the electric field is continuous in homogeneous media, this is not the case for the electric field at the interface between two materials with different permittivities. There are necessary conditions imposed on **E** and **D** to satisfy the fundamental relations at the interface, referred to as ***boundary conditions***.

To derive the boundary conditions for **E** and **D**, we consider an interface formed by two adjoining dielectrics with permittivities ε_1 and ε_2, as shown in Fig. 3.27. First, the circulation of **E** around a rectangular loop *abcda* is computed by assuming that the electric field intensities $\mathbf{E}_1$ and $\mathbf{E}_2$ are constant along the upper and lower sides of the loop, respectively. These represent the electric fields just above and just below the interface, respectively, as loop height Δh approaches zero. The circulation of **E** around the loop should be zero in order to conform to the irrotational nature of the electric field. That is,

$$\oint_{abcda} \mathbf{E} \cdot d\mathbf{l} = \int_{a}^{b} \mathbf{E} \cdot d\mathbf{l} + \int_{b}^{c} \mathbf{E} \cdot d\mathbf{l} + \int_{c}^{d} \mathbf{E} \cdot d\mathbf{l} + \int_{d}^{a} \mathbf{E} \cdot d\mathbf{l}$$

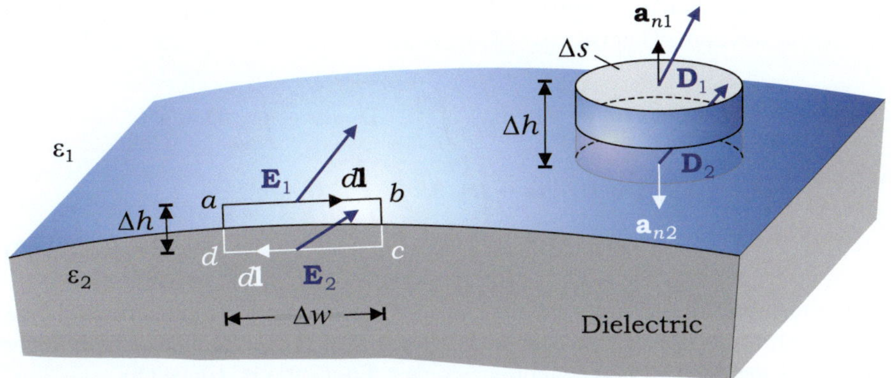

Fig. 3.27 Interface between two different dielectrics with permittivities ε_1 and ε_2

$$= (E_{1t})\,\Delta w + \int_b^c \mathbf{E} \cdot d\mathbf{l} + (-E_{2t})\,\Delta w + \int_d^a \mathbf{E} \cdot d\mathbf{l}$$

$$= 0 \tag{3.89}$$

where the subscript t represents the tangential component. As the height of the rectangle approaches zero, the line integrals along the left and right sides, bc and da, vanish and Eq. (3.89) is reduced to

$$(E_{1t})\,\Delta w + (-E_{2t})\,\Delta w = 0$$

where E_{1t} and E_{2t} are the tangential components of the electric fields on the opposite sides of the interface. Therefore, the boundary condition for $\mathbf{E}$ is given by

$$\boxed{E_{1t} = E_{2t}} \quad [\text{V/m}] \tag{3.90}$$

The tangential component of $\mathbf{E}$ *is continuous across the interface between the two dielectrics.* Inserting Eq. (3.81) into Eq. (3.90) leads to the boundary condition for $\mathbf{D}$, that is,

$$\frac{D_{1t}}{\varepsilon_1} = \frac{D_{2t}}{\varepsilon_2} \tag{3.91}$$

The tangential component of $\mathbf{D}$ is *not* continuous across the interface.

Next, we take as the Gaussian surface a circular cylinder with cross section Δs and height Δh, divided into two halves by the interface, as shown in Fig. 3.27. We apply Gauss's law to the cylinder surface by assuming that the electric flux densities $\mathbf{D}_1$ and $\mathbf{D}_2$ are constant over the top and bottom surfaces of the cylinder, respectively. They represent the electric flux densities just above and just below the interface, respectively, as the cylinder height Δh approaches zero. The integral of $\mathbf{D}$ over the cylinder surface should be equal to the net charge enclosed within the cylinder as Δh approaches zero. That is,

$$\oint_S \mathbf{D} \cdot d\mathbf{s} = \int_{top} \mathbf{D} \cdot d\mathbf{s} + \int_{bottom} \mathbf{D} \cdot d\mathbf{s} + \int_{side} \mathbf{D} \cdot d\mathbf{s}$$

$$= D_{1n}\,\Delta s - D_{2n}\,\Delta s + \int_{side} \mathbf{D} \cdot d\mathbf{s}$$

$$= \rho_s\,\Delta s \tag{3.92}$$

where the subscript n denotes the normal component, and ρ_s is the surface charge density on the interface. Because the surface integral of $\mathbf{D}$ over the side of the cylinder vanishes as the cylinder height approaches zero, Eq. (3.92) is reduced to

$$D_{1n}\,\Delta s - D_{2n}\,\Delta s = \rho_s\,\Delta s$$

where D_{1n} and D_{2n} are the normal components of the electric flux densities on the opposite sides of the interface. Therefore, the boundary condition for $\mathbf{D}$ is given by

$$\boxed{D_{1n} - D_{2n} = \rho_s} \quad [\mathrm{C/m^2}] \tag{3.93}$$

The surface charge density ρ_s is always enclosed by the cylinder, even if the cylinder height approaches zero; however, this is not the case for volume charge. If there is no surface charge on the interface, we have

$$\boxed{D_{1n} = D_{2n}} \tag{3.94}$$

The normal component of $\mathbf{D}$ *is continuous across the interface between the two different dielectrics if there is no surface charge at the interface.* Inserting Eq. (3.81) into Eq. (3.94) leads to the following boundary condition for $\mathbf{E}$:

$$\varepsilon_1 E_{1n} = \varepsilon_2 E_{2n} \tag{3.95}$$

The normal component of $\mathbf{E}$ is *not* continuous across the interface.

Example 3.16 If $\mathbf{E}_1$ and $\mathbf{E}_2$ represent the electric field intensities on the opposite sides of the interface between the two dielectrics with permittivities ε_1 and ε_2, respectively, as shown in Fig. 3.28, express E_2 and θ_2 in terms of E_1, θ_1, ε_1, and ε_2.

Solution

The normal and tangential components of $\mathbf{E}_1$ at the interface are

$$E_{1n} = E_1 \cos\theta_1$$

$$E_{1t} = E_1 \sin\theta_1$$

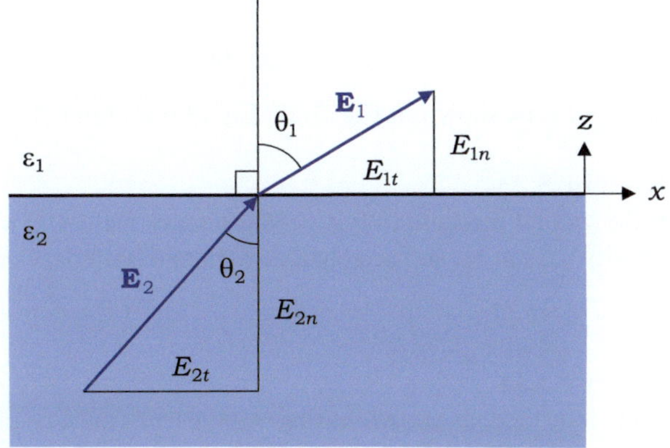

Fig. 3.28 Interface between two adjoining dielectrics

The tangential component of $\mathbf{E}$ is continuous across the interface. Thus,

$$E_{2t} = E_{1t} = E_1 \sin \theta_1 \tag{3.96a}$$

Dielectrics cannot carry free charges; thus, there is no practical way to accumulate surface charges on a dielectric-dielectric interface. Thus, the normal component of $\mathbf{D}$ is continuous across the interface. Thus,

$$\begin{aligned} D_{2n} &= D_{1n} \\ &= \varepsilon_1 E_{1n} = \varepsilon_1 E_1 \cos \theta_1 \end{aligned} \tag{3.96b}$$

With $D_{2n} = \varepsilon_2 E_{2n}$, Eq. (3.96b) becomes

$$E_{2n} = \frac{\varepsilon_1}{\varepsilon_2} E_1 \cos \theta_1 \tag{3.96c}$$

Combining Eqs. (3.96a) and (3.96c) yields

$$E_2 = \sqrt{(E_{2t})^2 + (E_{2n})^2} = E_1 \sqrt{\sin^2 \theta_1 + (\varepsilon_1/\varepsilon_2)^2 \cos^2 \theta_1}$$

Next, from Eqs. (3.96a) and (3.96c), the angle between $\mathbf{E}_2$ and the surface normal is obtained as

$$\theta_2 = \tan^{-1} \frac{E_{2t}}{E_{2n}} = \tan^{-1} \left[\frac{\varepsilon_2}{\varepsilon_1} \tan \theta_1 \right] \tag{3.96d}$$

Thus,

$$\frac{\tan \theta_1}{\tan \theta_2} = \frac{\varepsilon_1}{\varepsilon_2} \tag{3.96e}$$

If $\theta_1 > \theta_2$, as shown in Fig. 3.28, Eq. (3.96e) predicts $\varepsilon_1 > \varepsilon_2$.

Exercise 3.33

If the second medium in Fig. 3.28 is a lossy dielectric with a uniform volume charge density ρ_v, but no surface charge, write the boundary condition for D_n.

Ans. $D_{1n} = D_{2n}$.

Exercise 3.34

If the interface, as shown in Fig. 3.28, carries surface charge density ρ_s, express the boundary conditions for **E** and **D** in vector notation using unit vector $\mathbf{a}_z$.

Ans. $(\varepsilon_1 \mathbf{E}_1 - \varepsilon_2 \mathbf{E}_2) \cdot \mathbf{a}_z = \rho_s$ and $(\mathbf{E}_1 - \mathbf{E}_2) \times \mathbf{a}_z = 0$.

Exercise 3.35

If the interface, as shown in Fig. 3.28, carries no surface charges, show that the following boundary conditions for the electric potential fully conform to those for **E** and **D**:

$$V_1(z = 0^+) = V_2(z = 0^-) \text{ and } \varepsilon_1 \left.\frac{\partial V_1}{\partial z}\right|_{z=0^+} = \varepsilon_2 \left.\frac{\partial V_2}{\partial z}\right|_{z=0^-}.$$

Ans. Eqs. (3.90) and (3.94).

Review Questions

RQ 3.30	Define the electric dipole moment and electric polarization.	[(3.65)(3.67)]
RQ 3.31	Sate the relationship between electric polarization and polarization charge.	[(3.69)(3.72)]
RQ 3.32	Define **D** in the material medium.	[(3.76)]
RQ 3.33	State Gauss's law in material media.	[(3.77)]
RQ 3.34	What is the relationship between **P** and **E** in material media?	[(3.79)]
RQ 3.35	What is the relationship between **D** and **E** in material media?	[(3.81)]
RQ 3.36	Is ε_r always greater than unity in simple media under static conditions?	[(3.76)(3.79)(3.81)]
RQ 3.37	What is the internal electric field?	[Fig. 3.24]
RQ 3.38	Explain how the electric field lines are modified in the presence of a dielectric.	[Fig. 3.22]
RQ 3.39	State the boundary conditions for E_t and D_n at the interface between the two dissimilar dielectrics.	[(3.90)(3.94)]

RQ 3.40   Do electric field lines reflect from an interface?   [Fig. 3.28]

RQ 3.41   Which of the boundary conditions for **E** and **D** is   [(3.94)]
           related to the conservation of the electric flux?

3.6 Perfect Conductor in Static Electric Field

Conductors are electrically neutral under normal conditions although they contain a large number of free electrons. This is because the positively charged ionized lattice atoms balance the negative charges of free electrons. In the absence of an externally applied electric field, the free electrons collide with impurities or imperfections in the lattice and scatter in random directions; thus, there is no net motion of the free electrons. However, in the presence of an electric field, free electrons gain a net speed during the mean time between collisions and constitute a conduction current. Conductivity is a measure of the ease with which free electrons produce conduction current under the influence of an electric field. A perfect conductor has an infinite conductivity ($\sigma = \infty$), whereas a perfect dielectric has zero conductivity ($\sigma = 0$).

When charges are injected into a perfect conductor, the repelling force of the like charges immediately drives the charge to the surface, causing an accumulation of charges, irrespective of whether the charge is uniform on the surface. Therefore, the net charge should be distributed over the conductor surface, such that there is no electric field in the interior. Otherwise, the electric field generates an infinite conduction current that in turn redistributes the charge instantaneously until there is no electric field inside. Similarly, when an uncharged conductor is placed in an electric field, the free electrons are driven in the direction opposite to that of the external field, leaving positively charged ionized atoms on the opposite side. The induced surface charges produce an electric field of their own in such a way as to cancel out the external field in the interior. Therefore, there must be no electric field or net charge inside a perfect conductor. That is,

$$\boxed{\mathbf{E} = 0} \tag{3.97}$$

$$\boxed{\rho_v = 0} \tag{3.98}$$

where ρ_v is the net volume charge density. These are two basic properties of perfect conductors. Third, any net charge resides on the surface of the conductor. Fourth, the electric field of the net charge is always perpendicular to the conductor surface, pointing outwards. Fifth, the conductor surface is an equipotential surface.

Following the same procedure used for the dielectric-dielectric interface, we obtain the boundary conditions for **E** and **D** on the surface of a perfect conductor. Consider a perfect conductor residing in free space, as illustrated in Fig. 3.29. The circulation of **E** around the rectangular loop *abcda* must be zero. That is,

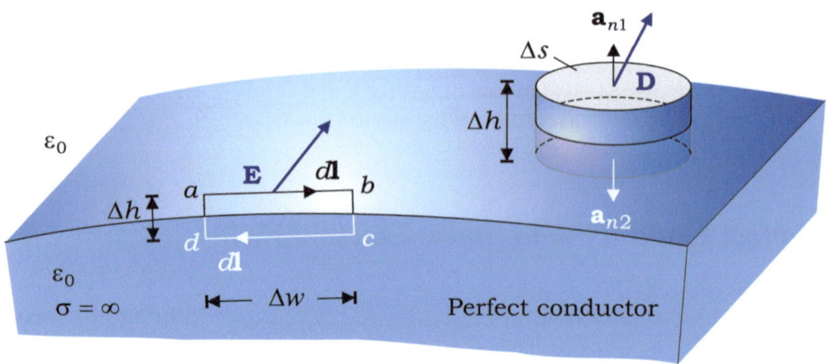

Fig. 3.29 Perfect conductor residing in free space

$$\oint_{abcda} \mathbf{E} \cdot d\mathbf{l} = \int_a^b \mathbf{E} \cdot d\mathbf{l} + \int_b^c \mathbf{E} \cdot d\mathbf{l} + \int_c^d \mathbf{E} \cdot d\mathbf{l} + \int_d^a \mathbf{E} \cdot d\mathbf{l}$$

$$= (E_{1t})\,\Delta w + \int_b^c \mathbf{E} \cdot d\mathbf{l} + 0 + \int_d^a \mathbf{E} \cdot d\mathbf{l}$$

$$= 0 \tag{3.99}$$

where the subscript t denotes the tangential component. The line integral along
the bottom side, cd, is zero, because $\mathbf{E}$ is zero inside the perfect conductor. The line
integrals along the left and right sides, da and bc, vanish as the height Δh approaches
zero. Under these conditions, Eq. (3.99) is reduced to

$$(E_{1t})\,\Delta w = 0$$

Consequently, on the surface of a perfect conductor,

$$\boxed{E_t = 0} \tag{3.100}$$

The tangential component of $\mathbf{E}$ *on the surface of a perfect conductor is always zero.*
In view of Eqs. (3.97) and (3.100), it follows that no work is done to move a charge
around on or in a perfect conductor. *A perfect conductor is an equipotential body.*

Next, Gauss's law is applied to a circular cylinder with its lower half immersed
in the conductor, as shown in Fig. 3.29. The integral of $\mathbf{D}$ over the bounding surface
of the cylinder is equal to the surface charge enclosed by the cylinder. That is,

$$\oint_S \mathbf{D} \cdot d\mathbf{s} = \int_{top} \mathbf{D} \cdot d\mathbf{s} + \int_{bottom} \mathbf{D} \cdot d\mathbf{s} + \int_{side} \mathbf{D} \cdot d\mathbf{s}$$

$$= D_{1n}\Delta s + 0 + \int_{side} \mathbf{D} \cdot d\mathbf{s}$$

$$= \rho_s \Delta s \tag{3.101}$$

where the subscript n denotes the normal component. The surface integral over the bottom plate of the cylinder is zero because $\mathbf{E} = 0 = \mathbf{D}$ in the conductor. As $\Delta h \to 0$, the surface integral over the side surface vanishes, but the surface charge $\rho_s \Delta s$ is always enclosed in the cylinder. Thus, Eq. (3.101) is reduced to

$$D_{1n} \Delta s = \rho_s \Delta s$$

Therefore, on the surface of a perfect conductor,

$$\boxed{D_n = \rho_s} \qquad (3.102)$$

The normal component of $\mathbf{D}$ *on the surface of a perfect conductor is equal to the surface charge density.*

Example 3.17 A conducting sphere of radius a with a net charge $+Q$ is enclosed by an uncharged conductive spherical shell of radii b and c, as shown in Fig. 3.30a. (a) Determine ρ_s induced on the shell, and identify its source. (b) Repeat part (a) when the shell is grounded, as shown in Fig. 3.30b.

Solution

From the spherical symmetry, we expect $\mathbf{D} = D_R(R)\,\mathbf{a}_R$ everywhere.

(a) In region $a < R \leq b$, from Gauss's law,

$$4\pi R^2 D_R = +Q$$

Thus, on the spherical surface at $R = b$,

$$\mathbf{D} = D_R\,\mathbf{a}_R = \frac{Q}{4\pi b^2}\,\mathbf{a}_R \qquad (3.103a)$$

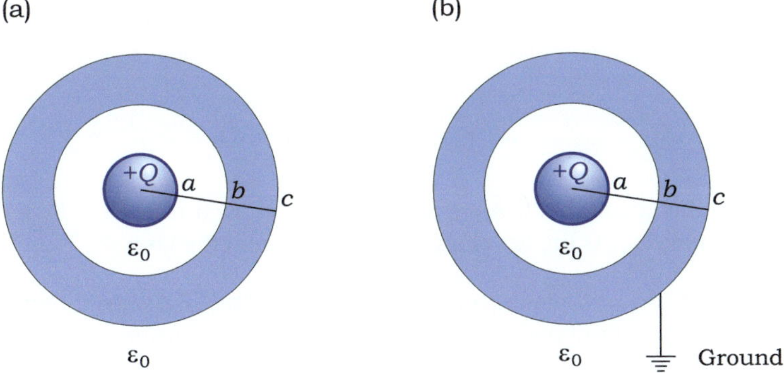

Fig. 3.30 Conducting sphere enclosed by a conductive spherical shell

Using Eq. (3.102), along with $\mathbf{a}_n = -\mathbf{a}_R$, the surface charge density at $R = b$ is obtained as

$$\rho_{s1} = -D_R = -\frac{Q}{4\pi b^2} \quad (\text{at } R = b) \tag{3.103b}$$

In region $R \geq c$, the net charge enclosed in the Gaussian surface is simply $+Q$, because the outer shell is uncharged and electrically neutral. From Gauss's law,

$$4\pi R^2 D_R = +Q$$

Thus, on the spherical surface at $R = c$,

$$\mathbf{D} = D_R \, \mathbf{a}_R = \frac{Q}{4\pi c^2} \, \mathbf{a}_R \tag{3.103c}$$

Using Eq. (3.102), along with $\mathbf{a}_n = \mathbf{a}_R$, the surface charge density at $R = c$ is obtained as

$$\rho_{s2} = D_R = \frac{Q}{4\pi c^2} \quad (\text{at } R = c) \tag{3.103d}$$

The negative charge ρ_{s1} arises from the free electrons of the outer conductor, whereas the positive charge ρ_{s2} originates from the ionized lattice atoms.

Note that the two charges $+Q$ and ρ_{s1} jointly produce $\mathbf{E} = 0$ inside the outer conductor.

(b)   The outer conductor is at zero potential because it is grounded and there are no electric field lines between the outer conductor and the ground. Thus, $\mathbf{E}$ is zero in the region $R \geq c$. It then follows that

$$D_n = 0 = \rho_{s2} \quad (\text{at } R = c) \tag{3.103e}$$

In region $a < R \leq b$, $\mathbf{D}$ is the same as that in part (a). Therefore, the surface charge density on the inner surface of the shell is given by

$$\rho_{s1} = -\frac{Q}{4\pi b^2} \quad (\text{at } R = b) \tag{3.103f}$$

The two charges $+Q$ and ρ_{s1} jointly produce $\mathbf{E} = 0$ in region $R > b$. The surface charge ρ_{s1} is the net charge transferred from the ground.

Example 3.18 The $x = 0$ plane corresponds to the surface of a semi-infinite dielectric. A long conducting cylinder is half-immersed in the dielectric, carrying a net charge of $+Q$ [C/m], as shown in Fig. 3.31. The electric field of the form $\mathbf{E} = E_o \mathbf{a}_\rho / \rho$ is taken as a trial solution for $\rho \geq a$, as it satisfies $\nabla \cdot \mathbf{D} = 0$ and $\nabla \times \mathbf{E} = 0$.

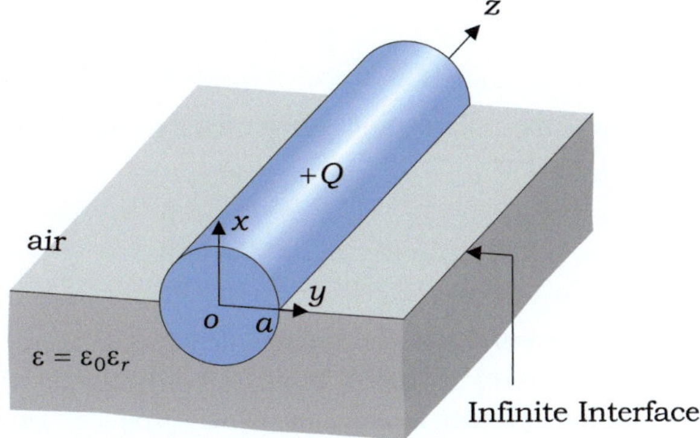

Fig. 3.31 Long conducting cylinder half-buried in dielectric

(a)   Does it satisfy the boundary conditions?

(b)   Determine E_o and the net and polarization surface charge densities, ρ_s and ρ_{ps}, on the cylinder.

(c)   Check whether the sum of ρ_s and ρ_{ps} is uniform over the cylinder surface to justify the trial solution.

Solution

(a)   The trial electric field is perpendicular to the conductor surface and continuous across the air–dielectric interface at $x = 0$.

(b)   From Gauss's law, for $\rho \geq a$,

$$2\pi\rho\frac{1}{2}\left[\varepsilon_0\frac{E_o}{\rho} + \varepsilon_0\varepsilon_r\frac{E_o}{\rho}\right] = Q$$

This yields

$$\mathbf{E} = \frac{Q}{\varepsilon_0(\varepsilon_r + 1)\pi\rho}\,\mathbf{a}_\rho \quad (\rho \geq a) \tag{3.104a}$$

Using Eq. (3.104a), the density of free surface charge on the cylinder for $x > 0$ is obtained as

$$\rho_s = D_\rho = \frac{Q}{(\varepsilon_r + 1)\pi a} \quad (x > 0) \tag{3.104b}$$

Similarly, the density of free surface charge on the cylinder for $x < 0$ is obtained as

$$\rho_s = D_\rho = \frac{\varepsilon_r Q}{(\varepsilon_r + 1)\pi a} \quad (x < 0) \tag{3.104c}$$

At the conductor–dielectric interface at $\rho = a$, from Eq. (3.104a),

$$\mathbf{P} = \mathbf{D} - \varepsilon_0 \mathbf{E} = \frac{(\varepsilon_r - 1)\, Q}{(\varepsilon_r + 1)\, \pi a}\, \mathbf{a}_\rho \qquad (3.104\text{d})$$

Thus, the polarization surface charge density on the cylinder for $x < 0$ is obtained as

$$\rho_{ps} = \mathbf{P} \cdot (-\mathbf{a}_\rho) = \frac{-(\varepsilon_r - 1)\, Q}{(\varepsilon_r + 1)\, \pi a} \qquad (3.104\text{e})$$

(c)   The sum of the surface charge densities in Eqs. (3.104c) and (3.104e) is equal to that in Eq. (3.104b). In other words, the density of the total surface charge on the cylinder is uniform, which produces an electric field with cylindrical symmetry, as implied in the trial solution. Thus, the trial solution is justified. The electric field in Eq. (3.104a) is a unique solution in accordance with Helmholtz's theorem; its divergence and curl vanish for $\rho \geq a$ and it satisfies the boundary conditions.

Exercise 3.36
Point charge q is held at rest above an infinitely large conducting slab. Determine the total charges induced on (a) the top, and (b) the bottom surfaces of the slab.

Ans. (a) $-q$, (b) $+q$.

Exercise 3.37
In what ways are the induced charges in Exercise 3.36 distributed on the surfaces?

Ans. $\mathbf{E}$ is normal to the surface and zero inside.

Exercise 3.38
Point charge q is at the center of a spherical cavity formed in an uncharged conducting sphere (Fig. 3.32). Find (a) ρ_{sa} and ρ_{sb}, and (b) $\mathbf{E}$ for $R \geq b$.

Ans. (a) $\rho_{sa} = -q/4\pi a^2$, $\rho_{sb} = q/4\pi b^2$, (b) $\mathbf{E} = q\, \mathbf{a}_R / 4\pi\varepsilon_0 R^2$.

Fig. 3.32 Spherical cavity inside the conducting sphere

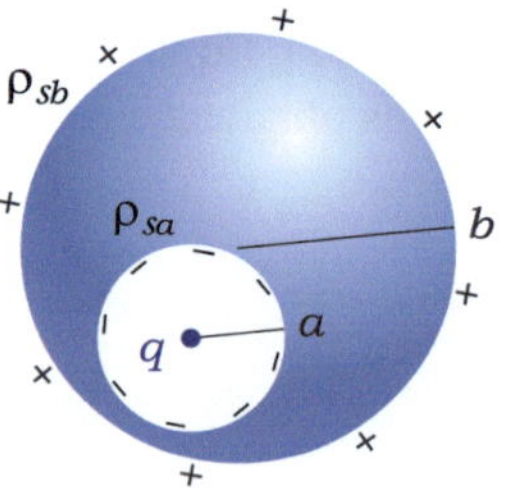

Review Questions

RQ 3.42 State electrostatic properties of a perfect conductor.   [(3.97)(3.98)]

RQ 3.43 What are the boundary conditions for **E** and **D** on the   [(3.100)(3.102)]
surface of a perfect conductor?

RQ 3.44 Explain how net charges are neutralized inside a   [Fig. 3.32]
conductor.

3.7 Electrostatic Potential Energy

The electric potential at a point in space is defined as the work done in bringing a unit charge from infinity to the point against the electric field present in the region. When a charge is placed at a point in an electric field, the product of the charge and the electric potential is the potential energy of the charge. To maintain the charge in position, the electric force acting on the charge must be balanced by an external mechanical force. If we release our hold on to the charge, the potential energy is transformed into kinetic energy, which accelerates the charge and returns it to infinity. By extending this concept to a system of charges, the work done to assemble the charge distribution is stored as the potential energy of the system.

For a collection of point charges residing in free space, the potential energy can be obtained by computing the work done to bring the individual charges from infinity to their predetermined positions and adding all the expended energies. For example, consider a system with three point-charges in free space. When the first charge q_1 is placed at its point, no energy is expended because there is no electric field exerting a force on the charge. That is,

$$W_1 = 0 \tag{3.105a}$$

When the second charge q_2 is brought from infinity to its prearranged position near q_1 against the electric field of q_1, the work done is given by

$$W_2 = q_2 \left[\frac{q_1}{4\pi\varepsilon_0 \mathcal{R}_{2-1}} \right]$$

where the term in brackets is the electric potential at the point of q_2 due to q_1. Using $\mathcal{R}_{2-1} = |\mathbf{r}_2 - \mathbf{r}_1| = \mathcal{R}_{1-2}$, W_2 can be rearranged as

$$W_2 = \frac{1}{2}\frac{q_2 q_1}{4\pi\varepsilon_0 \mathcal{R}_{2-1}} + \frac{1}{2}\frac{q_1 q_2}{4\pi\varepsilon_0 \mathcal{R}_{1-2}} \tag{3.105b}$$

By following the same procedure, the energy expended in bringing the third charge q_3 to its prearranged position can be expressed as

$$W_3 = q_3 \left[\frac{q_1}{4\pi\varepsilon_0 \mathcal{R}_{3-1}} \right] + q_3 \left[\frac{q_2}{4\pi\varepsilon_0 \mathcal{R}_{3-2}} \right]$$

$$= \frac{1}{2} \left[\frac{q_3 q_1}{4\pi\varepsilon_0 \mathcal{R}_{3-1}} + \frac{q_3 q_2}{4\pi\varepsilon_0 \mathcal{R}_{3-2}} \right] + \frac{1}{2} \left[\frac{q_1 q_3}{4\pi\varepsilon_0 \mathcal{R}_{1-3}} + \frac{q_2 q_3}{4\pi\varepsilon_0 \mathcal{R}_{2-3}} \right] \quad (3.105c)$$

Therefore, the potential energy of the system of three charges is expressed as

$$W_1 + W_2 + W_3$$

$$= \frac{q_1}{2} \left[\frac{q_2}{4\pi\varepsilon_0 \mathcal{R}_{1-2}} + \frac{q_3}{4\pi\varepsilon_0 \mathcal{R}_{1-3}} \right] + \frac{q_2}{2} \left[\frac{q_1}{4\pi\varepsilon_0 \mathcal{R}_{2-1}} + \frac{q_3}{4\pi\varepsilon_0 \mathcal{R}_{2-3}} \right]$$

$$+ \frac{q_3}{2} \left[\frac{q_1}{4\pi\varepsilon_0 \mathcal{R}_{3-1}} + \frac{q_2}{4\pi\varepsilon_0 \mathcal{R}_{3-2}} \right]$$

$$= \frac{1}{2} (q_1 V_1 + q_2 V_2 + q_3 V_3) \quad (3.106)$$

where V_1 is the electric potential at the point of q_1 due to the two other charges, and so on.

By extending this procedure to N point charges, the total energy stored in the system of N point charges can be expressed as

$$\boxed{W_E = \frac{1}{2} \sum_{i=1}^{N} q_i V_i} \quad [\text{J}] \quad (3.107)$$

Again, V_i is the electric potential at the point of q_i due to all the other charges.

Subsequently, the energy stored in a continuous charge distribution can be obtained in the same manner. For a volume charge of density ρ_v, the charged region is first subdivided into N differential volumes. Substituting q_i in Eq. (3.107) with incremental charge $\rho_v \Delta v_i$ and taking the limit as $N \to \infty$ and $\Delta v \to 0$, the stored energy is obtained as follows:

$$W_E = \lim_{\substack{N \to \infty \\ \Delta v \to 0}} \frac{1}{2} \sum_{i=1}^{N} \rho_v \Delta v_i V_i \quad (3.108)$$

The right-hand side of Eq. (3.108) is, by definition, the integral of $\rho_v V$ over the volume occupied by the charge. Therefore, the ***electrostatic potential energy*** of volume charge density ρ_v is expressed as

$$\boxed{W_E = \frac{1}{2} \int_{\mathcal{V}} \rho_v V \, dv} \quad [\text{J}] \quad (3.109)$$

where $\mathcal{V}$ is the volume occupied by the charge, and V is the electric potential.

It is often more convenient to express the potential energy in terms of $\mathbf{E}$ and $\mathbf{D}$, with no knowledge of ρ_v. Applying Gauss's law to Eq. (3.109) leads to

$$W_E = \frac{1}{2} \int_{\mathcal{V}} (\nabla \cdot \mathbf{D}) V \, dv \qquad (3.110)$$

Using the vector identity $\nabla \cdot (V\mathbf{D}) = V(\nabla \cdot \mathbf{D}) + \mathbf{D} \cdot (\nabla V)$, Eq. (3.110) can be rewritten as

$$\begin{aligned}
W_E &= \frac{1}{2} \int_{\mathcal{V}} \nabla \cdot (V\mathbf{D}) \, dv - \frac{1}{2} \int_{\mathcal{V}} \mathbf{D} \cdot (\nabla V) \, dv \\
&= \frac{1}{2} \oint_{\mathcal{S}} V\mathbf{D} \cdot d\mathbf{s} + \frac{1}{2} \int_{\mathcal{V}} \mathbf{D} \cdot \mathbf{E} \, dv
\end{aligned} \qquad (3.111)$$

where the divergence theorem is used along with $\mathbf{E} = -\nabla V$.

Although $\mathcal{S}$ in Eq. (3.111) is the bounding surface of $\mathcal{V}$, it may be arbitrarily large only if it encloses all charges. For simplicity, let it be a sphere of radius R. As R tends to infinity, the terms in the integrand V, $|\mathbf{D}|$, and $|d\mathbf{s}|$ vary as $1/R$, $1/R^2$, and R^2, respectively. Therefore, the closed-surface integral vanishes as $R \to \infty$, and the electrostatic potential energy is given by

$$\boxed{W_E = \frac{1}{2} \int_{\mathcal{V}} \mathbf{D} \cdot \mathbf{E} \, dv} \qquad [\mathrm{J}] \qquad (3.112)$$

where the volume $\mathcal{V}$ encompasses the entire space of $\mathbf{E}$. Using the constitutive relation $\mathbf{D} = \varepsilon \mathbf{E}$ in Eq. (3.112) gives

$$\boxed{W_E = \frac{1}{2} \int_{\mathcal{V}} \varepsilon E^2 \, dv} \qquad [\mathrm{J}] \qquad (3.113)$$

In view of Eq. (3.112), the ***electrostatic energy density*** is defined as

$$\boxed{w_e = \frac{1}{2} \mathbf{D} \cdot \mathbf{E}} \qquad [\mathrm{J/m^3}] \qquad (3.114)$$

which is given in units of joules per cubic meter.

Example 3.19 Determine the potential energy of the volume charge of a uniform density ρ_v assembled into a spherical shell of radii a and b in free space, as shown in Fig. 3.33.

Solution

Let us compute the work done by individually bringing and stacking thin spherical layers of charges. First, the total charge within a hollow sphere of outer radius R_1 is obtained as

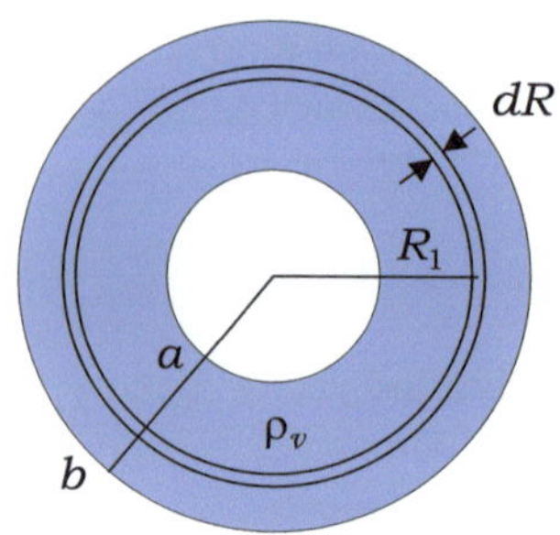

Fig. 3.33 Spherical shell formed by uniform volume charge

$$Q = \rho_v \frac{4\pi}{3}\left(R_1^3 - a^3\right)$$

Assuming that there is no charge in the region $R > R_1$, the electric potential at $R = R_1$ is the same as if Q is concentrated at the center, that is,

$$V = \frac{Q}{4\pi\varepsilon_0 R_1} = \frac{1}{4\pi\varepsilon_0 R_1}\left[\rho_v \frac{4\pi}{3}\left(R_1^3 - a^3\right)\right] \tag{3.115a}$$

Next, the total charge in a thin spherical layer with radius R_1 and thickness dR is calculated as

$$dq = \rho_v 4\pi R_1^2\, dR \tag{3.115b}$$

The energy expended in bringing the charge dq from infinity to its position can be obtained using Eqs. (3.115a) and (3.115b) as

$$dW = V\, dq = \left[\rho_v^2 \frac{4\pi}{3\varepsilon_0} R_1\left(R_1^3 - a^3\right)\right] dR$$

The total energy expended in assembling the volume charge is, therefore,

$$W_E = \int dW = \rho_v^2 \frac{4\pi}{3\varepsilon_0} \int_{R=a}^{R=b} R\left(R^3 - a^3\right) dR$$

$$= \rho_v^2 \frac{4\pi}{3\varepsilon_0}\left[\frac{1}{5}(b^5 - a^5) - a^3 \frac{1}{2}(b^2 - a^2)\right] \tag{3.115c}$$

If a is set to zero in Eq. (3.115c), the potential energy of a uniformly charged sphere of radius b is given in the form kb^5 with constant k. If the principle of superposition was valid, the potential energy of the hollow sphere would have been in the form of $k\left(b^5 - a^5\right)$, but not so. Although the electric potential obeys the principle of superposition, electrostatic potential energy *does not*.

Example 3.20 Two parallel conducting plates of area A are spaced d apart and carry net charges $+Q$ and $-Q$ (Fig. 3.34). The gap is half-filled with a dielectric of ε_r.

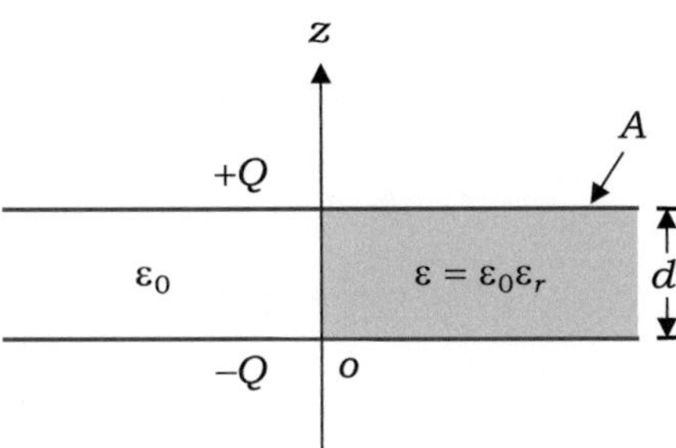

Fig. 3.34 Two parallel plates with net charges $+Q$ and $-Q$

Assuming that d is much smaller than the linear dimensions of the plate, determine (a) W_E, (b) W_E if the gap is emptied, and (c) explain why the former is less than the latter.

Solution

If the fringing effects of $\mathbf{E}$ at the edges are ignored, the proper trial solution is $\mathbf{E} = -E_o\,\mathbf{a}_z$ with a constant E_o in the area between the plates and $\mathbf{E} = 0$ elsewhere. It satisfies the boundary conditions such that $\mathbf{E}$ is normal to the conductor surface and continuous across the air-dielectric interface.

(a) Applying Gauss's law to the upper plate yields

$$\frac{A}{2}\varepsilon_0 E_o + \frac{A}{2}\varepsilon_0\varepsilon_r E_o = +Q$$

Thus,

$$\mathbf{E} = -\frac{2Q}{A\varepsilon_0(\varepsilon_r + 1)}\,\mathbf{a}_z \equiv -E_o\,\mathbf{a}_z$$

Subsequently, the electrostatic potential energy is given by

$$W_E = \frac{1}{2}\frac{Ad}{2}\varepsilon_0 E_o^2 + \frac{1}{2}\frac{Ad}{2}\varepsilon_0\varepsilon_r E_o^2 = \frac{Q^2 d}{A\varepsilon_0(\varepsilon_r + 1)} \tag{3.116a}$$

(b) Setting $\varepsilon_r = 1$ in Eq. (3.116a), the potential energy with the dielectric removed can be obtained as

$$W_E = \frac{Q^2 d}{2A\varepsilon_0} \tag{3.116b}$$

(c) The electrostatic potential energy was expended to produce electric polarization in the dielectric.

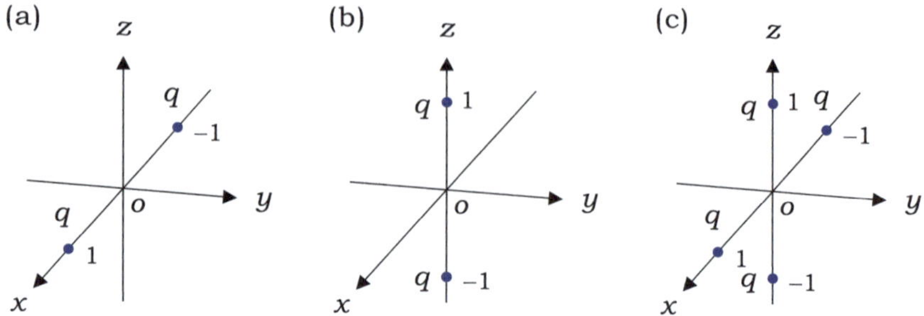

Fig. 3.35 Three systems of point charges

Exercise 3.39

Determine the potential energy when charge q [C] is uniformly distributed over a spherical surface of radius a.

Ans. $W_E = q^2/8\pi\varepsilon_0 a$.

Exercise 3.40

Identical point charges of $+q$ [C] are arranged as shown in Fig. 3.35. Determine W_E for each configuration. Is W_E in (c) equal to the sum of those in (a) and (b)?

Ans. (a) (b) $q^2/8\pi\varepsilon_0$, (c) $q^2(2\sqrt{2}+1)/4\pi\varepsilon_0$, (d) No.

Exercise 3.41

The two point-charges in Fig. 3.35a have the same mass m. If they are released from rest and move in opposite directions, find the velocity at greater distances.

Ans. $v = \pm q\,\mathbf{a}_x/(8\pi\varepsilon_0 m)^{1/2}$.

Review Questions

RQ 3.45	What is the physical significance of the qV?	[(3.107)]
RQ 3.46	Express the potential energy of the volume charge.	[(3.109)]
RQ 3.47	Express the potential energy in terms of field quantities.	[(3.112)]
RQ 3.48	Define the electrostatic energy density.	[(3.114)]
RQ 3.49	Does the potential energy obey the superposition principle?	[Fig. 3.35]

3.8 Electrostatic Boundary-Value Problems

When the charge distribution is given in a region of space, Coulomb's law, Gauss's law, and the concept of electric potential can be applied to determine the electric field, electric flux density, and electric potential. In many practical situations in which the charge distribution and/or electric potential are specified at the boundaries, straightforward application of one of these laws is not possible. For example, if two very long parallel conducting cylinders are maintained at constant potentials, it is difficult to determine the electric field between them and the distribution of surface charges on the cylinders. In this section, we focus on Poisson's and Laplace's equations, which allow us to determine V and $\mathbf{E}$ in a given region from the charge distribution and/or the electric potential at the boundaries. Electrostatic problems involving boundary values are called ***boundary-value problems***.

3.8.1 Poisson's and Laplace's Equations

The derivation of Poisson's equation is simple. We begin with two fundamental relations governing the static electric field. That is,

$$\nabla \cdot \mathbf{D} = \rho_v \tag{3.117}$$

$$\nabla \times \mathbf{E} = 0 \tag{3.118}$$

If we limit our discussion to homogeneous, linear, and isotropic media, the constitutive relation for $\mathbf{E}$ and $\mathbf{D}$ is simply

$$\mathbf{D} = \varepsilon \mathbf{E} \tag{3.119}$$

where the permittivity ε is a constant and is independent of both the magnitude and direction of $\mathbf{E}$. Let us recall the relationship between $\mathbf{E}$ and V, that is,

$$\mathbf{E} = -\nabla V \tag{3.120}$$

Inserting Eqs. (3.119) and (3.120) into Eq. (3.117) gives

$$-\nabla \cdot \varepsilon \nabla V = \rho_v \tag{3.121}$$

Because ε is constant in simple media, it can be taken outside the divergence operator such that

$$\boxed{\nabla^2 V = -\frac{\rho_v}{\varepsilon}} \tag{3.122}$$

This is known as ***Poisson's equation***. The Laplacian operator ∇^2 denotes the divergence of the gradient of a scalar field. If there is no net volume charge in a given region, Poisson's equation is reduced to ***Laplace's equation***. That is,

$$\boxed{\nabla^2 V = 0} \tag{3.123}$$

In many practical problems, the method of separation of variables can significantly facilitate the solution to Laplace's equation. It resolves the second-order partial differential equation into three second-order ordinary differential equations, each of which requires two independent boundary values to determine two integration constants. In a region with no volume charges, Laplace's equation is solved for electric potential by assuming that the point, line, and surface charges in the region are boundary values.

Laplace's equation in Cartesian coordinates is expressed as

$$\boxed{\frac{\partial^2 V}{\partial x^2} + \frac{\partial^2 V}{\partial y^2} + \frac{\partial^2 V}{\partial z^2} = 0} \quad \text{(Cartesian)} \tag{3.124a}$$

In other coordinate systems, the Laplacian of V is obtained by taking the divergence of the gradient of V, that is,

$$\boxed{\frac{1}{\rho}\frac{\partial}{\partial \rho}\left(\rho \frac{\partial V}{\partial \rho}\right) + \frac{1}{\rho^2}\left(\frac{\partial^2 V}{\partial \phi^2}\right) + \frac{\partial^2 V}{\partial z^2} = 0} \quad \text{(cylindrical)} \tag{3.124b}$$

$$\boxed{\frac{1}{R^2}\frac{\partial}{\partial R}\left(R^2 \frac{\partial V}{\partial R}\right) + \frac{1}{R^2 \sin\theta}\frac{\partial}{\partial \theta}\left(\sin\theta \frac{\partial V}{\partial \theta}\right) + \frac{1}{R^2 \sin^2\theta}\frac{\partial^2 V}{\partial \phi^2} = 0} \quad \text{(spherical)}$$

$$\tag{3.124c}$$

Exercise 3.42
Show that the following electric potentials satisfy Laplace's equation:
(a) $V = e^{-2x}\cos(2y)$, (b) $V = R\cos\theta$, and (c) $V = \ln[\tan(\theta/2)]$.

Exercise 3.43
Which of the following is incorrect?
(a) $\nabla^2 R^{-1} = 0$, (b) $\nabla^2(R-2)^{-1} = 0$, (c) $\nabla^2[(x-2)^2 + y^2 + z^2]^{-1/2} = 0$.

Ans. (b).

3.8.2   Uniqueness Theorem

From calculus, we know that a second-order ordinary differential equation has a general solution with two integration constants. If the constants can be determined from the given boundary values, then the particular solution is unique in a given region. Similarly, there is a **uniqueness theorem** for Poisson's equation, which is a second-order partial differential equation for three variables. *The solution to*

Poisson's equation is the only solution in the given region, if it satisfies the prescribed conditions for V at the entire boundary. The uniqueness theorem allows us to solve Poisson's equation in an intuitive manner such that we make a reasonable guess at the answer that satisfies both Poisson's equation and the boundary condition to obtain a unique solution.

To verify the uniqueness theorem, suppose V_1 and V_2 are two solutions of Poisson's equation in a region $\mathscr{R}$ with volume $\mathcal{V}$ and bounding surface $\mathcal{S}$ such that

$$\nabla^2 V_1(\mathbf{r}) = -\frac{\rho_v}{\varepsilon} \tag{3.125a}$$

$$\nabla^2 V_2(\mathbf{r}) = -\frac{\rho_v}{\varepsilon} \tag{3.125b}$$

If V_1 and V_2 satisfy the same boundary condition, it follows that

$$V_1(\mathbf{r}_s) = V_2(\mathbf{r}_s) = V_s \tag{3.125c}$$

where $\mathbf{r}_s$ is the position vector of a point on the boundary surface, and V_s represents the boundary value.

Next, we consider a scalar field defined as

$$\tilde{V}(\mathbf{r}) \equiv V_1(\mathbf{r}) - V_2(\mathbf{r}) \tag{3.126a}$$

Based on Eqs. (3.125a)–(3.125c), we can construct another boundary value problem, for which $\tilde{V}(\mathbf{r})$ is a unique solution, that is,

$$\nabla^2 \tilde{V}(\mathbf{r}) = 0 \tag{3.126b}$$

$$\tilde{V}(\mathbf{r}_s) = 0 \tag{3.126c}$$

Substituting $\tilde{V}(\mathbf{r})$ and $\nabla \tilde{V}(\mathbf{r})$ for A and $\mathbf{V}$ in $\nabla \cdot (A\mathbf{V}) = A(\nabla \cdot \mathbf{V}) + \mathbf{V} \cdot (\nabla A)$, which is a vector identity, we obtain

$$\nabla \cdot (\tilde{V} \, \nabla \tilde{V}) = \tilde{V}(\nabla \cdot \nabla \tilde{V}) + (\nabla \tilde{V}) \cdot (\nabla \tilde{V})$$

Integrating both sides of this over volume $\mathcal{V}$ gives

$$\int_{\mathcal{V}} \nabla \cdot (\tilde{V} \, \nabla \tilde{V}) \, dv = \int_{\mathcal{V}} \tilde{V}(\nabla \cdot \nabla \tilde{V}) \, dv + \int_{\mathcal{V}} (\nabla \tilde{V}) \cdot (\nabla \tilde{V}) \, dv \tag{3.127}$$

The first term on the right side of Eq. (3.127) vanishes because of the zero Laplacian of $\tilde{V}$, as shown in Eq. (3.126b). Applying the divergence theorem and the boundary condition in Eq. (3.126c) to the left-hand side of Eq. (3.127) gives

$$\int_V \nabla \cdot (\tilde{V}\,\nabla\tilde{V})\,dv = \oint_S (\tilde{V}\,\nabla\tilde{V}) \cdot d\mathbf{s} = 0$$

Therefore, Eq. (3.127) is reduced to

$$\int_V (\nabla\tilde{V}) \cdot (\nabla\tilde{V})\,dv = 0$$

As the integrand is always positive, $\nabla\tilde{V} = 0$ at all points in V. Thus, $\tilde{V}(\mathbf{r})$ is a constant in the given region:

$$\tilde{V}(\mathbf{r}) = V_1(\mathbf{r}) - V_2(\mathbf{r}) = C \tag{3.128}$$

where C is a constant. Because Eq. (3.128) is true for all points in the given region, including the boundary, it is evident from Eq. (3.126c) that $C = 0$. Thus,

$$V_1(\mathbf{r}) = V_2(\mathbf{r})$$

This result confirms the uniqueness theorem.

As the uniqueness theorem holds true irrespective of the volume charge in a given region, the uniqueness theorem for Laplace's equation was also verified. If a solution to Laplace's equation can be obtained through trial and error to satisfy the given boundary condition, then it is the only solution in the given region.

Exercise 3.44
Two infinite parallel plates at $y = \pm a$ are both maintained at $V = 0$. The air gap is partitioned using a strip with a width slightly less than $2a$ in the $x = 0$ plane, which is maintained at $V = V_o$. Show that the following $V(x, y)$ can be a unique solution in the air gap for $x > 0$:

$$V(x, y) = \sum_{n=1,3,5..}^{\infty} c_n e^{-n\pi x/2a} \cos(n\pi y/2a).$$

Ans. $\nabla^2 V = 0$ in the gap and V satisfies the boundary condition $V(x, y = \pm a) = 0$. For $x = 0$, c_n are the Fourier coefficients of a rectangular function of width $2a$ and height V_o. Furthermore, $V = 0$ for $x = \infty$. Thus, the given V satisfies all boundary conditions.

3.8.3   Examples of Boundary-Value Problems

Example 3.21 Two parallel conducting plates are separated by a distance d and maintained at $V = 0$ and $V = V_o$, as shown in Fig. 3.36. When the gap is filled with a volume charge of density $\rho_v = \rho_o \sin(\pi z/d)$ [C/m^3], determine (a) V in the space between the plates, and (b) ρ_s induced on the plates.

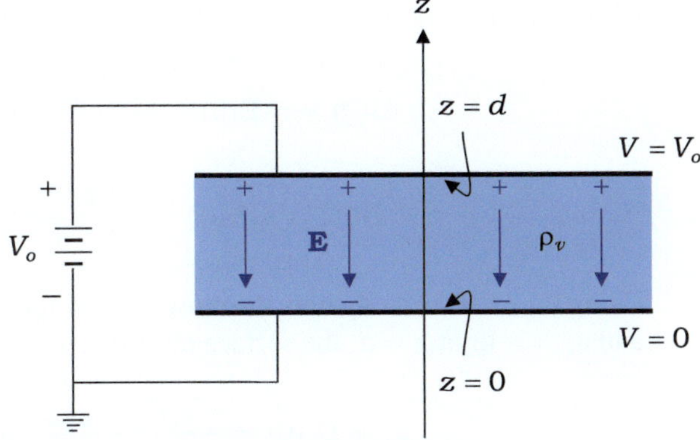

Fig. 3.36 Two parallel plates filled with volume charge

Solution

(a) By ignoring the fringing effects at the edges, the given system can be assumed to be a section of two infinite parallel plates over which V is independent of x and y. In this case, Poisson's equation is reduced to

$$\frac{d^2 V}{dz^2} = -\frac{\rho_o}{\varepsilon_0} \sin\left(\frac{\pi}{d}z\right)$$

Upon integrating both sides of this with respect to z twice, we obtain

$$V = \frac{\rho_o}{\varepsilon_0}\left(\frac{d}{\pi}\right)^2 \sin\left(\frac{\pi}{d}z\right) + c_1 z + c_2 \tag{3.129a}$$

where c_1 and c_2 are integration constants. Applying the boundary condition to Eq. (3.129a) results in

$$V(z = 0) = 0 = c_2$$

$$V(z = d) = V_o = c_1 d$$

Thus, V between the plates is obtained as

$$V = \frac{\rho_o}{\varepsilon_0}\left(\frac{d}{\pi}\right)^2 \sin\left(\frac{\pi}{d}z\right) + \frac{V_o}{d}z \tag{3.129b}$$

(b) From Eq. (3.129b), $\mathbf{E}$ between the plates is given by

$$\mathbf{E} = -\nabla V = -\left[\frac{\rho_o d}{\varepsilon_0 \pi}\cos\left(\frac{\pi}{d}z\right) + \frac{V_o}{d}\right]\mathbf{a}_z$$

On the conducting surfaces at $z = 0$ and $z = d$, the electric flux densities are

$$\mathbf{D}(0) = \varepsilon_0\,\mathbf{E}(0) = \varepsilon_0\left[-\frac{\rho_o d}{\varepsilon_0\,\pi} - \frac{V_o}{d}\right]\mathbf{a}_z$$

$$\mathbf{D}(d) = \varepsilon_0\,\mathbf{E}(d) = \varepsilon_0\left[\frac{\rho_o d}{\varepsilon_0\,\pi} - \frac{V_o}{d}\right]\mathbf{a}_z$$

Using Eq. (3.102), along with the outward unit normal vectors, $\mathbf{a}_n = \mathbf{a}_z$ at $z = 0$ and $\mathbf{a}_n = -\mathbf{a}_z$ at $z = d$, the surface charge densities are obtained as

$$\rho_s = D_z(0) = -\frac{\rho_o d}{\pi} - \frac{\varepsilon_0 V_o}{d} \quad (\text{at } z = 0) \tag{3.129c}$$

$$\rho_s = -D_z(d) = -\frac{\rho_o d}{\pi} + \frac{\varepsilon_0 V_o}{d} \quad (\text{at } z = d) \tag{3.129d}$$

The surface charge densities in Eqs. (3.129c) and (3.129d) are unequal in magnitude. The term $\pm\varepsilon_0 V_o/d$ is associated with the potential difference V_o. Meanwhile, the additional surface charge $-\rho_o d/\pi$ cancels the effect of the gap charge of $2\rho_o d/\pi$ [C] and ensures that $\mathbf{E} = 0$ in the interior of the conducting plate.

Example 3.22 Two semi-infinite conducting planes take wedge form, making angles ϕ_1 and ϕ_2 with the x-axis. They are separated by a very small insulating air gap and maintained at potentials V_1 and V_2, as shown in Fig. 3.37. Determine V in regions (a) $\phi_1 \le \phi \le \phi_2$, and (b) $\phi_2 \le \phi \le (\phi_1 + 2\pi)$.

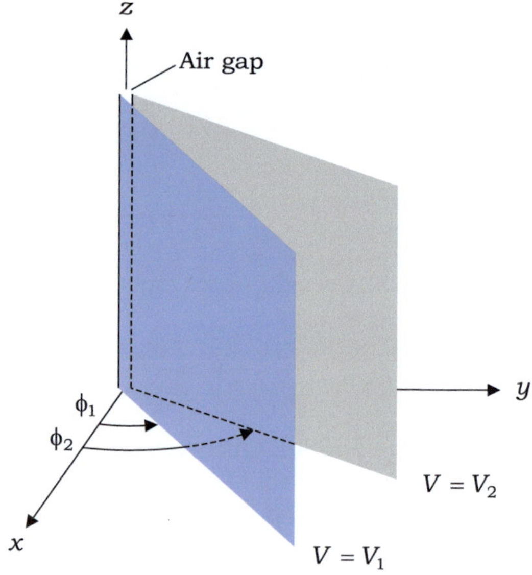

Fig. 3.37 Two semi-infinite conducting plates in wedge form

Solution

In view of the fact that the $\phi = \phi_1$ and $\phi = \phi_2$ planes are equipotential surfaces, our first guess at the solution is that any half-plane with constant ϕ is an equipotential surface. For this trial solution, V is independent of ρ and z. Thus, Laplace's equation in cylindrical coordinates is reduced to

$$\nabla^2 V = \frac{1}{\rho^2}\frac{\partial^2 V}{\partial \phi^2} = 0$$

Upon integrating both sides with respect to ϕ twice, we obtain

$$V = c_1 \phi + c_2 \quad (c_1 \text{ and } c_2, \text{ constants}) \tag{3.130a}$$

(a) In the region $\phi_1 \leq \phi \leq \phi_2$, boundary conditions are applied to Eq. (3.130a) to obtain

$$V(\phi_1) = V_1 = c_1\phi_1 + c_2$$

$$V(\phi_2) = V_2 = c_1\phi_2 + c_2$$

Solving these for c_1 and c_2 yields

$$c_1 = \frac{V_1 - V_2}{\phi_1 - \phi_2} \text{ and } c_2 = V_1 - \frac{V_1 - V_2}{\phi_1 - \phi_2}\phi_1$$

Thus,

$$V = \frac{V_1 - V_2}{\phi_1 - \phi_2}(\phi - \phi_1) + V_1 \quad (\phi_1 \leq \phi \leq \phi_2) \tag{3.130b}$$

(b) In the region $\phi_2 \leq \phi \leq (\phi_1 + 2\pi)$, boundary conditions are applied to the general solution given in Eq. (3.130a) to obtain

$$V(\phi_2) = V_2 = c_1\phi_2 + c_2$$

$$V(\phi_1 + 2\pi) = V_1 = c_1(\phi_1 + 2\pi) + c_2$$

Solving these for c_1 and c_2 yields

$$c_1 = \frac{V_2 - V_1}{\phi_2 - \phi_1 - 2\pi} \text{ and } c_2 = V_2 - \frac{V_2 - V_1}{\phi_2 - \phi_1 - 2\pi}\phi_2$$

Thus,

$$V = \frac{V_2 - V_1}{\phi_2 - \phi_1 - 2\pi}(\phi - \phi_2) + V_2 \quad (\phi_2 \le \phi \le (\phi_1 + 2\pi)) \tag{3.130c}$$

The electric potentials in Eqs. (3.130b) and (3.130c) are unique solutions in their respective regions because they satisfy Laplace's equation and the boundary conditions.

Example 3.23 An uncharged conducting sphere of radius a is introduced into a uniform electric field $\mathbf{E}_o = E_o \mathbf{a}_z$, as shown in Fig. 3.38. Surface charge ρ_s should be induced on the sphere to ensure that $\mathbf{E} = 0$ in the interior. At a distance, ρ_s may appear as an electric dipole that produces V, as expressed in Eq. (3.61d).

(a) Form a trial solution for V and find a unique solution.
(b) Determine ρ_s on the sphere.

Solution

(a) We take as a trial solution

$$V = -E_o R \cos\theta + c_1 \cos\theta / R^2 \tag{3.131a}$$

where the first term on the right-hand side is the electric potential due to the uniform external field ($z = R \cos\theta$), and the second term is the electric potential due to an equivalent electric dipole with a constant c_1. This trial solution satisfies Laplace's equation for spherical coordinates.

The conductor surface is an equipotential surface. Applying the boundary condition $V(R = a) = V_o$ (a constant) to Eq. (3.131a) yields

$$c_1 = V_o \frac{a^2}{\cos\theta} + E_o a^3$$

Therefore, the electric potential outside the sphere is

Fig. 3.38 Uncharged spherical conductor placed in an otherwise uniform electric field

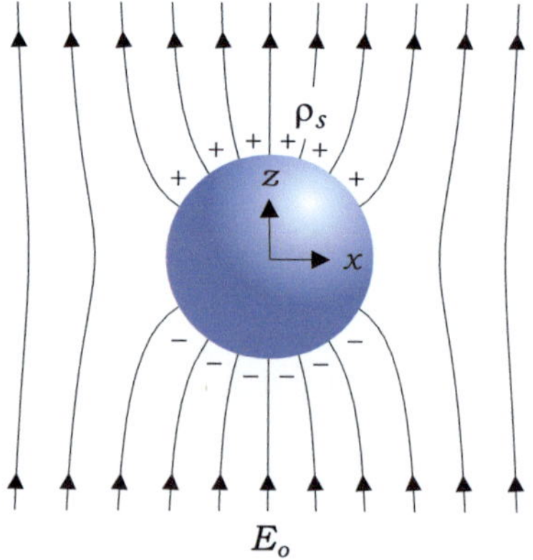

$$V = -E_o R \cos\theta + E_o a^3 \frac{\cos\theta}{R^2} + V_o \frac{a^2}{R^2}$$

Because the sphere has no net charge, $V(R = a) = 0$, which leads to $V_o = 0$. Therefore,

$$V = -E_o R \cos\theta + E_o a^3 \frac{\cos\theta}{R^2} \qquad (3.131b)$$

This is the only solution to the problem because it satisfies Laplace's equation and boundary conditions.

(b) At $R = a$, the electric field intensity is

$$\mathbf{E}(R = a) = -\nabla V|_{R=a} = 3E_o \cos\theta \, \mathbf{a}_R$$

The surface charge density is thus obtained as

$$\rho_s = D_n = 3\varepsilon_0 E_o \cos\theta \qquad (3.131c)$$

It is evident from Eq. (3.85) in Example 3.15 that ρ_s in Eq. (3.131c) produces a uniform field $\mathbf{E} = -E_o \mathbf{a}_z$ inside the sphere, which completely cancels out the external field. However, this modifies the external field outside the sphere, as shown in Fig. 3.38.

3.8.4 The Method of Images

The electric field lines of the electric dipole are symmetrical about an infinite plane in the middle (Fig. 3.39a), and are always perpendicular to the plane, making the plane an equipotential surface. As the middle plane extends to infinity, its electric potential becomes zero, which can be formally justified using Eq. (3.62). Let us digress briefly and consider a boundary value problem, in which the point charge $+q$ is located at a distance $d/2$ above an infinite conducting plane maintained at zero potential (Fig. 3.39b). According to potential theory, the electric field lines of the charge should terminate on the conductor at right angles. Here, we note that the behavior of the electric field lines on the conductor is exactly the same as that on the middle plane of the electric dipole (Fig. 3.39a). Because the solution to Laplace's equation is the only correct answer in the given region, if it satisfies the given boundary conditions, the electric field lines in Fig. 3.39b must be the same as those above the middle plane, as shown in Fig. 3.39a.

Instead of directly solving Laplace's equation, we can determine $\mathbf{E}$ and V of a charge above an infinite grounded conducting plane by simulating the boundary condition with an image charge. In other words, the field in a given region is computed from the combination of the charge and its image charge, with the conducting plane

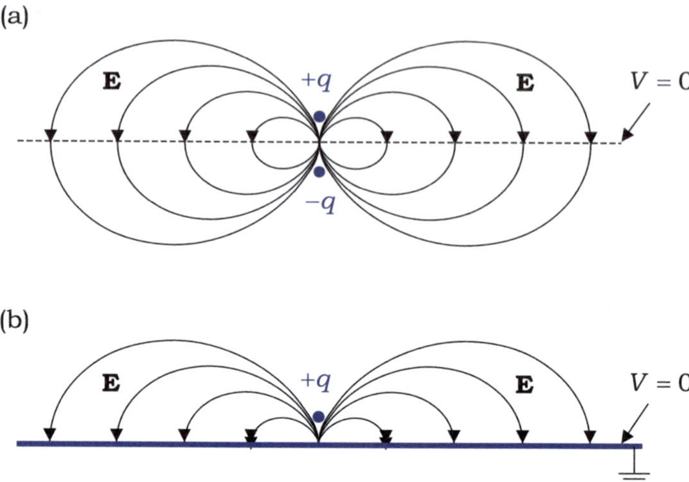

Fig. 3.39 **a** Dipole electric field. **b** Charge above the grounded conductor

removed. This method is called the ***method of images***. This can be extended to the continuous distribution of charges and equipotential bodies by invoking the principle of superposition (Fig. 3.40). The symmetrical charge distribution in the method of images greatly facilitates the solutions for $\mathbf{E}$ and V. It is important to note that the method of images is used to find the field only in the region where image charges are *not* present.

Example 3.24 A point charge q is at distance d above the $z = 0$ plane, which is perfectly conducting and is maintained at $V = 0$. Determine the surface charge density on the plane.

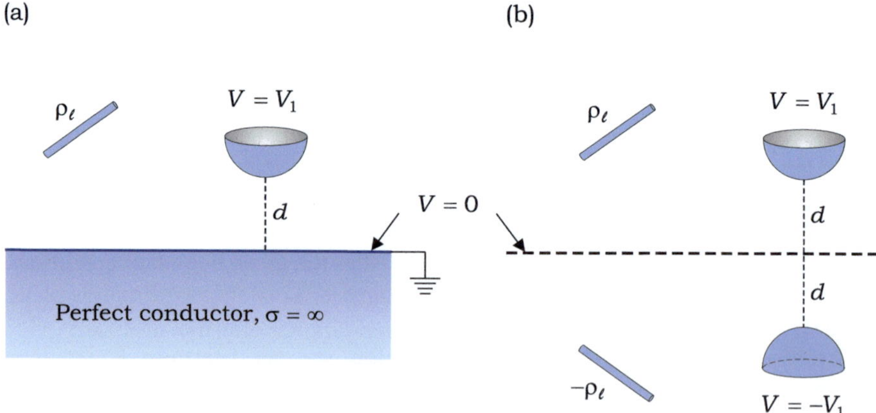

Fig. 3.40 **a** Boundary-value problem. **b** Method of images

Solution

Assuming an image charge $-q$ at $z = -d$, the electric potential in the region $z \geq 0$ is obtained as

$$V(x, y, z) = \frac{q}{4\pi\varepsilon_0}\left[\frac{1}{\sqrt{x^2 + y^2 + (z - d)^2}} - \frac{1}{\sqrt{x^2 + y^2 + (z + d)^2}}\right]$$

Thus, the surface charge density on the $z = 0$ plane is obtained as

$$\rho_s(x, y, 0) = -\varepsilon_0 \left.\frac{\partial V}{\partial z}\right|_{z=0} = -\frac{q}{2\pi}\frac{d}{(x^2 + y^2 + d^2)^{3/2}}$$

This can be rearranged as

$$\rho_s(x, y, 0) = 2\left[\frac{q}{4\pi(x^2 + y^2 + d^2)}\left\{\frac{x\,\mathbf{a}_x + y\,\mathbf{a}_y - d\,\mathbf{a}_z}{(x^2 + y^2 + d^2)^{1/2}}\right\}\right]\cdot\mathbf{a}_z$$
$$= 2\mathbf{D}(x, y, 0)\cdot\mathbf{a}_z \tag{3.132}$$

where the term in the braces is a unit vector of the distance vector from charge q to a point on the plane, the term in brackets is the electric flux density on the plane due to q alone, and $\mathbf{a}_z$ represents an outward unit normal to the conducting plane. In short, the surface charge density is twice the outward electric flux density on the conducting surface.

3.8.4.1 Line Images

Two infinitely long parallel line charges of opposite polarity produce electric field lines that are symmetrical about the middle plane. In this case, equipotential surfaces take the form of a circular cylinder, with their axes well separated from each other, but with their radii given such that equipotential surfaces of positive V enclose the positive line charge, and those of negative V enclose the negative line charge. These equipotential surfaces are very useful for solving boundary value problems involving cylindrical conductors, as well as straight-line conductors.

Consider two line charges of uniform densities ρ_ℓ and $-\rho_\ell$ that are a distance of $2d$ apart, as shown in Fig. 3.41. The electric potential at point p can be obtained using Eq. (3.52) as

$$V = \frac{\rho_\ell}{2\pi\varepsilon_0}\left[\ln\frac{\mathcal{R}_{1o}}{\mathcal{R}_1} - \ln\frac{\mathcal{R}_{2o}}{\mathcal{R}_2}\right] \tag{3.133}$$

where $\mathcal{R}_1$ and $\mathcal{R}_2$ are the perpendicular distances from field point p to the line charges, and $\mathcal{R}_{1o}$ and $\mathcal{R}_{2o}$ are the perpendicular distances from the zero-reference point to the line charges. We can set $\mathcal{R}_{1o} = \mathcal{R}_{2o}$ by assuming that the middle plane at $x = 0$ has

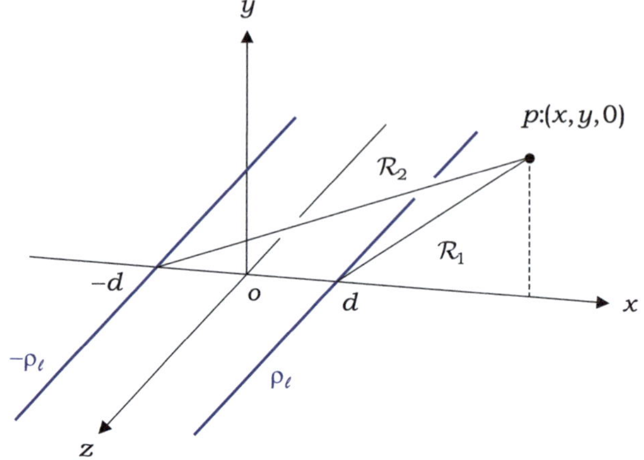

Fig. 3.41 Two parallel line charges of opposite polarity

a zero potential. Then, Eq. (3.133) becomes

$$V = \frac{\rho_\ell}{2\pi\varepsilon_0} \ln \frac{\mathcal{R}_2}{\mathcal{R}_1}$$

By expressing $\mathcal{R}_1$ and $\mathcal{R}_2$ in Cartesian coordinates, we have

$$V = \frac{\rho_\ell}{4\pi\varepsilon_0} \ln\left[\frac{(x+d)^2 + y^2}{(x-d)^2 + y^2}\right] \tag{3.134}$$

Let us rearrange Eq. (3.134) to define parameter K as follows:

$$\exp\left[\frac{4\pi\varepsilon_0 V}{\rho_\ell}\right] = \frac{(x+d)^2 + y^2}{(x-d)^2 + y^2} \equiv K \tag{3.135}$$

The second equality of Eq. (3.135) can be expressed as

$$(x+d)^2 + y^2 = K\left[(x-d)^2 + y^2\right]$$

Rearranging this leads to an expression for the equipotential surface of electric potential V, that is,

$$(x-\alpha)^2 + y^2 = \beta^2 \tag{3.136a}$$

with

$$\boxed{\alpha = d\,\frac{K+1}{K-1}} \tag{3.136b}$$

$$\boxed{\beta = \frac{2d\sqrt{K}}{K-1}} \tag{3.136c}$$

$$\boxed{K = \exp\left[\frac{4\pi\varepsilon_0 V}{\rho_\ell}\right]} \tag{3.136d}$$

From Eq. (3.136a), we can see that the equipotential surface of potential V is given by an infinitely long cylinder of radius β, with the axis positioned at $x = \alpha$.

From Eq. (3.136), it follows that

$$d^2 = \alpha^2 - \beta^2 \tag{3.137a}$$

$$\sqrt{K} = \frac{\alpha + d}{\beta} \tag{3.137b}$$

These relations are useful for solving certain boundary-value problems. For example, if two parallel conducting cylinders of radius β are separated by an axis-to-axis distance of 2α and maintained at potentials $-V$ and $+V$, the separation of the two equivalent line charges $2d$ can be determined using Eq. (3.137a), and the equivalent line charge density ρ_ℓ from Eqs. (3.137b) and (3.136d). Such line charges produce equipotential surfaces, two of which fit the given cylinders and satisfy boundary conditions. It is then straightforward to determine $\mathbf{E}$ and V outside the cylinders by using the two equivalent line charges.

Setting $V = 0$ in Eq. (3.136d) yields $K = 1$, and thus, $\alpha = \beta = \infty$, implying that the middle plane at $x = 0$ is at zero potential. Thus, the initial assumption is justified. For $V > 0$ (or $K > 1$), equipotential surfaces are formed in region $x > 0$ enclosing the positive line charge. Meanwhile, equipotential surfaces of $V < 0$ are formed in region $x < 0$, all enclosing the negative line charge.

When two infinitely long line charges of uniform density $\rho_\ell = \pm 1\,[\mathrm{nC/m}]$ are separated by 2 [m] in free space, equipotential surfaces are obtained using Eq. (3.136a) and drawn to scale in the xy-plane, as shown in Fig. 3.42.

If a boundary-value problem involves conductors in the form of infinite lines, planes, and/or cylinders, as illustrated in Fig. 3.42, Eq. (3.136) can be used to best advantage.

Example 3.25 For a grounded conducting plane coincident with the $x = 0$ plane, an infinite conducting cylinder of radius $\beta_1 = 0.1\,[\mathrm{m}]$ is located at distance $\alpha_1 = 0.125\,[\mathrm{m}]$, as shown in Fig. 3.43a. If the cylinder is at potential $V_1 = 40\,[\mathrm{V}]$, determine V outside the conductor.

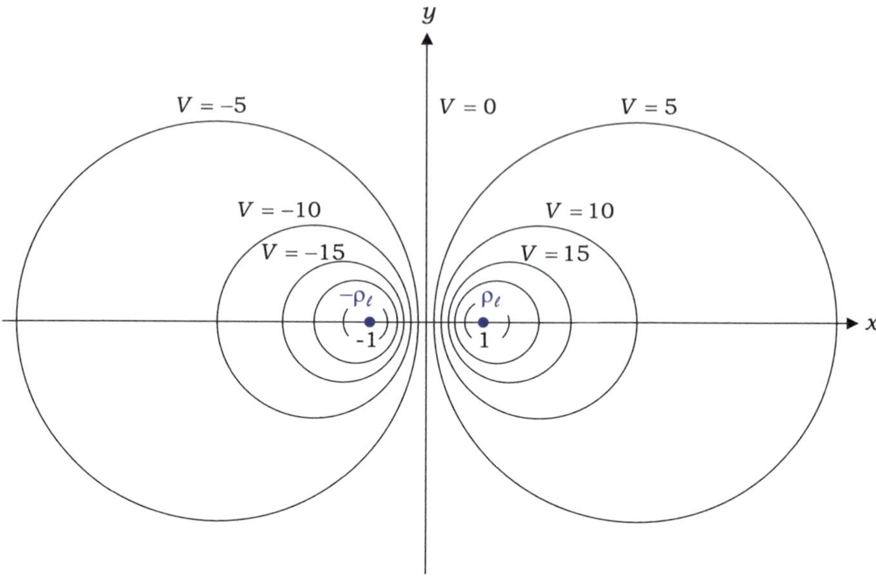

Fig. 3.42 Equipotential surfaces due to two parallel line charges

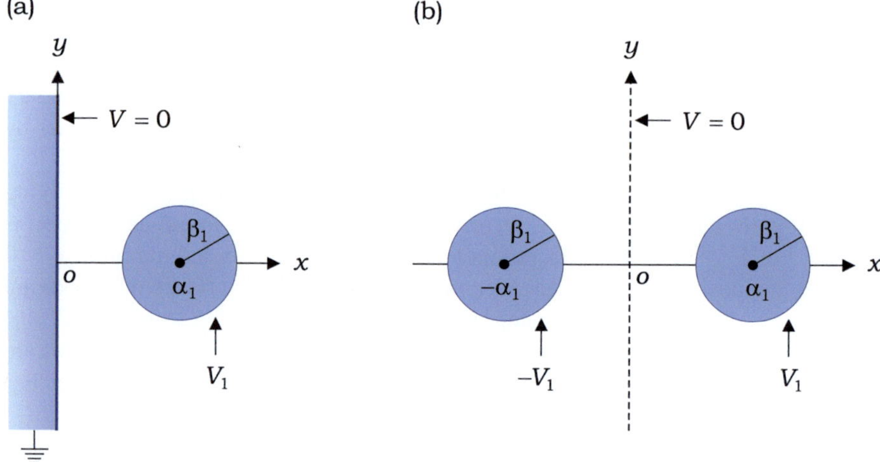

Fig. 3.43   **a** Infinite conducting cylinder parallel to grounded conducting plane. **b** Method of images

Solution

The grounded plane is replaced with the image of the cylinder maintained at potential $-V_1$, as shown in Fig. 3.43b. We look for equivalent line charges, whose equipotential surfaces have the same shape as the cylinder surfaces.

From Eq. (3.137a), the half-distance between two equivalent line charges is

$$d = \sqrt{\alpha_1^2 - \beta_1^2} = 0.75 \,[\text{m}]$$

From Eq. (3.137b), $K = 4$, and from Eq. (3.136d),

$$\rho_\ell = \frac{4\pi\varepsilon_0 V_1}{\ln K_1} = 3.21 \,[\text{nC/m}]$$

The electric potential outside the cylinder is obtained from Eq. (3.134) as

$$V = 28.85\left[\ln \frac{(x + d)^2 + y^2}{(x - d)^2 + y^2}\right] \,[\text{V}] \qquad (3.138)$$

Surface charges are induced on the cylinder and grounded conductor, and their values can be obtained using Eq. (3.138). These charges cause the electric field in the conductor to vanish.

Example 3.26 A line charge of density $-3.0 \,[\text{nC/m}]$ and a conducting cylinder with a net charge of $3.0 \,[\text{nC/m}]$ are parallel to each other, as shown in Fig. 3.44. If the cylinder radius is $\beta = 0.1 \,[\text{m}]$ and the separation is $\ell = 0.5 \,[\text{m}]$, determine the location of the image line charge, and then V outside the cylinder.

Solution

In view of the fact that the two conductors carry an equal charge but opposite polarity, and the cylinder is an equipotential object, we approach this problem with the method of images. The cylinder is replaced with an equivalent line charge of density ρ_ℓ placed at $x = d$, which is yet to be determined.

By setting $\alpha + d = 0.5$ and $\beta = 0.1$ in Eq. (3.137b), we obtain $K = 25$. Using this in Eq. (3.136b), we obtain $\alpha = 0.26$ and $d = 0.24$. Thus, the image line charge is at a distance 0.48 [m] from the line charge.

The electric potential outside the cylinder is obtained using Eq. (3.134) as

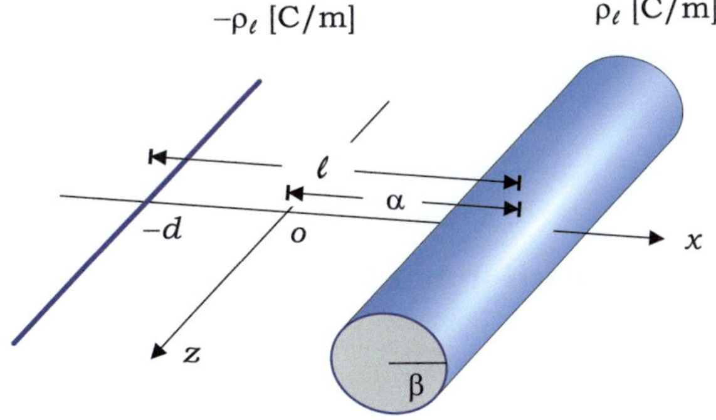

Fig. 3.44 Long line charge in parallel with a long charged cylinder

$$V = 26.96\left[\ln \frac{(x+0.24)^2 + y^2}{(x-0.24)^2 + y^2}\right] \qquad (3.139)$$

As a check, we calculate the total flux of ρ_ℓ crossing the $x = 0$ plane per unit length along the z-axis, such that

$$\Psi = \int_{-\infty}^{\infty} \varepsilon_0(-\nabla V|_{x=0}) \cdot (-\mathbf{a}_x)\, dy$$

$$= 26.96\varepsilon_0 \int_{-\infty}^{\infty} \frac{0.96}{(0.24)^2 + y^2}\, dy = 3.0\,[\text{nC/m}]$$

The total flux is equal to the line charge density, ρ_ℓ.

Exercise 3.45

Point charge q is at distance d from the grounded conducting plane, as shown in Fig. 3.39b. Find the (a) force on q, and (b) potential energy of the system.

Ans. (a) $F_e = q^2/16\pi\varepsilon_0 d^2$, (b) $W_E = -q^2/16\pi\varepsilon_0 d$.

Exercise 3.46

Does any of the electric field lines shown in Fig. 3.42 start at ρ_ℓ and follow a circular path to $-\rho_\ell$?

Ans. Yes. A circle of radius d centered at the origin crosses equipotential surfaces at right angles, as shown in Eq. (3.137a).

Review Questions

RQ 3.50	Write Poisson's and Laplace's equations.	[(3.122)(3.123)]
RQ 3.51	Explain why the usefulness of Poisson's and Laplace's equations is limited to simple media.	[(3.121)]
RQ 3.52	Write Laplace's equation for cylindrical and spherical systems.	[(3.124b,c)]
RQ 3.53	State the uniqueness theorem for the electric potential.	[(3.125c)]
RQ 3.54	How many independent boundary conditions are needed for a unique determination of V in a one-dimensional space?	[(3.129a)]
RQ 3.55	What are the underlying assumptions for the method of images?	[Fig. 3.39]

3.9   Capacitor and Capacitance

Any two conducting objects can form a *capacitor* regardless of their size and shape, if they are separated by a dielectric. A capacitor can store energy in the electric field induced in the dielectric by charges accumulated on the conductors. Let us consider two electrically neutral conductors embedded in a dielectric with permittivity ε, as shown in Fig. 3.45. In the thought experiment, we took some free electrons from conductor 1 and transferred them to conductor 2. Consequently, conductor 2 is negatively charged and conductor 1 is positively charged because of the ionized host atoms. The two conductors now have net charges of equal amounts, but opposite polarities. In practical situations, charge separation can be accomplished using a DC voltage source connected to conductors.

Excess charges should always reside on the conductor surface such that there is no electric field inside. Conductors are thus equipotential objects. The net charges give rise to an electric field $\mathbf{E}$ in the dielectric and a potential difference V between the conductors. Further charge separation certainly increases the surface charge density but leaves its spatial distribution unchanged. Otherwise, any change in the charge distribution directly leads to a nonzero electric field in the conductor. It follows that if the net charge is increased by a factor of k, then ρ_s, $\mathbf{E}$, and V increase everywhere by the same factor. This can be explained by the linear relationships $\rho_s = D_n = \varepsilon E_n$ and $\mathbf{E} = -\nabla V$. Consequently, the ratio of the net charge to the potential difference between the two conductors is constant for a given capacitor.

Charge separation is performed against an electric field already established in the dielectric. Thus, energy was expended for charge accumulation on the conductors. The work done is stored as the potential energy of the capacitor. For a fixed net charge on the conductor, the electric field in the space between the conductors strongly depends on the dielectric properties characterized by ε and the geometry of the capacitor, such as the size, shape, and separation of the conductors.

The ***capacitance*** of a capacitor is defined as the amount of net charge on the conductor required to build a potential difference of $1V$ between conductors. That

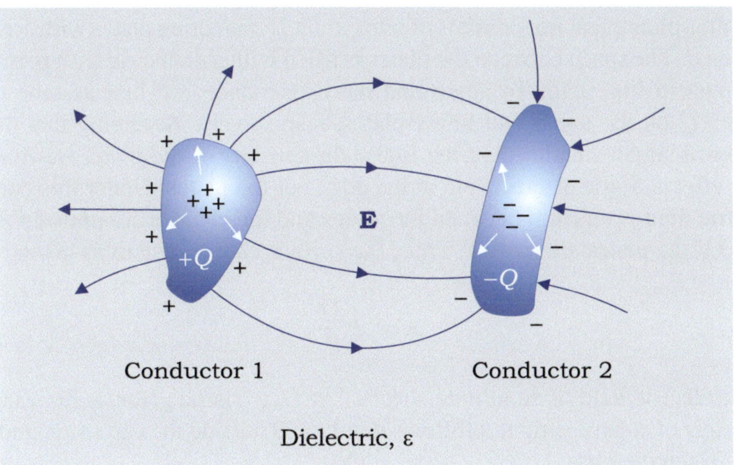

Fig. 3.45 Two conductors embedded in a dielectric

is,

$$\boxed{C = \frac{Q}{V}} \quad [\text{F}] \tag{3.140}$$

The capacitance is measured in farads [F], which is equivalent to the coulombs per volt. Again, it is important to note that the capacitance is independent of the total charge Q and the potential difference V because of the linear relationship between Q and V.

The capacitance between the two conductors can be obtained using Eq. (3.140) and taking the following steps:

1.   Assume that charges $+Q$ and $-Q$ are on two conductors.
2.   Choose an appropriate coordinate system.
3.   Find $\mathbf{E}$ of Q using Coulomb's law, Gauss's law, or other methods.
4.   Calculate V from the negative line integral of $\mathbf{E}$.
5.   Determine C from Q/V.

Alternatively,

1.   Assume that the two conductors are at electric potentials V_1 and V_2.
2.   Solve Laplace's equation for V between the conductors.
3.   Find $\mathbf{E}$ from the negative gradient of V.
4.   Calculate the surface charge density, $\rho_s = \varepsilon E_n$, and total charge Q.
5.   Determine C from $Q/(V_1 - V_2)$.

3.9.1   Parallel-Plate Capacitor

The parallel-plate capacitor consists of two parallel conducting plates with area S and separation d. The space between the plates is filled with a dielectric with permittivity ε, as shown in Fig. 3.46. To determine the capacitance, we first assume charges $+Q$ and $-Q$ on the upper and lower plates, respectively. Assuming that the plate separation is much smaller than the linear dimensions of the plate, we ignore the fringing effects of the electric field at the edges of the plates. Under this condition, the electric field is uniform between the plates and the charges are uniform over the plates, as if the plates are infinite. Thus, the surface charge density is given by

$$\rho_s = \pm \frac{Q}{S}$$

From the electric field of an infinite sheet of surface charge given in Eq. (3.38) and the principle of superposition, it follows that $\mathbf{E} = 0$ outside the capacitor, and, in the space between the plates,

$$\mathbf{E} = \frac{-\rho_s}{\varepsilon} \mathbf{a}_z \tag{3.141}$$

Next, the potential difference between the plates is calculated as

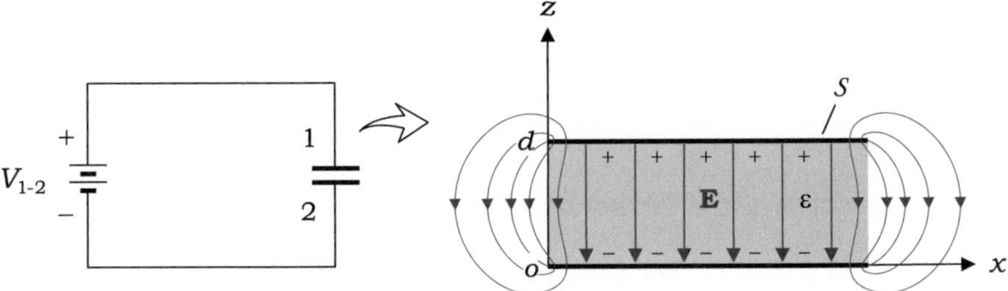

Fig. 3.46 Parallel-plate capacitor

$$V_{1-2} = -\int_{2}^{1} \mathbf{E} \cdot d\mathbf{l} = \frac{\rho_s}{\varepsilon} d$$

The capacitance of the parallel-plate capacitor is therefore

$$\boxed{C = \frac{Q}{V_{1-2}} = \frac{\varepsilon S}{d}} \qquad [\text{F}] \qquad (3.142)$$

The capacitance is directly proportional to the surface area and permittivity of the dielectric but inversely proportional to the plate separation.

Upon inserting Eq. (3.141) into Eq. (3.112), the electrostatic potential energy stored in the capacitor can be obtained as

$$W_E = \frac{1}{2}\int_{V} \mathbf{D} \cdot \mathbf{E}\, dv = \frac{\varepsilon}{2}\int_{V} E^2 dv$$

$$= \frac{\rho_s^2}{2\varepsilon} S d = \frac{1}{2}\left(\frac{\varepsilon S}{d}\right)\left[\frac{\rho_s d}{\varepsilon}\right]^2$$

where the term in parentheses is the capacitance, and the term in brackets is the potential difference across the capacitor. Thus, the electrostatic energy can be expressed as

$$\boxed{W_E = \frac{1}{2}C V^2} \qquad [\text{J}] \qquad (3.143)$$

The stored energy is proportional to the capacitance and square of the potential difference.

If the conducting plates are enlarged while the total charge remains the same, the two quantities ρ_s and $\mathbf{E}$ are reduced by the same factor. In this case, further charge separation is required for a potential difference of $1V$ across the capacitor; thus, the capacitance increases. Next, if the dielectric is replaced by one with a larger ε, $\mathbf{E}$ in the dielectric is reduced, and additional charge separation is required for a potential difference of $1V$. Thus, the capacitance increases.

Even if the plate separation is reduced, there is no change in $\mathbf{E}$ in the dielectric; however, the integration path for V is shortened. Thus, more charge separation is required for a build-up of $1V$, which increases capacitance.

3.9.1.1 The Principle of Virtual Displacement

Coulomb's law allows for the determination of the electrostatic force acting on a charged body caused by another body. The law of energy conservation can provide another method for computing electrostatic force, called *the principle of virtual displacement*.

Let us consider a parallel-plate capacitor with surface area S and an air gap of thickness d, as shown in Fig. 3.46. The battery is disconnected when the capacitor is charged to a potential difference of V_o with charges $+Q$ and $-Q$ on the upper and lower plates, respectively. The electric force acting on the upper plate is denoted as $\mathbf{F}$. If the upper plate is displaced by a differential distance $d\mathbf{l}$ because of $\mathbf{F}$, called *virtual displacement*, the differential work done by the system is

$$-dW_E = \mathbf{F} \cdot d\mathbf{l} \tag{3.144}$$

where the negative sign denotes that work has been done by the system at the expense of the electrostatic energy stored in the capacitor. From calculus, the differential of W_E can be expressed in terms of the gradient of W_E such as

$$dW_E = \nabla W_E \cdot d\mathbf{l} \tag{3.145}$$

Comparing Eqs. (3.144) and (3.145) leads to

$$\boxed{\mathbf{F} = -\nabla W_E} \tag{3.146}$$

The force on the upper plate is equal to the negative gradient of the electrostatic energy stored in the capacitor. Inserting Eq. (3.143) into Eq. (3.146) and using $Q = CV$, the force on the upper plate is calculated as

$$\mathbf{F} = -\frac{dW_E}{dz}\,\mathbf{a}_z = \frac{Q^2}{2C^2}\frac{dC}{dz}\,\mathbf{a}_z$$
$$= -\frac{V_o^2}{2}\frac{\varepsilon_0 S}{d^2}\,\mathbf{a}_z \tag{3.147}$$

where charge Q is assumed to be constant.

Next, if the capacitor is connected to the battery with voltage V_o, then Eq. (3.144) should be modified to include the work done by the battery such that

$$-dW_E + V_o\,dQ = \mathbf{F} \cdot d\mathbf{l} \tag{3.148}$$

where dQ is the differential charge supplied by the battery to maintain potential V_o across the capacitor during plate displacement. Setting $\mathbf{F} \cdot d\mathbf{l} = F\,dz$ in Eq. (3.148), we obtain

$$F = -\frac{dW_E}{dz} + V_o\frac{dQ}{dz} = -\frac{1}{2}V_o^2\frac{dC}{dz} + V_o^2\frac{dC}{dz} \tag{3.149a}$$

$$= \frac{1}{2}V_o^2\frac{dC}{dz} \tag{3.149b}$$

where we have used Eqs. (3.142) and (3.143), and assumed that the potential difference V_o is a constant. Using $C = \varepsilon_0 S/d$, Eq. (3.149b) can be rewritten as

$$\mathbf{F} = -\frac{V_o^2}{2}\frac{\varepsilon_0 S}{d^2}\,\mathbf{a}_z \tag{3.150}$$

Two results in Eqs. (3.147) and (3.150) are identical. However, in the latter case, the work has been done by the battery.

3.9.2 Examples of Capacitors

Example 3.27 A coaxial capacitor of length ℓ consists of two concentric cylindrical conductors with radii a and b (Fig. 3.47). The space between the conductors is filled with a dielectric with permittivity ε. Ignoring fringing effects at the edges, determine the capacitance by assuming (a) charges $\pm Q$, and (b) potentials $\pm V_1$ on the conductors.

Solution

(a) We begin by assuming that charges $+Q$ and $-Q$ are on the inner and outer conductors, respectively. If the fringing effect is ignored, the $\mathbf{E}$ in the gap is the same as that if the capacitor is infinitely long. Based on symmetry considerations, we expect the electric field to be $\mathbf{E} = E_\rho(\rho)\,\mathbf{a}_\rho$ everywhere.

In region $a < \rho < b$, from Gauss's law,

Fig. 3.47 Coaxial capacitor

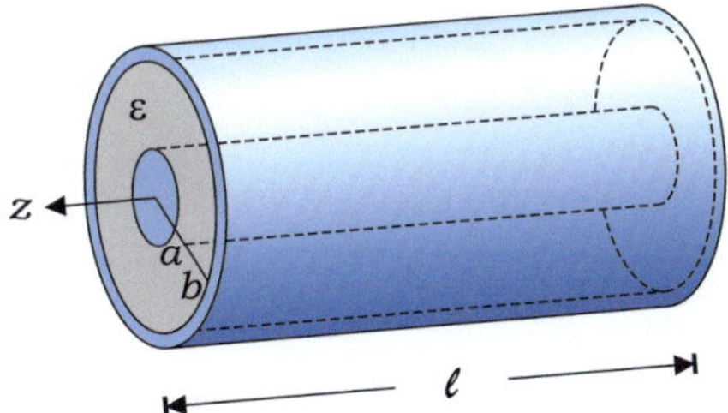

$$\varepsilon E_\rho 2\pi\rho\ell = Q$$

The electric field in the gap is thus

$$\mathbf{E} = E_\rho\,\mathbf{a}_\rho = \frac{Q}{2\pi\varepsilon\rho\ell}\mathbf{a}_\rho$$

The potential difference between the two conductors is given by

$$V_{a-b} = -\int_{\rho=b}^{\rho=a}\mathbf{E}\cdot d\mathbf{l} = -\int_{\rho=b}^{\rho=a}\frac{Q}{2\pi\varepsilon\rho\ell}\mathbf{a}_\rho\cdot\left(\mathbf{a}_\rho d\rho\right) = \frac{Q}{2\pi\varepsilon\ell}\ln\left(\frac{b}{a}\right)$$

Thus, the capacitance of the coaxial capacitor is

$$C = \frac{Q}{V_{a-b}} = \frac{2\pi\varepsilon\ell}{\ln(b/a)}\quad[\text{F}]\tag{3.151}$$

(b)   We assume potentials $+V_1$ and $-V_1$ on the inner and outer conductors, respectively. Subsequently, we form a trial solution in which any cylindrical surface centered on the z-axis is an equipotential surface. Subsequently, Laplace's equation in cylindrical coordinates is reduced to

$$\frac{1}{\rho}\frac{\partial}{\partial\rho}\left(\rho\frac{\partial V}{\partial\rho}\right) = 0$$

A general solution is written with constants of integration c_1 and c_2 as

$$V = c_1\ln\rho + c_2$$

By applying the boundary conditions for V to the general solution, we have

$$V_1 = c_1\ln a + c_2\quad(\text{at }\rho = a)$$

$$-V_1 = c_1\ln b + c_2\quad(\text{at }\rho = b)$$

Solving the above equations for c_1 and c_2 yields

$$V = V_1 - 2V_1\frac{\ln(\rho/a)}{\ln(b/a)}$$

Taking the negative gradient of V yields

$$\mathbf{E} = \frac{2V_1}{\ln(b/a)}\frac{1}{\rho}\mathbf{a}_\rho$$

$$\mathbf{D} = \varepsilon \mathbf{E} = \frac{2\varepsilon V_1}{\ln(b/a)} \frac{1}{\rho} \mathbf{a}_\rho$$

On the cylindrical surface at $\rho = a$, the surface charge density ρ_s and the total surface charge Q are obtained as follows:

$$\rho_s = D_\rho = \frac{2\varepsilon V_1}{\ln(b/a)} \frac{1}{a}$$

$$Q = 2\pi a \ell \rho_s = \frac{4\pi\varepsilon V_1 \ell}{\ln(b/a)}$$

Thus, the capacitance is

$$C = \frac{Q}{2V_1} = \frac{2\pi\varepsilon\ell}{\ln(b/a)} \quad [\mathrm{F}] \tag{3.152}$$

The results of Eqs. (3.151) and (3.152) are equal.

Example 3.28 Determine the capacitance of an isolated conducting sphere of radius a residing in free space.

Solution

First, we determine the capacitance of a spherical capacitor consisting of two concentric spheres with radii a and b (Fig. 3.48), and then make the outer sphere infinitely large. Assuming that charges $+Q$ and $-Q$ are on the spheres and considering spherical symmetry, we expect the electric field to be of the form $\mathbf{E} = E_R(R)\,\mathbf{a}_R$ everywhere.

In region $a < R < b$, by invoking Gauss's law, we obtain $D_R 4\pi R^2 = Q$. Thus,

$$\mathbf{E} = E_R \mathbf{a}_R = \frac{Q}{4\pi\varepsilon_0 R^2} \mathbf{a}_R$$

Fig. 3.48 Isolated sphere

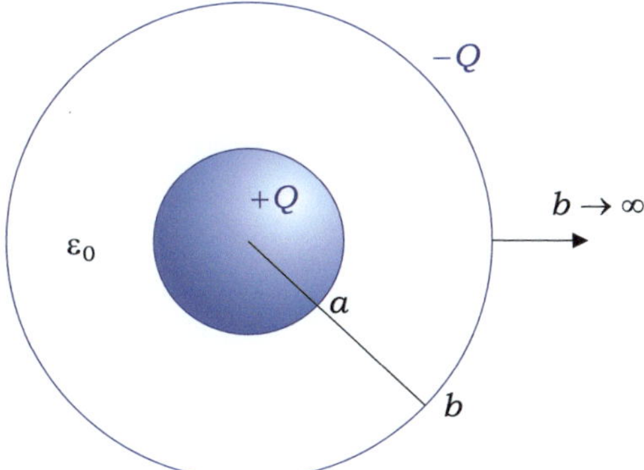

The potential difference between the two spheres is given by

$$V_{a-b} = -\int_{R=b}^{R=a} \mathbf{E} \cdot d\mathbf{l} = -\int_{R=b}^{R=a} \frac{Q}{4\pi\varepsilon_0 R^2}\mathbf{a}_R \cdot (\mathbf{a}_R \, dR) = \frac{Q}{4\pi\varepsilon_0}\left(\frac{1}{a} - \frac{1}{b}\right)$$

Thus, the capacitance of the spherical capacitor is

$$C = \frac{Q}{V_{a-b}} = 4\pi\varepsilon_0\left(\frac{1}{a} - \frac{1}{b}\right)^{-1} \quad [\text{F}] \tag{3.153a}$$

Taking the limit as $b \to \infty$, the capacitance of an isolated sphere is obtained as

$$C = 4\pi\varepsilon_0 a \quad [\text{F}] \tag{3.153b}$$

Example 3.29 Two parallel conducting wires of radius β_1 and length ℓ are separated by a center-to-center distance $2\alpha_1$ in free space (Fig. 3.49). Ignoring the edge effect, determine the capacitance between the wires.

Solution

Assuming potentials V_1 and $-V_1$ on the two wires, we solve this boundary value problem using the method of images depicted in Fig. 3.43b. From Eq. (3.136d), the equivalent line charge density is found to be

$$\rho_\ell = \frac{4\pi\varepsilon_0 V_1}{\ln K_1}$$

The total charge on the wire is

$$Q = \rho_\ell \ell = \frac{4\pi\varepsilon_0 V_1}{\ln K_1}\ell \tag{3.154a}$$

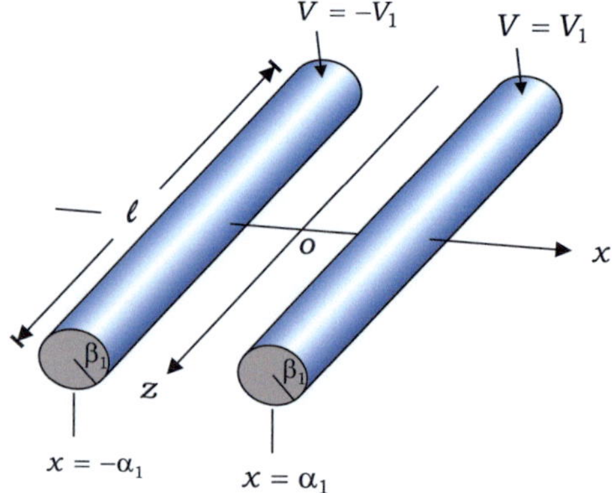

Fig. 3.49 Two parallel conducting wires

Inserting Eq. (3.137b), along with Eq. (3.137a), into Eq. (3.154a) gives

$$C = \frac{Q}{2V_1} = \frac{\pi\varepsilon_0\ell}{\ln\left[\alpha_1 + \sqrt{\alpha_1^2 - \beta_1^2}\right] - \ln\beta_1} \qquad (3.154b)$$

Using $\ln[a + (a^2 - 1)^{1/2}] = \cosh^{-1} a$ for $a > 1$, the capacitance is express as

$$C = \frac{\pi\varepsilon_0\ell}{\cosh^{-1}(\alpha_1/\beta_1)} \quad [\text{F}] \qquad (3.154c)$$

In addition, under the condition $\alpha_1 \gg \beta_1$, the capacitance can be expressed as

$$C = \frac{\pi\varepsilon_0\ell}{\ln(2\alpha_1/\beta_1)} \quad [\text{F}] \qquad (3.154d)$$

Exercise 3.47
For a parallel-plate capacitor, derive its capacitance from the boundary condition at the conducting plate, that is, $D_n = \rho_s$.

Ans. $\frac{\rho_s}{D_n} = \frac{Q/\mathcal{S}}{\varepsilon V/d} = 1$, $C = \frac{Q}{V} = \frac{\varepsilon\mathcal{S}}{d}$.

Exercise 3.48
Show that the capacitance of the coaxial capacitor given in Eq. (3.151) is reduced to that of a parallel-plate capacitor for $b/a \approx 1$.
[Hint: $\ln(b/a) \approx (b/a) - 1$.]

Exercise 3.49
If a conducting sphere of radius a carries net charge Q in free space, determine (a) the electrostatic energy of the system W_E, and (b) the capacitance from W_E.

Ans. (a) $W_E = Q^2/8\pi\varepsilon_0 a$, (b) $C = 4\pi\varepsilon_0 a$.

Review Questions

RQ 3.56	Define capacitance.	[(3.140)]
RQ 3.57	Express the capacitance of the parallel-plate, coaxial, and spherical capacitors.	[(3.142)(3.151) (3.153a)]
RQ 3.58	Does the capacitance depend on the dielectric strength?	[(3.142)]
RQ 3.59	Express the electrostatic energy stored in the capacitor.	[(3.143)]
RQ 3.60	Explain the principle of virtual displacement.	[(3.146)]

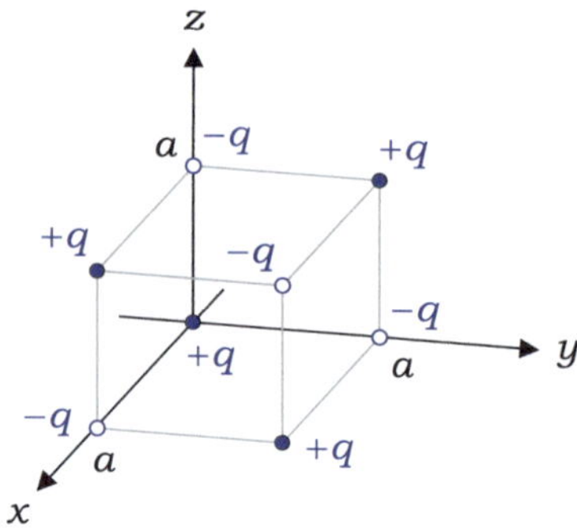

Fig. 3.50 Eight point charges at the corners of a cube (Problem 3.3)

3.10 Problems

Coulomb's law

3.1 Three point charges are arranged along the x-axis in free space such that $q_1 = +Q$ [C] at $x = 1$, an unknown charge q_2 at $x = 2$, and $q_3 = 4Q$ [C] at $x = 3$. Determine the value of q_2 necessary to make the force on q_1 vanish.

3.2 Point charge q_1 is fixed at point $(0, 2, 0)$, whereas the second charge q_2 is allowed to move along the x-axis. If they carry the same charge, locate q_2 to make the force on q_2 directed along the $+x$-axis maximum.

3.3 Positive charges alternate with negative charges at the eight corners of a cube of side a, as shown in Fig. 3.50. Determine the force acting on the charge at the origin.

3.4 Two point charges, $q_1 = 15$ [nC] and $q_2 = 10$ [nC], are positioned at $z = 2$ and $z = -3$, respectively, on the z-axis. In addition, the $z = 0$ plane represents an infinite sheet of surface charges of density $\rho_s = -40$ [pC/m^2]. Determine the electric forces acting on q_1 and q_2.

Electric field

3.5 An infinitely long line charge of density ρ_ℓ is parallel to the y-axis passing through point $(2, 3, 1)$ in free space. Determine $\mathbf{E}$ at the origin.

3.6 The line charge of density $\rho_{\ell o}$ extends from $z = z_1$ to $z = z_2$ in free space, as shown in Fig. 3.51. Express $\mathbf{E}$ at a distance ρ from the charge on the $z = 0$ plane in terms of the interior angles of the triangle formed by $\rho_{\ell o}$ and p.

3.7 Two infinitely long parallel line charges of density ρ_ℓ pass through a circle of radius a in the $z = 0$ plane, as shown in Fig. 3.52. These are mirror images of each other about the $y = 0$ plane. Find $\mathbf{E}$ at point $(a, 0, 0)$, and show that $\mathbf{E}$ is independent of ϕ_1. [Ans. $\mathbf{E} = \mathbf{a}_x \rho_\ell / 2\pi\varepsilon_0 a$.]

3.8 The uniform surface charge density ρ_s is confined to an infinitely long strip defined by $-a \leq x \leq a$, $y = 0$, and $-\infty < z < \infty$ in free space. To find $\mathbf{E}$ at point $(0, b, 0)$, first divide the charge into many infinitely long strips of width dx and then add the contributions from all equivalent line charges. [Hint: $\int (x^2 + a^2)^{-1} dx = a^{-1} \tan^{-1}(x/a)$.]

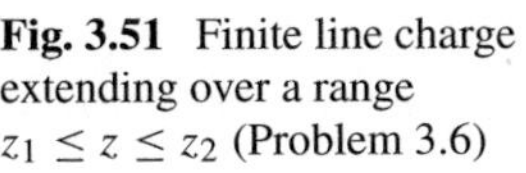

Fig. 3.51 Finite line charge extending over a range $z_1 \le z \le z_2$ (Problem 3.6)

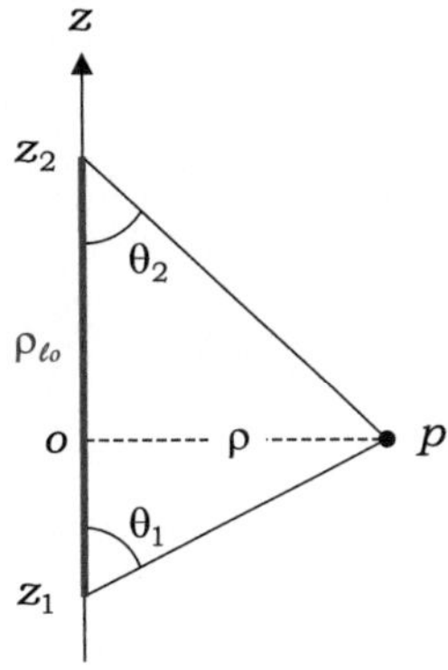

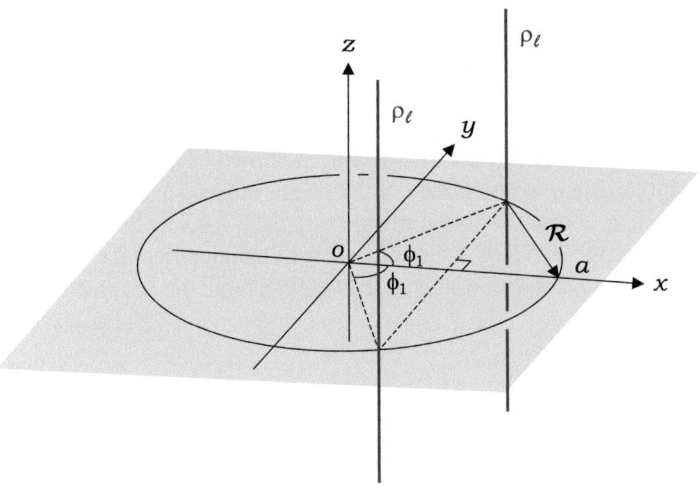

Fig. 3.52 Two parallel line charges (Problems 3.7 and 3.10)

3.9   A cylinder of radius a extends between $z = -b$ and $z = b$ with its axis along the z-axis and has a uniform surface charge density ρ_s. To determine $\mathbf{E}$ on the z-axis, first divide the charge into many circular strips of width dz, and then add the contributions from all equivalent circular line charges.

3.10   An infinitely long cylinder of radius a with its axis along the z-axis has a uniform surface charge density ρ_s. To determine $\mathbf{E}$ on the surface of the cylinder, divide the charge into many infinitely long strips of width $a d\phi$ and use the result in Problem 3.7.

3.11   The net charge Q [C] is uniformly distributed over a disk of radius a in the $z = 0$ plane centered at the origin. At a distance b above the origin, find

    (a)   electric field $\mathbf{E}$,

    (b)   the distance beyond which $\mathbf{E}$ approximates, with an error less than 1%, $\mathbf{E}_1$ obtained as if Q is concentrated at the center, and

(c) the distance within which $\mathbf{E}$ approximates, with an error less than 1%, $\mathbf{E}_2$ obtained as if the disk is infinite in extent, carrying the same surface charge density.

3.12 An infinite plane is defined by $6x + 3y + 2z = 12$, and carries a uniform surface charge density ρ_s in free space. Determine $\mathbf{E}$ at p:(1, 1, 1).

3.13 The uniform volume charge density ρ_v forms a hemispherical shell with an inner radius of 1 and outer radius of 2, as shown in Fig. 2.30. Determine $\mathbf{E}$ at the origin.

3.14 The electric field of an electric dipole placed at the origin in free space is given by $\mathbf{E}(\mathbf{r}) = (qd/4\pi\varepsilon_0 R^3)(2\cos\theta\,\mathbf{a}_R + \sin\theta\,\mathbf{a}_\theta)$. Find an equation for the electric field lines in the $\phi = 0$ plane. [Hint: $dR/(Rd\theta) = E_R/E_\theta$.]

Electric flux

3.15 An infinitely long and uniform line charge of density 20 [nC/m] lies along the x-axis. Find the net electric flux flowing out of a hypothetical sphere of radius 10 [cm] centered at the origin.

3.16 If a point charge $+q$ [C] is placed at the origin, what is the total electric flux through the triangle in the first octant, as shown in Fig. 3.53?

3.17 A point charge q is positioned at the origin in free space. Determine the total electric flux across the $z = a$ plane.

3.18 A point charge of 5 [μC] is at the center of a dielectric cube of 40 [cm] on the side. If the permittivity of the cube is ε, determine the net electric flux flowing out of the cube.

Gauss's law

3.19 Two parallel conducting plates with the same thickness t carry charges $+Q$ and $-Q$ (Fig. 3.54). Ignoring the edge effect, the charge is assumed to be uniformly distributed. Thus, one might assume that the charge is equally divided between the top and bottom surfaces of each plate (see the inset).

(a) To determine exactly what is wrong with this guess, determine $\mathbf{E}$ everywhere, including the interior of the conductor.

Fig. 3.53 Point charge at the origin and triangle in the first octant (Problem 3.16)

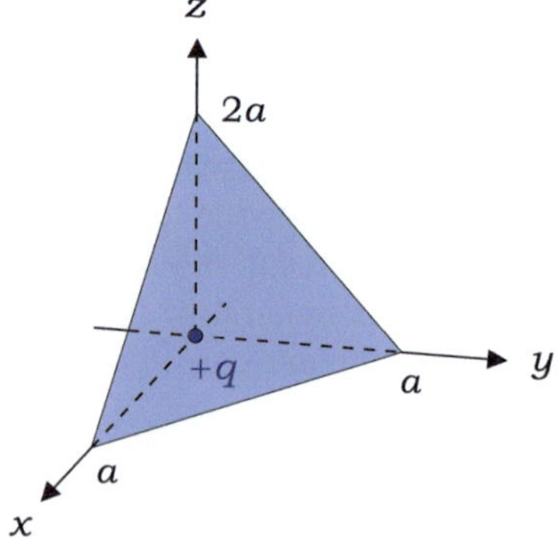

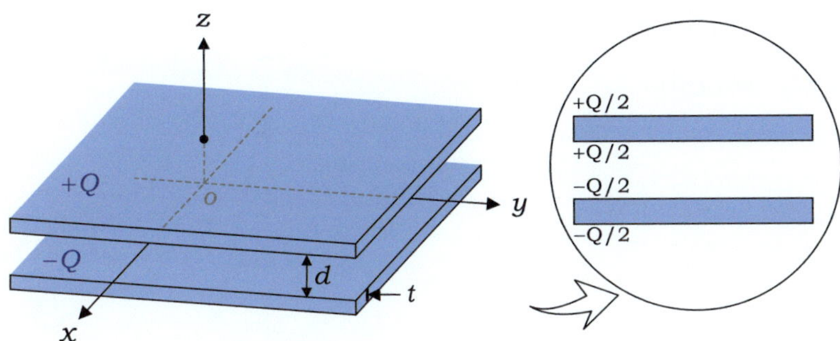

Fig. 3.54 Parallel plates with opposite charges (Problem 3.19)

   (b)   What is the correct charge distribution?

3.20 For the volume charge density $\rho_v = e^{-|z|}$ residing in free space, determine **D** everywhere.

3.21 The uniform volume charge of density ρ_v forms a cylindrical shell of radii a and b $(a < b)$, with the axis coincident with the z-axis in free space. Determine **D** everywhere.

3.22 When two concentric spheres of radii a and b $(a < b)$ are centered at the origin and carry uniform surface charges of densities ρ_{s1} and ρ_{s2}, respectively, determine **D** everywhere.

3.23 A sphere of radius 0.1 [m] with surface charge density 5 [μC/m^2] is centered at point $(0, 0, -2)$. In addition, an infinitely long line with charge density 0.1 [μC/m] is parallel to the x-axis passing through point $(0, 0, 2)$. Determine **D** at p:$(0, 1, 0)$.

3.24 For the electric flux density $\mathbf{D} = \mathbf{a}_R/(4\pi\varepsilon_0 R)$, determine the volume charge density and total charge within the space between two concentric spheres of radii a and b $(b > a)$ centered at the origin using Gauss's law in point form.

Electric potential

3.25 A square of side 2 is centered at the origin in the $z = 0$ plane, with its two corners at $(-1, -1, 0)$ and $(1, 1, 0)$. For vector field $\mathbf{K} = \mathbf{a}_x + 2x^2\mathbf{a}_y$, show that the closed-line integral of **K** around the square is zero. Is **K** a conservative field?

3.26 For the two vector fields $\mathbf{E}_1 = e^{-|z|}\mathbf{a}_x$ and $\mathbf{E}_2 = \mathbf{a}_\theta/R^2$ $(R > 0)$, determine whether they can possibly represent static electric fields in free space by evaluating their curls.

3.27 In the presence of an electric field $\mathbf{E} = y\,\mathbf{a}_x + x\,\mathbf{a}_y$ in free space, determine the work done in carrying a charge of 3 [μC] from point p_1:$(3, 4, 0)$ to point p_2:$(0, 1, 0)$ along (a) a parabola defined by $y = (x - 1)^2$, and (b) a straight line defined by $y = x + 1$.

3.28 Two charges, q and $-q$, are located at points $(3, 2, \sqrt{3})$ and $(5, 3, -\sqrt{2})$, respectively, in free space. Determine V at p:$(4, 1, 0)$ if $V = 0$ at the origin.

Fig. 3.55 Line charge in the form of a half-circle (Problems 3.30 and 3.31)

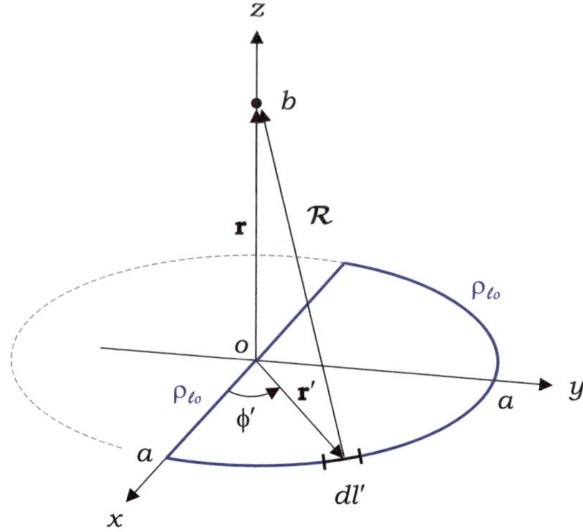

3.29   The net charge of Q [C] is equally divided into N point charges equally spaced along a circle of radius a in the $z = 0$ plane, with the center at the origin. Determine V on the x-axis.

3.30   A uniform line charge with density $\rho_{\ell o}$ forms a closed half-circle with radius a in the $z = 0$ plane, as shown in Fig. 3.55. Find V at distance b from the origin along the z-axis.

3.31   For the line charge shown in Fig. 3.55, explain why the electric field on the z-axis cannot be obtained by taking the negative gradient of the electric potential on the z-axis, that is, $\mathbf{E}(0,\ 0,\ z) \neq -\nabla V(0,\ 0,\ z)$.

3.32   Two infinitely long and thin cylinders of radii 0.2 [m] and 0.4 [m] are coaxial with the z-axis in free space. When the inner conductor carries a uniform surface charge of $3\,[\mu\text{C/m}^2]$, and the outer conductor is connected to the ground, determine V everywhere.

3.33   A perfectly conducting sphere with radius a and net charge Q is surrounded by a spherical dielectric with permittivity ε (Fig. 3.56). Determine V everywhere.

3.34   For the volume charge of density ρ_v formed into a sphere of radius a in free space, determine the electric potential inside the sphere.

Fig. 3.56 A sphere with charge Q is embedded in a dielectric sphere (Problem 3.33)

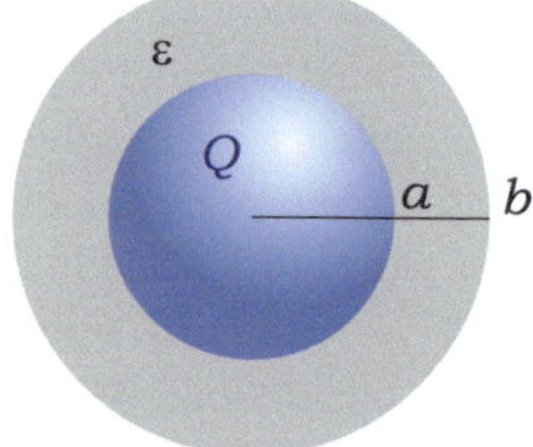

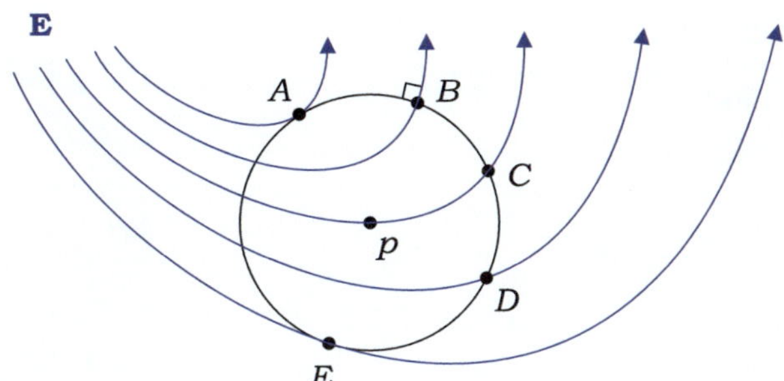

Fig. 3.57 Circle drawn in electric field lines (Problem 3.38)

3.35 Locate the zero-reference point for the electric potential that is defined by $V = 8\ln[\{(x+2)^2 + (y+3)^2\}^{1/2}/\{(x-2)^2 + (y-3)^2\}^{1/2}]$.

3.36 For a potential field $V = xy^2 - 2yz + 3$, express the equipotential line in the $z = 0$ plane that passes through $p{:}(2, 1, 0)$.

3.37 Start with the electric potential of an electric dipole given in Eq. (3.62) to find an expression for the equipotential surface in the $x = 0$ plane.

3.38 A circle is drawn around point p in the presence of electric field lines (Fig. 3.57). (a) Is p at a lower potential than those at points A and E? (b) Which point (B, C, or D) has the lowest potential?

Electric polarization

3.39 A point charge of q [C] is placed at the center of a dielectric sphere with radius a and dielectric constant ε_r. Find

  (a) **D**, **E**, and **P** inside the sphere,
  (b) polarization charge densities ρ_{ps} and ρ_{pv}, and
  (c) show that $Q_{ps} = (\varepsilon_r - 1)q/\varepsilon_r$ and $Q_{pv} = 0$.

3.40 When the dielectric sphere in Problem 3.39 has a net charge q at the center, the charge Q_{ps} on the surface should be balanced by $-Q_{ps}$ at the center. Show that the existence of $-Q_{ps}$ accounts for ε_r of the dielectric.

3.41 A uniform line charge of density ρ_ℓ coincides with the z-axis and is surrounded by a hollow cylinder with inner radius b, outer radius c, and dielectric constant ε_r. Determine ρ_{ps} at $\rho = b$ and c.

3.42 For a dielectric sphere with ε_r placed in an external field $\mathbf{E}_o = 10\,\mathbf{a}_z$, the internal field is measured as $\mathbf{E}_i = 7.5\,\mathbf{a}_z$. Use yet-to-be-determined ε_r to express (a) electric polarization, (b) polarization surface charge density, and (c) polarization field. (d) Show that $\varepsilon_r = 2$.

3.43 Two concentric spheres carry equal but opposite charges, as shown in Fig. 3.58. When the lower half of the gap is filled with a dielectric with ε_r, find

  (a) **E** for $a \le R \le b$,
  (b) ρ_{ps} on $\mathcal{S}_{a2}$ and $\mathcal{S}_{b2}$, and
  (c) ρ_s on $\mathcal{S}_{a1}$ and $\mathcal{S}_{a2}$ caused by $+Q$.
  (d) Check whether **E** in part (a) is a unique solution.

Fig. 3.58 Half-filled
concentric spheres (Problem
3.43)

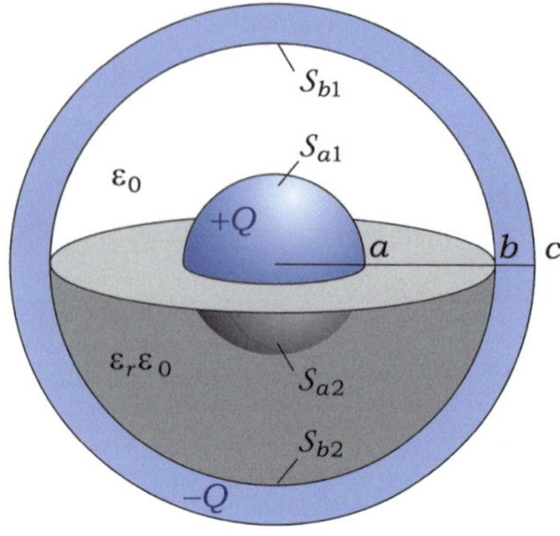

Boundary condition

3.44   The $z = 0$ plane is the interface between the two dielectrics: $\varepsilon_r = 2$ for $z > 0$
and $\varepsilon_r = 3$ for $z < 0$. If $\mathbf{E}_1 = 3\mathbf{a}_x - 2\mathbf{a}_y + 6\mathbf{a}_z$ in the upper region, find

(a) $\mathbf{E}_2$ in the lower region,
(b)   polarization surface charge density, and
(c)   polarization fields $\mathbf{E}_{ps1}$ and $\mathbf{E}_{ps2}$ in the two regions.
(d)   Show that $\mathbf{E}_1 - \mathbf{E}_{ps1} = \mathbf{E}_2 - \mathbf{E}_{ps2}$, which is equal to the external field.

3.45   If the electric potential is given by $V = -y - 3z$ in free space ($z < 0$),
determine V in the dielectric with $\varepsilon_r = 3$, occupying the region $z \geq 0$.

3.46   Determine the normal component of $\mathbf{E}$ on the semi-infinite dielectric with
$\varepsilon_r = 3$ carrying a polarization surface charge of $\rho_{ps} = 2\,[\text{nC/m}^2]$.

3.47   The $z = 0$ and $z = 0.2$ [m] planes carry charges of equal amounts but opposite polarities of $-4.5\,[\text{nC/m}^2]$ and $4.5\,[\text{nC/m}^2]$, respectively. If the region
$0 < z \leq 0.1$ [m] is filled with a dielectric with $\varepsilon_r = 1.5$ and the $z = 0$
plane is maintained at $V = 0$, determine

(a) $\mathbf{D}, \mathbf{E}$, and $\mathbf{P}$ for $0 < z \leq 0.2$, and
(b)   total charge densities at the $z = 0, 0.1$, and 0.2 [m] planes.

3.48   For a dielectric sphere placed in an external field, as in Problem 3.42, determine
the electric field immediately outside the sphere.

3.49   When an infinite dielectric slab with thickness d and dielectric constant $\varepsilon_r = 1.732$ is introduced into a uniform external field $\mathbf{E}_o$, the internal field is found
to be $\mathbf{E}_2 = 5\mathbf{a}_x + 5\mathbf{a}_z$, as shown in Fig. 3.59. Determine

(a) ρ_{ps} on the surface, and
(b) $\mathbf{E}_1$ immediately outside the slab.
(c)   Explain why $\mathbf{E}_o$ must be equal to $\mathbf{E}_1$ unlike other general cases.

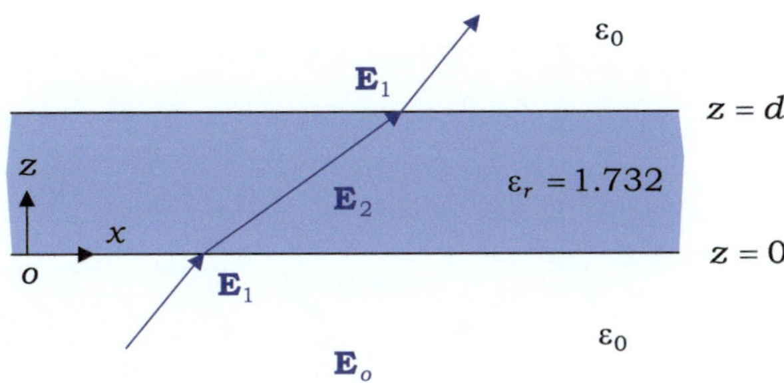

Fig. 3.59 Dielectric slab in an electric field (Problem 3.49)

3.50 Two infinite parallel plates with surface charges of equal amounts but opposite polarities, $\rho_s = \pm 5.77\varepsilon_0$ [C/m²], are placed in an external electric field $\mathbf{E}_o = 5\mathbf{a}_x + 5\sqrt{3}\,\mathbf{a}_z$, as shown in Fig. 3.60.

(a) Determine $\mathbf{E}_i$ in the gap.

(b) When the gap is filled with a dielectric of ε_r and the charged plates are removed, the electric fields remain unchanged. Determine ε_r.

3.51 When the point charge q is above the dielectric-dielectric interface, the mirror-image charge assumed in the second dielectric can significantly facilitate the solution of the electric fields in both the regions (Fig. 3.61). In this case, we can start with

$$\mathbf{E}_o + \mathbf{E}_1 = \frac{q}{4\pi\varepsilon_1}\left[\frac{\mathbf{r}-\mathbf{r}'}{|\mathbf{r}-\mathbf{r}'|^{3/2}} + A\frac{\mathbf{r}-\mathbf{r}''}{|\mathbf{r}-\mathbf{r}''|^{3/2}}\right] \quad (z \geq 0)$$

$$\mathbf{E}_2 = \frac{q}{4\pi\varepsilon_2}B\frac{\mathbf{r}-\mathbf{r}'}{|\mathbf{r}-\mathbf{r}'|^{3/2}} \quad (z \leq 0)$$

Show that

$$A = \frac{\varepsilon_1 - \varepsilon_2}{\varepsilon_1 + \varepsilon_2}, \quad B = \frac{2\varepsilon_2}{\varepsilon_1 + \varepsilon_2}, \quad \text{and} \quad \rho_{ps} = 2\varepsilon_0 A\left[\mathbf{E}_o|_{z=o} \cdot (-\mathbf{a}_z)\right]$$

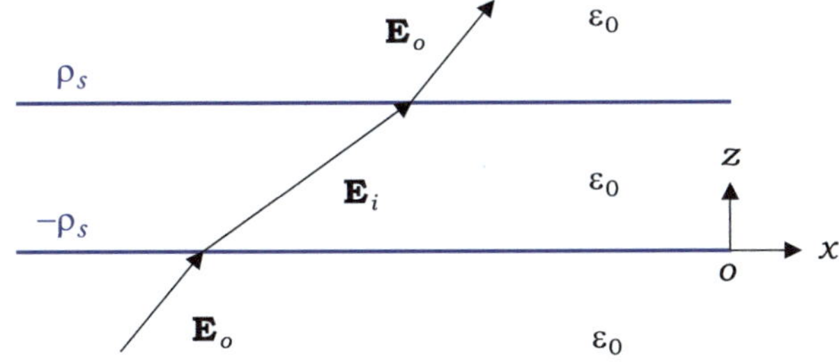

Fig. 3.60 Parallel surface charges in an external field (Problem 3.50)

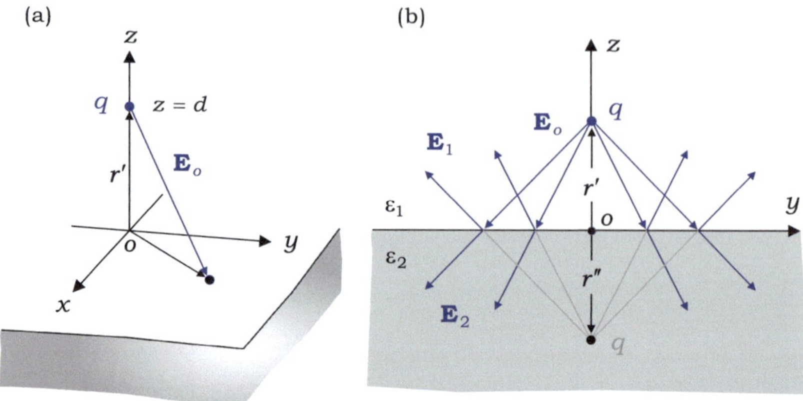

Fig. 3.61 Point charge q above the dielectric-dielectric interface and its mirror-image charge (Problems 3.51 and 3.52)

3.52 A point charge q in free space is at a distance d above a semi-infinite dielectric with permittivity ε_2, as shown in Fig. 3.61a. Determine the electric force acting on q.

Electrostatic potential energy

3.53 Three identical point charges of 5 [nC] are located at the corners of a right triangle in free space, defined by (1, 0, 0), (0, 0, 0), and (0, 1, 0), where all distances are in meters. Determine the potential energy of the system of three charges.

3.54 A sphere of radius a in free space has uniform surface charge density ρ_{so}. (a) Determine $\mathbf{E}$ everywhere and V at $R = a$. Compute the potential energy using (b) $\mathbf{E}$, (c) V, and (d) the work done to bring in infinitesimal surface charges from infinity to form the sphere.

3.55 A sphere of radius a is filled with a uniform volume charge of density ρ_v in free space. Compute the potential energy using an electric field.

Boundary value problems

3.56 A spherical conductor, described by $x^2 + y^2 + (z - 2)^2 = 4$, is maintained at a potential of 10 [V] in free space. Three trial solutions are formed to satisfy the boundary conditions. Which is the correct solution?

(a) $V = 20\{x^2 + y^2 + (z - 2)^2\}^{-1/2}$,
(b) $V = 40\{x^2 + y^2 + (z - 2)^2\}^{-1}$, and
(c) $V = 160\{x^2 + y^2 + (z - 2)^2\}^{-2}$.

3.57 Given an electric potential $V = V_o e^{-2R}/R$ in free space, compute the total charge in the entire space.

3.58 Find a general solution to Laplace's equation for a problem with (a) spherical symmetry, and (b) both cylindrical and translational symmetry along the z-axis.

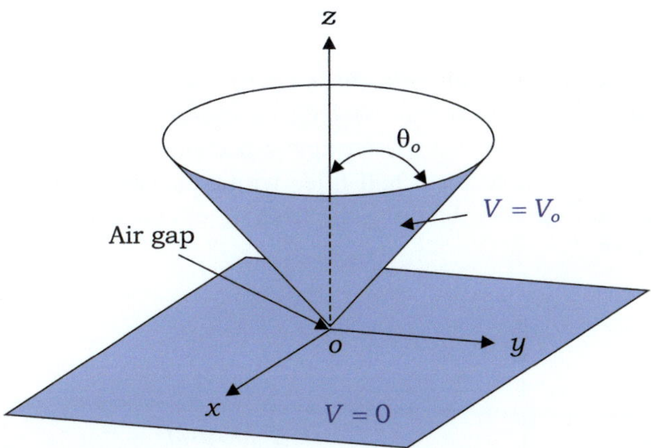

Fig. 3.62 Infinite conducting plane and cone (Problem 3.60)

3.59 Two infinite parallel conducting planes are defined by $2x + y + 3z = 6$ and $2x + y + 3z = 12$, and maintained at $V = 10\,[\text{V}]$ and $V = 0\,[\text{V}]$, respectively. Determine V everywhere.

3.60 A conical surface is maintained at $V = V_o$ and insulated from a grounded conducting plane with a small air gap, as shown in Fig. 3.62. When these surfaces are infinitely large and perfectly conducting, determine V in the region of $\theta_o \leq \theta \leq \pi/2$.

3.61 The space between two concentric spheres of radii a and b $(b > a)$ is filled with a dielectric with permittivity ε. If a point charge $+q$ is placed at the center and the inner and outer spheres are maintained at potentials V_1 and V_2, respectively, determine V everywhere.

Method of images

3.62 An infinite conducting plane at $z = 0$ is grounded, whereas two infinite line charges of density 3 [nC/m] cross each other at $z = 1$ [m] on the z-axis parallel to the x- and y-axes. Determine $\mathbf{E}$ at $z = 2$ [m] above the origin.

3.63 An infinite grounded conducting plane coincides with the $z = 0$ plane. Two identical point charges of q [C] are situated on the z-axis, one at $z = 1$ [m] and the other at $z = -2$ [m]. Determine $\mathbf{E}$ at $p{:}(0, 1, 1)$.

3.64 An infinite grounded conducting plane is bent at $90°$ and the point charge q is situated in the xy-plane, as shown in Fig. 3.63. Using the method of images, determine V in the first quadrant and ρ_s on the surface.

3.65 Two long parallel conducting cylinders of radius 0.1 [m] are separated by a center-to-center distance of 1 [m] and maintained at potentials -100 [V] and 100 [V], as shown in Fig. 3.64. Determine the equivalent line charges that produce the same V outside the cylinders.

Capacitance

3.66 A parallel-plate capacitor with surface area $\mathcal{S}$ and separation d is half-filled with a dielectric with permittivity ε_1, as shown in Fig. 3.65. Ignoring the fringing effects at the edges, determine capacitance.

3.67 A parallel-plate capacitor is half-filled with a dielectric with permittivity ε_1, as shown in Fig. 3.66. Ignoring the fringing effects at the edges, determine capacitance.

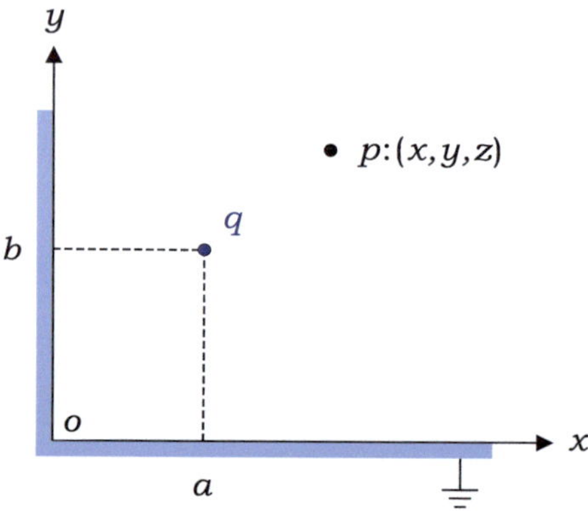

Fig. 3.63 An infinite conducting plane bent at 90° and a point charge (Problem 3.64)

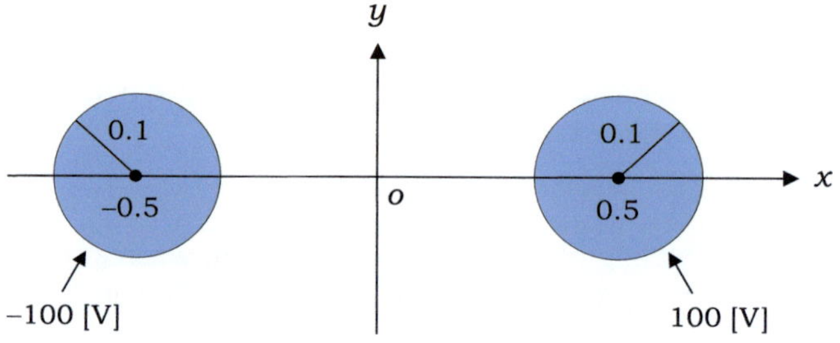

Fig. 3.64 Cross section of two parallel cylinders (Problems 3.65 and 3.70)

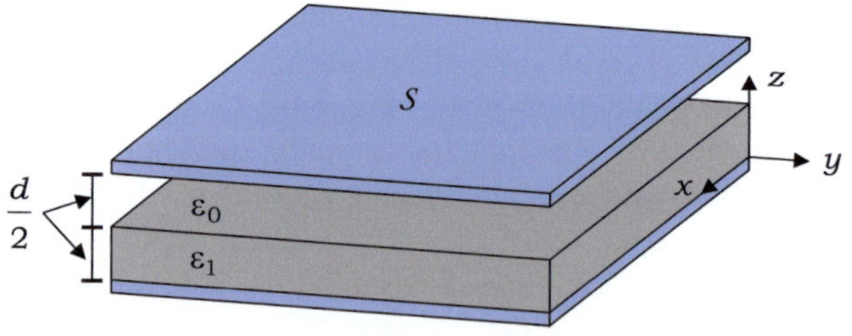

Fig. 3.65 Parallel-plate capacitor (Problem 3.66)

Fig. 3.66 Parallel-plate
capacitor (Problem 3.67)

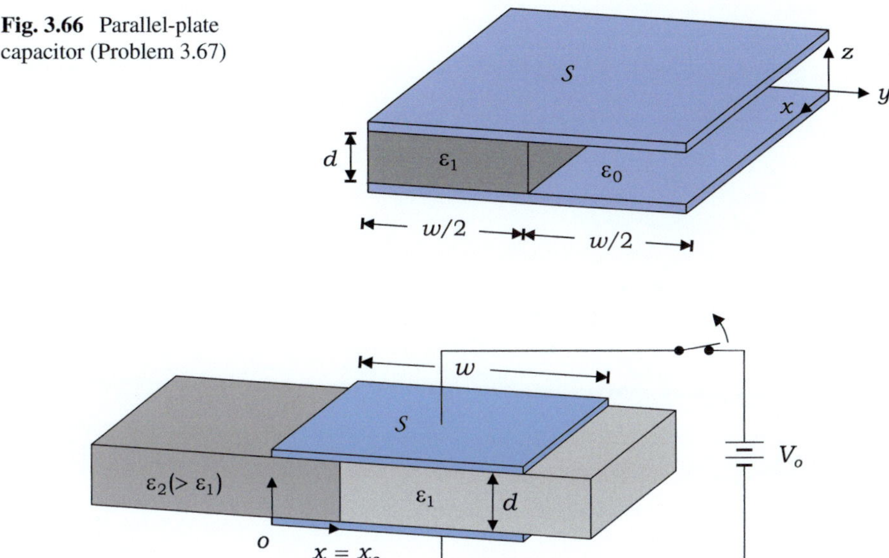

Fig. 3.67 Parallel-pate capacitor with two dielectric slabs (Problem 3.69)

3.68 A parallel-plate capacitor with plate area S and spacing d is filled with an inhomogeneous dielectric with $\varepsilon = \varepsilon_1 \exp(-z)$ and placed on the $z = 0$ plane. Ignoring the fringing effects at the edges, determine capacitance.

Principle of virtual displacement

3.69 A parallel-plate capacitor has two adjoining dielectric slabs, as shown in Fig. 3.67, and is charged to V_o while the slabs are held in place. When the switch is opened, the slab with permittivity ε_2 is completely pulled into the capacitor. Find (a) the force on the slab immediately after the switch is opened, and (b) the final voltage across the plates.

3.70 When cylinders, as shown in Fig. 3.64, are fully charged to ± 100 [V], the battery is disconnected, leaving net charges ± 2.43 [nC/m] on the cylinders. (a) Determine the potential energy per unit length. (b) If the two cylinders are moved a distance dx away from each other, find the change in the electric potential, and (c) the force on the cylinder.

Chapter 4
Steady Electric Current

In Chap. 3, we focused our discussion on static electric charges that were fixed in position and held constant over time. Otherwise, we assumed that the charges were instantly relaxed to a steady state. However, electric charges can move under the influence of an electric field and constitute a ***conduction current*** in the conductor. Meanwhile, the motion of electric charges in vacuum gives rise to a ***convection current***. According to the basic circuit theory, readers should be familiar with the conduction current in a simple electric circuit. Ohm's law states that the voltage across a resistor is equal to the product of the current and the resistance. However, the convection current is not proportional to the voltage across the electrodes in vacuum, and thus, is not governed by Ohm's law. According to the principle of charge conservation, electric charge cannot be created or destroyed. This principle manifests itself as ***the equation of continuity*** in electromagnetics and Kirchhoff's current law in the circuit theory.

In an electric circuit, the current flowing through a wire is defined as the charge passing through a reference point per unit time. In electromagnetics, we are more concerned with ***volume current density***, which is defined as the charge passing through a reference point per unit area per unit time. When current is confined to a surface, it is described by the ***surface current density***, which is defined as the charge passing through a reference point per unit width per unit time.

When time-varying fields are introduced in Chap. 6, it is shown that the time rate of charge of $\mathbf{D}$ gives rise to a ***displacement current density***. This is an equivalent current involving no electric charge, but behaves in exactly the same way as a conduction or convection current in generating time-varying magnetic fields.

4.1 Convection Current

Let us start by considering the case shown in Fig. 4.1, where the volume charge of density ρ_v [C/m^3] moves with velocity $\boldsymbol{v}$ in free space and passes through slanting plane $\mathcal{S}$. The charge crossing an incremental area Δs in time Δt is given by

$$\Delta Q = \rho_v |\boldsymbol{v}| \Delta t \, \Delta s \cos\theta \tag{4.1}$$

where $\Delta s \cos\theta$ is the projection of Δs onto a plane perpendicular to $\boldsymbol{v}$. Using $\cos\theta = \mathbf{a}_v \cdot \mathbf{a}_s$, where $\mathbf{a}_v$ is the unit vector in the direction of $\boldsymbol{v}$, and $\mathbf{a}_s$ is the unit normal to Δs, Eq. (4.1) can be rearranged as

$$\Delta Q = \rho_v \boldsymbol{v} \cdot \Delta \mathbf{s} \, \Delta t \tag{4.2}$$

where $\boldsymbol{v} = |\boldsymbol{v}| \mathbf{a}_v$ and $\Delta \mathbf{s} = \Delta s \, \mathbf{a}_s$.

Therefore, the incremental current across Δs is expressed as

$$\Delta I = \frac{\Delta Q}{\Delta t} = \rho_v \boldsymbol{v} \cdot \Delta \mathbf{s} \equiv \mathbf{J} \cdot \Delta \mathbf{s} \tag{4.3}$$

The volume current density, or simply the current density, is defined as

$$\boxed{\mathbf{J} \equiv \rho_v \boldsymbol{v}} \quad [\text{A/m}^2] \tag{4.4}$$

where ρ_v is the volume charge density, and $\boldsymbol{v}$ is the velocity of the flow of the volume charge. The volume current density is expressed in amperes per square meter, [A/m^2], or in coulombs per square meter per second, [C/m$^2 \cdot$ s]. Note that $\mathbf{J}$ is a vector whose unit vector points in the direction of the current flow, and its magnitude represents

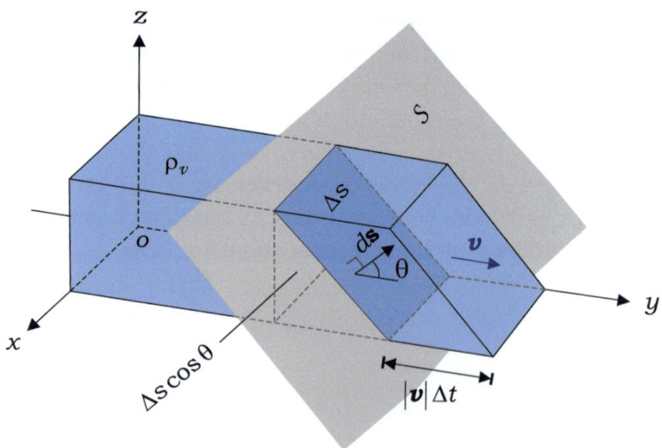

Fig. 4.1 Volume charge of density ρ_v moving at velocity $\boldsymbol{v}$

the charge crossing a unit area per unit time. It is important to note that a unit area should be taken in the plane perpendicular to $\mathbf{J}$.

The total current passing through surface $\mathcal{S}$ is given by the surface integral of $\mathbf{J}$ over $\mathcal{S}$. That is,

$$\boxed{I = \int_{\mathcal{S}} \mathbf{J} \cdot d\mathbf{s}} \quad [\text{A}] \tag{4.5}$$

Current I is scalar, whereas current density $\mathbf{J}$ is a vector that generally varies from point to point in space, and thus forms a vector field in a given region.

Example 4.1 A hot cathode in the $y = 0$ plane is maintained at zero potential, whereas the anode in the $y = d$ plane is maintained at potential V_o. Electrons are injected into the space between the electrodes by thermionic emission and accelerated toward the anode in a vacuum, as shown in Fig. 4.2. Assuming that the current density in the gap is uniform in the steady state, determine the volume charge density and electric potential in the gap.

Solution

The acceleration of an electron with mass m and charge e in the gap is given by $a = eE/m$, where E is the electric field. Rewriting this using the relations $v = dy/dt$, $a = dv/dt$, and $E = -dV/dy$, where v is the electron velocity, and V is the electric potential, we have

$$v \, dv = -\frac{e}{m} dV \tag{4.6a}$$

Integrating both sides of Eq. (4.6a) yields

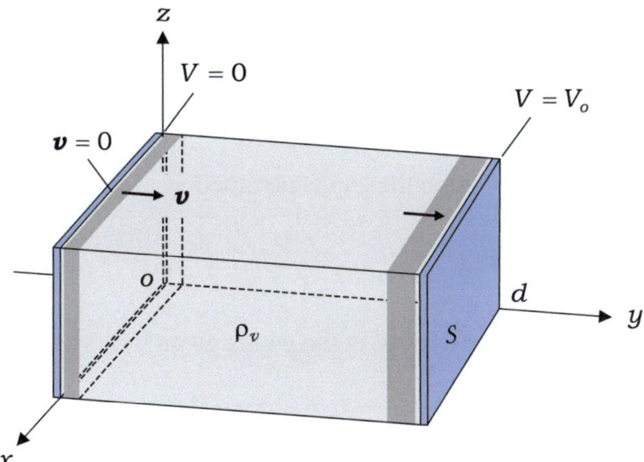

Fig. 4.2 Electron cloud accelerated between the two electrodes at $y = 0$ and $y = d$

$$\frac{1}{2}mv^2 = -eV \tag{4.6b}$$

where the constant of integration is set to zero, because V and v are zero at $y = 0$. Next, Poisson's equation for the gap is reduced to

$$\frac{d^2 V}{dy^2} = -\frac{\rho_v}{\varepsilon_0} \tag{4.6c}$$

Inserting Eq. (4.6b) into Eq. (4.6c) and using $\mathbf{J} = \rho_v v\, \mathbf{a}_y$, we obtain

$$\frac{d^2 V}{dy^2} = \frac{|J|}{\varepsilon_0\sqrt{2|e|/m}} V^{-1/2} \equiv AV^{-1/2} \tag{4.6d}$$

where A is a constant, and both J and e are negative values. Multiplying both sides of Eq. (4.6d) by $G \equiv dV/dy$ and integrating with respect to y, we obtain

$$\int G\frac{dG}{dy}\,dy = \int AV^{-1/2}\frac{dV}{dy}\,dy \quad \rightarrow \quad \frac{1}{2}G^2 = 2AV^{1/2}$$

Thus,

$$V^{-1/4}\frac{dV}{dy} = 2\sqrt{A} \tag{4.6e}$$

Integrating both sides of Eq. (4.6e) with respect to y yields

$$V = \left[\frac{3}{2}\sqrt{A}\right]^{4/3} y^{4/3} \tag{4.6f}$$

By applying the boundary condition to Eq. (4.6f), that is, $V = V_o$ at $y = d$, we obtain

$$V = V_o\left(\frac{y}{d}\right)^{4/3} \tag{4.6g}$$

Subsequently, the electric field in the gap is obtained as

$$\mathbf{E} = -\nabla V = -\frac{4}{3}\frac{V_o}{d}\left(\frac{y}{d}\right)^{1/3}\mathbf{a}_y \tag{4.6h}$$

From Gauss's law, the charge density in the gap is given by

$$\rho_v = -\frac{4}{9}\frac{\varepsilon_0 V_o}{d^2}\left(\frac{y}{d}\right)^{-2/3} \tag{4.6i}$$

In steady state, the current density $\mathbf{J}$ should be constant along the y-axis; otherwise, electrons accumulate over time at points where $\nabla \cdot \mathbf{J}$ is positive. This can be confirmed by inserting Eqs. (4.6b, f, i) into $\mathbf{J} = \rho_v \upsilon\, \mathbf{a}_y$. Furthermore, $\mathbf{J}$ is not proportional to $\mathbf{E}$, as is evident from Eq. (4.6h); $\mathbf{E}$ is a function of position, and $\mathbf{J}$ is constant. The convection current does not follow Ohm's law.

Exercise 4.1

How is the voltage across the electrodes related to the convection current flowing between them?

Ans. $V_o \sim I^{2/3}$.

Exercise 4.2

Given $\mathbf{J} = 4R \sin\theta\, \mathbf{a}_\phi$, determine the current flowing across a square of side unity with two sides coincident with the y- and z-axes.

Ans. $I = 2$ [A].

Review Questions

RQ 4.1	Relate current density and volume charge density.	[(4.4)]
RQ 4.2	Express the current in terms of the current density.	[(4.5)]
RQ 4.3	Does a steady current always require a steady charge flow?	[Fig. 4.2]

4.2 Conduction Current and Ohm's Law

Good conductors have extremely high conductivities. Solid copper is a typical example of this type. Cu atoms with a single valence electron are arranged at regular sites in a face-centered cubic crystal structure. The loosely bound valence electrons easily detach from the host atoms and form a sea of electrons called conduction electrons or free electrons. The electrostatic force between the negatively charged electron cloud and positively charged copper ions is the origin of metallic bonding. Initiated by the thermal energy of the conductor, free electrons migrate from atom to atom and, in the course of the migration, collide with crystal defects, impurities, and mostly with vibrating atomic lattice. In the absence of an externally applied electric field, the free electrons move in random directions, exhibiting no net displacement. However, an electric field can cause electrons to accelerate in one direction before colliding with a lattice. In endless cycles of acceleration and randomizing collisions, free electrons gain a constant velocity, called the **drift velocity**, and form a steady current, called the conduction current, which flows through the conductor.

Under the influence of an applied electric field $\mathbf{E}$, a conduction electron accelerates to velocity υ in the conductor before colliding with the lattice. Expressed mathematically,

$$\upsilon = \upsilon_o + \frac{e\mathbf{E}}{m_e} t_o \quad [\text{m/s}] \tag{4.7}$$

where v_o is the velocity of the electron immediately after collision, t_o is the period before the electron collides with the lattice, and e and m are the electron charge and mass, respectively. The average velocity of conduction electrons can be expresses as

$$\overline{v} = \frac{1}{N} \sum_{i=1}^{N} \left[v_{oi} + \frac{e\mathbf{E}}{m} t_{oi} \right] = \frac{e\mathbf{E}}{m} \overline{t}_o \quad [\text{m/s}] \tag{4.8}$$

The average of the initial velocities v_{oi} is zero because the electron velocity immediately after the collision is arbitrary. The average time $\overline{t}_o$ is the **mean time between collisions**, and the average velocity $\overline{v}$ is the drift velocity. It is evident from Eq. (4.8) that the drift velocity is directly proportional to the electric field in the conductor, that is,

$$\boxed{v = -\mu \mathbf{E}} \quad [\text{m/s}] \tag{4.9}$$

where μ is the **electron mobility** measured in square meters per volt per second. For simplicity, the bar at the top of v is omitted. Typical values of μ are $0.0032\,[\text{m}^2/\text{V} \cdot \text{s}]$ for copper and $0.0056\,[\text{m}^2/\text{V} \cdot \text{s}]$ for silver.

Substituting Eq. (4.9) into Eq. (4.4) leads to the **conduction current density**, that is,

$$\mathbf{J} = -\rho_v \mu \mathbf{E} \quad [\text{A/m}^2] \tag{4.10}$$

where ρ_v is the volume charge density of conduction electrons, which is given by

$$\boxed{\rho_v = ne} \tag{4.11}$$

where n is the volume density of the conduction electrons or the number of electrons per unit volume, and $e = -1.602 \times 10^{-19}\,[\text{C}]$ is the electron charge. Even if $\mathbf{J}$ in Eq. (4.10) has a negative sign, $\mathbf{J}$ is always parallel to $\mathbf{E}$ in metallic conductors because ρ_v is negative.

Inserting Eq. (4.11) into Eq. (4.10) leads to the **point form of Ohm's law**, that is,

$$\boxed{\mathbf{J} = \sigma \mathbf{E}} \quad [\text{A/m}^2] \tag{4.12}$$

The **conductivity** of the metallic conductor is defined as

$$\boxed{\sigma = -ne\mu} \quad [\text{S/m}] \tag{4.13}$$

This is measured in siemens per meter [S/m] or amperes per volt per meter [A/V $\cdot$ m]. In Eq. (4.13), n is the density of the conduction electrons and μ is the electron mobility. Note that σ is always positive, because the electron charge is negative. The conductivity is almost constant over a wide range of current densities and electric fields. However, it increases almost linearly with decreasing temperature in the room temperature region for metallic conductors. Typical values are $\sigma = 5.80 \times 10^7$ [S/m] for copper and $\sigma = 6.17 \times 10^7$ [S/m] for silver. A **perfect conductor** is a material with $\sigma = \infty$, whereas a **perfect dielectric** is a material with $\sigma = 0$.

Semiconductors contain two types of charge carriers, electrons and holes. They both contribute to the conductivity such that

$$\sigma = n_e |e| \mu_e + n_h |e| \mu_h \quad \text{[S/m]} \tag{4.14}$$

where $|e|$ is the absolute value of the electron charge, n_e and n_h are the electron and hole volume densities, respectively, and μ_e and μ_h are the electron and hole mobilities, respectively. The conductivities of intrinsic semiconductors increase with increasing temperature, whereas those of the metallic conductors decrease. This is because the number of charge carriers increases with the temperature in the intrinsic semiconductor, whereas the electron mobility decreases in the metallic conductor. In semiconductors, the conductivity is in the range $10\text{--}10^{-10}$ [S/m]. For example, at room temperature, $n_e = n_h = 1.0 \times 10^{16}$ [m^{-3}], $\mu_e = 0.14$ [m^2/V · s], and $\mu_h = 0.045$ [m^2/V · s] for intrinsic silicon, and for Ge, $n_e = n_h = 2.3 \times 10^{19}$ [m^{-3}], $\mu_e = 0.39$ [m^2/V · s], and $\mu_h = 0.19$ [m^2/V · s].

Example 4.2 For a copper wire with diameter 2 [mm], conductivity $\sigma = 5.8 \times 10^7$ [S/m], and mobility $\mu = 0.0032$ [m^2/V · s], determine

(a)   the density of conduction electrons, and
(b)   the drift velocity at a current of 25 [A] in wire.

Solution

(a)   From Eq. (4.13),

$$n = \frac{\sigma}{|e|\mu} = \frac{5.8 \times 10^7}{(1.6 \times 10^{-19})(0.0032)} = 1.13 \times 10^{29} \ \text{[m}^{-3}\text{]} \tag{4.15a}$$

(b)   The current density in the wire is

$$J = \frac{\text{current}}{\text{area}} = \frac{25}{\pi \, (10^{-3})^2} = 7.96 \times 10^6 \ \text{[A/m}^2\text{]}$$

The drift velocity is obtained using Eq. (4.4) as

$$\upsilon = \frac{J}{\rho_v} = \frac{J}{n|e|}$$

$$= \frac{7.96 \times 10^6}{1.13 \times 10^{29} \ \times 1.6 \times 10^{-19}} = 4.4 \times 10^{-4} \ \text{[m/s]} \tag{4.15b}$$

For a 25 [A] current flowing in the wire, conduction electrons move with a drift velocity of a mere 0.44 [mm/s]. This value is remarkably small compared with the velocity of a current pulse, which is comparable to the speed of light in vacuum, $c = 3 \times 10^8$ [m/s].

Exercise 4.3
What can be used to increase the conduction current in a copper wire? (a) electric field, (b) drift velocity, (c) mobility, (d) conductivity, (e) density of free electrons, and (f) mean time between collisions.

Ans. (a), (b).

Exercise 4.4
Compare the densities of conduction electrons in solid copper and silver using known values of σ and μ.

Ans. $n^{Cu}/n^{Ag} = 1.65$.

Exercise 4.5
For intrinsic silicon at 300 K, compare the contributions of electrons and holes to the conductivity.

Ans. $\sigma_e/\sigma_h = 3.1$.

Review Questions

RQ 4.4	Why do conduction electrons move with constant velocity in metallic conductors, even though they are accelerated by an electric field?	[(4.8)]
RQ 4.5	Define the mobility of conduction electrons.	[(4.9)]
RQ 4.6	Write the point form of Ohm's law.	[(4.12)]
RQ 4.7	What makes the conductivity to vary from one conductor to another?	[(4.13)]
RQ 4.8	What makes Ag to have a larger σ than Cu (n, μ, or both)?	[(4.13)]
RQ 4.9	What distinguishes conduction current from convection current?	[(4.13)]
RQ 4.10	What material parameters are used for classifying materials as conductors, insulators, or semiconductors?	[(4.14)]

4.3   The Equation of Continuity

The principle of charge conservation states that electric charge is conserved. Electric charges can neither be created nor destroyed and only equal amounts of positive and negative charges can be generated by separation or lost by recombination. In discharged conductors, the negative charge of the free electrons is balanced by the positive charge of the ionized atoms such that the net charge is zero. Charge conservation is the law of nature that cannot be derived from other principles. This is manifested in the equation of continuity discussed in this section.

The total current flowing through closed surface S can be obtained by integrating current density $\mathbf{J}$ over the surface such that

$$I = \oint_{\mathcal{S}} \mathbf{J} \cdot d\mathbf{s} \tag{4.16}$$

For simplicity, we assume that the current is due to the positive charges in motion, instead of electrons. If the net outward current across a closed surface is nonzero, the total charge enclosed by the surface should decrease, according to the principle of charge conservation. Expressed mathematically,

$$I = -\frac{dQ}{dt} = -\frac{d}{dt} \int_{\mathcal{V}} \rho_v \, dv \tag{4.17}$$

where I is the net outward current, Q is the total charge inside $\mathcal{S}$, ρ_v is the volume charge density, and $\mathcal{V}$ is the volume bounded by $\mathcal{S}$. Combining Eqs. (4.16) and (4.17) yields

$$\oint_{\mathcal{S}} \mathbf{J} \cdot d\mathbf{s} = -\frac{d}{dt} \int_{\mathcal{V}} \rho_v \, dv \tag{4.18}$$

The net outward current flowing across a closed surface is equal to the time rate of charge decrease in the region enclosed by the surface. Using the divergence theorem, Eq. (4.18) can be rewritten as

$$\int_{\mathcal{V}} \nabla \cdot \mathbf{J} \, dv = \int_{\mathcal{V}} \left(-\frac{d\rho_v}{dt} \right) dv \tag{4.19}$$

where the time derivative is taken inside the volume integral because $\mathcal{V}$ is independent of time. Here, $\mathcal{V}$ may be arbitrary only if it encloses all charges. Therefore, the integrands in Eq. (4.19) should be equal at every point in $\mathcal{V}$, that is,

$$\nabla \cdot \mathbf{J} = -\frac{\partial \rho_v}{\partial t} \tag{4.20}$$

This equation is referred to as the equation of continuity. The current density $\mathbf{J}$ may represent either the conduction or convection current or both.

Under static conditions, the charge density is independent of time. Thus, the equation of continuity is reduced to

$$\nabla \cdot \mathbf{J} = 0 \tag{4.21}$$

By taking the integral of Eq. (4.21) over a volume $\mathcal{V}$ and applying the divergence theorem, we obtain

$$\oint_{\mathcal{S}} \mathbf{J} \cdot d\mathbf{s} = 0 \tag{4.22a}$$

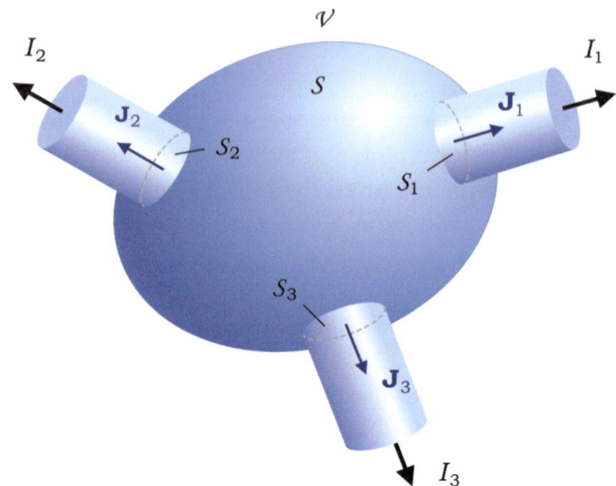

Fig. 4.3 Kirchhoff's current law

If surface $\mathcal{S}$ consists of many elemental surfaces, the closed-surface integral in Eq. (4.22a) can be broken into integrals over the elemental surfaces such that

$$\int_{\mathcal{S}_1} \mathbf{J} \cdot d\mathbf{s} + \int_{\mathcal{S}_2} \mathbf{J} \cdot d\mathbf{s} + \int_{\mathcal{S}_3} \mathbf{J} \cdot d\mathbf{s} + \ldots = 0 \qquad (4.22b)$$

Each term on the left-hand side of Eq. (4.22b) represents the current flowing through the elemental surface (Fig. 4.3). Rewriting Eq. (4.22b) in terms of the current leads to **Kirchhoff's current law**, that is,

$$\sum_i I_i = 0 \qquad (4.23)$$

This also indicates that the sum of all currents flowing out of the junction is zero.

4.3.1   Relaxation Time

If a net charge is introduced into a conductor, it should be distributed on the surface such that $\rho_v = 0$ and $\mathbf{E} = 0$ inside the conductor. The time required for the charge to move away from a given point in the conductor can be computed using the continuity equation. We begin by placing a volume charge with a density of ρ_v inside a good conductor with permittivity ε. Gauss's law dictates that the charge produces an electric field such that

$$\nabla \cdot \mathbf{E} = \frac{\rho_v}{\varepsilon} \qquad (4.24)$$

Using the point form of Ohm's law, $\mathbf{J} = \sigma\mathbf{E}$, Eq. (4.24) can be rewritten as

$$\boxed{\nabla \cdot \mathbf{J} = \frac{\sigma}{\varepsilon}\rho_v} \tag{4.25}$$

where conductivity σ is constant in a homogenous, linear, and isotropic material.
  Comparing Eqs. (4.25) and (4.20) yields

$$\frac{\partial \rho_v}{\partial t} + \frac{\sigma}{\varepsilon}\rho_v = 0 \tag{4.26}$$

Solving Eq. (4.26) for ρ_v with the initial condition $\rho_v(t = 0) = \rho_o$ results in

$$\boxed{\rho_v(t) = \rho_o\, e^{-(\sigma/\varepsilon)t}} \tag{4.27}$$

As the excess charge spreads in every direction in the conductor, the initial charge ρ_o decreases exponentially with time and reduces to $1/e$ (or 36.8%) of its initial value in a ***relaxation time***, defined as

$$\boxed{\tau = \frac{\varepsilon}{\sigma}} \quad \text{[s]} \tag{4.28}$$

For example, using $\sigma = 5.80 \times 10^7$ [S/m] and $\varepsilon \cong \varepsilon_0 = 8.854 \times 10^{-12}$ [F/m] in Eq. (4.28), the relaxation time for copper is computed as $\tau = 1.53 \times 10^{-19}$ [s], which is probably the shortest time one may encounter in electromagnetics.
  It should be noted that the relaxation time is *not* the time required for excess charge to reach the conductor surface. Multiplying both sides of Eq. (4.25) by the differential volume Δv shows that the relaxation of charge $\rho_v \Delta v$ over time ε/σ on the right-hand side of the equation is equal to the total current flowing out of volume Δv on the left-hand side. This implies that the relaxation time is associated with the magnitude of the induced current pulse, and not its velocity. For the sake of argument, we assume that this current pulse has the form of a rectangle traveling in the $\mathbf{a}_R$-direction, for which ε/σ is the width in the time dimension, and $\rho_v \Delta v\, \sigma/\varepsilon$ is the height. Although this current pulse travels through the conductor at a speed comparable to that at $c = 3 \times 10^8$ [m/s] (the speed of the electric field in conductors), the conduction electrons within the rectangle must move at a drift velocity of the order of 10^{-4} [m/s]. For example, a group of conduction electrons 1 [m] ahead of the pulse will begin to move $1/c$ [s] later. A steady state can be reached only when the current pulse reaches the conductor surface and delivers the excess charge to the surface.

Exercise 4.6

Find the relaxation time in amber for which the dielectric constant is 2.7 and the resistivity is 5×10^{14} [$\Omega \cdot$ m].

Ans. 3.3 h.

Exercise 4.7

Find the ratio between the relaxation times of copper and silver.

Ans. $\tau_{Cu}/\tau_{Ag} = 1.06$.

Review Questions

RQ 4.11	State the equation of continuity.	[(4.20)]
RQ 4.12	State the equation of continuity under static conditions.	[(4.21)]
RQ 4.13	State Kirchhoff's current law.	[(4.23)]
RQ 4.14	Define the relaxation time of the conductor.	[(4.28)]
RQ 4.15	Does the relaxation time determine the current pulse width?	[(4.28)]
RQ 4.16	Explain why a greater ε leads to a longer relaxation time.	[Fig. 3.24]

4.4 Steady Currents at the Interface

The steady current density constitutes a vector field in a given region, and can be uniquely determined if its divergence and curl are specified in accordance with Helmholtz's theorem. The equation of continuity given in Eq. (4.21) defines the divergence of $\mathbf{J}$ such that

$$\boxed{\nabla \cdot \mathbf{J} = 0} \tag{4.29}$$

It is important to note that Eq. (4.29) is independent of the material that occupies the region under consideration.

Next, by combining the fundamental relation, $\nabla \times \mathbf{E} = 0$, and the point form of Ohm's law, $\mathbf{J} = \sigma\mathbf{E}$, we obtain

$$\boxed{\nabla \times \frac{\mathbf{J}}{\sigma} = 0} \tag{4.30}$$

where the conductivity σ may not be taken outside the curl operator because it may vary with position. Furthermore, σ may undergo an abrupt change across the interface between the two conductors.

Note that Eqs. (4.29) and (4.30) constitute two fundamental relationships for steady current. They can be converted to an integral form by applying the divergence and Stokes's theorems, as follows:

$$\oint_{\mathcal{S}} \mathbf{J} \cdot d\mathbf{s} = 0 \tag{4.31a}$$

$$\oint_{\mathcal{C}} (\mathbf{J}/\sigma) \cdot d\mathbf{l} = 0 \tag{4.31b}$$

The closed surface S in Eq. (4.31a) and closed loop C in Eq. (4.31b) are independent of each other, and the conductivity σ in Eq. (4.31b) cannot be taken outside the integral symbol if loop C traverses the materials with different conductivities.

The same procedure used for the boundary conditions for $\mathbf{E}$ and $\mathbf{D}$ can be followed to obtain the boundary conditions for $\mathbf{J}$ at the interface between the two conductors with conductivities σ_1 and σ_2. That is,

$$\boxed{J_{1n} = J_{2n}} \tag{4.32a}$$

$$\boxed{\frac{J_{1t}}{\sigma_1} = \frac{J_{2t}}{\sigma_2}} \tag{4.32b}$$

where subscripts n and t denote the normal and tangential components, respectively. The normal component of $\mathbf{J}$ is continuous across the interface, whereas the tangential component of $\mathbf{J}$ is discontinuous.

Example 4.3 The $z = 0$ plane is the interface between two lossy dielectrics with dissimilar permittivities and conductivities, as shown in Fig. 4.4. For $\mathbf{J}_1 = b\,\mathbf{a}_y + c\,\mathbf{a}_z$ [A/m^2] in region $z \leqslant 0$, determine

(a) $\mathbf{J}_2$ in region $z \geqslant 0$,
(b) $\mathbf{E}_1$ and $\mathbf{E}_2$, and
(c)   surface charge density induced at interface.

Solution

(a)   We begin with the general form of $\mathbf{J}_2 = J_{2y}\,\mathbf{a}_y + J_{2z}\,\mathbf{a}_z$ for $z \geqslant 0$. By applying the boundary conditions to $\mathbf{J}_1$ and $\mathbf{J}_2$, we obtain

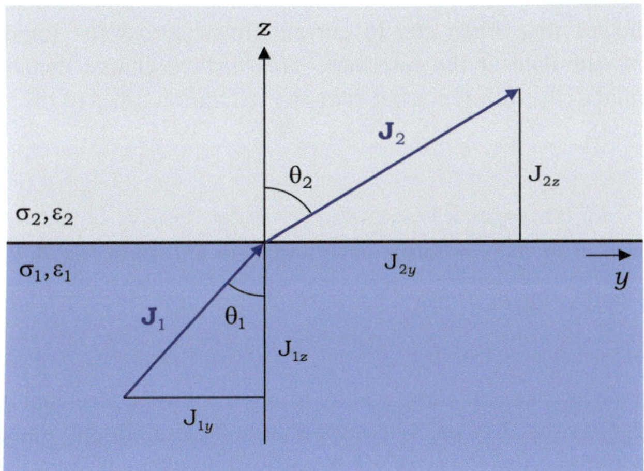

Fig. 4.4   Steady currents in two adjacent lossy media

$$J_{2y} = \frac{\sigma_2}{\sigma_1} b \quad \text{(tangential component)}$$

$$J_{2z} = c \quad \text{(normal component)}$$

Combining these components yields

$$\mathbf{J}_2 = \frac{\sigma_2}{\sigma_1} b\, \mathbf{a}_y + c\, \mathbf{a}_z$$

(b) From the point form of Ohm's law,

$$\mathbf{E}_1 = \frac{\mathbf{J}_1}{\sigma_1} = \frac{b}{\sigma_1} \mathbf{a}_y + \frac{c}{\sigma_1} \mathbf{a}_z$$

$$\mathbf{E}_2 = \frac{\mathbf{J}_2}{\sigma_2} = \frac{b}{\sigma_1} \mathbf{a}_y + \frac{c}{\sigma_2} \mathbf{a}_z$$

(c) The normal components of $\mathbf{D}$ at the interface are

$$D_{1z} = \varepsilon_1 E_{1z} = \varepsilon_1 \frac{c}{\sigma_1}$$

$$D_{2z} = \varepsilon_2 E_{2z} = \varepsilon_2 \frac{c}{\sigma_2}$$

The normal component of $\mathbf{D}$ at the interface induces surface chares such as

$$\rho_s = D_{2z} - D_{1z} = c\left(\frac{\varepsilon_2}{\sigma_2} - \frac{\varepsilon_1}{\sigma_1}\right) \quad [\text{C/m}^2] \tag{4.33}$$

This indicates that when steady current flows across the interface, surface charges accumulate at the interface. The surface charge density is directly proportional to the difference between the relaxation times of the two adjoining materials.

Exercise 4.8

Express the ratio of σ_1 to σ_2 in terms of θ_1 and θ_2 as shown in Fig. 4.4.

Ans. $\sigma_1/\sigma_2 = \tan\theta_1 / \tan\theta_2$.

Exercise 4.9

For a long wire with a steady current lying along the z-axis, a current density in the form of $\mathbf{J} = J_\phi(\rho)\, \mathbf{a}_\phi + J_z(\rho)\, \mathbf{a}_z$ is expected from the cylindrical, translational, and anti-rotational symmetries. Show that $\mathbf{J}$ is uniform inside.

Ans. $\nabla \times (\mathbf{J}/\sigma) = 0$; thus, $\mathbf{J}$ is constant and uniform.

Review Questions

RQ 4.17   What are the two governing equations for **J**?   [(4.29)(4.30)]

RQ 4.18   State the boundary conditions for **J**.   [(4.32a,b)]

RQ 4.19   Which component of **D** is responsible for surface charges   [(4.33)]
induced at the conductor-conductor interface?

4.5 The Charge Relaxation Method

When a steady current flows through a long and straight wire, the two fundamental relations in Eqs. (4.29) and (4.30), along with symmetry considerations, show that the electric field is uniform inside and parallel to the surface of the conductor. Moreover, the boundary conditions for the electric field dictate that the electric field is nonzero outside the wire and that surface charges should exist on the conductor surface. The question then arises as to what causes the surface charge to remain intact when there is no current flowing towards the conductor surface. The charge relaxation method allows us to interpret the steady conduction current as the relaxation of electrode charges at regular intervals of relaxation time flowing along the electric flux of the electrode charge.

Let us consider the case in which a steady current starts from an electrode and flows towards the interface between two dissimilar conductors, as shown in Fig. 4.5. For simplicity, we assume that a point electrode is positioned at $(0, -a, b)$ in Cartesian coordinates, and that the $y = 0$ plane is the interface between two conductors with conductivities σ_1 and σ_2. The point electrode is assumed to maintain a net charge of q [C] at all times despite the fact that charges are released from the electrode at a fixed rate of $q(\varepsilon_1/\sigma_1)^{-1}$ [C/s] to form a steady current of such an amount, where ε_1/σ_1 is the relaxation time in conductor 1. When the electrode charge induces electric fields and steady currents in both conductors, the boundary condition and equation of continuity require that the tangential component of **E** and normal component of **J** are both continuous across the interface. The use of an image charge can significantly facilitate the fulfilment of these requirements at the interface.

We begin by assuming image charge q at point $(0, a, b)$ in the second conductor. The electric flux in conductor 1 is assumed to be a linear combination of the electric flux from the source and image charges. Meanwhile, the electric flux in conductor 2 is assumed to be proportional to the electric flux of the source charge only, ignoring the presence of the image charge. That is,

$$\mathbf{D}_1 = \frac{q}{4\pi}\left[\frac{x\mathbf{a}_x + (y+a)\mathbf{a}_y + (z-b)\mathbf{a}_z}{[x^2 + (y+a)^2 + (z-b)^2]^{3/2}} + A\frac{x\mathbf{a}_x + (y-a)\mathbf{a}_y + (z-b)\mathbf{a}_z}{[x^2 + (y-a)^2 + (z-b)^2]^{3/2}}\right]$$

$$\equiv \mathbf{D}_1^s + \mathbf{D}_1^i \tag{4.34a}$$

$$\mathbf{D}_2 = \frac{q}{4\pi}B\frac{x\mathbf{a}_x + (y+a)\mathbf{a}_y + (z-b)\mathbf{a}_z}{[x^2 + (y+a)^2 + (z-b)^2]^{3/2}}$$

$$= B\mathbf{D}_1^s \tag{4.34b}$$

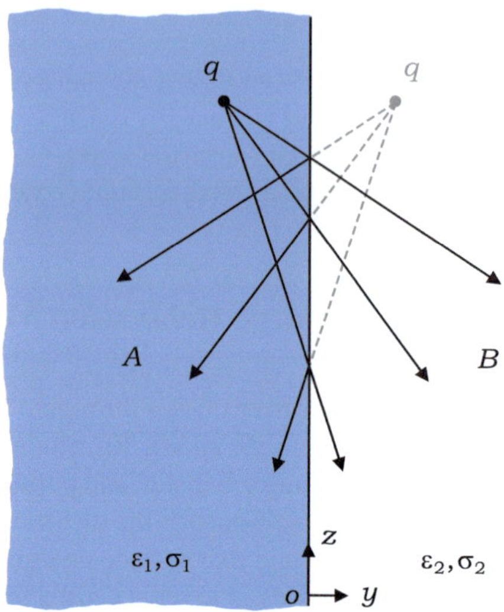

Fig. 4.5 Point electrode with charge q and its image charge (gray point) induce steady currents in the two adjoining conductors during the process of charge relaxation

where A and B are constants yet to be determined, and $\mathbf{D}_1^s$ and $\mathbf{D}_1^i$ are the electric flux densities in conductor 1, referring to the first and second terms in the brackets, respectively. Here, the superscripts s and i denote the source and image charges, respectively. The electric flux lines in Eq. (4.34) are schematically drawn for both conductors, as shown in Fig. 4.5. The electric flux may be viewed as if it flows from the source charge towards the interface and reflects off it.

Next, at the interface at $y = 0$, the normal components of $\mathbf{D}_1$ and $\mathbf{D}_2$ can be expressed in terms of the normal component of $\mathbf{D}_1^s$, which is the electric flux density due to the source charge only, ignoring the image charge. In a more general form, the normal components of the total electric flux densities at the opposite sides of the interface are expressed as

$$D_{1n} = \mathbf{D}_1\big|_{y=0} \cdot \mathbf{a}_n = (1 - A)D_{1n}^s \tag{4.35a}$$

$$D_{2n} = \mathbf{D}_2\big|_{y=0} \cdot \mathbf{a}_n = B D_{1n}^s \tag{4.35b}$$

where n denotes the normal component, and $\mathbf{a}_n = \mathbf{a}_y$ is the unit normal to the interface directed away from the source charge. Similarly, the tangential components of $\mathbf{D}_1$ and $\mathbf{D}_2$ can be expressed in a more general form as

$$D_{1t} = (1 + A)D_{1t}^s \tag{4.36a}$$

$$D_{2t} = B D_{1t}^s \tag{4.36b}$$

where t denotes the tangential component.

To make Eq. (4.35) satisfy the boundary conditions for $\mathbf{J}$, the electric flux density is first converted into the current density by multiplying it with σ/ε. Enforcing the normal component of $\mathbf{J}$ to be continuous across the interface results in

$$\frac{\sigma_1}{\varepsilon_1}(1 - A) = \frac{\sigma_2}{\varepsilon_2}B \tag{4.37a}$$

In a similar procedure as above, enforcing the tangential component of $\mathbf{E}$ in Eq. (4.36) to be continuous across the interface results in

$$\frac{1}{\varepsilon_1}(1 + A) = \frac{1}{\varepsilon_2}B \tag{4.37b}$$

Solving these equations yields

$$\boxed{A = \frac{1 - \sigma_2/\sigma_1}{1 + \sigma_2/\sigma_1}} \tag{4.38a}$$

$$\boxed{B = 2\frac{\varepsilon_2/\varepsilon_1}{1 + \sigma_2/\sigma_1}} \tag{4.38b}$$

As Gauss's law states, surface charges should be present across the interface, such as $\rho_s = D_{2n} - D_{1n}$. From Eqs. (4.35a) and (4.35b), the induced surface charge density is expressed as

$$\boxed{\rho_s = (A + B - 1)D_{1n}^s} \tag{4.39}$$

where D_{1n}^s is the normal component of $\mathbf{D}_1^s$ at the interface, which is produced only by the source charge and is directed away from the source. The so-called "electric flux coefficients" A and B refer to Eqs. (4.38a) and (4.38b). Using Eq. (4.35b), Eq. (4.39) can be rewritten as

$$\boxed{\rho_s = \frac{(A + B - 1)}{B}D_{2n}} \tag{4.40}$$

where D_{2n} is the normal component of $\mathbf{D}_2$ at the interface, and is directed away from the source charge.

As the image charge successfully prescribed the charge distribution at the interface, it is dismissed from its role in fulfilling the boundary conditions for $\mathbf{J}$ and $\mathbf{E}$. It is now considered that the electric fields in the two conductors are caused by the electrode and surface charges, with the image charge removed. The electric field $\mathbf{E}_1^i = \mathbf{D}_1^i/\varepsilon_1$ in Eq. (4.34a) was initially assumed to be caused by the image charge. However, it is now the electric field of the surface charge expressed by Eq. (4.39) or Eq. (4.40). From symmetry considerations, the electric field of ρ_s for $y \geqslant 0$ is a mirror image of $\mathbf{E}_1^i$, denoted by $\overline{\mathbf{E}}_1^i$, which can be obtained by simply replacing the term $y - a$ in $\mathbf{E}_1^i$ with $y + a$ for $\overline{\mathbf{E}}_1^i$. Therefore, in region $y \geqslant 0$, the total electric field is the sum of the electric fields of the source and surface charges, that is, $\mathbf{E}_1^s(y \geq 0) + \overline{\mathbf{E}}_1^i(y \geq 0)$, which can be readily shown to be exactly the same as $\mathbf{E}_2$ or $\mathbf{D}_2/\varepsilon_2$, as expressed by Eq. (4.34b).

As an example, a two-dimensional model of a finite wire is constructed with a slab conductor with width $2a$ and height $2c$, extending between $-\infty$ and ∞ along the x-axis, as shown in Fig. 4.6. Two infinitely long parallel filaments are embedded in the slab for two ideal sources, through which a steady current enters and exits the conductor. The two filaments are connected to a battery at infinity and are assumed to be perfectly conductive, so that the distribution of charges on the filaments is instantaneous during the flow of steady current from one filament to the other in the $-z$-direction, sustaining constant net charges at all times. Using this structure, the electric fields, surface charge densities, and equipotential surfaces in the yz-plane are computed using numerical methods.

Assuming the electrode charge to be $\pm 2\pi$ [mC/m] and setting $a = 1$ [cm], $b = 9$ [cm], $c = 10$ [cm], $\varepsilon_1 = \varepsilon_0$, and $\sigma_2 = 0$, the induced surface charge is computed. In the transient state, the total electric field due to the electrode and surface charges is first computed using Coulomb's law at points on the conductor surface. Next, the induced surface charge is calculated from the electric field using Eq. (4.39). This procedure is repeated until there is little change in the distribution of the surface charges.

The steady-state distribution of the surface charges is plotted against the distance from the top center of the slab, as shown in Fig. 4.7a. The surface charge peaks sharply at the corners of the conductor but varies almost linearly with distance in the middle of the conductor. The electric fields and equipotential surfaces are obtained

Fig. 4.6 Two-dimensional model of a finite wire carrying a steady current in the $-z$-direction

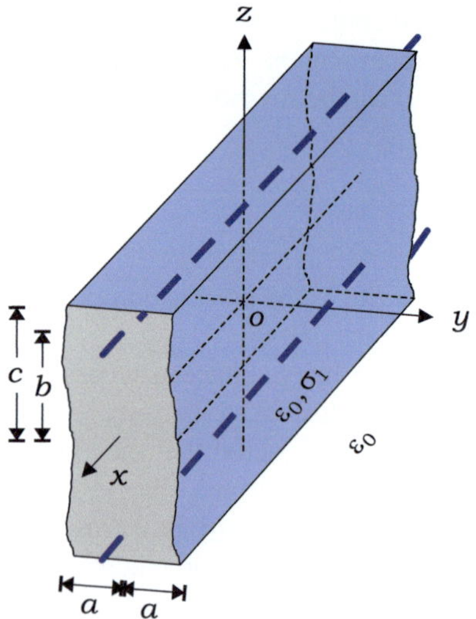

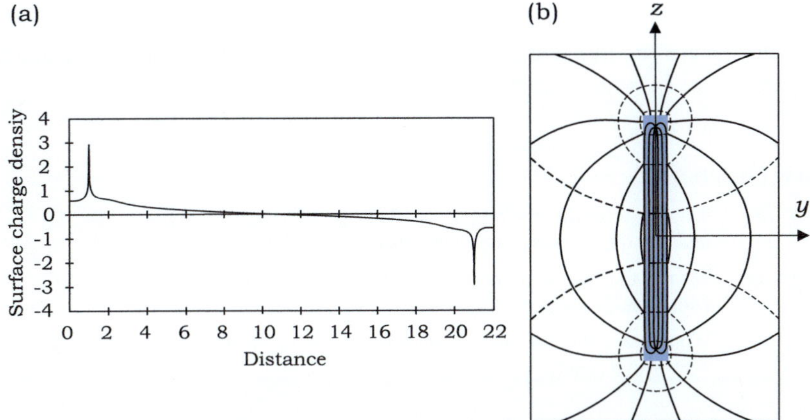

Fig. 4.7   **a** Surface charge density in units of $[C/m^2]$ versus distance in units of $[cm]$, starting from the top center of the slab to the bottom center. **b** Electric field lines, solid lines; Equipotential lines, dotted lines

from the steady-state electrode and surface charges and drawn in the yz-plane, as shown in Fig. 4.7b. The electric field is uniform inside the conductor, except in the region close to the electrode.

Example 4.4 When the current density is $\mathbf{J}_1 = b\,\mathbf{a}_y + c\,\mathbf{a}_z \,[A/m^2]$ for $z \leqslant 0$, as shown in Fig. 4.4, determine the surface charge density at the interface using Eq. (4.39).

Solution

The normal component of $\mathbf{D}_1$ at the interface is

$$D_{1z} = \frac{\varepsilon_1}{\sigma_1} c \tag{4.41a}$$

In view of Eq. (4.34a), the electric flux in Eq. (4.41a) must be caused by a source at infinity and surface charge at the interface. In addition, Eq. (4.35a) shows that the normal component of $\mathbf{D}$ at the interface, D_{1z}, is given by

$$D_{1z} = (1 - A)D_{1n}^s \tag{4.41b}$$

Combining Eqs. (4.41a) and (4.41b) and using the result in Eq. (4.39), the surface charge density is obtained as

$$\rho_s = \frac{A + B - 1}{1 - A} \frac{\varepsilon_1}{\sigma_1} c = c\left(\frac{\varepsilon_2}{\sigma_2} - \frac{\varepsilon_1}{\sigma_1} \right) \tag{4.41c}$$

The two results in Eqs. (4.33) and (4.41c) are equal.

Exercise 4.10

Determine the electric flux coefficients A and B at the conductor-air interface, where $\sigma_2 = 0$ and $\varepsilon_1 = \varepsilon_2 = \varepsilon_0$, as shown in Fig. 4.5.

Ans. $A = 1$ and $B = 2$.

Review Questions

RQ 4.20　What justifies that the relaxation of charge in the period of the relaxation time constitutes a conduction current?　　[(4.25)]

RQ 4.21　What is the source of charge piled up on a conducting wire with a steady current?　　[(4.39)]

RQ 4.22　What makes the electric field uniform inside a conducting wire with a steady current?　　[Fig. 4.7]

4.6 Resistance

A conductive body with a finite conductivity is called a **resistor**. When a voltage is applied across the terminals of a resistor, an electric field is established in the conductor and a conduction current flows between the terminals. In the transient state, the current flows towards the conductor surface and causes charge accumulation. In the steady state, the electric field is determined by both electrode and surface charges. As there is no net charge within the current-carrying conductors, the electric field should satisfy the fundamental relations $\nabla \cdot \mathbf{D} = 0$ and $\nabla \times \mathbf{E} = 0$ in the interior. Laplace's equation must be solved for analytical solutions of the electric field and the steady current. However, the boundary conditions for the electric potential and charge distribution on the conductor surface cannot be determined easily. Nevertheless, we can always use the charge relaxation method in Sect. 4.5 to obtain numerical solutions. Moreover, in many situations, using the steady current condition, in which the current flows parallel to the conductor surface, and considering the boundary condition, in which the electric field is perpendicular to the electrode surface at equipotential, the electric field is assumed to be uniform along the current-carrying wire.

For a straight resistor carrying steady current, as shown in Fig. 4.8, the internal electric field is uniform and parallel to the resistor surface; otherwise, the charge accumulates on the surface over time. This uniform electric field satisfies the two fundamental relations and the boundary conditions at the end faces. In this case, the total current through a cross section is

$$I = \int_{S} \mathbf{J} \cdot d\mathbf{s} = JS = \sigma ES \tag{4.42}$$

where σ is the conductivity, and S is the cross-sectional area. From the definition of potential difference, the relationship between the voltage across the terminals and the electric field inside the resistor is expressed as

$$V_{a-b} = -\int_{b}^{a} \mathbf{E} \cdot d\mathbf{l} = E\ell \tag{4.43}$$

Fig. 4.8 Straight resistor

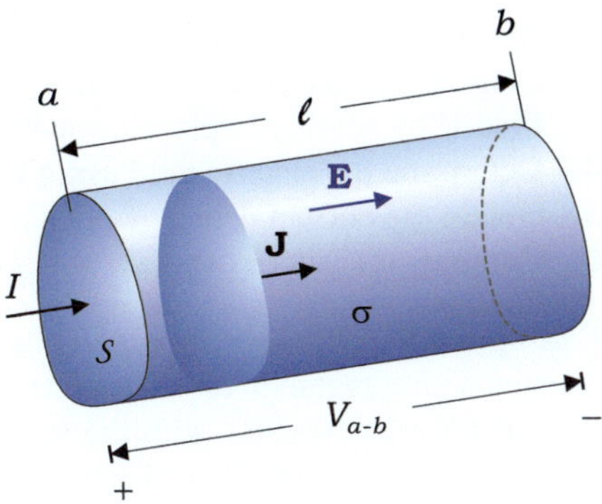

where ℓ is the resistor length. Combining Eqs. (4.42) and (4.43) yields the following voltage-current relationship:

$$\boxed{R = \frac{V_{a-b}}{I} = \frac{\ell}{\sigma S}} \quad [\Omega] \tag{4.44}$$

The ratio between the voltage and current is referred to as the **resistance**, denoted by R, and is measured in ohms $[\Omega]$. Resistance can also be expressed in terms of **resistivity** ρ such that

$$\boxed{R = \frac{\rho\ell}{S}} \quad [\Omega] \tag{4.45}$$

Resistivity is the reciprocal of conductivity, that is, $\rho = 1/\sigma$. It is measured in units of ohm meters $[\Omega \cdot m]$. The resistivity ρ should not be confused with the radial distance in cylindrical coordinates ρ or with the line charge density ρ_ℓ.

From Eq. (4.44), **Ohm's law** is derived as

$$\boxed{V = RI} \quad [V] \tag{4.46}$$

where the subscript $a-b$ is omitted for simplicity. Ohm's law states that the voltage across a resistor is equal to the product of the resistance and the current flowing in the resistor.

The **conductance** G is the reciprocal of resistance R, that is,

$$G = \frac{1}{R} = \frac{\sigma S}{\ell} \quad [S] \tag{4.47}$$

The conductance is measured in units of siemens $[S]$, whereas the conductivity is measured in units of siemens per meter $[S/m]$.

The resistance of a homogenous conductive body can be obtained by solving the boundary value problem. The procedure for determining resistance is as follows:

1. Assume a potential difference V_o across the conductor terminals.
2. Choose the coordinate system.
3. Find V by solving Laplace's equation.
4. Determine the electric field using $\mathbf{E} = -\nabla V$.
5. Obtain the current density from $\mathbf{J} = \sigma\mathbf{E}$ and total current I.
6. Compute resistance using $R = V_o/I$.

Example 4.5 A conductive material with conductivity σ is formed into a half-ring resistor with a rectangular cross section (Fig. 4.9a). In addition, the same material is used for a bar resistor with the same cross section and volume as the half-ring (Fig. 4.9b). Determine the resistance of the two resistors.

Solution

(a) Considering the geometry of the half ring, we assume that the plane of constant ϕ is an equipotential surface. Let $V = V_o$ at $\phi = \pi$, and $V = 0$ at $\phi = 2\pi$. Under these conditions, Laplace's equation is reduced to

$$\frac{d^2 V}{d\phi^2} = 0$$

This has a general solution, with constants c_1 and c_2, of the form

$$V = c_1\,\phi + c_2 \tag{4.48a}$$

By applying the boundary conditions to Eq. (4.48a), we obtain

$$V(\phi = \pi) = V_o = \pi c_1 + c_2$$

$$V(\phi = 2\pi) = 0 = 2\pi c_1 + c_2$$

Solving these equations for c_1 and c_2 yields

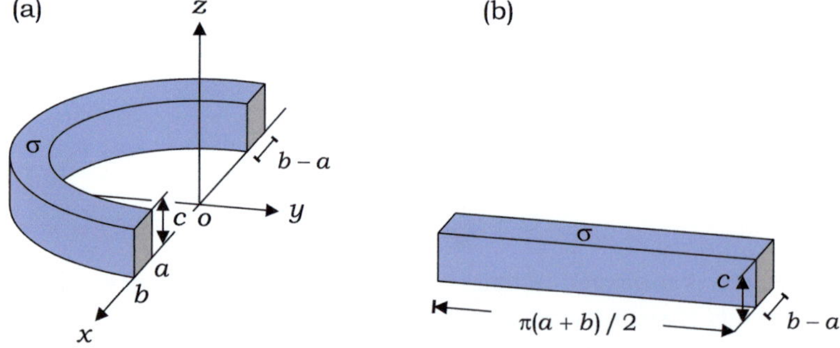

Fig. 4.9 A half-ring and a bar have the same cross section and volume

$$V = -\frac{V_o}{\pi}\phi + 2V_o$$

Thus, the electric field and current density inside are

$$\mathbf{E} = -\nabla V = \frac{V_o}{\rho\pi}\mathbf{a}_\phi \qquad (4.48b)$$

$$\mathbf{J} = \sigma\mathbf{E} = \frac{\sigma V_o}{\rho\pi}\mathbf{a}_\phi \qquad (4.48c)$$

The electric field in Eq. (4.48b) ensures that the two end plates are equipotential surfaces and satisfies another boundary condition in which $\mathbf{E}$ is parallel to the conductor surface. Therefore, this is a unique solution.

The total current in the conductor is

$$I = \int_S \mathbf{J}\cdot d\mathbf{s} = \int_{\rho=a}^{\rho=b}\int_{z=0}^{z=c} \frac{\sigma V_o}{\rho\pi}\mathbf{a}_\phi \cdot (dz\,d\rho\,\mathbf{a}_\phi)$$

$$= \frac{\sigma V_o}{\pi}c\ln(b/a) \qquad (4.48d)$$

Thus, the resistance of the half-ring is

$$R_1 = \frac{V_o}{I} = \frac{\pi}{\sigma c\ln(b/a)} \qquad (4.48e)$$

(b) A uniform internal electric field is a unique solution for this boundary value problem. Therefore, using Eq. (4.44), the resistance of the bar is obtained as

$$R_2 = \frac{\ell}{\sigma S} = \frac{\pi\,(a+b)}{2\sigma c\,(b-a)} \qquad (4.48f)$$

Using numerical methods, it can be shown that $R_2 > R_1$.

Exercise 4.11

The copper wire (No. 10 AWG) is listed as having a diameter of 2.588 [mm] and the resistance of 3.277 [Ω/km]. Verify the listed resistance.

Exercise 4.12

If a conducting wire of length ℓ is stretched to 2ℓ in length while the volume is kept the same, determine the new resistance.

Ans. $R_{new} = 4R$.

Review Questions

RQ 4.23	Express resistance in terms of conductivity.	[(4.44)]
RQ 4.24	State Ohm's law.	[(4.46)]
RQ 4.25	Distinguish between resistance and resistivity.	[(4.45)]
RQ 4.26	Distinguish between conductance and conductivity.	[(4.47)]

4.7 Power Dissipation and Joule's Law

When a conductive body is placed in an electric field, the free electrons of the conductor accelerate during the meantime between collisions. The electrons soon collide with the atomic lattice and are scattered in random directions. In the course of acceleration and collision, the electrons gain kinetic energy from the electric field and release it as thermal energy. From the perspective of the electric field, some of the potential energy of the electric field is converted into kinetic energy of free electrons and dissipated as heat.

From circuit theory, the electrical power is given by the product of voltage and current. Nevertheless, it is shown in the present section that electrical power can be expressed using the general relationship between the energy and force. Suppose that the electric field $\mathbf{E}$ exerts a Coulomb force on a free electron, causing it to move by differential distance $d\mathbf{l}$. If the displacement is performed in a short period dt, the power delivered to the electron is given by

$$p = \frac{e\mathbf{E} \cdot d\mathbf{l}}{dt} = e\mathbf{E} \cdot \boldsymbol{v} \quad [\text{W}] \tag{4.49}$$

where e is the electron charge, and $\boldsymbol{v} = d\mathbf{l}/dt$ is the drift velocity of the electron. Note that $\boldsymbol{v}$ is opposite in direction to $\mathbf{E}$. Next, the total power delivered to the electrons within a differential volume dv is computed as

$$\begin{aligned} dP &= (ndv)p = (ndv)\,e\mathbf{E} \cdot \boldsymbol{v} \\ &= \mathbf{E} \cdot \mathbf{J}\,dv \quad [\text{W}] \end{aligned} \tag{4.50}$$

where n is the number density of electrons, and $\mathbf{J}$ is the current density. The total power dissipated in volume $\mathcal{V}$ is therefore

$$\boxed{P = \int_{\mathcal{V}} \mathbf{E} \cdot \mathbf{J}\,dv} \quad [\text{W}] \tag{4.51}$$

This is known as **Joule's law**. The power is measured in watts [W] or joules per second.

In view of Eq. (4.51), the volume power density is defined as

$$\frac{dP}{dv} = \mathbf{E} \cdot \mathbf{J} \quad [\text{W/m}^3] \tag{4.52}$$

This is known as the point form of Joule's law.

Joule's law in Eq. (4.51) can be transformed into a more common form: $P = VI$. Consider a straight conducting wire in which the differential volume is given by $dv = dl\,ds$, where dl is the differential length along the wire, and ds is the differential area in the cross section. Because the electric field is parallel to the wire and uniform over the cross section, Eq. (4.51) can be rewritten as

$$P = \int_{\mathcal{V}} \mathbf{E} \cdot \mathbf{J}\,dv = \left(\int_{\mathcal{L}} E\,dl\right)\left(\int_{\mathcal{S}} J\,ds\right) = VI \quad [\text{W}] \tag{4.53}$$

Substituting Ohm's law, $V = IR$, into Eq. (4.53) leads to a familiar expression for ohmic power loss, that is,

$$\boxed{P = I^2 R} \quad [\text{W}] \tag{4.54}$$

This is the power dissipated in the resistance R.

Example 4.6 A slow-blow type of silver fuse is designed to disconnect the circuit in 10 [s] at twice the current rating of the fuse. Metallic silver with density 10.49 [g/cm^3] has a resistivity 1.59×10^{-8} [$\Omega \cdot$ m], specific heat 0.233 [J/g $\cdot$ K], and melting point of 1235 K. Assuming that the coefficients are independent of temperature, determine the diameter of the silver fuse rated at a DC current of 20 [A].

Solution

Let the silver fuse have cross section $\mathcal{S}$ [m^2] and length ℓ [m]. The thermal energy required to raise the fuse temperature from 293 K (room temperature) to 1235 K (melting point) is

$$W_T = 0.233 \left[\frac{\text{J}}{\text{g} \cdot \text{K}}\right] \times 10.49 \left[\frac{\text{g}}{\text{cm}^3}\right] \times (\mathcal{S} \times \ell \times 10^6)\,[\text{cm}^3] \times (1235 - 293)\,[\text{K}]$$

$$= (\mathcal{S}\ell) \times 2.30 \times 10^9\,[\text{J}]$$

The electrical energy expended in the resistance R in 10 [s] for a dc-current of 40 [A] is

$$W_E = I^2 R \times 10\,[\text{s}] = 40^2 \times \frac{1.59 \times 10^{-8}\ell}{\mathcal{S}} \times 10 = 2.54 \times 10^{-4}\frac{\ell}{\mathcal{S}}\,[\text{J}]$$

Equating W_T with W_E yields

$$\mathcal{S}^2 = (\pi D^2/4)^2 = 1.10 \times 10^{-13}$$

Thus, the diameter is

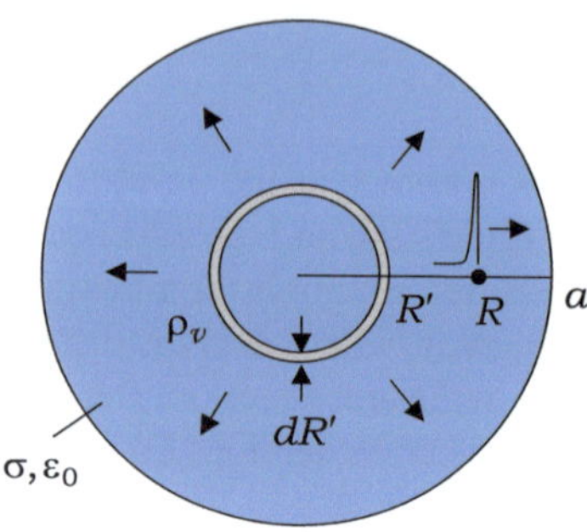

Fig. 4.10 Relaxation of a net charge produces a spherical current pulse that travels at c in the conductive sphere

$$D = 0.65[\text{mm}].$$

Example 4.7 A point charge of Q [C] is placed at the center of a conductive sphere with conductivity σ at time $t = 0$. As the charge relaxes, it forms a spherical shell of thickness $(\varepsilon_0/\sigma)\, c$, which is the product of the relaxation time and speed of light in the conductor (Fig. 4.10). (a) Express $\mathbf{J}$ and $\mathbf{E}$ in the sphere as functions of R and t. (b) Determine the total energy expended in the sphere. (c) Show that the result in (b) is equal to the difference between the potential energies of the charge at $t = 0$ and surface charge at $t = \infty$.

Solution

(a) The charge relaxes over time as

$$q(t) = Q\, e^{-(\sigma/\varepsilon_0)t}$$

From $I = -dq/dt$, the current density at a sphere of radius R' is given by

$$\mathbf{J}(R') = \mathbf{a}_R \frac{I}{4\pi R'^2} = \mathbf{a}_R \frac{\sigma Q}{4\pi\varepsilon_0 R'^2} e^{-(\sigma/\varepsilon_0)t'}$$

where t' denotes the time required for the current pulse to travel a distance R'. However, this expression only shows the variation of the current over time at a fixed position R'. In other words, the traveling of the current pulse was not incorporated. As a matter of fact, at an advanced position R, the current pulse can be observed at a later time, delayed by R/c [s]. Therefore, in general, the current density in the sphere is expressed as

$$\mathbf{J}(R, t) = \mathbf{a}_R \frac{\sigma Q}{4\pi\varepsilon_0 R^2} e^{-(\sigma/\varepsilon_0)[t - R/c]} \quad (R/c \le t < \infty) \tag{4.55a}$$

At a fixed position, R_o, Eq. (4.55a) describes the temporal variation of the current across the sphere of radius R_o as time advances, starting from $t = R_o/c$. Meanwhile, at an instant in time $t_o = R_o/c$, Eq. (4.55a) describes the spatial variation of the current in region $R \leqslant R_o$, which clearly tails off towards the center of the sphere.

Finally, from the point form of Ohm's law, the electric field in the sphere is obtained as

$$\mathbf{E}(R, t) = \frac{\mathbf{J}}{\sigma} = \mathbf{a}_R \frac{Q}{4\pi\varepsilon_0 R^2} e^{-(\sigma/\varepsilon_0)[t - R/c]} \quad (R/c \leq t < \infty) \qquad (4.55b)$$

(b) The total energy expended in the conductor can be expressed as

$$W = \int_T \int_V \mathbf{E} \cdot \mathbf{J}\, dv\, dt$$

where t ranges from 0 to ∞, and V is the volume of the spherical current pulse of thickness $(\varepsilon_0/\sigma)c$ [m]. When the current pulse passes through a sphere of radius R, the energy expended in a thin spherical shell of volume $4\pi R^2 dR$ is given by

$$\Delta W = \int_{t=R/c}^{t=\infty} (4\pi R^2 dR)\, \sigma \left(\frac{Q}{4\pi\varepsilon_0 R^2} \right)^2 e^{-2(\sigma/\varepsilon_0)[t - R/c]}\, dt$$

$$= \frac{Q^2}{8\pi\varepsilon_0 R^2} dR$$

where Eqs. (4.55a) and (4.55b) are used. Because the tail of the current pulse actually extends to $t = \infty$, the upper limit of the integration should be $t = \infty$, not $t = \varepsilon_0/\sigma$. The lower limit $t = R/c$ corresponds to the instant when the pulse arrives at position R.

Next, integrating ΔW over the volume of the conducting sphere yields

$$W = \int_{R=\delta}^{R=a} \Delta W$$

$$= \lim_{\delta \to 0} \frac{Q^2}{8\pi\varepsilon_0} \left(\frac{1}{\delta} - \frac{1}{a} \right) \qquad (4.55c)$$

where δ is used to exclude a singular point at the center.

(c) At $t = 0$, the charge is concentrated at the center. With the help of Exercise 3.39, the potential energy of the point charge Q at the center can be expressed as

$$W_1 = \lim_{\delta \to 0} \frac{1}{2} \frac{Q^2}{4\pi\varepsilon_0 \delta} \qquad (4.55d)$$

At steady state ($t = \infty$), the charge is uniformly distributed over the surface of the sphere, and its potential energy is given by

$$W_2 = \frac{1}{2}\frac{Q^2}{4\pi\varepsilon_0 a} \tag{4.55e}$$

The difference, $W_1 - W_2$, is equal to the total energy expended in conductor W, as expressed in Eq. (4.55c).

Exercise 4.13
For the copper wire (No. 10 AWG) rated at 30 [A] (see Exercise 4.11), determine the maximum power dissipated in the wire per kilometer.

Ans. 2.95 [kW].

Exercise 4.14
Two voltage sources, 110 [V] and 220 [V], deliver the same power to the load through identical low-loss wires. Determine the ratios between (a) the currents, and (b) the powers dissipated in the wires.

Ans. (a) $I_{110}/I_{220} = 2$, (b) $P_{110}/P_{220} = 4$.

Exercise 4.15
For two resistors connected (a) in parallel, and (b) in series, is the total dissipated power equal to the sum of the powers dissipated in the individual resistors?

Ans. (a) Yes, (b) Yes.

Review Questions

RQ 4.27	State Joule's law.	[(4.51)]
RQ 4.28	Express ohmic power loss.	[(4.53)(4.54)]
RQ 4.29	Why is the higher voltage in power transmission lines advantageous?	[Exercise 4.14]

4.8 Analogy Between D and J

Permittivity is characteristic of dielectrics, whereas conductivity is characteristic of conductors. When a material with permittivity ε and conductivity σ has an electric field inside, the electric flux density is related to the electric field through permittivity and the current density is related to the electric field through conductivity. That is,

$$\mathbf{D} = \varepsilon \mathbf{E} \tag{4.56a}$$

$$\mathbf{J} = \sigma \mathbf{E} \tag{4.56b}$$

In the same manner that the closed-surface integral of $\mathbf{D}$ is equal to the total enclosed charge Q, the closed-surface integral of $\mathbf{J}$ is equal to the total outward current I. Expressed mathematically,

$$\oint_{S} \mathbf{D} \cdot d\mathbf{s} = Q \tag{4.57a}$$

$$\oint_{S} \mathbf{J} \cdot d\mathbf{s} = I \tag{4.57b}$$

In addition, the boundary conditions dictate that both $\mathbf{D}$ and $\mathbf{J}$ terminate at right angles on the conductor surface. In view of these considerations, we conclude that $\mathbf{D}$, ε, and Q are analogous to $\mathbf{J}$, σ, and I, respectively.

We consider two conductors embedded in a lossy dielectric connected to a dc-voltage source, as illustrated in Fig. 4.11. From the analogy between $\mathbf{D}$ and $\mathbf{J}$, the relationship between the capacitance and leakage resistance of the conductors can be derived, and is particularly useful for determining the capacitance of a given system from a known resistance, and vice versa.

As stated before, the potential difference between the conductors is given by

$$V_{a-b} = -\int_{b}^{a} \mathbf{E} \cdot d\mathbf{l} \tag{4.58}$$

Using Eqs. (4.57b) and (4.58), the leakage resistance between the conductors can be expressed as

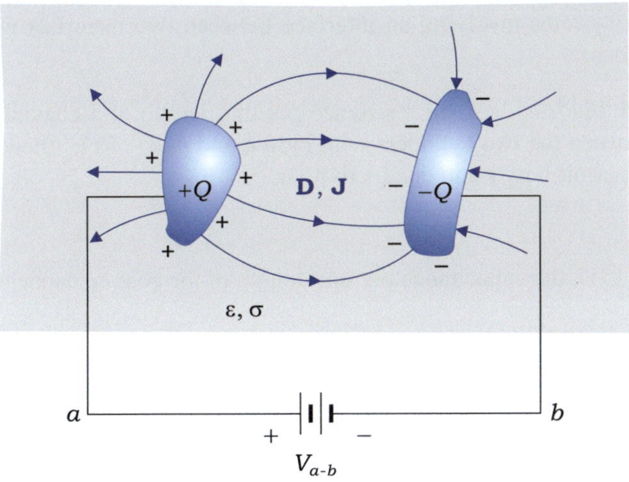

Fig. 4.11 Analogy between $\mathbf{D}$ and $\mathbf{J}$

$$R = \frac{V_{a-b}}{I} = \frac{-\int_b^a \mathbf{E} \cdot d\mathbf{l}}{\oint_S \mathbf{J} \cdot d\mathbf{s}} \tag{4.59}$$

Next, using Eqs. (4.57a) and (4.58), the capacitance between the conductors can be expressed as

$$C = \frac{Q}{V_{a-b}} = \frac{\oint_S \mathbf{D} \cdot d\mathbf{s}}{-\int_b^a \mathbf{E} \cdot d\mathbf{l}} \tag{4.60}$$

In a simple medium, for which ε and σ are constants, taking the product of Eq. (4.59) and Eq. (4.60), together with Eq. (4.56), we obtain

$$\boxed{RC = \frac{\varepsilon}{\sigma}} \tag{4.61}$$

The product of R and C in a two-conductor system is equal to the relaxation time of the surrounding material. This relation can be conveniently used to determine the leakage resistance from a known capacitance, and vice versa. With the relation $R = 1/G$, Eq. (4.61) can be rewritten as

$$\boxed{\frac{C}{G} = \frac{\varepsilon}{\sigma}} \tag{4.62}$$

It is important to note that the relationships in Eqs. (4.61) and (4.62) may not be applicable to systems involving an interface between two materials with different relaxation times.

Example 4.8 Find the leakage resistance per unit length of a coaxial capacitor if the space between the two cylinders with radii a and b $(a < b)$ is filled with a lossy dielectric of permittivity ε and conductivity σ.

Solution

From Eq. (3.151), the capacitance per unit length of the coaxial capacitor is

$$C = \frac{2\pi\varepsilon}{\ln(b/a)} \quad [\text{F/m}]$$

With the help of Eq. (4.61), the leakage resistance is obtained as

$$R = \frac{\varepsilon}{\sigma} \frac{1}{C} = \frac{\ln(b/a)}{2\pi\sigma} \quad [\Omega] \tag{4.63}$$

This is exactly what can be obtained by following the procedure for R described in Sect. 4.6.

Exercise 4.16

A parallel-plate capacitor of 0.01 [μF] is filled with lossy dielectric of $\varepsilon_r = 2.5$ and $\sigma = 10^{-3}$ [S/m]. Ignoring the fringing effect, determine the leakage resistance.

Ans. 2.2 [Ω].

Review Questions

RQ 4.30   What makes the field lines of **D** and **J** look the same?   [(4.56)]

RQ 4.31   Explain the analogy between **D** and **J**.   [Fig. 4.11]

RQ 4.32   Relate the capacitance to the leakage resistance.   [(4.61)]

4.9 Problems

Current and current density

4.1 Given the current density $\mathbf{J} = 0.3\mathbf{a}_y + 0.4\mathbf{a}_z$ [A/cm²], determine the total current flowing through a half-cylinder defined by $\rho = 2$ [cm], $0 \leq \phi \leq \pi$, and $0 \leqslant z \leqslant 2$ [cm], as shown in Fig. 4.12.

4.2 For the current density $\mathbf{J} = 20e^{-10\rho^2}\mathbf{a}_z$ [A/m²], determine the total current flowing through a section of a sphere, defined by $R = 0.4$ [m], $0 \leq \theta \leq 50^o$, and $0 \leq \phi \leq 2\pi$, as shown in Fig. 4.13.

4.3 Repeat Problem 4.2 for current density $\mathbf{J} = 6\cos(\rho^2)\,\mathbf{a}_z$ [A/m²]. [Hint: Equation of continuity.]

4.4 A copper wire with a diameter of 3 [mm] carries a steady current of 1.6 [A]. Find the number of conduction electrons crossing a plane at an angle of 45° to the wire axis per second.

Conductivity

Fig. 4.12 Half-cylinder (Problem 4.1)

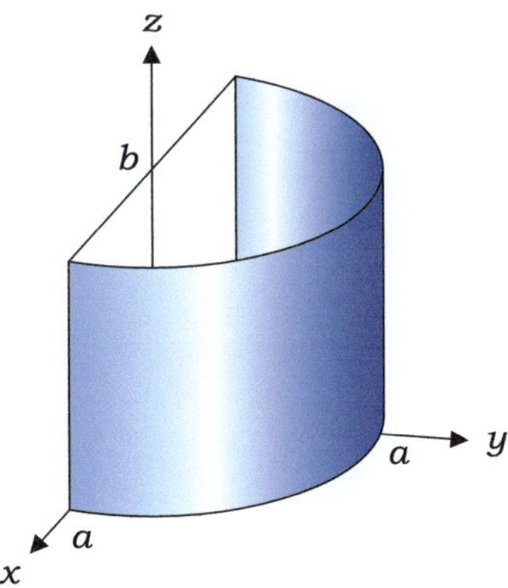

Fig. 4.13 A section of a sphere (Problems 4.2 and 4.3)

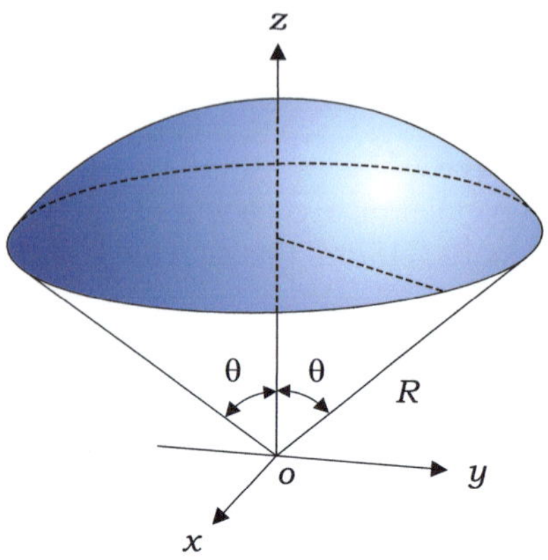

4.5   A 10 [mm]-long intrinsic silicon bar has a rectangular cross section of 4 [mm] × 2 [mm], for which $\mu_e = 0.14\,[\mathrm{m^2/V \cdot s}]$, $\mu_h = 0.045\,[\mathrm{m^2/V \cdot s}]$, and $n_e = n_h = 1.0 \times 10^{16}\,[\mathrm{m^{-3}}]$ at 300 K. When the voltage drop between the two ends of the bar is 5 [V], determine

    (a)   the conductivity,
    (b)   the total current,
    (c)   the resistance, and
    (d)   the power dissipated in the bar.

4.6   When intrinsic silicon is doped with donors, the electron density increases by 10% at 300 K. Compute the increase in conductivity using the known material parameters.

Resistance

4.7   A wire with a circular cross section of radius a has conductivity σ_1 and is coated with a material with conductivity σ_2 up to $\rho = b$ (Fig. 4.14). Determine b such that an equal current flows into the core and coating layers.

Fig. 4.14 Coated wire (Problems 4.7 and 4.14)

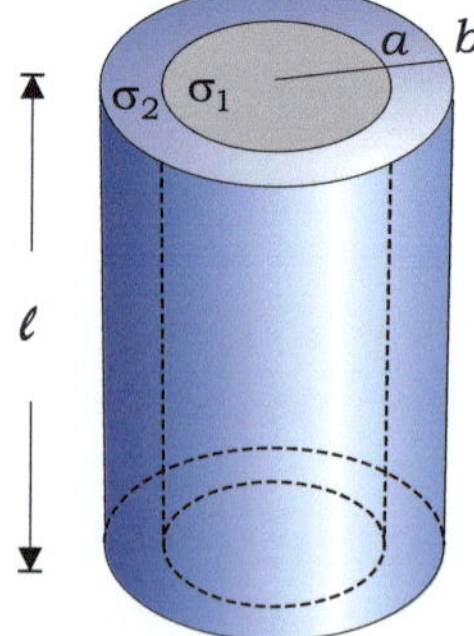

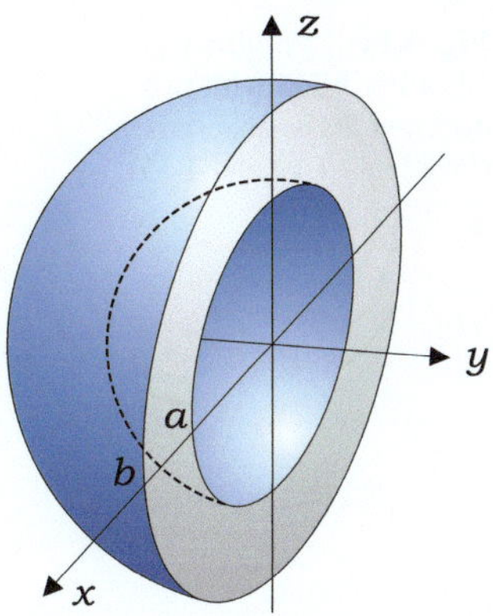

Fig. 4.15 Hemispherical shell made of ohmic material (Problem 4.10)

4.8 A 10 [cm] long coaxial resistor consists of two conducting cylinders with radii $a = 5$ [cm] and $b = 6$ [cm]. When the gap is filled with graphite with $\sigma = 7 \times 10^4$ [S/m], determine the resistance between the cylinders by solving Laplace's equation, ignoring the edge effect.

4.9 Repeat Problem 4.8, assuming that the inner and outer cylinders are at potentials of 10 [V] and 0 [V], respectively, by directly solving $\nabla \cdot \mathbf{E} = 0$ in the gap.

4.10 An ohmic medium with conductivity σ is formed into a hemispherical shell with radii a and b, as shown in Fig. 4.15. Ignoring the edge effect, determine the resistance between the two hemispherical surfaces.

4.11 A carbon composition is formed into a cylinder with a diameter of 2.4 [mm] and length of 6.7 [mm]. If it has resistance 1 [MΩ], determine the conductivity of the material.

4.12 The conductivity of copper depends on the temperature T in degrees Celsius in such a way that $\sigma = \sigma_o[1 + \alpha(T - T_o)]^{-1}$, where $\sigma_o = 5.8 \times 10^7$ [S/m], $T_o = 20^o$C, and $\alpha = 0.0039$ [C^{-1}]. Determine the resistance of a 10 [m] long copper wire with a diameter of 0.2 [mm] at (a) $T = -30^o$C, and (b) $T = 40^o$C.

4.13 A cylindrical capacitor consists of two coaxial cylinders with a gap filled with two lossy dielectrics (Fig. 4.16). When it is connected to a voltage source, V_o [V], ignoring the edge effect, determine (a) the current density, (b) ρ_s at the $\rho = a$, $\rho = b$, and $\rho = c$ surfaces, and (c) the resistance.

4.14 For the coated wire, as shown in Fig. 4.14, determine b such that the two regions have the same resistance.

4.15 If the two ends of the double-layered cylinder, as shown in Fig. 4.16, are capped with conducting disks and the conducting cylinders at $\rho = a$ and $\rho = c$ are removed, determine the resistance between the two disks.

4.16 Two parallel conducting plates with area S coincide with the $z = d$ and $z = 0$ planes, and are maintained at potentials $V = V_o$ and $V = 0$, respectively. If the gap is filled with an inhomogeneous material with conductivity $\sigma = \sigma_o[1 + (z/d)]^{1/2}$, ignoring the edge effect, determine the current density, electric field, and resistance between the plates.

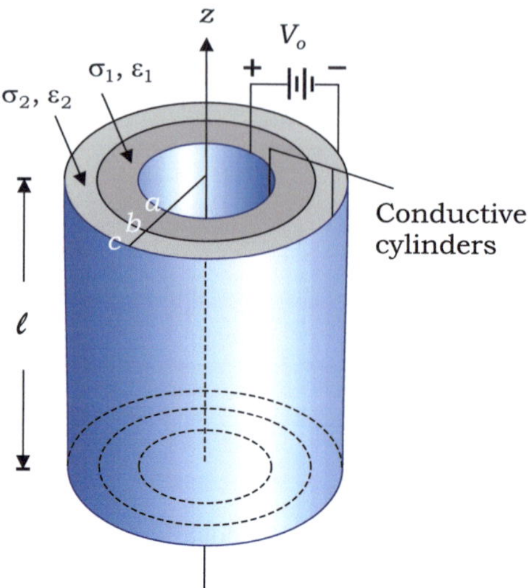

Fig. 4.16 Cylindrical capacitor filled with two different lossy materials (Problems 4.13 and 4.15)

4.17 A conic object with conductivity σ has two spherical ends coated with perfectly conductive material (Fig. 4.17). The steady current flowing between the end surfaces conforms to the cone such that **J** is normal to the end surface and parallel to the conic surface. Determine (a) the radii of curvature of the end surfaces, and (b) the resistance between the two end surfaces.

Boundary conditions

4.18 The $z = 0$ plane is the interface between the two lossy dielectrics: σ_1 and ε_1 for $z < 0$, and σ_2 and ε_2 for $z > 0$. For $\mathbf{J}_1 = 2\mathbf{a}_y + 3\mathbf{a}_z$ [A/m^2] in region $z < 0$, find (a) $\mathbf{J}_2$ in region $z > 0$, (b) **E** everywhere, and (c) ρ_s at the interface.

Fig. 4.17 Truncated cone with steady current (Problem 4.17)

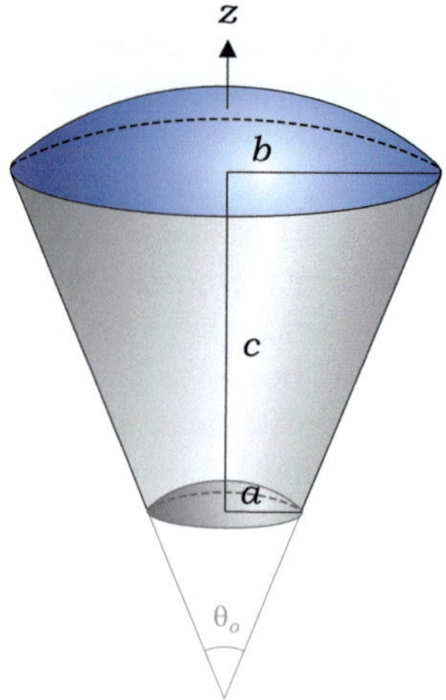

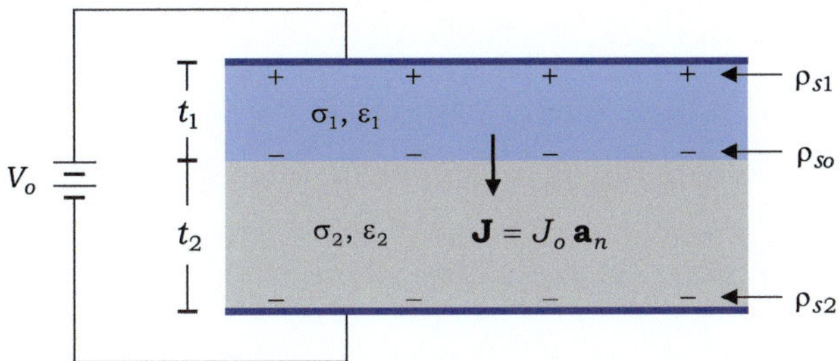

Fig. 4.18 Parallel plates filled with two lossy dielectrics (Problems 4.19, 4.24, and 4.25)

4.19 The space between the two parallel perfectly conducting plates is filled with two lossy dielectrics, as shown in Fig. 4.18. Ignoring the fringing effect, determine (a) $\mathbf{J}$ and $\mathbf{E}$ in the gap from the applied voltage V_o, and (b) three surface charge densities, ρ_{so}, ρ_{s1}, and ρ_{s2}.

Charge relaxation

4.20 A wire with conductivity σ and permittivity ε is in contact with the thin leads at the ends. The two point-contacts maintain equal but opposite charges, $\pm q$ [C], at all times because the battery is connected to the leads. Determine the current flowing through the wire.

4.21 If the net charge of q [C/m^3] is reduced to $0.01q$ in 1 [ns] in a material with $\varepsilon_r = 10$, determine the conductivity of the material.

4.22 The relaxation of net charge can be a measure of the performance of an insulator. Which is the better insulator: distilled water with $\varepsilon_r = 81$ and $\sigma = 10^{-3}$ [S/m], or dry sand with $\varepsilon_r = 4$ and $\sigma = 10^{-4}$ [S/m]?

4.23 A copper sphere with a radius of 0.2 [m] is charged at 1 [C] and connected to the ground through a thin Nichrome wire with a radius of 1 [mm] and conductivity 6.70×10^5 [S/m]. Assuming that the charge is uniformly distributed on the sphere during discharge, determine the transient current in the wire.

4.24 Starting with $\mathbf{J} = J_o\,\mathbf{a}_n$ in the space between the two parallel plates, as shown in Fig. 4.18, determine the surface charge density at the interface, ρ_{so}, using Eq. (4.40).

4.25 The conductive plates, as shown in Fig. 4.18, have a finite conductivity of $\sigma_0(< \sigma_1)$ and permittivity ε_1. Starting with $\mathbf{J} = J_o\,\mathbf{a}_n$ flowing between the planes, determine ρ_{s1} on the upper plate using Eq. (4.40), and check that the plate is negatively charged, even if it is at a higher potential than the lower plate.

Chapter 5
Magnetostatics

In Chap. 3, we studied electrostatics, which are primarily concerned with the electric field and electric flux density in various types of dielectrics caused by various static charge distributions. In this chapter, we focus on magnetostatics that deals with the magnetic field and magnetic flux density in various types of magnetic materials caused by various steady current distributions. As the rules of electrostatics are based on Coulomb's law and the principle of superposition, magnetostatic rules are based on the Biot-Savart law and the superposition principle. An analogy exists between the electric and magnetic fields. Just as the electric flux density and electric field are related by the constitutive relation $\mathbf{D} = \varepsilon\mathbf{E}$, where ε is the permittivity, the magnetic flux density and magnetic field are related by the constitutive relation $\mathbf{B} = \mu\mathbf{H}$, where μ is the ***permeability***. As in the case of electrostatics, a static magnetic field can be uniquely determined by its divergence and curl. Although such an analogy can help us to comprehend the concept of magnetostatics to a certain degree, we should not rely entirely on this analogy when studying magnetostatics. For example, $\mathbf{B}$ is responsible for the magnetic force exerted on the electric charge in motion, whereas $\mathbf{E}$ is responsible for the electric force exerted on the charge. In the same way as $\mathbf{D}$ is defined in dielectrics, $\mathbf{H}$ is defined in magnetic materials such that it is connected only to free currents and is thus independent of the material. From the point of view of physics, it would be a convincing argument that $\mathbf{B}$ and $\mathbf{H}$ are the respective counterparts of $\mathbf{E}$ and $\mathbf{D}$.

Magnetism was discovered by ancients, who observed that pieces of lodestone could attract small pieces of iron. A lodestone is a piece of naturally magnetized magnetite, which is believed to be caused by a strong magnetic field around a lightning bolt. Lodestones were found near the ancient city of Magnesia, and the word 'magnet' was derived from the Greek word 'magnesian stone'. Magnetism was thought to be independent of electricity until 1819 when Hans Christian Oersted observed that the needle of a compass was deflected in the vicinity of a current-carrying wire. Oersted's observations established a link between magnetic and electric fields in that an electric field produces a current in the wire, and the current generates a magnetic field around the wire. However, this link is incomplete, because static magnetic fields

Y. H. Lee, *Introduction to Engineering Electromagnetics*,
https://doi.org/10.1007/978-3-031-28659-9_5

cannot generate static electric fields. In Chap. 6, we show that a time-varying $\mathbf{B}$ can induce a time-varying $\mathbf{E}$, which completes the link between magnetic and electric fields.

Our discussion on static magnetic fields begins with the Biot-Savart law, which states that a steady current produces a static magnetic field. Next, we explore Ampere's circuital law, which is a special case of the Biot-Savart law, just as Gauss's law is a special case of Coulomb's law. Ampere's circuital law is particularly useful when the geometry under consideration has certain symmetry. The next topic of discussion is the magnetization vector, which is essential for explaining the interaction between an external magnetic field and magnetic material. The magnetic field intensity $\mathbf{H}$ is redefined in magnetic materials such that Ampere's law can be expressed in terms of free currents alone. Using the two fundamental relations for static magnetic fields, that is, $\nabla \cdot \mathbf{B} = 0$ and $\nabla \times \mathbf{H} = \mathbf{J}$, we discuss the boundary conditions for $\mathbf{B}$ and $\mathbf{H}$ at the interface between two different magnetic materials. We extend our discussion to the inductance, magnetic energy, and torque.

5.1 Lorentz Force Equation

From Chap. 3, we recall that an electric charge q, either at rest or in motion in an electric field $\mathbf{E}$, experiences an electric force expressed as

$$\mathbf{F}_e = q\,\mathbf{E} \quad [\text{N}] \tag{5.1}$$

Just as the electric field is defined as the electric force on a unit charge, the ***magnetic flux density*** may be defined as the magnetic force exerted on a unit charge moving with a unity velocity. From experimental observations, it was deduced that an electric charge q moving at velocity $\boldsymbol{v}$ in magnetic flux density $\mathbf{B}$ experiences a magnetic force expressed as

$$\mathbf{F}_m = q\,\boldsymbol{v} \times \mathbf{B} \quad [\text{N}] \tag{5.2}$$

The magnetic flux density is given in units of tesla [T] or webers per square meter $[\text{Wb/m}^2]$.

If charge q moves with velocity $\boldsymbol{v}$ in the presence of both electric field $\mathbf{E}$ and magnetic flux density $\mathbf{B}$, the total force exerted on the charge is given by

$$\boxed{\mathbf{F} = \mathbf{F}_e + \mathbf{F}_m = q(\mathbf{E} + \boldsymbol{v} \times \mathbf{B})} \quad [\text{N}] \tag{5.3}$$

This is known as the ***Lorentz force equation***. It is important to note that the electric force acts on the electric charge regardless of whether the charge is moving, whereas the magnetic force involves only electric charge in motion. Because $\mathbf{F}_m$ is always perpendicular to the direction of motion of the charge, there is no expenditure of magnetic energy even though the charge may be displaced by the magnetic force. In other words, there is no transfer of magnetic energy stored in **B** to the kinetic energy of the charge. *The magnetic force does no work.* It cannot alter the speed of the charge but changes the direction of motion of the charge.

Example 5.1 In the presence of a uniform magnetic flux density $\mathbf{B} = B_o \mathbf{a}_z$, an electron crosses the x-axis at velocity $\mathbf{v} = v_o \mathbf{a}_y$ at time $t = 0$. Determine the radius of the circle that the electron traces for $t \geq 0$.

Solution

The magnetic force on the electron is

$$\mathbf{F}_m = e\,\mathbf{v} \times \mathbf{B} \qquad (e < 0) \tag{5.4a}$$

$\mathbf{F}_m$ is always perpendicular to $\mathbf{v}$ in the xy-plane. Nevertheless, the acceleration of the electron of mass m is

$$\mathbf{a} = \mathbf{F}_m / m \tag{5.4b}$$

This is the time rate of $\mathbf{v}$, that is, $\mathbf{a} = d\mathbf{v}/dt$. If the angular separation between $\mathbf{v}$ and $\mathbf{v}'$ is $\Delta\theta$, as illustrated in Fig. 5.1, then, by definition,

$$|\mathbf{a}| = \lim_{\Delta t \to 0}\left|\frac{\mathbf{v}' - \mathbf{v}}{\Delta t}\right| = \lim_{\Delta t \to 0}\frac{\Delta\theta\, v_o}{\Delta t} \qquad (|\mathbf{v}| = v_o)$$

Thus,

$$\frac{|\mathbf{a}|}{v_o} = \frac{\Delta\theta}{\Delta t} \equiv \omega \tag{5.4c}$$

where ω is the angular velocity, which is also defined as $\omega\, r_o = v_o$. Using Eqs. (5.4a) and (5.4b) on the left-hand side of Eq. (5.4c) and the relation $\omega = v_o/r_o$ on the right-hand side, we obtain

$$r_o = \frac{m v_o}{|e| B_o} \tag{5.4d}$$

Exercise 5.1

Determine the electric field $\mathbf{E}$ applied for $t \geq 0$ to make the electron in Fig. 5.1, along with **B**, move in a straight line.

Ans. $\mathbf{E} = -v_o B_o \mathbf{a}_x$.

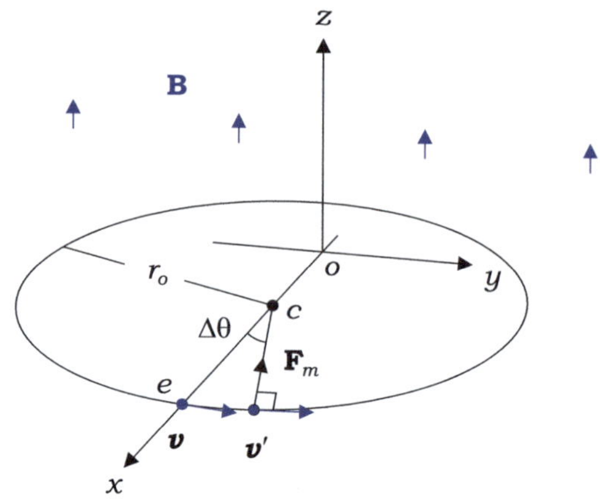

Fig. 5.1 An electron is in circular motion in uniform **B**

Exercise 5.2

If an electric field $\mathbf{E} = E_o \mathbf{a}_z$ is applied to the electron, as shown in Fig. 5.1, in addition to **B**, what trace is formed by the electron?

Ans. Helix.

Review Questions

RQ 5.1   What would lead to the statement that energy is never transferred   [(5.2)]
from the magnetic field to the charge in motion?

RQ 5.2   State Lorentz force equation.   [(5.3)]

5.2   The Biot-Savart Law

Previously, magnetic flux density was introduced as a vector field that exerted a force on the charge in motion. We now shift our focus to the magnetic field, which is a vector field induced by steady current. The ***Biot-Savart law*** states that the differential magnetic field due to a steady current I flowing in the line segment $d\mathbf{l}'$ is given by

$$\boxed{d\mathbf{H} = \frac{I\,d\mathbf{l}' \times \mathbf{a}_\mathcal{R}}{4\pi \mathcal{R}^2}} \quad [\text{A/m}] \tag{5.5}$$

where $\mathbf{a}_\mathcal{R}$ is the unit vector in the direction of distance vector $\mathcal{R}$ from the position of $I\,d\mathbf{l}'$ (source point) to the position of $d\mathbf{H}$ (field point). A conducting wire of negligible thickness is referred to as a ***filament*** throughout the text. The filamentary current I in Eq. (5.5) is given in amperes [A] and the magnetic field is measured in units of amperes per meter [A/m].

   Although the cross product in Eq. (5.5) follows the right-hand rule, the magnetic field itself does not involve any rotation. The cross product is simply a mathematical

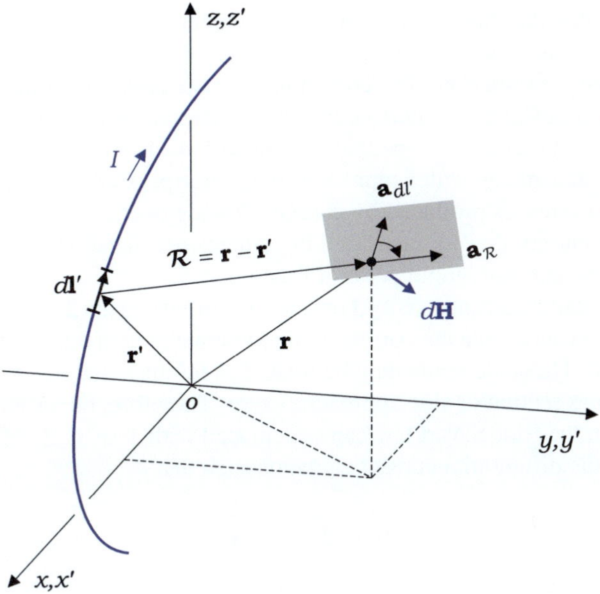

Fig. 5.2   Differential magnetic field at $\mathbf{r}$ due to a differential current element at $\mathbf{r}'$

symbol that represents a direction normal to the plane containing two vectors $d\mathbf{l}'$ and $\mathbf{a}_{\mathcal{R}}$, and a magnitude proportional to the sine of the angle between the vectors. It is important to note that $d\mathbf{l}' \times \mathbf{a}_{\mathcal{R}}$ represents a vector quantity observed at the field point, although it contains a vector at the source point, $d\mathbf{l}'$. The cross product is more readable if it is rearranged as $|d\mathbf{l}'|\mathbf{a}_{dl'} \times \mathbf{a}_{\mathcal{R}}$, where the unit vector $\mathbf{a}_{dl'}$ can be displaced to the field point for the cross product (Fig. 5.2).

The magnetic field is inversely proportional to the square of the distance between the differential current element and the field point, just as the electric field is inversely proportional to the square of the distance between the charge and the field point. Moreover, both the fields depend linearly on the source. However, in contrast with the electric field directed along the distance vector from the source to the field point, the magnetic field is always perpendicular to the distance vector.

The principle of superposition allows us to obtain the magnetic field of a steady current I flowing in a filament using the vector sum of the contributions from all the individual current elements that comprise the filamentary current. That is,

$$\boxed{\mathbf{H} = \int_{C'} \frac{I\, d\mathbf{l}' \times \mathbf{a}_{\mathcal{R}}}{4\pi \mathcal{R}^2}} \quad [\mathrm{A/m}] \tag{5.6}$$

where C' denotes the path of integration or filament with current I. The path of integration is a mathematical term, whereas the filament is an object in the physical world. The positive direction of C' corresponds to the direction of the current flow.

As discussed in Chap. 4, when a current is generated by the flow of charges in a three-dimensional region, it is specified by the volume current density measured in amperes per square meter, which represents the charge passing through a unit area per unit time. If a current is produced by charges flowing over a surface, it is described by the surface current density measured in amperes per meter, which represents the charges crossing per unit length per unit time.

When the volume current density $\mathbf{J}$ or surface current density $\mathbf{J}_s$ is given in a region of space, we first subdivide the corresponding volume or area into many elements of volume or area. These elements may be treated as infinitesimal line segments in the limit as the cross section or side approaches zero. Therefore, the differential current element $I\,d\mathbf{l}'$ in the Biot-Savart law can be replaced with $\mathbf{J}'\,dv'$ or $\mathbf{J}_s'\,ds'$. Equivalence exists among the differential current elements such that

$$\boxed{I\,d\mathbf{l}' = \mathbf{J}'\,dv' = \mathbf{J}_s'\,ds'} \tag{5.7}$$

where the differential volume dv' and differential area ds' are centered at the source point with position vector $\mathbf{r}'$ as the differential length $d\mathbf{l}'$. Using Eq. (5.7) in Eq. (5.6), the resulting magnetic field can be expressed as

$$\boxed{\mathbf{H} = \int_{\mathcal{V}'} \frac{\mathbf{J}' \times \mathbf{a}_{\mathcal{R}}\,dv'}{4\pi\mathcal{R}^2}} \quad [\text{A/m}] \tag{5.8}$$

$$\boxed{\mathbf{H} = \int_{\mathcal{S}'} \frac{\mathbf{J}_s' \times \mathbf{a}_{\mathcal{R}}\,ds'}{4\pi\mathcal{R}^2}} \quad [\text{A/m}] \tag{5.9}$$

Again, $\mathcal{R}$ and $\mathbf{a}_{\mathcal{R}}$ are the magnitude and unit vector of $\mathcal{R} = \mathbf{r} - \mathbf{r}'$, respectively. Note that Eqs. (5.6), (5.8), and (5.9) employ a mixed-coordinate system, in which the primed and unprimed coordinates completely overlap. When integration is performed over the primed coordinates, the quantities given in the unprimed coordinates are treated as constants. The cross product is evaluated at the field point defined in the unprimed coordinates.

Example 5.2 Determine $\mathbf{H}$ of an infinitely long filament lying along the z-axis and carrying a steady current I, as shown in Fig. 5.3.

Solution

From symmetry considerations, field point p is assumed to be on the $z = 0$ plane in cylindrical coordinates without loss of generality.

The distance vector is expressed in terms of the base vectors at p such that

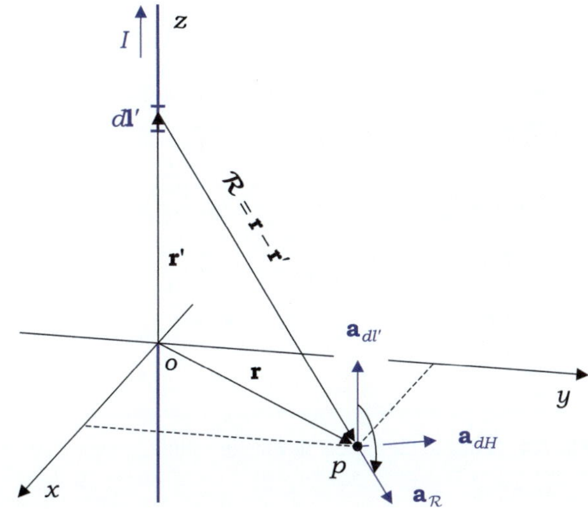

Fig. 5.3 An infinitely long filamentary current I

$$\mathcal{R} = \mathbf{r} - \mathbf{r}' = \rho\,\mathbf{a}_\rho - z'\mathbf{a}_{z'}$$
$$= \rho\,\mathbf{a}_\rho - z'\mathbf{a}_z$$

The differential length vector along the line current is

$$d\mathbf{l}' = dz'\mathbf{a}_z$$

From the Biot-Savart law,

$$\mathbf{H} = \int_{z'=-\infty}^{z'=+\infty} \frac{I\,dz'\mathbf{a}_z \times \left(\rho\,\mathbf{a}_\rho - z'\mathbf{a}_z\right)}{4\pi\mathcal{R}^3}$$

$$= \frac{I\,\mathbf{a}_\phi}{4\pi} \int_{z'=-\infty}^{z'=+\infty} \frac{\rho\,dz'}{(\rho^2 + z'^2)^{3/2}} = \mathbf{a}_\phi \frac{I\rho}{4\pi} \frac{z'/\rho^2}{\sqrt{\rho^2 + z'^2}}\Bigg|_{z'=-\infty}^{z'=+\infty} \qquad (5.10)$$

where $\mathbf{a}_\phi$ is taken outside the integral sign because it is independent of z'.

The magnetic field due to an infinitely long filamentary current I is therefore

$$\boxed{\mathbf{H} = \frac{I}{2\pi\rho}\mathbf{a}_\phi} \qquad \text{[A/m]} \qquad (5.11)$$

The direction of **H** *and the direction of flow of I are related by the right-hand rule.* The right thumb points in the direction of I when the fingers follow the direction of **H**.

Let us now explore the symmetry of the problem to see what it can tell us about **H**. The infinitely long filamentary current exhibits cylindrical symmetry (no change

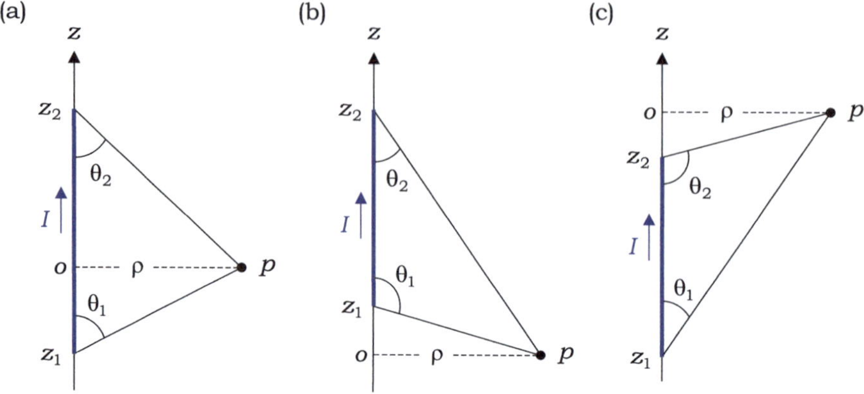

Fig. 5.4 **H** due to the current flowing in a finite straight filament

after rotation about the z-axis) and translational symmetry in the z-direction (no change after displacement along the z-axis). Accordingly, the resultant **H** should be independent of the ϕ- and z-coordinates. Moreover, the cross product $d\mathbf{l}' \times \mathbf{a}_R$ always points in the $\mathbf{a}_\phi$-direction. As a result, **H** has only a ϕ-component that depends only on ρ, as is evident from Eq. (5.11).

Here, we derive a formula that is useful for determining **H** of a finite filamentary current extending between $z = z_1$ and $z = z_2$ (Fig. 5.4). At a distance ρ from the filament in the $z = 0$ plane, **H** is computed as

$$\mathbf{H} = \mathbf{a}_\phi \frac{I}{4\pi\rho} \left(\left. \frac{z'}{\sqrt{\rho^2 + z'^2}} \right|_{z'=z_1}^{z'=z_2} \right) = \mathbf{a}_\phi \frac{I}{4\pi\rho} \left(\frac{z_2}{\sqrt{\rho^2 + z_2^2}} - \frac{z_1}{\sqrt{\rho^2 + z_1^2}} \right) \quad (5.12)$$

Using two interior angles of triangle $\triangle p z_1 z_2$, θ_1 and θ_2, the magnetic field at a distance ρ from the filamentary current can be expressed as

$$\boxed{\mathbf{H} = \frac{I}{4\pi\rho}[\cos\theta_1 + \cos\theta_2]\mathbf{a}_\phi} \quad [\text{A/m}] \quad (5.13)$$

This holds true regardless of the positions of z_1 and z_2 with respect to the $z = 0$ plane. The direction of **H** is determined by the right-hand rule: when the right thumb points in the direction of I, the fingers rotate in the direction of **H**.

It should be noted that **H** in Eq. (5.13) is non-irrotational, that is, $\nabla \times \mathbf{H} \neq 0$, meaning that it does not represent a magnetic field in real life. However, Eq. (5.13) is useful for determining the contribution of a segment of a closed current loop or

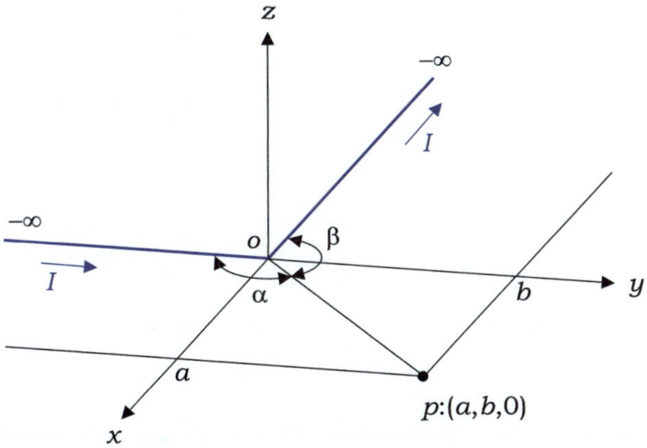

Fig. 5.5   An L shaped filament carrying steady current I

filamentary current that extends to infinity. The same is true for $d\mathbf{H}$ in Eq. (5.5), indicating that *there are no magnetic point sources* (see Sect. 5.3).

Example 5.3   Determine $\mathbf{H}$ at point p:$(a, b, 0)$ when the steady current I flows in an infinitely long filament bent into an L shape, as shown in Fig. 5.5.

Solution

For a semi-infinite segment along the y-axis, setting $\theta_1 = 0$, $\theta_2 = \alpha$, and $\rho = a$ in Eq. (5.13) gives

$$\mathbf{H}_1 = \frac{I}{4\pi a}[1 + \cos\alpha](-\mathbf{a}_z)$$

Similarly, for a semi-infinite segment along the x-axis, setting $\theta_1 = \beta$, $\theta_2 = 0$, and $\rho = b$ gives

$$\mathbf{H}_2 = \frac{I}{4\pi b}[\cos\beta + 1](-\mathbf{a}_z)$$

The directions of $\mathbf{H}_1$ and $\mathbf{H}_2$ were determined using the right-hand rule. Next, from Fig. 5.5,

$$\cos\alpha = -\sin(\alpha - 90°) = -\frac{b}{\sqrt{a^2 + b^2}}$$

$$\cos\beta = -\sin(\beta - 90°) = -\frac{a}{\sqrt{a^2 + b^2}}$$

Therefore, the magnetic field at point p is given by

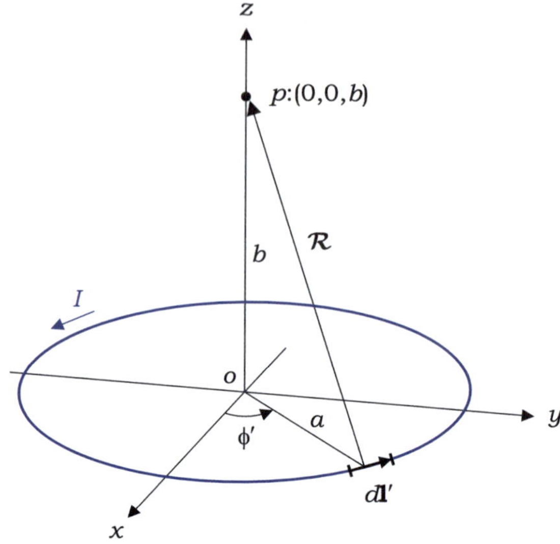

Fig. 5.6 A circular loop carrying steady current I

$$\mathbf{H} = \mathbf{H}_1 + \mathbf{H}_2 = \frac{I}{4\pi a}\left[1 - \frac{b}{\sqrt{a^2 + b^2}}\right](-\mathbf{a}_z) + \frac{I}{4\pi b}\left[-\frac{a}{\sqrt{a^2 + b^2}} + 1\right](-\mathbf{a}_z)$$

$$= \mathbf{a}_z \frac{I}{4\pi ab}\left[\sqrt{a^2 + b^2} - (a + b)\right]$$

Example 5.4 A filament is formed into a circle of radius a in the xy-plane, carrying a steady current I, as shown in Fig. 5.6. Determine $\mathbf{H}$ at $p{:}(0, 0, b)$.

Solution

In cylindrical coordinates,

$$dl' = a\,d\phi'\mathbf{a}_{\phi'}$$

$$\mathcal{R} = \mathbf{r} - \mathbf{r}' = b\,\mathbf{a}_z - a\,\mathbf{a}_{\rho'}$$

Inserting the above expressions into Eq. (5.6) gives

$$\mathbf{H} = \int_{C'} \frac{I\,dl' \times \mathbf{a}_{\mathcal{R}}}{4\pi \mathcal{R}^2} = \int_{\phi'=0}^{\phi'=2\pi} \frac{I a\,d\phi'\mathbf{a}_{\phi'} \times (b\,\mathbf{a}_z - a\,\mathbf{a}_{\rho'})}{4\pi \mathcal{R}^3} \tag{5.14a}$$

This can be rearranged as

$$\mathbf{H} = \frac{I ab}{4\pi(a^2 + b^2)^{3/2}} \int_{\phi'=0}^{\phi'=2\pi} \mathbf{a}_{\rho'}\,d\phi' + \frac{I a^2\,\mathbf{a}_z}{4\pi(a^2 + b^2)^{3/2}} \int_{\phi'=0}^{\phi'=2\pi} d\phi' \tag{5.14b}$$

Referring to the first integral on the right side of Eq. (5.14b), increasing ϕ' from 0 to 2π provides unit vector $\mathbf{a}_{\rho'}$ with a complete turn; therefore, the integral disappears.

At a distance b above a circular loop with radius a and steady current I, the magnetic field is given by

$$\mathbf{H} = \frac{Ia^2}{2(a^2 + b^2)^{3/2}}\mathbf{a}_z \qquad [\text{A/m}] \qquad (5.15)$$

Exercise 5.3
Check that the curl of the magnetic field of an infinitely long filamentary current in Eq. (5.11) is zero.

Ans. $\nabla \times \mathbf{H} = 0$ for $\rho > 0$.

Exercise 5.4
A square of side a lies in the xy-plane with the center at the origin and carries current I flowing counterclockwise, as viewed from above. Find $\mathbf{H}$ at the origin.

Ans. $\mathbf{H} = \mathbf{a}_z I2\sqrt{2}/(\pi a)$.

Exercise 5.5
For a current-carrying filament, as illustrated in Fig. 5.4, show that $\mathbf{H}$ at point p is the sum of the $\mathbf{H}$'s due to the segments $\overline{z_1 z_3}$ and $\overline{z_3 z_2}$, where $z_1 < z_3 < z_2$.

Ans. $\cos\theta + \cos(\pi - \theta) = 0$.

Exercise 5.6
The two dots in the following figures represent two parallel filaments carrying equal current. From the magnetic field lines shown, determine the direction of the current.

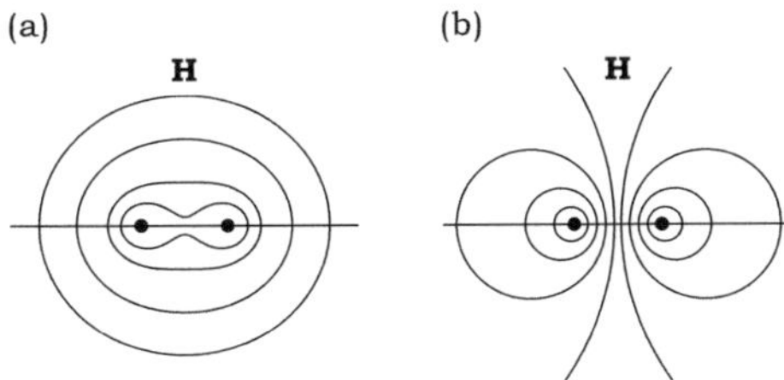

Ans. Two currents flow (a) in the same direction, and (b) in opposite directions.

Review Questions

RQ 5.3   What are the uses of primed and unprimed coordinates in the   [Fig. 5.2]
          expression of the Biot-Savart law?

RQ 5.4   Is $d\mathbf{l}' \times \mathbf{a}_{\mathcal{R}}$ a vector specific to a point in space?   [Fig. 5.2]

RQ 5.5   Does the principle of superposition apply to a magnetic field?   [(5.6)]

RQ 5.6   Define differential current elements for the volume and   [(5.7)]
          surface currents.

RQ 5.7   In what way does $\mathbf{H}$ of an infinitely long filamentary current   [(5.11)]
          vary with distance?

RQ 5.8   Express $\mathbf{H}$ of the finite filamentary current.   [(5.13)]

RQ 5.9   In what way does $\mathbf{H}$ vary with the distance above the current   [(5.15)]
          loop at large distances?

5.3 Ampere's Circuital Law

Ampere's circuital law is analogous to Gauss's law for electricity in the sense that it allows us to determine the $\mathbf{H}$ of a current distribution with a certain symmetry by solving a simple algebraic equation. Ampere's circuital law is a special case of the Biot-Savart law. The former is derived from the latter, as discussed in Sect. 5.5. Just as the point form of Gauss's law relates the divergence of $\mathbf{D}$ to the distribution of volume charge, the point form of Ampere's circuital law relates the curl of $\mathbf{H}$ to the distribution of volume current.

Ampere's circuital law states that *the line integral of* $\mathbf{H}$ *around a closed path is equal to the current enclosed in that path.* Expressed mathematically,

$$\boxed{\oint_{C} \mathbf{H} \cdot d\mathbf{l} = I}$$
(5.16)

The direction of travel on the closed path C and the direction of flow of the current I are governed by the right-hand rule; the right thumb points in the direction of I when the fingers follow the positive direction on C.

It is important to clarify the meaning of "the current enclosed by a loop". If a conducting wire passes through the loop surface, the current flowing in the wire is the current enclosed by the loop regardless of whether the wire is perpendicular to the loop surface.

Example 5.5 Show that the closed-line integrals of $\mathbf{H}$ around loops C_1, C_2, and C_3, as shown in Fig. 5.7, are equal to the filamentary current I flowing along the z-axis.

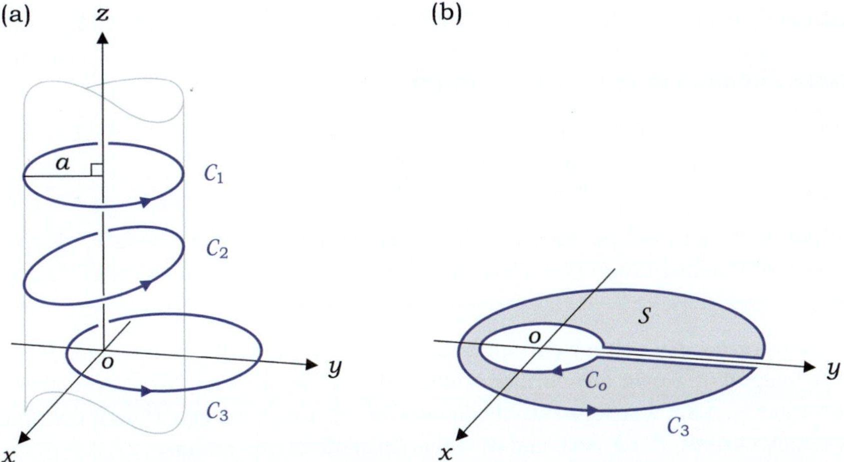

Fig. 5.7 **a** Loops C_1 and C_2 formed around a circular cylinder of radius a. **b** Open surface $\mathcal{S}$ with a tiny through-hole has C_o (circle centered at the origin) and C_3 (arbitrary loop) as the boundaries as the through-hole shrinks to zero

Solution

The line integrals of the magnetic field, as given in Eq. (5.11), around loops C_1 and C_2 are as follows:

$$\oint_{C_1} \mathbf{H} \cdot (a\,d\phi\,\mathbf{a}_\phi) = \frac{I}{2\pi} \int_{\phi=0}^{\phi=2\pi} d\phi = I \tag{5.17a}$$

$$\oint_{C_2} \mathbf{H} \cdot (a\,d\phi\,\mathbf{a}_\phi + dz\,\mathbf{a}_z) = \frac{I}{2\pi} \int_{\phi=0}^{\phi=2\pi} d\phi = I \tag{5.17b}$$

Consider an open surface $\mathcal{S}$, as shown in Fig. 5.7b, whose boundary C consists of loops C_o and C_3, and the edges of a through-hole. As surface $\mathcal{S}$ excludes the filamentary current, from Stokes's theorem and the result of Exercise 5.3, the closed-line integral of $\mathbf{H}$ vanishes, that is,

$$\oint_C \mathbf{H} \cdot d\mathbf{l} = \int_{\mathcal{S}} (\nabla \times \mathbf{H}) \cdot d\mathbf{s} = 0 \tag{5.17c}$$

As the through-hole approaches zero, the integral in Eq. (5.17c) can be split into two closed-line integrals as follows:

$$\oint_C \mathbf{H} \cdot d\mathbf{l} = \oint_{C_3} \mathbf{H} \cdot d\mathbf{l} + \oint_{C_o} \mathbf{H} \cdot d\mathbf{l} = 0 \tag{5.17d}$$

Because C_o is a circle centered at the origin, the line integral of $\mathbf{H}$ around C_o is equal to $-I$, as expected from Eq. (5.17a), where the negative sign accounts for the reverse direction of travel along C_o in the right-hand rule. Thus,

$$\oint_{C_3} \mathbf{H} \cdot d\mathbf{l} = -\oint_{C_o} \mathbf{H} \cdot d\mathbf{l} = I \tag{5.17e}$$

If filamentary current I passes through the loop surface, it is the current enclosed by the loop or the integration path, regardless of the shape and orientation of the loop.

In the presence of volume current density $\mathbf{J}$, the current enclosed by loop C is equal to the integral of $\mathbf{J}$ over any surface bounded by C. Consider a closed hemisphere as shown in Fig. 5.8a, whose surface is denoted by S_o. From the equation of continuity for steady current, $\nabla \cdot \mathbf{J} = 0$, and the divergence theorem, we have

$$\int_{\mathcal{V}} \nabla \cdot \mathbf{J} \, dv = \oint_{S_o} \mathbf{J} \cdot d\mathbf{s} = 0 \tag{5.18}$$

The closed-surface integral can be split into two surface integrals as follows:

$$\oint_{S_o} \mathbf{J} \cdot d\mathbf{s} = \int_{\substack{hemi- \\ sphere}} \mathbf{J} \cdot d\mathbf{s} + \int_{disk} \mathbf{J} \cdot d\mathbf{s} = 0 \tag{5.19}$$

where $d\mathbf{s}$ is directed away from the volume. If the positive direction of $d\mathbf{s}$ is taken in the general direction of $\mathbf{J}$, then,

$$\int_{\substack{hemi- \\ sphere}} \mathbf{J} \cdot d\mathbf{s} = \int_{disk} \mathbf{J} \cdot d\mathbf{s} = I \tag{5.20}$$

where I is the total current that passes through the surface. This indicates that the currents through any surface with the same contour are identical. Next, with the base plate removed, the hemisphere constitutes an open surface with loop C as a contour, as shown in Fig. 5.8b. This leads to the conclusion that *the current enclosed by a loop is equal to the current flowing through any surface bounded by that loop.*

Ampere's circuital law can be expressed in terms of the volume current density $\mathbf{J}$ as follows:

$$\boxed{\oint_C \mathbf{H} \cdot d\mathbf{l} = \int_S \mathbf{J} \cdot d\mathbf{s}} \tag{5.21}$$

where S is the surface bounded by closed loop C. The directions of $d\mathbf{l}$ along C and $d\mathbf{s}$ on S are related by the right-hand rule, in which the right thumb points in the direction of $d\mathbf{s}$ when the fingers advance in the direction of $d\mathbf{l}$. Ampere's circuital law states that *the line integral of $\mathbf{H}$ around a loop is equal to the current flowing*

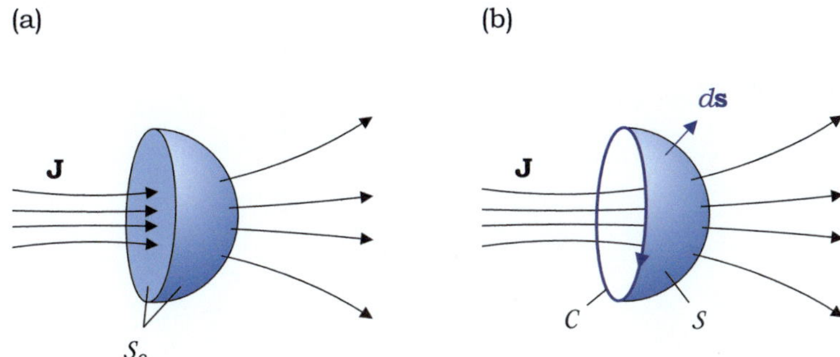

Fig. 5.8 The current enclosed by C is equal to the current flowing through S

through a surface bounded by the loop. Applying Stokes's theorem to the left-hand side of Eq. (5.21) yields

$$\int_{S} \nabla \times \mathbf{H} \cdot d\mathbf{s} = \int_{S} \mathbf{J} \cdot d\mathbf{s} \tag{5.22}$$

As surface S may be arbitrary, the two integrands of Eq. (5.22) should be equal at every point on S. Therefore, Ampere's circuital law is expressed in point form as

$$\boxed{\nabla \times \mathbf{H} = \mathbf{J}} \tag{5.23}$$

Note that Eq. (5.23) is one of two fundamental relations governing the static magnetic field. However, under time-varying conditions, Ampere's circuital law should be modified to accommodate the electromagnetic induction observed by Faraday (see Chap. 6).

Ampere's circuital law, which is expressed by Eq. (5.21), is useful for determining the magnetic field if the path of integration C is chosen in such a way that $\mathbf{H}$ *is constant and tangential to C, or perpendicular to C otherwise.* This path is called the ***Amperian path***.

Example 5.6 An infinitely long filament lying along the z-axis carries a steady current, I, which flows in the $+z$-direction (Fig. 5.9). Determine $\mathbf{H}$ using Ampere's circuital law.

Solution

The line current has cylindrical and translational symmetries. Moreover, the Biot-Savart law indicates that $d\mathbf{H}$ always points in the direction of $\mathbf{a}_\phi$. Therefore, the magnetic field should have the form of $\mathbf{H} = H_\phi(\rho)\,\mathbf{a}_\phi$. In view of this, we take a circle centered on the z-axis as the Amperian path.

Applying Ampere's circuital law to an Amperian circle of radius ρ yields

Fig. 5.9 Infinitely long filamentary current and proper Amperian path

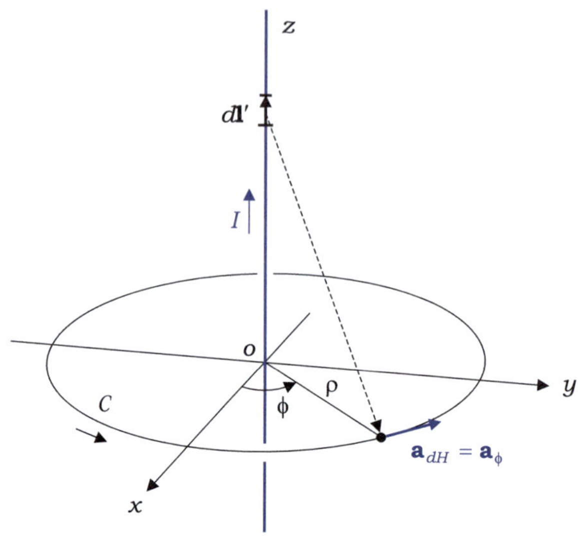

$$\oint \mathbf{H} \cdot d\mathbf{l} = \int_{\phi=0}^{2\pi} H_\phi(\rho)\,\mathbf{a}_\phi \cdot \left(\rho\,d\phi\,\mathbf{a}_\phi\right) = H_\phi(\rho)\,2\pi\rho$$

$$= I$$

Thus,

$$\mathbf{H} = H_\phi(\rho)\,\mathbf{a}_\phi = \frac{I}{2\pi\rho}\mathbf{a}_\phi \tag{5.24}$$

This result is the same as that obtained using the Biot-Savart law, as shown in Eq. (5.11).

Example 5.7 An infinitely long coaxial cable consists of an inner conductor with radius a and an outer conductor with inner radius b and outer radius c, as shown in Fig. 5.10. A steady current I flows in the opposite direction in the two conductors and is uniformly distributed in the cross section. Determine $\mathbf{H}$ everywhere.

Solution

A coaxial cable with a uniform current has cylindrical and translational symmetries such that the resultant $\mathbf{H}$ is independent of ϕ and z. The current in the inner conductor can be viewed as the sum of the many filamentary currents. Because the magnetic field of a long filamentary current is circumferential, a pair of filamentary currents at (ρ_1, ϕ_1) and $(\rho_1, -\phi_1)$ produces $\mathbf{H}$ in the direction of $\mathbf{a}_\phi$, as shown in Fig. 5.10b. The same is true for the current in the outer conductor. Therefore, the magnetic field is expected to be in the form of $\mathbf{H} = H_\phi(\rho)\,\mathbf{a}_\phi$ everywhere if it is not zero.

In the region $0 < \rho \le a$, taking a circle of radius ρ as the Amperian path, we have

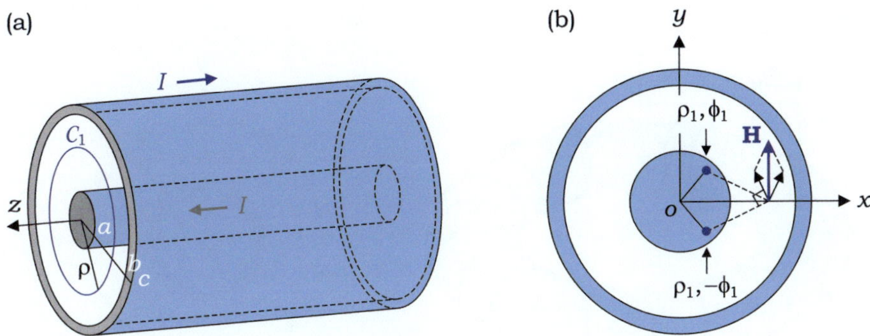

Fig. 5.10 Coaxial cable carrying steady current

$$\oint_C \mathbf{H} \cdot d\mathbf{l} = \int_{\phi=0}^{2\pi} H_\phi(\rho)\,\mathbf{a}_\phi \cdot (\rho\,d\phi\,\mathbf{a}_\phi) = 2\pi\rho H_\phi(\rho) \qquad (5.25a)$$

The current enclosed by the path is equal to $\rho^2 I/\pi a^2$. Thus,

$$\mathbf{H} = H_\phi \mathbf{a}_\phi = \frac{I\rho}{2\pi a^2}\mathbf{a}_\phi \qquad (0 < \rho \leq a) \qquad (5.25b)$$

In other regions, the result of Eq. (5.25a) can also be used for the closed-line integral of $\mathbf{H}$, because a circle is still a proper Amperian path. However, the current enclosed by the path is different in different regions such that

$$I \qquad\qquad (a \leq \rho \leq b)$$

$$I\frac{c^2 - \rho^2}{c^2 - b^2} \qquad (b \leq \rho \leq c)$$

$$0 \qquad\qquad (\rho \geq c)$$

It then follows that

$$\mathbf{H} = \frac{I}{2\pi\rho}\mathbf{a}_\phi \qquad (a \leq \rho \leq b) \qquad (5.25c)$$

$$\mathbf{H} = \frac{I}{2\pi\rho}\frac{c^2 - \rho^2}{c^2 - b^2}\mathbf{a}_\phi \qquad (b \leq \rho \leq c) \qquad (5.25d)$$

$$\mathbf{H} = 0 \qquad (\rho \geq c) \qquad (5.25e)$$

The magnetic field vanishes outside the cable, as expressed in Eq. (5.25e). In general, the zero circulation of $\mathbf{H}$ does not necessarily mean that $\mathbf{H}$ is zero at every point in the loop. However, in this case, $\mathbf{H}$ is always constant and tangential to the loop. Thus,

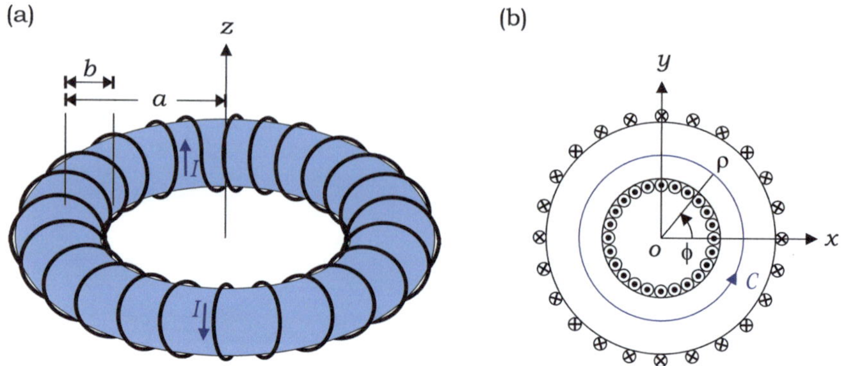

Fig. 5.11 Toroidal coil

the zero circulation of **H** directly leads to **H** $= 0$ in any loop, and subsequently, **H** vanishes at every point in a given region.

Example 5.8 A toroidal coil has N turns closely wound on an air core with a mean radius a, and carries a steady current I, as shown in Fig. 5.11. Determine **H** everywhere.

Solution

The toroidal coil has rotational symmetry about the z-axis; therefore, the resulting **H** must be independent of ϕ. Next, if the current I is reversed, causality dictates that the direction of **H** is reversed at every point in space. The reversal of I can be achieved by turning the toroid upside down. However, in this case, the ρ-component of the resulting **H** remains unchanged; thus, **H** cannot have a ρ-component. Based on the symmetry considerations thus far, **H** is expected to be of the form $\mathbf{H} = H_\phi(\rho, z)\,\mathbf{a}_\phi + H_z(\rho, z)\,\mathbf{a}_z$. Next, we consider the circulation of **H** around a rectangle in the $\phi = \phi_1$ plane, along which the differential length vector is $d\mathbf{l} = d\rho\,\mathbf{a}_\rho$ or $d\mathbf{l} = dz\,\mathbf{a}_z$. Thus,

$$\oint_C \mathbf{H} \cdot d\mathbf{l} = \oint_C \left(H_\phi \mathbf{a}_\phi + H_z \mathbf{a}_z \right) \cdot d\mathbf{l}$$

$$= \int_{z=z_1}^{z=z_2} H_z(\rho_1, z)\, dz - \int_{z=z_1}^{z=z_2} H_z(\rho_2, z)\, dz \qquad (5.26a)$$

The integration path is assumed to have sides at $z = z_1$, $z = z_2$, $\rho = \rho_1$, and $\rho = \rho_2$, such that the toroid penetrates through the surface of the rectangle. However, no current passes through the loop surface, because the current flows only in the plane of constant ϕ for an ideal toroidal coil. In other words, no current is enclosed in the loop, and the circulation of **H**, as shown in Eq. (5.26a), vanishes. This can only be fulfilled if H_z is independent of ρ everywhere. At large distances, the toroid appears to be a point that leads to **H** $= 0$ at infinity, which implies that H_z should vanish everywhere. Finally, $\mathbf{H}$ is expected to be of the form $\mathbf{H} = H_\phi(\rho, z)\,\mathbf{a}_\phi$ if it is not zero.

Inside the toroid, taking a circle of radius ρ as the Amperian path and using $d\mathbf{l} = \rho\,d\phi\,\mathbf{a}_\phi$, we have

$$\oint_C \mathbf{H} \cdot d\mathbf{l} = \rho H_\phi \int_0^{2\pi} d\phi = 2\pi\rho H_\phi \tag{5.26b}$$

As loop C encloses a total current of NI, the magnetic field is obtained as

$$\mathbf{H} = \frac{NI}{2\pi\rho}\,\mathbf{a}_\phi \quad \text{(inside)} \tag{5.26c}$$

Outside the toroid, the Amperian path encloses no current. Thus,

$$\mathbf{H} = 0 \quad \text{(outside)} \tag{5.26d}$$

Exercise 5.7
A long cylinder with radius a and an axis coincident with the z-axis carries a uniform surface current $\mathbf{J}_s = J_o\,\mathbf{a}_z$ [A/m]. Determine $\mathbf{H}$ everywhere.

Ans. $\mathbf{H} = 0$ for $\rho < a$, and $\mathbf{H} = \mathbf{a}_\phi J_o a/\rho$ for $\rho > a$.

Exercise 5.8
What is the functional form of $\mathbf{H}$ inside a toroidal coil with a rectangular cross section?

Ans. $\mathbf{H} = H_\phi(\rho)\,\mathbf{a}_\phi$.

Exercise 5.9
If the rectangle bounds the region defined by $0 \le x \le 1$, $0 \le y \le 2$, and $z = 0$ in the magnetic field $\mathbf{H} = y\,\mathbf{a}_x + 2x\,\mathbf{a}_y$, determine the enclosed current using (a) Eq. (5.16), and (b) Eq. (5.23).

Ans. (a) $\oint_C \mathbf{H} \cdot d\mathbf{l} = 2$, (b) $\int_S \nabla \times \mathbf{H} \cdot d\mathbf{s} = 2$.

Review Questions

RQ 5.10   State Ampere's circuital law.   [(5.16)]

RQ 5.11   What is the meaning of the current enclosed by a loop?   [Figs. 5.7,5.8]

RQ 5.12   What determines the positive direction of C in Eq. (5.16)?   [Fig. 5.8]

RQ 5.13   In what ways are the directions of $d\mathbf{l}$ and $d\mathbf{s}$ related?   [Fig. 5.8]

RQ 5.14   How is the Amperian path selected?   [Fig. 5.9]

RQ 5.15   Is it possible to apply Ampere's circuital law to a current   [(5.13)]
          element of finite extent?

RQ 5.16   Under what conditions does the zero circulation of **H** around   [Example 5.7]
           a loop directly lead to **H** = 0 at every point in the loop?

RQ 5.17   Write the point form of Ampere's circuital law.   [(5.23)]

5.4 Magnetic Flux Density

Just as **D** is related to **E** by the constitutive relation $\mathbf{D} = \varepsilon_0 \mathbf{E}$ in free space, the magnetic flux density **B** is related to **H** in free space by the constitutive relation,

$$\boxed{\mathbf{B} = \mu_0 \mathbf{H}} \quad \text{[T] or [Wb/m}^2\text{]} \tag{5.27}$$

which is measured in teslas [T] or in webers per square meter [Wb/m^2]. The proportionality constant μ_0 is referred to as the ***permeability of free space***, and has a value of $\mu_0 = 4\pi \times 10^{-7}$ [H/m].

The ***magnetic flux*** passing through a surface is given by the integral of **B** over the surface, that is,

$$\boxed{\Phi = \int_{\mathcal{S}} \mathbf{B} \cdot d\mathbf{s}} \quad \text{[Wb]} \tag{5.28}$$

which is in webers. The dot product in Eq. (5.28) is to convert the differential area $|d\mathbf{s}|$ in $\mathcal{S}$ into an equivalent area in the cross section of **B** because the quantity $|\mathbf{B}|$ represents the magnetic flux passing through a unit area in a plane perpendicular to **B**.

From electrostatics, we recall that electric flux lines always begin at positive charges and end at negative charges. Accordingly, the net outward electric flux through a closed surface is nonzero if the surface encloses a net positive charge. In contrast, magnetic flux lines are always closed by themselves because there are no isolated magnetic charges or no isolated magnetic poles. One might try to break up a permanent magnet into minute pieces on an atomic scale to separate the north and south poles but only find that each piece still has a north and south pole.

According to the Biot-Savart law, as expressed in Eq. (5.5), it is apparent that a differential current element produces magnetic field lines in the form of concentric circles. The total magnetic field is given by the vector sum of the contributions of all individual current elements, as shown in Eq. (5.6). Therefore, it is a continuous function of position and its field lines close upon themselves.

Gauss's law of magnetism states that *the net outward magnetic flux through a closed surface is zero*. That is,

$$\boxed{\oint_{\mathcal{S}} \mathbf{B} \cdot d\mathbf{s} = 0} \tag{5.29}$$

This can be viewed as an expression for the nonexistence of isolated magnetic charges and *the law of conservation of magnetic flux*. By applying the divergence theorem to Eq. (5.29), the *point form of Gauss's law for magnetism* is obtained as

$$\boxed{\nabla \cdot \mathbf{B} = 0} \tag{5.30}$$

It should be noted that Eqs. (5.23) and (5.30) are two fundamental relations for the static magnetic field.

In Chap. 3, we have defined the curl of $\mathbf{E}$ and the divergence of $\mathbf{D}$, which are essential for the unique determination of the static electric field. For future reference, we will collect the fundamental relations for static electric and static magnetic fields as follows:

$$\boxed{\begin{aligned} \nabla \times \mathbf{E} &= 0 \\ \nabla \cdot \mathbf{D} &= \rho_v \\ \nabla \times \mathbf{H} &= \mathbf{J} \\ \nabla \cdot \mathbf{B} &= 0 \end{aligned}} \tag{5.31}$$

With the help of the divergence and Stokes's theorems, these relations can be expressed in integral form as follows:

$$\boxed{\begin{aligned} \oint_C \mathbf{E} \cdot d\mathbf{l} &= 0 \\ \oint_S \mathbf{D} \cdot d\mathbf{s} &= \int_V \rho_v \, dv \\ \oint_C \mathbf{H} \cdot d\mathbf{l} &= \int_S \mathbf{J} \cdot d\mathbf{s} \\ \oint_S \mathbf{B} \cdot d\mathbf{s} &= 0 \end{aligned}} \tag{5.32}$$

There is no coupling between the static electric and magnetic fields. However, time-varying electric and magnetic fields become coupled such that a time-varying electric field generates a time-varying magnetic field, and vice versa. Under time-varying conditions, the two divergence equations in Eq. (5.31) still remain valid. However, the two curl equations must be modified to accommodate Faraday's electromagnetic induction and Maxwell's concept of displacement current density.

Example 5.9 A very thin spherical shell of radius a is connected to two straight wires, as shown in Fig. 5.12. The filamentary current I becomes a surface current with $\mathbf{J}_s = J_o \mathbf{a}_\theta$ on the sphere. By forming an appropriate trial solution and applying Helmholtz's theorem, determine $\mathbf{H}$ everywhere.

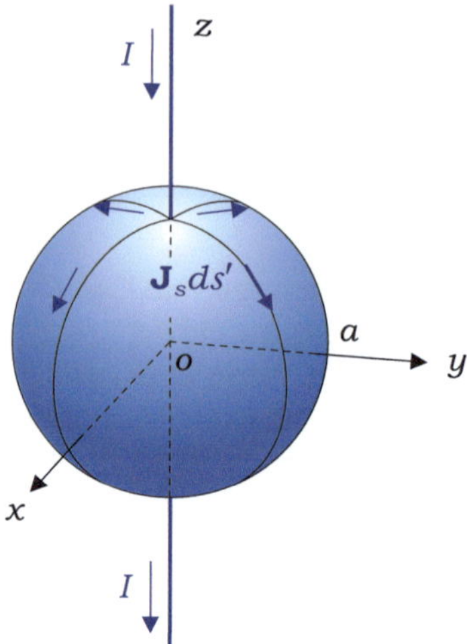

Fig. 5.12 Thin spherical shell is connected to two straight wires

Solution

At far distances, the given object would appear just as an infinitely long filamentary current I. Thus, outside the sphere, we choose as a trial solution

$$\mathbf{H} = -\frac{I}{2\pi\rho}\mathbf{a}_\phi \tag{5.33a}$$

A circular Amperian path encloses no current inside the sphere. Thus, $\mathbf{H}$ is zero inside the sphere. Outside the sphere, Eq. (5.33a) satisfies the two fundamental relations,

$$\nabla \cdot \mathbf{B} = 0 \text{ and } \nabla \times \mathbf{H} = 0 \tag{5.33b}$$

If the current is uniform in the $\mathbf{a}_\phi$-direction on the sphere, then the surface current density is given by $\mathbf{J}_s = I\,\mathbf{a}_\theta/(2\pi a \sin\theta)$. Subsequently, the boundary condition for $\mathbf{H}$ requires $\mathbf{H} = -|\mathbf{J}_s|\mathbf{a}_\phi$ on the sphere (Eq. (5.96)). As a matter of fact, the trial solution in Eq. (5.33a) produces the same tangential $\mathbf{H}$ on the sphere, that is, $\mathbf{H} = -I\,\mathbf{a}_\phi/(2\pi a \sin\theta)$. In other words, the trial solution satisfies the boundary condition for $\mathbf{H}$ as well as the two fundamental relations. Therefore, the given trial solution is the only solution such that

$$\mathbf{H} = -\frac{I}{2\pi\rho}\,\mathbf{a}_\phi \quad \text{(outside)} \tag{5.33c}$$

$$\mathbf{H} = 0 \quad \text{(inside)} \tag{5.33d}$$

where $\rho = a \sin\theta$ is used in the solution.

Review Questions

RQ 5.18	In what way are $\mathbf{B}$ and $\mathbf{H}$ related in free space?	[(5.27)]
RQ 5.19	Define the magnetic flux.	[(5.28)]
RQ 5.20	State Gauss's law for magnetism.	[(5.29)(5.30)]
RQ 5.21	What are the mathematical expressions for the observation that magnetic flux lines are always closed?	[(5.29)(5.30)]
RQ 5.22	Express the law of conservation of the magnetic flux.	[(5.29)(5.30)]
RQ 5.23	What are the two fundamental relations for static magnetic fields?	[(5.31)(5.32)]
RQ 5.24	If the trial solution satisfies $\nabla \cdot \mathbf{B} = 0$ and $\nabla \times \mathbf{H} = 0$ in the source-free region, is this the only solution?	[Fig. 5.12]

5.5 Vector Magnetic Potential

In Chap. 3, we saw that the irrotational nature of the electric field was expressed by $\nabla \times \mathbf{E} = 0$ and allowed us to define the electric potential through the relationship, $\mathbf{E} = -\nabla V$. We also observed that the electric potential is much easier to handle than the electric field because of its scalar nature. Therefore, the electric potential was conveniently used to determine the electric field by first calculating V from a given charge distribution and then taking the negative gradient of V. In the same way as the electric potential, the *scalar magnetic potential* V_m can be defined through the relation $\mathbf{H} = -\nabla V_m$. However, V_m is a many-valued function because of the nonconservative nature of $\mathbf{H}$. For example, in the presence of $\mathbf{H}$ caused by an infinitely long filamentary current I, the scalar magnetic potential is given by $V_m = -I\phi/(2\pi)$. However, this has many values at a point in space, as ϕ increases beyond 2π. From the relationship $\nabla \times \mathbf{H} = \nabla \times (-\nabla V_m) = \mathbf{J}$, it is evident that V_m is useful only in regions with no volume current. In this section, we focus on the *vector magnetic potential* $\mathbf{A}$, which is useful for determining $\mathbf{B}$ regardless of whether the given region contains currents.

We begin with the solenoidal nature of $\mathbf{B}$, expressed as $\nabla \cdot \mathbf{B} = 0$. Considering the vector identity $\nabla \cdot (\nabla \times \mathbf{U}) = 0$, the vector magnetic potential is defined as

$$\boxed{\mathbf{B} = \nabla \times \mathbf{A}} \tag{5.34}$$

The unit of $\mathbf{A}$ is weber per meter [Wb/m]. If the vector magnetic potential is obtained from a given current distribution, the magnetic flux density is simply given by the curl of the vector magnetic potential.

The relationship between the vector magnetic potential and volume current density can be obtained from the Biot-Savart law. First, the Biot-Savart law is rewritten for the magnetic flux density such as

$$\mathbf{B} = \frac{\mu_0}{4\pi} \int_{C'} \frac{I\,d\mathbf{l}' \times \boldsymbol{\mathcal{R}}}{\mathcal{R}^3} \tag{5.35}$$

Borrowing from Eq. (2.41),

$$\nabla \frac{1}{\mathcal{R}} = -\frac{\boldsymbol{\mathcal{R}}}{\mathcal{R}^3}$$

where $\boldsymbol{\mathcal{R}} = \mathbf{r} - \mathbf{r}'$ is the distance vector. Inserting this into Eq. (5.35) yields

$$\mathbf{B} = -\frac{\mu_0 I}{4\pi} \int_{C'} d\mathbf{l}' \times \left(\nabla \frac{1}{\mathcal{R}} \right) \tag{5.36}$$

Setting $f = 1/\mathcal{R}$ and $\mathbf{U} = d\mathbf{l}'$ in the identity $\nabla \times (f\mathbf{U}) = (\nabla f) \times \mathbf{U} + f(\nabla \times \mathbf{U})$ and using it in Eq. (5.36), we obtain

$$\mathbf{B} = \frac{\mu_0 I}{4\pi} \int_{C'} \left[\nabla \times \left(\frac{d\mathbf{l}'}{\mathcal{R}} \right) - \frac{1}{\mathcal{R}} (\nabla \times d\mathbf{l}') \right] = \nabla \times \int_{C'} \frac{\mu_0 I\,d\mathbf{l}'}{4\pi \mathcal{R}} \tag{5.37}$$

where $\nabla \times d\mathbf{l}' = 0$ because the curl operates only on the unprimed coordinates, and the other curl operator is taken outside the line integral because it is independent of the path C'. Comparison of Eq. (5.37) with Eq. (5.34) leads to an expression for the vector magnetic potential, that is,

$$\boxed{\mathbf{A} = \frac{\mu_0}{4\pi} \int_{C'} \frac{I\,d\mathbf{l}'}{\mathcal{R}}} \quad \text{[Wb/m]} \quad \text{(line current)} \tag{5.38a}$$

where C' denotes the path of the line current I.

When the current is distributed over volume $\mathcal{V}$ or across surface $\mathcal{S}$, the vector magnetic potential can be obtained by substituting the equivalent current element in Eq. (5.7) into Eq. (5.38a) such that

$$\boxed{\mathbf{A} = \frac{\mu_0}{4\pi} \int_{\mathcal{V}'} \frac{\mathbf{J}'}{\mathcal{R}}\,dv'} \quad \text{[Wb/m]} \quad \text{(volume current)} \tag{5.38b}$$

$$\boxed{\mathbf{A} = \frac{\mu_0}{4\pi} \int_{\mathcal{S}'} \frac{\mathbf{J}'_s}{\mathcal{R}}\,ds'} \quad \text{[Wb/m]} \quad \text{(surface current)} \tag{5.38c}$$

where $\mathcal{R}$ is the magnitude of the distance vector. We note that the expression for $\mathbf{A}$ is much simpler than that for the Biot-Savart law because it has no cross product of the two vectors. The vector magnetic potential discussed thus far may be a multi-valued function, because $\nabla \cdot \mathbf{A}$ has not yet been specified. For example, there is no change in $\mathbf{B}$ even if $\mathbf{A} + \nabla \varphi$ is used instead of $\mathbf{A}$ in equation $\mathbf{B} = \nabla \times \mathbf{A}$. It is convenient to set $\varphi = 0$ such that $\nabla \cdot \mathbf{A} = 0$ under static conditions (see the next section).

The vector magnetic potential has a physical significance. The closed-line integral of $\mathbf{A}$ is equal to the magnetic flux passing through the surface that is bounded by the integration path. The magnetic flux through surface $\mathcal{S}$ is given by

$$\Phi = \int_{\mathcal{S}} \mathbf{B} \cdot d\mathbf{s} \tag{5.39}$$

By substituting $\mathbf{B} = \nabla \times \mathbf{A}$ and applying Stokes's theorem to Eq. (5.39), we obtain

$$\Phi = \int_{\mathcal{S}} (\nabla \times \mathbf{A}) \cdot d\mathbf{s} = \oint_{C} \mathbf{A} \cdot d\mathbf{l} \tag{5.40}$$

Combining Eqs. (5.39) and (5.40) yields

$$\oint_{C} \mathbf{A} \cdot d\mathbf{l} = \int_{\mathcal{S}} \mathbf{B} \cdot d\mathbf{s} \tag{5.41}$$

The circulation of $\mathbf{A}$ around the loop is equal to the magnetic flux enclosed by the loop.

5.5.1   Ampere's Circuital Law from the Biot-Savart Law

The point form of Ampere's circuital law can be derived from the Biot-Savart law using the vector magnetic potential. First, the curl of $\mathbf{H}$ in free space is rewritten using $\mathbf{B} = \mu_0 \mathbf{H}$ and $\mathbf{B} = \nabla \times \mathbf{A}$ as follows:

$$\nabla \times \mathbf{H} = \frac{1}{\mu_0} \nabla \times \mathbf{B} = \frac{1}{\mu_0} \nabla \times \nabla \times \mathbf{A} \tag{5.42}$$

Applying the vector identity $\nabla \times \nabla \times \mathbf{A} = \nabla(\nabla \cdot \mathbf{A}) - \nabla^2 \mathbf{A}$ to Eq. (5.42) yields

$$\nabla \times \mathbf{H} = \frac{1}{\mu_0}\left[\nabla(\nabla \cdot \mathbf{A}) - \nabla^2 \mathbf{A}\right] \tag{5.43}$$

At this point, we briefly digress and show that the divergence of the vector magnetic potential is always zero. First, taking the divergence of both sides of Eq. (5.38b), we obtain

$$\nabla \cdot \mathbf{A} = \frac{\mu_0}{4\pi} \int_{\mathcal{V}'} \nabla \cdot \left(\frac{\mathbf{J}'}{\mathcal{R}}\right) dv' \tag{5.44}$$

where the divergence operator is brought inside the volume integral because it is independent of $\mathcal{V}'$. Next, it can be shown that the volume integral in Eq. (5.44) vanishes. Let us consider the following vector identities:

$$
\nabla \cdot \left(\frac{\mathbf{J}'}{\mathcal{R}} \right) = \left(\nabla \frac{1}{\mathcal{R}} \right) \cdot \mathbf{J}' + \frac{1}{\mathcal{R}} \nabla \cdot \mathbf{J}'
\tag{5.45a}
$$

$$
\nabla' \cdot \left(\frac{\mathbf{J}'}{\mathcal{R}} \right) = \left(\nabla' \frac{1}{\mathcal{R}} \right) \cdot \mathbf{J}' + \frac{1}{\mathcal{R}} \nabla' \cdot \mathbf{J}'
\tag{5.45b}
$$

Note that $\nabla \cdot \mathbf{J}' = 0$ because ∇ is independent of the primed coordinates, and $\nabla' \cdot \mathbf{J}' = 0$ because of the equation of continuity under static conditions. With the identity $\nabla(1/\mathcal{R}) = -\nabla'(1/\mathcal{R})$, Eqs. (5.45a) and (5.45b) can be combined to yield

$$
\nabla \cdot \left(\frac{\mathbf{J}'}{\mathcal{R}} \right) = -\nabla' \cdot \left(\frac{\mathbf{J}'}{\mathcal{R}} \right)
\tag{5.46}
$$

Inserting Eq. (5.46) into Eq. (5.44) and invoking the divergence theorem,

$$
\nabla \cdot \mathbf{A} = -\frac{\mu_0}{4\pi} \int_{\mathcal{V}'} \nabla' \cdot \left(\frac{\mathbf{J}'}{\mathcal{R}} \right) dv' = -\frac{\mu_0}{4\pi} \oint_{\mathcal{S}'} \frac{\mathbf{J}' \cdot d\mathbf{s}'}{\mathcal{R}}
\tag{5.47}
$$

Here, volume $\mathcal{V}'$ may be chosen in such a way that it includes $\mathbf{J}'$ and an empty space surrounding the current. In this case, no current passes through surface $\mathcal{S}'$, which is the boundary surface of the volume. Thus, the closed-surface integral on the right side of Eq. (5.47) vanishes. Therefore,

$$
\boxed{\nabla \cdot \mathbf{A} = 0}
\tag{5.48}
$$

The vector magnetic potential is a solenoidal field.

Subsequently, using Eq. (5.48), Eq. (5.43) is reduced to

$$
\nabla \times \mathbf{H} = -\frac{1}{\mu_0} \nabla^2 \mathbf{A}
\tag{5.49}
$$

The curl of $\mathbf{H}$ is linearly proportional to the Laplacian of $\mathbf{A}$.

Next, we focus our attention on the Laplacian of $\mathbf{A}$ of the source current $\mathbf{J}$. The Laplacian of $\mathbf{A}$ is a vector consisting of Laplacians of three scalar components such that

$$
\nabla^2 \mathbf{A} = (\nabla^2 A_x)\, \mathbf{a}_x + (\nabla^2 A_y)\, \mathbf{a}_y + (\nabla^2 A_z)\, \mathbf{a}_z
\tag{5.50}
$$

Inserting $\mathbf{A}$, as expressed in Eq. (5.38b), into Eq. (5.50) and collecting the terms with unit vector $\mathbf{a}_x$, we obtain

$$
\nabla^2 A_x = \frac{\mu_0}{4\pi} \int_{\mathcal{V}'} J_x' \left(\nabla^2 \frac{1}{\mathcal{R}} \right) dv'
\tag{5.51}
$$

where ∇^2 is brought inside the integral sign, because it is independent of the primed coordinates. Similar expressions can be used for other components.

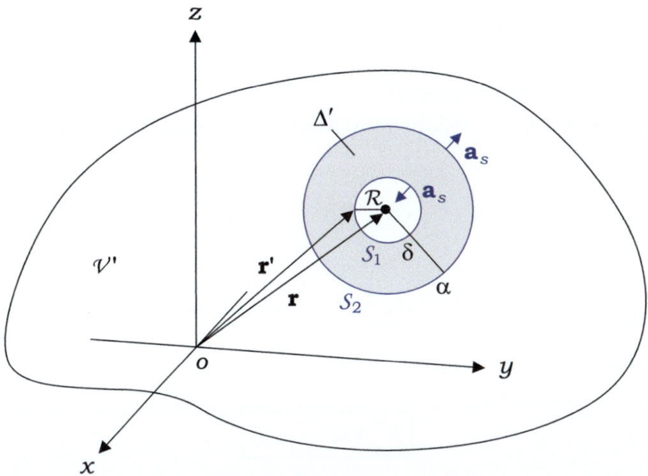

Fig. 5.13 Infinitesimal volume Δ' between spheres $\mathcal{S}_1$ and $\mathcal{S}_2$ excludes the field point at the center

To evaluate the volume integral in Eq. (5.51), we consider a small region surrounding the field point $\mathbf{r}$. As source point $\mathbf{r}'$ approaches the field point, term $\nabla^2(1/\mathcal{R})$ in Eq. (5.51) increases rapidly compared with the source current J_x' and becomes infinite for $\mathbf{r}' = \mathbf{r}$. This suggests that the volume integral can be performed only over an infinitesimal volume Δ centered at the field point. The exclusion of a single point from a given region does not affect the volume integral over that region; therefore, the source point at $\mathbf{r}' = \mathbf{r}$ is excluded from the volume integral. In other words, the volume integral is performed in a region without a point at which $\nabla^2(1/\mathcal{R})$ becomes infinite. To accommodate this exclusion, a new infinitesimal volume Δ' is introduced, which is a hollow sphere with outer radius α and inner radius δ, as shown in Fig. 5.13. The outer radius is sufficiently small such that J_x' may be assumed to be constant within Δ' and yet sufficiently large such that $\nabla^2(1/\mathcal{R}) \approx 0$ at the outer surface of Δ'. By setting $J_x'(\mathbf{r}') = J_x'(\mathbf{r})$ and bringing it outside the volume integral, we have

$$\nabla^2 A_x = \frac{\mu_0 J_x}{4\pi} \int_{\Delta'} \left(\nabla^2 \frac{1}{\mathcal{R}}\right) dv' = \frac{\mu_0 J_x}{4\pi} \lim_{\delta \to 0} \oint_{\mathcal{S}_1 + \mathcal{S}_2} \left(\nabla \frac{1}{\mathcal{R}}\right) \cdot d\mathbf{s}' \qquad (5.52)$$

where the identity $\nabla^2 U = \nabla \cdot \nabla U$ and the divergence theorem are used. Surface $\mathcal{S}_1 + \mathcal{S}_2$ corresponds to the bounding surface of Δ'. Because $\nabla(1/\mathcal{R})$ is negligible at the outer surface $\mathcal{S}_2$, we are left with a closed-surface integral over the inner surface $\mathcal{S}_1$. Using the relation $\nabla(1/\mathcal{R}) = -(\mathbf{a}_\mathcal{R}/\mathcal{R}^2)$ in Eq. (5.52), where $\mathbf{a}_\mathcal{R}$ is directed along the unit normal to the inner sphere $\mathbf{a}_s$, we have

$$\nabla^2 A_x = -\frac{\mu_0 J_x}{4\pi} \lim_{\delta \to 0} \oint_{\mathcal{S}_1} \frac{\mathbf{a}_\mathcal{R}}{\mathcal{R}^2} \cdot d\mathbf{s}' = -\frac{\mu_0 J_x}{4\pi} \lim_{\delta \to 0} \left[\frac{4\pi\delta^2}{\delta^2}\right]$$

$$= -\mu_0 J_x \tag{5.53}$$

where $4\pi\delta^2$ represents the area of $\mathcal{S}_1$.

Following the same procedure used for A_x, the Laplacians of A_y and A_z are obtained as follows:

$$\nabla^2 A_y = -\mu_0 J_y \tag{5.54a}$$

$$\nabla^2 A_z = -\mu_0 J_z \tag{5.54b}$$

Finally, the **vector Poisson's equation** is expressed as

$$\boxed{\nabla^2 \mathbf{A} = -\mu_0 \mathbf{J}} \tag{5.55}$$

Next, inserting Eq. (5.55) into Eq. (5.49) leads to the point form of Ampere's circuital law,

$$\nabla \times \mathbf{H} = \mathbf{J} \tag{5.56}$$

As stated previously, the Biot-Savart law relates the current element $I\,d\mathbf{l}'$ at the source point to the differential magnetic field $d\mathbf{H}$ at the field point. Meanwhile, the point form of Ampere's circuital law relates the curl of $\mathbf{H}$ at a point in space to the current density $\mathbf{J}$ given at the point, and thus is particularly useful for describing local effects in magnetostatics.

Example 5.10 An infinite sheet of current with uniform density $\mathbf{J}_s = J_o \mathbf{a}_y$ coincides with the $z = 0$ plane (Fig. 5.14). Find the magnetic field along the z-axis by determining the vector magnetic potential.

Solution

From Eq. (5.38c), the vector magnetic potential $d\mathbf{A}$ at field point p due to the current element $\mathbf{J}_s\,ds'$ at source point p' is given by

$$d\mathbf{A} = \frac{\mu_0 J_o \mathbf{a}_y}{4\pi \mathcal{R}} ds'$$

The total vector magnetic potential at p is

$$\mathbf{A} = \int_{x'=-\infty}^{\infty} \int_{y'=-\infty}^{\infty} \frac{\mu_0 J_o \mathbf{a}_y}{4\pi \mathcal{R}} dx' dy' \tag{5.57a}$$

Subsequently, the magnetic flux density is obtained as follows:

$$\mathbf{B} = \nabla \times \mathbf{A} = \frac{\mu_0 J_o}{4\pi} \int_{x'=-\infty}^{\infty} \int_{y'=-\infty}^{\infty} \nabla \times \frac{\mathbf{a}_y}{\mathcal{R}} dx' dy' \tag{5.57b}$$

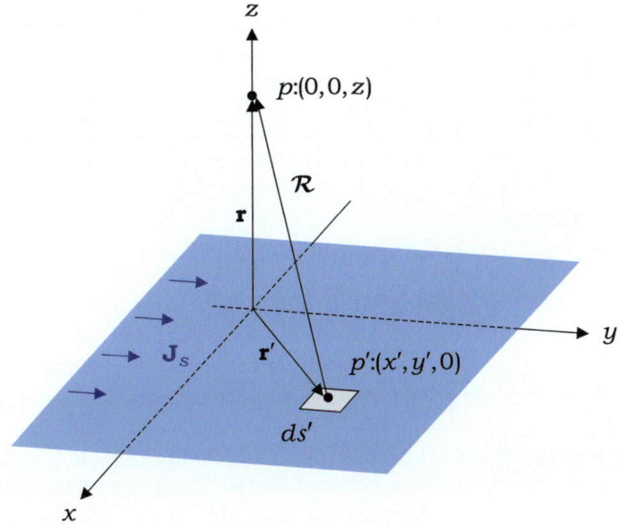

Fig. 5.14 Infinite plane carrying a uniform surface current

Using $\mathcal{R} = [(x - x')^2 + (y - y')^2 + z^2]^{1/2}$, the cross product is performed as follows:

$$\nabla \times \frac{\mathbf{a}_y}{\mathcal{R}} = \begin{vmatrix} \mathbf{a}_x & \mathbf{a}_y & \mathbf{a}_z \\ \dfrac{\partial}{\partial x} & \dfrac{\partial}{\partial y} & \dfrac{\partial}{\partial z} \\ 0 & \left[(x - x')^2 + (y - y')^2 + z^2\right]^{-1/2} & 0 \end{vmatrix}$$

$$= \left[z\,\mathbf{a}_x - (x - x')\,\mathbf{a}_z\right]\left[(x - x')^2 + (y - y')^2 + z^2\right]^{-3/2} \tag{5.57c}$$

Inserting Eq. (5.57c) into Eq. (5.57b) yields

$$\mathbf{B} = \frac{\mu_0 J_o}{4\pi} \int_{x'=-\infty}^{\infty} \int_{y'=-\infty}^{\infty} \frac{\left[z\,\mathbf{a}_x - (x - x')\,\mathbf{a}_z\right] dx'dy'}{\left[(x - x')^2 + (y - y')^2 + z^2\right]^{3/2}} \tag{5.57d}$$

Because the integration interval ranges from $-\infty$ to ∞, terms $(x - x')$ and $(y - y')$ can be replaced by $-x'$ and $-y'$, respectively. Thus,

$$\mathbf{B} = \frac{\mu_0 J_o}{4\pi} \int_{x'=-\infty}^{\infty} \int_{y'=-\infty}^{\infty} \frac{\left[z\,\mathbf{a}_x + x'\,\mathbf{a}_z\right] dx'dy'}{\left[(x')^2 + (y')^2 + z^2\right]^{3/2}} \tag{5.57e}$$

The second term in the numerator is an odd function of x'; thus, it does not contribute to $\mathbf{B}$. Integrating the first term in cylindrical coordinates yields

$$\mathbf{B} = \frac{\mu_0 J_o z\, \mathbf{a}_x}{4\pi} \int_{\phi'=0}^{2\pi} \int_{\rho'=0}^{\infty} \frac{\rho'\, d\rho'\, d\phi'}{\left[(\rho')^2 + z^2\right]^{3/2}} = \frac{\mu_0 J_o z\, \mathbf{a}_x}{2|z|}$$

Therefore, the answer is

$$\mathbf{B} = +\frac{1}{2}\mu_0 J_o\, \mathbf{a}_x, \quad \mathbf{H} = +\frac{1}{2} J_o\, \mathbf{a}_x \quad (z > 0) \tag{5.57f}$$

$$\mathbf{B} = -\frac{1}{2}\mu_0 J_o\, \mathbf{a}_x, \quad \mathbf{H} = -\frac{1}{2} J_o\, \mathbf{a}_x \quad (z < 0) \tag{5.57g}$$

It is important to note that even if the field point is located along the z-axis, it is defined by (x, y, z), and not $(0, 0, z)$, in the expression for $\mathcal{R}$ in Eq. (5.57c). Otherwise, the curl of $(\mathbf{a}_y/\mathcal{R})$ becomes trivial because the derivative of $(\mathbf{a}_y/\mathcal{R})$ with respect to x or y along the z-axis is zero.

Example 5.11 An infinitely long wire of radius a is concentric with the z-axis and carries uniform current $\mathbf{J} = J_o\, \mathbf{a}_z$ [A/m^2]. Determine $\mathbf{A}$ everywhere.

Solution

From symmetry considerations, the vector magnetic potential is expected to have the form $\mathbf{A} = A_z(\rho)\, \mathbf{a}_z$. Accordingly, the vector Poisson's equation is reduced to

$$\nabla^2 A_z(\rho) = \frac{1}{\rho}\frac{\partial}{\partial \rho}\left(\rho \frac{\partial A_z(\rho)}{\partial \rho}\right) = -\mu_0 J_o$$

Solving the equation yields a general solution,

$$A_z(\rho) = -\mu_0 J_o \frac{1}{4}\rho^2 + c_1 \quad (\rho \le a) \tag{5.58a}$$

$$A_z(\rho) = c_2 \ln \rho + c_3 \quad (\rho \ge a) \tag{5.58b}$$

Subsequently, from $\nabla \times \mathbf{A} = \mu_0 \mathbf{H}$,

$$H_\phi = \frac{J_o \rho}{2} \quad (\rho \le a) \tag{5.58c}$$

$$H_\phi = -\frac{c_2}{\mu_0 \rho} \quad (\rho \ge a) \tag{5.58d}$$

Applying the boundary condition for $\mathbf{H}$ to the surface at $\rho = a$ yields

$$c_2 = -\frac{\mu_0 J_o a^2}{2}$$

The boundary condition for $\mathbf{A}$ at $\rho = a$ is $A_z(a^-) = A_z(a^+)$, which can be derived from Eq. (5.34) by noting that the closed-line integral of $\mathbf{A}$ along a rectangle straddling the interface should vanish as the loop surface decreases to zero. Applying this to Eqs. (5.58a) and (5.58b) yields

$$c_1 = \frac{1}{4}\mu_0 J_o a^2 \text{ and } c_3 = \frac{1}{2}\mu_0 J_o a^2 \ln a$$

The answer is therefore

$$\mathbf{A} = -\frac{1}{4}\mu_0 J_o (\rho^2 - a^2)\,\mathbf{a}_z \qquad (\rho \le a) \tag{5.58e}$$

$$\mathbf{A} = -\frac{1}{2}\mu_0 J_o a^2 \ln(\rho/a)\,\mathbf{a}_z \qquad (\rho \ge a) \tag{5.58f}$$

Exercise 5.10
A circular loop of radius a is centered at the origin in the $z = 0$ plane. If it carries a steady current I, determine $\mathbf{A}$ along the z-axis.

Ans. 0.

Exercise 5.11
For an infinite current-sheet, as shown in Fig. 5.14, find a general solution of $\mathbf{A}$ by solving (a) Eq. (5.38c), and (b) Eq. (5.55). [Hint: Translational symmetry, $\mathbf{J} = 0$.]

Ans. (a) $\mathbf{A} = A_y(z)\,\mathbf{a}_y$, (b) $\mathbf{A} = (c_1 z + c_2)\,\mathbf{a}_y$.

Exercise 5.12
Derive the boundary conditions for $\mathbf{A}$ at the interface using Eqs. (5.34) and (5.48).

Ans. The tangential and normal components of $\mathbf{A}$ are continuous.

Review Questions

RQ 5.25	Define the vector magnetic potential.	[(5.34)]
RQ 5.26	Express the relationship between $\mathbf{A}$ and its source.	[(5.38a,b,c)]
RQ 5.27	Express the vector Poisson's equation.	[(5.55)]
RQ 5.28	Which of the three expressions in Eqs. (5.38a)–(5.38c) is a solution to the vector Poisson's equation?	[(5.38b)]
RQ 5.29	What are the two fundamental relations for $\mathbf{A}$?	[(5.34)(5.48)]
RQ 5.30	State the boundary conditions for $\mathbf{A}$.	[Exercise 5.12]

5.6 Magnetic Dipole

A ***magnetic dipole*** is a small loop that carries steady current. This is the simplest magnetic model of an atom, whereas the electric dipole is the simplest electric model of an atom. The magnetic dipole provides an easy way to analyze the interaction between a material medium and a magnetic field, and plays a major role in studying the magnetic properties of materials. The magnetic dipole is characterized by the ***magnetic dipole moment***, which is the product of the loop current and loop area.

To explore the magnetic dipole, we consider a circular loop of radius a centered at the origin in the $z = 0$ plane carrying a steady current I, as shown in Fig. 5.15. Without loss of generality, we assume that field point p is in the yz-plane. Point p' is the source point at which differential current element $I\,d\mathbf{l}'$ is positioned. Repeating Eq. (5.38a) for convenience, the vector magnetic potential at point p is given by

$$\mathbf{A} = \frac{\mu_0 I}{4\pi} \oint_{C'} \frac{d\mathbf{l}'}{\mathcal{R}}$$

In a mixed-coordinate system, we expand the vectors $\mathbf{r}$, $\mathbf{r}'$, and $\mathcal{R}$ using the base vectors of the Cartesian system but express their scalar components in terms of spherical coordinates such that

$$\mathbf{r} = R \sin\theta \, \mathbf{a}_y + R \cos\theta \, \mathbf{a}_z \tag{5.59a}$$

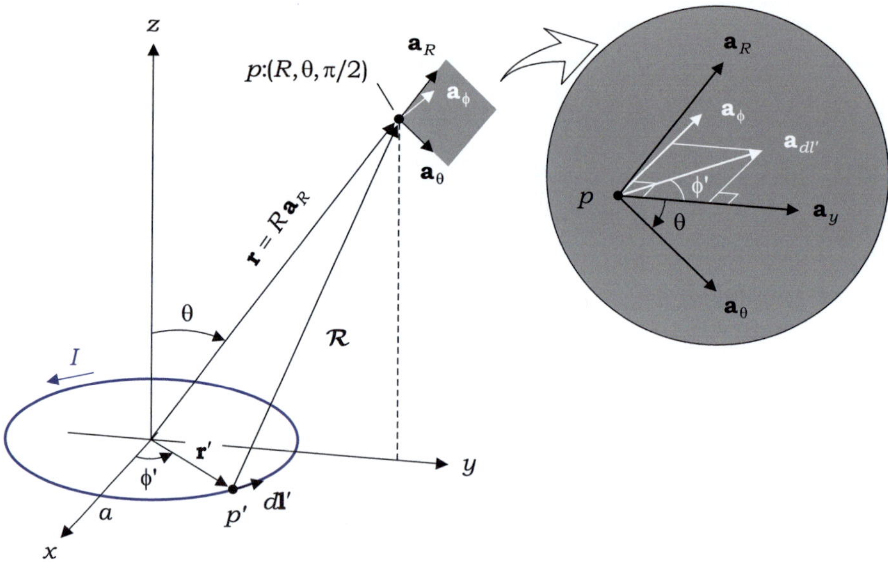

Fig. 5.15 Magnetic dipole

$$\mathbf{r'} = a\cos\phi'\,\mathbf{a}_x + a\sin\phi'\,\mathbf{a}_y \tag{5.59b}$$

$$\mathcal{R}^{-1} = \left|\mathbf{r} - \mathbf{r'}\right|^{-1} = [R^2 + a^2 - 2aR\sin\theta\sin\phi']^{-1/2} \tag{5.59c}$$

At large distances ($R \gg a$), with the help of binomial expansion, the inverse distance term, $\mathcal{R}^{-1}$, can be approximated as

$$
\begin{aligned}
\mathcal{R}^{-1} &= \frac{1}{R}\left[1 + \frac{a^2}{R^2} - \frac{2a}{R}\sin\theta\sin\phi'\right]^{-1/2} \\
&\cong \frac{1}{R}\left(1 + \frac{a}{R}\sin\theta\sin\phi'\right)
\end{aligned}
\tag{5.60}
$$

where higher-order terms such as $(a/R)^2$ and $(a/R)^3$ are ignored.

The differential length vector at point p' along the loop is $d\mathbf{l'} = a\,d\phi'\,\mathbf{a}_{\phi'}$. This can be expressed in terms of the base vectors at point p as follows:

$$d\mathbf{l'} = a\,d\phi'\left(\cos\phi'\sin\theta\,\mathbf{a}_R + \cos\phi'\cos\theta\,\mathbf{a}_\theta + \sin\phi'\,\mathbf{a}_\phi\right) \tag{5.61}$$

Substituting Eqs. (5.60) and (5.61) into Eq. (5.38a), we obtain

$$
\begin{aligned}
\mathbf{A} &= \frac{\mu_0 I}{4\pi}\int_{\phi'=0}^{2\pi}\frac{1}{R}\left(1 + \frac{a}{R}\sin\theta\sin\phi'\right)a\,d\phi'\left(\cos\phi'\sin\theta\,\mathbf{a}_R\right. \\
&\qquad\qquad\qquad + \left.\cos\phi'\cos\theta\,\mathbf{a}_\theta + \sin\phi'\,\mathbf{a}_\phi\right) \\
&= \frac{\mu_0 I}{4\pi}\frac{a}{R}\mathbf{a}_\phi\int_{\phi'=0}^{2\pi}\left(\frac{a}{R}\sin\theta\sin\phi'\right)\sin\phi'\,d\phi' \\
&= \frac{\mu_0 I a^2}{4R^2}\sin\theta\,\mathbf{a}_\phi
\end{aligned}
\tag{5.62}
$$

In vector notation, the vector magnetic potential of a magnetic dipole with magnetic dipole moment $\mathbf{m}$ can be expressed as

$$\boxed{\mathbf{A} = \frac{\mu_0}{4\pi}\frac{\mathbf{m}\times\mathbf{a}_R}{R^2}} \quad [\text{Wb/m}] \tag{5.63}$$

where the magnetic dipole moment is defined as

$$\boxed{\mathbf{m} = I\pi a^2\mathbf{a}_z = IS\,\mathbf{a}_z} \quad [\text{A}\cdot\text{m}^2] \tag{5.64}$$

The magnetic dipole moment is a vector whose magnitude is the product of the loop current and area, and its unit vector is normal to the loop surface in accordance with the right-hand rule that the right thumb points in the direction of $\mathbf{m}$ when the fingers follow the direction of I in the loop.

By taking the curl of $\mathbf{A}$ in Eq. (5.63) in spherical coordinates, $\mathbf{B}$ of the magnetic dipole moment is given by

$$\boxed{\mathbf{B} = \frac{\mu_0\, m}{4\pi R^3}(2\cos\theta\,\mathbf{a}_R + \sin\theta\,\mathbf{a}_\theta)} \quad [\text{T}] \qquad (5.65)$$

It can be shown that $\mathbf{B}$ in Eq. (5.65) satisfies the two fundamental relations $\nabla \cdot \mathbf{B} = 0$ and $\nabla \times \mathbf{H} = 0$. We see from Eq. (5.65) that $\mathbf{B}$ of a magnetic dipole is analogous to $\mathbf{E}$ of an electric dipole. The expression for $\mathbf{B}$ is identical to that of $\mathbf{E}$ in Eq. (3.63) if m and μ_0 are replaced by p and $1/\varepsilon_0$, respectively. It should be noted that Eqs. (3.63) and (5.65) both represent the far-field patterns valid under the conditions $R \gg d$ and $R \gg a$, respectively.

The electric field lines of an electric dipole and magnetic field lines of a magnetic dipole can be obtained using numerical methods without relying on binomial expansion. They are drawn to scale, as shown in Fig. 5.16.

Example 5.12 A long cylinder with radius a and surface current $\mathbf{J}_s = J_o\,\mathbf{a}_\phi$ [A/m] is centered along the z-axis. Starting by assuming that it is a stack of identical current loops of radius a, find the vector magnetic potential and magnetic flux density everywhere.

Solution

At large distances from the z-axis, $\mathbf{A}$ of a loop current $J_o\,dz$ placed in the $z' = 0$ plane can be obtained using Eq. (5.62) as

$$\mathbf{A} = \frac{\mu_0 J_o\,dz\,a^2}{4}\frac{\rho}{(\rho^2 + z^2)^{3/2}}\,\mathbf{a}_\phi \qquad (\rho \gg a) \qquad (5.66\text{a})$$

Adding the contributions of all current loops placed at different z'-planes results in

$$\begin{aligned}
\mathbf{A}_2 &= \frac{\mu_0 J_o\,a^2\rho}{4}\,\mathbf{a}_\phi \int_{z'=-\infty}^{z'=\infty} \frac{dz'}{[\rho^2 + (z' - z)^2]^{3/2}} \qquad (\rho \gg a) \qquad (5.66\text{b})\\
&= \frac{\mu_0 J_o\,a^2}{2\rho}\,\mathbf{a}_\phi
\end{aligned}$$

This will be used as a trial solution for $\mathbf{A}$ in region $\rho \geq a$. Next, as a trial solution for $\mathbf{A}$ in the region $\rho \leq a$, we choose

$$\mathbf{A}_1 = \left(c_2\,\rho^2 + c_1\,\rho + c_o\right)\mathbf{a}_\phi \qquad (\rho \leq a) \qquad (5.66\text{c})$$

Here, c_o, c_1, and c_2 are constants that must be determined. $\mathbf{B}$ is then computed as

$$\mathbf{B}_1 = \nabla \times \mathbf{A}_1 = \left(3c_2\,\rho + 2c_1 + c_o\,\rho^{-1}\right)\mathbf{a}_z \qquad (\rho \leq a) \qquad (5.66\text{d})$$

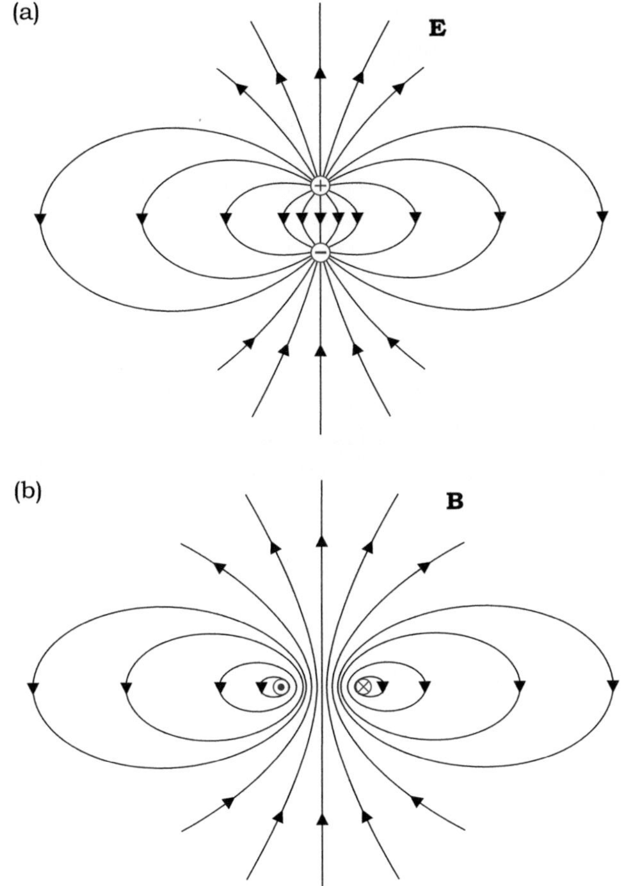

Fig. 5.16 Near-field patterns of **a** electric and **b** magnetic dipoles

$$\mathbf{B}_2 = \nabla \times \mathbf{A}_2 = 0 \qquad (\rho \geq a) \tag{5.66e}$$

Enforcing Eq. (5.66d) to satisfy the fundamental relations for magnetic fields, we obtain

$$\nabla \cdot \mathbf{B}_1 = 0 \tag{5.66f}$$

$$\nabla \times \mathbf{H}_1 = -\frac{1}{\mu_0}\left(3c_2 - c_o\,\rho^{-2}\right)\mathbf{a}_\phi \tag{5.66g}$$

As there is no free current inside the solenoid, $\nabla \times \mathbf{H}_1 = 0$. Thus,

$$c_2 = c_o = 0$$

Next, applying the boundary condition for $\mathbf{A}$, that is, $\mathbf{A}_1(\rho = a) = \mathbf{A}_2(\rho = a)$, to Eqs. (5.66b) and (5.66c), we obtain

$$c_1 = \frac{1}{2}\mu_0 J_o$$

Therefore, the answer is

$$\mathbf{A}_1 = \frac{\mu_0 J_o \rho}{2}\,\mathbf{a}_\phi \text{ and } \mathbf{B}_1 = \mu_0 J_o \mathbf{a}_z \qquad (\rho \le a) \tag{5.66h}$$

$$\mathbf{A}_2 = \frac{\mu_0 J_o a^2}{2\rho}\,\mathbf{a}_\phi \text{ and } \mathbf{B}_2 = 0 \qquad (\rho \ge a) \tag{5.66i}$$

Furthermore, the magnetic fields in Eqs. (5.66h) and (5.66i) satisfy the boundary conditions for $\mathbf{H}$ at $\rho = a$, that is, $H_2 - H_1 = J_o$.

Exercise 5.13

For a magnetic dipole moment $\mathbf{m} = m\,\mathbf{a}_z$ placed at the origin, in what direction is $|\mathbf{B}|$ at its maximum?

Ans. $+z$- and $-z$-directions.

Exercise 5.14

For an electron circulating around a circle as shown in Fig. 5.1, determine the line current in the loop and the magnetic dipole moment.

Ans. $I = |e|\omega/2\pi$, $\mathbf{m} = -\mathbf{a}_z m v_o^2 /2B$.

Exercise 5.15

Write a differential equation for the magnetic flux lines in the yz-plane at large distances from the magnetic dipole located at the origin.

Ans. $\dfrac{dz}{dy} = \dfrac{B_z}{B_y} = \dfrac{2\cos^2\theta - \sin^2\theta}{3\sin\theta\cos\theta}$, where $\theta = \tan^{-1}(y/z)$.

Review Questions

RQ 5.31	What is a magnetic dipole?	[Fig. 5.15]
RQ 5.32	Define the magnetic dipole moment.	[(5.64)]
RQ 5.33	In what ways do the far-field patterns of the electric and magnetic dipoles vary in space?	[(3.63)(5.65)]
RQ 5.34	Sketch the near-field patterns of the electric and magnetic dipoles.	[Fig. 5.16]

5.7 Magnetic Materials

The simplest atomic model consists of a positively charged nucleus and negatively charged electrons orbiting around it. The orbiting electrons form a small current loop that creates a magnetic dipole moment in the atom. The electrons and nucleus also spin about their axes, constituting additional magnetic dipole moments rooted in the quantum theory. However, the effect of the nuclear spin is three orders of magnitude smaller than that of the orbiting electron or electron spin, and is thus neglected in most cases.

The magnetic properties of a material are determined by the interaction of the lattice atoms with an external magnetic field. Under normal circumstances, orbital magnetic moments are randomly oriented and cancel out. However, an externally applied magnetic field may exert either centripetal or centrifugal force on the orbiting electrons. However, the electron orbit is quantized and cannot be changed. Therefore, the electron velocity must be varied in order to balance the magnetic force. This change causes each atom to gain another orbital magnetic moment, which is always opposite to the applied magnetic field. This is the magnetization process of **diamagnetic materials**, in which the material is **magnetized**.

To further examine magnetization in diamagnetic materials, let us consider two identical atoms with opposite magnetic moments, $\mathbf{m} = \pm m_o \mathbf{a}_z$, placed in a uniform magnetic field $\mathbf{B} = B_o \mathbf{a}_z$, as shown in Fig. 5.17. If the magnetic moment is caused by an electron moving in a circle of radius r_o in the xy-plane, the equation of motion can be expressed in cylindrical coordinates as

$$\frac{-q^2}{4\pi\varepsilon_0 r_o^2}\,\mathbf{a}_\rho \pm (v_o + \Delta v)q\,B_o\,\mathbf{a}_\rho + m_e\frac{(v_o + \Delta v)^2}{r_o}\,\mathbf{a}_\rho = 0 \qquad (5.67)$$

where Δv is the change in electron velocity. The first two terms on the left-hand side of Eq. (5.67) are the electric and magnetic forces on the electron, respectively,

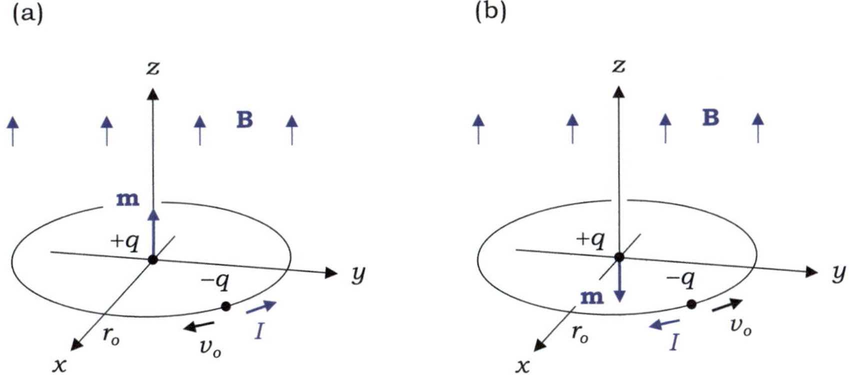

Fig. 5.17 Electron orbiting around the nucleus

and the third term is the centrifugal force on the electron with mass m_e. The electron motion is clockwise for $+$ sign and counterclockwise for $-$ sign, as viewed from above (see Fig. 5.17).

Because Δv is caused by $\mathbf{B}$, it should be zero when $B_o = 0$. Using this and assuming $\Delta v \ll v_o$, Eq. (5.67) can be reduced to

$$\pm q v_o B_o \mathbf{a}_\rho = -m_e \frac{2 v_o \Delta v}{r_o} \mathbf{a}_\rho$$

Thus,

$$\Delta v = \mp q B_o \frac{r_o}{2 m_e} \tag{5.68}$$

where the negative sign indicates that the electron velocity and magnetic moment decreased, as shown in Fig. 5.17a. In contrast, the positive sign indicates that the magnetic moment increases because of $\mathbf{B}$, as shown in Fig. 5.17b. Nevertheless, the incremental magnetic moment is always opposite in direction to $\mathbf{B}$. In other words, in the presence of an external magnetic field, *the net magnetic moment of the diamagnetic materials is antiparallel to the applied field.*

Diamagnetic materials such as gold, silver, copper, lead, and silicon have negative magnetic susceptibilities on the order of -10^{-5} and relative permeabilities that are slightly less than unity. Superconductors are perfect diamagnetic materials with susceptibility $\chi_m = -1$, and thus, zero **relative permeability**. No magnetic field is allowed inside the superconductor. The superconductor completely expels the magnetic field, and can thus be levitated. All materials exhibit diamagnetism, even if the diamagnetic effect is overwhelmed by other stronger magnetic effects in some materials.

Diamagnetism is mainly due to the orbital magnetic moment of atoms with even numbers of electrons; electrons exist in pairs with opposite spin directions in accordance with the Pauli exclusion principle. In contrast, ***paramagnetism*** is caused by the orbital and spin magnetic moments of atoms with odd numbers of electrons. In paramagnetic materials, the two magnetic moments do not completely cancel each other, and atoms have permanent magnetic moments. Under normal conditions, the magnetic moments are randomly oriented, such that there is no net magnetic moment in the material. However, in the presence of an externally applied magnetic field, the permanent magnetic moments align with the field, resulting in a net magnetic moment, which, in turn, increases the total magnetic field in the material. However, as discussed in Chap. 3, aligned electric dipole moments decrease the total electric field in the material. Paramagnetic materials, such as air, aluminum, platinum, titanium, and tungsten, have magnetic susceptibilities on the order of 10^{-5} and relative permeabilities that are slightly greater than unity. For most practical purposes, we can set $\mu = \mu_0$ for diamagnetic and paramagnetic materials and consider them as nonmagnetic materials.

The magnetization of diamagnetic and paramagnetic materials can be maintained only in the presence of an external magnetic field. Meanwhile, *ferromagnetic materials* can sustain magnetization without an external field. The atoms in ferromagnetic materials have large permanent magnetic moments because of the dominant effect of electron spin. Such magnetic dipoles have a strong tendency to align with their neighbors, which originates from quantum mechanics. The complete alignment of all magnetic dipoles is possible in microscopic regions called *magnetic domains*, where each domain is fully magnetized. Under normal conditions, the magnetic domains are randomly oriented with respect to the magnetization and yield no net magnetic moment. However, in the presence of an external magnetic field, the magnetic domains oriented toward the external field grow at the expense of nearby domains and produce a large net magnetic moment. If the external field is sufficiently strong, one domain will have overwhelming dominance over other domains. In this case, the material is said to be saturated.

Ferromagnetic materials such as iron, cobalt, and nickel have very large relative permeabilities ranging from 250 to 5000, and their alloys such as permalloy and mumetal have relative permeabilities as large as 10^5. Unlike diamagnetic and paramagnetic materials, ferromagnetic materials are nonlinear materials in that the relative permeability depends not only on the magnetic field intensity, but also on the past magnetic states of the material.

5.7.1 *Magnetization and Magnetization Current*

According to the atomic model of matter, a magnetized material may be considered an aggregate of discrete magnetic dipoles positioned at the corresponding lattice points in free space. For the macroscopic properties of a magnetic material, we define *magnetization* $\mathbf{M}$ as the magnetic dipole moment per unit volume. Expressed mathematically,

$$\mathbf{M} = \lim_{\Delta v \to 0} \frac{1}{\Delta v} \sum_{i=1}^{n\Delta v} \mathbf{m}_i \quad [\text{A/m}] \tag{5.69}$$

where n is the number density of the lattice atoms or magnetic dipoles, and $\mathbf{m}_i$ are the magnetic dipole moments of the atoms within an incremental volume Δv. Note that $\mathbf{M}$ is a continuous function of position, whereas $\mathbf{m}_i$ are discrete vectors positioned at separate points in space. Magnetization is measured in amperes per meter. A magnetic material with nonzero $\mathbf{M}$ is said to be *magnetized*.

Magnetization induces currents on the surface of the material. To investigate this, we consider an interface between a magnetized material in region $z \leq 0$ and the free space above, or air with little error, as shown in Fig. 5.18. We take a square loop with side a carrying steady current I as the magnetic dipole induced in the material; its center is denoted by a black dot. In this figure, the magnetic dipoles are aligned parallel to a unit vector $\mathbf{a}_m$ and make an angle θ with the unit normal to the interface

$\mathbf{a}_n$. This implies that the material is fully magnetized by the magnetic field applied in the direction of $\mathbf{a}_m$. This is because the magnetic field, magnetization, and magnetic dipole moments are all parallel to $\mathbf{a}_m$ in the material.

For the sake of argument, the material is sliced into hypothetical layers of thickness $(a/2)\sin\theta$. A side view of the magnetic dipole is shown in Fig. 5.18a, where a small circle with a dot (or cross) indicates the direction of current flowing on the upper (or lower) side of the square loop. As can be seen from Fig. 5.18a, the magnetic dipoles, with their centers placed in layers 1 and 3, contribute currents to layer 2. However, the two currents completely cancel each other. Similarly, no net current is induced in the region below layer 2. In contrast, the magnetic dipoles with their centers placed in layers 1 and 2 contribute to the currents flowing in air and layer 1. These net currents are considered to be surface currents induced by magnetization.

The front view of the magnetic dipole is shown in Fig. 5.18b. Because the square loops are tilted away from the viewer, they appear as rectangles of size $a \times a\sin\theta$. As stated earlier, only magnetic dipoles with centers in layers 1 and 2 contribute to the

(a)

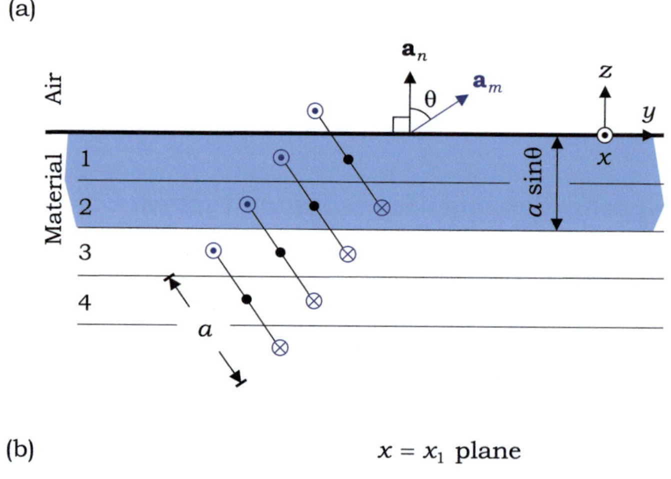

(b)

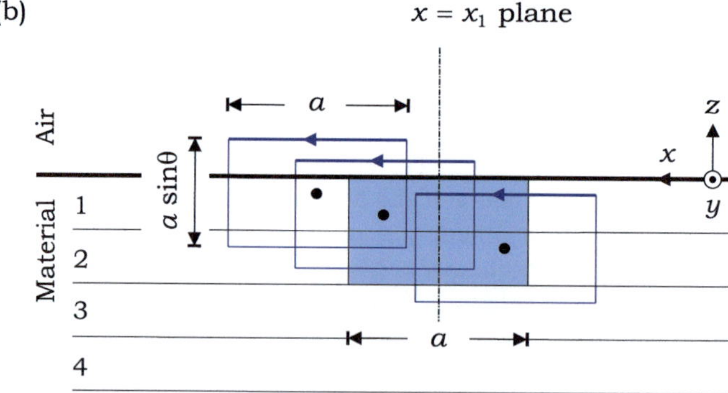

Fig. 5.18 Magnetization surface current. The magnetic dipoles are represented by square loops: **a** Sideview, and **b** Front view

surface current. However, only those with their centers in the shaded area, as shown in Fig. 5.18b, can contribute to the current flowing through the reference plane at $x = x_1$. In view of these considerations, the density of the induced surface current is defined as the current flowing through the reference plane per unit length along the y-axis. That is,

$$J_{ms} = In(a^2 \sin \theta) = n|\mathbf{m}| \sin \theta \tag{5.70}$$

where n is the number density of the magnetic dipoles, and the term in parentheses is the size of the shaded area in Fig. 5.18b. Using the magnetization defined by $\mathbf{M} = n\,\mathbf{m}$ and replacing sine with the cross product of the unit vectors $\mathbf{a}_m$ and $\mathbf{a}_n$, the ***magnetization surface current density*** is expressed as

$$\boxed{\mathbf{J}_{ms} = \mathbf{M} \times \mathbf{a}_n} \quad [\text{A/m}] \tag{5.71}$$

Here, $\mathbf{M}$ is the magnetization at the surface of the material, and $\mathbf{a}_n$ is the unit normal to the surface pointing out of the material.

Magnetization may also induce a volume current inside the material, which is characterized by the ***magnetization current density*** $\mathbf{J}_m$ [A/m^2]. Let us consider the interface between a material with magnetization $\mathbf{M}$ in region $z \leq 0$ and the free space above it, as shown in Fig. 5.19. Circular loop $\mathcal{C}$ is drawn on the surface in the presence of magnetization surface currents. If there is a net surface current crossing and flowing out of the circle, another current should emerge in the loop area from the interior in accordance with the continuity equation. This current is referred to as the ***magnetization current***. Both the magnetization and magnetization surface currents originate from magnetic dipoles; the current in a square loop, or magnetic dipole, contributes to $\mathbf{J}_m$ in the interior and $\mathbf{J}_{ms}$ on the surface of the material, as illustrated in Fig. 5.19.

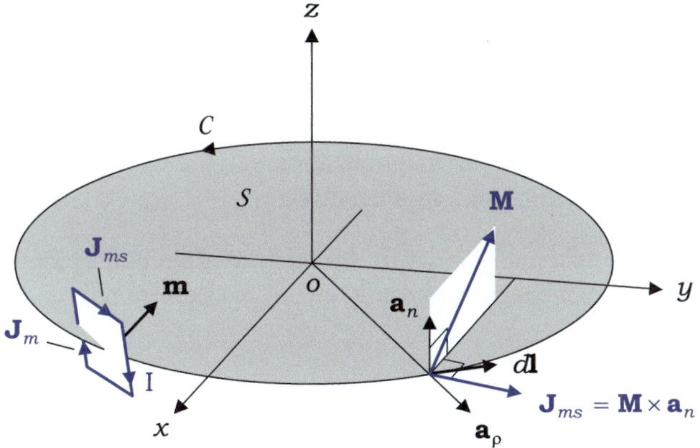

Fig. 5.19 Magnetization volume and surface current densities $\mathbf{J}_m$ and $\mathbf{J}_{ms}$

We are now ready to obtain an expression for $\mathbf{J}_m$. Referring to the circle drawn on the surface of the material or the $z = 0$ plane (Fig. 5.19), the differential current passing through the differential length $|d\mathbf{l}|$ along the circle is expressed as

$$dI_{ms} = \mathbf{J}_{ms} \cdot \mathbf{a}_\rho |d\mathbf{l}| = (\mathbf{M} \times \mathbf{a}_n) \cdot \mathbf{a}_\rho |d\mathbf{l}| \tag{5.72}$$

where the unit vector $\mathbf{a}_\rho$ is normal to $d\mathbf{l}$ directed along the circle, whereas the unit vector $\mathbf{a}_n$ is normal to the loop surface and directed along the $+z$-direction. Applying a scalar triple product, $(\mathbf{A} \times \mathbf{B}) \cdot \mathbf{C} = \mathbf{A} \cdot (\mathbf{B} \times \mathbf{C})$, to Eq. (5.72), we obtain

$$dI_{ms} = \mathbf{M} \cdot \left(\mathbf{a}_n \times \mathbf{a}_\rho\right)|d\mathbf{l}| = \mathbf{M} \cdot d\mathbf{l} \tag{5.73}$$

where $\mathbf{a}_n \times \mathbf{a}_\rho = \mathbf{a}_\phi$ is directed along the circle. Thus, the net surface current flowing out of the circle can be expressed as

$$I_{ms} = \oint_C \mathbf{M} \cdot d\mathbf{l} = \int_S (\nabla \times \mathbf{M}) \cdot d\mathbf{s} \quad [\text{A}] \tag{5.74}$$

where Stokes's theorem is used, and surface S is bounded by circle C. The continuity equation indicates that the total surface current I_{ms} is equal to the volume current emerging from the interior and enclosed by C. That is,

$$I_{ms} = \int_S \mathbf{J}_m \cdot d\mathbf{s} \quad [\text{A}] \tag{5.75}$$

Comparing Eqs. (5.74) and (5.75) leads to the conclusion that the two integrands should be the same at every point on the surface of the material. Thus, the magnetization current density is expressed as

$$\boxed{\mathbf{J}_m = \nabla \times \mathbf{M}} \quad [\text{A/m}^2] \tag{5.76}$$

If there is no free volume current in a simple medium, it follows that $\nabla \times \mathbf{H} = 0$, and thus $\nabla \times \mathbf{M} = 0$ in the material, because $\mathbf{M}$ and $\mathbf{H}$ are linearly related to each other, as shown in Eq. (5.87). This implies that no magnetization current is present in the material. Nevertheless, the magnetization surface current is not always zero on the boundary surface for nonzero magnetization.

Example 5.13 A permanent magnet in the form of a circular cylinder of radius a and height b has uniform magnetization $\mathbf{M} = M_o \mathbf{a}_z$. Determine $\mathbf{J}_m$ and $\mathbf{J}_{ms}$.

Solution

From Eq. (5.76),

$$\mathbf{J}_m = \nabla \times (M_o \mathbf{a}_z) = 0$$

From Eq. (5.71),

$$\begin{aligned}
\mathbf{J}_{ms} &= \mathbf{M} \times \mathbf{a}_n \\
&= M_o\,\mathbf{a}_z \times \mathbf{a}_z = 0 \qquad \text{(top)} \\[6pt]
\mathbf{J}_{ms} &= M_o\,\mathbf{a}_z \times (-\mathbf{a}_z) = 0 \quad \text{(bottom)} \\[6pt]
\mathbf{J}_{ms} &= M_o\,\mathbf{a}_z \times \mathbf{a}_\rho = M_o\,\mathbf{a}_\phi \; \text{(side)}
\end{aligned} \tag{5.77}$$

The results show that the permanent magnet can be replaced with a cylindrical current sheet of radius a and height b with a surface current density of $\mathbf{J}_s = M_o\,\mathbf{a}_\phi$ with respect to the magnetic field.

Example 5.14 Find the vector magnetic potential of the magnetization $\mathbf{M}$.

Solution

In a magnetized material with nonzero $\mathbf{M}$, the magnetic moment in the differential volume dv' is equal to $\mathbf{M}dv'$. Using this in Eq. (5.63), the differential vector magnetic potential is given by

$$d\mathbf{A} = \frac{\mu_0}{4\pi} \frac{dv'\,\mathbf{M} \times \mathbf{a}_{\mathcal{R}}}{\mathcal{R}^2} \tag{5.78a}$$

Using the identity given in Eq. (2.41) and the relation $\nabla(1/\mathcal{R}) = -\nabla'(1/\mathcal{R})$, this can be rewritten as

$$d\mathbf{A} = \frac{\mu_0}{4\pi}\mathbf{M} \times \left(\nabla' \frac{1}{\mathcal{R}} \right) dv' \tag{5.78b}$$

Using the identity $\nabla \times (f\,\mathbf{U}) = f(\nabla \times \mathbf{U}) - \mathbf{U} \times (\nabla f)$ and integrating $d\mathbf{A}$ in Eq. (5.78b) over a given volume $\mathcal{V}'$, we obtain

$$\mathbf{A} = \frac{\mu_0}{4\pi} \int_{\mathcal{V}'} \left[\frac{1}{\mathcal{R}} \nabla' \times \mathbf{M} - \nabla' \times \left(\frac{\mathbf{M}}{\mathcal{R}} \right) \right] dv' \tag{5.78c}$$

Let $\mathbf{G}$ be a constant in the identity $\nabla \cdot (\mathbf{G} \times \mathbf{K}) = \mathbf{K} \cdot (\nabla \times \mathbf{G}) - \mathbf{G} \cdot (\nabla \times \mathbf{K})$ to obtain $\nabla \cdot (\mathbf{G} \times \mathbf{K}) = -\mathbf{G} \cdot (\nabla \times \mathbf{K})$. Keeping the second integral of Eq. (5.78c) in mind, we set $\mathbf{K} = \mathbf{M}/\mathcal{R}$ in this equation and take the volume integrals of both sides to obtain

$$\begin{aligned}
\int_{\mathcal{V}'} \mathbf{G} \cdot \nabla' \times \left(\frac{\mathbf{M}}{\mathcal{R}} \right) dv' &= -\int_{\mathcal{V}'} \nabla' \cdot \left(\mathbf{G} \times \frac{\mathbf{M}}{\mathcal{R}} \right) dv' \\[6pt]
&= -\oint_{\mathcal{S}'} \left(\mathbf{G} \times \frac{\mathbf{M}}{\mathcal{R}} \right) \cdot d\mathbf{s}'
\end{aligned}$$

$$= -\oint_{S'} \mathbf{G} \cdot \frac{\mathbf{M}}{\mathcal{R}} \times d\mathbf{s}' \tag{5.78d}$$

where we have used Stokes's theorem and the scalar triple product. The constant $\mathbf{G}$ can come outside the integral, and thus, Eq. (5.78d) is reduced to

$$-\int_{V'} \nabla' \times \left(\frac{\mathbf{M}}{\mathcal{R}}\right) dv' = \oint_{S'} \frac{\mathbf{M}}{\mathcal{R}} \times d\mathbf{s}' \tag{5.78e}$$

Using Eq. (5.78e) in Eq. (5.78c) yields

$$\mathbf{A} = \frac{\mu_0}{4\pi} \int_{V'} \frac{1}{\mathcal{R}} \nabla' \times \mathbf{M} \, dv' + \frac{\mu_0}{4\pi} \oint_{S'} \frac{1}{\mathcal{R}} \mathbf{M} \times \mathbf{a}_n \, ds' \tag{5.78f}$$

where $\mathbf{a}_n$ denotes an outward unit normal to the surface.

Furthermore, with the help of Eqs. (5.71) and (5.76), Eq. (5.78f) can be written as

$$\mathbf{A} = \frac{\mu_0}{4\pi} \int_{V'} \frac{\mathbf{J}'_m}{\mathcal{R}} dv' + \frac{\mu_0}{4\pi} \oint_{S'} \frac{\mathbf{J}'_{ms}}{\mathcal{R}} ds' \tag{5.79}$$

Comparing Eq. (5.79) with Eqs. (5.38b, c) leads to the conclusion that $\mathbf{J}_m$ and $\mathbf{J}_{ms}$ produce $\mathbf{A}$, and thus $\mathbf{B}$, in the same manner as the free currents $\mathbf{J}$ and $\mathbf{J}_s$ in free space.

Exercise 5.16

A material with $\mathbf{M} = M_o \mathbf{a}_z$ forms a sphere of radius a. Find $\mathbf{J}_{ms}$ on the surface.

Ans. $(M_o \sin\theta)\, \mathbf{a}_\phi$ [A/m].

Review Questions

RQ 5.35	Explain diamagnetism, paramagnetism, and ferromagnetism.	[Fig. 5.17]
RQ 5.36	Define magnetization.	[(5.69)]
RQ 5.37	Relate $\mathbf{J}_{ms}$ and $\mathbf{J}_m$ to magnetization.	[(5.71)(5.76)]
RQ 5.38	What are the units of $\mathbf{J}_{ms}$ and $\mathbf{J}_m$?	[(5.71)(5.76)]
RQ 5.39	Does $\mathbf{J}_m = 0$ in a magnetic material mean $\mathbf{M} = 0$ inside it?	[(5.76)]
RQ 5.40	Is it true that $\mathbf{J}_{ms}$ is not always zero on the boundary surface of a magnetized material even if $\mathbf{J}_m$ is zero in the interior?	[Example 5.13]

5.7.2 *Permeability*

The magnetization and internal magnetic field of a magnetized material can be understood by analogy with the electric polarization and internal electric field of a dielectric material, respectively. When a magnetic material is placed in an external magnetic field, the induced magnetization produces a magnetization current in the material, which, in turn, generates a magnetic field called the magnetization field. The external magnetic field and magnetization field constitute the ***internal magnetic field***. Here, we redefine $\mathbf{H}$ such that it is related only to the free current provided by turning on the current source.

In view of the fact that both the free and magnetization currents generate magnetic fields in the material, by invoking Ampere's law, we can write

$$\nabla \times \frac{\mathbf{B}}{\mu_0} = \mathbf{J} + \mathbf{J}_m$$

$$= \mathbf{J} + \nabla \times \mathbf{M} \tag{5.80}$$

where $\mathbf{J}$ is the free current density, which may be the conduction current, convection current, or both, and $\mathbf{J}_m$ is the magnetization current density. The constant μ_0 in Eq. (5.80) indicates that both $\mathbf{J}$ and $\mathbf{J}_m$ are placed in free space in accordance with the atomic model of matter, which states that a magnetic material is an aggregate of magnetic dipoles in free space. Although the magnetization surface current $\mathbf{J}_{ms}$ can also produce a magnetic field, it is excluded from Eq. (5.80) because it cannot describe local effects; it only relates to the magnetic field at points away from it. Instead, $\mathbf{J}_{ms}$ constitutes the boundary condition for a unique solution in a given region. Equation (5.80) can be rearranged as

$$\nabla \times \left(\frac{\mathbf{B}}{\mu_0} - \mathbf{M} \right) = \mathbf{J} \tag{5.81}$$

We are now ready to redefine $\mathbf{H}$ in the magnetic material such that

$$\mathbf{H} \equiv \frac{\mathbf{B}}{\mu_0} - \mathbf{M} \tag{5.82}$$

Thus,

$$\boxed{\mathbf{B} = \mu_0(\mathbf{H} + \mathbf{M})} \quad [\text{T}] \tag{5.83}$$

The newly defined $\mathbf{H}$ allows us to write the point form of Ampere's law in terms of the free current $\mathbf{J}$ alone, that is,

$$\boxed{\nabla \times \mathbf{H} = \mathbf{J}} \quad [\text{A/m}^2] \tag{5.84}$$

It is important to note that Ampere's law in Eq. (5.84) always holds true *irrespective of the material*. Next, taking the surface integrals of both sides of Eq. (5.84) over surface $\mathcal{S}$, we have

$$\int_{\mathcal{S}} (\nabla \times \mathbf{H}) \cdot d\mathbf{s} = \int_{\mathcal{S}} \mathbf{J} \cdot d\mathbf{s} \tag{5.85}$$

Applying Stokes's theorem to the left-hand side of Eq. (5.85) and noting that the right side represents the total current flowing through $\mathcal{S}$, Ampere's circuital law is expressed as

$$\boxed{\oint_{\mathcal{C}} \mathbf{H} \cdot d\mathbf{l} = I} \quad [\text{A}] \tag{5.86}$$

where loop $\mathcal{C}$ corresponds to the perimeter of the surface $\mathcal{S}$. The definition of $\mathbf{H}$ in Eq. (5.82) makes it possible for Ampere's circuital law to always be true, regardless of the magnetization volume and surface currents that may be present in the material.

The magnetization depends on the *total* magnetic field in the material, or the internal magnetic field, such that

$$\boxed{\mathbf{M} = \chi_m \mathbf{H}} \quad [\text{A/m}] \tag{5.87}$$

where χ_m is the **magnetic susceptibility**, which is the same at all points in homogeneous materials, independent of the magnitude of $\mathbf{H}$ in linear materials, and independent of the direction of $\mathbf{H}$ in isotropic materials.

Inserting Eq. (5.87) into Eq. (5.83) leads to the constitutive relation that relates $\mathbf{H}$ to $\mathbf{B}$. That is,

$$\boxed{\mathbf{B} = \mu_0(1 + \chi_m)\mathbf{H} = \mu_0 \mu_r \mathbf{H} = \mu \mathbf{H}} \quad [\text{T}] \tag{5.88}$$

Here, μ is the permeability measured in henrys per meter [H/m], μ_0 is the permeability of free space, and μ_r is the relative permeability, which is a dimensionless constant for a simple medium.

From Eq. (5.88), the relationship between χ_m and μ is obtained as

$$\chi_m = \frac{\mu}{\mu_0} - 1 \tag{5.89}$$

This is useful in solving practical problems.

Magnetic materials can be classified into three groups according to their relative permeabilities:

$$\mu_r \leq 1 \quad (\text{diamagnetic}) \tag{5.90a}$$

$$\mu_r \geq 1 \quad \text{(paramagnetic)} \tag{5.90b}$$

$$\mu_r \gg 1 \quad \text{(ferromagnetic)} \tag{5.90c}$$

The typical values of χ_m are in the order of -10^{-6} to -10^{-4} for diamagnetic materials, 10^{-5} to 10^{-3} for paramagnetic materials, and 10^1 to 10^5 for ferromagnetic materials.

In summary, Ampere's circuital law and Gauss's law of magnetism constitute two fundamental relations for static magnetic fields. They allow us to uniquely determine the magnetic field in a given region with the help of appropriate boundary conditions regardless of the material medium that may fill the region. The fundamental relations can be expressed in integral form as follows:

$$\boxed{\oint_C \mathbf{H} \cdot d\mathbf{l} = I} \tag{5.91a}$$

$$\boxed{\oint_S \mathbf{B} \cdot d\mathbf{s} = 0} \tag{5.91b}$$

$$\boxed{\nabla \times \mathbf{H} = \mathbf{J}} \tag{5.91c}$$

$$\boxed{\nabla \cdot \mathbf{B} = 0} \tag{5.91d}$$

$\mathbf{B}$ and $\mathbf{H}$ are related by the constitutive relation such that

$$\mathbf{B} = \mu \mathbf{H} \tag{5.92}$$

with the permeability μ.

Example 5.15 A long filamentary current I oriented along the z-axis is enclosed by a long hollow cylinder with permeability μ_1, as shown in Fig. 5.20. Determine

(a) $\mathbf{H}$ and $\mathbf{B}$ everywhere, and
(b) $\mathbf{J}_m$ and $\mathbf{J}_{ms}$.

Solution

From symmetry considerations, the magnetic field is expected to be of the form $\mathbf{H} = H_\phi(\rho)\,\mathbf{a}_\phi$, for which the proper Amperian path is a circle centered on the z-axis.

(a) Ampere's circuital law is independent of material. Thus,

$$\mathbf{H} = \frac{I}{2\pi\rho}\,\mathbf{a}_\phi \quad \text{(everywhere)}$$

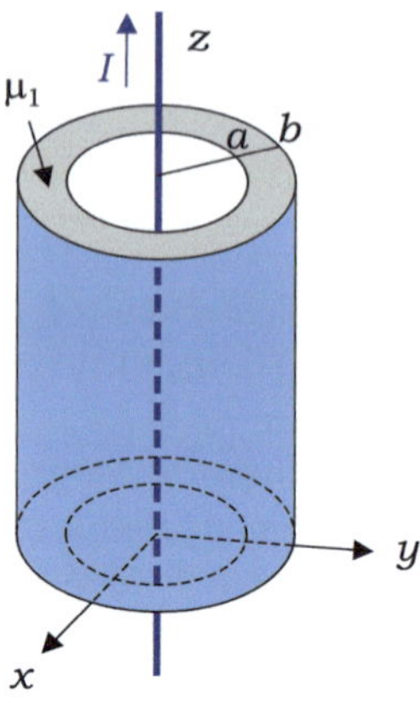

Fig. 5.20 Filamentary current I enclosed by a hollow cylinder with μ

From the constitutive law,

$$\mathbf{B} = \mu_0\mathbf{H} = \frac{I\mu_0}{2\pi\rho}\,\mathbf{a}_\phi \qquad (0 < \rho < a) \qquad\qquad (5.93a)$$

$$\mathbf{B} = \mu_1\mathbf{H} = \frac{I\mu_1}{2\pi\rho}\,\mathbf{a}_\phi \qquad (a \le \rho \le b) \qquad\qquad (5.93b)$$

$$\mathbf{B} = \mu_0\mathbf{H} = \frac{I\mu_0}{2\pi\rho}\,\mathbf{a}_\phi \qquad (\rho > b) \qquad\qquad (5.93c)$$

(b)   The magnetization of the hollow cylinder can be obtained from Eq. (5.87) as

$$\mathbf{M} = \chi_m\mathbf{H} = \left(\frac{\mu_1}{\mu_0} - 1\right)\frac{I}{2\pi\rho}\,\mathbf{a}_\phi \qquad\qquad (5.93d)$$

The curl of $\mathbf{M}$ in Eq. (5.93d) vanishes. Thus,

$$\mathbf{J}_m = 0$$

On the surface of the cylinder, by setting $\rho = a$ or $\rho = b$ in Eq. (5.93d), the magnetization is obtained as

$$\mathbf{M}_a = \left(\frac{\mu_1}{\mu_0} - 1\right)\frac{I}{2\pi a}\,\mathbf{a}_\phi \qquad (\rho = a) \qquad\qquad (5.93e)$$

$$\mathbf{M}_b = \left(\frac{\mu_1}{\mu_0} - 1\right)\frac{I}{2\pi b}\,\mathbf{a}_\phi \qquad (\rho = b) \qquad\qquad (5.93f)$$

The outward unit normal to the cylinder surface is $\mathbf{a}_n = -\mathbf{a}_\rho$ at $\rho = a$, and $\mathbf{a}_n = \mathbf{a}_\rho$ at $\rho = b$. Thus, the magnetization surface current density is

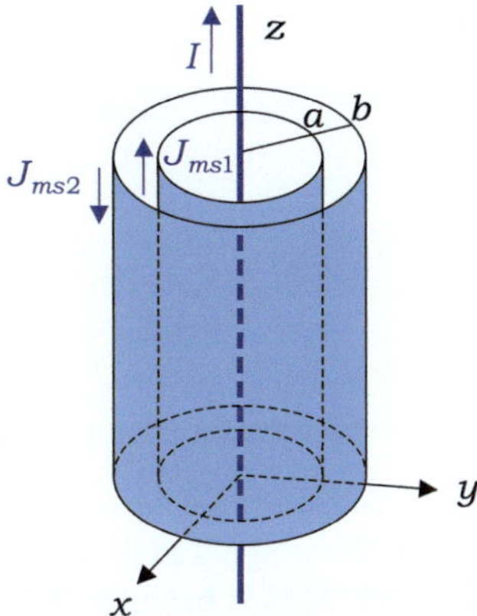

Fig. 5.21 Concentric cylinders with surface currents in free space

$$\mathbf{J}_{ms1} = \mathbf{M}_a \times (-\mathbf{a}_\rho) = \left(\frac{\mu_1}{\mu_0} - 1\right)\frac{I}{2\pi a}\,\mathbf{a}_z \qquad (\rho = a) \qquad (5.93\text{g})$$

$$\mathbf{J}_{ms2} = \mathbf{M}_b \times \mathbf{a}_\rho = -\left(\frac{\mu_1}{\mu_0} - 1\right)\frac{I}{2\pi b}\,\mathbf{a}_z \qquad (\rho = b) \qquad (5.93\text{h})$$

We note that the magnetization surface currents on the two cylindrical surfaces are the same in magnitude, but opposite in direction.

The magnetic field in the material is traced only to the filamentary and magnetization surface currents because $\mathbf{J}_m$ is zero.

Example 5.16 The magnetization surface currents $\mathbf{J}_{ms1}$ and $\mathbf{J}_{ms2}$ in Example 5.15 may be assumed to reside in free space with the hollow cylinder removed (Fig. 5.21). Find $\mathbf{B}$ everywhere and compare it with Example 5.15.

Solution

Considering the symmetry of the given system, we begin with the magnetic flux density of the form $\mathbf{B} = B_\phi(\rho)\,\mathbf{a}_\phi$.

Ampere's circuital law yields

$$\oint_C (\mathbf{B}/\mu_0) \cdot d\mathbf{l} = I \qquad (0 < \rho < a) \qquad (5.94\text{a})$$

$$\oint_C (\mathbf{B}/\mu_0) \cdot d\mathbf{l} = I + 2\pi a J_{ms1} \qquad (a \le \rho \le b) \qquad (5.94\text{b})$$

$$\oint_C (\mathbf{B}/\mu_0) \cdot d\mathbf{l} = I + 2\pi a J_{ms1} - 2\pi b J_{ms2} \qquad (\rho > b) \qquad (5.94\text{c})$$

By inserting $\mathbf{J}_{ms1}$ and $\mathbf{J}_{ms2}$, as given in Eqs. (5.93g) and (5.93h), into the above equations, we obtain

$$\mathbf{B} = \frac{I\mu_0}{2\pi\rho}\,\mathbf{a}_\phi \qquad (0 < \rho < a) \qquad\qquad (5.94\text{d})$$

$$\mathbf{B} = \frac{I\mu_1}{2\pi\rho}\,\mathbf{a}_\phi \qquad (a \le \rho \le b) \qquad\qquad (5.94\text{e})$$

$$\mathbf{B} = \frac{I\mu_0}{2\pi\rho}\,\mathbf{a}_\phi \qquad (\rho > b) \qquad\qquad (5.94\text{f})$$

The results are the same as those for Example 5.15.

Ampere's circuital law, expressed by Eq. (5.86), cannot be applied to this case, in which the magnetization and magnetization surface currents are directly taken into account. This is because $\mathbf{H}$ is related only to free currents.

A short magnetic cylinder with large permeability is placed in a uniform magnetic field, as shown in Fig. 5.22a. At first glance, one might think that the induced magnetization is uniform in the cylinder, conforming to a uniform external field. However, they must have considered the wrong thing. Even if uniform magnetization induces a uniform magnetization surface current, the resulting magnetization field is not uniform because of the finite extent of surface current and edge effects. In fact, the magnetization itself is not uniform at first because of the nonuniform internal field. For simplicity, we assumed that the magnetization surface current was in the form of a circle, and computed the magnetic field lines inside and outside the cylinder using numerical methods. The results are plotted on a scale, as shown in Fig. 5.22b. The figure shows that the magnetic field lines are concentrated in the interior of the cylinder, strongly suggesting that a structure composed of a magnetic material with high permeability can be used to confine the magnetic field to its interior and direct flow.

Example 5.17 An infinitely long cylinder with radius a has uniform permanent $\mathbf{M} = M_o\mathbf{a}_z$, as shown in Fig. 5.23a. A small section was removed from the cylinder and used as the permanent magnet. The magnetic flux lines of the remaining segments were computed numerically, as shown in Fig. 5.23b. Sketch field lines of $\mathbf{B}$ and $\mathbf{H}$ for the permanent magnet.

Solution

The permanent magnetization $\mathbf{M} = M_o\mathbf{a}_z$ leads to $\mathbf{B} = \mu_0\mathbf{M}$ and $\mathbf{H} = 0$ inside the infinite cylinder and $\mathbf{B} = \mathbf{H} = 0$ outside, as shown in Fig. 5.23a. According to the principle of superposition, the magnetic flux lines of the permanent magnet are $\mathbf{B}_1 = \mathbf{B} - \mathbf{B}_2$ in the region $\rho < a$, and $\mathbf{B}_1 = -\mathbf{B}_2$ in the region $\rho > a$, as shown in Fig. 5.24. The magnetic flux lines are closed, but change abruptly on the cylindrical surface of the permanent magnet.

Although $\mathbf{H}$ is zero inside the original cylinder, it is nonzero inside the permanent magnet. The principle of superposition provides $\mathbf{B} = \mathbf{B}_1 + \mathbf{B}_2 = \mu_0\mathbf{M}$ in the region

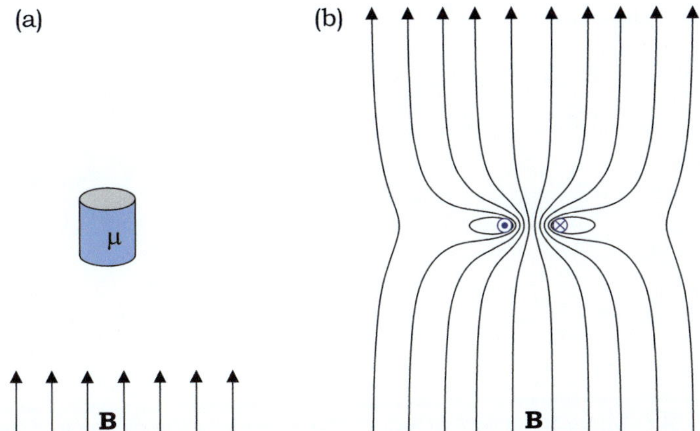

Fig. 5.22 **a** Magnetic cylinder with large μ placed in uniform **B**. **b** Magnetic field lines calculated by assuming $\mathbf{J}_{ms}$ to be a circular loop current

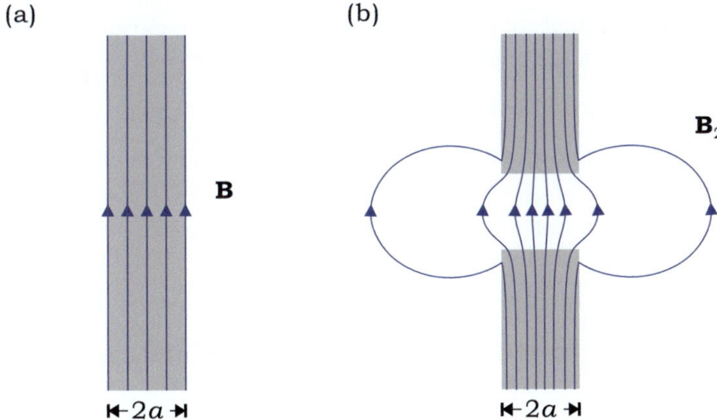

Fig. 5.23 Magnetic flux lines. **a** Infinite cylinder with $\mathbf{M} = M_o\,\mathbf{a}_z$. **b** Same cylinder with a small section removed

Fig. 5.24 Magnetic flux lines of a permanent magnet with $\mathbf{M} = M_o\,\mathbf{a}_z$

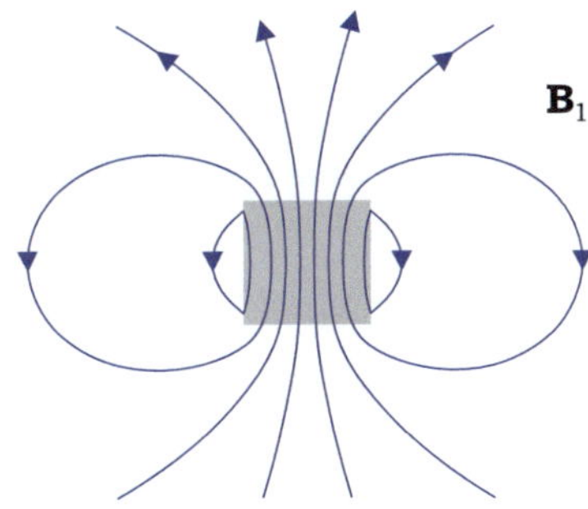

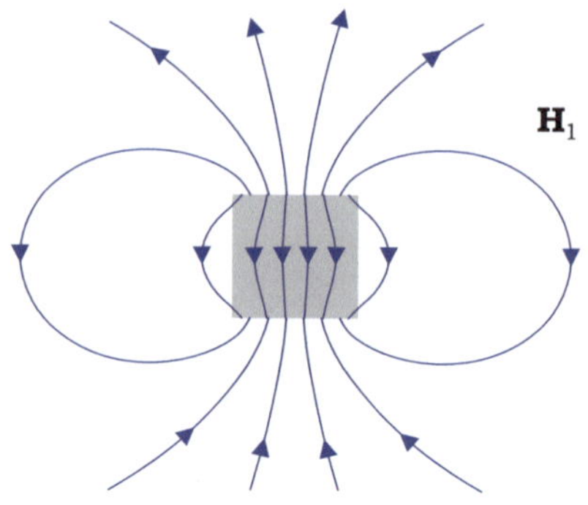

Fig. 5.25 H-field lines of a permanent magnet with $\mathbf{M} = M_o\,\mathbf{a}_z$

of the original cylinder that corresponds to the interior of the permanent magnet. By definition, $\mathbf{H}_1 = \mathbf{B}_1/\mu_0 - \mathbf{M}$ inside the permanent magnet. Combining these two equations yields $\mathbf{H}_1 = -\mathbf{B}_2/\mu_0$ inside the permanent magnet. The exterior of the permanent magnet is in free space. Thus, $\mathbf{H}_1 = \mathbf{B}_1/\mu_0$, as shown in Fig. 5.25.

Note that the **H**-field lines of the permanent magnet do not close upon themselves! They appear to start at the upper end and end at the lower end.

Exercise 5.17
Determine μ_r of the linear material if $\mathbf{M}$ is 10 times as large as internal $\mathbf{H}$.

Ans. $\mu_r = 11$.

Exercise 5.18
For $|\mathbf{B}| = 0.02$ [T] in solid Nickel with $\mu_r = 250$, determine χ_m, H, and M inside.

Ans. $\chi_m = 249$, $H = 63.7$ [A/m], and $M = 15.9 \times 10^3$ [A/m].

Exercise 5.19
An infinitely long cylinder of radius a is coaxial with the z-axis. If it has a uniform permanent magnetization $\mathbf{M} = M_o\,\mathbf{a}_z$, determine $\mathbf{B}$ everywhere.

Ans. $\mathbf{B} = \mu_0 M_o\,\mathbf{a}_z$ inside, and $\mathbf{B} = 0$ outside.

Exercise 5.20
An infinitely long cylinder with susceptibility χ_m carries a uniform surface current $\mathbf{J}_s = J_o\,\mathbf{a}_\phi$. Determine $\mathbf{B}$ everywhere.

Ans. $\mathbf{B} = \mu_0(\chi_m + 1)J_o\,\mathbf{a}_z$ inside, and $\mathbf{B} = 0$ outside.

Review Questions

RQ 5.41	Define **H** in matter.	[(5.82)]
RQ 5.42	Relate the magnetic flux density to magnetization.	[(5.83)]
RQ 5.43	Relate **H** to the free current in the magnetic material.	[(5.84)]
RQ 5.44	Write Ampere's circuital law for a magnetic material.	[(5.86)]
RQ 5.45	Define χ_m, μ_r, and μ.	[(5.87)–(5.89)]
RQ 5.46	What are the three types of magnetic materials?	[(5.90a,b,c)]
RQ 5.47	Is there any way to make a short magnetic cylinder have uniform magnetization using an external magnetic field?	[Fig. 5.22]
RQ 5.48	A surface current $\mathbf{J}_s = \sin\theta\, \mathbf{a}_\phi$ on the sphere produces a uniform **H** inside. Does this imply that a spherical material can be uniformly magnetized by a uniform external field?	[Exercise 5.16, Problem 5.27]

5.7.3 Hysteresis of Ferromagnetic Materials

An externally applied magnetic field can strongly magnetize ferromagnetic materials. The material exhibits unique magnetic properties, which are mainly due to the formation of magnetized domains. The relationship between **B** and **H** is highly nonlinear such that the relative permeability strongly depends on the internal magnetic field. Moreover, the material can retain a fair amount of magnetization even after the applied field is removed.

The atoms of a ferromagnetic material have a large permanent magnetic moment that originates from the electron spin moment. Magnetic moments tend to align with each other over small regions, called ***magnetic domains***, because of the strong interatomic force (Fig. 5.26a). Magnetic domains have various shapes and sizes ranging from micrometers to millimeters in linear dimensions. Each magnetic domain is fully magnetized such that all the constituent magnetic moments point in the same direction, producing uniform magnetization. In the absence of an external magnetic field, the magnetic domains are randomly oriented and yield no net magnetic moment. However, under the influence of an externally applied field, magnetic domains with a magnetization vector parallel to the applied field grow at the expense of their neighbors, resulting in a large net magnetic moment (Fig. 5.26b). When the external field is removed, the magnetic domains cannot be broken again into pieces with the original orientations. However, residual magnetization remains in the material. This phenomenon is referred to as ***magnetic hysteresis***. The magnetic state at the present time depends on the past magnetic state of the material. However, above the Curie temperature, ferromagnetic materials behave like paramagnetic materials, because their permanent magnetic moments are randomly oriented and do not form magnetic domains. For example, the Curie temperature of Fe is 770° C.

The magnetic behavior of a ferromagnetic material can be well described by the $B{-}H$ ***magnetization curve***, which is a plot of the magnetic flux density B measured in the material versus the applied magnetic field intensity H, as shown in Fig. 5.27. Let

(a)   (b)

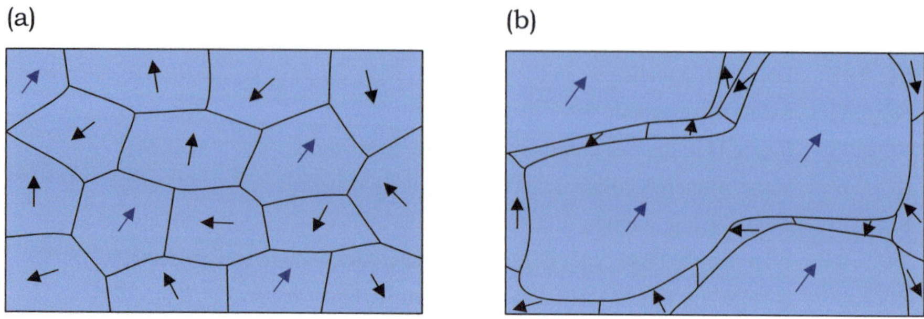

Fig. 5.26 Magnetic domains in ferromagnetic materials. **a** No external **H** is present. **b** Strong external **H** is applied along the blue arrow

us begin with a completely demagnetized ferromagnetic material. First, we increase the external magnetic field from zero to H_1, which can be accomplished simply by placing the material inside a solenoidal coil and increasing the current in the coil. The magnetic flux density then increases from zero to B_1 in the material, and its level follows the dotted line shown in Fig. 5.27a, which is called the ***initial magnetization curve***. If the magnetic field is decreased from H_1 to zero, the magnetic flux density in the material does not go to zero, but its level follows the upper path of the $B-H$ curve and remains at a nonzero value, B_r, called the ***residual flux density***. The phenomenon in which magnetization lags behind the field is called hysteresis, meaning "to lag" in Greek. To ensure that the residual flux density vanishes from the material, a magnetic field H_c, called ***coercive field intensity***, is applied in the opposite direction. Note that both B_r and H_c depend on the maximum field intensity H_1.

If the external magnetic field varies periodically between two maximum values $\pm H_1$, then B traces a closed loop per cycle of H in the $B-H$ plot, which is referred to as the ***hysteresis loop***. The path taken by B for H to increase from $-H_1$ to H_1 is different from that for H to change in the opposite way. The magnetization depends not only on the applied magnetic field but also on past magnetic states. The shape and size of the hysteresis loop depend on the material and maximum applied field. The hysteresis loop is a direct consequence of the nonlinear relationship between B and H.

In a ferromagnetic material, the permeability is equal to the ratio B/H in the $B-H$ curve, which is not necessarily tangent to the curve and is a nonlinear function of H. The area of the hysteresis loop corresponds to the energy dissipated in the form of heat per unit volume of material per cycle of H. Hard ferromagnetic materials have wide hysteresis loops, as shown in Fig. 5.27a, whereas soft ferromagnetic materials exhibit narrow hysteresis loops, as shown in Fig. 5.27b. A permanent magnet is possible because of the residual flux density that remains in the ferromagnetic material. Hard ferromagnetic materials with a larger coercive field intensity produce better permanent magnets; they are hardly demagnetized by an external magnetic field of moderate strength.

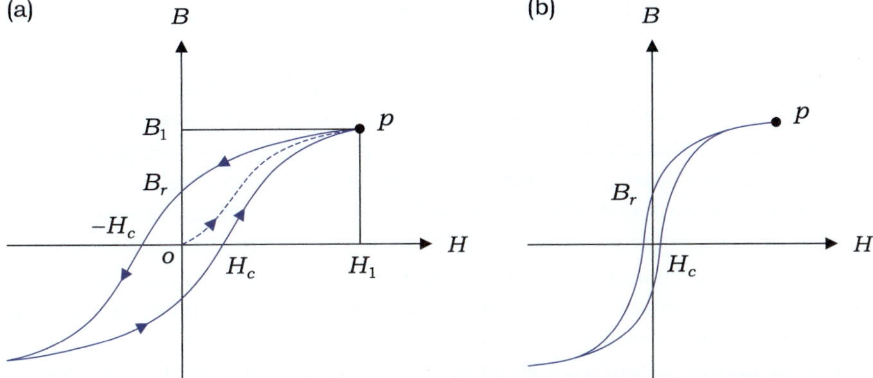

Fig. 5.27 Typical hysteresis loops for **a** hard, and **b** soft ferromagnetic materials

When the applied magnetic field is sufficiently strong to enforce the total alignment of the magnetic dipole moments with the field, the material is said to be *saturated*. A further increase in H would cause no change in the magnetization. Nevertheless, the internal magnetic flux increases beyond saturation because the external magnetic field provides an additional magnetic flux via the relationship $\mathbf{B} = \mu_0 \mathbf{H}$.

Exercise 5.21
An iron metal with number density 8.5×10^{28} [m^{-3}] has a saturation magnetization of 7.9×10^5 [A/m]. Find the magnetic dipole moment of each atom.

Ans. 9.3×10^{-24} [A $\cdot$ m^2].

Exercise 5.22
At saturation point p in Fig. 5.27b, the tangent to the curve is $7°$ to the H-axis, while the line from the origin to p is $45°$. Determine μ_r at p.

Ans. $\mu_r = \tan 45° / \tan 7° = 8.1$ (The numerator is μ and the denominator is μ_0).

Exercise 5.23
Which of the following statements is true of ferromagnetic materials? (a) μ_r is constant, (b) the hysteresis loop becomes a straight line beyond saturation, and (c) the area of the hysteresis loop is directly related to energy loss.

Ans. (b) and (c).

Review Questions

RQ 5.49	What is the magnetic domain?	[Fig. 5.26]
RQ 5.50	Explain the Curie temperature.	[Fig. 5.26]
RQ 5.51	What is the hysteresis loop?	[Fig. 5.27]
RQ 5.52	Define the residual flux density and coercive field intensity.	[Fig. 5.27]
RQ 5.53	What is a hard ferromagnetic material?	[Fig. 5.27]

5.7.4 *Magnetic Boundary Conditions*

Gauss's law of magnetism is one of the two fundamental relations for static magnetic fields, and is directly related to the law of conservation of magnetic flux. The net outward magnetic flux through any closed surface is always zero regardless of the material medium. Ampere's circuital law is another fundamental relation that states that the closed-line integral of the magnetic field is equal to the free current enclosed by the path. This holds true regardless of the material. Therefore, these relations must be satisfied even at the interface between two dissimilar materials. We are now ready to derive the boundary conditions of **B** and **H** at such an interface following the same procedure used for the boundary conditions for **E** and **D**.

First, we apply Ampere's circuital law to a rectangular loop *abcda*, extending the same distance above and below the interface (Fig. 5.28), and write

$$\oint_C \mathbf{H} \cdot d\mathbf{l} = H_{1t}\,\Delta w - H_{2t}\,\Delta w + \int_b^c \mathbf{H} \cdot d\mathbf{l} + \int_d^a \mathbf{H} \cdot d\mathbf{l}$$

$$= J_{sn}\,\Delta w \tag{5.95}$$

where subscript t denotes the tangential component, and J_{sn} represents the surface current density perpendicular to the loop surface. The line integrals along the left and right sides of the loop vanish as the loop height approaches zero. Therefore, the boundary condition for the tangential component of **H** is given by

$$\boxed{H_{1t} - H_{2t} = J_{sn}} \tag{5.96}$$

The tangential component of **H** is discontinuous across the interface when surface current exists at the interface. A more general expression for the boundary condition can be obtained using a unit normal to the interface, $\mathbf{a}_{1-2}$, which is directed from medium 2 to medium 1 (Fig. 5.29). That is,

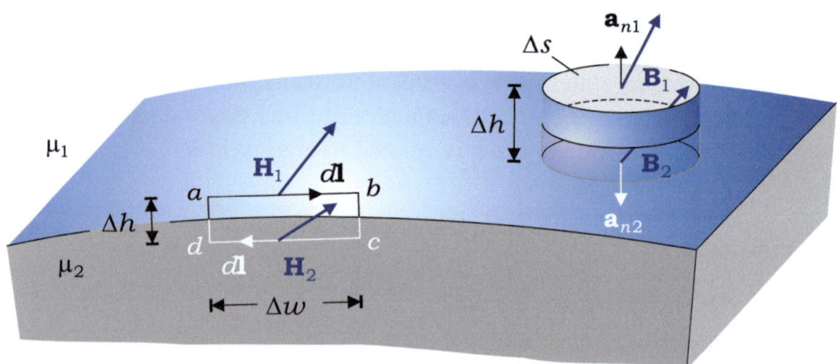

Fig. 5.28 Interface between two magnetic materials with μ_1 and μ_2

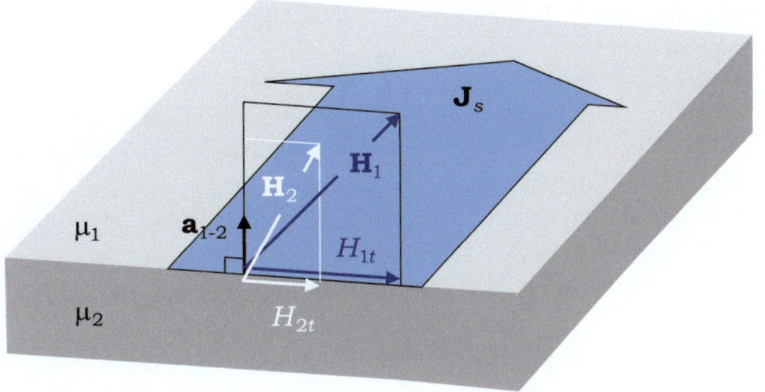

Fig. 5.29 Boundary conditions for **H** at interface carrying surface current

$$\boxed{\mathbf{a}_{1-2} \times (\mathbf{H}_1 - \mathbf{H}_2) = \mathbf{J}_s} \tag{5.97}$$

When two adjoining materials have finite conductivities, a volume current can flow across the interface, as discussed in Chap. 4. In this case, even if the current leaves surface charges at the interface, no surface current flows over the interface. With no surface current over the interface, the boundary condition for the tangential component of **H** is reduced to

$$\boxed{H_{1t} = H_{2t}} \tag{5.98}$$

The tangential component of **H** *is continuous across the interface.* By applying the constitutive relation, $\mathbf{B} = \mu\mathbf{H}$, to Eq. (5.98), the boundary condition for tangential component of **B** is obtained as

$$\frac{B_{1t}}{\mu_1} = \frac{B_{2t}}{\mu_2} \tag{5.99}$$

The tangential component of **B** is discontinuous across the interface.

Next, Gauss's law of magnetism is applied to a cylinder buried half and half in two regions, as shown in Fig. 5.28. Following the same procedure used for **D**, the boundary condition for the normal component of **B** is obtained as

$$\boxed{B_{1n} = B_{2n}} \tag{5.100}$$

The normal component of **B** *is continuous across the interface.* By invoking the constitutive relation to Eq. (5.100), the boundary condition for the normal component of **H** is obtained as

$$\mu_1 H_{1n} = \mu_2 H_{2n} \tag{5.101}$$

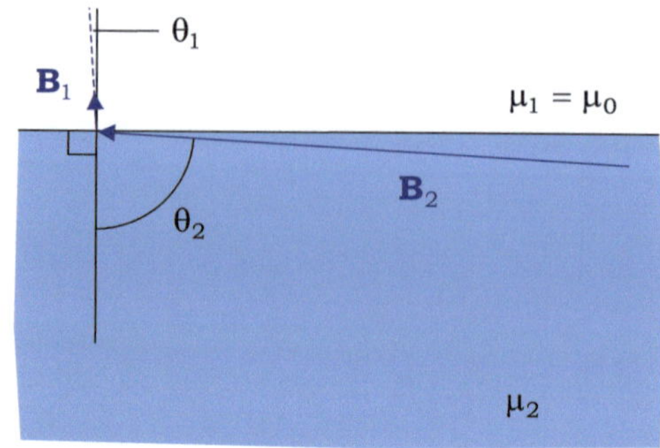

Fig. 5.30 Interface between free space and iron metal

The normal component of **H** is discontinuous across the interface.

Example 5.18 Consider an interface between free space and iron metal with $\mu = 1000\mu_0$. In free space, $\mathbf{B}_1$ makes an angle $\theta_1 = 1.0°$ with the surface normal (Fig. 5.30). Determine (a) the direction of $\mathbf{B}_2$ in the metal, and (b) the ratio B_2/B_1.

Solution

(a) From the boundary conditions for **H** and **B**,

$$\frac{1}{\mu_2} B_{2t} = \frac{1}{\mu_0} B_1 \sin(1°) \tag{5.102a}$$

$$B_{2n} = B_1 \cos(1°) \tag{5.102b}$$

In iron metal,

$$\tan\theta_2 = \frac{B_{2t}}{B_{2n}} = \frac{(\mu_2/\mu_0)B_1 \sin(1°)}{B_1 \cos(1°)} = 17.46$$

Thus, $\theta_2 = 86.7°$

(b) Combining Eqs. (5.102a) and (5.102b) yields

$$B_2 = \sqrt{(B_{2t})^2 + (B_{2n})^2} = B_1\sqrt{(1000)^2 \sin^2(1°) + \cos^2(1°)}$$

Thus, $B_2/B_1 = 17.48$.

Example 5.19 A very long filament carrying a steady current I_o penetrates a dielectric sphere of radius a with a permeability $\mu = \mu_0\mu_r$, as shown in Fig. 5.31. Verify using Helmholtz's theorem that the magnetic field expressed by $\mathbf{H} = (I_o/2\pi\rho)\,\mathbf{a}_\phi$ is a unique solution everywhere, except for $\rho = 0$.

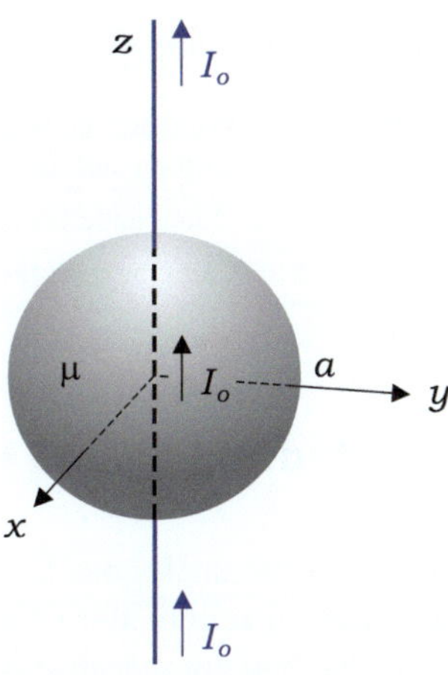

Fig. 5.31 A long filamentary current penetrates the dielectric sphere

Solution

The trial solution for **H** satisfies two fundamental relations, $\nabla \times \mathbf{H} = 0$ and $\nabla \cdot \mathbf{B} = 0$, inside and outside the sphere, and vanishes at infinity. The given **H** is tangential to the surface of the sphere and continuous across the surface, thereby satisfying boundary conditions. Therefore, this is a unique solution for region $\rho > 0$.

If the dielectric was in the form of a cube, the given **H** could not be a solution to the problem because it could not satisfy the boundary conditions for **H** and **B** at the surface of the cube.

Exercise 5.24
Relate the two permeabilities in Fig. 5.30 to the tilt angles of the two **B**'s.

Ans. $\tan\theta_1 / \tan\theta_2 = \mu_1/\mu_2$.

Exercise 5.25
Which boundary condition is related to the conservation of magnetic flux?

Ans. $B_{1n} = B_{2n}$.

Exercise 5.26
Characterize **B** on the surface of a material with permeability close to infinity.

Ans. **B** is normal to the surface.

Exercise 5.27
If uniform external **B** in Fig. 5.22a induced a uniform **M** in the cylinder, while **B** remained unchanged outside, what else would go wrong?

Ans. The boundary condition for H_t cannot be satisfied.

Review Questions

RQ 5.54   What are the boundary conditions for **H** at the interface   [(5.98)–(5.101)]
          with no surface current?

RQ 5.55   Is magnetic flux always conserved across an interface?   [(5.100)]

RQ 5.56   Can magnetic flux lines bend abruptly at the interface?   [(5.99)(5.100)]

5.8   Inductance and Inductor

Inductors are analogous to capacitors in that they store energy in a magnetic field, whereas capacitors store energy in an electric field. A single-wire turn may be the simplest inductor, because typical inductors consist of many turns of wires. It is well known from the previous discussion that the magnetic flux lines due to the loop current always pass through the loop surface and form closed lines around it. As capacitance is defined as the ratio of the charge accumulated on the conductor to the potential difference between the conductors, inductance is defined as the ratio of the magnetic flux linking the loop to the current flowing in the loop.

Let us consider a conducting loop C_1 with a steady current I_1 and second loop C_2 situated in the neighborhood of the first loop, as shown in Fig. 5.32. When the current in C_1 produces a magnetic flux $\mathbf{B}_1$, some fluxes pass through the surface bounded by C_2. Here, the mutual flux Φ_{12} is defined as the magnetic flux produced by current I_1 and linked with C_2 or passed through surface S_2. Expressed mathematically

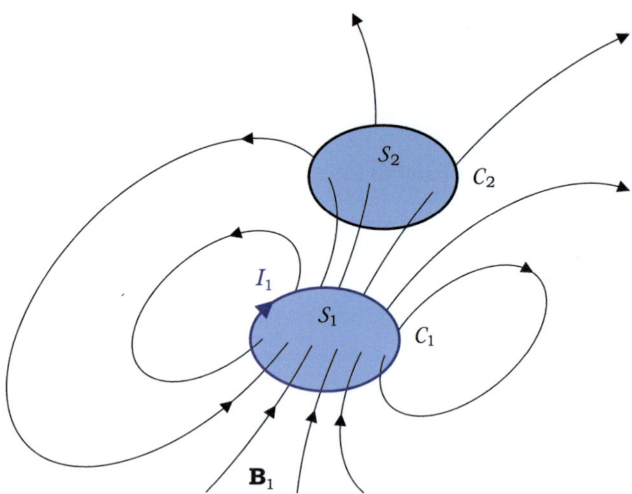

Fig. 5.32 Magnetic coupling between two conducting loops

Fig. 5.33 Magnetic flux linkage with a coiled loop

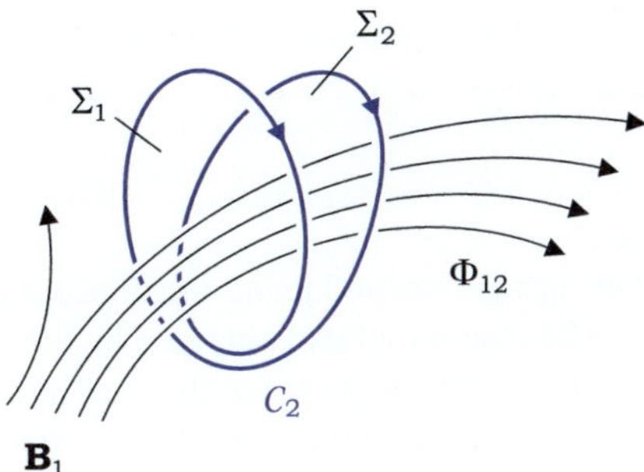

$$\Phi_{12} = \int_{\mathcal{S}_2} \mathbf{B}_1 \cdot d\mathbf{s} \tag{5.103}$$

If loop C_2 has N_2 turns of the wire, **_magnetic flux linkage_** Λ_{12} is defined as

$$\Lambda_{12} = N_2 \Phi_{12} \tag{5.104}$$

where Φ_{12} denotes the magnetic flux linked to a single turn in C_2. In our notation, the subscript "12" denotes something "from 1 to 2," whereas subscript "1–2" denotes something "from 2 to 1".

The magnetic flux linked to loop C_2 with multiple turns is equal to the total magnetic flux passing through the entire surface of the uncoiled loop. Suppose that the loop has two wire turns, and Σ_1 and Σ_2 are surfaces bounded by individual turns facing the same direction, as shown in Fig. 5.33. If the loop is uncoiled into a simple loop, the total loop surface is $\mathcal{S} = \Sigma_1 + \Sigma_2$ and the magnetic flux Φ_{12} passes through the uncoiled loop surface twice in the same direction. Therefore, the magnetic flux linked to the loop is $2\Phi_{12}$. In general, the magnetic flux linkage is equal to the product of the magnetic flux linking a single turn and number of turns.

The two neighboring loops shown in Fig. 5.32 are magnetically coupled through mutual flux. To describe magnetic coupling, the **_mutual inductance_** between the loops is defined as

$$\boxed{M_{12} = \frac{\Lambda_{12}}{I_1} = \frac{N_2}{I_1} \int_{\mathcal{S}_2} \mathbf{B}_1 \cdot d\mathbf{s}} \quad [\text{H}] \tag{5.105}$$

where N_2 is the number of turns in loop C_2, $\mathcal{S}_2$ is the surface bounded by a single turn of C_2, and $\mathbf{B}_1$ is the magnetic flux density caused by the current flowing in loop C_1. The mutual inductance is measured in henrys [H] and should not be confused with magnetization $\mathbf{M}$. In simple media, the permeability is a constant that is independent of the magnitude and direction of the magnetic field. If this is the case, $\mathbf{B}_1$ is linearly proportional to the current I_1 flowing in loop C_1 according to Ampere's law, $\nabla \times (\mathbf{B}/\mu) = \mathbf{J}$. Therefore, the mutual inductance is independent of current.

Similarly, mutual inductance M_{21} is defined as the ratio of the magnetic flux linking loop C_1 to the current flowing in loop C_2. In linear media,

$$\boxed{M_{12} = M_{21}} \quad [\text{H}] \tag{5.106}$$

This can be verified using the concept of mutual energy described in Sect. 5.9.

The magnetic flux generated by the current flowing in loop C_1 is linked to the loop and is referred to as magnetic flux linkage Λ_{11}. The **self-inductance** or **inductance** of a loop is defined as the ratio of the magnetic flux linking the loop to the current flowing through it. That is,

$$\boxed{L_1 = \frac{\Lambda_{11}}{I_1} = \frac{N_1}{I_1} \int_{S_1} \mathbf{B}_1 \cdot d\mathbf{s}} \quad [\text{H}] \tag{5.107}$$

where N_1 is the number of turns in loop C_1, and S_1 is the surface enclosed by a single turn of C_1. The inductance is measured in henrys [H]. Even if the inductance is independent of the current, it is highly dependent on the geometry of the conducting structure and permeability of the surrounding medium.

For a given inductor, the inductance can be determined taking the following steps:

1. Assume current I in the conductor.
2. Choose a coordinate system considering symmetry.
3. Find $\mathbf{B}$ from I using the Biot-Savart law or Ampere's law.
4. Determine the magnetic flux linkage, $\Lambda = N\Phi = N \int \mathbf{B} \cdot d\mathbf{s}$.
5. Find L from $L = \Lambda/I$.

The magnetic flux linkage with a current-carrying loop is illustrated in Fig. 5.34. The magnetic flux lines due to the current form closed lines around the loop. The inductance of the loop is equivalent to the ratio between the number of flux lines and loop current. If the loop has other shapes, the number of flux lines will not be the same and the loop will have a different inductance.

Fig. 5.34 Magnetic flux linkage with a current-carrying loop

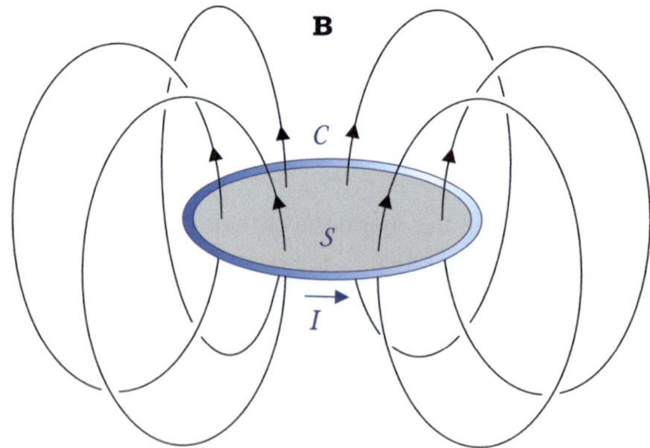

Example 5.20 Determine the mutual inductance between a very long straight wire and a rectangular loop residing in free space, as illustrated in Fig. 5.35.

Solution

If a steady current I is assumed in the straight wire, $\mathbf{B}$ is induced such that

$$\mathbf{B} = \frac{\mu_0 I}{2\pi\rho}\mathbf{a}_\phi$$

The magnetic flux linking the loop is

$$\Lambda_{12} = \Phi_{12} = \int_{S_2} \mathbf{B} \cdot d\mathbf{s} = \int_{\rho=a}^{\rho=b} \frac{\mu_0 I}{2\pi\rho}\mathbf{a}_\phi \cdot \left(h\, d\rho\, \mathbf{a}_\phi\right)$$

$$= \frac{\mu_0 I h}{2\pi}\ln\frac{b}{a} \tag{5.108a}$$

Therefore, the mutual inductance is

$$M_{12} = \frac{\Lambda_{12}}{I} = \frac{\mu_0 h}{2\pi}\ln\frac{b}{a} \quad [\text{H}] \tag{5.108b}$$

It would be mathematically demanding to determine the mutual inductance M_{21} by assuming current I in the loop and computing the flux linkage with the straight wire. Nevertheless, M_{21} is equal to M_{12}.

Example 5.21 Find the self-inductance per unit length of a very long solenoid consisting of N turns of wire per unit length wound tightly on an air core of radius a.

Solution

Let us first examine the geometry for symmetry:

(1) The solenoid has cylindrical and translational symmetries that lead to the magnetic flux density being independent of ϕ and z such that

$$\mathbf{B} = B_\rho(\rho)\,\mathbf{a}_\rho + B_\phi(\rho)\,\mathbf{a}_\phi + B_z(\rho)\,\mathbf{a}_z$$

(2) $\mathbf{B}$ should point in the opposite direction everywhere if current I is switched, which is equivalent to a vertical flip of both the solenoid and $\mathbf{B}$. However, the ρ-component, $B_\rho(\rho)\,\mathbf{a}_\rho$, remains the same after the flip, and thus, should be excluded from $\mathbf{B}$.

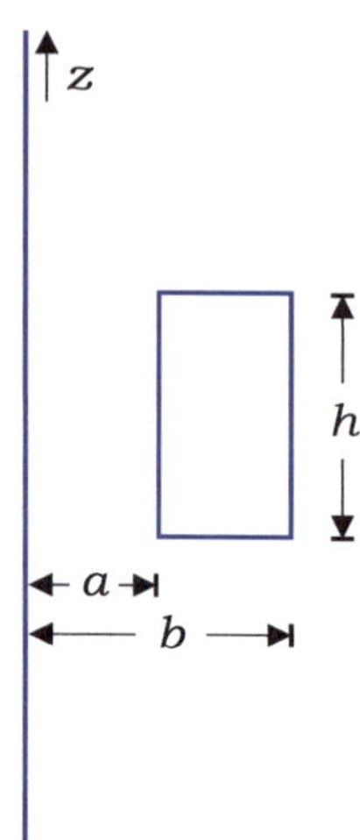

Fig. 5.35 A very long straight wire and rectangular loop in its vicinity

(3) Because the current flows only in the $\mathbf{a}_\phi$-direction in an ideal solenoid, no current flows through loop $\mathcal{C}$, which is placed in the xy-plane (Fig. 5.36). Applying Ampere's circuital law to loop $\mathcal{C}$ leads to $B_\phi = 0$, as is evident from

$$\oint_{\mathcal{C}} \mathbf{H} \cdot d\mathbf{l} = \int_{\phi=0}^{2\pi} B_\phi(\rho)\, \mathbf{a}_\phi \cdot (\rho d\phi\, \mathbf{a}_\phi) = 0$$

Thus far, $\mathbf{B}$ has been expected to assume the form $\mathbf{B} = B_z(\rho)\, \mathbf{a}_z$. Applying Ampere's circuital law to loops $\mathcal{C}_1$ and $\mathcal{C}_2$, as shown in Fig. 5.36, ensures that $\mathbf{B}$ is independent of ρ or constant everywhere ($\mathbf{B} = B_o\, \mathbf{a}_z$). This is because the Amperian path encloses no currents. At large distances, the solenoid appears as a simple straight line with no current flowing; thus, $\mathbf{B}$ should vanish at infinity. Therefore, $\mathbf{B}$ must be zero at each point outside the solenoid.

Next, applying Ampere's circuital law to loop $\mathcal{C}_3$, we have

$$h H_z = NIh$$

Thus, in the interior of the solenoid,

$$\mathbf{H} = NI\, \mathbf{a}_z \tag{5.109a}$$

$$\mathbf{B} = \mu_0 NI\, \mathbf{a}_z \tag{5.109b}$$

The total magnetic flux through the cross section of the solenoid is

$$\Phi = B\pi a^2 = \mu_0 NI \pi a^2$$

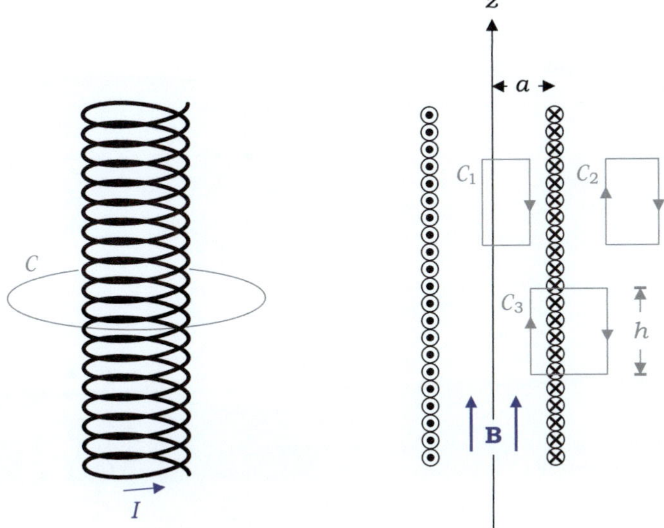

Fig. 5.36 Very long solenoid with many turns of wire tightly wound

The magnetic flux linkage per unit length of the solenoid is

$$\Lambda = N\Phi = \mu_0 N^2 I \pi a^2$$

Therefore, the self-inductance per unit length of the solenoid is

$$L = \frac{\Lambda}{I} = \mu_0 N^2 \pi a^2 \quad [\text{H/m}] \tag{5.109c}$$

Example 5.22 A long coaxial cable consists of a solid inner conductor of radius a and conductive cylindrical sheet of radius b, as shown in Fig. 5.37. Assuming that the steady current I constitutes a uniform volume current in the inner conductor and a uniform surface current in the outer conductor, determine the inductance per unit length of the cable.

Solution

Based on symmetry considerations, $\mathbf{B}$ is expected to have the form $B_\phi(\rho)\,\mathbf{a}_\phi$ everywhere.

From Ampere's circuital law,

$$\mathbf{B} = 0 \quad (\rho > b) \tag{5.110a}$$

$$\mathbf{B} = \frac{\mu_0 I}{2\pi\rho}\mathbf{a}_\phi \quad (a < \rho < b) \tag{5.110b}$$

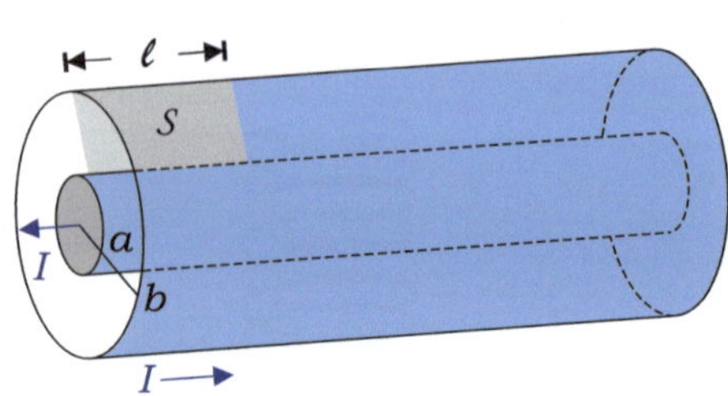

Fig. 5.37 Coaxial cable consisting of a solid conductor and conductive cylindrical sheet

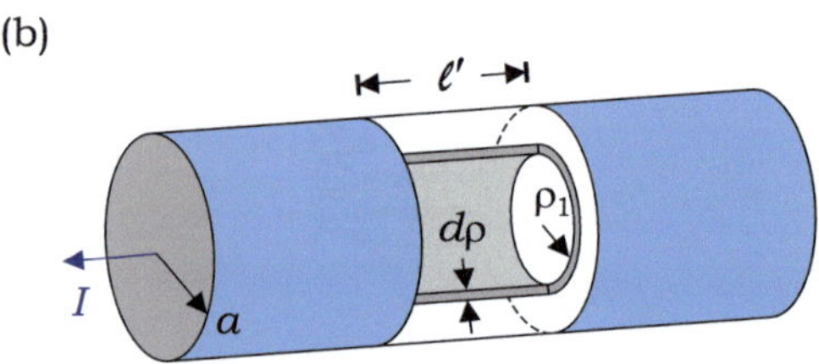

$$\mathbf{B} = \frac{\mu_0 \rho I}{2\pi a^2}\mathbf{a}_\phi \qquad (0 \le \rho \le a) \qquad\qquad (5.110c)$$

(1) In region $a < \rho < b$, the magnetic flux through the surface $\mathcal{S}$, as shown in Fig. 5.37, is given by

$$\Phi = \int_{\mathcal{S}} \mathbf{B} \cdot d\mathbf{s} = \int_{\rho=a}^{\rho=b} \frac{\mu_0 I}{2\pi\rho}\ell d\rho = \frac{\mu_0 I \ell}{2\pi}\ln\frac{b}{a}$$

Therefore, the magnetic flux linking the inner conductor per unit length is

$$\Lambda_1 = \frac{\Phi}{\ell} = \frac{\mu_0 I}{2\pi}\ln\frac{b}{a} \qquad\qquad (5.110d)$$

(2) In region $0 \le \rho \le a$, we consider a circular pipe with radius $\rho_1 \le b$, length ℓ', and thickness $d\rho$, as shown in Fig. 5.37b. The magnetic flux confined to the thin pipe wall can be obtained from Eq.(5.110c) as

$$d\Phi = B_\phi \ell' d\rho = \frac{\mu_0 \rho_1 I}{2\pi a^2}\ell' d\rho \qquad\qquad (5.110e)$$

This was caused by the current flowing in the enclosure with $\rho \le \rho_1$, that is, $I_1 = \rho_1^2 I/a^2$. Let us remember that the inductance requires a magnetic flux linkage with the total current I. Therefore, $d\Phi$ must be converted to an equivalent magnetic flux $d\Phi'$ linked to I. As discussed in Sect. 5.9, the magnetic energy is given by the product of the magnetic flux linkage and current multiplied by $1/2$.

Accordingly, the magnetic energy within the thin pipe wall can be expressed as $dW_m = \frac{1}{2}I_1 d\Phi = \frac{1}{2}(\rho_1^2 I/a^2)\, d\Phi \equiv \frac{1}{2}I\, d\Phi'$. Using this relationship, the differential magnetic flux linkage per unit length is obtained as

$$d\Lambda = \frac{d\Phi'}{\ell'} = \frac{d\Phi}{\ell'}\frac{\rho_1^2}{a^2} = \frac{\mu_0 \rho_1^3 I}{2\pi a^4}\, d\rho$$

Subsequently, the internal magnetic flux linking the inner conductor per unit length is computed as

$$\Lambda_2 = \int d\Lambda = \int_{\rho=0}^{\rho=a} \frac{\mu_0 \rho^3 I}{2\pi a^4}\, d\rho = \frac{\mu_0 I}{8\pi} \tag{5.110f}$$

Finally, the magnetic flux linkage per unit length of the coaxial cable is the sum of Λ_1 and Λ_2, and the inductance per unit length is

$$L = \frac{\Lambda_1 + \Lambda_2}{I} = \frac{\mu_0}{2\pi}\ln\frac{b}{a} + \frac{\mu_0}{8\pi} \quad [\text{H/m}] \tag{5.110g}$$
$$\equiv L_{ex} + L_{in}$$

where L_{ex} is the **external inductance**, and L_{in} is the **internal inductance**.

Example 5.23 Two long parallel wires of equal radius a are separated by an axis-to-axis distance b ($b \gg a$) in free space, as shown in Fig. 5.38. When they carry an equal but opposite current that is uniformly distributed in the cross section, compute the external and internal inductances per unit length.

Solution

Fig. 5.38 Two parallel wires carrying equal but opposite currents

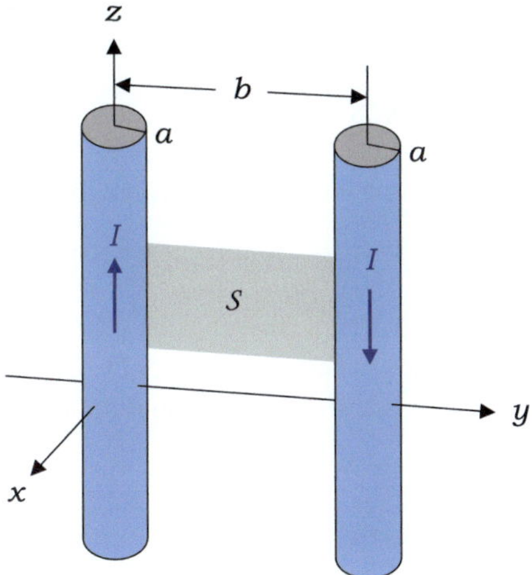

In the shaded area $\mathcal{S}$, specified by $x = 0$, $a < y < (b - a)$, and unity height, the magnetic fields of the two currents are superimposed to yield

$$\mathbf{B} = \mathbf{B}_1 + \mathbf{B}_2 = -\frac{\mu_0 I}{2\pi y}\mathbf{a}_x - \frac{\mu_0 I}{2\pi(b-y)}\mathbf{a}_x$$

The magnetic flux linkage per unit length is

$$\Lambda_{ex} = \int_{\mathcal{S}} (\mathbf{B}_1 + \mathbf{B}_2) \cdot d\mathbf{s} = \frac{\mu_0 I}{2\pi} \int_{y=a}^{y=b-a} \left[\frac{1}{y} + \frac{1}{b-y}\right] dy$$

$$\cong \frac{\mu_0 I}{\pi} \ln \frac{b}{a}$$

Therefore, the external inductance per unit length is

$$L_{ex} = \frac{\Lambda_{ex}}{I} = \frac{\mu_0}{\pi} \ln \frac{b}{a} \tag{5.111a}$$

Under the condition $b \gg a$, in the interior of a conductor, $\mathbf{B}$ of the current flowing in the other conductor can be ignored. Therefore, the internal inductance per unit length of the wire is the same as in Eq. (5.110f). For the two-wire line, the internal inductance per unit length is thus

$$L_{in} = \frac{\mu_0}{4\pi} \quad [\text{H/m}] \tag{5.111b}$$

Finally, the inductance per unit length of the two-wire line is

$$L = L_{ex} + L_{in} = \frac{\mu_0}{\pi}\left[\ln \frac{b}{a} + \frac{1}{4}\right] \quad [\text{H/m}] \tag{5.111c}$$

Exercise 5.28
If a coaxial cable, as shown in Fig. 5.37, is made of conductors with permeability μ_1 and insulator with μ_2, determine the external and internal inductances.

Ans. $L_{ex} = (\mu_2/2\pi) \ln(b/a)$, $L_{in} = \mu_1/8\pi$.

Exercise 5.29
The toroid, as shown in Fig. 5.11, has $a \gg b$, for which $\mathbf{B}$ is assumed to be uniform in the cross section. Determine the inductance.

Ans. $L = \mu_0 N^2 b^2/2a$.

Exercise 5.30
For the two current-carrying loops below, the mutual inductances M_{12} and M_{21} are zero for various reasons. Explain.

Ans. $M_{12} = 0$, no **B** is induced by C_1; $M_{21} = 0$, the loop area of C_1 is zero.

Review Questions

RQ 5.57   What is meant by the magnetic flux linkage?   [(5.104),Fig. 5.34]

RQ 5.58   Define mutual and self-inductances.   [(5.105)(5.107)]

RQ 5.59   Explain external and internal inductances.   [(5.110g)]

5.9 Magnetic Energy

In Sect. 3.7, we observed that the electrostatic potential energy of a charge distribution is equal to the work done in moving individual charges from infinity to predetermined points, and that energy is stored in an electric field with energy density $w_e = \varepsilon E^2/2$. Energy is also expended when sending a current into a conducting loop and producing a magnetic field. The energy is stored in a magnetic field at an energy density of $w_m = \mu H^2/2$. As the magnetic field builds up around the current-carrying loop, an electric field is induced in the loop to oppose the current in accordance with Faraday's law. Work should be done against this electric field to maintain a constant current and generate a magnetic field.

5.9.1 Magnetic Energy in Inductor

An inductor is a conducting device that has self-inductance and stores energy in its magnetic field. We know from circuit theory that for an *ac*-current ii flowing in an inductor, the voltage across it is given by $v = L \, di/dt$. This $v - i$ relationship can be derived from Faraday's law, as described in Chap. 6. However, it is currently used only to obtain an expression for the magnetic energy stored in an inductor. Hereafter, the time-varying quantities are denoted by letters in the script font.

We begin by considering a conducting loop with inductance L. As the current i is increased from zero to a constant value I in the loop, the current induces not only the magnetic flux linking the loop but also an opposing voltage v in the loop, as specified by Faraday's law of induction. The product of the voltage and current vi represents the power delivered by the source to the loop or the energy stored in the magnetic field per unit time. Thus, the total magnetic energy stored in the inductor is expressed as

$$W_m = \int vi \, dt = \int i \, L \frac{di}{dt} \, dt = L \int_{i=0}^{i=I} i \, di$$

Thus,

$$\boxed{W_m = \frac{1}{2} L I^2} \quad [\text{J}] \tag{5.112}$$

By substituting $L = \Lambda/I = N\Phi/I$ into Eq. (5.112), the magnetic energy of the inductor can also be expressed as

$$\boxed{W_m = \frac{1}{2}\Lambda I = \frac{1}{2} N I \Phi} \quad [\text{J}] \tag{5.113}$$

where I is the steady current, N is the number of turns, Φ is the magnetic flux linkage with a single turn, and Λ is the magnetic flux linkage in the loop.

Next, we consider two neighboring loops, C_1 and C_2, which carry steady currents I_1 and I_2, respectively, as shown in Fig. 5.39. The magnetic flux linking each loop is

$$\Lambda_1 = N_1(\Phi_{11} + \Phi_{21}) \tag{5.114a}$$

$$\Lambda_2 = N_2(\Phi_{12} + \Phi_{22}) \tag{5.114b}$$

where N_1 and N_2 are the numbers of turns in loops C_1 and C_2, respectively. In these equations, Φ_{12} is the magnetic flux of the current in C_1 linked to loop C_2, and Φ_{21} is the magnetic flux of the current in C_2 linked to loop C_1. Similarly, Φ_{11} and Φ_{22} are the magnetic fluxes of the currents in C_1 and C_2, respectively, and they are linked to their loops. Inserting Eqs. (5.114a, b) into Eq. (5.113), the magnetic energy stored in the two-loop system is expressed as

Fig. 5.39 Two neighboring loops with currents flowing in the same direction (Shading is used to show the loop surface)

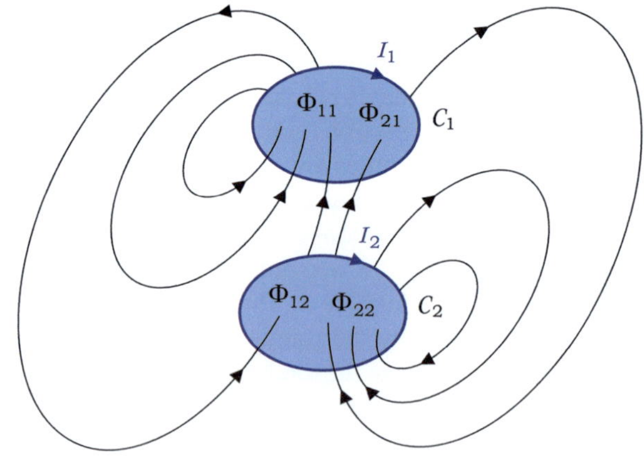

$$W_m = \frac{1}{2}(\Lambda_1 I_1 + \Lambda_2 I_2) = \frac{1}{2}N_1 I_1(\Phi_{11} + \Phi_{21}) + \frac{1}{2}N_2 I_2(\Phi_{12} + \Phi_{22})$$

where I_1 and I_2 are the currents in C_1 and C_2, respectively. With the definitions $L_1 = N_1 \Phi_{11}/I_1$, $L_2 = N_2 \Phi_{22}/I_2$, $M_{12} = N_2 \Phi_{12}/I_1$, and $M_{21} = N_1 \Phi_{21}/I_2$, and the relation $M_{12} = M_{21}$, the magnetic energy stored in the two-loop system is given by

$$\boxed{W_m = \frac{1}{2}L_1 I_1^2 + \frac{1}{2}L_2 I_2^2 \pm M_{12} I_1 I_2} \qquad \text{[J]} \qquad (5.115)$$

The positive sign in front of M_{12} is used for the case in which I_1 and I_2 flow in the same direction and the negative sign otherwise.

To examine the physical meanings of the terms on the right-hand side of Eq. (5.115), we start with two loops with no currents at time $t = 0$. First, the current i_2 is increased from zero to I_2 in loop C_2, while maintaining the current in loop C_1 at zero. In this case, the work done by the source connected to C_2 is given by the second term on the right-hand side of Eq. (5.115). Meanwhile, no work has been done in loop C_1, even if an *emf* (electromotive force) due to the mutual flux Φ_{21} may have arisen and then vanish (*emf* is induced by the time derivative of Φ_{21}). Second, while maintaining the current in C_2 at I_2, current i_1 is increased from zero to I_1 in loop C_1. The work done by the source connected to C_1 is equal to the first term on the right-hand side of Eq. (5.115). However, during the buildup of I_1, the time-varying Φ_{12} induces a voltage v_{12} in C_2 in such a way as to oppose the change in the magnetic flux linking C_2, which is $\Phi_{22} + \Phi_{12}$. Additional work must be done by the source connected to C_2 to maintain the current at I_2. With the relationship $v_{12} = M_{12}\, di_1/dt$, the additional work done is computed as

$$W_{12} = \int v_{12} I_2\, dt = M_{12} I_2 \int_{i_1=0}^{i_1=I_1} di_1 = M_{12} I_1 I_2 \qquad (5.116)$$

which is equal to the third term on the right side of Eq. (5.115). If I_1 and I_2 flow in opposite directions, Φ_{12} tends to reduce the magnetic flux linking C_2, and v_{12} is induced in such a way as to increase the current in C_2; thus, the additional work done is negative, $W_{12} < 0$.

Even if we started our calculation of W_m by increasing i_1 from zero to I_1 while maintaining $i_2 = 0$, we would obtain the same result as that in Eq. (5.115), except that M_{12} was replaced by M_{21}. In other words, the magnetic energy of the given system is the same, regardless of whether we start with i_1 or i_2. Thus, the relationship $M_{12} = M_{21}$ is verified.

5.9.2 Magnetic Energy in Terms of Magnetic Field

It would be more convenient to express the magnetic energy in terms of field quantities **B** and **H**, just as the electrostatic energy is expressed in terms of **D** and **E**. Let us start by considering a rectangular loop C' carrying a steady current I in the $z = 0$ plane, as shown in Fig. 5.40. The magnetic field lines pass through surface S', bounded by C', and form closed lines. We construct a closed loop, C, along one of the magnetic field lines, which is used as the Amperian path. Subsequently, upon applying Ampere's circuital law to loop C, we have

$$\oint_C \mathbf{H} \cdot d\mathbf{l} = NI \tag{5.117}$$

where N is the number of turns in loop C', and $d\mathbf{l}$ is the differential length vector along loop C, which is always parallel to both **H** and **B** because loop C corresponds to a magnetic flux line in a simple medium.

The total magnetic flux enclosed by loop C' is given by

$$\Phi = \int_{S'} \mathbf{B}' \cdot d\mathbf{s}'$$

where prime denotes the quantity measured on surface S' in the $z = 0$ plane. Next, from Eq. (5.113), the magnetic energy of the current I in loop C' is

$$W_m = \frac{1}{2} NI \int_{S'} \mathbf{B}' \cdot d\mathbf{s}' \tag{5.118}$$

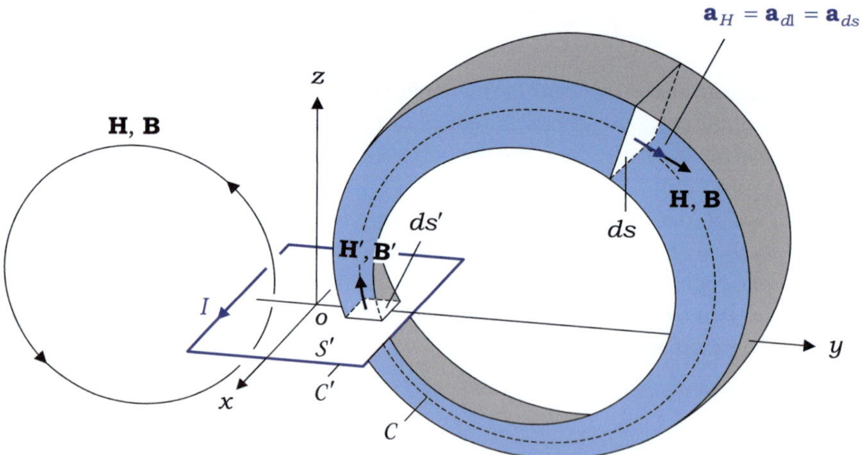

Fig. 5.40 A bundle of magnetic field lines in the form of a ring

Upon substituting Eq. (5.117) into Eq. (5.118), we have

$$W_m = \frac{1}{2}\left[\oint_C \mathbf{H}\cdot d\mathbf{l}\right]\left[\int_{S'}\mathbf{B}'\cdot d\mathbf{s}'\right] = \int_{S'}\frac{1}{2}\oint_C\left[(\mathbf{B}'\cdot d\mathbf{s}')\,\mathbf{H}\cdot d\mathbf{l}\right]$$

$$\equiv \int_{S'} dW \tag{5.119}$$

The quantities on surface S', such as $d\mathbf{s}'$ and $\mathbf{B}'$, are independent of the quantities along loop C, such as $d\mathbf{l}$ and $\mathbf{H}$. Thus, the closed-line integral is taken inside the surface integral in Eq. (5.119), and the integrand of the surface integral is denoted by dW.

As shown in Fig. 5.40, a bundle of magnetic field lines that pass through the differential area $d\mathbf{s}'$ in S' takes the form of a ring in a three-dimensional space. From the conservation of magnetic flux, it follows that

$$d\Phi = \mathbf{B}'\cdot d\mathbf{s}' = \mathbf{B}\cdot d\mathbf{s} \tag{5.120}$$

where $d\mathbf{s}$ represents the cross-sectional area of the ring. The differential area vector $d\mathbf{s}'$ always points in the $+z$-direction on S', whereas the magnitude and direction of $d\mathbf{s}$ vary with position along the ring. However, $\left|d\mathbf{s}'\right|$ and $|d\mathbf{s}|$ are the cross-sectional areas through which the same magnetic flux passes. Inserting Eq. (5.120) into Eq. (5.119) gives

$$dW = \frac{1}{2}\oint_C\left[(\mathbf{B}\cdot d\mathbf{s})\,\mathbf{H}\cdot d\mathbf{l}\right]$$

Because $\mathbf{B}$, $\mathbf{H}$, $d\mathbf{l}$, and $d\mathbf{s}$ are all tangential to loop C, dW can be rewritten as

$$dW = \frac{1}{2}\oint_C\left[(\mathbf{B}\cdot\mathbf{H})\,d\mathbf{s}\cdot d\mathbf{l}\right] = \oint_C\frac{1}{2}(\mathbf{B}\cdot\mathbf{H})\,ds\,dl \tag{5.121}$$

By identifying term $ds\,dl$ with the differential volume of the ring, we conclude that the right-hand side of Eq. (5.121) represents the integration of $(\mathbf{B}\cdot\mathbf{H})/2$ over the volume of the ring. Subsequently, the surface integral in Eq. (5.119) is equivalent to the integration of $(\mathbf{B}\cdot\mathbf{H})/2$ over the entire space, where the magnetic field is nonzero. The magnetic flux lines passing through the surface S' fill the entire space. Therefore, the magnetic energy can be expressed as

$$\boxed{W_m = \frac{1}{2}\int_V \mathbf{B}\cdot\mathbf{H}\,dv}\qquad [\text{J}] \tag{5.122}$$

In a simple medium, where μ is a constant, Eq. (5.122) becomes

$$W_m = \frac{\mu}{2}\int_V H^2\,dv = \frac{1}{2\mu}\int_V B^2\,dv\qquad [\text{J}] \tag{5.123}$$

where the volume integral is conducted over the entire space. In view of Eq. (5.122), the magnetic energy density is defined as $(\mathbf{B} \cdot \mathbf{H})/2$, measured in joules per cubic meter.

If the magnetic energy of the conducting device is computed using Eq. (5.123), the inductance can be obtained using Eq. (5.112) as follows:

$$\boxed{L = \frac{2W_m}{I^2}} \quad \text{[J]} \tag{5.124}$$

Example 5.24 For the coaxial cable, as shown in Fig. 5.37, find the inductance per unit length of the cable by determining the magnetic energy per unit length.

Solution

Inserting Eq. (5.110b) into Eq. (5.122) yields the magnetic energy stored in region $a < \rho < b$ such that

$$W_{m1} = \frac{1}{2\mu_0} \int_{\rho=a}^{\rho=b} \int_{\phi=0}^{\phi=2\pi} \left[\frac{\mu_0 I}{2\pi\rho} \right]^2 \rho \, d\rho \, d\phi = \frac{\mu_0 I^2}{4\pi} \ln \frac{b}{a} \tag{5.125a}$$

Similarly, inserting Eq. (5.110c) into Eq. (5.122) yields the magnetic energy stored in region $0 \le \rho \le a$, as follows:

$$W_{m2} = \frac{1}{2\mu_0} \int_{\rho=0}^{\rho=a} \int_{\phi=0}^{\phi=2\pi} \left[\frac{\mu_0 \rho I}{2\pi a^2} \right]^2 \rho \, d\rho \, d\phi = \frac{\mu_0 I^2}{16\pi} \tag{5.125b}$$

From Eq. (5.124), the inductance per unit length of the cable is

$$L = \frac{2}{I^2}(W_{m1} + W_{m2}) = \frac{\mu_0}{2\pi} \ln \frac{b}{a} + \frac{\mu_0}{8\pi} \quad \text{[H/m]} \tag{5.125c}$$

This result is the same as that in Eq. (5.110g).

Example 5.25 Two very long solenoids are concentric with the z-axis and have N_1 and N_2 turns per unit length wound on air cores with radii of a and b, respectively, as shown in Fig. 5.41. Find the self and mutual inductances by determining (a) the magnetic flux linkage, and (b) the magnetic energy.

Solution

The solenoid produces a uniform $\mathbf{H}$ in its interior without an external magnetic field.

For the inner solenoid,

$$\begin{aligned} \mathbf{H}_1 &= N_1 I_1 \, \mathbf{a}_z \quad (0 \le \rho < a) \\ &= 0 \qquad\qquad (\rho > a) \end{aligned}$$

Fig. 5.41 Two concentric solenoids placed in free space

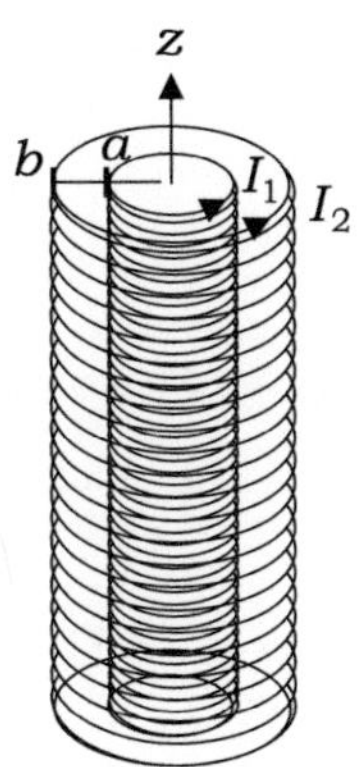

For the outer solenoid,

$$\mathbf{H}_2 = N_2 I_2 \, \mathbf{a}_z \quad (0 \le \rho < b)$$
$$= 0 \qquad (\rho > b)$$

(a) These magnetic fields provide the following magnetic flux per unit length:

$$\Phi_{11} = \mu_0 N_1 I_1 \pi a^2, \ \ \Phi_{12} = \mu_0 N_1 I_1 \pi a^2$$

$$\Phi_{22} = \mu_0 N_2 I_2 \pi b^2, \ \ \Phi_{21} = \mu_0 N_2 I_2 \pi a^2$$

The self-inductances per unit length are

$$L_1 = \frac{N_1 \Phi_{11}}{I_1} = \mu_0 N_1^2 \pi a^2 \ \text{and} \ L_2 = \frac{N_2 \Phi_{22}}{I_2} = \mu_0 N_2^2 \pi b^2 \qquad (5.126a)$$

The mutual inductances per unit length are

$$M_{12} = \frac{N_2 \Phi_{12}}{I_1} = \mu_0 N_1 N_2 \pi a^2 \ \text{and} \ M_{21} = \frac{N_1 \Phi_{21}}{I_2} = \mu_0 N_1 N_2 \pi a^2 \quad (5.126b)$$

Note that $M_{12} = M_{21}$.

(b) The total magnetic energy per unit length is equal to the sum of those in regions $a < \rho < b$ and $0 \le \rho < a$, that is,

$$W_m = \frac{\mu_0}{2}(N_2 I_2)^2 \pi (b^2 - a^2) + \frac{\mu_0}{2}(N_1 I_1 + N_2 I_2)^2 \pi a^2$$

$$= \frac{1}{2}\mu_0 \pi \left[(N_1 I_1)^2 a^2 + (N_2 I_2)^2 b^2 + 2 N_1 N_2 I_1 I_2 a^2 \right] \qquad (5.126c)$$

Comparing Eq. (5.126c) with Eq. (5.115) gives

$$L_1 = \mu_0 N_1^2 \pi a^2 \text{ and } L_2 = \mu_0 N_2^2 \pi b^2 \tag{5.126d}$$

$$M_{12} = \mu_0 N_1 N_2 \pi a^2 \tag{5.126e}$$

The results are the same as those in Part (a).

Exercise 5.31

By what factor does W_m increase if the (a) radius, (b) length, (c) current, (d) number of turns, and (e) μ of the solenoid core are increased by a factor of two?

Ans. (a) 4, (b) 2, (c) 4, (d) 4, (e) 2.

Review Questions

RQ 5.60	Express magnetic energy in terms of inductance.	[(5.112)]
RQ 5.61	Express magnetic energy in terms of flux linkage.	[(5.113)]
RQ 5.62	Express magnetic energy in terms of field quantities.	[(5.122)]
RQ 5.63	Express the magnetic energy of a two-loop system.	[(5.115)]

5.10   Magnetic Force and Torque

The Lorentz force equation states that charge q moving at velocity $\boldsymbol{v}$ in magnetic flux density $\mathbf{B}$ experiences magnetic force $\mathbf{F}_m = q\,\boldsymbol{v} \times \mathbf{B}$. Because the current-carrying wire necessarily involves the motion of conduction electrons, the wire experiences magnetic force even if it is fixed in position in a magnetic field.

5.10.1   *Magnetic Force on Current-Carrying Conductor*

Consider a segment of a current-carrying wire with cross-sectional area $\mathcal{S}$ and length $|d\mathbf{l}|$ placed in a magnetic field $\mathbf{B}$. If the conduction electrons of density n $[\mathrm{m}^{-3}]$ move at a velocity $\boldsymbol{v}$ [m/s] in the wire, the magnetic force exerted on the segment is

$$
\begin{aligned}
d\mathbf{F}_m &= en\mathcal{S}|d\mathbf{l}|\,\boldsymbol{v} \times \mathbf{B} \\
&= -en\mathcal{S}|\boldsymbol{v}|\,d\mathbf{l} \times \mathbf{B}
\end{aligned} \tag{5.127}
$$

where e is the electron charge given by -1.6×10^{-19} [C], and $en\mathcal{S}|d\mathbf{l}|$ represents the total charge in a given segment. Because $d\mathbf{l}$ is assumed to be parallel to the direction of the current, it is opposite in direction to electron velocity $\boldsymbol{v}$. With the current density $\mathbf{J} = en\boldsymbol{v}$ and total current $I = |\mathbf{J}|\mathcal{S}$, Eq. (5.127) can be written as

$$dF_m = \mathbf{J} \times \mathbf{B}\, dv \tag{5.128a}$$

$$d\mathbf{F}_m = I\, d\mathbf{l} \times \mathbf{B} \tag{5.128b}$$

where $dv = \mathcal{S}|d\mathbf{l}|$. Thus, the magnetic force exerted on a conducting device with volume current density $\mathbf{J}$ or filamentary current I is given by

$$\boxed{\mathbf{F}_m = \int_{\mathcal{V}} \mathbf{J} \times \mathbf{B}\, dv} \quad [\text{N}] \tag{5.129a}$$

$$\boxed{\mathbf{F}_m = I \oint_{\mathcal{C}} d\mathbf{l} \times \mathbf{B}} \quad [\text{N}] \tag{5.129b}$$

where $d\mathbf{l}$ points in the direction of the flow of I, which is taken as the direction of travel along loop $\mathcal{C}$.

When two current-carrying wires reside in free space, which are not necessarily parallel to each other, the first wire experiences a magnetic force in the presence of the magnetic flux produced by the current in the second wire, and vice versa. First, $\mathbf{B}$ at point $\mathbf{r}_1$ on wire $\mathcal{C}_1$, caused by current I_2 in wire $\mathcal{C}_2$, is obtained by invoking the Biot-Savart law such that

$$\mathbf{B}_{1-2} = \frac{\mu_0 I_2}{4\pi} \oint_{\mathcal{C}_2} \frac{d\mathbf{l}_2 \times \mathbf{a}_{\mathcal{R}}}{\mathcal{R}^2} \tag{5.130}$$

where $\mathbf{a}_{\mathcal{R}}$ is a unit vector directed from source point $\mathbf{r}_2$ to field point $\mathbf{r}_1$. The magnetic force on wire $\mathcal{C}_1$ with current I_1 residing in $\mathbf{B}_{1-2}$ is thus

$$\mathbf{F}_1 = I_1 \oint_{\mathcal{C}_1} d\mathbf{l}_1 \times \mathbf{B}_{1-2} \tag{5.131}$$

Upon inserting Eq. (5.130) into Eq. (5.131), we obtain

$$\boxed{\mathbf{F}_1 = \frac{\mu_0}{4\pi} I_1 I_2 \oint_{\mathcal{C}_1} \oint_{\mathcal{C}_2} \frac{d\mathbf{l}_1 \times (d\mathbf{l}_2 \times \mathbf{a}_{\mathcal{R}})}{\mathcal{R}^2}} \quad [\text{N}] \tag{5.132}$$

It may not be straightforward to evaluate the double line-integral in Eq. (5.132). In many cases, it is more convenient to divide the integral into two parts, as expressed in Eqs. (5.130) and (5.131), and evaluate them separately.

Example 5.26 A transmission line consists of two very long parallel wires separated by a distance d [m] in free space, carrying a steady current I [A] in opposite directions, as shown in Fig. 5.42. Determine the magnetic force per unit length for each line.

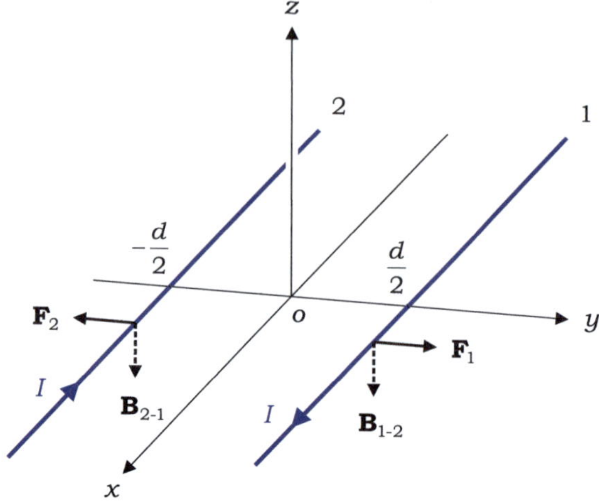

Fig. 5.42 Two parallel wires with steady currents flowing in opposite directions

Solution

From Ampere's circuital law, **B** on wire 1 due to current I in wire 2 is

$$\mathbf{B}_{1-2} = -\frac{\mu_0 I}{2\pi d}\mathbf{a}_z$$

The differential length vector on wire 1 is

$$d\mathbf{l}_1 = dx\,\mathbf{a}_x$$

Thus, the magnetic force per unit length of wire 1 is given by

$$\mathbf{F}_1 = I \oint_{C_1} d\mathbf{l}_1 \times \mathbf{B}_{1-2} = I \int_{x=0}^{x=1} dx\,\mathbf{a}_x \times \left(-\frac{\mu_0 I}{2\pi d}\mathbf{a}_z\right)$$

$$= \frac{\mu_0 I^2}{2\pi d}\mathbf{a}_y \tag{5.133a}$$

The magnetic force is a mutual force such that the magnetic force on wire 2 is simply

$$\mathbf{F}_2 = -\mathbf{F}_1 = -\frac{\mu_0 I^2}{2\pi d}\mathbf{a}_y \tag{5.133b}$$

Exercise 5.32

A circular loop is centered at the origin in the xy-plane, carrying a steady current I in a uniform field $\mathbf{B} = B_o\,\mathbf{a}_z$. Determine the net force acting on the loop.

Ans. Zero.

5.10.2 *Magnetic Force and Virtual Work*

In the previous section, the Lorentz force equation was used to determine the magnetic force exerted on a moving charge or current-carrying conductor placed in a magnetic field. We now introduce an alternative method for determining the magnetic force called the method of virtual displacement. This method is based on the fundamental relationship between energy and force, that is, energy is equal to the line integral of the force.

Consider an electromagnet as shown in Fig. 5.43, where the current in the coil induces a magnetic flux in the core, gap, and armature (moving piece). If the fringing effects at the edges are ignored, then the magnetic field is uniform in the gap. Suppose we apply an external force **F** to move the armature downward by a small distance. The work done is stored as magnetic energy in the incremental gap volume. Conversely, the attractive magnetic force on the armature can be determined by assuming a "virtual" displacement of the armature in the upward direction, and computing the magnetic energy contained in the decremental gap volume. Note that the method of virtual displacement assumes that **B** remains unchanged even if the gap size may be changed.

If the attractive force of the electromagnet, $\mathbf{F}_m$, causes the armature to move upward by a small distance $d\mathbf{l}$, work must be done at the expense of the magnetic energy of the system. Expressed mathematically,

$$\mathbf{F}_m \cdot d\mathbf{l} = -dW_m \tag{5.134}$$

where $-dW_m$ represents the decrease in the magnetic energy stored in the gap. From calculus, the differential of a scalar quantity can be expressed in terms of its gradient as $dW = (\nabla W) \cdot d\mathbf{l}$. Using this relationship, Eq. (5.134) can be written as

Fig. 5.43 Electromagnet and armature

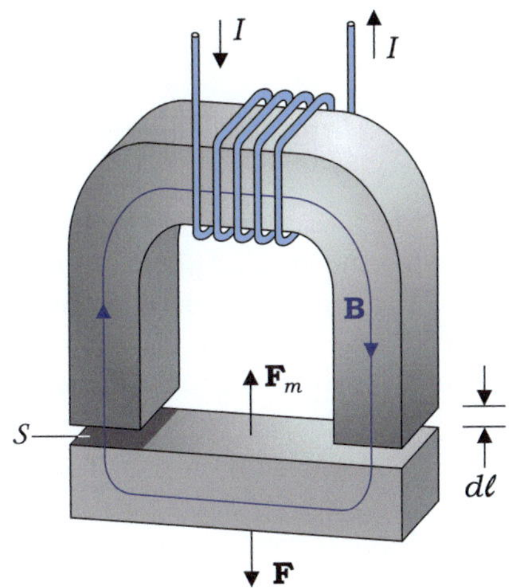

$$\mathbf{F}_m \cdot d\mathbf{l} = -(\nabla W_m) \cdot d\mathbf{l} \tag{5.135}$$

Because $d\mathbf{l}$ is arbitrary in magnitude, Eq. (5.135) directly leads to

$$\boxed{\mathbf{F}_m = -\nabla W_m} \quad [\text{N}] \tag{5.136}$$

The attractive force on the armature is equal to the negative gradient of the magnetic energy stored in the gap.

Example 5.27 For the electromagnet, as shown in Fig. 5.43, express the magnetic force of the electromagnet $\mathbf{F}_m$ in terms of $\mathbf{B}$.

Solution

For a virtual displacement of the armature by a distance $d\ell$ in the upward direction, the magnetic energy in the gap is reduced by

$$dW_m = -2\left[\frac{B^2}{2\mu_0}\mathcal{S}\,d\ell\right] \tag{5.137a}$$

Comparing Eqs. (5.137a) and (5.134), along with $d\mathbf{l} = d\ell\,\mathbf{a}_z$, we obtain

$$\mathbf{F}_m = \frac{B^2\mathcal{S}}{\mu_0}\,\mathbf{a}_z \quad [\text{N}] \tag{5.137b}$$

Exercise 5.33
Referring to Fig. 5.43, show that the tensile force per unit area of the surface $\mathcal{S}$ is equal to the magnetic energy density in the gap.

Ans. $P = |\mathbf{F}_m|/2\mathcal{S} = B^2/2\mu_0\ [\text{N/m}^2]$.

5.10.3 Magnetic Torque

A rigid body can rotate about its pivot axis by applying force. **Moment arm r** is defined as the distance vector from the pivot axis to the point of application of the force. The angular acceleration of the body depends on the length of the moment arm and force normal to the moment arm. **Torque** is defined as the cross product of the moment arm and applied force. That is,

$$\boxed{\mathbf{T} = \mathbf{r} \times \mathbf{F}} \quad [\text{N} \cdot \text{m}] \tag{5.138}$$

where $\mathbf{F}$ is assumed to lie in the plane perpendicular to the pivot axis, similar to moment arm $\mathbf{r}$. Otherwise, the orientation of the pivot axis is determined by the direction of $\mathbf{T}$. The unit of torque is newton-meter.

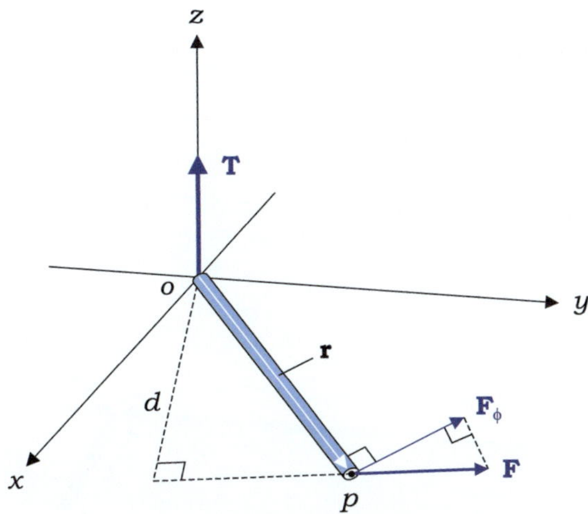

Fig. 5.44 Torque **T** acting on the rigid bar

For example, with reference to Fig. 5.44, the solid bar extends from the origin to point p in the xy-plane and pivots about the z-axis because of the external force **F** applied to p. From Eq. (5.138), the magnitude of the torque is

$$T = r\,F_\phi = F\,d \qquad (5.139)$$

where r is the length of the moment arm, F_ϕ is the component of **F** perpendicular to **r**, and d is the ***effective moment arm*** or perpendicular distance from the origin to **F**. The direction of the torque obeys the right-hand rule: the right thumb points in the direction of the torque when the fingers follow the direction of rotation of the body.

An externally applied magnetic field may exert torque on the current-carrying loop and cause it to rotate about its pivot axis until the magnetic dipole moment of the loop aligns with the magnetic field. Consider a small rectangular loop carrying steady current I, as shown in Fig. 5.45. The loop is constrained to rotate about the x-axis in the presence of a uniform magnetic flux density **B** applied along the y-axis. At time $t = 0$, the magnetic dipole moment of loop **m** is at an angle α with respect to the y-axis. Magnetic forces are exerted on the four sides of the loop according to the Lorentz force equation. The magnetic forces on the top and bottom sides, $\mathbf{F}_1$ and $\mathbf{F}_2$, exert torque on the loop. However, the magnetic forces on the left and right sides, $\mathbf{F}_3$ and $\mathbf{F}_4$, are either parallel or antiparallel to the pivot axis and thus do not contribute. Therefore,

$$\mathbf{T} = -d(F_1 + F_2)\,\mathbf{a}_x \qquad (5.140)$$

where d is the effective moment arm at the top and bottom of the loop. From Eq. (5.129b), the magnitudes of $\mathbf{F}_1$ and $\mathbf{F}_2$ are

$$F_1 = F_2 = I\,w\,B \qquad (5.141)$$

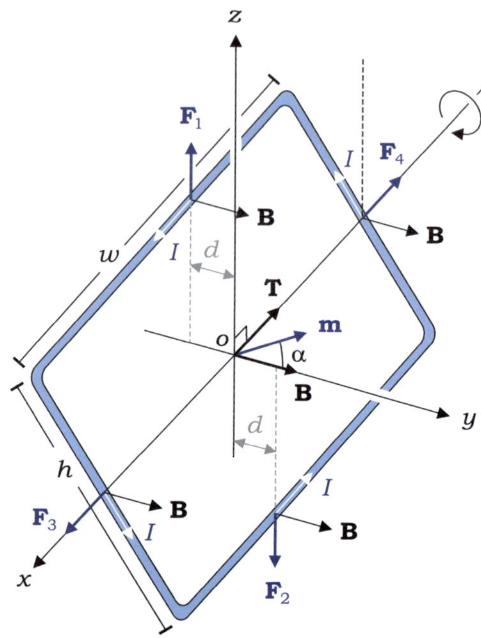

Fig. 5.45 Rectangular loop with current I pivoting about the x-axis in **B**

where w is the width of the rectangular loop. Inserting Eq. (5.141) and the relation $d = (h/2) \sin \alpha$ into Eq. (5.140), the torque on the loop is obtained as

$$\mathbf{T} = -hwIB \sin \alpha \, \mathbf{a}_x = -mB \sin \alpha \, \mathbf{a}_x \qquad (5.142)$$

where m is the magnitude of the magnetic dipole moment given by $\mathbf{m} = hwI \, \mathbf{a}_n$. In vector notation, the torque can be expressed using the cross product of the magnetic dipole moment and magnetic flux density. That is,

$$\boxed{\mathbf{T} = \mathbf{m} \times \mathbf{B}} \quad [\text{N} \cdot \text{m}] \qquad (5.143)$$

The torque tends to rotate the current loop such that the magnetic dipole moment aligns with the applied magnetic field.

Example 5.28 A closed loop in the xy-plane carries current I in the presence of uniform **B**, as shown in Fig. 5.46. Show that the torque on the loop pivoting about the origin is always given by $\mathbf{T} = \mathbf{m} \times \mathbf{B}$ regardless of the loop shape.

Solution

At a point on the loop, the magnetic field $\mathbf{B} = B_1 \mathbf{a}_x + B_2 \mathbf{a}_y + B_3 \mathbf{a}_z$ exerts a force $d\mathbf{F} = I d\mathbf{l} \times \mathbf{B}$, where $d\mathbf{l} = dx \, \mathbf{a}_x + dy \, \mathbf{a}_y$, such that

Fig. 5.46 Current-carrying loop of an arbitrary shape in uniform **B**

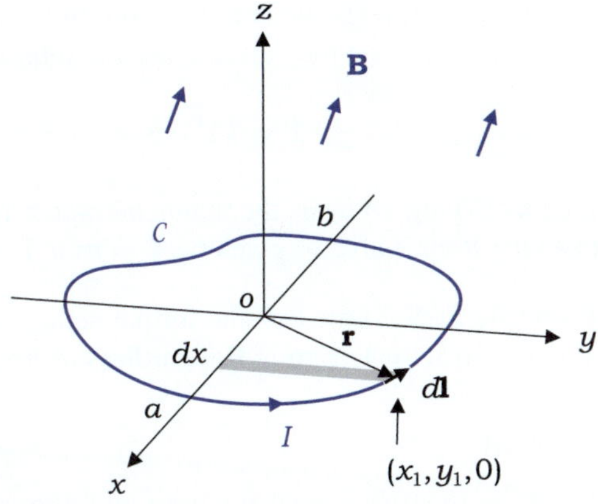

$$dF = I \begin{vmatrix} \mathbf{a}_x & \mathbf{a}_y & \mathbf{a}_z \\ dx & dy & 0 \\ B_1 & B_2 & B_3 \end{vmatrix} = I \left[dy\,B_3\,\mathbf{a}_x - dx\,B_3\,\mathbf{a}_y + (dx\,B_2 - dy\,B_1)\,\mathbf{a}_z \right] \qquad (5.144a)$$

Using $d\mathbf{T} = \mathbf{r} \times d\mathbf{F}$ and the position vector $\mathbf{r} = x\,\mathbf{a}_x + y\,\mathbf{a}_y$, the differential torque is computed as

$$d\mathbf{T} = \mathbf{r} \times (I\,d\mathbf{l} \times \mathbf{B})$$

$$= I \begin{vmatrix} \mathbf{a}_x & \mathbf{a}_y & \mathbf{a}_z \\ x & y & 0 \\ dy\,B_3 & -dx\,B_3 & (dx\,B_2 - dy\,B_1) \end{vmatrix}$$

$$= I \left[(y\,dx\,B_2 - y\,dy\,B_1)\,\mathbf{a}_x - (x\,dx\,B_2 - x\,dy\,B_1)\,\mathbf{a}_y + (-x\,dx\,B_3 - y\,dy\,B_3)\,\mathbf{a}_z \right]$$
$$(5.144b)$$

The total torque can be obtained by integrating $d\mathbf{T}$ around the loop. In this case, the integration of xdx or ydy around the loop vanishes because the integration path is closed, as shown below:

$$\oint_C x\,dx = \frac{x^2}{2}\Big|_{x=a}^{x=b} + \frac{x^2}{2}\Big|_{x=b}^{x=a} = 0, \text{ and similarly, } \oint_C y\,dy = 0 \qquad (5.144c)$$

In contrast, the integration of ydx around the loop gives the negative of the loop area A, that is,

$$\oint_C y\,dx = \int_{x=a}^{x=b} y\,dx + \int_{x=b}^{x=a} y\,dx = -A \qquad (5.144d)$$

where the first integral in the middle gives the loop area for $y > 0$ multiplied by -1 because dx is negative, and the second gives the loop area for $y < 0$ multiplied by

-1 because y is negative in this region. In contrast, the integration of $x\,dy$ around the loop is equal to the loop area. Thus, the total torque on the loop is given by

$$\mathbf{T} = IA\big[-B_2\,\mathbf{a}_x + B_1\,\mathbf{a}_y\big] = IA\,\mathbf{a}_z \times \mathbf{B} \qquad (5.144e)$$

By identifying $IA\,\mathbf{a}_z$ as the magnetic dipole moment $\mathbf{m}$, the torque on the current-carrying loop is always given by $\mathbf{T} = \mathbf{m} \times \mathbf{B}$ regardless of the loop shape.

Example 5.29 Show that the torque acting on the loop, as shown in Fig. 5.46, is always given by $\mathbf{T} = \mathbf{m} \times \mathbf{B}$ regardless of the position of the pivot point.

Solution

Let $\boldsymbol{\mathcal{R}}$ be the distance vector from an arbitrary pivot point to the origin. Then, the differential torque due to $d\mathbf{F}_m$ at the position vector $\mathbf{r}$ is

$$d\mathbf{T} = (\boldsymbol{\mathcal{R}} + \mathbf{r}) \times d\mathbf{F}_m = (\boldsymbol{\mathcal{R}} + \mathbf{r}) \times (I\,d\mathbf{l} \times \mathbf{B}) \qquad (5.145a)$$

where $d\mathbf{F} = I\,d\mathbf{l} \times \mathbf{B}$. Using BAC-CAB rule, one of the two cross products on the right side of Eq. (5.145a) can be written as

$$\boldsymbol{\mathcal{R}} \times (I\,d\mathbf{l} \times \mathbf{B}) = I\,d\mathbf{l}\,(\boldsymbol{\mathcal{R}} \cdot \mathbf{B}) - I\mathbf{B}\,(\boldsymbol{\mathcal{R}} \cdot d\mathbf{l}) \qquad (5.145b)$$

Because I, $\boldsymbol{\mathcal{R}}$, and $\mathbf{B}$ are constants, the integrations of $d\mathbf{l}$ and $\boldsymbol{\mathcal{R}} \cdot d\mathbf{l}$ around a closed path vanish. For example, if $\boldsymbol{\mathcal{R}}$ is parallel to the x-axis and the loop crosses the axis at x_1 and x_2, the line integral of $\boldsymbol{\mathcal{R}} \cdot d\mathbf{l}$ from x_1 to x_2 along the loop is equal to the length between x_1 and x_2, and the line integral along the rest of the loop yields the negative of the former result. Therefore, $d\mathbf{T}$ can be reduced to Eq. (5.144b), and the torque is given by $\mathbf{T} = \mathbf{m} \times \mathbf{B}$. This result indicates that the torque is the same, even if the pivot point is located outside the loop.

Exercise 5.34
If a small loop of area 0.5 [cm^2] carrying a steady current 200 [mA] experiences a maximum torque of 1 [μN $\cdot$ m], determine $|\mathbf{B}|$.

Ans. 0.1 [Wb/m^2].

Exercise 5.35
When a square and circular loop of the same perimeter carry an equal current in uniform $\mathbf{B}$, determine the ratio between the torques on the loops.

Ans. $T_{square}/T_{circle} = \pi/4$.

Fig. 5.47   A charge moving around the filamentary current (Problem 5.2)

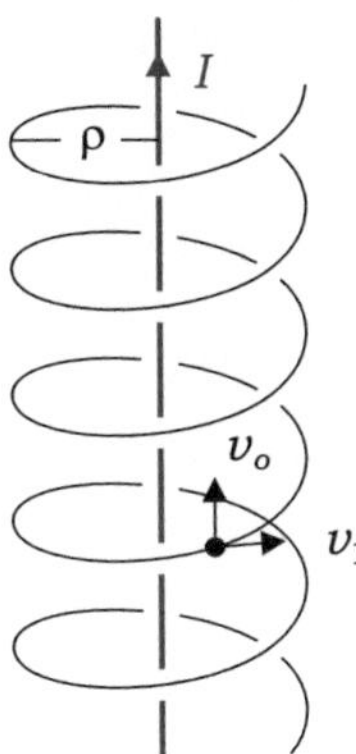

Review Questions

RQ 5.64   Express the magnetic force on the current-carrying wire placed in a uniform magnetic field.   [(5.129b)]

RQ 5.65   Express magnetic force in terms of magnetic energy.   [(5.136)]

RQ 5.66   Write the relationship between torque and force.   [(5.138)]

RQ 5.67   Define the moment arm and effective moment arm.   [Fig. 5.44]

RQ 5.68   Express the torque on the magnetic dipole placed in **B**.   [(5.143)]

RQ 5.69   Does $\mathbf{T} = 0$ on a current-carrying loop mean $\mathbf{F}_m = 0$ at every point in the loop placed in **B**?   [Fig. 5.46]

5.11   Problems

Lorentz force equation

5.1   In a uniform magnetic flux density $\mathbf{B} = B_o\,\mathbf{a}_z$, point charge q moves at velocity $v = a\,\mathbf{a}_x + b\,\mathbf{a}_y$ at time $t = 0$. Determine the electric field required to move the charge along a straight line for $t \geq 0$.

5.2   A particle with charge q and mass m traces a helix of radius ρ around infinite filamentary current I, as shown in Fig. 5.47. If the particle rotates at angular velocity ω, determine its vertical speed v_o. [Hint: Centripetal acceleration, $\mathbf{a} = -\rho\omega^2\,\mathbf{a}_\rho$.]

5.3   A long conducting bar is oriented along the y-axis in a uniform field $\mathbf{B} = B_o\,\mathbf{a}_z$, carrying a current I [A] uniformly distributed over the rectangular cross section (Fig. 5.48). If the current is caused by the holes of density n [m^{-3}] in motion, determine

(a)   the drift velocity of the hole, and
(b)   the electric field $\mathbf{E}_H$ induced in the bar by magnetic force.

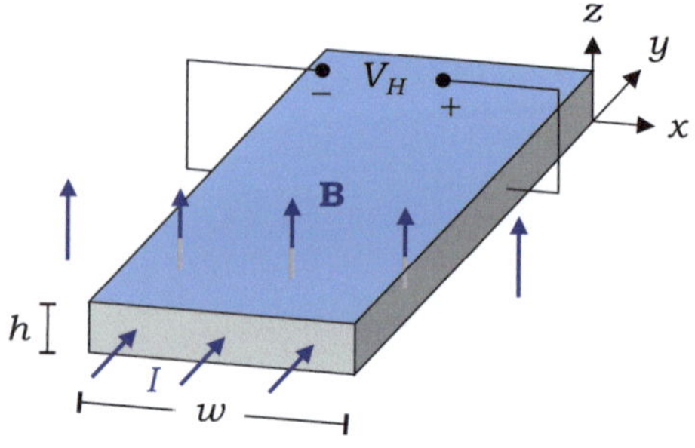

Fig. 5.48 Hall voltage
(Problem 5.3)

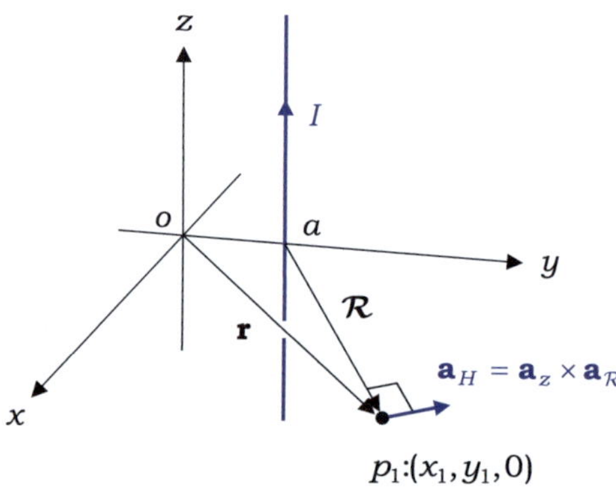

Fig. 5.49 A filamentary
current parallel to the z-axis
(Problem 5.4)

The Biot-Savart law

5.4 An infinite filament crosses the y-axis at $y = a$ and carries a steady current I in the $+z$-direction (Fig. 5.49). Determine $\mathbf{H}$ at point $p_1:(x_1, y_1, 0)$. [Hint: The current I along the z-axis produces $\mathbf{H} = I\,\mathbf{a}_\phi/2\pi\rho$.]

5.5 Three infinitely long parallel filaments carry the same current I, as shown in Fig. 5.50. Determine $\mathbf{H}$ at point p due to (a) the center filament, and (b) the two side filaments. (c) Determine filament separation, a, for which the two $\mathbf{H}$'s in parts (a) and (b) are equal.

5.6 A wire is formed into an equilateral triangle of side a in the xy-plane, with the center at the origin (Fig. 5.51). When a steady current I flows counterclockwise, as viewed from above, determine $\mathbf{H}$ at the origin.

5.7 Determine $\mathbf{H}$ at the origin caused by the steady current I flowing in a closed loop consisting of a semicircle and rectangle, as shown in Fig. 5.52.

5.8 A right triangle carries a steady current I, as shown in Fig. 5.53. When point p is positioned along a line perpendicular to the hypotenuse, determine $\mathbf{H}$ at p for $t = 2$ and $a = 1$.

5.9 A square loop centered at the origin in the xy-plane carries a steady current I, as shown in Fig. 5.54. Determine $\mathbf{H}$ at points (a) p, and (b) q.

Fig. 5.50 Three parallel filamentary currents (Problem 5.5)

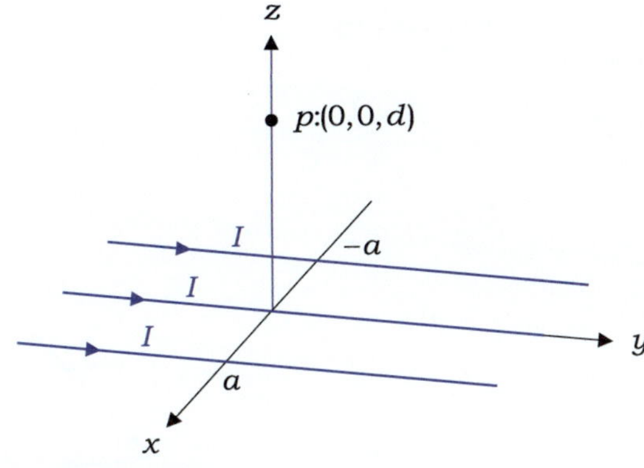

Fig. 5.51 Equilateral triangle centered at the origin (Problem 5.6)

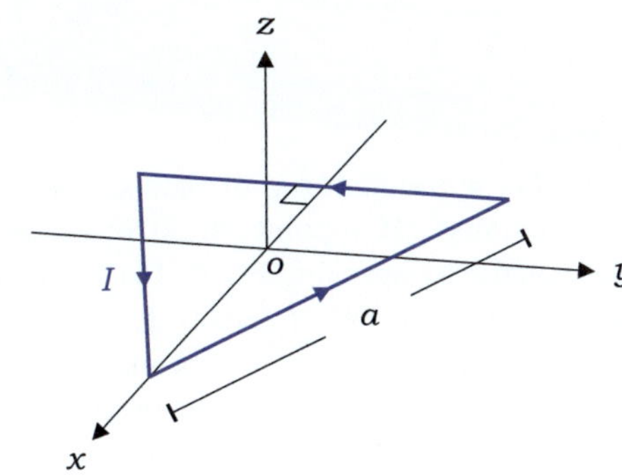

Fig. 5.52 Current-carrying loop (Problems 5.7 and 5.47)

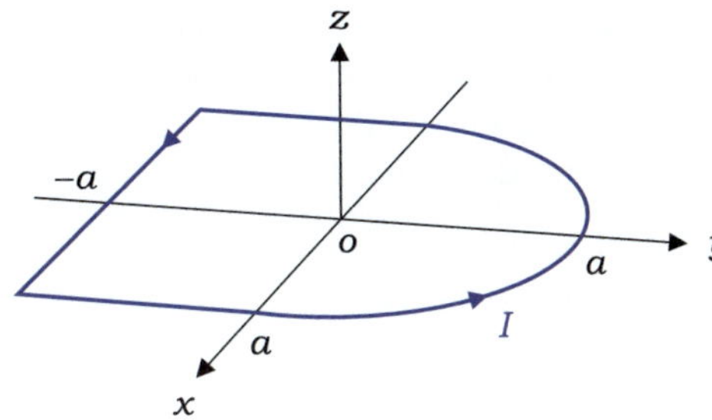

Fig. 5.53 Right triangle carrying current I (Problem 5.8)

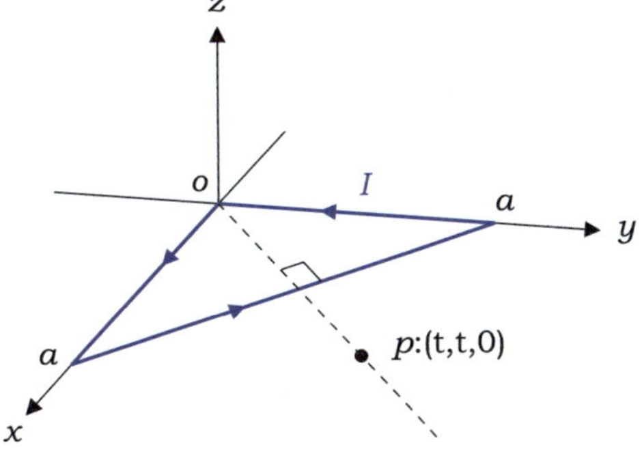

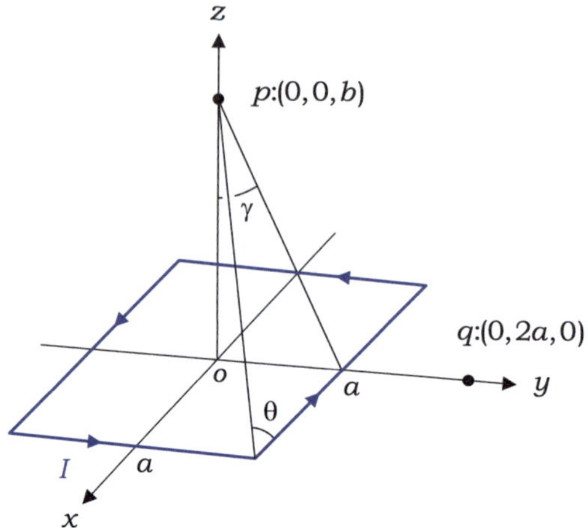

Fig. 5.54 Current-carrying square loop (Problem 5.9)

5.10 For a circular loop of radius a with steady current I, as shown in Fig. 5.55, express $d\mathbf{H}$ at point p:(0, 0, b) caused by the current element $I\,dl$ in terms of a and b by applying Eq. (5.13) to triangle $\triangle pmn$. Determine $\mathbf{H}$ at p.

5.11 A surface current density $\mathbf{J}_s = J_o\,\mathbf{a}_y$ is confined to a region defined by $-a \leq x \leq a$ and $-\infty < y < \infty$ in the $z = 0$ plane, as shown in Fig. 5.56. Determine $\mathbf{H}$ at distance b above the origin by invoking the Biot-Savart law.

5.12 The current strip shown in Fig. 5.56 can be broken into many narrow strips of width dx carrying line current $J_o dx$. Start with $\mathbf{H}$ of an infinite filamentary current to determine $\mathbf{H}$ at point p.

5.13 A uniform surface current with $\mathbf{J}_s = J_o\,\mathbf{a}_\phi$ [A/m] is confined to a circular band of radius a and thickness c, as shown in Fig. 5.57. Determine $\mathbf{H}$ at distance b from the origin.

Fig. 5.55 Circular loop with steady current (Problem 5.10)

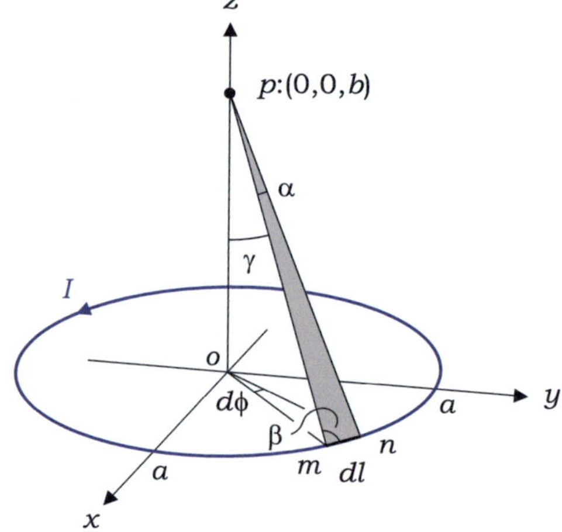

Fig. 5.56 Surface current in an infinite strip (Problems 5.11 and 5.12)

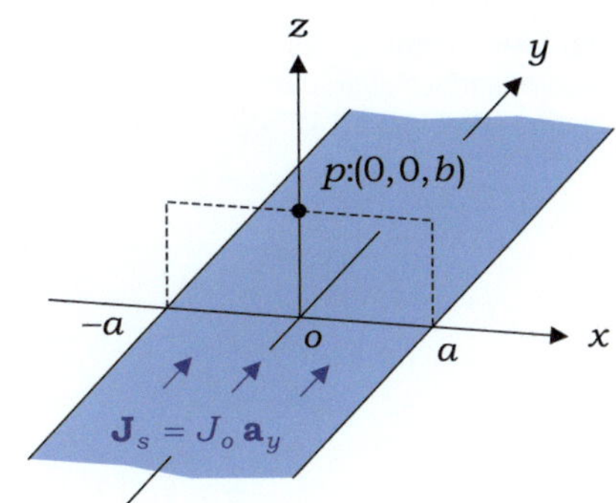

Fig. 5.57 Circular band with uniform surface current (Problem 5.13)

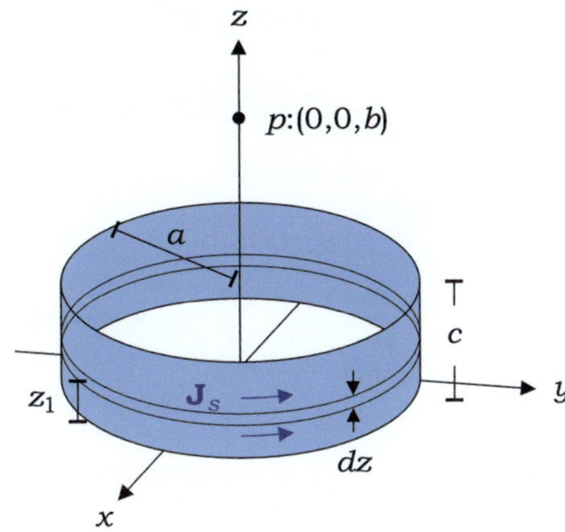

5.14   An annulus with a uniform surface charge density ρ_s [C/m^2] rotates about the z-axis at angular velocity ω [rad/s] (Fig. 5.58). Determine

   (a)   the surface current density in the annulus,
   (b)   **H** at point p by using the Biot-Savart law, and
   (c)   **H** at point p by applying Eq. (5.15) to the narrow circular strips of width $d\rho$ that comprise the annulus.

5.15   Verify that the curl of **H** in Eq. (5.6) vanishes if the path of integration or the current-carrying filament is either a closed or infinite line.
   [Hint: $\nabla \times (\mathbf{A} + \mathbf{B}) = \mathbf{A}(\nabla \cdot \mathbf{B}) - \mathbf{B}(\nabla \cdot \mathbf{A}) + (\mathbf{B} \cdot \nabla)\mathbf{A} - (\mathbf{A} \cdot \nabla)\mathbf{B}.$]

Ampere's law

5.16   A uniform surface current with $\mathbf{J}_s = J_o\,\mathbf{a}_y$ coincides with the $z = 0$ plane (Fig. 5.59). Show that symmetry considerations can predict $\mathbf{H} = H_x(z)\,\mathbf{a}_x$ for $z > 0$ and $\mathbf{H} = -H_x(|z|)\,\mathbf{a}_x$ for $z < 0$. Find **H** everywhere.

5.17   A uniform volume current $\mathbf{J} = J_o\,\mathbf{a}_y$ [A/m^2] forms an infinite slab, defined by $-d \le z \le d$ and $-\infty < (x, y) < \infty$, as shown in Fig. 5.60. Using a general

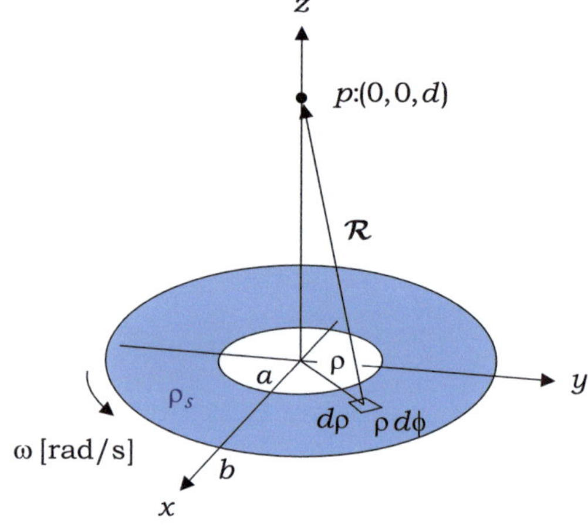

Fig. 5.58 Annulus with uniform surface charge in rotation (Problem 5.14)

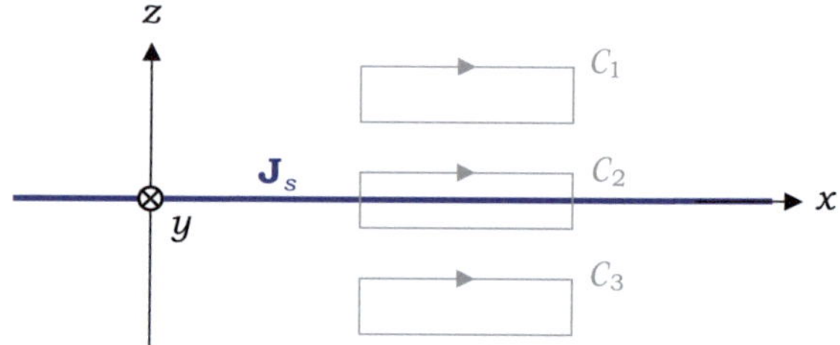

Fig. 5.59 An infinite current sheet and suggested Amperian paths (Problem 5.16)

solution, $\mathbf{H} = H_x(z)\,\mathbf{a}_x$ for $z > 0$ and $\mathbf{H} = -H_x(|z|)\,\mathbf{a}_x$ for $z < 0$, find $\mathbf{H}$ everywhere.

Magnetic field, magnetic flux density, and unique solution

5.18 Can the following vector fields possibly represent the magnetic fields in free space?

(a) $\mathbf{K} = \frac{1}{x^2+y^2}(-y\,\mathbf{a}_x + x\,\mathbf{a}_y)$,

(b) $\mathbf{L} = \dfrac{1}{\rho\sqrt{\rho^2+1}}\,\mathbf{a}_\phi$,

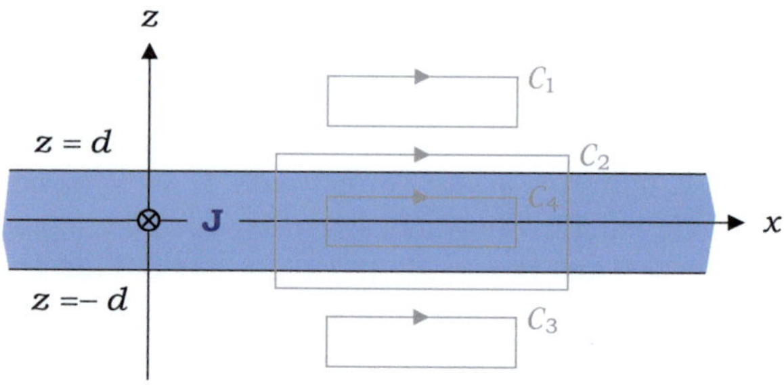

Fig. 5.60 Uniform volume current in the form of an infinite slab and suggested Amperian paths (Problem 5.17)

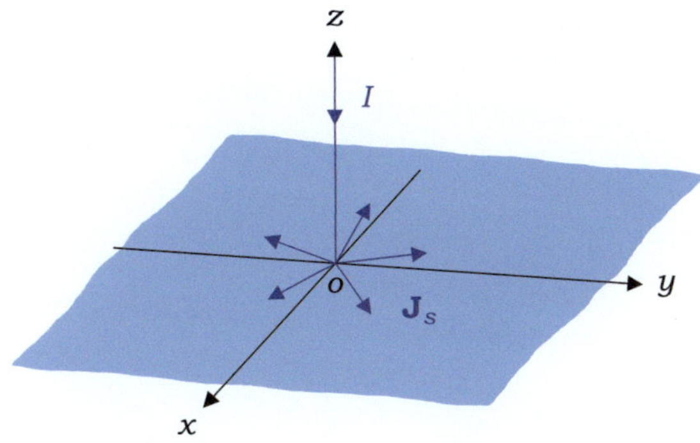

Fig. 5.61 A filamentary current becomes a surface current (Problem 5.21)

 (c) $\mathbf{N} = \frac{\sin\theta}{R^2}\,\mathbf{a}_\phi$, and

 (d) $\mathbf{O} = \frac{1}{R^3}(2\cos\theta\,\mathbf{a}_R + \sin\theta\,\mathbf{a}_\theta)$.

5.19   A long straight wire of radius a is oriented along the z-direction and carries a steady current, I, uniformly distributed over the cross section. (a) Determine the current density $\mathbf{J}$ and then $\mathbf{H}$ using Ampere's circuital law. (b) Show $\nabla \times \mathbf{H} = \mathbf{J}$ in the conductor.

5.20   For the toroidal coil, as shown in Fig. 5.11, starting with a trial solution $\mathbf{H} = H_\phi(\rho,\phi,z)\,\mathbf{a}_\phi$ everywhere, (a) show that H_ϕ is independent of ϕ and z using two fundamental relations for $\mathbf{H}$, and (b) find $\mathbf{H}$ everywhere using Ampere's circuital law and the boundary condition.

5.21   The filamentary current I flowing in the $-z$-direction becomes the surface current $\mathbf{J}_s = J_o(\rho)\,\mathbf{a}_\rho$ in the $z = 0$ plane, as shown in Fig. 5.61. Taking $\mathbf{H} = -I\,\mathbf{a}_\phi/(2\pi\rho)$ above the sheet as the trial solution, (a) verify that this is a unique solution, and (b) find $\mathbf{H}$ below the sheet.

Fig. 5.62 Conductive can connected to two straight wires (Problem 5.22)

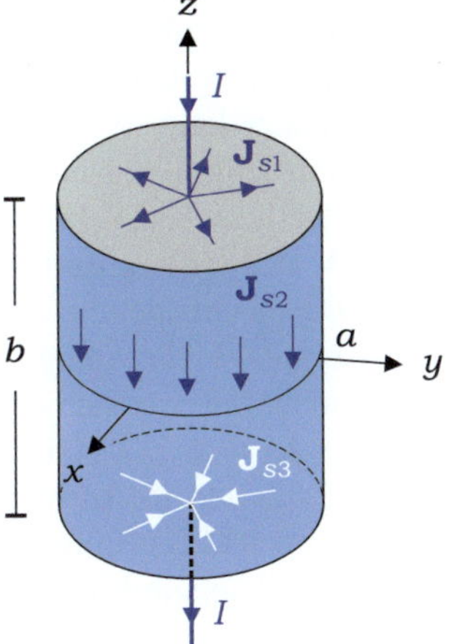

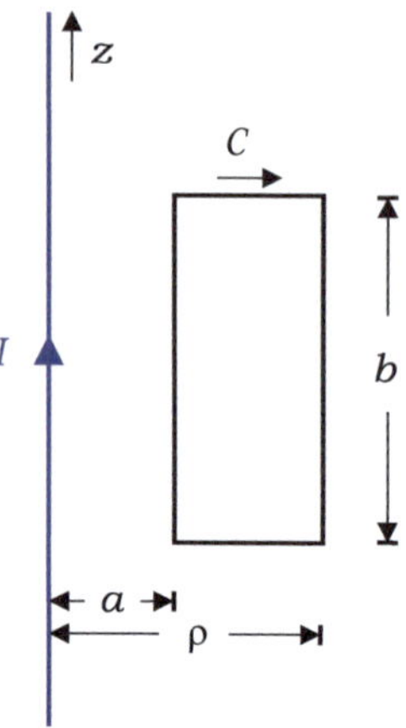

Fig. 5.63 Infinitely long
line current (Problem 5.24)

5.22   A closed can is connected to two very long straight wires, as shown in Fig. 5.62.
Line current I becomes the surface current on the can. Starting with the trial
solution $\mathbf{H} = -I\,\mathbf{a}_\phi/(2\pi\rho)$ in the exterior,

   (a)   find the surface current densities $\mathbf{J}_{s1}(\rho)$, $\mathbf{J}_{s2}$, and $\mathbf{J}_{s3}(\rho)$,
   (b)   show that $\mathbf{H}$ satisfies the boundary condition on the can surface, and
   (c)   check whether $\mathbf{H}$ is the only solution.

Vector magnetic potential

5.23   A straight wire extending between $z = -a$ and $z = a$ carries a steady current, I,
flowing in the $+z$-direction. Find the vector magnetic potential $\mathbf{A}$ in cylindrical
coordinates and then $\mathbf{H}$ from $\mathbf{A}$ outside the wire.

5.24   For an infinitely long line current, I, find $\mathbf{A}$ by taking the line integral of $\mathbf{A}$
around a rectangular loop and setting $\mathbf{A}(\rho = a) = 0$ (Fig. 5.63).

5.25   When steady current I flows along the z-axis, use its $\mathbf{B}$ in solving the equation
$\mathbf{B} = \nabla \times \mathbf{A}$ for $\mathbf{A}$.

5.26   A very long solenoid with its axis coincident with the z-axis has N turns per
unit length and is closely wound on an air core of radius a. If it carries a
steady current I, (a) show that symmetry considerations lead to a general form
$\mathbf{A} = A_\phi(\rho)\,\mathbf{a}_\phi$, and (b) find $\mathbf{A}$ everywhere by using $\mathbf{B} = \mu_0 N I\,\mathbf{a}_z$ inside and
$\mathbf{B} = 0$ outside.

5.27   A sphere with radius a and uniform surface charge density ρ_s [C/m^2] rotates
about the z-axis at ω [rad/s] (Fig. 5.64). A trial solution for $\mathbf{A}$ can be formed
as $\mathbf{A}_2 = \mathbf{a}_\phi c_2 R^{-2}\sin\theta$ (outside) and $\mathbf{A}_1 = \mathbf{a}_\phi c_1 R \sin\theta$ (inside) based on Eq.
(5.62) and the boundary condition for $\mathbf{A}$. (a) Find $\mathbf{A}$ and $\mathbf{B}$ everywhere, (b)
show that $\mathbf{B}$ is uniform inside the sphere, and (c) show that the current density
on the sphere is proportional to $\mathbf{B} \times \mathbf{a}_n$ with unit surface normal $\mathbf{a}_n$.

Fig. 5.64 Sphere with ρ_s rotates about the z-axis (Problems 5.27 and 5.48)

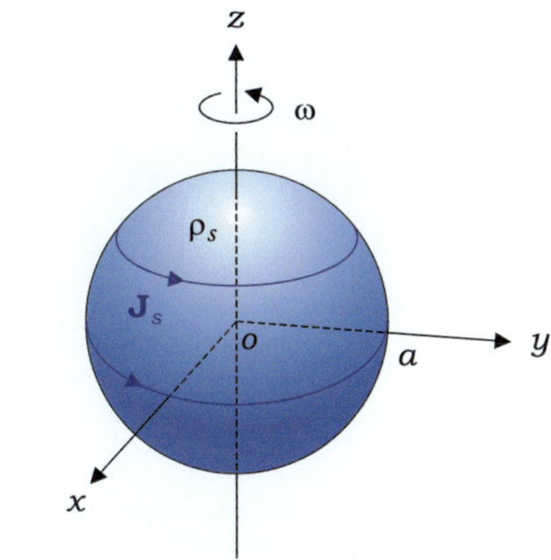

Fig. 5.65 Coaxial cable filled with two dissimilar materials (Problems 5.30 and 5.43)

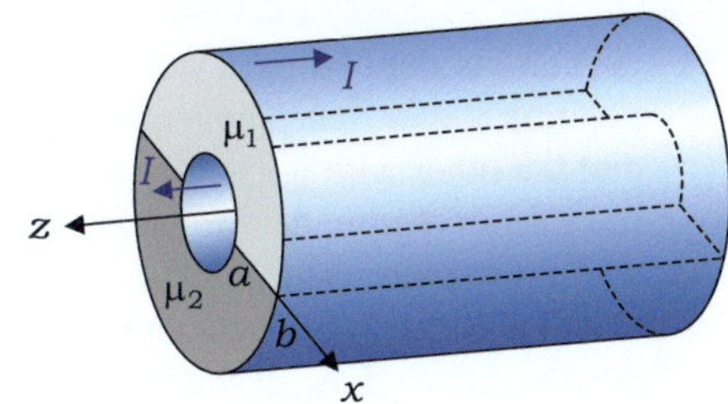

Magnetization and permeability

5.28 When the internal field is $B = 1.2\,[\text{T}]$ in a block of nickel with molar mass 58.69 [g/mol], density 8.91 [g/cm^3], and $\mu_r = 250$, determine the magnitudes of (a) magnetization, and (b) magnetic dipole moment of each atom.

5.29 A large magnetic slab with finite thickness is placed on the xy-plane. The magnetic field is $\mathbf{H} = 1.02 \times 10^5 \mathbf{a}_z\,[\text{A/m}]$ immediately below the slab, and the magnetization is $\mathbf{M} = 10^5 \mathbf{a}_z\,[\text{A/m}]$ inside. Determine μ of the material.

5.30 The space between the two long and thin coaxial cylinders is filled with two dissimilar materials (μ_1 for $0 < \phi < \pi$ and μ_2 for $\pi < \phi < 2\pi$) (Fig. 5.65). When current I flows in opposite directions in the cylinders, starting with a trial solution, $\mathbf{B} = B_\phi \mathbf{a}_\phi$ in the gap and $\mathbf{B} = 0$ for $\rho < a$, determine

(a) $\mathbf{B}$ and $\mathbf{H}$ in the region between the cylinders,

(b) $\mathbf{J}_s$ and $\mathbf{J}_{ms}$ at $\rho = a$, and

(c) verify that $\mathbf{B}$ is a unique solution.

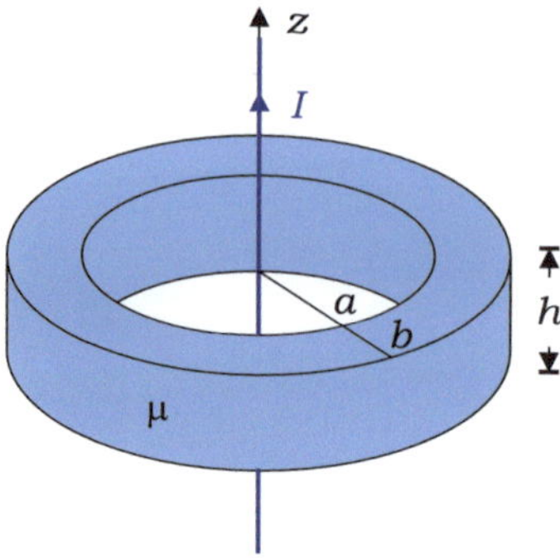

Fig. 5.66 Filamentary current along the axis of a toroid (Problem 5.33)

Magnetic boundary condition

5.31   A plane is defined by $\mathbf{k} \cdot \mathbf{r} = 1$, where $\mathbf{k} = 2\mathbf{a}_x + 2\mathbf{a}_y + \mathbf{a}_z$, and $\mathbf{r}$ is the position vector. It separates two materials: one with $\mu_1 = 10\,\mu_0$ for $\mathbf{k} \cdot \mathbf{r} > 1$, and the other with $\mu_2 = 20\,\mu_0$ for $\mathbf{k} \cdot \mathbf{r} < 1$. If $\mathbf{H}_1 = 12\mathbf{a}_x + 24\mathbf{a}_y + 21\mathbf{a}_z$ in the former region, determine $\mathbf{H}_2$ in the latter region.

5.32   The $z = 0$ plane is the interface between a magnetic material occupying region $z \leq 0$ and the free space above, and carries a uniform magnetization surface current $\mathbf{J}_{ms} = J_o\mathbf{a}_x$. If the magnetic flux density in the material is $\mathbf{B} = B_o\mathbf{a}_y$, determine $\mathbf{B}$ in free space.

5.33   In the presence of a steady current I flowing along the z-axis, a toroid with a rectangular cross section and permeability μ surrounds the line current, as shown in Fig. 5.66. Determine $\mathbf{B}$ everywhere.

5.34   An infinite magnetic slab with thickness d and permeability μ lies on the $z = 0$ plane, as shown in Fig. 5.67. When $\mathbf{B}_1$ makes an angle θ_1 with the z-axis, express the following quantities in terms of B_1, θ_1, and μ:

(a) B_2 and θ_2,
(b) B_3 and θ_3, and
(c) $\mathbf{J}_{ms}$ in the $z = 0$ and $z = d$ planes.

Inductance and magnetic energy

5.35   A toroidal coil has N turns closely wound around the toroid with a rectangular cross section and permeability μ (Fig. 5.68). Find the inductance.

5.36   A toroidal coil has N turns wound closely around two adjoining toroids of the same shape but with different permeabilities, as shown in Fig. 5.69. Find the inductance.

5.37   A toroidal coil has N turns wound closely around the two halves of the toroid with different permeabilities (Fig. 5.70). Ignoring the edge effects in the gap, which is infinitesimal, find (a) the magnetic energy of the coil, and (b) the inductance.

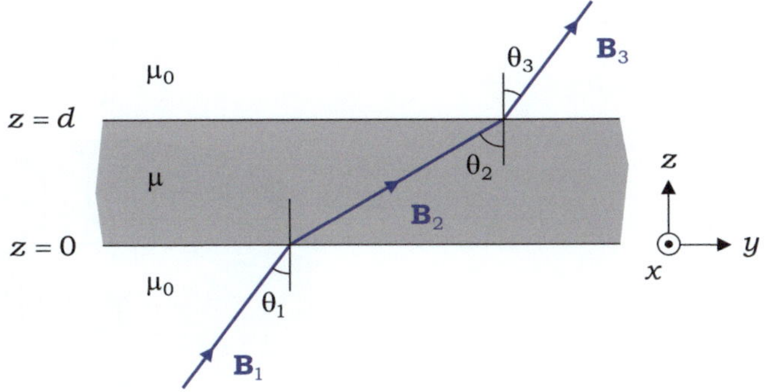

Fig. 5.67 Infinite magnetic slab lying on the $z = 0$ plane (Problem 5.34)

Fig. 5.68 Toroidal coil
(Problem 5.35)

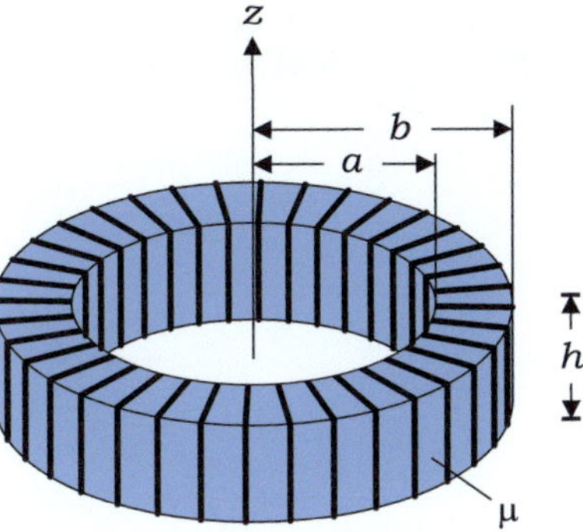

Fig. 5.69 Toroidal coil
around two adjoining toroids
(Problem 5.36)

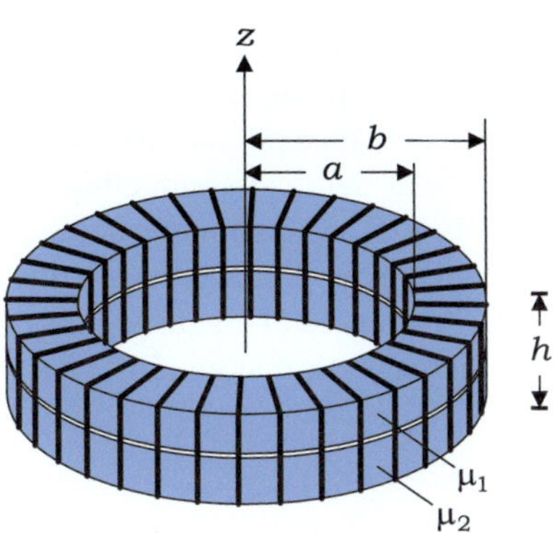

5.38   A solenoid has N turns per unit length and is tightly wound around a cylinder
made of two different materials, as shown in Fig. 5.71. When it carries a steady
current I, ignoring the edge effect, find

(a)   **B** and **H** in the two core regions,
(b)   the magnetic energy to determine the inductance per unit length, and
(c)   the flux linkage to determine the inductance per unit length.

Fig. 5.70 Toroidal coil
around two halves of a toroid
(Problem 5.37)

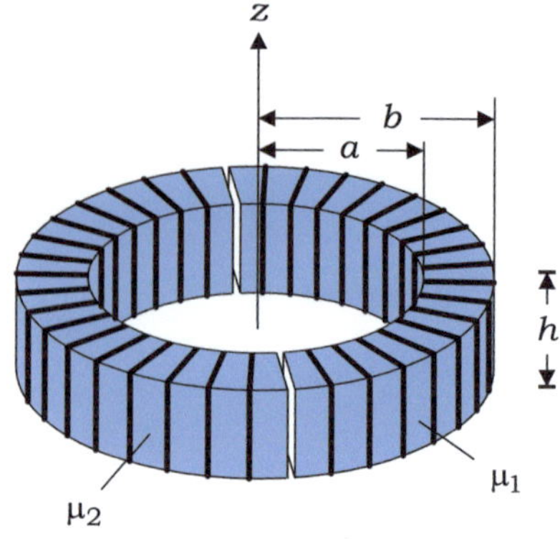

Fig. 5.71 Solenoidal coil
around coaxial cylinders
(Problem 5.38)

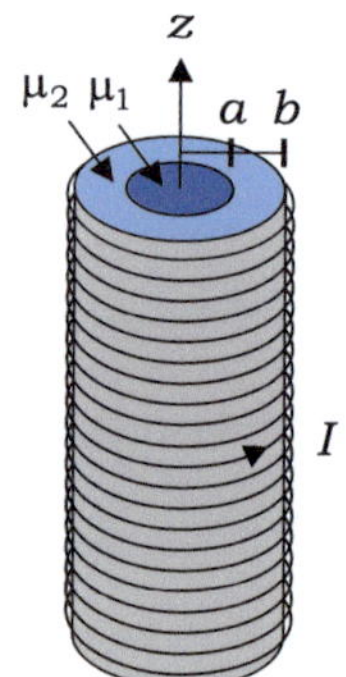

5.39   The space between the two thin coaxial cylinders is filled with two different
       magnetic materials, as shown in Fig. 5.72. Ignoring the edge effect, find the
       inductance per unit length.

5.40   Two identical conducting loops are arranged along the z-axis, as shown in
       Fig. 5.73. Assuming $b \gg a$, and using the vector magnetic potential of a
       magnetic dipole, show that the mutual inductance is $M_{12} = \pi\mu_0 a^4/2b^3$.

5.41   A solenoid of length ℓ has n_T turns closely wound around an air core and stores
       a magnetic energy of W_m because of current I. Find the magnetic energy stored

Fig. 5.72 Coaxial cable
filled with two dissimilar
materials (Problem 5.39)

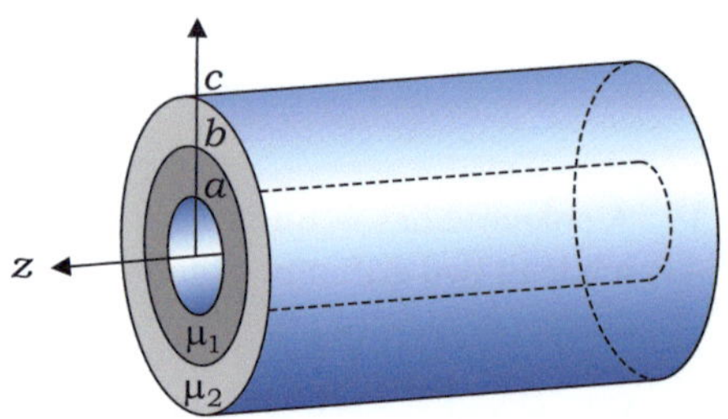

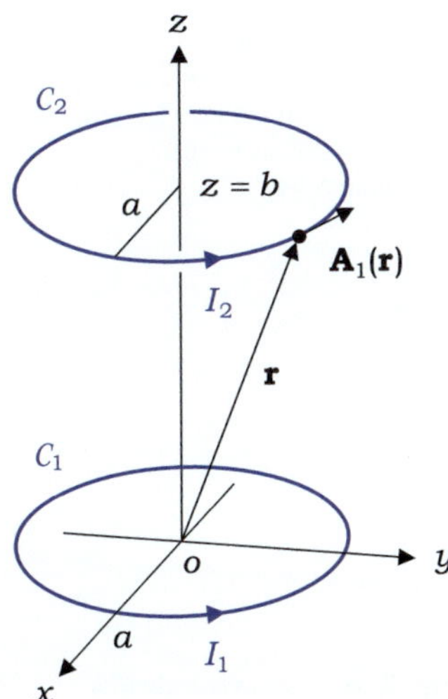

Fig. 5.73 Two identical current-carrying loops (Problems 5.40 and 5.46)

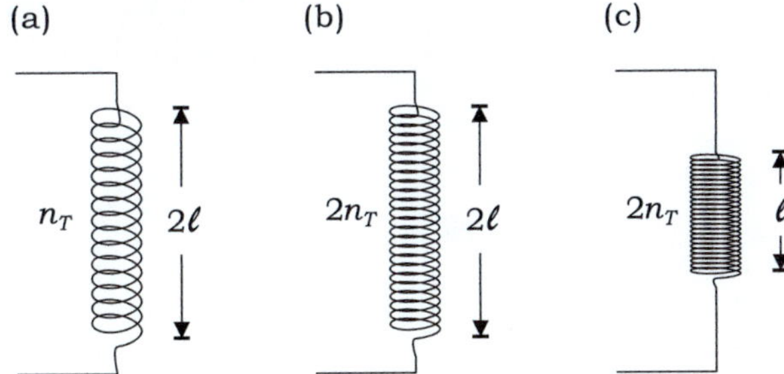

Fig. 5.74 Solenoids with the same radius and current (Problem 5.41)

in similar solenoids with the same radius and current, but different ℓ and/or n_T (Fig. 5.74), ignoring the stray fields.

5.42 The coaxial cable, shown in Fig. 5.10 has radii $a = 2\,[\text{cm}]$, $b = 4\,[\text{cm}]$, and $c = 5\,[\text{cm}]$. Assuming steady current I to be uniformly distributed in the cross section of each conductor, find the internal inductance of the outer conductor per unit length to compare it with that of the inner conductor.

5.43 If the coaxial cable shown in Fig. 5.65 carries a steady current I, find (a) the stored energy, and (b) the inductance per unit length.

Magnetic force and torque

5.44 Referring to the two-conductor system shown in Fig. 5.75. Determine the magnetic force exerted on (a) the rectangular loop, and (b) the right isosceles triangle.

5.45 A long solenoid has N turns per unit length and is closely wound around two iron rods separated by a small air gap ($d \ll a$), as shown in Fig. 5.76. Ignoring the edge effect, determine the force between the two rods.

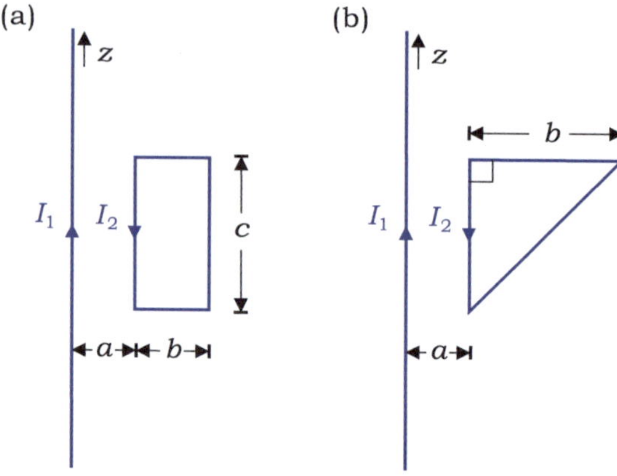

Fig. 5.75 A long filamentary current and a closed loop in close proximity to each other (Problem 5.44)

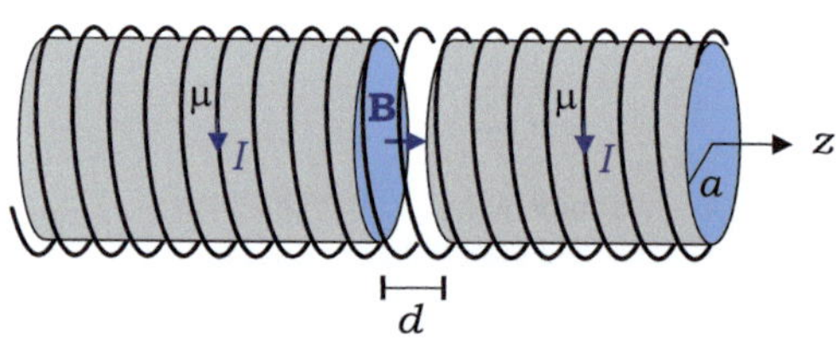

Fig. 5.76 Two iron rods in a solenoid (Problem 5.45)

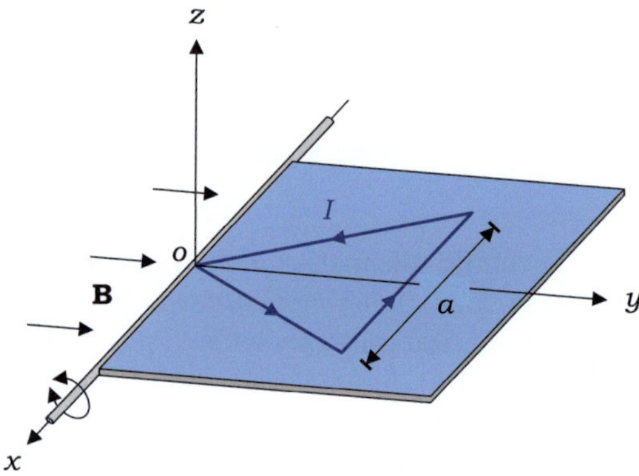

Fig. 5.77 Circuit board with trace placed in **B** (Problem 5.51)

5.46 For the two-loop system, shown in Fig. 5.73, assume $b \gg a$ to determine $\mathbf{F}_m$ on $\mathcal{C}_2$ due to I_1 using

    (a) Eq. (5.129b), and

    (b) the method of virtual displacement, together with M_{12}.

5.47 If the current-carrying loop, as shown in Fig. 5.52, resides in a uniform field $\mathbf{B} = B_2\,\mathbf{a}_y + B_3\,\mathbf{a}_z$, determine the torque on the loop about the origin.

5.48 For a sphere with uniform surface charge density ρ_s [C/m^2] rotating at ω [rad/s], as shown in Fig. 5.64, determine the magnetic dipole moment.

5.49 A finite solenoid with radius a and N turns is centered along the z-axis. When it carries a steady current I in the ϕ-direction in a uniform external field $\mathbf{B} = B_o\,\mathbf{a}_x$, determine the torque at the center.

5.50 Two magnetic dipoles with the same magnetic moment $\mathbf{m} = m_o\,\mathbf{a}_z$ are positioned at the origin and point $p{:}(1, \sqrt{2}, 1)$. Determine the torque on the magnetic dipole positioned at p.

5.51 A printed circuit board is constrained to pivot about the x-axis, as shown in Fig. 5.77. When steady current I flows in the trace of the form of an equilateral triangle placed in uniform $\mathbf{B} = B_o\,\mathbf{a}_y$, determine the torque.

Chapter 6
Time-Varying Fields and Maxwell's Equations

Until now, we have devoted ourselves to static electric and magnetic fields that are time invariant. In summary, we observed that the electric field and electric flux density of the charge distribution are related by the constitutive relation $\mathbf{D} = \varepsilon\mathbf{E}$, where permittivity ε is a measure of the electrical properties of the material. We learned that the irrotational nature of $\mathbf{E}$ ($\nabla \times \mathbf{E} = 0$) originates from the principle of energy conservation and that Gauss's law ($\nabla \cdot \mathbf{D} = \rho_v$) is a direct consequence of Coulomb's law. The divergence and curl equations constitute two fundamental relations for static electric fields, in the sense that, along with proper boundary conditions, they enable us to uniquely determine the electric field in a given region. Similarly, the magnetic field and magnetic flux density of the steady current distribution are related by $\mathbf{B} = \mu\mathbf{H}$, where the permeability μ is a measure of the magnetic properties of the material. The solenoidal nature of $\mathbf{B}$ ($\nabla \cdot \mathbf{B} = 0$) directly follows from the closed nature of magnetic flux lines, and Ampere's circuital law ($\nabla \times \mathbf{H} = \mathbf{J}$) is based on the Biot-Savart law. These divergence and curl equations constitute two fundamental relations for static magnetic fields.

There is no mutual relationship between the static electric and magnetic fields, although they may coexist in a conductor in such a way that a static electric field generates a steady current that in turn induces a static magnetic field around it. However, static magnetic fields do not generate static electric fields. Under time-varying conditions, the electric and magnetic fields are coupled to form an electromagnetic wave that propagates in a given medium.

In this chapter, we focus on the time-varying electric and magnetic fields. The concept of wave motion is discussed in Chap. 7 and the propagation of electromagnetic waves in various media is discussed in Chap. 8. To avoid confusion, bold script letters such as $\mathscr{E}$ and $\mathscr{H}$ denote the time-varying electric and magnetic fields, respectively. Other time-varying quantities, such as time-varying current density (vector) and volume charge density (scalar), are denoted by bold italic and italic letters $\boldsymbol{J}$ and ρ_v, respectively.

Time-varying electromagnetic fields differ from static electric and magnetic fields in several important respects, apart from the fact that they vary with time. First,

Y. H. Lee, *Introduction to Engineering Electromagnetics*,
https://doi.org/10.1007/978-3-031-28659-9_6

time-varying electromagnetic fields are generated by accelerated charges or electric currents, which vary over time. Second, time-varying electric and magnetic fields are coupled such that a time-varying electric field induces a time-varying magnetic field, and vice versa. In the same way that a static field is governed by its divergence and curl, time-varying electric and magnetic fields are governed by their divergence and curl. Gauss's laws for electricity and magnetism remain valid under time-varying conditions: $\nabla \cdot \mathscr{D} = \rho_v$ and $\nabla \cdot \mathscr{B} = 0$. However, the two curl equations for $\mathscr{E}$ and $\mathscr{H}$ must be modified to conform to Faraday's electromagnetic induction and incorporate the concept of the displacement current density introduced by Maxwell. The four equations for the curl of $\mathscr{E}$, curl of $\mathscr{H}$, divergence of $\mathscr{D}$, and divergence of $\mathscr{B}$ are **Maxwell's equations**. These are the fundamental relations that time-varying electromagnetic fields must satisfy at all times.

6.1 Faraday's Law

Michael Faraday observed that an **electromotive force** (*emf*) is induced in a wire loop when a permanent magnet moves near the loop or when the loop moves near the magnet. The electromotive force is the voltage induced in the loop. However, this was named inappropriately because it had dimensions of energy. Therefore, we prefer the *emf*. Faraday's law of electromagnetic induction states that *the induced emf in a wire loop is equal to the negative time rate of increase of the magnetic flux linking the loop.*

The *emf* induced in a loop of N turns is expressed mathematically as

$$\boxed{emf = -\frac{d\Lambda}{dt} = -N\frac{d\Phi}{dt}} \quad \text{[V]} \tag{6.1}$$

where Λ is the magnetic flux linkage with the loop, and Φ is the magnetic flux enclosed by a single turn. Furthermore, *emf* is defined as the integral of the induced electric field along the loop. That is,

$$\boxed{emf = \oint_C \mathscr{E} \cdot d\mathbf{l}} \quad \text{[V]} \tag{6.2}$$

where $\mathscr{E}$ is the induced electric field, and path C represents the wire loop. Although *emf* and $\mathscr{E}$ in Eq. (6.2) are induced by a time-varying magnetic flux, they may or may not vary over time. Unlike static electric fields, which are irrotational and conservative, the closed-line integral of $\mathscr{E}$ is nonzero. Thus, $\mathscr{E}$ is a nonconservative field and cannot be obtained from the negative gradient of the electric potential.

Combining Eqs. (6.1) and (6.2) leads to a more general expression for the induced *emf* such that

$$emf = \oint_C \mathscr{E} \cdot d\mathbf{l} = -\frac{d}{dt} \int_S \mathscr{B} \cdot d\mathbf{s} \qquad (6.3)$$

If the loop has many turns, the closed-line integral should be conducted around each turn of the loop, and the surface integral should be performed over all surfaces enclosed by all the turns of the loop. The directions of $d\mathbf{l}$ along C and $d\mathbf{s}$ on S are governed by the right-hand rule in which the right thumb points in the direction of $d\mathbf{s}$ when the four fingers rotate in the direction of $d\mathbf{l}$.

The negative sign in Eq. (6.3) is compatible with Lenz's law, which states that *emf* is generated in such a way that it induces a current in the loop and thus produces a magnetic flux to oppose the change in the magnetic flux linking the loop. Surprisingly, Eq. (6.3) is valid even for an imaginary loop.

The electromagnetic induction experiment is illustrated in Fig. 6.1. In Fig. 6.1a, the permanent magnet moves toward the loop with open ends so that the magnetic flux linking the loop increases with time. According to Faraday's and Lentz's laws, electric field $\mathscr{E}$ and current i are induced in the loop in the clockwise direction. Because of the open ends, the current accumulates positive charges at terminal 1 and negative charges at terminal 2, resulting in a terminal voltage v_{1-2} that balances the *emf* and prevents further current flow in the loop. If resistance R is connected across the terminals, the ohmic current i flows from terminal 1 to terminal 2 through the resistor, which constitutes a conduction current flowing clockwise in the loop.

The polarity of the induced *emf* often confuses the electromagnetic induction. Note that the polarity of the *emf* is determined solely by the positive direction of C. A positive *emf* indicates that the induced $\mathscr{E}$ is directed along $d\mathbf{l}$ or the positive direction of C. For example, if the positive direction of loop C in Fig. 6.1a is taken

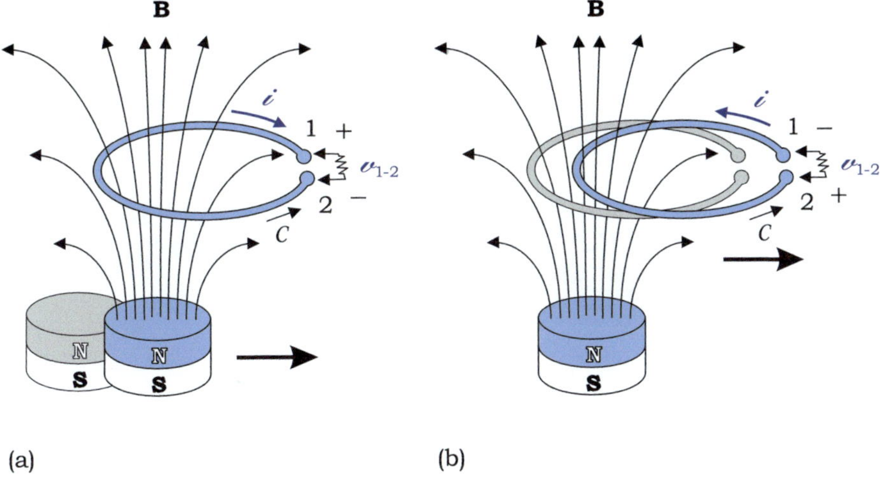

Fig. 6.1 Electromagnetic induction. The induced *emf* gives rise to current i, and thus, terminal voltage v_{1-2}: **a** $v_{1-2} > 0$ and *emf* < 0, **b** $v_{1-2} < 0$ and *emf* > 0

in the opposite direction, *emf* has opposite polarity. However, there is no change in the terminal voltage v_{1-2}.

Exercise 6.1
In a uniform and constant $\partial\mathscr{B}/\partial t$, a wire of finite length is formed into (a) a single loop, and (b) a smaller loop with two turns. Compare two *emf*s.

Ans. The *emf* in (a) is twice as large as that in (b).

6.1.1 Transformer emf

If an incremental *emf* is induced by an incremental magnetic flux $\Delta\Phi = \mathscr{B} \cdot \Delta\mathbf{s}$, in accordance with Eq. (6.3), there are three cases in which the time derivative of $\Delta\Phi$ is nonzero:

(1) The loop is stationary in a time-varying field: $d(\Delta\Phi)/dt = (d\mathscr{B}/dt) \cdot \Delta\mathbf{s}$.
(2) The loop moves in a static magnetic field: $d(\Delta\Phi)/dt = \mathscr{B} \cdot d(\Delta\mathbf{s})/dt$.
(3) The loop moves in a time-varying field: $d(\Delta\Phi)/dt = d(\mathscr{B} \cdot \Delta\mathbf{s})/dt$.

The **transformer emf** is the one that is induced in a stationary loop placed in a time-varying magnetic field, as in Case (1). Because $d\mathbf{s}$ is independent of time, the transformer *emf* can be expressed as

$$\boxed{emf_t = \oint_C \mathscr{E} \cdot d\mathbf{l} = -\int_S \frac{\partial\mathscr{B}}{\partial t} \cdot d\mathbf{s}} \quad [\text{V}] \tag{6.4}$$

where subscript t denotes the transformer *emf*. The time derivative is moved inside the integral sign and changed to a partial derivative. Applying Stokes's theorem to the middle term in Eq. (6.4) leads to

$$\int_S (\nabla \times \mathscr{E}) \cdot d\mathbf{s} = -\int_S \frac{\partial\mathscr{B}}{\partial t} \cdot d\mathbf{s} \tag{6.5}$$

Here, surface S can be arbitrarily selected only if it is bounded by loop C. Therefore, the two integrands in Eq. (6.5) should be the same at every point on S. This leads to the **point form of Faraday's law**, that is,

$$\boxed{\nabla \times \mathscr{E} = -\frac{\partial\mathscr{B}}{\partial t}} \tag{6.6}$$

As we will see later, Eq. (6.6) is a member of Maxwell's equations.

Example 6.1 A circular loop of radius a is stationary in the xy-plane in the presence of a magnetic field with $\mathscr{B}(t) = \mathbf{a}_z B_o \cos(\omega t)$, as shown in Fig. 6.2. Determine the *emf* induced in the loop.

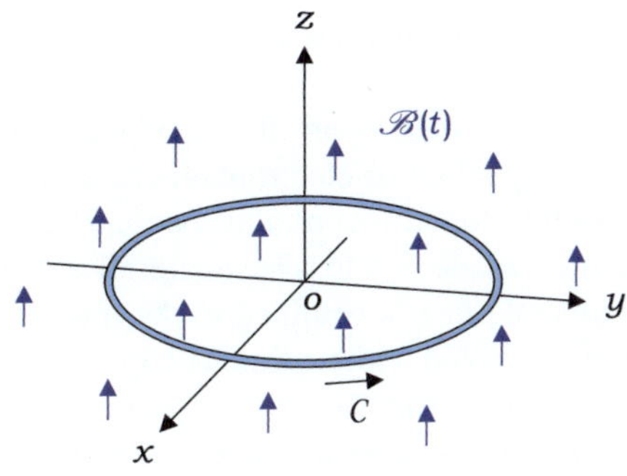

Fig. 6.2 Stationary circular loop in a time-varying magnetic field

Solution

The counterclockwise direction is taken as the positive direction of the loop. Then, $d\mathbf{s}$ should be directed along the $+z$-direction in accordance with the right-hand rule. The magnetic flux linking the loop is

$$\Phi = \int_{S} \mathscr{B} \cdot d\mathbf{s} = \int_{\phi=0}^{\phi=2\pi} \int_{\rho=0}^{\rho=a} \mathbf{a}_z B_o \cos(\omega t) \cdot (\mathbf{a}_z \rho \, d\rho \, d\phi)$$
$$= \pi B_o a^2 \cos(\omega t)$$

Thus, the transformer *emf* induced in the loop is

$$emf_t = -\frac{d\Phi}{dt} = \pi B_o a^2 \omega \sin(\omega t) \tag{6.7}$$

For $emf_t > 0$, the induced electric field is directed along the positive direction of C or in the counterclockwise direction. The flux linkage and transformer *emf* are normalized to unity and plotted against t, as shown in Fig. 6.3. The shaded area represents the time interval in which Φ decreases with time, and thus emf_t is positive. The induced *emf* always lags behind the flux linkage by $90°$ in time phase.

Exercise 6.2

When a circular loop of radius a and a square loop of side b are separated in $\mathscr{B}(t) = \mathbf{a}_z B_o \cos \omega t$, determine a/b for the same (a) *emf*, and (b) $\mathscr{E}$ in the loops.

Ans. (a) 0.56, (b) 0.5.

Fig. 6.3 Normalized flux linkage and transformer *emf* plotted against time

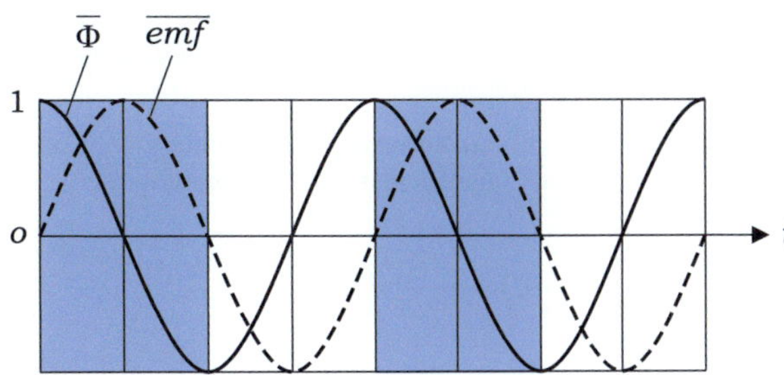

6.1.1.1   Ideal Transformer

A transformer is an alternating current (AC) device that operates according to Faraday's law of electromagnetic induction. The transformer consists of two coils wound around a common magnetic core, as shown in Fig. 6.4. An ideal magnetic core is made of a lossless material with infinite permeability ($\mu = \infty$) such that the magnetic flux is confined to the interior of the core, and the two coils are magnetically coupled with no flux leakage. Transformers can transform voltage, current, and impedance from the primary circuit to the secondary circuit.

Let us consider the case in which the primary coil with N_1 turns is connected to the ac-voltage source $v_1(t)$ and the secondary coil with N_2 turns is connected to the load resistance R_L. Voltage v_1 is responsible for the magnetic flux Φ established in the magnetic core, which is then linked to both the primary and secondary coils. According to Faraday's law, emfs are induced across the two coils as follows:

$$emf_1 = -N_1 \frac{d\Phi}{dt} \equiv -v_1 \tag{6.8a}$$

$$emf_2 = -N_2 \frac{d\Phi}{dt} \equiv -v_2 \tag{6.8b}$$

(a)

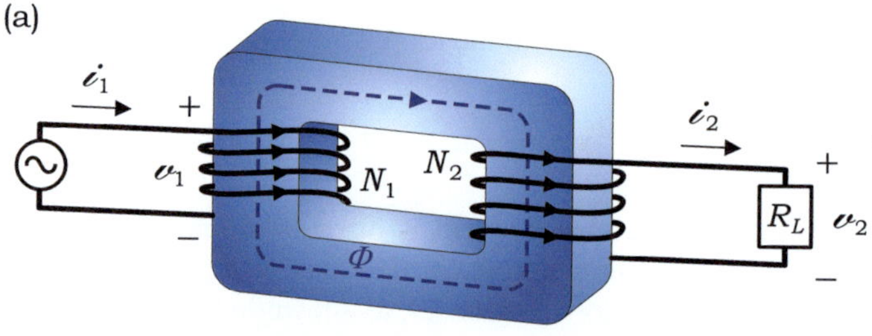

(b)

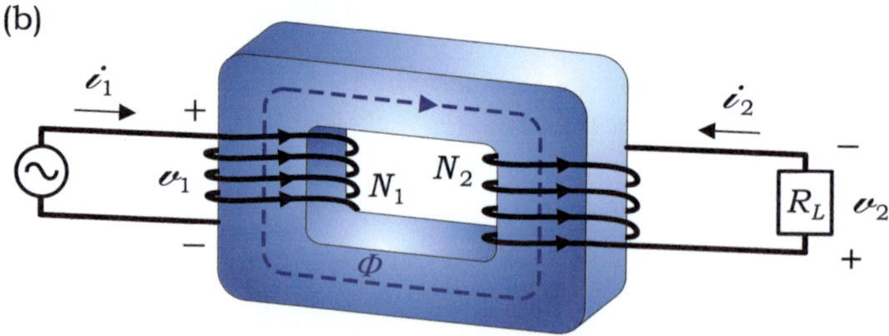

Fig. 6.4   Two identical transformers. The secondary coils are wound in opposite directions, and thus v_2 and i_2 in **b** are opposite in polarity to those in **a**

where v_1 and v_2 are the terminal voltages of the primary and secondary coils, respectively. It should be noted that the polarity of the terminal voltage is opposite to that of *emf*. For example, as illustrated in Fig. 6.4a, the arrow in the second coil indicates the direction of the electric field due to the *emf* in the coil, which is generally directed upward. Meanwhile, the electric field in the load caused by the terminal voltage is directed downward. In addition, the terminal voltage of the primary coil due to emf_1 should be equal to the source voltage v_1. Otherwise, the discrepancy causes Φ to increase or decrease until balance is reached. Simultaneously, emf_2 provides a load voltage v_2 across R_L and gives rise to a current i_2 in the secondary circuit, following Ohm's law.

The power dissipated in the primary circuit should be equal to that in the secondary circuit, based on the principle of energy conservation, such that

$$P_1 = v_1 i_1$$
$$= P_2 = v_2 i_2 \tag{6.9}$$

Combining Eqs. (6.8a), (6.8b) and (6.9) leads to

$$\frac{v_1}{v_2} = \frac{i_2}{i_1} = \frac{N_1}{N_2} \equiv \alpha \tag{6.10}$$

which is known as ***turns ratio***. The voltage ratio, v_1/v_2, is proportional to the turns ratio, whereas the current ratio, i_1/i_2, is inversely proportional.

A transformer can transform not only the voltage and current, but also the impedance of the circuit. In the primary circuit, the transformer can be regarded as an equivalent load with impedance Z_1, that is,

$$Z_1 = \frac{v_1}{i_1} = \frac{(N_1/N_2)\, v_2}{(N_2/N_1)\, i_2} \tag{6.11}$$

With the load impedance, $Z_2 = v_2/i_2$, the effective impedance of the transformer seen by the source in the primary circuit is given by Eq. (6.11) as

$$Z_1 = \left(\frac{N_1}{N_2}\right)^2 Z_2 \tag{6.12}$$

The impedance transformation involves the square of the turns ratio.

Example 6.2 When a current source with $i_1(t) = I_1 \cos(\omega t)$ is connected to the primary coil with an inductance L_1, as shown in Fig. 6.4a, current i_2 flows in the secondary coil with inductance L_2. Find (a) v_1, v_2, and i_2, and (b) the time-averaged powers dissipated in the primary and secondary circuits.

Solution

If the secondary circuit is opened, the primary coil can be considered a simple inductor. From the definition of inductance, that is, $L_1 i_1 = N_1 \Phi_o$, the magnetic flux in the core and the voltage in the primary circuit are given by

$$\Phi_o = \frac{L_1 I_1}{N_1} \cos(\omega t) \tag{6.13a}$$

$$v_o = N_1 \frac{d\Phi_o}{dt} = -L_1 I_1 \omega \sin(\omega t) \tag{6.13b}$$

Subsequently, when the secondary coil is terminated with a load R_L to close the secondary circuit, the magnetic flux Φ_o is linked to the secondary coil. The flux linkage induces voltage v_2 across the load resistance and current i_2 in the secondary circuit such as

$$v_2 = N_2 \frac{d\Phi_o}{dt} = -\frac{N_2}{N_1} L_1 I_1 \omega \sin(\omega t) \tag{6.13c}$$

$$i_2 = \frac{v_2}{R_L} = -\frac{N_2}{N_1} \frac{L_1 I_1 \omega}{R_L} \sin(\omega t) \tag{6.13d}$$

Then, current i_2 generates an additional magnetic flux Φ' in the core in such a way as to oppose the change in Φ_o. That is,

$$\Phi' = \frac{L_2 i_2}{N_2} = -\frac{1}{N_1} \frac{L_1 L_2 I_1 \omega}{R_L} \sin(\omega t) \tag{6.13e}$$

This additional flux induces an *emf* in the primary coil. Thus, the additional terminal voltage is expressed as

$$v' = N_1 \frac{d\Phi'}{dt} = -\frac{L_1 L_2 I_1 \omega^2}{R_L} \cos(\omega t) \tag{6.13f}$$

Because the magnetic flux Φ' opposes the change in Φ_o, the total terminal voltage in the primary circuit is given by

$$v_1 = v_o - v' = -L_1 I_1 \omega \sin(\omega t) + \frac{L_1 L_2 I_1 \omega^2}{R_L} \cos(\omega t) \tag{6.13g}$$

Next, the time-averaged powers dissipated in the primary and secondary circuits can be calculated as

$$P_{av1} = \frac{1}{T} \int i_1 v_1 \, dt = \frac{L_1 L_2 I_1^2 \omega^2}{2 R_L} \tag{6.13h}$$

$$P_{av2} = \frac{1}{T} \int i_2 v_2 \, dt = \left(\frac{N_2}{N_1}\right)^2 \frac{L_1^2 I_1^2 \omega^2}{2R_L} \tag{6.13i}$$

where $i_1(t) = I_1 \cos(\omega t)$, as given in the problem. From $L_2/L_1 = (N_2/N_1)^2$, it is evident that the results of Eqs. (6.13h) and (6.13i) are equal, thereby satisfying the principle of energy conservation.

Exercise 6.3

For the transformer in Example 6.2, (a) find the mutual flux Φ_{12}, and (b) express v_2 in terms of Φ_{12}, which must be the same as that in Eq. (6.13c).

Ans. (a) $\Phi_{12} = (L_1 I_1/N_1) \cos(\omega t)$, (b) $v_2 = N_2 (d\Phi_{12}/dt)$.

6.1.2 Motional emf

A **motional emf** is induced in a conducting loop in motion in a static magnetic field. For example, if a wire loop rotates in a uniform static **B** or moves rectilinearly in a nonuniform static **B**, a motional *emf* is induced in the loop. The motional *emf* is expressed mathematically as

$$\boxed{emf_m = \oint_C \mathscr{E} \cdot d\mathbf{l} = -\frac{d}{dt} \int_S \mathbf{B} \cdot d\mathbf{s}} \quad [\text{V}] \tag{6.14}$$

As an example, we consider a rectangular loop moving into a uniform static field $\mathbf{B} = B_o \mathbf{a}_z$ (for $y \geq y_o$) at constant velocity $\boldsymbol{v} = v_o \mathbf{a}_y$, as shown in Fig. 6.5. At time $0 \leq t \leq y_o/v_o$, a motional *emf* is induced in the loop such that

$$emf_m = -\frac{d}{dt} \int_S \mathbf{B} \cdot d\mathbf{s}$$

$$= -\frac{d}{dt}(B_o x_o v_o t) = -B_o x_o v_o \tag{6.15}$$

where x_o and y_o are the loop side lengths. $d\mathbf{s}$ is assumed to be directed along the $+z$-direction, and thus, the counterclockwise direction is the positive direction of the loop. The negative *emf* in Eq. (6.15) indicates that the rotation of the induced electric field and current is clockwise in the loop.

The motional *emf* is directly related to the magnetic force exerted on conduction electrons that move with the wire. Let us focus on a single conduction electron moving at a velocity $\boldsymbol{v}$ in a static **B** field. From the Lorentz force equation, the magnetic force on the electron is given by

$$\mathbf{F}_m = e\boldsymbol{v} \times \mathbf{B} \tag{6.16}$$

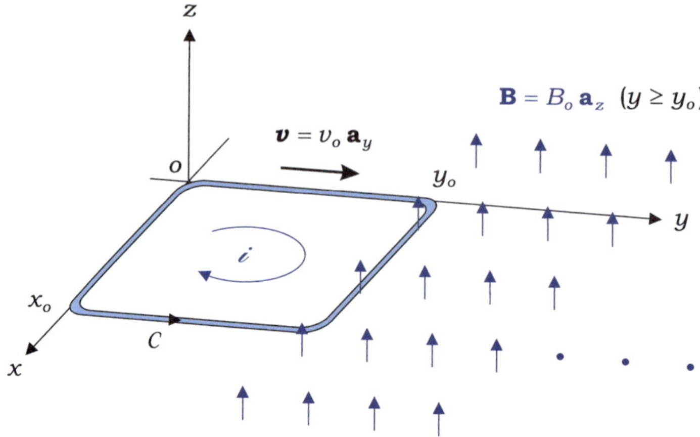

Fig. 6.5 A loop moves into a static magnetic field

where e denotes the electron charge of -1.6×10^{-19} [C]. In the wire, the magnetic force per unit charge is referred to as the ***motional electric field*** and is expressed as

$$\boxed{\mathscr{E}_m = \frac{\boldsymbol{F}_m}{e} = \boldsymbol{v} \times \mathbf{B}} \tag{6.17}$$

By integrating $\mathscr{E}_m$ around the wire loop, the motional *emf* can be expressed as

$$\boxed{emf_m = \oint_C \mathscr{E}_m \cdot d\mathbf{l} = \oint_C (\boldsymbol{v} \times \mathbf{B}) \cdot d\mathbf{l}} \quad [\mathrm{V}] \tag{6.18}$$

where $\boldsymbol{v}$ is the velocity of the wire loop, and $\mathbf{B}$ is the static magnetic flux density.

To see how Eq. (6.18) can be used to obtain motional *emf*, we revisit the rectangular loop shown in Fig. 6.5. As is evident from the figure, only the right-hand side of the loop contributes to emf_m. Specifically, there are no contributions from the top and bottom sides because $\mathscr{E}_m$ is perpendicular to the wire and no contributions from the left side because it is still in the region of zero magnetic field for $0 < t < y_o/v_o$. Thus, from Eq. (6.18), the motional *emf* is

$$
\begin{aligned}
emf_m &= \int_{x=x_o}^{x=0} (\boldsymbol{v} \times \mathbf{B}) \cdot d\mathbf{l} \\
&= \int_{x=x_o}^{x=0} v_o B_o\, \mathbf{a}_x \cdot \mathbf{a}_x\, dx = -B_o x_o v_o
\end{aligned} \tag{6.19}
$$

where $d\mathbf{l} = dx\,\mathbf{a}_x$. The result in Eq. (6.19) is the same as that in Eq. (6.15) obtained by Faraday's law.

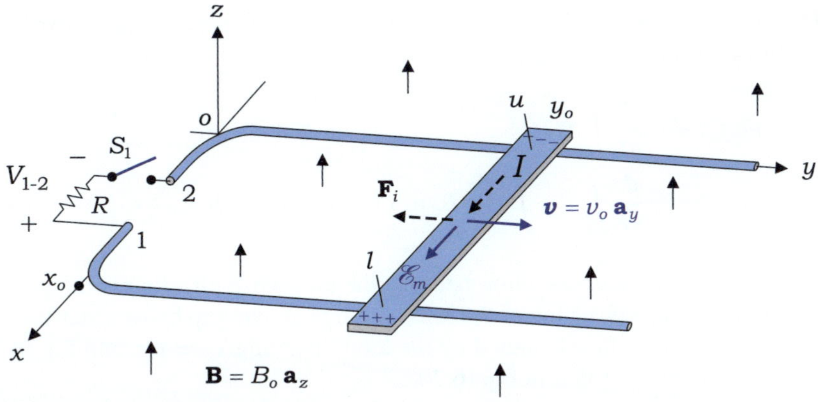

Fig. 6.6 Sliding bar moving over a pair of conducting rails

Next, we consider a classical example of motional *emf*, as illustrated in Fig. 6.6, where a conducting bar is allowed to slide over a pair of conducting rails. Let us examine the case in which the bar moves at a constant velocity $v = v_o\mathbf{a}_y$ in a uniform field $\mathbf{B} = B_o\mathbf{a}_z$ because of the external force applied to the bar. If switch S_1 is open, then the induced motional *emf* appears as the terminal voltage. Using Eq. (6.18), the motional *emf* induced in the sliding bar can be expressed as

$$
\begin{aligned}
emf_m &= \int_u^l (v \times \mathbf{B}) \cdot d\mathbf{l} \\
&= \int_{x=0}^{x=x_o} v_o B_o\,\mathbf{a}_x \cdot \mathbf{a}_x\,dx = v_o B_o x_o
\end{aligned}
\tag{6.20}
$$

where u and l denote the upper and lower contact points of the sliding bar, respectively. The clockwise direction is taken as the positive direction of the closed path, which consists of the sliding bar, rails, and the gap between terminals 1 and 2. The positive *emf* in Eq. (6.20) implies that the induced $\mathscr{E}_m$ is in the $+x$-direction in the bar. Thus, the terminal voltage is given by

$$
V_{1-2} = v_o B_o x_o
\tag{6.21}
$$

A positive value of V_{1-2} indicates that terminal 1 has a higher potential than terminal 2.

In the transient state, the motional electric field in the sliding bar causes a conduction current to flow in the bar in the same direction as the motional electric field, which then causes the accumulation of negative charges at point u and positive charges at point l, until the electric field of the charges balances the motional electric field. Because the rails are equipotential objects, terminal 1 is certainly at a higher potential than terminal 2.

The motional *emf* can also be obtained directly from Faraday's law, as expressed in Eq. (6.14). That is,

$$emf_m = -\frac{d}{dt} \int_S \mathbf{B} \cdot d\mathbf{s}$$

$$= -\frac{d}{dt} \int_{y=0}^{y=y_o} \int_{x=0}^{x=x_o} B_o \, \mathbf{a}_z \cdot (-dx\,dy \, \mathbf{a}_z) = B_o x_o \frac{dy_o}{dt} \qquad (6.22)$$

where $d\mathbf{s}$ points in the $-z$-direction because the clockwise direction is the positive direction of the loop consisting of the rails, sliding bar, and gap between terminals 1 and 2, and S is the surface bounded by the loop. Inserting $y_o = v_o t$ into Eq. (6.22) yields the same result as that in Eq. (6.20).

Next, consider the case in which switch S_1 is closed while the bar moves at the same velocity as before because of an external force. In this case, the motional *emf* induced in the bar causes the current to flow along a closed path comprising the rail, sliding bar, and resistor R. As a sliding bar with a nonzero current moves in a magnetic field, it must experience a magnetic force directed opposite to v, as specified by Eq. (5.129b); thus, the energy is expended when moving the bar. The magnetic force on the bar is

$$\mathbf{F}_i = I \int_C d\mathbf{l} \times \mathbf{B}$$

$$\qquad (6.23)$$

$$= I \int_{x=0}^{x=x_o} B_o(-\mathbf{a}_y)\, dx = -I B_o x_o \, \mathbf{a}_y \quad [\text{N}]$$

where $\mathbf{B} = B_o \, \mathbf{a}_z$, and $d\mathbf{l} = dx \, \mathbf{a}_x$. Here, the clockwise direction is taken as the positive direction of the loop, conforming to the direction of the current.

Because the magnetic force in Eq. (6.23) opposes the motion of the bar, mechanical power is required to move the bar at a constant velocity v, that is,

$$P_{me} = -\mathbf{F}_i \cdot v = I B_o x_o v_o \quad [\text{W}] \qquad (6.24)$$

where $-\mathbf{F}_i$ is the mechanical force required to balance the magnetic force.

Next, using V_{1-2} given in Eq. (6.21), the electric power dissipated in the resistor can be computed as

$$P_{el} = I V_{1-2} = I B_o x_o v_o \quad [\text{W}] \qquad (6.25)$$

The two powers in Eqs. (6.24) and (6.25) are equal, thereby satisfying the principle of energy conservation.

Example 6.3 A rectangular wire loop connected to a battery carries a steady current I_2 in the vicinity of a long filamentary current I_1, as shown in Fig. 6.7.

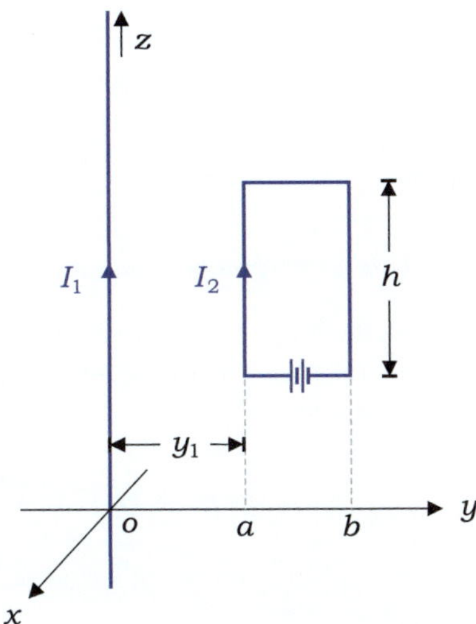

Fig. 6.7 Two current-carrying conductors

(a) Starting with W_m in Eq. (5.115) and using the principle of virtual displacement, find the force on the loop, $\mathbf{F}_m$, and the work done in moving the loop by a distance dy in the $-y$-direction.

(b) Unknown forces cause the loop to move at $v = -v_o\mathbf{a}_y$ for $y_1 = a$. Find the *emf* in the loop using Faraday's law and the power expended by the battery to maintain the current at I_2 during displacement by dy in the $-y$-direction.

(c) Show that the expended power in (b) is consistent with the work done in (a).

(d) For the loop in motion, as described in part (b), invoke the Lorentz force equation to show that the magnetic force on the loop, $\overline{\mathbf{F}}_m$, is directed along v and $\overline{\mathbf{F}}_m = -\mathbf{F}_m$.

Solution

(a) The mutual inductance between two conductors is given by Eq. (5.108):

$$M_{12} = \frac{\mu_0 h}{2\pi} \ln\left(\frac{y_1 + b - a}{y_1}\right) \tag{6.26a}$$

The magnetic energy of the system W_m is given by Eq. (5.115). Because the self-inductances of the two conductors are independent of the space variable y_1, by invoking the principle of virtual displacement, the magnetic force on the loop is obtained as

$$\mathbf{F}_m = -\nabla W_m\big|_{y=a}$$

$$= -\mathbf{a}_y\left[\frac{d}{dy}(M_{12}I_1 I_2)\right]_{y=a} = \mathbf{a}_y I_1 I_2 \frac{\mu_0 h}{2\pi}\left(\frac{1}{a} - \frac{1}{b}\right) \tag{6.26b}$$

where y is used instead of y_1, for notational simplicity.

Moving the loop by a distance dy in the $-y$-direction can be accomplished by applying a counter force $-\mathbf{F}_m$. Thus, the work done is

$$\Delta W = -\mathbf{F}_m \cdot (-dy\,\mathbf{a}_y) = I_1 I_2 \frac{\mu_0 h}{2\pi} \left(\frac{1}{a} - \frac{1}{b} \right) dy \tag{6.26c}$$

This corresponds to an increase in the system energy.

(b) As the loop moves closer to the filamentary current because of an external force, the magnetic flux linkage increases with time, and an *emf* is induced in the loop such that

$$\begin{aligned} emf &= -\frac{d\Phi}{dt} \\ &= -\frac{d}{dt}(M_{12} I_1) = -I_1 \frac{dM_{12}}{dy} \frac{dy}{dt} \\ &= -I_1 \left[\frac{\mu_0 h}{2\pi} \left(\frac{1}{a} - \frac{1}{b} \right) \right] v_o \end{aligned} \tag{6.26d}$$

where $dy/dt = -v_o$. The negative sign in Eq. (6.26d) signifies that *emf* induces a current flowing counterclockwise or in the direction opposite to the flow of I_2.

To maintain the loop current at I_2, work is done by the battery. The power supplied by the battery is

$$\Delta P = |emf| \times I_2 = I_1 I_2 \left[\frac{\mu_0 h}{2\pi} \left(\frac{1}{a} - \frac{1}{b} \right) \right] \left| \frac{dy}{dt} \right| \tag{6.26e}$$

(c) ΔP in Eq. (6.26e) is equal to ΔW of Eq. (6.26c) divided by Δt, implying that the magnetic energy of the system originates from the battery.

(d) The magnetic flux density of I_1 is

$$\mathbf{B} = \frac{\mu_0 I_1}{2\pi\rho} \mathbf{a}_\phi$$

The Lorentz forces on the left and right sides of the loop are

$$\mathbf{F}_{m,l} = I_2 h\, \mathbf{a}_z \times \mathbf{B} = I_2 h \frac{\mu_0 I_1}{2\pi a}(-\mathbf{a}_y)$$

$$\mathbf{F}_{m,r} = I_2 h \frac{\mu_0 I_1}{2\pi b}\, \mathbf{a}_y$$

Because the forces on the top and bottom sides cancel out, the net Lorentz force on the loop is

$$\overline{\mathbf{F}}_m = \mathbf{F}_{m,l} + \mathbf{F}_{m,r} = -\frac{\mu_0 h I_1 I_2}{2\pi}\left[\frac{1}{a} - \frac{1}{b}\right]\mathbf{a}_y \qquad (6.26f)$$

The magnetic force, $\overline{\mathbf{F}}_m$, is directed along $-\mathbf{a}_y$ and parallel to $\boldsymbol{v}$. Furthermore, $\overline{\mathbf{F}}_m = -\mathbf{F}_m$. From these results, we can conclude that the magnetic force $\overline{\mathbf{F}}_m$ causes the loop to move in the $-\mathbf{a}_y$-direction. When the battery current I_2 increases, the Lorentz force $\overline{\mathbf{F}}_m$ increases, and the loop is pulled towards the filament, resulting in an increase in the magnetic energy of the system. Therefore, it is evident that magnetic force and energy originate from the current flowing through the conductor. However, the magnetic field did no work.

Exercise 6.4
What types of *emfs* are induced in the loops in Figs. 6.1a and 6.1b, respectively.

Ans. (a) Transformer *emf*, (b) Motional *emf*.

Exercise 6.5
If the magnetic flux linking the loop in Fig. 6.1a varies as $\Phi = \Phi_o[1 - \cos(\omega t)]$, determine (a) *emf*, and (b) i in load resistance R.

Ans. (a) $emf = -\omega\Phi_o \sin(\omega t)$, (b) $i = (\omega\Phi_o/R)\sin(\omega t)$.

Exercise 6.6
By what factor does the dissipated power increase if the load resistance R in Fig. 6.6 is doubled?

Ans. 0.5.

6.1.3 Loop Moving in Time-Varying Magnetic Field

We now consider a more general case, in which a closed wire loop moves or rotates in a time-varying magnetic field. In this case, the total *emf* is the sum of the transformer and motional *emf*s, that is,

$$\boxed{\begin{aligned} emf &= \oint_C \mathscr{E} \cdot d\mathbf{l} \\ &= -\int_S \frac{\partial \mathscr{B}}{\partial t} \cdot d\mathbf{s} + \oint_C (\boldsymbol{v} \times \mathscr{B}) \cdot d\mathbf{l} \end{aligned}} \qquad [\text{V}] \qquad (6.27)$$

where C represents the closed loop moving with velocity $\boldsymbol{v}$ in $\mathscr{B}$, and S is the surface bounded by C. The directions of $d\mathbf{s}$ and $d\mathbf{l}$ follow the right-hand rule. The surface integral in Eq. (6.27) represents the transformer *emf*, which is evaluated by assuming that the loop is fixed in position. The second closed-line integral in Eq. (6.27) represents the motional *emf*, which is evaluated by assuming that $\mathscr{B}$ is constant. The total *emf* is generally expressed by Eq. (6.3).

Applying Stokes's theorem to Eq. (6.27) results in

$$\nabla \times \mathscr{E} = -\frac{\partial \mathscr{B}}{\partial t} + \nabla \times (\boldsymbol{v} \times \mathscr{B}) \tag{6.28}$$

The first term on the right side of Eq. (6.28) originates from the transformer *emf*, whereas the second term originates from the motional *emf*.

Example 6.4 A rectangular wire loop rotates about the x-axis at angular velocity ω_o in a time-varying field $\mathscr{B} = B\sin(\omega t)\,\mathbf{a}_y$, as shown in Fig. 6.8. At time $t = t_1$, a unit normal to the loop surface makes an angle $\varphi = \omega_o t_1$ with the $+y$-axis. Find the (a) transformer, (b) motional, and (c) total electromotive forces.

Solution

(a) The total magnetic flux linking the loop with dimensions $a \times b$ is

$$\Phi = \int \mathscr{B} \cdot d\mathbf{s} = B\sin(\omega t)\,\mathbf{a}_y \cdot (ab\,\mathbf{a}_s)$$

$$= abB\sin(\omega t)\cos\varphi \tag{6.29a}$$

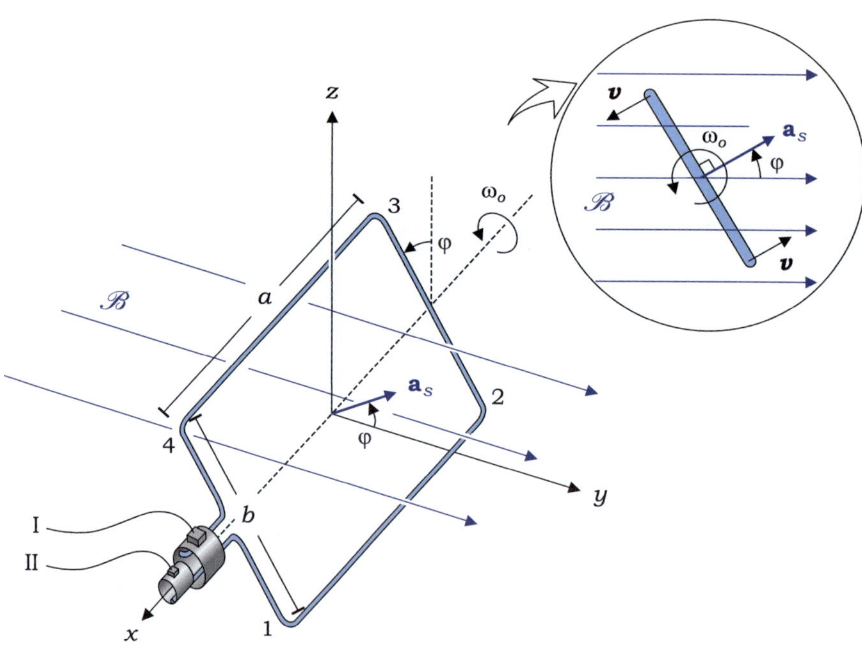

Fig. 6.8 Rectangular wire loop rotating in a time-varying magnetic field

The transformer *emf* is obtained by taking the negative time derivative of Φ, while maintaining a constant φ, such that

$$emf_t = -\left.\frac{d\Phi}{dt}\right|_{\varphi=constant}$$

$$= -\omega ab B \cos(\omega t) \cos \varphi \qquad (6.29b)$$

As can be seen from Fig. 6.8, the direction of $\mathbf{a}_s$ indicates that the loop is traversed counterclockwise for the closed-line integral of $\mathscr{E}$ to obtain emf_t. Therefore, a positive value of emf_t indicates that terminal *I* has a higher potential than terminal *II*.

(b) The motional *emf* is obtained using Eq. (6.18) with $\mathscr{B}$ held constant such that

$$emf_m = \oint_C (\boldsymbol{v} \times \mathscr{B}) \cdot d\mathbf{l}$$

$$= \int_1^2 \left[\omega_o\left(\tfrac{1}{2}b\right)\mathbf{a}_s \times \mathbf{a}_y B \sin(\omega t_1)\right] \cdot (dx\, \mathbf{a}_x)$$

$$+ \int_3^4 \left[\omega_o\left(\tfrac{1}{2}b\right)(-\mathbf{a}_s) \times \mathbf{a}_y B \sin(\omega t_1)\right] \cdot (dx\, \mathbf{a}_x) \qquad (6.29c)$$

where $\boldsymbol{v} = \omega_o(b/2)(\pm\mathbf{a}_s)$ on the bottom (positive sign) and top (negative sign) sides of the loop, as shown in the inset of Fig. 6.8. There is no contribution from the left or right side of the loop because the vector $\boldsymbol{v} \times \mathscr{B}$ is perpendicular to the side.

Thus,

$$emf_m = \omega_o ab B \sin(\omega t_1) \sin \varphi \qquad (6.29d)$$

(c) In the general case, the instantaneous magnetic flux linking the loop is obtained by setting $\varphi = \omega_o t$ in Eq. (6.29a), that is,

$$\Phi = ab B \sin(\omega t) \cos(\omega_o t) \qquad (6.29e)$$

Therefore, the total *emf* at $t = t_1$ is

$$emf = -\left.\frac{d\Phi}{dt}\right|_{t=t_1}$$

$$= -\omega ab B \cos(\omega t_1) \cos(\omega_o t_1) + \omega_o ab B \sin(\omega t_1) \sin(\omega_o t_1) \qquad (6.29f)$$

Here, ω should not be confused with ω_o: ω is the angular frequency of $\mathscr{B}$ and ω_o is the angular velocity of the loop. The result in Eq. (6.29f) is equal to the sum of those in Eqs. (6.29b) and (6.29d) if $\omega_o t_1$ is set to φ.

Exercise 6.7

In Fig. 6.6, the bar moving at $\boldsymbol{v} = \upsilon_o \mathbf{a}_y$ crosses the $y = 0$ line at $t = 0$ in the presence of $\mathscr{B} = \mathbf{a}_z B_o \cos(\omega_o t - y)$. Find the transformer and motional *emfs*.

Ans. $emf_t = x_o B_o \omega_o [\cos(\omega_o t - \upsilon_o t) - \cos(\omega_o t)], emf_m = -\upsilon_o x_o B_o \cos(\omega_o t - \upsilon_o t)$.

Review Questions

RQ 6.1	State Faraday's law of electromagnetic induction.	[(6.1)]
RQ 6.2	State Lenz's law.	[(6.1)]
RQ 6.3	What is meant by a negative *emf*?	[Fig. 6.1]
RQ 6.4	Express transformer *emf*.	[(6.4)]
RQ 6.5	Express the point form of Faraday's law.	[(6.6)]
RQ 6.6	What is turns ratio?	[(6.10)]
RQ 6.7	Express the impedance transformation in an ideal transformer.	[(6.12)]
RQ 6.8	Express the motional *emf*.	[(6.18)]
RQ 6.9	What is the motional electric field intensity?	[(6.17)]
RQ 6.10	Define the curl of the time-varying electric field.	[(6.28)]

6.2 Displacement Current Density

As we saw in the previous section, time-varying electric fields are nonconservative. Under time-varying conditions, the irrotational nature of the static electric field, expressed as $\nabla \times \mathbf{E} = 0$, must be modified to $\nabla \times \mathscr{E} = -\partial \mathscr{B}/\partial t$ to conform to the Faraday electromagnetic induction. Similarly, Ampere's circuital law, expressed as $\nabla \times \mathbf{H} = \mathbf{J}$, should be modified to incorporate Maxwell's hypothesis of the displacement current density under time-varying circumstances.

Ampere's circuital law is inconsistent with the equation of continuity under time varying conditions, which is expressed in point form as

$$\nabla \times \mathbf{H} = \mathbf{J} \tag{6.30}$$

By taking the divergence of both sides of Eq. (6.30), we have

$$\nabla \cdot \nabla \times \mathbf{H} = \nabla \cdot \mathbf{J} \tag{6.31}$$

The left side of Eq. (6.31) vanishes because of the vector identity $\nabla \cdot \nabla \times \mathbf{U} = 0$, and the equation is reduced to

$$\nabla \cdot \mathbf{J} = 0$$

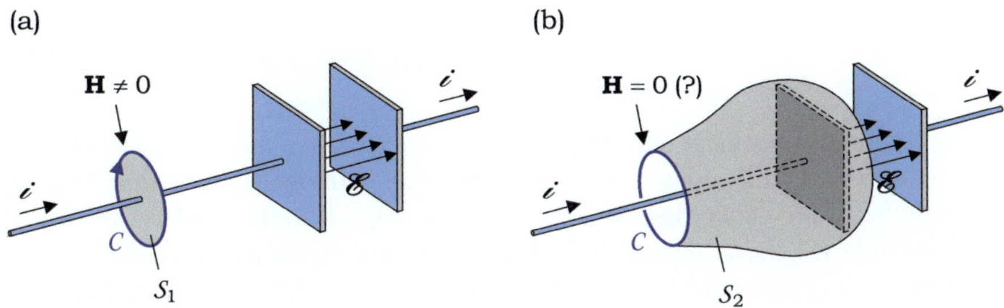

(a)

(b)

Fig. 6.9 Parallel-plate capacitor is charged by increasing the conduction current i. Two surfaces S_1 and S_2 have the same contour C

It is evident from the equation of continuity, $\nabla \cdot \boldsymbol{J} = -\partial \rho_v / \partial t$, that Ampere's circuital law in Eq. (6.30) holds true only under static conditions.

Ampere's circuital law gives rise to a contradiction when applied to a parallel-plate capacitor charged by an increasing lead current i, as illustrated in Fig. 6.9. Let us construct a closed loop C around the capacitor lead and consider S_1 and S_2 as two independent loop surfaces, each bounded by C. Surface S_1 is flat, as shown in Fig. 6.9a, whereas surface S_2 resembles an untied balloon whose surface passes between the capacitor plates, as shown in Fig. 6.9b. Applying Ampere's circuital law to loop C with loop surface S_1, we observe that the line integral of **H** around C is nonzero because the conduction current i passes through the loop surface. In contrast, if we apply Ampere's circuital law to loop C with loop surface S_2, no conduction current actually passes through S_2, and we are confronted with the contradictory conclusion that the line integral of **H** around the same loop is zero.

Conduction currents are not the only source of time-varying magnetic fields. When the conduction current increases in the parallel-plate capacitor, a time-varying magnetic field is also induced in the space between the capacitor plates even if there is no conduction current flowing in the gap. When the capacitor is charged to Q [C], the electric field in the gap is

$$\mathscr{E} = \frac{Q}{\varepsilon_0 A} \, \mathbf{a}_z \tag{6.32}$$

where A is the area of the plates perpendicular to the z-axis, air is used as the insulator, and the fringing effects at the edges are ignored. By taking the time derivatives of both sides of Eq. (6.32) and using the relation $i = dQ/dt$, we have

$$\frac{\partial(\varepsilon_0 \mathscr{E})}{\partial t} = \frac{i}{A} \, \mathbf{a}_z \tag{6.33}$$

From the right-hand side of Eq. (6.33), it is evident that the term on the left-hand side has the dimensions of the current density. Therefore, the **_displacement current density_** $\boldsymbol{J}_d$ is defined as the time derivative of the electric flux density $\mathscr{D}$.

$$\boxed{J_d \equiv \frac{\partial \mathscr{D}}{\partial t}} \quad [\text{A/m}^2] \tag{6.34}$$

Although the displacement current density does not involve any motion of electric charges, it behaves like conduction and convection currents as far as time-varying magnetic fields are concerned. In a parallel-plate capacitor, the conduction current is responsible for the magnetic field induced around perfectly conducting leads, whereas the displacement current density is solely responsible for the magnetic field in the gap. Conduction and displacement currents can coexist in lossy dielectrics or conductors with finite conductivities, both of which contribute to the magnetic field of the material. However, in good conductors, the effect of the displacement current is ignored compared with that of the conduction current.

Under time-varying conditions, the point form of Ampere's law is modified to incorporate the displacement current density as follows:

$$\boxed{\nabla \times \mathscr{H} = J + \frac{\partial \mathscr{D}}{\partial t}} \tag{6.35}$$

This equation is referred to as **generalized Ampere's law**, or simply Ampere's law. The current density, J, in Eq. (6.35) may represent conduction current, convection current, or both, which vary with time. By applying Stokes's theorem to Eq. (6.35), the integral form of Ampere's law is given by

$$\boxed{\oint_C \mathscr{H} \cdot d\mathbf{l} = \int_S \left(J + \frac{\partial \mathscr{D}}{\partial t} \right) \cdot d\mathbf{s}} \tag{6.36}$$

Let us now check if Eq. (6.35) conforms to the equation of continuity. By taking the divergence of both sides of Eq. (6.35), we have

$$\nabla \cdot \nabla \times \mathscr{H} = \nabla \cdot J + \frac{\partial}{\partial t} (\nabla \cdot \mathscr{D}) \tag{6.37}$$

The divergence and time derivative can be interchanged in the rightmost part of Eq. (6.37), because they are mutually orthogonal. Then, with the vector identity $\nabla \cdot \nabla \times \mathbf{U} = 0$ and Gauss's law $\nabla \cdot \mathscr{D} = \rho_v$, Eq. (6.37) is reduced to

$$0 = \nabla \cdot J + \frac{\partial \rho_v}{\partial t} \tag{6.38}$$

which is the equation of continuity.

Example 6.5 An AC voltage source with $v(t) = V_o \sin(\omega t)$ is connected to a parallel-plate capacitor with a plate area S and separation d. Ignoring fringing effects, show that the displacement current in the lossless dielectric with permittivity ε filling the gap is equal to the conduction current in the perfectly conductive lead.

Solution

Ignoring the fringing effects and assuming that the upper plate has a higher potential than the lower plate, the electric flux density in the gap is given as

$$\mathscr{D} = -\mathbf{a}_z \varepsilon \frac{V_o}{d} \sin(\omega t)$$

The displacement current density is thus

$$\mathbf{J}_d = \frac{\partial \mathscr{D}}{\partial t} = -\mathbf{a}_z \omega \varepsilon \frac{V_o}{d} \cos(\omega t)$$

The total displacement current in the $-z$-direction is

$$i_d = \int_S \mathbf{J}_d \, ds = S \omega \varepsilon \frac{V_o}{d} \cos(\omega t)$$
$$= C V_o \omega \cos(\omega t) \qquad\qquad (6.39a)$$

where $C = \varepsilon S / d$. From circuit theory, the conduction current in the lead is

$$i_c = C \frac{d}{dt} v(t) = C V_o \omega \cos(\omega t) \qquad\qquad (6.39b)$$

The two results in Eqs. (6.39a) and (6.39b) are equal.

Exercise 6.8
Does the displacement current density depend on the dielectric filling the parallel-plate capacitor?

Ans. No.

Exercise 6.9
A coaxial capacitor of unity length has a capacitance C. If the voltage across the inner and outer conductors is $v = V_o \cos(\omega t)$, ignoring the edge effect, determine the displacement current density in the gap.

Ans. $\mathbf{J}_d = -\mathbf{a}_\rho \omega C V_o \sin(\omega t)/(2\pi\rho)$ [A/m^2].

Review Questions

RQ 6.11	What is the unit of $\partial \mathscr{D}/\partial t$?	[(6.34)]
RQ 6.12	What is the significance of displacement current density?	[(6.35)]
RQ 6.13	What physical laws justify the displacement current density?	[(6.38)]

6.3 Maxwell's Equations

In 1873, James Clerk Maxwell published a unified theory of electricity and magnetism by formulating the previously known experimental results of Coulomb, Gauss, Ampere, Faraday, and others and by incorporating the concept of displacement current density. This theory comprises four fundamental relations, called Maxwell's equations, in which any electromagnetic field should satisfy at all times under time-varying conditions, regardless of the material medium. Maxwell's equations can be expressed in either integral or differential form. The integral form is advantageous for describing underlying physical concepts, whereas the differential or point form is advantageous for specifying electromagnetic quantities at each point in a given region of space.

Maxwell's equations in point form are as follows:

$$\boxed{\nabla \times \mathscr{E} = -\frac{\partial \mathscr{B}}{\partial t}} \tag{6.40a}$$

$$\boxed{\nabla \times \mathscr{H} = \boldsymbol{J} + \frac{\partial \mathscr{D}}{\partial t}} \tag{6.40b}$$

$$\boxed{\nabla \cdot \mathscr{D} = \rho_v} \tag{6.40c}$$

$$\boxed{\nabla \cdot \mathscr{B} = 0} \tag{6.40d}$$

These equations may be individually referred to as Faraday's, Ampere's, Gauss's, and Gauss's laws for magnetism.

Auxiliary equations are essential in solving electromagnetic problems. The relations between $\mathscr{D}$ and $\mathscr{E}$ and between $\mathscr{B}$ and $\mathscr{H}$ are called constitutive relations:

$$\mathscr{D} = \varepsilon \mathscr{E} \tag{6.41a}$$

$$\mathscr{B} = \mu \mathscr{H} \tag{6.41b}$$

where ε and μ are the permittivity and permeability, respectively, of the material. The conduction and convection current densities are defined as

$$\boldsymbol{J} = \sigma \mathscr{E} \tag{6.42a}$$

$$\boldsymbol{J} = \rho_v \boldsymbol{v} \tag{6.42b}$$

where σ is the conductivity, ρ_v is the volume charge density, and $\boldsymbol{v}$ is the charge velocity. The total force exerted on charge q moving in $\mathscr{E}$ and $\mathscr{B}$ is specified by the Lorentz force equation.

$$F = q(\mathscr{E} + \boldsymbol{v} \times \mathscr{B})\tag{6.43}$$

where $\boldsymbol{v}$ is the charge velocity. Under time-varying conditions, the equation of continuity is given by

$$\nabla \cdot \boldsymbol{J} = -\frac{\partial \rho_v}{\partial t}\tag{6.44}$$

which is based on the conservation of the electric charge.

Example 6.6 Given $\mathscr{E} = E_o\mathbf{a}_x \cos(\omega t - kz)$ and $\mathscr{H} = E_o\mathbf{a}_y \sqrt{\varepsilon_0/\mu_0}\,\cos(\omega t - kz)$ in free space, along with $k = \omega\sqrt{\mu_0\varepsilon_0}$, show that $\mathscr{E}$ and $\mathscr{H}$ satisfy Maxwell's equations, and thus constitute genuine electromagnetic fields.

Solution

Substitution of $\mathscr{E}$ and $\mathscr{H}$ into Eq. (6.40a) gives

$$kE_o\mathbf{a}_y \sin(\omega t - kz) = \omega E_o\mathbf{a}_y \sqrt{\mu_0\varepsilon_0}\,\sin(\omega t - kz)\tag{6.45a}$$

Substitution of $\mathscr{E}$ and $\mathscr{H}$ into Eq. (6.40b), with $\boldsymbol{J} = 0$ (free space), gives

$$-kE_o\mathbf{a}_x \sqrt{\varepsilon_0/\mu_0}\,\sin(\omega t - kz) = -\omega\varepsilon_0 E_o\mathbf{a}_x \sin(\omega t - kz)\tag{6.45b}$$

It is evident from the relation $k = \omega\sqrt{\mu_0\varepsilon_0}$ that Eqs. (6.45a) and (6.45b) are satisfied at all times.

From Eq. (6.40c), with $\rho_v = 0$ (free space), we obtain

$$\nabla \cdot \mathscr{D} = \nabla \cdot [\varepsilon_0 E_o\mathbf{a}_x \cos(\omega t - kz)] = 0\tag{6.45c}$$

From Eq. (6.40d),

$$\nabla \cdot \left[\mu_0 E_o\mathbf{a}_y \sqrt{\varepsilon_0/\mu_0}\,\cos(\omega t - kz)\right] = 0\tag{6.45d}$$

The $\mathscr{E}$ and $\mathscr{H}$ given in the problem satisfy Maxwell's equations, and thus constitute the electromagnetic fields in free space. It can be shown that even if the argument of the cosine is changed to $\omega t + kz$, the new $\mathscr{E}$ and $\mathscr{H}$ also constitute the electromagnetic fields in free space.

Exercise 6.10

Which of the following fields represents a time-varying electric or magnetic field in free space?

(a) $\mathscr{E} = \mathbf{a}_x E_o e^{(\omega t - kz)}$, (b) $\mathscr{H} = \mathbf{a}_\phi (I/2\pi\rho) \cos(\omega t)$, (c) $\mathscr{E} = \mathbf{a}_z E_o \sin(\omega t - kz)$, (d) $\mathscr{E} = \mathbf{a}_z E_o \cos(ky) \cos(\omega t)$, (e) $\mathscr{H} = \mathbf{a}_z H_o \rho^{-1} \cos(\omega t - k\rho)$.

Ans. (a) and (d) if $(k/\omega)^2 = \varepsilon_0\mu_0$.

6.3.1 Maxwell's Equations in Integral Form

Maxwell's equations in point form given in Eqs. (6.40a)–(6.40d) can be converted into an integral form using the divergence and Stokes's theorems as follows:

$$\oint_C \mathscr{E} \cdot d\mathbf{l} = -\int_S \frac{\partial \mathscr{B}}{\partial t} \cdot d\mathbf{s} \tag{6.46a}$$

$$\oint_C \mathscr{H} \cdot d\mathbf{l} = \int_S \left[\mathbf{J} + \frac{\partial \mathscr{D}}{\partial t} \right] \cdot d\mathbf{s} \tag{6.46b}$$

$$\oint_S \mathscr{D} \cdot d\mathbf{s} = \int_V \rho_v \, dv \tag{6.46c}$$

$$\oint_S \mathscr{B} \cdot d\mathbf{s} = 0 \tag{6.46d}$$

The directions of $d\mathbf{l}$ and $d\mathbf{s}$ are governed by the right-hand rule, in which $d\mathbf{s}$ on the surface S points in the direction of the right thumb when the fingers follow $d\mathbf{l}$ along the loop C.

6.3.2 Electromagnetic Boundary Conditions

Following the same procedure as that used for the boundary conditions for static fields $\mathbf{E}$, $\mathbf{D}$, $\mathbf{H}$, and $\mathbf{B}$, we can obtain the boundary conditions for $\mathscr{E}$, $\mathscr{D}$, $\mathscr{H}$, and $\mathscr{B}$ at the interface between the two dissimilar materials. The boundary conditions for the tangential components of $\mathscr{E}$ and $\mathscr{H}$ are obtained from Faraday's law and Ampere's law, respectively, whereas those for the normal components of $\mathscr{D}$ and $\mathscr{B}$ are obtained from Gauss's law and Gauss's law for magnetism, respectively.

Because the line integral of $\mathscr{E}$ or $\mathscr{H}$ is conducted around a closed loop extending across the interface as the loop surface shrinks to zero, neither the term $\partial \mathscr{B}/\partial t$ in Faraday's law nor the term $\partial \mathscr{D}/\partial t$ in Ampere's law affects the boundary conditions for the tangential components of $\mathscr{E}$ and $\mathscr{H}$. Therefore,

$$\mathscr{E}_{1t} = \mathscr{E}_{2t} \tag{6.47a}$$

$$\mathscr{H}_{1t} - \mathscr{H}_{2t} = J_s \tag{6.47b}$$

where t denotes the tangential component, and J_s represents the surface current flowing over the interface in the direction normal to the loop surface.

The boundary conditions for the normal components of $\mathscr{D}$ and $\mathscr{B}$ are

$$\boxed{\mathscr{D}_{1n} - \mathscr{D}_{2n} = \rho_s} \tag{6.47c}$$

$$\boxed{\mathscr{B}_{1n} = \mathscr{B}_{2n}} \tag{6.47d}$$

where n denotes the normal component, and ρ_s represents the surface charge density at the interface.

Two frequently encountered interfaces are those between two lossless dielectrics and between a lossless dielectric and perfect conductor. As lossless dielectrics are characterized by zero conductivity ($\sigma = 0$), the interface between two lossless dielectrics carries no surface charges and no surface currents, that is,

$$\rho_s = J_s = 0 \tag{6.48}$$

The boundary conditions at the interface between the two lossless dielectrics are

$$\boxed{\mathscr{E}_{1t} = \mathscr{E}_{2t}} \tag{6.49a}$$

$$\boxed{\mathscr{H}_{1t} = \mathscr{H}_{2t}} \tag{6.49b}$$

$$\boxed{\mathscr{D}_{1n} = \mathscr{D}_{2n}} \tag{6.49c}$$

$$\boxed{\mathscr{B}_{1n} = \mathscr{B}_{2n}} \tag{6.49d}$$

This indicates that the tangential and normal components of the electromagnetic field are continuous across the interface.

A perfect conductor has infinite conductivity ($\sigma = \infty$). In most practical situations, good conductors such as silver, copper, gold, and aluminum with high conductivities on the order of $\sigma \sim 10^7$ [S/m] may be considered perfect conductors. The infinite conductivity is related to the unique characteristic of a perfect conductor, that is, a zero electric field in the interior. Consequently, any net charge that a perfect conductor may have should reside on the surface only and any current that a perfect conductor may carry should flow over the surface only. The relationship between $\mathscr{E}$ and $\mathscr{H}$ ensures that the magnetic field also vanishes inside a perfect conductor (Fig. 6.10). Therefore,

$$\mathscr{E}_2 = 0 = \mathscr{D}_2 \tag{6.50a}$$

$$\mathscr{H}_2 = 0 = \mathscr{B}_2 \tag{6.50b}$$

where subscript 2 denotes region 2 occupied by a perfect conductor. The boundary conditions at the interface between the lossless dielectric in region 1 and perfect conductor in region 2 are as follows:

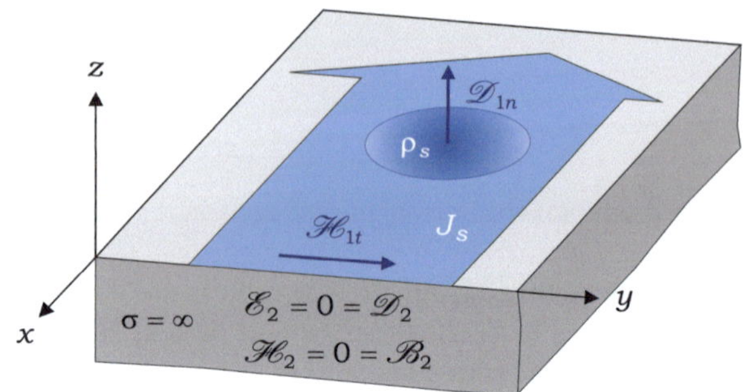

Fig. 6.10 Interface between the perfect conductor and lossless dielectric

$$\boxed{\mathscr{E}_{1t} = 0} \tag{6.51a}$$

$$\boxed{\mathscr{H}_{1t} = J_s} \tag{6.51b}$$

$$\boxed{\mathscr{D}_{1n} = \rho_s} \tag{6.51c}$$

$$\boxed{\mathscr{B}_{1n} = 0} \tag{6.51d}$$

where t and n denote tangential and normal components, respectively. It should be noted that a unit surface normal is taken away from the conductor. For example, $\mathscr{D}_{1n}$ is the normal component of $\mathscr{D}$ on the conductor surface, and is directed away from it.

Example 6.7 In free space above a perfect conductor ($z \leq 0$), the electric field is given by

$$\mathscr{E} = a(\sin\theta\,\mathbf{a}_y + \cos\theta\,\mathbf{a}_z)\cos(ky\cos\theta - kz\sin\theta - \omega t)$$
$$+ a(-\sin\theta\,\mathbf{a}_y + \cos\theta\,\mathbf{a}_z)\cos(ky\cos\theta + kz\sin\theta - \omega t)$$

and the magnetic field is directed along the $+x$-direction. Determine the surface charge and current densities in the $z = 0$ plane.

Solution

In the $z = 0$ plane, we have

$$\mathscr{D} = \varepsilon_0\mathscr{E}(z = 0) = \mathbf{a}_z\varepsilon_0 2a\cos\theta\,\cos(ky\cos\theta - \omega t)$$

The induced surface charge density is

$$\rho_s = \mathscr{D}_n = \varepsilon_0 2a\cos\theta\cos(ky\cos\theta - \omega t) \tag{6.52a}$$

Inserting $\mathscr{H} = H_x(\mathbf{r}, t)\,\mathbf{a}_x$ into Faraday's law gives

$$\mathbf{a}_y \frac{\partial H_x}{\partial z} - \mathbf{a}_z \frac{\partial H_x}{\partial y} = \frac{\partial \mathcal{D}}{\partial t}$$

At $z = 0$, $\mathcal{D}$ only has a z-component. Thus,

$$-\frac{\partial H_x}{\partial y} = \omega \varepsilon_0 2a \cos \theta \sin(ky \cos \theta - \omega t) \tag{6.52b}$$

Integrating Eq. (6.52b) over y yields

$$H_x = \frac{\omega}{k} \varepsilon_0 2a \cos(ky \cos \theta - \omega t)$$

Thus, the induced surface current density is

$$\mathbf{J}_s = \mathbf{a}_z \times \mathcal{H}_t = \mathbf{a}_y \frac{\omega}{k} \varepsilon_0 2a \cos(ky \cos \theta - \omega t) \tag{6.52c}$$

The surface charge and current densities in Eqs. (6.52a) and (6.52c) are related by $\mathbf{J}_s = \rho_s \mathbf{v}_p$. This shows that the surface charge density propagates with velocity $\mathbf{v}_p = \omega/(k \cos \theta)$, but not the surface charge itself, which can be explained by the wave concept in Chap. 7.

Exercise 6.11

The two magnetic fields, $\mathcal{H}_1 = \mathbf{a}_x H_o \cos(\omega t - kz)$ and $\mathcal{H}_2 = \mathbf{a}_x H_o \cos(\omega t + kz)$, coexist above a perfect conductor occupying the region $z \le 0$. Find $\mathbf{J}_s$ at $z = 0$.

Ans. $\mathbf{J}_s = \mathbf{a}_y 2H_o \cos(\omega t)$ [A/m].

Review Questions

RQ 6.14	Write Maxwell's equations in point form.	[(6.40a,b,c,d)]
RQ 6.15	Write Maxwell's equations in integral form.	[(6.46a,b,c,d)]
RQ 6.16	What makes the unified theory of electricity and magnetism possible?	[(6.34)]
RQ 6.17	Is the equation of continuity incorporated into Maxwell's equations?	[(6.40b)]
RQ 6.18	Write the boundary conditions for $\mathcal{E}$ and $\mathcal{H}$.	[(6.47a,b)]
RQ 6.19	Explain why $\mathcal{E}$ and $\mathcal{H}$ both vanish in perfect conductors.	[(6.50a,b)]
RQ 6.20	What are the electromagnetic boundary conditions on the surface of a perfect conductor?	[(6.51a,b,c,d)]

6.4   Retarded Potential

In Chaps. 3 and 5, we saw that the electric potential of the volume charge can be expressed by Eq. (3.51a), and the vector magnetic potential of the steady volume current is expressed as Eq. (5.38b), which are repeated as follows:

$$V(\mathbf{r}) = \int_{\mathcal{V}'} \frac{\rho_v(\mathbf{r}')}{4\pi\varepsilon\mathcal{R}} \, dv' \quad [\text{V}] \tag{6.53a}$$

$$\mathbf{A}(\mathbf{r}) = \int_{\mathcal{V}'} \frac{\mu\mathbf{J}(\mathbf{r}')}{4\pi\mathcal{R}} \, dv' \quad [\text{Wb/m}] \tag{6.53b}$$

where $\mathcal{R} = |\mathbf{r} - \mathbf{r}'|$ is the distance between the field and source point.

When the source charge and current vary with time, one might be tempted to rewrite the above equations by replacing $\rho_v(\mathbf{r}')$ and $\mathbf{J}(\mathbf{r}')$ with $\rho_v(\mathbf{r}', t)$ and $\mathbf{J}(\mathbf{r}', t)$, respectively, for time-varying potentials. However, the negative gradient or curl of such a function cannot constitute time-varying $\mathscr{E}$ or $\mathscr{B}$ because it does not satisfy Maxwell's equations. In fact, the potential cannot respond instantaneously at large distances from the source; it takes time for an electromagnetic field to reach a distant field point and for the effect of the source to appear at that point. In other words, a change in the source at point $\mathbf{r}'$ is felt at point $\mathbf{r}$ at a later time, delayed by $\mathcal{R}/\upsilon$, where $\mathcal{R}$ is the distance between the two points, and υ is the speed of propagation of electromagnetic fields. The potentials $V(\mathbf{r}, t_1)$ and $\mathbf{A}(\mathbf{r}, t_1)$ at time $t = t_1$ are caused by the source given at an earlier time $t = t_1 - \mathcal{R}/\upsilon$, that is, $\rho_v(\mathbf{r}', t_1 - \mathcal{R}/\upsilon)$ and $\mathbf{J}(\mathbf{r}', t_1 - \mathcal{R}/\upsilon)$. Expressed mathematically,

$$V(\mathbf{r}, t) = \int_{\mathcal{V}'} \frac{\rho_v(t - \mathcal{R}/\upsilon)}{4\pi\varepsilon\mathcal{R}} \, dv' \quad [\text{V}] \tag{6.54a}$$

$$\mathbf{A}(\mathbf{r}, t) = \int_{\mathcal{V}'} \frac{\mu\mathbf{J}(t - \mathcal{R}/\upsilon)}{4\pi\mathcal{R}} \, dv' \quad [\text{Wb/m}] \tag{6.54b}$$

which are called the ***retarded scalar potential*** and ***retarded vector potential***, respectively. In free space, the speed of light is $\upsilon = c = 1/\sqrt{\varepsilon_0\mu_0}$, where ε_0 is the permittivity and μ_0 is the permeability of free space. This retarded potential is useful for solving radiation problems, as discussed in Chap. 11.

Starting with the definition of the vector magnetic potential, we derive useful formulas for the retarded potential. Just as the vector magnetic potential is defined such that it follows the solenoidal nature of magnetic flux density and the null identity $\nabla \cdot (\nabla \times \mathbf{U}) = 0$, the time-varying vector magnetic potential A is defined as

$$\mathscr{B} = \nabla \times \boldsymbol{A} \tag{6.55}$$

Inserting Eq. (6.55) into Faraday's law gives

$$\nabla \times \mathscr{E} = -\frac{\partial}{\partial t}(\nabla \times \boldsymbol{A}) \tag{6.56}$$

By interchanging the time derivative and curl on the right-hand side of Eq. (6.56), we write

$$\nabla \times \left(\mathscr{E} + \frac{\partial \boldsymbol{A}}{\partial t} \right) = 0 \tag{6.57}$$

Considering the null identity $\nabla \times (\nabla W) = 0$, the ***time-varying scalar potential*** V is defined as

$$\mathscr{E} + \frac{\partial \boldsymbol{A}}{\partial t} = -\nabla V \tag{6.58}$$

or

$$\boxed{\mathscr{E} = -\nabla V - \frac{\partial \boldsymbol{A}}{\partial t}} \tag{6.59}$$

Note that Eq. (6.59) is reduced to the well-known relation $\boldsymbol{E} = -\nabla V$ under static conditions.

Next, inserting Eq. (6.55) into Ampere's law in Eq. (6.40b), we obtain

$$\nabla \times \nabla \times \boldsymbol{A} = \mu \boldsymbol{J} + \mu \frac{\partial \mathscr{D}}{\partial t} \tag{6.60}$$

By applying the identity $\nabla \times \nabla \times \boldsymbol{U} = \nabla(\nabla \cdot \boldsymbol{U}) - \nabla^2 \boldsymbol{U}$ and substituting Eq. (6.59) into Eq. (6.60), we obtain

$$\nabla^2 \boldsymbol{A} - \mu\varepsilon \frac{\partial^2 \boldsymbol{A}}{\partial t^2} = -\mu \boldsymbol{J} + \nabla \left(\nabla \cdot \boldsymbol{A} + \mu\varepsilon \frac{\partial V}{\partial t} \right) \tag{6.61}$$

where $\mathscr{D} = \varepsilon\mathscr{E}$ is used.

For a unique determination of the vector field $\boldsymbol{A}$, the specification of its curl and divergence is required by Helmholtz's theorem. The curl of $\boldsymbol{A}$ is already given by Eq. (6.55). To be consistent with the divergence of $\boldsymbol{A}$ in Eq. (5.48), we let

$$\boxed{\nabla \cdot \boldsymbol{A} = -\mu\varepsilon \frac{\partial V}{\partial t}} \tag{6.62}$$

which is called the Lorentz condition for the potentials. Then Eq. (6.61) is reduced to the ***inhomogeneous wave equation for the vector magnetic potential***:

$$\boxed{\nabla^2 \boldsymbol{A} - \mu\varepsilon \frac{\partial^2 \boldsymbol{A}}{\partial t^2} = -\mu \boldsymbol{J}} \tag{6.63}$$

Under static conditions, Eq. (6.62) reduces to $\nabla \cdot \mathbf{A} = 0$, as shown in Eq. (5.48), and Eq. (6.63) reduces to the vector Poisson's equation, as shown in Eq. (5.55).

Substitution of Eq. (6.59) into Gauss's law yields

$$\frac{\rho_v}{\varepsilon} = \nabla \cdot \left(-\nabla V - \frac{\partial \mathbf{A}}{\partial t} \right)$$

$$= -\nabla^2 V - \frac{\partial}{\partial t}(\nabla \cdot \mathbf{A}) \tag{6.64}$$

By substituting the Lorentz condition in Eq. (6.62) into Eq. (6.64), the **inhomogeneous wave equation for the scalar potential** is given as

$$\boxed{\nabla^2 V - \mu\varepsilon\frac{\partial^2 V}{\partial t^2} = -\frac{\rho_v}{\varepsilon}} \tag{6.65}$$

Under static conditions, Eq. (6.65) is reduced to Poisson's equation. We see from Eqs. (6.63) and (6.65) that the wave equations for $\mathbf{A}$ and V are decoupled from each other through the Lorentz condition. Moreover, a remarkable symmetry exists between the two equations.

For $\rho_v = 0 = \mathbf{J}$, Eqs. (6.63) and (6.65) are reduced to homogeneous differential wave equations for $\mathbf{A}$ and V with general solutions of the forms $U(t \pm \mathcal{R}/\upsilon)$ and $W(t \pm \mathcal{R}/\upsilon)$, respectively. These solutions represent waves propagating at velocity $\upsilon = 1/\sqrt{\mu\varepsilon}$, as discussed in Chap. 7.

Example 6.8 Find the retarded vector magnetic potential at point p due to an incremental current element with linear dimension h and current $i(t) = I_o \cos(\omega t)$ located at the origin in free space, as shown in Fig. 6.11.

Solution

Rewriting Eq. (6.54b) for the line current, along with $c = 1/\sqrt{\varepsilon_0\mu_0}$, we have

$$\mathbf{A}(\mathbf{r}, t) = \int_{\mathcal{L}'} \frac{\mu_0}{4\pi\mathcal{R}} i(t - \mathcal{R}/c)\, \mathbf{a}_z\, dl' \tag{6.66a}$$

If h is assumed to be negligibly small, the radial distance R in spherical coordinates approximates $\mathcal{R}$. Therefore,

$$\mathbf{A}(\mathbf{r}, t) = \frac{\mu_0 h}{4\pi R} I_o \cos[\omega(t - R/c)]\, \mathbf{a}_z \tag{6.66b}$$

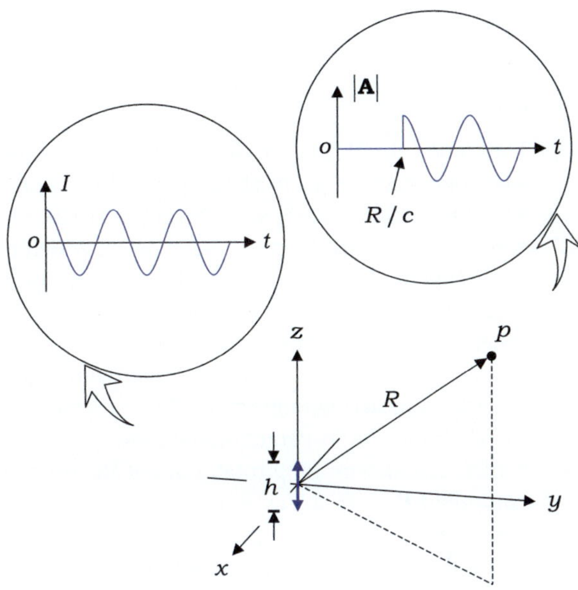

Fig. 6.11 Retarded potential due to a current element at the origin

The vector magnetic potential at a distance R from the origin is retarded by R/c in time with respect to the current at the origin. The retardation time is the travel time of the electromagnetic field propagating from the source to the field point.

Exercise 6.12

Explain why $\mathscr{H}$ cannot be obtained from Eq. (5.8) simply by replacing $\mathbf{J}'$ with $\mathbf{J}'(t - \mathcal{R}/\upsilon)$.

Ans. Displacement current density should be accounted for.

Exercise 6.13

Is the second term on the right side of Eq. (6.59) directly related to Faraday's electromagnetic induction?

Ans. Yes, the circulations of $\mathbf{A}$ and $\mathscr{E}$ are Φ and emf, respectively.

Review Questions

RQ 6.21	Explain how $\mathscr{E}$ and $\mathscr{B}$ are obtained from V and A.	[(6.59)(6.55)]
RQ 6.22	What is the Lorentz condition for the potentials?	[(6.62)]
RQ 6.23	Write inhomogeneous wave equations for V and A.	[(6.65)(6.63)]
RQ 6.24	Are the V and A propagating waves under time-varying conditions?	[(6.65)(6.63)]

6.5 Problems

Faraday's law

6.1 The same current flows in two identical square loops of side a with resistance R when one is connected to a battery with V_o and the other is placed in $\mathscr{B} = B_o t\,\mathbf{a}_z$, which increases linearly with time (Fig. 6.12). Express B_o in terms of V_o and a.

6.2 A rectangular wire loop is connected to resistor R and placed in the yz-plane in a magnetic field $\mathscr{B} = \mathbf{a}_x B_o \sin(\pi y/2a)\sin(\omega t)$ (Fig. 6.13). Determine the induced *emf* and current in the loop.

6.3 When a magnetic dipole with magnetic moment $\mathbf{m} = \mathbf{a}_z m_o \cos \omega t$ is placed at the origin, find (a) $\mathscr{B}$ and A everywhere under quasi-static conditions, ignoring the time retardation, and (b) *emf* in a circular loop of radius a placed in the $z = d$ plane with its center on the z-axis.

6.4 A rectangular wire loop rotates about its left edge at angular velocity ω_o in a static field $\mathbf{B} = B_o \mathbf{a}_x$, as shown in Fig. 6.14. Find *emf* using (a) the motional electric field, and (b) the general formula for Faraday's law.

6.5 A wavy wire lying in the xy-plane rotates about the z-axis at angular velocity ω_o in static field $\mathbf{B} = B_o \mathbf{a}_z$, as shown in Fig. 6.15. Show that the motional *emf* is independent of the wire shape.

(a)   (b)

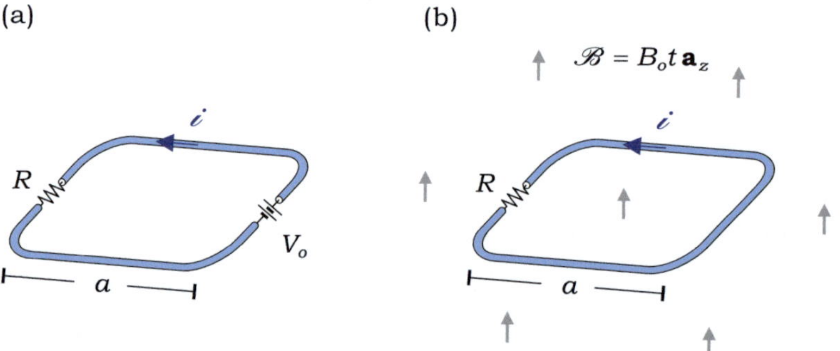

Fig. 6.12 Two loops have the same resistance and current (Problem 6.1)

Fig. 6.13 Rectangular wire loop (Problem 6.2)

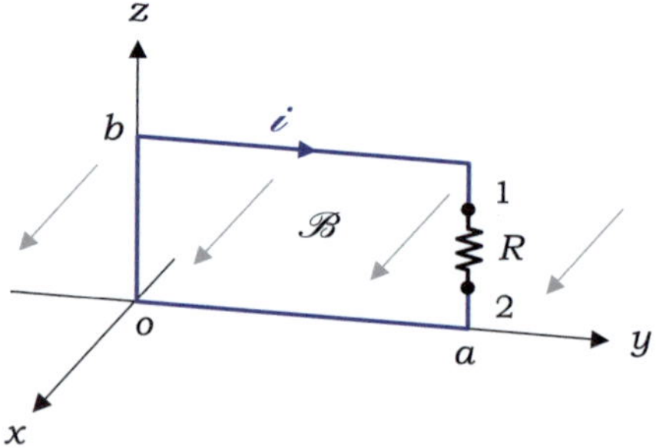

Fig. 6.14 Wire loop rotates
in static **B** (Problem 6.4)

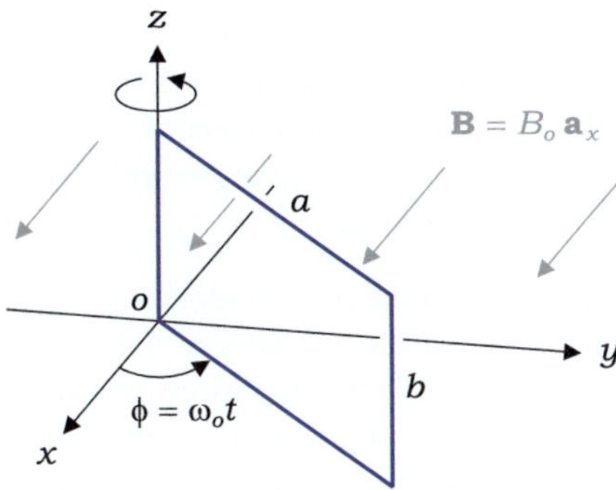

Fig. 6.15 Wavy wire in the
xy-plane rotates in static **B**
(Problem 6.5)

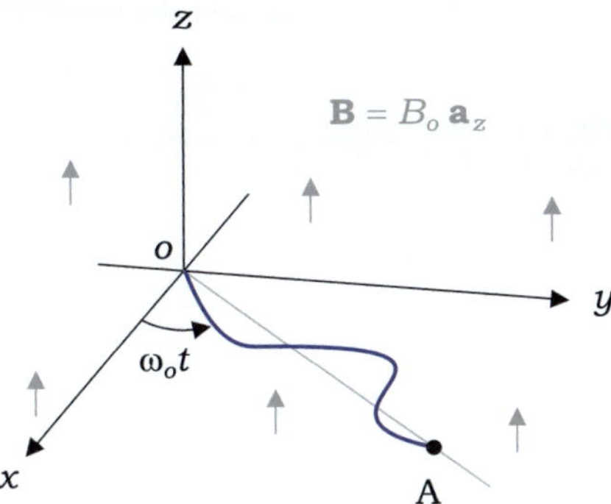

6.6   When a steady current I flows through two parallel rails and a sliding bar in the presence of a static field $\mathbf{B} = B_o\,\mathbf{a}_z$, as shown in Fig. 6.16, the bar moves at velocity $\boldsymbol{v} = v_o\,\mathbf{a}_y$. Determine

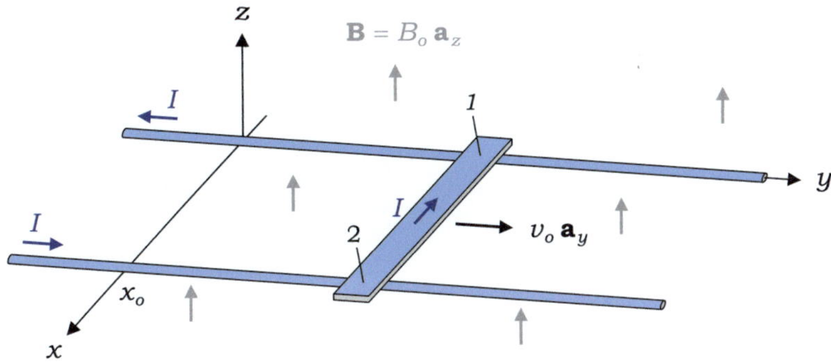

Fig. 6.16 A rail gun (Problem 6.6)

(a) the motional *emf*,
(b) the electric power dissipated in the circuit using $P_{el} = I \times emf$,
(c) the magnetic force, $\mathbf{F}_m$, on the bar,
(d) the mechanical power dissipated in the bar using $P_m = \mathbf{F}_m \cdot v$, and
(e) show that $P_{el} = P_m$.

6.7 A rectangular wire loop with internal resistance $R\,[\Omega]$ moves at velocity $v = v_o\,\mathbf{a}_y$ in the vicinity of a current-carrying wire because of an external force (Fig. 6.17). For the loop, determine (a) *emf*, (b) current, (c) magnetic force, (d) mechanical power dissipated, and (e) electric power dissipated.

6.8 A conductive disk of radius a rotates at angular velocity ω_o in a uniform static field $\mathbf{B} = B_o\,\mathbf{a}_z$, as shown in Fig. 6.18. Find the *emf* between sliding contacts 1 and 2 assuming that the conductive shaft is negligibly thin.

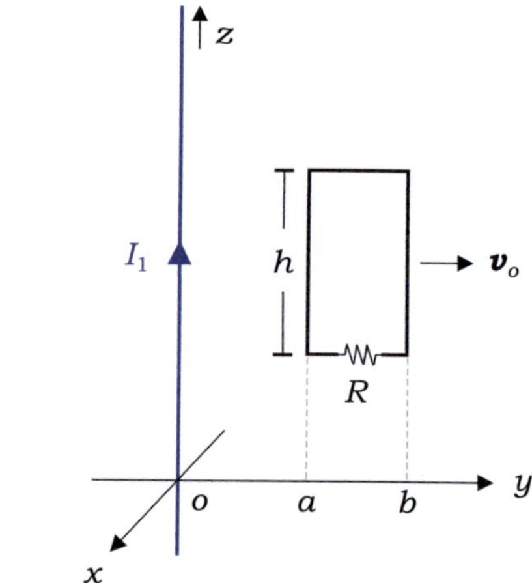

Fig. 6.17 Wire loop in motion in the vicinity of the current-carrying wire (Problem 6.7)

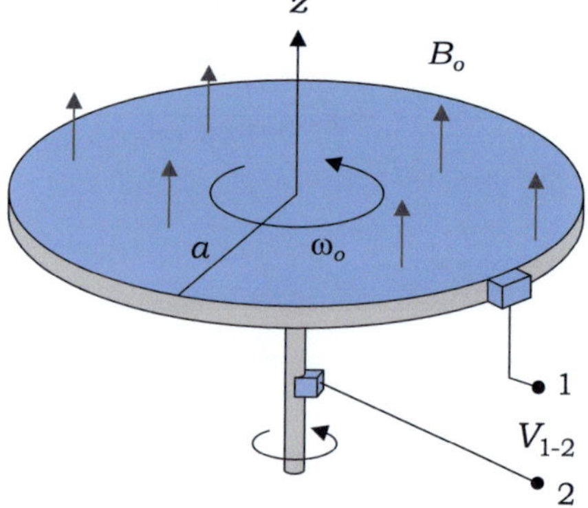

Fig. 6.18 Conductive disk rotating in static **B** field (Problem 6.8)

Fig. 6.19 Conductive disk
in a time-varying field
(Problem 6.9)

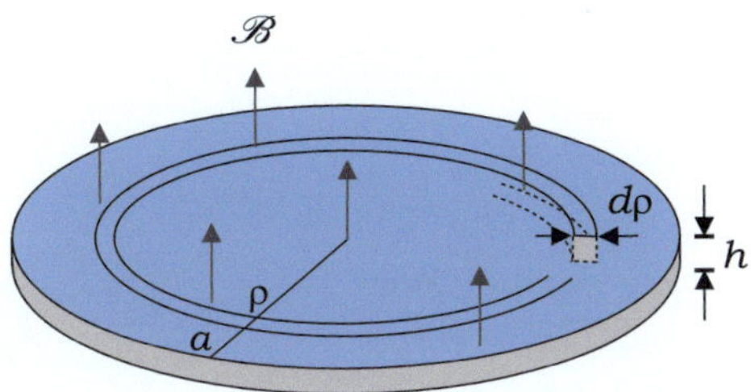

6.9 A thin disk of thickness h with conductivity σ lies on the xy-plane in a uniform time-varying field $\mathscr{B} = \mathbf{a}_z B_o \cos(\omega t)$, as shown in Fig. 6.19. Find

   (a) the *emf* induced in a circle of radius ρ on the disk, and
   (b) the total time-averaged power dissipated in the disk.

Displacement current

6.10 For an electric field $\mathscr{E} = E_o \mathbf{a}_x e^{-\alpha z} \cos(\omega t - kz)$ present in a conductor with conductivity σ and permittivity ε, determine

   (a) the conduction current density $\mathbf{J}_c$,
   (b) the displacement current density $\mathbf{J}_d$, and
   (c) the ratio between the amplitudes of $\mathbf{J}_c$ and $\mathbf{J}_d$.

6.11 At a frequency of 10 [GHz], what is the ratio $|\mathbf{J}_c/\mathbf{J}_d|$ in the copper wire for which $\varepsilon_r = 1$ and $\sigma = 5.8 \times 10^7$ [S/m]?

6.12 A coaxial capacitor of length ℓ is composed of two thin conductive cylinders of radii a and b ($a < b$), and the gap is filled with a dielectric with ε and μ_0. When it is connected to an AC voltage source $v = V_o \cos(\omega t)$, ignoring the edge effects and assuming quasi-static conditions, determine

   (a) the conduction current in the circuit, and
   (b) the displacement current in the gap. [Hint: Example 3.27.]

6.13 A capacitor is charged by a battery for $t \geq 0$ (Fig. 6.20). The current flows at $i_C = (V_o/R)e^{-t/RC}$ in the circuit, and charges build up and are uniformly distributed on plates of radius a under quasi-static conditions. Find (a) the surface charge density ρ_s on the plate, and (b) $\mathscr{D}_2$, $\mathbf{J}_d$, and $\mathscr{H}_2$ in the gap, ignoring the edge effects. Next, find (c) the surface current density $\mathbf{J}_s(\rho)$ on the plate, and (d) $\mathscr{H}_2$ in the gap assuming $\mathscr{H}_1 = (i_C/2\pi\rho)(-\mathbf{a}_\phi)$ for $z > 0$ and using the boundary conditions. (e) Compare two $\mathscr{H}_2$s in (b) and (d).

Maxwell's equations

6.14 Given a time-varying but spatially uniform field $\mathscr{E}(t) = E_o \mathbf{a}_x \cos(\omega t)$, show that it is not an electromagnetic field through the following steps:

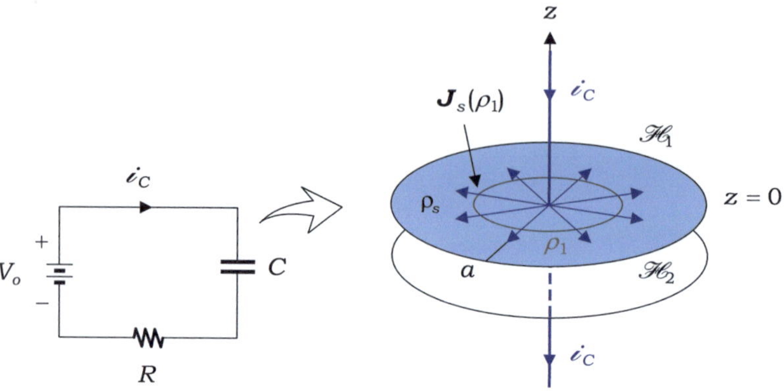

Fig. 6.20 Capacitor is charged by the battery (Problem 6.13)

(a)   Compute $\mathscr{H}$ from $\mathscr{E}$ using Faraday's law.
(b)   Compute $\mathscr{E}$ from $\mathscr{H}$ in (a) using Ampere's law.
(c)   Compare $\mathscr{E}$ in (b) with $\mathscr{E}$ given in the problem.

6.15   Given $\mathscr{E} = \mathbf{a}_x f(t - z\sqrt{\mu_0\varepsilon_0})$ in free space, where f is a smooth function in space and time, and thus differentiable, obtain $\mathscr{H}$ from $\mathscr{E}$ using Faraday's law, and show that both satisfy Ampere's law and the two Gauss's laws of Maxwell's equations.

6.16   In a lossless dielectric with ε and μ_0,

(a)   write Maxwell's equations in point form,
(b)   find $\mathscr{H}$ if $\mathscr{E} = (E_1\,\mathbf{a}_x + E_2\,\mathbf{a}_y)\cos(\omega t - kz)$, and
(c)   express k in terms of ω, ε and μ_0.

6.17   For an electromagnetic field in the form of $\mathscr{E} = \mathscr{E}_x(z, t)\,\mathbf{a}_x$ in free space, derive a second-order differential equation from Maxwell's equations that $\mathscr{E}$ should satisfy.

Electromagnetic boundary conditions

6.18   A lossless dielectric with $\varepsilon = n^2\varepsilon_0$ and μ_0 occupies region $z \geq 0$, where n is a constant. We have $\mathscr{E}_1 = E_t\,\mathbf{a}_x \cos(\omega t - zn\omega\sqrt{\mu_0\varepsilon_0})$ for $z \geq 0$ and $\mathscr{E}_0 = E_i\,\mathbf{a}_x \cos(\omega t - z\omega\sqrt{\mu_0\varepsilon_0}) + E_r\,\mathbf{a}_x \cos(\omega t + z\omega\sqrt{\mu_0\varepsilon_0})$ for $z < 0$ or free space. Show, in these regions, $\mathscr{H}_1 = n\sqrt{\varepsilon_0/\mu_0}\,E_t\,\mathbf{a}_y \cos(\omega t - zn\omega\sqrt{\mu_0\varepsilon_0})$ and $\mathscr{H}_o = \sqrt{\varepsilon_0/\mu_0}\{E_i\,\mathbf{a}_y \cos(\omega t - z\omega\sqrt{\mu_0\varepsilon_0}) - E_r\,\mathbf{a}_y \cos(\omega t + z\omega\sqrt{\mu_0\varepsilon_0})\}$. Express E_t and E_r in terms of E_i by invoking the boundary conditions.

6.19   The $z = 0$ plane is an interface between free space and a perfect conductor occupying region $z \leq 0$. Two electromagnetic fields coexist in free space, $\mathscr{E}_1 = E_1\,\mathbf{a}_x \cos(\omega t - k_y y - k_z z)$ and $\mathscr{E}_2 = E_2\,\mathbf{a}_x \cos(\omega t - k_y y + k_z z)$. Determine (a) the ratio E_1/E_2, (b) $\mathscr{H}$ in free space, and (c) the surface current density on the conductor.

6.20 Given a vector magnetic potential $\boldsymbol{A} = A_o \mathbf{a}_x \cos[\omega(t - z/c)]$ in free space, find $\mathscr{E}$ using

(a) $\mathscr{B} = \nabla \times \boldsymbol{A}$ and $\nabla \times \mathscr{H} = \partial \mathscr{D}/\partial t$, and

(b) $\nabla \cdot \boldsymbol{A} = -\mu_0 \varepsilon_0 (\partial V/\partial t)$ and $\mathscr{E} = -\nabla V - \partial \boldsymbol{A}/\partial t$.

6.21 Given $\boldsymbol{A} = A_o \mathbf{a}_z (1/R) \cos[\omega(t - R/c)]$ in free space, where $c = 1/\sqrt{\varepsilon_0 \mu_0}$, and R is the radial distance, show that $\mathscr{E} = -A_o \omega\, \mathbf{a}_\theta (\sin\theta/R) \sin[\omega(t - R/c)]$ and $\mathscr{H} = -A_o \omega \sqrt{\varepsilon_0/\mu_0}\, \mathbf{a}_\phi (\sin\theta/R) \sin[\omega(t - R/c)]$ at large distances.

Chapter 7
Wave Motion

7.1 One-Dimensional Wave

A wave is a disturbance of a continuous medium that propagates at a constant velocity with its shape unchanged. Mechanical waves in a stretched spring, water waves on the surface of a lake, and sound waves in air are good examples of traveling waves. There are two types of traveling waves. The wave in spring, as shown in Fig. 7.1a, is a ***longitudinal wave*** in which the medium or spring is displaced in the same direction as the direction of wave propagation. The wave shown in Fig. 7.1b is a ***transverse wave*** in which the medium is displaced perpendicular to the propagation direction. Although waves can transport energy from one point to another, the material medium itself cannot advance in space. Therefore, the waves can propagate at high speeds.

Let us consider the wave shown in Fig. 7.1b, which propagates along the horizontal axis or $+x$-axis. The spatial variation in the spring or disturbance of the medium occurs only along the x-axis. Thus, this wave is called a one-dimensional wave. A mathematical expression for the disturbance is called the ***wavefunction***, which is generally a function of position and time, such as

$$\psi = f(x, t) \tag{7.1}$$

where f is a smooth function of x and t, and is thus differentiable with respect to x and t. The wavefunction carries all information about the wave, such as the wave shape, propagation direction, and wave velocity.

Although a traveling wave can be well described by its wavefunction, it cannot be drawn on paper because it varies simultaneously with position and time. To avoid this difficulty, we take a detour. First, while holding the variable t constant, we draw ψ as a function of position, which is known as the ***wave profile***. This corresponds to taking a picture of a wave at a particular time. Second, while holding the variable x fixed, we draw ψ as a function of time.

Consider the two profiles of waves obtained at $t = 0$ and $t = t_1$, as shown in Fig. 7.2. They are defined by functions $f(x)$ and $\overline{f}(x)$ as follows:

© The Author(s), under exclusive license to Springer Nature Switzerland AG 2024

Y. H. Lee, *Introduction to Engineering Electromagnetics*,

https://doi.org/10.1007/978-3-031-28659-9_7

Fig. 7.1 Waves in a spring:
a Longitudinal wave, and **b**
Transverse wave

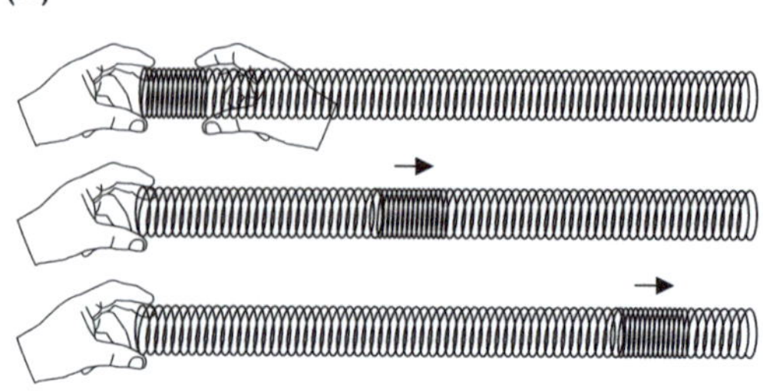

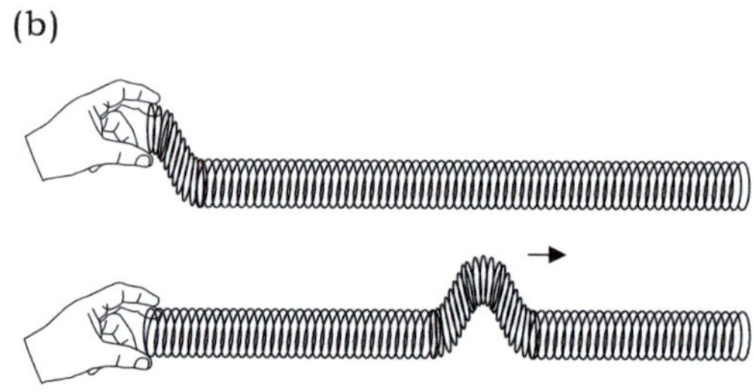

Fig. 7.2 Two wave profiles
obtained at $t = 0$ and $t = t_1$

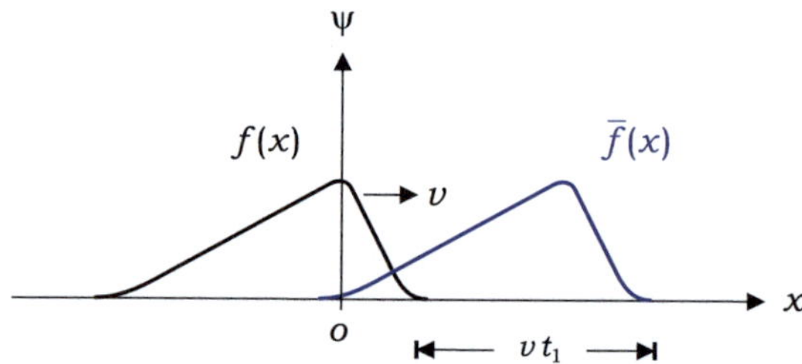

$$\psi(x, 0) = f(x) \tag{7.2a}$$

$$\psi(x, t_1) = \overline{f}(x) \tag{7.2b}$$

As function $\overline{f}$ is displaced to the right by vt_1 with respect to the function f, where v is the velocity of the wave, the wavefunction at time t_1 can be expressed as

$$\psi(x, t_1) = f(x - vt_1) \tag{7.3}$$

If t_1 is allowed to vary, Eq. (7.3) leads to a general expression for the wave, that is,

$$\boxed{\psi(x, t) = f(x - vt)} \tag{7.4}$$

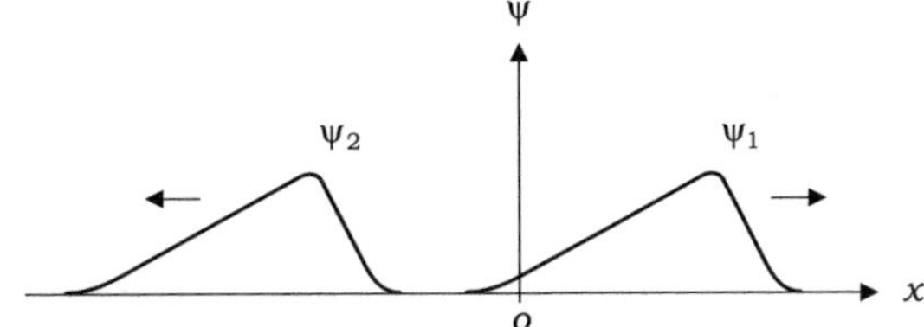

Fig. 7.3 Two waves
$\psi_1 = f(x - vt)$ and
$\psi_2 = f(x + vt)$ with the
same profile

The negative sign of the wavefunction indicates that the wave propagates in the $+x$-direction. If the wave propagates in the opposite direction, its wavefunction is expressed as

$$\boxed{\psi(x, t) = f(x + vt)}$$
(7.5)

It is noteworthy that the two waves expressed in Eqs. (7.4) and (7.5) have the same profile, but propagate in opposite directions, as illustrated in Fig. 7.3.

The ***differential wave equation*** was well known long before Maxwell and was used to describe the wave motion in different types of media. This equation is a second-order partial differential equation. As we will see in Sect. 7.3, Maxwell's equations can be combined into a three-dimensional differential wave equation. For now, we focus our attention on waves propagating in one-dimensional space. In this case, all the waves under consideration should satisfy the one-dimensional differential wave equation, that is,

$$\boxed{\frac{\partial^2 \psi}{\partial x^2} = \frac{1}{v^2} \frac{\partial^2 \psi}{\partial t^2}}$$
(7.6)

The wave velocity v is constant in a homogeneous and linear medium. Through direct substitution, we can show that the second-order differential wave equation in Eq. (7.6) has two independent solutions of the form

$$\psi = f(x \pm vt)$$
(7.7)

The solution can also be expressed as

$$\psi = f(vt \pm x)$$
(7.8)

It is important to note that f can be any function only if it is twice differentiable and has, as its argument, the special combination $x \pm vt$ or $vt \pm x$. Again, a positive sign is associated with a wave propagating in the $-x$-direction, whereas a negative sign is associated with a wave propagating in the $+x$-direction.

Example 7.1 Determine whether the following functions represent the traveling waves in homogeneous and linear media: (a) $\psi = \exp[-(x - 3t)^2]$, and (b) $\psi = \sin(2xt)$.

Solution

(a) The second derivatives of ψ with respect to x and t are

$$\frac{\partial^2 \psi}{\partial x^2} = -2e^{-(x-3t)^2} + 4(x-3t)^2 e^{-(x-3t)^2} \tag{7.9a}$$

$$\frac{\partial^2 \psi}{\partial t^2} = -18e^{-(x-3t)^2} + 36(x-3t)^2 e^{-(x-3t)^2} \tag{7.9b}$$

By taking the ratio between Eqs. (7.9a) and (7.9b), we obtain

$$v^2 = \frac{\partial^2 \psi}{\partial t^2} \bigg/ \frac{\partial^2 \psi}{\partial x^2} = 9$$

It is a traveling wave with a velocity $v = 3$.

(b) The second derivatives of ψ are

$$\frac{\partial^2 \psi}{\partial x^2} = -(2t)^2 \sin(2xt) \tag{7.9c}$$

$$\frac{\partial^2 \psi}{\partial t^2} = -(2x)^2 \sin(2xt) \tag{7.9d}$$

By taking the ratio between Eqs. (7.9c) and (7.9d), we obtain

$$\frac{\partial^2 \psi}{\partial t^2} \bigg/ \frac{\partial^2 \psi}{\partial x^2} = \left(\frac{x}{t}\right)^2 \neq \text{constant} \tag{7.9e}$$

This is not a traveling wave because the ratio in Eq. (7.9e), which corresponds to the wave velocity, is not a constant.

Exercise 7.1

Determine the propagation direction of the following waves: (a) $\psi = f(-x - t)$, (b) $\psi = 10 \sin^2(-z + 5t)$, and (c) $\psi = \cos(4z - 3t + 10)$.

Ans. (a) $-x$-, (b) $+z$-, (c) $+z$-directions.

Exercise 7.2

Show by direct substitution that the following functions satisfy the differential wave equation: (a) $\psi = f(-x - vt)$, (b) $\psi = f(t + x/v)$, and (c) $\psi = f(t - x/v)$.

Exercise 7.3

Is ψ a longitudinal or transverse wave?

(a) $\psi = 5\mathbf{a}_x \cos(2x - 3t)$, and (b) $\psi = 4\mathbf{a}_y \cos(-2x - 3t)$.

Ans. (a) Longitudinal wave, (b) Transverse wave.

Review Questions

RQ 7.1	Define longitudinal and transverse waves.	[Fig. 7.1]
RQ 7.2	Why is it that wavefunctions are smooth in space and time?	[(7.6)]
RQ 7.3	Explain how the direction of wave propagation is determined from the wavefunction.	[(7.7)]
RQ 7.4	What is it that distinguishes wavefunctions from ordinary functions?	[(7.6)]

7.1.1 Harmonic Wave

A harmonic wave has a sine or cosine curve as its profile. Conventionally, cosine functions are used for the harmonic waves. The harmonic wave is generally expressed as

$$\psi = A\cos[k(x - vt) + \varphi_o] = A\cos[kx - \omega t + \varphi_o] \qquad (7.10)$$

where A is the **amplitude**, k is the **propagation constant**, and ω is the **angular frequency**, which is equal to the frequency multiplied by 2π, that is, $\omega = 2\pi f$.

The entire argument for the cosine function in Eq. (7.10) is called the **phase**. The phase is generally a function of position and time such that

$$\varphi = kx - \omega t + \varphi_o \qquad (7.11)$$

where φ_o is the **initial phase**.

If both the input and output of an electromagnetic process are represented by harmonic waves with neither beginning nor end, then the system is considered to be in a **sinusoidal steady state**. Under this condition, the phase can be conveniently used to describe the present state of sinusoidal motion, which is represented by an angle in the range of 0 to 2π [rad]. For example, if the phase, as expressed by Eq. (7.11), is increased from 0 to 2π because of a change in the position, or x, the time-harmonic motion observed at a new position is said to be in phase with that at the previous position. It is important to note that crests are indistinguishable under sinusoidal steady-state conditions but only exhibit the same state of sinusoidal motion.

The **phase velocity** of a harmonic wave is the velocity at which the crest moves in the direction of the wave propagation. Let us consider a harmonic wave traveling with a phase velocity v_p in the $+x$-direction. The crest observed at point $x = x_1$ and time $t = t_1$ appears at a new position $x = x_1 + \Delta x$ and later time $t = t_1 + \Delta t$, where Δx and Δt are related by $\Delta x = v_p \Delta t$ with phase velocity v_p. The two wave motions observed at such conditions are in phase, and therefore

$$\varphi_1 = kx_1 - \omega t_1 + \varphi_o = k(x_1 + \Delta x) - \omega(t_1 + \Delta t) + \varphi_o \qquad (7.12)$$

This equation can be satisfied only if

$$k\Delta x - \omega \Delta t = 0 \qquad (7.13)$$

By inserting $\upsilon_p = \Delta x / \Delta t$ into Eq. (7.13), the phase velocity is expressed as

$$\boxed{\upsilon_p = \frac{\omega}{k}} \quad \text{[m/s]} \qquad (7.14)$$

Phase velocity is the ratio of the angular frequency to the propagation constant.

The harmonic wave repeats itself at regular intervals along the x-axis, called the **spatial period** or **wavelength** (denoted by λ), and at regular intervals along time, called the **temporal period** (denoted by τ). The wavelength is the distance between two neighboring crests of the wave, whereas the temporal period is the time taken for a second crest to pass through the same point in space after the first crest. At two different points separated by a wavelength, $x = x_1$ and $x = x_1 + \lambda$, the harmonic wave must exhibit the same temporal behavior. Therefore,

$$A\cos[k(x_1 + \lambda) - \omega t] = A\cos(kx_1 - \omega t) \qquad (7.15)$$

This leads to the relationship $k\lambda = 2\pi$. Similarly, the spatial patterns of the wave observed at two different instants separated by a temporal period, $t = t_1$ and $t = t_1 + \tau$, must be identical. Therefore,

$$A\cos[kx - \omega(t_1 + \tau)] = A\cos(kx - \omega t_1) \qquad (7.16)$$

This leads to the relationship $\omega\tau = 2\pi$. Consequently, the propagation constant k and angular frequency ω of the harmonic wave can be expressed as follows:

$$\boxed{k = \frac{2\pi}{\lambda}} \quad \text{[rad/m]} \qquad (7.17a)$$

$$\boxed{\omega = \frac{2\pi}{\tau} = 2\pi f} \quad \text{[rad/s]} \qquad (7.17b)$$

where λ is the wavelength, τ is the temporal period, and f is the frequency. Inserting Eqs. (7.17a, b) into Eq. (7.14) yields the following useful relationship:

$$\boxed{\upsilon_p = f\lambda} \quad \text{[m/s]} \qquad (7.18)$$

The phase velocity of a harmonic wave is equal to the product of frequency and wavelength.

To examine the relationship between the temporal and spatial variations of a harmonic wave, we consider a wave traveling with velocity v, as shown in Fig. 7.4, which was generated by a source at $x = 0$ and propagated along the x-axis for $0 \leq t \leq t_1$. The figure shows the spatial variation of the wave at time $t = t_1$. The vertical line on the rightmost end represents the leading edge of the wave. The temporal variations of the wave at points $x = x_1$ and $x = x_2$ are shown in the insets. Note that the wavefunction in the time domain is the reverse of that in the space domain because of the negative sign of the wavefunction, as shown in Eq. (7.10). For $x_1 = 3\lambda/8$ and $x_2 = 15\lambda/8$, the temporal variation observed at point x_2 is delayed by $1.5\lambda/v$ [s] with respect to that at point x_1, which corresponds to 1.5τ [s]. If a sinusoidal steady state is assumed, for which the leading edge is completely ignored, we say that the wave at point x_2 lags behind that at point x_1 by π radians in time phase, based on the fact that 1.5τ [s] corresponds to $1.5 \times 2\pi$ [rad].

Example 7.2 Given a harmonic wave $\psi = -4\cos[2\pi(0.2x - 3t)]$, determine the (a) amplitude, (b) direction of propagation, (c) wavelength, (d) temporal period, (e) frequency, and (f) phase velocity.

Solution

(a) 4
(b) $+x$-direction
(c) $k = 2\pi/\lambda = 2\pi \times 0.2 \quad \rightarrow \quad \lambda = 5\,[\text{m}]$
(d) $\omega = 2\pi/\tau = 2\pi \times 3 \quad \rightarrow \quad \tau = 0.33\,[\text{s}]$
(e) $\omega = 2\pi f = 2\pi \times 3 \quad \rightarrow \quad f = 3\,[\text{s}^{-1}]$

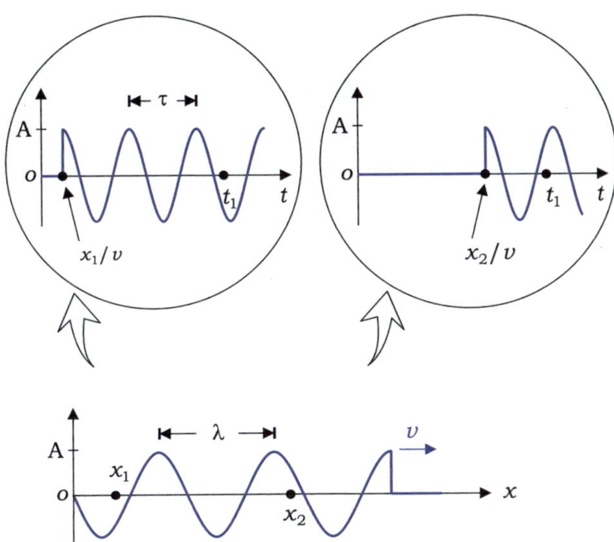

Fig. 7.4 Harmonic wave generated by the source at $x = 0$. The insets show the time behaviors at points $x = x_1$ and $x = x_2$

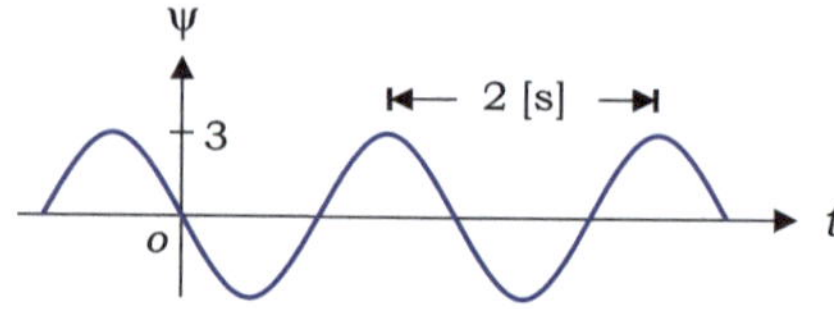

Fig. 7.5 Temporal variation of the harmonic wave at $x = 0$

(f) $\quad \upsilon_p = \omega/k = f\lambda \qquad \rightarrow \qquad \upsilon_p = 15\,[\text{m/s}]$

Exercise 7.4

When a harmonic wave travels in the $+x$-direction with $\upsilon_p = 10\,[\text{m/s}]$, its temporal variation at $x = 0$ is as shown in Fig. 7.5. Express its wavefunction.

Ans. $\psi = 3\cos(0.1\pi x - \pi t - \pi/2)$.

7.1.2 Harmonic Wave in Complex Form

When harmonic waves are expressed in terms of cosine functions, a heavy dose of trigonometric identities is required to handle the wavefunctions. This difficulty can be circumvented by using complex exponentials instead of cosine functions. Moreover, the use of complex exponentials is advantageous when dealing with phases, impedances, and superposition of harmonic waves.

By definition, a complex number $\hat{z}$ comprises real and imaginary parts,

$$\hat{z} = x + iy \tag{7.19}$$

where $i = \sqrt{-1}$, and x and y are real numbers. An exponential function $\exp(x)$ can be extended to a complex exponential function $\exp(\hat{z})$ using the **Euler formula**:

$$\boxed{e^{\pm i\theta} \equiv \cos\theta \pm i\sin\theta} \tag{7.20}$$

From Eq. (7.20), we have

$$\cos\theta = \text{Re}\left[e^{i\theta}\right] = \frac{1}{2}\left[e^{i\theta} + e^{-i\theta}\right] \tag{7.21a}$$

$$\sin\theta = \text{Im}\left[e^{i\theta}\right] = \frac{1}{2i}\left[e^{i\theta} - e^{-i\theta}\right] \tag{7.21b}$$

where $\text{Re}[.]$ and $\text{Im}[.]$ denote the real and imaginary parts of $e^{i\theta}$, respectively. For the cosine reference, the harmonic wave is expressed as

$$\psi(x, t) = A\cos(kx - \omega t + \varphi_o) = \text{Re}\left[Ae^{i(kx - \omega t + \varphi_o)}\right] \tag{7.22}$$

This is known as the **real instantaneous form** of harmonic waves. From this point, with the understanding that the actual wave is its real part, a harmonic wave is expressed simply as

$$\psi = A\,e^{i(kx - \omega t + \varphi_o)} \tag{7.23}$$

This is known as the **complex form** of harmonic waves. Under sinusoidal steady-state conditions, all calculations can be performed in a complex domain using harmonic waves in complex form. Finally, by taking the real part of the result, a solution in the real instantaneous form can be obtained.

A harmonic wave can be expressed in complex form as

$$\psi = \left[\left(A\,e^{i\varphi_o}\right)e^{ikx}\right]e^{-i\omega t} \equiv \left[\hat{A}\,e^{ikx}\right]e^{-i\omega t} \tag{7.24}$$

where $\hat{A}$ is the **complex amplitude** defined by

$$\hat{A} \equiv A\,e^{i\varphi_o} \tag{7.25}$$

Here, A is the amplitude, and φ_o is the phase angle. Throughout the text, a complex number is denoted by $\wedge$ at the top (called "caret" or "hat").

The frequency of a harmonic wave should remain the same, even if the wave propagates through different linear media. Therefore, for simplicity in notation, the time-harmonic term $e^{-i\omega t}$ is omitted in Eq. (7.24) and the harmonic wave is simply expressed as

$$\boxed{\psi = \hat{A}\,e^{ikx}} \tag{7.26}$$

If we are given a harmonic wave expressed in a time-independent complex form as in Eq. (7.26), we obtain its instantaneous value by taking the real part of the product of ψ and $e^{-i\omega t}$.

Exercise 7.5
Given that $\hat{A}_1 = 2\,e^{i2.353}$ and $\hat{A}_2 = 3.46\,e^{i3.924}$, determine the amplitude and phase of the sum $\hat{A}_1 + \hat{A}_2$.

Ans. $4.00\,e^{i3.400}$.

7.1.3 Harmonic Wave in Phasor Form

There is no change in the real instantaneous value given by Eq. (7.22) even if $-i$ is used instead of i in the complex form. The use of $-i$ is preferred when solving practical electromagnetic problems, although the use of i may offer a definite advantage in certain cases. To avoid confusion, it is customary to express $-i$ as $-j$, where $j = \sqrt{-1}$. Therefore, a harmonic wave can also be expressed as

$$\psi(x, t) = \hat{A}\,e^{j(\omega t - kx)} \tag{7.27}$$

As in the case of $e^{-i\omega t}$, the time-harmonic term $e^{j\omega t}$ is dropped from Eq. (7.27) with the understanding that the frequency of a harmonic wave is the same in different linear media. Accordingly, a harmonic wave can be expressed as

$$\boxed{\psi = \hat{A}\,e^{-jkx}} \tag{7.28}$$

This is known as the **phasor form** of a harmonic wave or a **phasor**. The phasor is a complex number whose modulus represents the amplitude of the time-harmonic oscillation and its phase angle represents the initial phase of the oscillation at time $t = 0$.

The instantaneous value can be obtained by taking the real part of the product of the phasor and $e^{j\omega t}$, that is,

$$\psi = \mathrm{Re}\left[\hat{A}\,e^{-jkx}e^{j\omega t}\right] = A\cos(\omega t - kx + \varphi_o) \tag{7.29}$$

where $\hat{A} = A\,e^{j\varphi_o}$ is the complex amplitude.

The two complex forms, given in Eqs. (7.26) and (7.28), have distinct advantages and disadvantages of their own. The complex form in Eq. (7.26) enables us to visualize the spatial variation of the wave at a particular time more easily. For example, the spatial variation at a later time $t_1 + \Delta t$ can be obtained simply by displacing the former wave variation by a distance $\upsilon\Delta t$ along the direction of wave propagation. In contrast, the phasor form in Eq. (7.28) is useful for specifying the time delays of waves at different points in space. A positive phase indicates a lead and a negative phase indicates a lag in the time phase. This also allows us to handle the impedances in a straightforward manner. For example, if the phase angle of the intrinsic impedance is positive, then the electric field leads the magnetic field in the time phase. In summary, for harmonic time-dependence $\exp(-i\omega t)$, the complex form of Eq. (7.26) describes the distribution of *space phases* at a fixed time. For harmonic time-dependence $\exp(j\omega t)$, the phasor in Eq. (7.28) specifies the *time phases* at different points in space.

It is important to remember that only one form should be used consistently throughout a given problem. As a matter of fact, the phasor form will be used extensively in the following chapters.

Example 7.3 A harmonic wave with propagation constant k reflects off the surface of a semi-infinite medium filling the region $x \geq 0$. The transmitted wave is also a harmonic wave, but with k' (Fig. 7.6). Derive the reflection and transmission coefficients of the incident wave using the boundary condition that the disturbance and its spatial derivative are both continuous across the interface.

Solution

In the region $x \leq 0$, the total disturbance is the sum of the disturbances of the incident and reflected waves, that is,

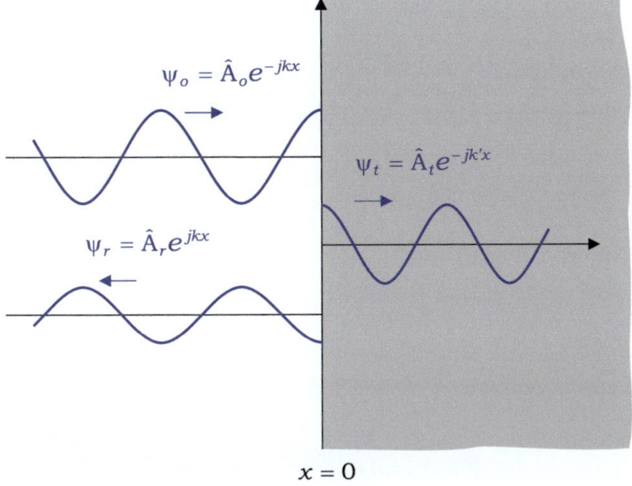

Fig. 7.6 Reflection and transmission of a harmonic wave

$$\psi = \psi_o + \psi_r = \hat{A}_o e^{-jkx} + \hat{A}_r e^{jkx}$$

At the interface at $x = 0$, from the boundary conditions,

$$\hat{A}_o + \hat{A}_r = \hat{A}_t$$

$$-k\hat{A}_o + k\hat{A}_r = -k'\hat{A}_t$$

By solving these equations for complex amplitudes $\hat{A}_r$ and $\hat{A}_t$, the reflection and transmission coefficients can be obtained as follows:

$$\boxed{\Gamma = \frac{\hat{A}_r}{\hat{A}_o} = \frac{k - k'}{k + k'}} \tag{7.30a}$$

$$\boxed{\tau = \frac{\hat{A}_t}{\hat{A}_o} = \frac{2k}{k + k'}} \tag{7.30b}$$

where the former represents the reflection coefficient and the latter represents the transmission coefficient.

Example 7.4 A harmonic wave with $\psi = A e^{-j0.5\pi x + j\omega t}$ propagates in the $+x$-direction in free space. At time $t = 0$, $\psi(x)$ is given as shown in Fig. 7.7.

(a) Plot ψ versus ωt for $t \geq 0$ at points $x = x_1$ and $x = x_2$.
(b) How much $\psi(x_2, t)$ lags behind $\psi(x_1, t)$ in time phase?

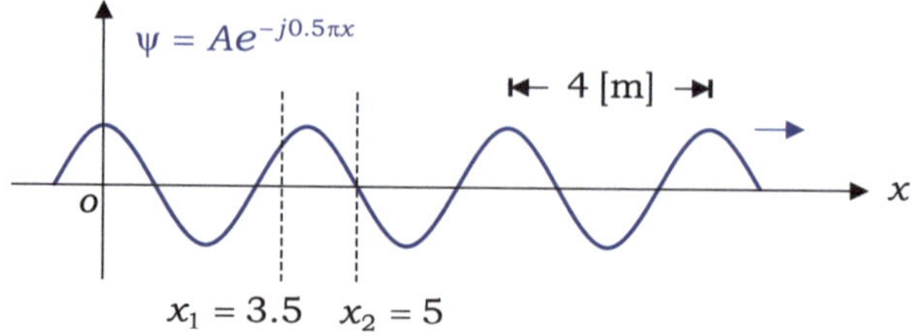

Fig. 7.7 Harmonic wave at time $t = 0$

Solution

(a) Disturbances at points $x = x_1$ and $x = x_2$ are

$$\psi(x_1) = A e^{-j0.5\pi \times 3.5} = A e^{-j1.75\pi}$$

$$\psi(x_2) = A e^{-j0.5\pi \times 5} = A e^{-j2.5\pi}$$

In real instantaneous form, the temporal variations at $x = x_1$ and $x = x_2$ are

$$\psi(x_1, t) = A \cos(\omega t - 1.75\pi)$$

$$\psi(x_2, t) = A \cos(\omega t - 2.5\pi)$$

These can be plotted as functions of ωt, as shown in Fig. 7.8.

(b) The phase difference is given by

$$\Delta\varphi = \varphi_2 - \varphi_1 = (\omega t - 2.5\pi) - (\omega t - 1.75\pi)$$
$$= -0.75\pi \qquad\qquad (7.31)$$

The negative sign in Eq. (7.31) indicates that $\psi(x_2, t)$ lags behind $\psi(x_1, t)$ by 0.75π [rad] in time phase.

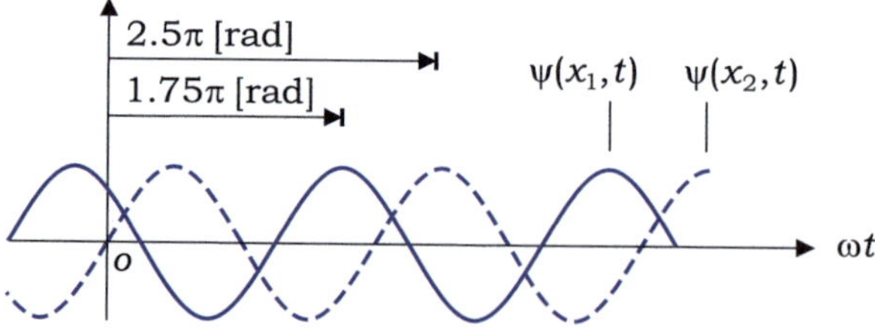

Fig. 7.8 Temporal variations at points $x = x_1$ and $x = x_2$

Exercise 7.6
Given phasor $\psi = 2\,e^{j4.3}$, determine its real instantaneous value.

Ans. $\psi = 2\cos(\omega t + 4.3)$.

Exercise 7.7
For the following pairs of phasors, determine the time phase by which ψ_2 lags behind ψ_1: (a) $\psi_1 = 2\,e^{j4.3}$ and $\psi_2 = 3\,e^{j2.1}$, (b) $\psi_1 = 4\,e^{j\pi}$ and $\psi_2 = -4\,e^{j1.2\pi}$, (c) $\psi_1 = 6\,e^{j0.7\pi}$ and $\psi_2 = 5\,e^{j1.8\pi}$, and (d) $\psi_1 = 2\,e^{-j0.6}$ and $\psi_2 = 2\,e^{-j1.8}$.

Ans. (a) 2.2 [rad], (b) 0.8π [rad], (c) 0.9π [rad], (d) 1.2 [rad].

Review Questions

RQ 7.5	Define the phase of the one-dimensional harmonic wave.	[(7.11)]
RQ 7.6	Express the phase velocity of the harmonic wave.	[(7.14)(7.18)]
RQ 7.7	Define the propagation constant of a wave and its unit.	[(7.17a)]
RQ 7.8	Write the Euler formula.	[(7.20)]
RQ 7.9	What is the complex amplitude?	[(7.26)]
RQ 7.10	What is the phasor form of the harmonic wave?	[(7.28)]
RQ 7.11	What is the time-phase?	[(7.28)]
RQ 7.12	Can the phasor determine wave velocity?	[(7.14)(7.28)]
RQ 7.13	How is the instantaneous value obtained from a phasor?	[(7.29)]

7.2 Plane Wave in Three-Dimensional Space

An extension of the differential wave equation in Eq. (7.6) into three dimensions leads to the following three-dimensional differential wave equation:

$$\nabla^2\psi = \frac{1}{\upsilon^2}\frac{\partial^2\psi}{\partial t^2} \tag{7.32}$$

where ∇^2 is the Laplacian operator, and υ is the wave velocity. It is worth noting that the propagation of a wave in three-dimensional space is governed by Eq. (7.32).

In this section, we demonstrate that the simplest solution of the differential wave equation in Eq. (7.32) is a **uniform plane wave**, for which the points of a constant phase form a "plane" in three-dimensional space, called the **phase front**, and the disturbance is constant or "uniform" over that plane. Because the energy of an actual wave gradually spreads out over its phase front as the wave advances, the wave profile cannot be strictly maintained in a three-dimensional space. A uniform plane wave has the same profile at every point in space and extends to infinity. Therefore, it is a mathematical object. However, a superposition of uniform plane waves can be made to form a wave of an arbitrary shape, satisfying the differential wave equation, in the same way that an arbitrary waveform in time domain can be considered as a superposition of time-harmonic functions by Fourier analysis.

A general solution of the three-dimensional differential wave equation can be expressed as

$$\psi = A_o \cos\left[\left(k_x x + k_y y + k_z z\right) \pm \omega t + \varphi_o\right] \qquad (7.33)$$

where φ_o is a constant phase. The solution in Eq. (7.33) can be justified by direct substitution. The two independent solutions in Eq. (7.33), one with positive sign and another with negative sign in front of ωt, can be linearly combined in such a way as to satisfy given boundary conditions. This particular solution must be unique in a given region of space in accordance with the uniqueness theorem. In many practical problems, we are primarily asked to determine the complex amplitude $A_o \exp(i\varphi_o)$ and the direction of **wavevector** $\mathbf{k} = k_x \mathbf{a}_x + k_y \mathbf{a}_y + k_z \mathbf{a}_z$ in a given region. The three-dimensional wave in Eq. (7.33) reduces to a one-dimensional wave, as shown in Eq. (7.10), if the wave is constrained to propagate along the x-axis.

The wave in Eq. (7.33) can be expressed in complex form as

$$\psi = \hat{A}\, e^{i\mathbf{k}\cdot\mathbf{r} \,\pm\, i\omega t} \qquad (7.34)$$

where the complex amplitude is generally expressed as $\hat{A} = A_o \exp(i\varphi_o)$, and wavevector $\mathbf{k}$ is expressed in Cartesian coordinates as

$$\mathbf{k} = k_x \mathbf{a}_x + k_y \mathbf{a}_y + k_z \mathbf{a}_z \qquad (7.35\text{a})$$

$$|\mathbf{k}| = \sqrt{k_x^2 + k_y^2 + k_z^2} = \frac{2\pi}{\lambda} \qquad (7.35\text{b})$$

where λ is the wavelength. The position vector $\mathbf{r}$ in Eq. (7.34) is expressed in Cartesian coordinates as

$$\mathbf{r} = x\,\mathbf{a}_x + y\,\mathbf{a}_y + z\,\mathbf{a}_z \qquad (7.36)$$

For a wave traveling in three-dimensional space, the spatial variation of the phase constitutes a scalar field that is a smooth function of position at an instant of time. Accordingly, the points of a constant phase define a smooth surface in space, which we refer to as a phase front or a **wavefront**. Let us explore the wave expressed by Eq. (7.34) for the phase front. The phase depends on the dot product of wavevector $\mathbf{k}$ and position vector $\mathbf{r}$, that is, $\varphi = \mathbf{k} \cdot \mathbf{r}$, where $\mathbf{k}$ is a constant vector, and $\mathbf{r}$ varies as a function of position. The spatial points in the same phase a can be determined by solving the following equation:

$$\mathbf{k} \cdot \mathbf{r} = a \qquad (7.37)$$

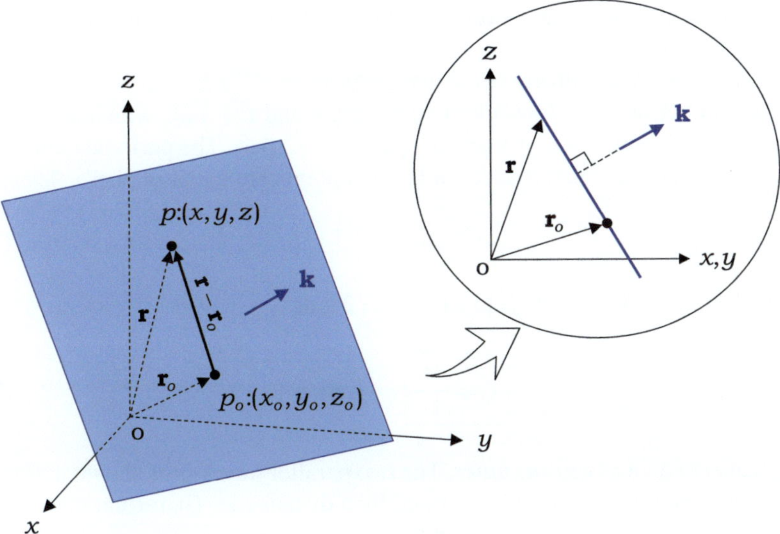

Fig. 7.9 Wavefront perpendicular to wavevector **k**

Fig. 7.10 Two wavefronts
separated by a wavelength

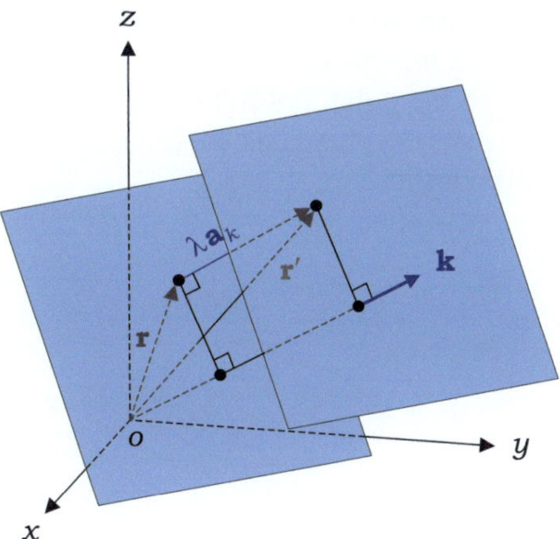

If $\mathbf{r}_o$ is a position vector satisfying Eq. (7.37), that is, $\mathbf{k} \cdot \mathbf{r}_o = a$, Eq. (7.37) can be rewritten as

$$\mathbf{k} \cdot (\mathbf{r} - \mathbf{r}_o) = 0 \tag{7.38}$$

The relationship in Eq. (7.38) can be illustrated graphically as shown in Fig. 7.9. The dot product in Eq. (7.38) ensures that vector $\mathbf{r} - \mathbf{r}_o$ is always perpendicular to constant vector **k**. Therefore, **r** refers to points on the plane perpendicular to **k**, including point

$\mathbf{r}_o$. In other words, the wave expressed in Eq. (7.34) has planar wavefronts, and thus is called a uniform plane wave.

The wavefront of a uniform plane wave repeats itself at regular wavelength intervals in the direction of $\mathbf{k}$. Consider two points, $\mathbf{r}$ and $\mathbf{r} + \lambda\mathbf{a}_k$, which are on a line oriented along the direction of $\mathbf{k}$ and at a distance λ apart. The periodic nature of the waves ensures that the disturbances at the two points are equal in phase. Therefore,

$$\hat{A}\,e^{i\mathbf{k}\cdot\mathbf{r}} = \hat{A}\,e^{i\mathbf{k}\cdot(\mathbf{r}+\lambda\mathbf{a}_k)} \tag{7.39}$$

The equality in Eq. (7.39) leads to $\lambda\mathbf{k}\cdot\mathbf{a}_k = 2\pi$ or

$$\boxed{k = \frac{2\pi}{\lambda}} \quad [\text{rad/m}] \tag{7.40}$$

which is also known as **wavenumber**. The two parallel wavefronts shown in Fig. 7.10 are both perpendicular to $\mathbf{k}$ and separated by a distance λ. Therefore, all the points on these wavefronts are in the same phase.

A three-dimensional view of a uniform plane wave is shown in Fig. 7.11, where the infinite parallel planes represent wavefronts of different phases that are perpendicular to wavevector $\mathbf{k}$ and move in unison in the direction of $\mathbf{k}$ as time progresses. A cosine curve can be used conveniently to represent these three-dimensional wavefronts. Each point on the cosine curve corresponds to an infinite planar wavefront that passes through the point and intersects the horizontal axis at a right angle. The abscissa represents the phase and the ordinate represents the magnitude of the disturbance at a given instant.

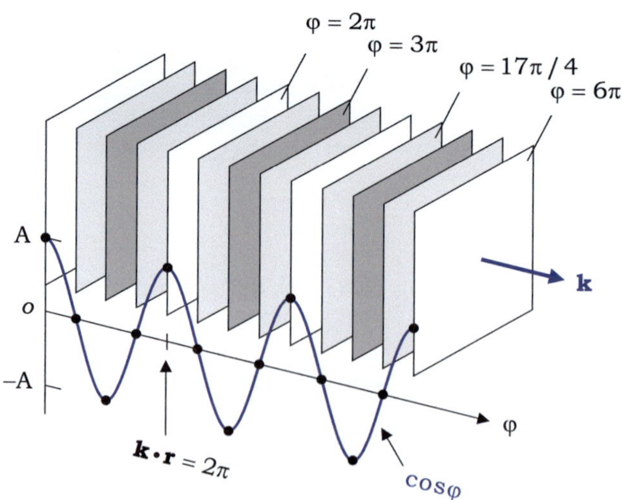

Fig. 7.11 Planar wavefronts moving in unison along the direction of wavevector $\mathbf{k}$

Fig. 7.12 Wavefronts of
uniform plane wave

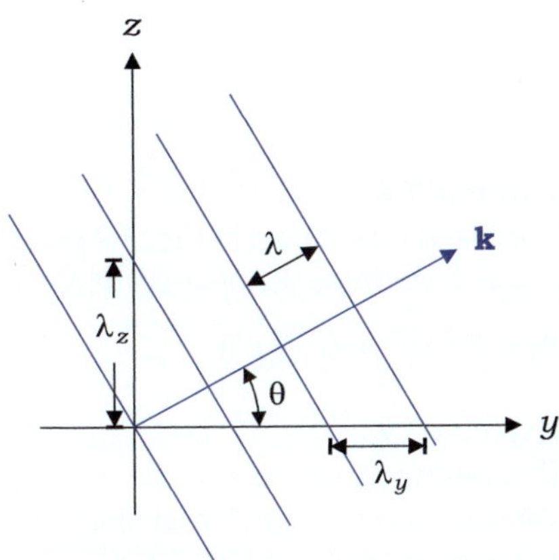

Example 7.5 A uniform plane wave with amplitude A propagates in the direction of **k**, which makes an angle θ with the y-axis in the yz-plane, as shown in Fig. 7.12. (a) Express the wavefunction in complex form, and (b) determine the spatial periods λ_y and λ_z measured along the y- and z-axes, respectively.

Solution

(a) The wavevector is expressed as

$$\mathbf{k} = \frac{2\pi}{\lambda}\left(\cos\theta\,\mathbf{a}_y + \sin\theta\,\mathbf{a}_z\right)$$

The wavefunction in complex form is

$$\psi = A\,e^{i\frac{2\pi}{\lambda}(y\cos\theta + z\sin\theta) - i\omega t} \tag{7.41a}$$

(b) From Eq. (7.41a), the phases at $y = y_1$ and $y = y_1 + \lambda_y$ are

$$\varphi_1 = \frac{2\pi}{\lambda}y_1\cos\theta - \omega t \quad \text{at} \quad y = y_1 \tag{7.41b}$$

$$\varphi_2 = \frac{2\pi}{\lambda}\left(y_1 + \lambda_y\right)\cos\theta - \omega t \quad \text{at} \quad y = y_1 + \lambda_y \tag{7.41c}$$

Because the phase difference is $\varphi_2 - \varphi_1 = 2\pi$, we have

$$\lambda_y = \frac{\lambda}{\cos\theta}$$

Following the same procedure as for λ_y, we obtain

$$\lambda_z = \frac{\lambda}{\sin \theta}$$

Exercise 7.8

The wavefront shown in Fig. 7.9 intercepts the coordinate axes at $x = a$, $y = b$, and $z = c$. Determine the phase difference between the wavefront and origin.

Ans. $ka^2(a^2 + b^2 + c^2)^{-1/2}$.

Exercise 7.9

If a uniform plane wave, as shown in Fig. 7.12, has an angular frequency ω, determine the phase velocity measured along the z-axis.

Ans. $\omega \lambda_z / 2\pi$.

Exercise 7.10

Given a uniform plane wave $4e^{-j(2x+2y+z)}$, determine the tilting angle of the wavefront with respect to the $z = 0$ plane.

Ans. $70.5°$.

Exercise 7.11

For a uniform plane wave with $\tilde{\mathbf{E}} = \sqrt{2}(\mathbf{a}_x + 2\mathbf{a}_y)\exp(-jkz)$, sketch the electric field vectors in the $z = 0$ plane at times (a) $\omega t = 0$, and (b) $\omega t = 3\pi/4$.

Ans.

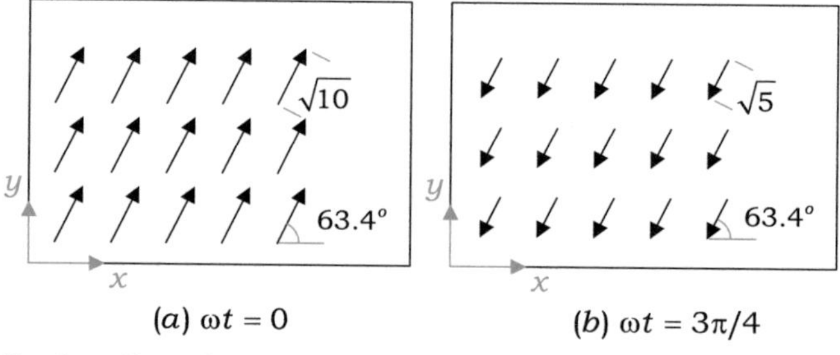

Review Questions

RQ 7.14	Write the three-dimensional differential wave equation.	[(7.32)]
RQ 7.15	Express a uniform plane wave.	[(7.33)(7.34)]
RQ 7.16	What can be learned about the wave from its wavevector?	[(7.35a,b)]
RQ 7.17	Explain how a cosine curve is used to describe a plane wave.	[Fig. 7.11]
RQ 7.18	Explain why the phase velocity of a plane wave depends on the direction of consideration.	[Fig. 7.12]

7.3 Electromagnetic Plane Wave

In free space, where there are no free charges or currents, Maxwell's equations are reduced to

$$\nabla \times \mathscr{E} = -\mu_0 \frac{\partial \mathscr{H}}{\partial t} \tag{7.42a}$$

$$\nabla \times \mathscr{H} = \varepsilon_0 \frac{\partial \mathscr{E}}{\partial t} \tag{7.42b}$$

$$\nabla \cdot \mathscr{E} = 0 \tag{7.42c}$$

$$\nabla \cdot \mathscr{H} = 0 \tag{7.42d}$$

where ε_0 and μ_0 are the permittivity and permeability of free space, respectively. Taking the curl on both sides of Eq. (7.42a) and using Eq. (7.42b), we obtain

$$\nabla \times \nabla \times \mathscr{E} = -\varepsilon_0 \mu_0 \frac{\partial^2 \mathscr{E}}{\partial t^2} \tag{7.43}$$

By applying the vector identity $\nabla \times \nabla \times \mathbf{U} = \nabla\nabla \cdot \mathbf{U} - \nabla^2 \mathbf{U}$ and using Eq. (7.42c) into Eq. (7.43), the three-dimensional differential wave equation is obtained as

$$\boxed{\nabla^2 \mathscr{E} = \varepsilon_0 \mu_0 \frac{\partial^2 \mathscr{E}}{\partial t^2}} \tag{7.44}$$

Because the three vector components of $\mathscr{E}$ are mutually exclusive, Eq. (7.44) can be decomposed into three scalar differential wave equations. In Cartesian coordinates,

$$\nabla^2 \mathscr{E}_x = \varepsilon_0 \mu_0 \frac{\partial^2 \mathscr{E}_x}{\partial t^2} \tag{7.45a}$$

$$\nabla^2 \mathscr{E}_y = \varepsilon_0 \mu_0 \frac{\partial^2 \mathscr{E}_y}{\partial t^2} \tag{7.45b}$$

$$\nabla^2 \mathscr{E}_z = \varepsilon_0 \mu_0 \frac{\partial^2 \mathscr{E}_z}{\partial t^2} \tag{7.45c}$$

These equations have three uniform plane waves as solutions:

$$\mathscr{E}_x(\mathbf{r}, t) = E_1 \cos(\mathbf{k}_1 \cdot \mathbf{r} - \omega t + \varphi_1) \tag{7.46a}$$

$$\mathscr{E}_y(\mathbf{r}, t) = E_2 \cos(\mathbf{k}_2 \cdot \mathbf{r} - \omega t + \varphi_2) \tag{7.46b}$$

$$\mathscr{E}_z(\mathbf{r}, t) = E_3 \cos(\mathbf{k}_3 \cdot \mathbf{r} - \omega t + \varphi_3) \qquad (7.46c)$$

where $\mathbf{k}_1$, $\mathbf{k}_2$, and $\mathbf{k}_3$ are wavevectors with the same magnitude $\omega\sqrt{\varepsilon_0\mu_0}$, $\mathbf{r}$ is the position vector, ω is the angular frequency, and φ_1, φ_2, and φ_3 are constant phases. Because the three components in Eqs. (7.46a)–(7.46c) should constitute a single wave, it is required that $\mathbf{k}_1 \cdot \mathbf{r} = \mathbf{k}_2 \cdot \mathbf{r} = \mathbf{k}_3 \cdot \mathbf{r}$ at every point in space, which only holds for $\mathbf{k}_1 = \mathbf{k}_2 = \mathbf{k}_3$. Therefore, the solution of the three-dimensional differential wave equation is given by a uniform plane wave, that is,

$$\boxed{\mathscr{E}(\mathbf{r}, t) = \mathrm{Re}\left[\hat{\mathbf{E}}_o\, e^{i\mathbf{k}\cdot\mathbf{r} - i\omega t}\right]} \qquad (7.47)$$

In general, $\hat{\mathbf{E}}_o = E_1 e^{i\varphi_1}\mathbf{a}_x + E_2 e^{i\varphi_2}\mathbf{a}_y + E_3 e^{i\varphi_3}\mathbf{a}_z$. For a linearly polarized wave, $\varphi_1 = \varphi_2 = \varphi_3 = \varphi_o$, as discussed in Chap. 8. In this case,

$$\hat{\mathbf{E}}_o = E_o e^{j\varphi_o}\, \mathbf{a}_E \qquad (7.48)$$

where E_o is the amplitude, φ_o is the phase angle, and $\mathbf{a}_E$ is the unit vector in the direction of the electric field. Note that $\hat{\mathbf{E}}_o$ is a vector with complex scalar components.

Although the wavefunction in Eq. (7.47) describes the spatiotemporal variation of a uniform plane wave in exactly the same way as that in Eq. (7.34), electromagnetic waves should be in the form of Eq. (7.47), where the electric field vector $\hat{\mathbf{E}}_o$ is perpendicular to the wavevector $\mathbf{k}$, as discussed later in the following section.

In free space, the electromagnetic plane wave propagates in the direction of $\mathbf{k}$ with the phase velocity given by

$$\upsilon_p = \frac{\omega}{k} = \frac{1}{\sqrt{\varepsilon_0\mu_0}} \cong 3 \times 10^8 \ \ [\mathrm{m/s}] \qquad (7.49)$$

Because the phase velocity is equal to the speed of light, c, the wavenumber in free space can be expressed as

$$\boxed{k = \frac{\omega}{c}} \qquad (7.50)$$

In material media, c should be replaced by υ_p, which is a distinct characteristic of the medium.

Following the same procedure as for $\mathscr{E}$, we can derive the differential wave equation for the magnetic field from Maxwell's equations, as follows:

$$\boxed{\nabla^2 \mathscr{H} = \varepsilon_0\mu_0 \frac{\partial^2 \mathscr{H}}{\partial t^2}} \qquad (7.51)$$

A uniform plane wave is the simplest solution to Eq. (7.51), which has the same form as Eq. (7.47), that is,

$$\mathscr{H}(\mathbf{r}, t) = \text{Re}\left[\hat{\mathbf{H}}_o\, e^{i\mathbf{k}\cdot\mathbf{r}\, -\, i\omega t}\right] \tag{7.52}$$

where $\hat{\mathbf{H}}_o$ is the complex amplitude. In view of the fact that $\mathscr{E}$ and $\mathscr{H}$ are related to each other by a constant of proportionality, that is, $|\hat{\mathbf{E}}_o/\hat{\mathbf{H}}_o| = \sqrt{\mu_0/\varepsilon_0}$ in free space, we only need to solve either one of Eqs. (7.44) and (7.51) and obtain the other field, if necessary, from Maxwell's equations.

Exercise 7.12
Show that Eq. (7.44) has another solution, $\mathscr{E}(\mathbf{r}, t) = \text{Re}[\hat{\mathbf{E}}_o \exp(i\mathbf{k}\cdot\mathbf{r} + i\omega t)]$.

Exercise 7.13
Find the ratio $|\hat{\mathbf{E}}_o/\hat{\mathbf{B}}_o|$ in free space.

Ans. c, speed of light in free space.

Exercise 7.14
What can be discovered regarding an electromagnetic plane wave propagating in free space from its wavevector?

Ans. Direction of propagation, wavelength, and frequency.

7.3.1 Transverse Electromagnetic Wave

If a uniform plane wave propagates in the $+z$-direction, the planar wavefront coincides with the xy-plane and the electric field is uniform over the wavefront at any instant. In this case, $\mathscr{E}$ is independent of x and y such that

$$\frac{\partial \mathscr{E}(\mathbf{r}, t)}{\partial x} = \frac{\partial \mathscr{E}(\mathbf{r}, t)}{\partial y} = 0 \tag{7.53}$$

Therefore, $\mathscr{E}$ can be expressed in Cartesian coordinates as

$$\mathscr{E}(\mathbf{r}, t) = \mathscr{E}_x(z, t)\,\mathbf{a}_x + \mathscr{E}_y(z, t)\,\mathbf{a}_y + \mathscr{E}_z(z, t)\,\mathbf{a}_z \tag{7.54}$$

This should satisfy Maxwell's equations. Subsequently, inserting Eq. (7.54) into Eq. (7.42c) yields

$$\nabla \cdot \mathscr{E} = \frac{\partial \mathscr{E}_z}{\partial z} = 0 \tag{7.55}$$

This indicates that $\mathscr{E}_z$ is constant throughout the entire space. However, a constant cannot be an electric field within the context of electromagnetics, in which both

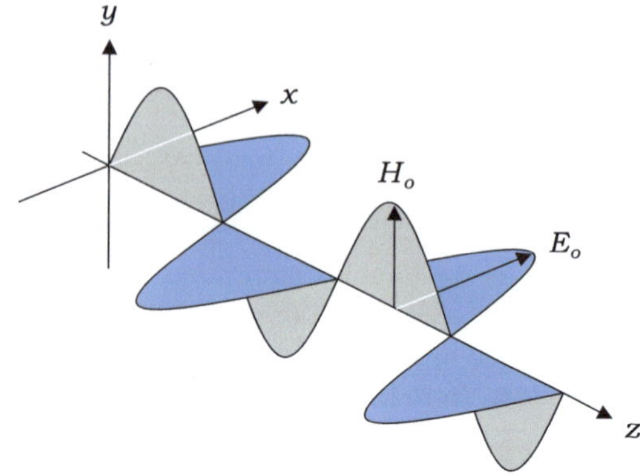

Fig. 7.13 Electromagnetic plane wave propagating in the $+z$-direction

electric and magnetic fields vary in space and time. Therefore,

$$\mathscr{E}_z = 0 \tag{7.56}$$

Subsequently, Eq. (7.54) is reduced to

$$\mathscr{E} = \mathscr{E}_x(z, t)\,\mathbf{a}_x + \mathscr{E}_y(z, t)\,\mathbf{a}_y \tag{7.57}$$

This shows that the electric field is perpendicular to the direction of wave propagation or $\mathbf{a}_z$.

Next, inserting Eq. (7.57) into Eq. (7.42a), we obtain

$$-\frac{\partial \mathscr{E}_y}{\partial z}\,\mathbf{a}_x + \frac{\partial \mathscr{E}_x}{\partial z}\,\mathbf{a}_y = -\mu_0 \frac{\partial \mathscr{H}_x}{\partial t}\,\mathbf{a}_x - \mu_0 \frac{\partial \mathscr{H}_y}{\partial t}\,\mathbf{a}_y - \mu_0 \frac{\partial \mathscr{H}_z}{\partial t}\,\mathbf{a}_z \tag{7.58}$$

The equality in Eq. (7.58) requires $\mathscr{H}_z$ to be constant over time. Because a constant cannot be a magnetic field, we have

$$\mathscr{H}_z = 0 \tag{7.59}$$

This indicates that the corresponding magnetic field is also perpendicular to the direction of wave propagation. The results in Eqs. (7.56) and (7.59) lead to the conclusion that *electromagnetic waves are transverse waves*.

The time-varying electric and magnetic fields of an electromagnetic wave are mutually orthogonal and perpendicular to the wavevector (Fig. 7.13). According to the right-hand rule, the right thumb points in the direction of $\mathbf{k}$ when the fingers rotate from $\mathscr{E}$ to $\mathscr{H}$, that is,

$$\boxed{\mathbf{a}_E \times \mathbf{a}_H = \mathbf{a}_k} \tag{7.60}$$

where $\mathbf{a}_E$, $\mathbf{a}_H$, and $\mathbf{a}_k$ are the unit vectors in the directions of $\mathscr{E}$, $\mathscr{H}$, and $\mathbf{k}$, respectively. The relation in Eq. (7.60) holds true at every point in space, at all times.

Example 7.6 For an electric field $\mathscr{E} = E_o\,\mathbf{a}_x \cos(kz - \omega t)$ in free space, find the associated magnetic field.

Solution

Inserting $\mathscr{E}$ into Eq. (7.42a) gives

$$\mathbf{a}_y E_o \frac{\partial}{\partial z}\cos(kz - \omega t) = -\mu_0 \frac{\partial}{\partial t}\left(\mathscr{H}_x\,\mathbf{a}_x + \mathscr{H}_y\,\mathbf{a}_y + \mathscr{H}_z\,\mathbf{a}_z\right) \tag{7.61a}$$

For equality to hold, $\mathscr{H}_x = \mathscr{H}_z = 0$. Thus, Eq. (7.61a) reduces to

$$kE_o \sin(kz - \omega t) = \mu_0 \frac{\partial \mathscr{H}_y}{\partial t} \tag{7.61b}$$

Integrating both sides of Eq. (7.61b) over t, and using $k = \omega\sqrt{\varepsilon_0 \mu_0}$, we obtain

$$\mathscr{H}_y = \sqrt{\frac{\varepsilon_0}{\mu_0}}\,E_o \cos(kz - \omega t) \tag{7.61c}$$

Thus, the associated magnetic field is

$$\mathscr{H} = \sqrt{\frac{\varepsilon_0}{\mu_0}}\,E_o\,\mathbf{a}_y \cos(kz - \omega t) \tag{7.61d}$$

where the constant of integration is set to zero, because it does not constitute an electromagnetic wave. Note that the expressions for $\mathscr{E}$ and $\mathscr{H}$ are identical except for their amplitudes. The amplitude of $\mathscr{H}$ differs from that of $\mathscr{E}$ by a factor of $\sqrt{\varepsilon_0/\mu_0}$, which is a real quantity. Therefore, the electric and magnetic fields are in phase in free space. Moreover, the electric and magnetic field vectors are perpendicular to each other in such a way that $\mathscr{E} \times \mathscr{H}$ points in the direction of wave propagation or the wavevector.

Exercise 7.15
Even if the time-harmonic function $\mathscr{E}(\mathbf{r}, t)$ satisfies Eq. (7.44), it may not necessarily represent an electromagnetic wave. Explain why.

Ans. Electromagnetic waves are transverse waves.

Exercise 7.16
For a uniform plane wave with a wavevector parallel to $\mathbf{a}_x + 2\mathbf{a}_y$ and $\mathscr{E}$ parallel to $\mathbf{a}_z$, determine the direction of $\mathscr{H}$.

Ans. $2\mathbf{a}_x - \mathbf{a}_y$.

Exercise 7.17

Explain why the following cannot be an electromagnetic wave in free space: (a) $\mathscr{E} = (3\mathbf{a}_x + 5\mathbf{a}_z)\cos(2x - 6 \times 10^8 t)$, and (b) $\mathscr{E} = 4\,\mathbf{a}_z\cos(2x + 3y - 3 \times 10^8 t)$.

Ans. (a) It is not a transverse wave, (b) Velocity is not the speed of light.

Review Questions

RQ 7.19	Write Maxwell's equations in free space.	[(7.42)]
RQ 7.20	Write the differential wave equation for $\mathscr{E}$ in free space.	[(7.44)]
RQ 7.21	Define the wavenumber.	[(7.40)(7.50)]
RQ 7.22	What makes $\mathscr{H}$ different from $\mathscr{E}$, even if they satisfy differential wave equations of the same form?	[(7.42a,b)]
RQ 7.23	Explain how the direction of $\mathbf{k}$ is determined from $\mathscr{E}$ and $\mathscr{H}$.	[(7.60)]
RQ 7.24	What is the proportionality constant between $|\mathscr{E}|$ and $|\mathscr{H}|$ in free space?	[(7.61d)]
RQ 7.25	Is it true that $\mathscr{E}$ and $\mathscr{H}$ in free space are always in phase?	[Fig. 7.13]

7.4 Problems

One-dimensional waves

7.1 Determine which of the following can represent a wave traveling in free space:

(a) $\psi = (2x - 3t)^2$,
(b) $\psi = \sin(2x^2 - 3t^2)$,
(c) $\psi = \cos(x^2 + t^2 + 2xt)$, and
(d) $\psi = e^{-3x}\cos(2t - x)$.

Harmonic waves

7.2 Given a traveling wave $\psi = e^{2+j1.2}e^{j100\pi t - j0.02\pi z}$ [cm], determine the following wave characteristics: (a) amplitude, (b) wavelength, (c) frequency, (d) phase velocity, and (e) propagation direction (SI units are used unless otherwise specified).

7.3 When a harmonic wave travels with phase velocity $v_p = 15$ [m/s] in the $+x$-direction, its profile taken at time $t = 0.4$ [s] is as shown in Fig. 7.14. Express the wavefunction using cosine reference.

7.4 A harmonic wave in free space is given by $\psi = A_o \exp(-jkx)$ for time dependence $\exp(j\omega t)$. The wave reflected off the interface at $x = x_3$ becomes a wave with $\psi' = A_o' \exp(jkx)$, as shown in Fig. 7.15. If the boundary condition requires $\psi'/\psi = -0.5$ at the interface, find

(a) A_o',

(b) $\psi_T = \psi + \psi'$ at $x = x_1$ and $x = x_2$, and

(c) time delay of $\psi_T(x_2)$ with respect to $\psi_T(x_1)$.

7.5 If the interface shown in Fig. 7.6 coincides with the $x = a$ plane, instead of the $x = 0$ plane, express $\hat{A}_r/A_o$ and $\hat{A}_t/A_o$ in terms of Γ, τ, and a.

7.6 Harmonic wave ψ_1 is normally incident on the dielectric slab, as shown in Fig. 7.16. The wave bounces back and forth inside the slab. The sum of all forward waves constitutes ψ_3 and the sum of all backward waves constitutes ψ_3'. The transmitted wave is denoted as ψ_4. Express the complex amplitudes in terms of A_1 using Eqs. (7.30a) and (7.30b) and the boundary condition, where ψ and $\partial\psi/\partial x$ are continuous across the interface.

7.7 When a harmonic wave travels at phase velocity $v_p = 10\,[\text{m/s}]$ in the $+x$-direction, its temporal behavior at $x = 4\,[\text{m}]$ is as shown in Fig. 7.17. Determine (a) angular frequency, and (b) propagation constant. Next, (c) express the wavefunction using cosine reference.

7.8 When a uniform plane wave with a wavelength of 18.57 [m] propagates in free space, the waves at points $(2, 4, 0)$ and $(1, 1, 0)$ lag behind the wave at the origin by 0.44π [rad] and 0.14π [rad] in time phase, respectively. Using the general expression $Ae^{-j\mathbf{k}\cdot\mathbf{r}+j\omega t}$, determine the wavevector.

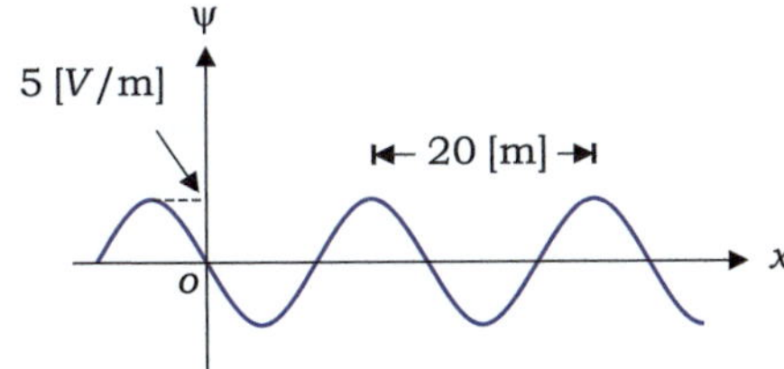

Fig. 7.14 Profile of harmonic wave at $t = 0.4$ [s] (Problem 7.3)

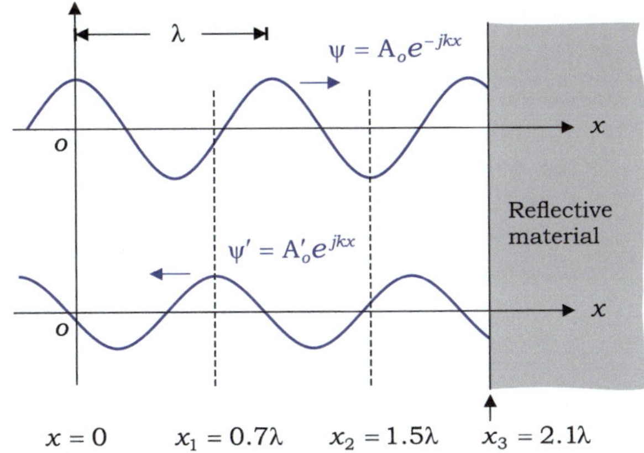

Fig. 7.15 Reflection of harmonic wave off the plane interface (Problem 7.4)

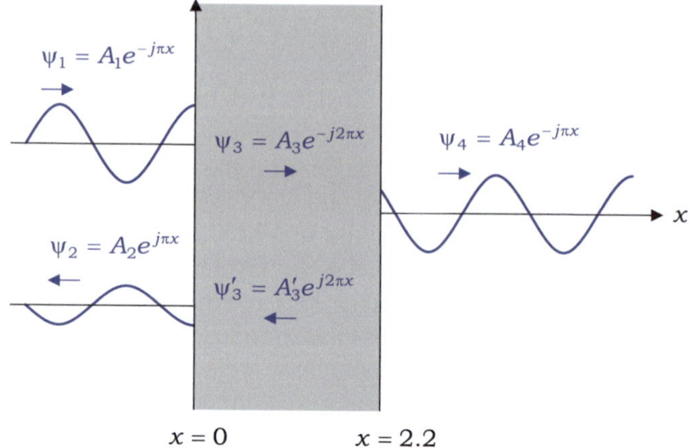

Fig. 7.16 Harmonic wave incident on a dielectric slab (Problem 7.6)

Fig. 7.17 Temporal behavior of harmonic wave (Problem 7.7)

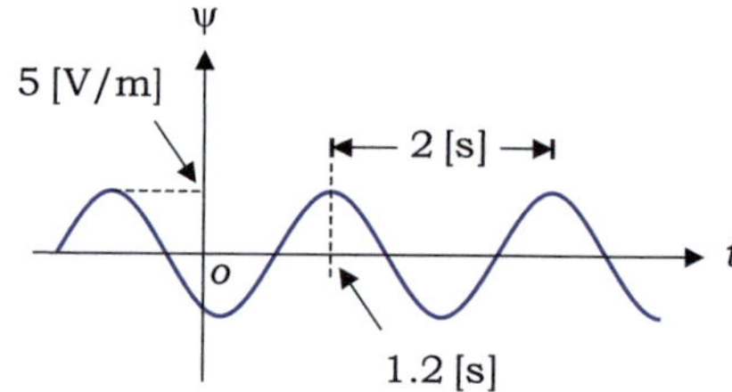

Plane waves

7.9   A uniform plane wave with a frequency of 60 [Hz] travelling in free space has wavevector $\mathbf{k} = 5\mathbf{a}_x + 3\mathbf{a}_y + \sqrt{2}\,\mathbf{a}_z$. Determine (a) phase velocity, and (b) phase velocity measured along the x-axis.

7.10   For a uniform plane wave with wavevector $\mathbf{k} = 10\big(\mathbf{a}_x + 2\mathbf{a}_y + 2\mathbf{a}_z\big)$, determine (a) the wavelength, and (b) the spatial period measured along the x-axis.

7.11   A uniform plane wave with wavelength $\lambda = 0.25\pi$ [m] travels away from the origin in free space. At time $t = t_o$, the wavefront intersects with the Cartesian axes at $x = 1/2$, $y = 1/\sqrt{3}$, and $z = 1/3$. For the time dependence $e^{-i\omega t}$, determine (a) the wavevector, and (b) the phase difference between the wavefront and the origin.

7.12   A planar wavefront intersects the Cartesian axes at $x = 3$, $y = 2\sqrt{3}$, and $z = 2$ at time $t = 0$. If it passes through the origin at $t = 0.2$ [s], determine (a) the direction of wave propagation, and (b) the phase velocity.

7.13   A uniform plane wave with $\psi = 5\exp[i40\pi(x+2y+2z) - i\omega t]$ travels at phase velocity $\upsilon_p = 50$ [m/s]. If the wavefront intersects the x-axis at $x = 0.5$ [m] at time $t = 0$, determine the x-intercept of the wavefront at time $t = 0.02$ [s].

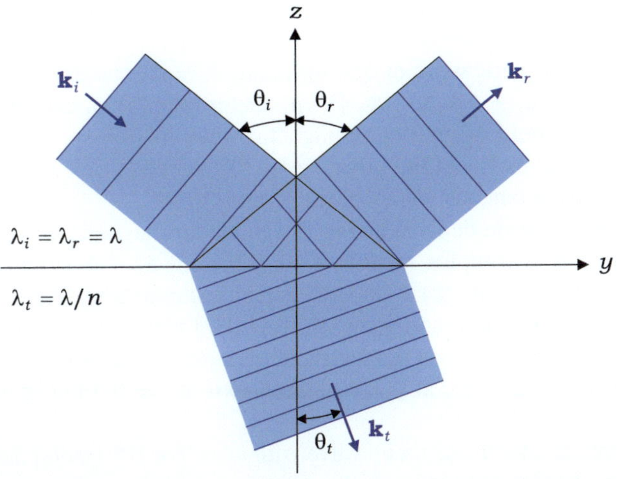

Fig. 7.18 Incident, reflected, and transmitted plane waves with wavevectors all lying in the yz-plane (Problem 7.14)

7.14 A uniform plane wave of wavelength λ is incident on an interface at $z = 0$ and is partly reflected from and partly transmitted across the interface, as shown in Fig. 7.18. The boundary conditions require that the spatial periods along the y-axis are the same. Show that $\theta_r = \theta_i$ and $\sin\theta_i = n\sin\theta_t$.

Electromagnetic plane waves

7.15 For each of the following functions, determine whether it can represent an electromagnetic wave propagating in free space. If not, explain why (The units of distance and time are meter and second, respectively).

(a) $\mathscr{E} = (2\mathbf{a}_x + 3\mathbf{a}_y)\sin(z - 3 \times 10^8 t)$,

(b) $\mathscr{E} = 10\,\mathbf{a}_z\cos(0.1x - 3 \times 10^6 t)$,

(c) $\mathscr{E} = 4\cos(3x + 9 \times 10^8 t)$,

(d) $\mathscr{E} = (3\mathbf{a}_x + \mathbf{a}_z)\sin(0.1z + 3 \times 10^7 t)$,

(e) $\mathscr{E} = 2\,\mathbf{a}_y\cos(100x + 3 \times 10^{10} t + 0.2)$,

(f) $\mathscr{E} = 6\,\mathbf{a}_z\cos(4x)\cos(12 \times 10^8 t)$, and

(g) $\mathscr{E} = 0.5\,\mathbf{a}_z\cos(3x - 4y + 15 \times 10^8 t)$.

7.16 Which of the following is a transverse electromagnetic wave propagating in free space?

(a) $\psi = (\mathbf{a}_x - i2\mathbf{a}_y)\,e^{i10z}$,

(b) $\psi = (1 + i0.5)\,\mathbf{a}_x e^{-i12.5z + i0.3}$,

(c) $\psi = 3e^{i1.4}\mathbf{a}_x e^{i(2y - 5z)}$,

(d) $\psi = (15\mathbf{a}_x + 12\mathbf{a}_y)\,e^{i(4x - 5y + 3z)}$,

(e) $\psi = (15\mathbf{a}_x - 12\mathbf{a}_y)\,e^{i(4x - 5y - 3z)}$,

(f) $\psi = (15\mathbf{a}_x + i12\mathbf{a}_y)\,e^{i(4x - 5y + 3z)}$, and

(g) $\psi = 15\mathbf{a}_x e^{i(4x-5y+3z)}$.

7.17 An electromagnetic wave in free space has $\mathscr{E} = \mathbf{E}_o \cos(3x + 4y - \omega t)$ [V/m] with $\mathbf{E}_o = 12\mathbf{a}_x + b\,\mathbf{a}_y + 10\mathbf{a}_z$ [V/m]. Determine (a) ω, and (b) $\mathbf{E}_o$.

7.18 The electric field of an electromagnetic wave in free space is expressed in complex form as $\mathbf{E} = (3\mathbf{a}_x + 4\mathbf{a}_y) \exp(i10^6 z)$ for the time dependence $e^{-i\omega t}$. Find the instantaneous values of (a) $\mathscr{E}$, and (b) $\mathscr{H}$.

7.19 For harmonic time-dependence $e^{-i\omega t}$, an electromagnetic wave in free space is expressed in complex form as $\mathbf{E} = [(1 + i2)\,\mathbf{a}_x + \mathbf{a}_y]\exp(i10^4 z - i0.8)$. Determine (a) ω, and (b) $\mathscr{E}$ and $\mathscr{H}$ in real instantaneous form.

7.20 An electromagnetic plane wave of 10 [MHz] propagates in the direction of $3\mathbf{a}_x + 4\mathbf{a}_y$ in free space. The electric field points in the z-direction and reaches a maximum of 2 [V/m] at $y = 6$ on the y-axis at $t = 0$. Find an expression for $\mathscr{E}(\mathbf{r}, t)$.

7.21 Given that an electromagnetic wave with $\omega = 9 \times 10^9$ [rad/s] has electric and magnetic field amplitudes of $20\mathbf{a}_z$ [V/m] and $(1/90\pi)(9\mathbf{a}_x - 12\mathbf{a}_y)$ [A/m], respectively, in free space, find $\mathscr{E}$ and $\mathscr{H}$.

Chapter 8
Time-Harmonic Electromagnetic Wave

Thus far, we have introduced the concept of wave motion, which is governed by the differential wave equation, and have shown that a uniform plane wave is the most general solution to the differential wave equation. We showed that Maxwell's equations can be combined to provide three-dimensional differential wave equations that time-varying electric and magnetic fields satisfy at all times. We also observed that the electromagnetic wave was a transverse wave and its frequency remained constant as it propagated through different linear media. Therefore, the phasor concept is particularly useful for describing the time-harmonic behavior of the electric and magnetic fields under sinusoidal steady-state conditions.

In this chapter, we focus our attention on the propagation of electromagnetic waves in low-loss dielectrics and good conductors under sinusoidal steady-state conditions. All waves are treated as single-frequency plane waves with harmonic time-dependence $e^{j\omega t}$. We begin with the phasor form of Maxwell's equations and discuss the changes in Maxwell's equations, and thus in the wavefunction, caused by the damping force and conduction electrons in lossy media. Material parameters such as intrinsic impedance, attenuation, and phase constants are introduced. Electromagnetic waves are transverse waves that carry electromagnetic power. The polarization vector and the concept of the Poynting vector are also discussed. Next, the boundary conditions for time-varying electric and magnetic fields are derived from Maxwell's equations, followed by the reflection and transmission coefficients of a wave obliquely incident on a plane interface. The Brewster and critical angles are explained.

8.1 The Phasor

Perfect dielectrics contain no free charges or currents (i.e., $\rho_v = 0$ and $\boldsymbol{J} = 0$). If the material is homogeneous, linear, and isotropic, the permittivity and permeability are constants that are independent of the direction and magnitude of the electric and magnetic fields. Under these conditions, Maxwell's equations become

$$\nabla \times \mathscr{E} = -\mu \frac{\partial \mathscr{H}}{\partial t} \tag{8.1a}$$

$$\nabla \times \mathscr{H} = \varepsilon \frac{\partial \mathscr{E}}{\partial t} \tag{8.1b}$$

$$\nabla \cdot \mathscr{E} = 0 \tag{8.1c}$$

$$\nabla \cdot \mathscr{H} = 0 \tag{8.1d}$$

where $\mathscr{E}$ and $\mathscr{H}$ are the real instantaneous values of the electric and magnetic fields, respectively, ε is the permittivity, and μ is the permeability.

Maxwell's equations can be combined to yield the three-dimensional differential wave equation for $\mathscr{E}$, that is,

$$\nabla^2 \mathscr{E} = \mu\varepsilon \frac{\partial^2 \mathscr{E}}{\partial t^2} \tag{8.2}$$

In a medium of infinite extent, which is said to be *unbounded*, a uniform plane wave is a general solution to the differential wave equation, and is expressed as

$$\mathscr{E}(\mathbf{r}, t) = \mathrm{Re}\left[\hat{\mathbf{E}}_o\, e^{-j\mathbf{k}\cdot\mathbf{r} + j\omega t}\right] \tag{8.3}$$

where $j = \sqrt{-1}$, and $\hat{\mathbf{E}}_o$ is a vector with complex scalar components. Hereafter, the complex numbers are denoted by the symbol ^ (call "hat"). As discussed in the previous chapter, the exponent in Eq. (8.3), or $-\mathbf{k}\cdot\mathbf{r}$, indicates that the phase front is an infinite plane perpendicular to wavevector $\mathbf{k}$. As $\hat{\mathbf{E}}_o$ is constant, the magnitude, direction, and phase of the electric field are constant and uniform over the phase front. Therefore, this wave is called a uniform plane wave. The time factor in Eq. (8.3), or $e^{j\omega t}$, indicates that $\mathscr{E}$ varies sinusoidally over time. Thus, this wave is also called a **time-harmonic wave**. The wavevector is defined as

$$\mathbf{k} = \omega\sqrt{\mu\varepsilon}\, \mathbf{a}_k \tag{8.4}$$

where ω is the angular frequency. Although the differential wave equation in Eq. (8.2) passes on information about the magnitude of $\mathbf{k}$ through ε and μ, it imposes no limit on the direction of $\mathbf{k}$. In an unbounded medium, the direction of $\mathbf{k}$ is solely determined by the source location.

When a uniform plane wave impinges on a plane interface between two adjoining media of different types, a part of the wave is reflected back into the first medium. Therefore, the total electric field in the first medium is given by the sum of those of the incident and reflected waves, that is,

$$\mathscr{E}(\mathbf{r}, t) = \mathrm{Re}\left[\left(\hat{\mathbf{E}}_o^i\, e^{-j\mathbf{k}_i\cdot\mathbf{r}} + \hat{\mathbf{E}}_o^r\, e^{-j\mathbf{k}_r\cdot\mathbf{r}}\right) e^{j\omega t}\right] \tag{8.5}$$

where i and r denote incident and reflected waves, respectively. It is important to note that, even if the direction of the reflected wave is opposite to that of the incident wave, the two waves are expressed in the same form. However, the reverse direction is taken care of by the negative sign implicit in wavevector $\mathbf{k}_r$. A direct substitution shows that the general solution given in Eq. (8.5) satisfies the differential wave equation only if $|\mathbf{k}_i| = |\mathbf{k}_r| = \omega\sqrt{\mu\varepsilon}$. The direction of $\mathbf{k}_i$ and amplitude $\hat{\mathbf{E}}_o^i$ are prescribed by the source of the incident wave; they are usually known in advance. In contrast, the direction of $\mathbf{k}_r$ and amplitude $\hat{\mathbf{E}}_o^r$ are as-yet-unknown constants, which can be determined later by applying the boundary conditions for $\mathscr{E}$ at the interface.

Rewriting Eq. (8.5) in a more general form, we have

$$\boxed{\mathscr{E}(\mathbf{r}, t) = \mathrm{Re}\left[\tilde{\mathbf{E}}(x, y, z)\, e^{j\omega t}\right]} \tag{8.6}$$

where the space coordinates are separated from the time variable through a time-independent function $\tilde{\mathbf{E}}(x, y, z)$ called the **electric field phasor**. It prescribes *the amplitude and phase of the time-harmonic oscillation of the electric field as a function of position*. This should not be confused with phasor amplitudes such as $\hat{\mathbf{E}}_o^i$ and $\hat{\mathbf{E}}_o^r$, which are constants. The real instantaneous value of the electric field can be obtained at any time by multiplying the phasor by time factor $e^{j\omega t}$ and taking the real part.

Following the same procedure used for $\mathscr{E}$, Maxwell's equations can be combined to yield a differential wave equation for $\mathscr{H}$, which has the same form as Eq. (8.2). The general solution for $\mathscr{H}$ has the same form as Eq. (8.6), that is,

$$\boxed{\mathscr{H}(\mathbf{r}, t) = \mathrm{Re}\left[\tilde{\mathbf{H}}(x, y, z)\, e^{j\omega t}\right]} \tag{8.7}$$

where $\tilde{\mathbf{H}}(x, y, z)$ is the magnetic field phasor.

Example 8.1 Show that phasor $\tilde{\mathbf{E}} = \mathbf{a}_y\, 2E_o \cos(k_1 x)\, e^{-jk_3 z}$ can be resolved into two uniform plane waves, and sketch their wavefronts in region $z \le 0$.

Solution

Using the Euler formula, the phasor can be rearranged as

$$\tilde{\mathbf{E}} = \mathbf{a}_y 2E_o \frac{1}{2}\left[e^{jk_1 x} + e^{-jk_1 x}\right] e^{-jk_3 z}$$

$$= \mathbf{a}_y E_o\, e^{-j(-k_1 x + k_3 z)} + \mathbf{a}_y E_o\, e^{-j(k_1 x + k_3 z)}$$

$$\equiv \tilde{\mathbf{E}}' + \tilde{\mathbf{E}}''$$

where $\tilde{\mathbf{E}}'$ and $\tilde{\mathbf{E}}''$ denote plane waves with wavevectors $\mathbf{k}' = -k_1\,\mathbf{a}_x + k_3\,\mathbf{a}_z$ and $\mathbf{k}'' = k_1\,\mathbf{a}_x + k_3\,\mathbf{a}_z$, respectively. As the two waves travel in the same medium, they have the

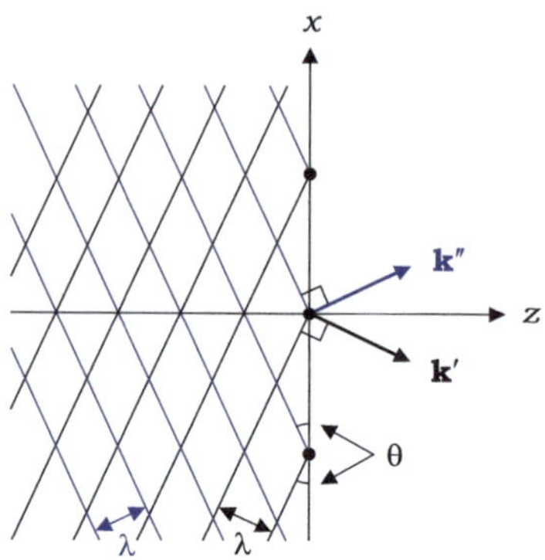

Fig. 8.1 Superposition of two uniform plane waves

same wavenumber $\left|\mathbf{k}'\right| = \left|\mathbf{k}''\right| = (k_1^2 + k_3^2)^{1/2}$. Therefore, the wavelength is given by $\lambda = 2\pi/(k_1^2 + k_3^2)^{1/2}$. The wavefronts make an angle of $\theta = \tan^{-1}(k_1/k_3)$ with respect to the x-axis, as shown in Fig. 8.1. We observe that constructive interference occurs at points on the x-axis marked with a dot, where $\tilde{\mathbf{E}} = \mathbf{a}_y 2E_o$, and also at the midpoint between two neighboring dots, where $\tilde{\mathbf{E}} = -\mathbf{a}_y 2E_o$.

Exercise 8.1
Find the real instantaneous value associated with phasor $\tilde{\mathbf{A}} = A_o(\mathbf{a}_x + j\,\mathbf{a}_y)$. Can this possibly represent an electromagnetic wave?

Ans. $\mathscr{A} = A_o\,\mathbf{a}_x \cos(\omega t) - A_o\,\mathbf{a}_y \sin(\omega t)$. No.

Exercise 8.2
Is it possible to express function $\mathscr{B} = B_o\,\mathbf{a}_x \cos(10t) \cos(15t - 30z)$ in phasor form? If not, why not?

Ans. No. It has two frequencies, $\omega = 5$ and $\omega = 25$.

8.1.1 Maxwell's Equations in Phasor Form

Under sinusoidal steady-state conditions, electric and magnetic fields and their sources vary sinusoidally over time. Therefore, Maxwell's equations can be conveniently expressed in terms of the phasors. First, substituting Eqs. (8.6) and (8.7) into Eq. (8.1a), we obtain

$$\nabla \times \mathrm{Re}\left[\tilde{\mathbf{E}}(x, y, z)\, e^{j\omega t}\right] = -\mu \frac{\partial}{\partial t} \mathrm{Re}\left[\tilde{\mathbf{H}}(x, y, z)\, e^{j\omega t}\right] \tag{8.8}$$

Because the curl, time derivative, and taking the real part are mutually exclusive operations, Eq. (8.8) can be written as

$$\mathrm{Re}\left[e^{j\omega t}\, \nabla \times \tilde{\mathbf{E}}(x, y, z)\right] = -\mu\, \mathrm{Re}\left[\tilde{\mathbf{H}}(x, y, z)\, j\omega\, e^{j\omega t}\right]$$

Given that the real instantaneous value of a quantity is simply the real part of the product of its phasor and the time factor $e^{j\omega t}$, the symbol Re[..] and the term $e^{j\omega t}$ are omitted in the above equation, and the equation is expressed in phasor form as

$$\nabla \times \tilde{\mathbf{E}} = -j\omega\mu\tilde{\mathbf{H}}$$

Following the same procedure as above, Maxwell's equations can be converted into phasor form. For a homogeneous, linear, and isotropic perfect dielectric, Maxwell's equations can be expressed in phasor form, as follows:

$$\boxed{\nabla \times \tilde{\mathbf{E}} = -j\omega\mu\tilde{\mathbf{H}}} \tag{8.9a}$$

$$\boxed{\nabla \times \tilde{\mathbf{H}} = j\omega\varepsilon\tilde{\mathbf{E}}} \tag{8.9b}$$

$$\boxed{\nabla \cdot \tilde{\mathbf{E}} = 0} \tag{8.9c}$$

$$\boxed{\nabla \cdot \tilde{\mathbf{H}} = 0} \tag{8.9d}$$

Again, the electric and magnetic field phasors $\tilde{\mathbf{E}}$ and $\tilde{\mathbf{H}}$ vary with space coordinates, but are independent of time. In general, these are vectors with complex scalar components.

Following the same procedure used for the differential wave equation in $\mathscr{E}$ or $\mathscr{H}$, Maxwell's equations in phasor form can be combined to form a second-order differential equation in $\tilde{\mathbf{E}}$ or $\tilde{\mathbf{H}}$. Taking the curl on both sides of Eq. (8.9a) and using Eq. (8.9b), we obtain

$$\nabla \times \nabla \times \tilde{\mathbf{E}} = -j\omega\mu\nabla \times \tilde{\mathbf{H}}$$
$$= \omega^2\mu\varepsilon\tilde{\mathbf{E}} \tag{8.10}$$

Applying the vector identity $\nabla \times \nabla \times \mathbf{U} = \nabla\nabla \cdot \mathbf{U} - \nabla^2\mathbf{U}$ to the left-hand side of Eq. (8.10) and using Eq. (8.9c) leads to the **vector Helmholtz equation**, that is,

$$\boxed{\nabla^2 \tilde{\mathbf{E}} + k^2 \tilde{\mathbf{E}} = 0} \tag{8.11}$$

The wavenumber k is defined as

$$\boxed{k = \omega\sqrt{\mu\varepsilon} = \frac{\omega}{\upsilon} = \frac{2\pi}{\lambda}} \quad \text{[rad/m]} \tag{8.12}$$

where ω is the angular frequency, μ is the permeability, ε is the permittivity, and λ is the wavelength. In the material, the phase velocity is given by

$$v = \frac{1}{\sqrt{\mu\varepsilon}} = \frac{c}{\sqrt{\mu_r \varepsilon_r}}$$

where μ_r and ε_r are the relative permeability and permittivity, respectively, and $c = 1/\sqrt{\mu_0 \varepsilon_0}$ is the speed of light in free space.

Exercise 8.3

Which of the following can be determined from the electric field phasor alone: (a) direction of wave propagation, (b) wave amplitude, (c) polarization state, (d) wavelength, (e) phase velocity, and (f) frequency.

Ans. (a)–(d).

Exercise 8.4

An electromagnetic wave in air with $\lambda = 0.4$ [m] is transmitted to a nonmagnetic dielectric, where it propagates with $\lambda = 0.2$ [m]. Determine ε_r of the material.

Ans. 4.

Review Questions

RQ 8.1 Write a general expression for the time-harmonic electric field   [(8.6)]
        in phasor form.

RQ 8.2 Write Maxwell's equations in free space in phasor form.   [(8.9a,b,c,d)]

RQ 8.3 Write the vector Helmholtz equation.   [(8.11)]

8.2 Waves in Homogeneous Medium

The permittivity and permeability of a homogeneous medium do not vary with the position, implying an infinite extent of the medium. As the medium has no boundaries or interfaces, a uniform plane wave can propagate without reflection or refraction.

8.2.1 Uniform Plane Wave in Lossless Dielectric

A lossless dielectric is a perfect medium in which electromagnetic waves propagate without any energy loss. In this case, wave propagation is governed by the permittivity and permeability of the material through the relation $k = \omega\sqrt{\varepsilon\mu}$. To learn how to use phasors to describe the propagation of harmonic waves in material media, we begin with the vector Helmholtz equation in Cartesian coordinates. That is,

$$\nabla^2\left(\tilde{E}_x\, \mathbf{a}_x + \tilde{E}_y\, \mathbf{a}_y + \tilde{E}_z\, \mathbf{a}_z\right) + k^2\left(\tilde{E}_x\, \mathbf{a}_x + \tilde{E}_y\, \mathbf{a}_y + \tilde{E}_z\, \mathbf{a}_z\right) = 0 \qquad (8.13)$$

where $\tilde{E}_x$, $\tilde{E}_y$, and $\tilde{E}_z$ are the three components of the electric field phasor. Because the unit vectors in Cartesian coordinates are mutually exclusive, Eq. (8.13) can be resolved into three scalar Helmholtz equations as follows:

$$\nabla^2 \tilde{E}_x + k^2 \tilde{E}_x = 0 \qquad (8.14a)$$

$$\nabla^2 \tilde{E}_y + k^2 \tilde{E}_y = 0 \qquad (8.14b)$$

$$\nabla^2 \tilde{E}_z + k^2 \tilde{E}_z = 0 \qquad (8.14c)$$

In the Cartesian coordinate system, Eq. (8.14a) can be written as

$$\frac{\partial^2 \tilde{E}_x}{\partial x^2} + \frac{\partial^2 \tilde{E}_x}{\partial y^2} + \frac{\partial^2 \tilde{E}_x}{\partial z^2} + k^2 \tilde{E}_x = 0 \qquad (8.15)$$

Through direct substitution, it can be shown that a uniform plane wave is a general solution to Eq. (8.15), that is,

$$\tilde{E}_x = \hat{E}_{ox}\, e^{-j(k_x x + k_y y + k_z z)} = \hat{E}_{ox}\, e^{-j\mathbf{k}\cdot\mathbf{r}} \qquad (8.16a)$$

where $\hat{E}_{ox}$ is the complex amplitude, $\mathbf{r}$ is the position vector, and $\mathbf{k}$ is the wavevector with $|\mathbf{k}| = k = (k_x^2 + k_y^2 + k_z^2)^{1/2}$. Similar procedures can be followed to obtain the y- and z-components of $\hat{\mathbf{E}}$ such that

$$\tilde{E}_y = \hat{E}_{oy}\, e^{-j\mathbf{k}\cdot\mathbf{r}} \qquad (8.16b)$$

$$\tilde{E}_z = \hat{E}_{oz}\, e^{-j\mathbf{k}\cdot\mathbf{r}} \qquad (8.16c)$$

The three components in Eqs. (8.16a, b, c) can be combined to form a solution of the vector Helmholtz equation only if they have the same wavevector; otherwise, they represent three independent waves. Consequently, in a lossless dielectric of infinite extent, a general solution of the vector Helmholtz equation is a uniform plane wave with wavevector $\mathbf{k}$. Although k or $|\mathbf{k}|$ in the vector Helmholtz equation is determined by the material, the direction of $\mathbf{k}$ is independent of the material, and is determined by the source direction.

The electric field phasor may take on different forms such as

$$\tilde{\mathbf{E}} = \tilde{E}_x \mathbf{a}_x + \tilde{E}_y \mathbf{a}_y + \tilde{E}_z \mathbf{a}_z = \left(\hat{E}_{ox} \mathbf{a}_x + \hat{E}_{oy} \mathbf{a}_y + \hat{E}_{oz} \mathbf{a}_z \right) e^{-j\mathbf{k}\cdot\mathbf{r}}$$

In a more compact form, it can be expressed as

$$\boxed{\tilde{\mathbf{E}} = \hat{\mathbf{E}}_o\, e^{-j\mathbf{k}\cdot\mathbf{r}}} \qquad (8.17)$$

The phasor amplitude $\hat{\mathbf{E}}_o$ provides information on the amplitude and direction of the time-harmonic oscillation of the electric field. The general form of $\hat{\mathbf{E}}_o$ is

$$\hat{\mathbf{E}}_o = \hat{E}_o \mathbf{a}_E = E_o e^{j\varphi_o} \mathbf{a}_E \tag{8.18}$$

where $\hat{E}_o$ is the complex amplitude, E_o is the amplitude, φ_o is the phase angle, and $\mathbf{a}_E$ is a unit vector in the direction of $\hat{\mathbf{E}}_o$. For a wave traveling in an infinite medium, the constant phase φ_o has no physical significance and is typically set to zero for simplicity.

The general expression for $\tilde{\mathbf{E}}$ in Eq. (8.17) contains three discrete vectors. It is important to distinguish one vector from the other. Wavevector $\mathbf{k}$ provides information on the wavelength and direction of the wave propagation. The position vector $\mathbf{r}$ defines the field point where the wave motion is observed as a function of time, and is used to specify the relative phase of the wave motion via $-\mathbf{k} \cdot \mathbf{r}$. Finally, the phasor amplitude $\hat{\mathbf{E}}_o$ describes the amplitude and direction of the time-harmonic electric field.

Although the uniform plane wave, expressed in Eq. (8.17), satisfies the vector Helmholtz equation, it is identified as an electromagnetic wave only after it is shown to satisfy Maxwell's equations, specifically the Gauss law. Inserting Eq. (8.17) into Eq. (8.9c) gives

$$\nabla \cdot \tilde{\mathbf{E}} = -j \left(\mathbf{k} \cdot \hat{\mathbf{E}}_o \right) e^{-j\mathbf{k} \cdot \mathbf{r}} = 0$$

This simply leads to

$$\mathbf{k} \cdot \hat{\mathbf{E}}_o = 0 \tag{8.19}$$

This indicates that $\hat{\mathbf{E}}_o$ is always perpendicular to the wavevector. In other words, $\hat{\mathbf{E}}_o$ lies in the wavefront perpendicular to the wavevector. Therefore, *electromagnetic waves are transverse waves.*

Next, substituting the electric field phasor given in Eq. (8.17) into Eq. (8.9a) gives

$$\nabla \times \tilde{\mathbf{E}} = -j\mathbf{k} \times \hat{\mathbf{E}}_o e^{-j\mathbf{k} \cdot \mathbf{r}}$$
$$= -j\omega\mu\tilde{\mathbf{H}} \tag{8.20}$$

By letting $k = \omega\sqrt{\mu\varepsilon}$ in Eq. (8.20), the relationship between the electric and magnetic field phasors is obtained as

$$\tilde{\mathbf{H}} = \frac{1}{\eta} \left(\mathbf{a}_k \times \tilde{\mathbf{E}} \right) \tag{8.21}$$

where $\mathbf{a}_k$ is a unit vector in the direction of the wavevector. For a medium with permeability μ and permittivity ε, the ***intrinsic impedance*** η is defined as

$$\boxed{\eta = \sqrt{\frac{\mu}{\varepsilon}}} \quad [\Omega] \tag{8.22}$$

This is a characteristic parameter of the medium.

Next, using Eq. (8.17) in Eq. (8.21) shows that the electric and magnetic fields are of the same form, and are coupled to each other such that

$$\tilde{\mathbf{H}} = \frac{1}{\eta}\left(\mathbf{a}_k \times \hat{\mathbf{E}}_o\right) e^{-j\mathbf{k}\cdot\mathbf{r}}$$
$$\equiv \hat{\mathbf{H}}_o\, e^{-j\mathbf{k}\cdot\mathbf{r}} \tag{8.23}$$

where the phasor amplitude $\hat{\mathbf{H}}_o$ is defined as

$$\hat{\mathbf{H}}_o = \frac{\hat{E}_o}{\eta}(\mathbf{a}_k \times \mathbf{a}_E)$$
$$\equiv \hat{H}_o\, \mathbf{a}_H \tag{8.24}$$

From Eq. (8.24) it is evident that the intrinsic impedance is the ratio between the complex amplitudes $\hat{E}_o$ and $\hat{H}_o$. Its value is measured in ohms $[\Omega]$. The intrinsic impedance of free space is $\eta_o = \sqrt{\mu_0/\varepsilon_0} = 120\pi \cong 377\,[\Omega]$. In lossless dielectrics, η is real. Thus, the magnetic field is always in phase with the electric field. In contrast, in lossy dielectrics, η is complex, and the magnetic field is generally out of phase with the electric field. Nevertheless, the three vectors $\hat{\mathbf{E}}_o$, $\hat{\mathbf{H}}_o$, and $\mathbf{k}$ are always perpendicular to each other and follow the right-hand rule. When the right fingers rotate from the electric field to the magnetic field vector, the thumb points in the direction of the wavevector. Expressed mathematically,

$$\boxed{\mathbf{a}_k = \mathbf{a}_E \times \mathbf{a}_H} \tag{8.25}$$

This leads to the conclusion that both the electric and magnetic field vectors lie in the wavefront perpendicular to the wavevector, confirming that electromagnetic waves are transverse waves.

The real instantaneous values of the electric and magnetic fields are obtained from their phasors as follows:

$$\mathscr{E}(\mathbf{r}, t) = \mathbf{a}_E\, E_o \cos(\omega t - \mathbf{k}\cdot\mathbf{r}) \tag{8.26a}$$

$$\mathscr{H}(\mathbf{r}, t) = \mathbf{a}_H\, \frac{E_o}{\eta}\cos(\omega t - \mathbf{k}\cdot\mathbf{r}) \tag{8.26b}$$

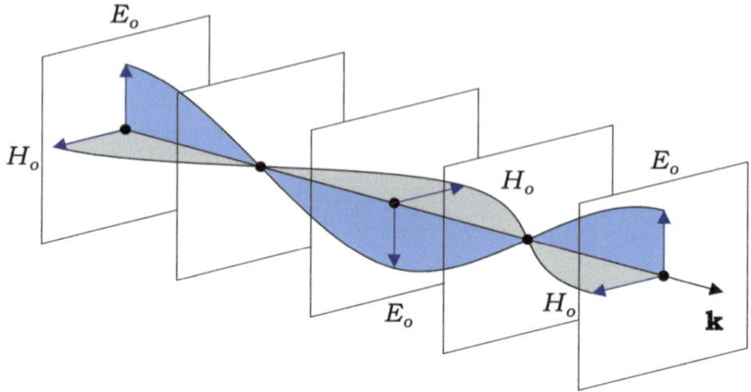

Fig. 8.2 Uniform plane wave in lossless media

where the phase angle of $\hat{\mathbf{E}}_o$ is set to zero for simplicity.

Again, in lossless dielectrics, the electric and magnetic fields are always in phase and propagate in the direction of $\mathbf{k}$, as illustrated in Fig. 8.2.

Example 8.2 An electromagnetic wave with electric field phasor $\tilde{\mathbf{E}} = \mathbf{a}_x E_o (1 + j) e^{-jkz}$ propagates in a dielectric with $\varepsilon_r = 4$ and $\mu_r = 1$. When its frequency is 2 [GHz], determine (a) η, and (b) k, and find (c) expressions for $\mathscr{E}$ and $\mathscr{H}$.

Solution

(a) From Eq. (8.22), the intrinsic impedance is

$$\eta = \sqrt{\frac{\mu_0}{\varepsilon_0 \varepsilon_r}} = 188.50 \ [\Omega]$$

(b) From Eq. (8.12),

$$k = \omega\sqrt{\varepsilon\mu} = \frac{2\pi \times 2 \times 10^9 \times 2}{3 \times 10^8} = 83.78 \ [\text{rad/m}]$$

(c) The given phasor can be rewritten as

$$\tilde{\mathbf{E}} = \mathbf{a}_x E_o \sqrt{2}\, e^{j\pi/4} e^{-jkz} \tag{8.27}$$

Its real instantaneous value is

$$\mathscr{E} = \text{Re}\left[\mathbf{a}_x E_o \sqrt{2}\, e^{j\pi/4} e^{-jkz} e^{j\omega t}\right]$$
$$= \mathbf{a}_x E_o \sqrt{2} \cos\left(1.26 \times 10^{10} t - 83.78z + \pi/4\right)$$

Using Eq. (8.27) in Eq. (8.21) yields

$$\tilde{\mathbf{H}} = \frac{1}{\eta}(\mathbf{a}_z \times \mathbf{a}_x) E_o \sqrt{2}\, e^{j\pi/4} e^{-jkz} = \frac{1}{\eta}\mathbf{a}_y E_o \sqrt{2}\, e^{j\pi/4} e^{-jkz}$$

The real instantaneous value is

$$\mathscr{H} = \mathbf{a}_y 7.50 \times 10^{-3} E_o \cos\left(1.26 \times 10^{10} t - 83.78z + \pi/4\right)$$

Exercise 8.5
Even if $\tilde{\mathbf{E}}$ and $\tilde{\mathbf{H}}$ satisfy the vector Helmholtz equations in the same form, they are not independent of each other. Explain why.

Ans. Eq. (8.21).

Exercise 8.6
Given an electric field phasor $\tilde{\mathbf{E}} = 4(2\mathbf{a}_x + j\mathbf{a}_y)\, e^{-j600z}$ in a medium with $\eta = 200$, determine the magnetic field phasor.

Ans. $\tilde{\mathbf{H}} = 0.02(-j\mathbf{a}_x + 2\mathbf{a}_y)\, e^{-j600z}$.

Review Questions

RQ 8.4	Express the uniform plane wave in phasor form.	[(8.17)]
RQ 8.5	Write a general expression for the amplitude of the electric field phasor.	[(8.18)]
RQ 8.6	Are electromagnetic waves transverse waves?	[(8.19)]
RQ 8.7	In what way is $\tilde{\mathbf{H}}$ related to $\tilde{\mathbf{E}}$?	[(8.21)]
RQ 8.8	Define intrinsic impedance.	[(8.22)]
RQ 8.9	In what way is $\mathbf{a}_k$ related to $\mathbf{a}_E$ and $\mathbf{a}_H$?	[(8.25)]

8.2.2 *Flow of Wave Power and Poynting Vector*

Chapters 3 and 5 show that energy can be stored in static electric and magnetic fields with energy densities of $(1/2)\,\mathbf{D}\cdot\mathbf{E}$ and $(1/2)\,\mathbf{B}\cdot\mathbf{H}$, respectively. Energy can also be stored in the time-varying electric and magnetic fields. In this case, the stored energy propagates in the direction of $\mathbf{k}$ with the same velocity as that of the electromagnetic wave. Accordingly, an electromagnetic wave can transfer energy between two points in space and deliver power to the load.

In this section, we are mainly concerned with the power density of electromagnetic waves and the flow of power in space. We start with Maxwell's two curl equations,

$$\nabla \times \mathscr{E} = -\frac{\partial \mathscr{B}}{\partial t} \tag{8.28a}$$

$$\nabla \times \mathscr{H} = \mathbf{J} + \frac{\partial \mathscr{D}}{\partial t} \tag{8.28b}$$

Consider a vector identity that always holds for any two smooth vector fields,

$$\nabla \cdot (\mathscr{E} \times \mathscr{H}) = \mathscr{H} \cdot (\nabla \times \mathscr{E}) - \mathscr{E} \cdot (\nabla \times \mathscr{H}) \tag{8.29}$$

Substituting Eqs. (8.28a, b) into Eq. (8.29) yields

$$-\nabla \cdot (\mathscr{E} \times \mathscr{H}) = \mathscr{E} \cdot \boldsymbol{J} + \mathscr{E} \cdot \frac{\partial \mathscr{D}}{\partial t} + \mathscr{H} \cdot \frac{\partial \mathscr{B}}{\partial t} \tag{8.30}$$

In simple media for which ε, μ, and σ are constant, the second and third terms on the right side of Eq. (8.30) can be rearranged as

$$\mathscr{E} \cdot \frac{\partial \mathscr{D}}{\partial t} = \frac{\varepsilon}{2} \frac{\partial}{\partial t}(\mathscr{E} \cdot \mathscr{E}) = \frac{1}{2} \frac{\partial}{\partial t}(\mathscr{D} \cdot \mathscr{E}) \tag{8.31a}$$

$$\mathscr{H} \cdot \frac{\partial \mathscr{B}}{\partial t} = \frac{\mu}{2} \frac{\partial}{\partial t}(\mathscr{H} \cdot \mathscr{H}) = \frac{1}{2} \frac{\partial}{\partial t}(\mathscr{B} \cdot \mathscr{H}) \tag{8.31b}$$

Inserting Eqs. (8.31a, b) back into Eq. (8.30) gives

$$\boxed{-\nabla \cdot (\mathscr{E} \times \mathscr{H}) = \mathscr{E} \cdot \boldsymbol{J} + \frac{1}{2} \frac{\partial}{\partial t}(\mathscr{D} \cdot \mathscr{E}) + \frac{1}{2} \frac{\partial}{\partial t}(\mathscr{B} \cdot \mathscr{H})} \tag{8.32}$$

Next, integrating both sides of Eq. (8.32) over a volume $\mathcal{V}$ and applying the divergence theorem, we obtain

$$-\oint_S (\mathscr{E} \times \mathscr{H}) \cdot d\mathbf{s} = \int_{\mathcal{V}} \mathscr{E} \cdot \boldsymbol{J}\,dv + \frac{\partial}{\partial t}\left[\int_{\mathcal{V}} \frac{1}{2}(\mathscr{D} \cdot \mathscr{E})dv\right] + \frac{\partial}{\partial t}\left[\int_{\mathcal{V}} \frac{1}{2}(\mathscr{B} \cdot \mathscr{H})dv\right] \tag{8.33}$$

The first term on the right side of Eq. (8.33) is the ohmic power loss in volume $\mathcal{V}$. The two brackets represent the energies stored in $\mathscr{E}$ and $\mathscr{H}$, respectively, for a given volume. Accordingly, the sum of the second and third terms on the right-hand side represents the time rate of increase of the total electromagnetic energy stored in the volume. Based on this understanding, we can conclude that the left-hand side of Eq. (8.33) is equal to the total instantaneous power flowing into the volume; part of the power is dissipated in the volume, and the rest is stored as electromagnetic energy, as shown on the right-hand side. Equation (8.33) is referred to as **Poynting's theorem**. Again, *the net power flowing into a given volume is equal to the sum of the ohmic power loss suffered in the volume and the time rate of increase of the energy stored in the volume.*

The net power flowing into volume $\mathcal{V}$ is obtained by integrating $\mathscr{E} \times \mathscr{H}$ over the bounding surface of $\mathcal{V}$, as shown in Eq. (8.33). Therefore, $\mathscr{E} \times \mathscr{H}$ is recognized as the **power density** measured in watts per square meter. For an electromagnetic wave with electric field $\mathscr{E}$ and magnetic field $\mathscr{H}$, the **Poynting vector S** is defined as

$$\boxed{\mathbf{S} = \mathscr{E} \times \mathscr{H}} \quad [\text{W/m}^2] \tag{8.34}$$

The Poynting vector is normal to both the electric and magnetic field vectors, and is thus parallel to the wavevector. In other words, the electromagnetic power flows along the direction of the wavevector.

Because the power density depends on $\mathscr{E}$ and $\mathscr{H}$, it varies rapidly over time. In practice, the time-averaged value is more important than the instantaneous one. Moreover, the relationship between $|\mathbf{S}|$ and $|\mathscr{E}|$ is nonlinear, because $|\mathbf{S}|$ is proportional to the square of $|\mathscr{E}|$. This is a rare case, in which we deal with a nonlinear quantity in electromagnetics. Therefore, the power density cannot be simply expressed as $\tilde{\mathbf{S}} = \tilde{\mathbf{E}} \times \tilde{\mathbf{H}}$ by using phasors. Before we can derive the correct phasor form, we digress briefly and describe how the amplitude E_o, complex amplitude $\hat{E}_o$, phasor amplitude $\hat{\mathbf{E}}_o$, and phasor $\tilde{\mathbf{E}}$ are used to express $\mathscr{E}$. In general, the time-harmonic electric field can be expressed as

$$
\begin{aligned}
\mathscr{E} &= \mathbf{a}_E E_o \cos(\omega t - \mathbf{k} \cdot \mathbf{r} + \varphi_e) \\
&= \frac{1}{2}\left[\mathbf{a}_E\left(E_o e^{j\varphi_e}\right)e^{j\omega t - j\mathbf{k}\cdot\mathbf{r}} + \mathbf{a}_E\left(E_o e^{-j\varphi_e}\right)e^{-j\omega t + j\mathbf{k}\cdot\mathbf{r}} \right] \\
&\equiv \frac{1}{2}\left[\left(\mathbf{a}_E \hat{E}_o\right)e^{-j\mathbf{k}\cdot\mathbf{r}}e^{j\omega t} + \left(\mathbf{a}_E \hat{E}_o^{*}\right)e^{j\mathbf{k}\cdot\mathbf{r}}e^{-j\omega t} \right] \\
&\equiv \frac{1}{2}\left[\hat{\mathbf{E}}_o e^{-j\mathbf{k}\cdot\mathbf{r}}e^{j\omega t} + \hat{\mathbf{E}}_o^{*} e^{j\mathbf{k}\cdot\mathbf{r}}e^{-j\omega t} \right] \\
&\equiv \frac{1}{2}\left[\tilde{\mathbf{E}}\, e^{j\omega t} + \tilde{\mathbf{E}}^{*} e^{-j\omega t} \right]
\end{aligned}
\tag{8.35}
$$

where $*$ denotes a complex conjugate. Similarly, the instantaneous magnetic field can be expressed as

$$
\mathscr{H} = \mathbf{a}_H H_o \cos(\omega t - \mathbf{k} \cdot \mathbf{r} + \varphi_h) = \frac{1}{2}\left[\tilde{\mathbf{H}}\, e^{j\omega t} + \tilde{\mathbf{H}}^{*} e^{-j\omega t} \right]
\tag{8.36}
$$

Subsequently, inserting Eqs. (8.35) and (8.36) into Eq. (8.34) yields

$$
\begin{aligned}
\mathbf{S} &= \frac{1}{2}\left[\tilde{\mathbf{E}}\, e^{j\omega t} + \tilde{\mathbf{E}}^{*} e^{-j\omega t} \right] \times \frac{1}{2}\left[\tilde{\mathbf{H}}\, e^{j\omega t} + \tilde{\mathbf{H}}^{*} e^{-j\omega t} \right] \\
&= \frac{1}{4}\left[\tilde{\mathbf{E}} \times \tilde{\mathbf{H}}\, e^{j2\omega t} + \tilde{\mathbf{E}}^{*} \times \tilde{\mathbf{H}}^{*} e^{-j2\omega t} + \tilde{\mathbf{E}} \times \tilde{\mathbf{H}}^{*} + \tilde{\mathbf{E}}^{*} \times \tilde{\mathbf{H}} \right]
\end{aligned}
\tag{8.37}
$$

In general, the average of a time-harmonic quantity $\mathbf{G}$ is defined as

$$
\langle \mathbf{G} \rangle = \frac{1}{T} \int_{-T/2}^{T/2} \mathbf{G}\, dt
$$

where T is the time period of $\mathbf{G}$. Accordingly, the ***time-averaged power density*** or ***time-averaged Poynting vector*** is defined as

$$\boxed{\langle \mathbf{S} \rangle = \frac{1}{2}\mathrm{Re}\left[\tilde{\mathbf{E}} \times \tilde{\mathbf{H}}^*\right]} \quad [\mathrm{W/m^2}] \tag{8.38}$$

where $\tilde{\mathbf{E}}$ and $\tilde{\mathbf{H}}$ are the electric and magnetic field phasors, respectively, and $*$ denotes the complex conjugate.

Because the quantities in the script font in Eq. (8.32) are time-harmonic functions, it is evident from trigonometry that the time averages of the second and third terms on the right-hand side of Eq. (8.32) vanish. Consequently,

$$-\nabla \cdot \langle \mathscr{E} \times \mathscr{H} \rangle = \langle \mathscr{E} \cdot \mathbf{J} \rangle \tag{8.39}$$

where $\langle \rangle$ denotes the time-averaged quantity. In phasor notation,

$$\boxed{-\nabla \cdot \langle \mathbf{S} \rangle = \frac{1}{2}\mathrm{Re}\left[\tilde{\mathbf{E}} \cdot \tilde{\mathbf{J}}^*\right]} \tag{8.40}$$

The negative divergence of the time-averaged Poynting vector is equal to the time-averaged ohmic power loss per unit volume.

Example 8.3 For an electromagnetic wave with $\mathscr{H}(\mathbf{r}, t) = \mathbf{a}_y E_o \sqrt{\varepsilon/\mu} \cos(\omega t - kz)$ and electric field $\mathscr{E}(\mathbf{r}, t) = \mathbf{a}_x E_o \cos(\omega t - kz)$, compute $\langle \mathbf{S} \rangle$ (a) using $\mathscr{E}$ and $\mathscr{H}$, and (b) using $\tilde{\mathbf{E}}$ and $\tilde{\mathbf{H}}$.

Solution

(a) From Eq. (8.34),

$$\mathbf{S} = \mathscr{E} \times \mathscr{H} = \mathbf{a}_z E_o^2 \sqrt{\varepsilon/\mu}\, \cos^2(\omega t - kz)$$

With the help of trigonometry, $\mathbf{S}$ can be rewritten as

$$\mathbf{S} = \mathbf{a}_z E_o^2 \sqrt{\varepsilon/\mu}\, \tfrac{1}{2}[1 + \cos(2\omega t - 2kz)]$$

Taking the time average of $\mathbf{S}$ yields

$$\langle \mathbf{S} \rangle = \mathbf{a}_z \frac{E_o^2}{2}\sqrt{\frac{\varepsilon}{\mu}} \tag{8.41a}$$

(b) In phasor form, the electric and magnetic fields are given by

$$\tilde{\mathbf{E}} = \mathbf{a}_x E_o e^{-jkz} \quad \text{and} \quad \tilde{\mathbf{H}} = \mathbf{a}_y E_o \sqrt{\varepsilon/\mu}\, e^{-jkz}$$

Using these in Eq. (8.38) yields

$$\langle \mathbf{S} \rangle = \tfrac{1}{2}\mathrm{Re}\left[\tilde{\mathbf{E}} \times \tilde{\mathbf{H}}^*\right] = \tfrac{1}{2}\mathrm{Re}\left[\left(\mathbf{a}_x E_o e^{-jkz}\right) \times \left(\mathbf{a}_y E_o \sqrt{\varepsilon/\mu}\, e^{-jkz}\right)^*\right]$$

$$= \mathbf{a}_z \frac{E_o^2}{2} \sqrt{\frac{\varepsilon}{\mu}} \qquad (8.41\text{b})$$

This result is the same as that in Eq. (8.41a).

Rewriting Eq. (8.41b), we have

$$\boxed{\langle \mathbf{S} \rangle = \mathbf{a}_z \left(\varepsilon \frac{E_o^2}{2} \right) \frac{1}{\sqrt{\varepsilon\mu}}} \qquad (8.41\text{c})$$

The time-averaged power density is equal to the product of the time-averaged energy density and phase velocity.

Exercise 8.7
For a wave with $\tilde{\mathbf{E}} = 10\,\mathbf{a}_x e^{-j600\,(3y+4z)}$ in free space, find the time-averaged power flowing through a unit area of the $z = 0$ plane.

Ans. 0.11 [W].

Exercise 8.8
Find the ratio of the energy densities stored in the electric and magnetic fields of an electromagnetic wave in lossless dielectrics.

Ans. 1.

Exercise 8.9
Show that the magnitude of $\mathbf{S}$ is equal to the sum of the energy densities of $\mathscr{E}$ and $\mathscr{H}$, multiplied by the phase velocity.

Ans. $(2\sqrt{\varepsilon\mu})^{-1}(\varepsilon|\mathscr{E}|^2 + \mu|\mathscr{H}|^2)$ is equal to $|\mathbf{S}| = |\mathscr{E}|^2/\eta$.

Review Questions

RQ 8.10	State Poynting's theorem in words.	[(8.33)]
RQ 8.11	Express Poynting vector in terms of $\mathscr{E}$ and $\mathscr{H}$.	[(8.34)]
RQ 8.12	Express the time-averaged power density in terms of $\tilde{\mathbf{E}}$ and $\tilde{\mathbf{H}}$.	[(8.38)]

8.2.3 *Wave Polarization*

Electromagnetic waves are transverse waves in which the electric field vectors lie in a plane perpendicular to the direction of wave propagation. The electric field of a time-harmonic wave reverses direction every half of its temporal period. In general, the direction changes continuously over time in the transverse plane. The ***polarization*** of a wave describes the manner in which the electric field vector varies with time in the transverse plane. If the tip of the electric field vector moves back and forth along

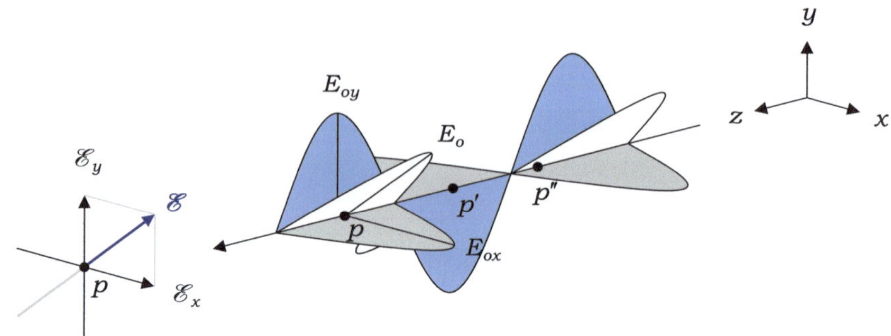

Fig. 8.3 Linearly polarized wave and its component waves

a straight line as time passes, the wave becomes ***linearly polarized*** (see Fig. 8.3). If the locus traced by the tip is a circle, the wave is ***circularly polarized*** (see Figs. 8.4 and 8.5). In general, the tip follows an elliptical track in the transverse plane. As far as the polarization of a wave is concerned, we only need to trace the electric field vector because the magnetic field is directly related to the electric field through Eq. (8.21).

8.2.3.1 Linear Polarization

When a uniform plane wave propagates in the $+z$-direction, its electric field vector lies in the xy-plane, and its electric field phasor is generally expressed as

$$\tilde{\mathbf{E}} = \hat{\mathbf{E}}_o e^{-jkz} = \left(\hat{E}_{ox}\, \mathbf{a}_x + \hat{E}_{oy}\, \mathbf{a}_y \right) e^{-jkz} \tag{8.42}$$

where $\hat{E}_{ox}$ and $\hat{E}_{oy}$ are the complex amplitudes of the x- and y-components, respectively.

If the amplitudes are real (i.e., $\hat{E}_{ox} = E_{ox}$ and $\hat{E}_{oy} = E_{oy}$), the instantaneous electric field is given by

$$\begin{aligned} \mathscr{E} &= \mathrm{Re}\left[\left(E_{ox}\, \mathbf{a}_x + E_{oy}\, \mathbf{a}_y \right) e^{-jkz} e^{j\omega t} \right] \\ &= \left(E_{ox}\, \mathbf{a}_x + E_{oy}\, \mathbf{a}_y \right) \cos(\omega t - kz) \end{aligned} \tag{8.43}$$

At a fixed point with $z = z_1$, the electric field of this wave oscillates along a straight line parallel to $E_{ox}\, \mathbf{a}_x + E_{oy}\, \mathbf{a}_y$, which is called the wave polarization vector. Thus, this wave is said to be linearly polarized in the direction of $E_{ox}\, \mathbf{a}_x + E_{oy}\, \mathbf{a}_y$. In this case, the wave may be considered as the sum of two uniform plane waves: one is linearly polarized in the direction of $\mathbf{a}_x$, and the other is linearly polarized in the direction of $\mathbf{a}_y$, as shown in Fig. 8.3. As the amplitudes E_{ox} and E_{oy} are both real, the two component waves are in phase, such that the two cosine curves cross the z-axis at the same point.

Example 8.4 Linearly polarized plane wave with $\tilde{\mathbf{E}} = E_o(\mathbf{a}_x + \sqrt{3}\,\mathbf{a}_y)\,e^{-jkz}$ travels in free space. Determine (a) the angle between the electric field vector and the x-axis, and (b) the time-averaged power density.

Solution

(a) The rotation angle of the wave-polarization vector, $\mathbf{a}_x + \sqrt{3}\,\mathbf{a}_y$, with respect to the x-axis is

$$\theta = \tan^{-1}(\sqrt{3}) = 60^o$$

(b) From Eq. (8.21), the magnetic field phasor is

$$\tilde{\mathbf{H}} = \sqrt{\varepsilon_0/\mu_0}\,\mathbf{a}_z \times E_o\left(\mathbf{a}_x + \sqrt{3}\,\mathbf{a}_y\right)e^{-jkz} = E_o\sqrt{\varepsilon_0/\mu_0}\left(\mathbf{a}_y - \sqrt{3}\,\mathbf{a}_x\right)e^{-jkz}$$

From Eq. (8.38),

$$\langle \mathbf{S}\rangle = \tfrac{1}{2}\mathrm{Re}\left[E_o\left(\mathbf{a}_x + \sqrt{3}\,\mathbf{a}_y\right)e^{-jkz} \times \left\{ E_o\sqrt{\varepsilon_0/\mu_0}\left(\mathbf{a}_y - \sqrt{3}\,\mathbf{a}_x\right)e^{-jkz}\right\}^*\right]$$
$$= \tfrac{1}{2}E_o^2\sqrt{\varepsilon_0/\mu_0}\left(\mathbf{a}_x + \sqrt{3}\,\mathbf{a}_y\right) \times \left(\mathbf{a}_y - \sqrt{3}\,\mathbf{a}_x\right)$$
$$= \tfrac{1}{2}E_o^2\sqrt{\varepsilon_0/\mu_0}\,(1 + 3)\,\mathbf{a}_z$$

This result indicates that the time-averaged power density of the wave is equal to the sum of those of the two component waves.

Exercise 8.10
Are the following waves linearly polarized? (a) $\tilde{\mathbf{E}} = 4(\mathbf{a}_x e^{j0.2} - \mathbf{a}_y e^{j0.2})\,e^{-j30z}$, (b) $\tilde{\mathbf{E}} = 2(\mathbf{a}_x e^{j0.3} + \mathbf{a}_y e^{-j0.3})\,e^{-j40z}$, (c) $\tilde{\mathbf{E}} = [\mathbf{a}_x \sin(30^o) + \mathbf{a}_y \cos(30^o)]\,e^{-j50z}$, (d) $\tilde{\mathbf{E}} = (3\mathbf{a}_x + j\mathbf{a}_y)\,e^{j60z}$, (e) $\tilde{\mathbf{E}} = (3 + j)\,\mathbf{a}_x\,e^{j70z}$, (f) $\tilde{\mathbf{E}} = (\mathbf{a}_x + \mathbf{a}_y e^{j\pi})\,e^{-j80z}$.

Ans. (a) Yes, (b) No, (c) Yes, (d) No, (e) Yes, (f) Yes.

Exercise 8.11
For a linearly polarized wave, as shown in Fig. 8.3, sketch the electric field vector in the transverse planes crossing points p, p', and p'' at an instant in time.

Ans.

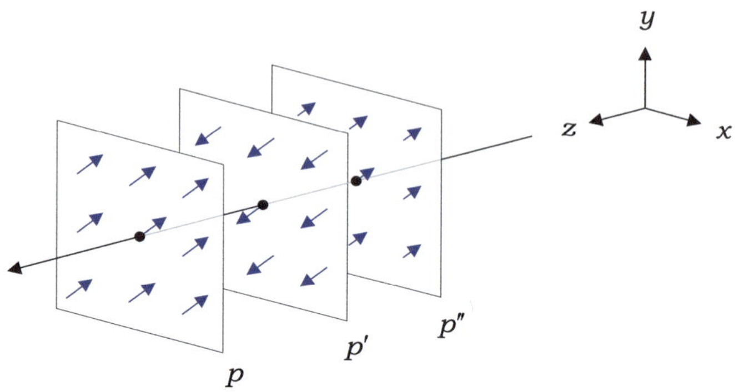

8.2.3.2 Circular Polarization

We now consider the case when the two complex amplitudes are equal in magnitude (i.e., $|\hat{E}_{ox}| = |\hat{E}_{oy}| = E_o$) but the y-component lags behind the x-component by 90° in the time dimension. This wave is said to be ***right-hand circularly polarized*** for the reason that will become evident shortly. If the wave travels in the $+z$-direction, its electric field phasor can be expressed as

$$\tilde{\mathbf{E}} = \left(E_o\mathbf{a}_x - jE_o\mathbf{a}_y\right)e^{-jkz} = \left(E_o\mathbf{a}_x + E_o e^{-j\pi/2}\mathbf{a}_y\right)e^{-jkz} \qquad (8.44)$$

Subsequently, the instantaneous electric field is given by

$$\mathscr{E}(z,t) = \mathbf{a}_x E_o \cos(\omega t - kz) + \mathbf{a}_y E_o \cos(\omega t - kz - \pi/2) \qquad (8.45)$$

Next, we examine the time-harmonic behavior of the electric field vector in the transverse plane, or simply in the $z = 0$ plane. In this case, the expression for $\mathscr{E}$ is reduced to

$$\mathscr{E}(0,t) = \mathbf{a}_x E_o \cos(\omega t) + \mathbf{a}_y E_o \sin(\omega t) \qquad (8.46)$$

For example, $\mathscr{E}(0, t = 0) = E_o\mathbf{a}_x$ and $\mathscr{E}(0, t = \pi/2\omega) = E_o\mathbf{a}_y$ at two different instants, but the magnitude is the same at all times, such that $|\mathscr{E}(0,t)| = E_o$. This leads to the conclusion that the electric field vector rotates in a counterclockwise direction over time in the xy-plane and its tip traces a circle, as shown in Fig. 8.4. The electric field vector completes a turn as ωt increases from zero to 2π. This wave is referred to as a right-hand circularly polarized wave. Polarization handedness is determined by the sense of rotation of the electric field vector with reference to the direction of wave propagation. In the present case, the right thumb points in the direction of wave propagation when the fingers follow the rotation of the electric field vector.

A right-hand circularly polarized wave traveling in the $+z$-direction is illustrated in Fig. 8.4. The y-component wave lags behind the x-component wave by $90°$ in time phase. In other words, the peak of the y-component wave arrives at point p a quarter of the temporal period later than that of the x-component wave. At one instant, the y-component curve appears to be displaced backward (i.e., in the $-z$-direction) by a quarter of the wavelength with respect to the x-component curve. As the two component waves propagate in the same direction with the same phase velocity, the phase relationship between them is maintained at all times at any point in space. Because the phase difference is relative, the y-component wave can be said to lead the x-component wave by $270°$ in time phase.

If the y-component leads the x-component by $90°$ in time phase and has the same amplitude as the x-component, this wave is called a ***left-hand circularly polarized wave***. In other words, when the left fingers follow the rotation of the electric field vector, the thumb points in the direction of wave propagation. For this wave, the electric field phasor is expressed as

$$\tilde{\mathbf{E}} = \left(E_o\mathbf{a}_x + jE_o\mathbf{a}_y\right)e^{-jkz} = \left(E_o\mathbf{a}_x + E_o e^{j\pi/2}\mathbf{a}_y\right)e^{-jkz} \qquad (8.47)$$

The instantaneous electric field is

$$\mathscr{E}(z, t) = \mathbf{a}_x E_o \cos(\omega t - kz) + \mathbf{a}_y E_o \cos(\omega t - kz + \pi/2) \qquad (8.48)$$

Next, setting $z = 0$ in Eq. (8.48), we examine the time-harmonic behavior of the electric field vector in the transverse plane. Given that

$$\mathscr{E}(0, t) = \mathbf{a}_x E_o \cos(\omega t) - \mathbf{a}_y E_o \sin(\omega t) \qquad (8.49)$$

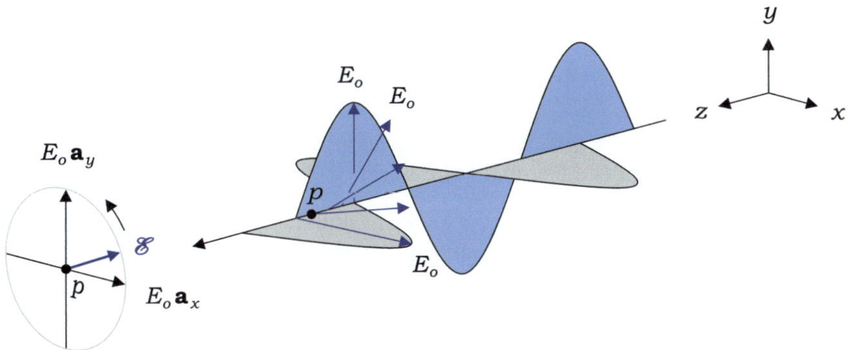

Fig. 8.4 Right-hand circularly polarized wave and its component waves. The blue arrow represents the electric field vector traveling along the z-axis

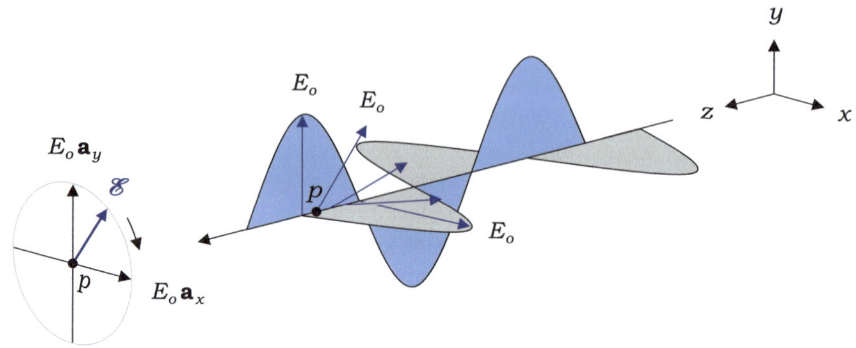

Fig. 8.5 Left-hand circularly polarized wave and its component waves. The blue arrow represents the electric field vector traveling along the z-axis

it follows that $\mathscr{E}(0,0) = E_o\,\mathbf{a}_x$ and $\mathscr{E}(0, t = \pi/2\omega) = -E_o\,\mathbf{a}_y$, as an example. This observation, along with $|\mathscr{E}(0, t)| = E_o$, leads us to conclude that the electric field vector rotates clockwise in the xy-plane, and its tip leaves a circular trace as it completes a turn for every 2π of ωt.

A left-hand circularly polarized wave traveling in the $+z$-direction is illustrated in Fig. 8.5. Because the y-component *leads* the x-component by $90°$ in time phase, the peak of the y-component reaches point p one-quarter of the temporal period *earlier* than the x-component. At one instant, the y-component curve appears to be displaced *forward* (i.e., in the $+z$-direction) by one-fourth of the wavelength relative to the x-component curve.

Example 8.5 Resolve a linearly polarized wave expressed as $\tilde{\mathbf{E}}_1 = E_o\,\mathbf{a}_x e^{-jkz}$ into two circularly polarized waves.

Solution

By adding and subtracting the term $(jE_o/2)\,\mathbf{a}_y$ on the right-hand side of the expression, we obtain

$$\tilde{\mathbf{E}}_1 = \frac{1}{2}\left[\left(E_o\,\mathbf{a}_x - jE_o\,\mathbf{a}_y\right) + \left(E_o\,\mathbf{a}_x + jE_o\,\mathbf{a}_y\right)\right]e^{-jkz} \tag{8.50}$$

The first term in brackets represents a right-hand circularly polarized wave, whereas the second term represents a left-hand circularly polarized wave.

Conversely, if two circularly polarized waves of equal amplitude have electric field vectors rotating in opposite directions and crossing the y-axis at the same instant, the sum of these waves constitutes a wave linearly polarized along the y-axis.

Exercise 8.12

Determine the polarization handedness of these circularly polarized waves: (a) $\tilde{\mathbf{E}} = (j\,\mathbf{a}_x + \mathbf{a}_y)\,e^{-j30z}$, (b) $\tilde{\mathbf{E}} = 2(j\,\mathbf{a}_y - \mathbf{a}_z)\,e^{-j40x}$, and (c) $\tilde{\mathbf{E}} = 3(\mathbf{a}_x - j\,\mathbf{a}_y)\,e^{j50z}$.

Ans. (a) Right-hand, (b) Left-hand, (c) Left-hand.

8.2.3.3 Elliptical Polarization

An electromagnetic wave is ***elliptically polarized*** if it is not linearly or circularly polarized. The tip of the electric field vector traces an ellipse in the transverse plane over time. Linear or circular polarization is a special case of elliptical polarization. The electric field phasor of an elliptically polarized wave is generally expressed as

$$\tilde{\mathbf{E}} = \left(E_{ox}\,\mathbf{a}_x + E_{oy}\,e^{j\varphi}\,\mathbf{a}_y \right) e^{-jkz} \tag{8.51}$$

where the amplitudes E_{ox} and E_{oy} are positive real quantities, and the phase angle φ determines the phase difference between the two components. The real instantaneous value is

$$\mathscr{E} = E_{ox}\,\mathbf{a}_x\,\cos(\omega t - kz) + E_{oy}\,\mathbf{a}_y\,\cos(\omega t - kz + \varphi) \tag{8.52}$$

It is evident from Eq. (8.52) that the electric field vector changes its magnitude and direction as time progresses. In the transverse plane at $z = 0$, the two components of $\mathscr{E}$ are given by

$$\mathscr{E}_x = E_{ox}\,\cos(\omega t) \tag{8.53a}$$

$$\mathscr{E}_y = E_{oy}\,\cos(\omega t + \varphi) \tag{8.53b}$$

This indicates that the trace of the tip of $\mathscr{E}$ is inscribed in a rectangle with sides $2E_{ox}$ and $2E_{oy}$ centered at the origin. At time $t = 0$, we have $\mathscr{E}_x = E_{ox}$ and $\mathscr{E}_y = E_{oy}\cos\varphi$, and the tip of $\mathscr{E}$ is located on the right side of the rectangle for $0 < \varphi < \pi$. At a later time, $\omega t = \pi/2$, we have $\mathscr{E}_x = 0$ and $\mathscr{E}_y = -E_{oy}\sin\varphi$, and the tip of $\mathscr{E}$ is located on the negative y-axis. It is evident that the vector $\mathscr{E}$ rotates clockwise in the xy-plane, and the wave is said to be ***left-hand elliptically polarized***. Moreover, the elongated trace is slanted to the right for $0 < \varphi < \pi/2$ and to the left for $\pi/2 < \varphi < \pi$ (see Exercise 8.13). If the phase angle is in the range of $-\pi < \varphi < 0$, the wave is ***right-hand elliptically polarized***.

The shape of the polarization ellipse depends on the E_{oy}/E_{ox} ratio and phase angle φ. The ellipse in Fig. 8.6 has a major axis of $2E'_{ox}$ and minor axis of $2E'_{oy}$. For now, we are interested in the rotation angle of the major axis with the horizontal axis, θ. It should be noted that $\theta \neq \tan^{-1}(E_{oy}/E_{ox})$. Let us begin with a left-hand elliptically polarized wave with an elliptical locus, as shown in Fig. 8.6. In this case, the electric field can be expressed in the primed coordinate system as

$$\mathscr{E}_x{}' = E'_{ox}\,\cos(\omega t + \varphi') \tag{8.54a}$$

$$\mathscr{E}_y{}' = -E'_{oy}\,\sin(\omega t + \varphi') \tag{8.54b}$$

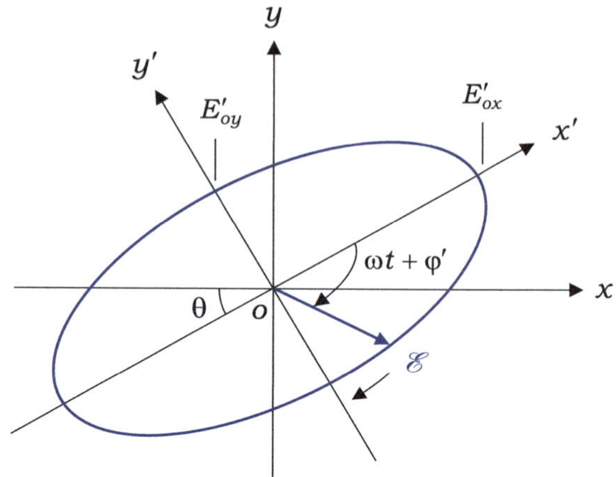

Fig. 8.6 Elliptical trace of the electric field vector, which is rotated by angle θ

where the negative sign indicates that the electric field vector rotates clockwise in the $x'y'$-plane. The terms E'_{ox}, E'_{oy}, and φ' in Eqs. (8.54a) and (8.54b) can be expressed in terms of E_{ox}, E_{oy}, and φ, as given in Eqs. (8.53a) and (8.53b), by taking the coordinate transformation of Eqs. (8.53a) and (8.53b) into the primed system. That is,

$$\mathscr{E}'_x = E_{ox} \cos(\omega t) \cos\theta + E_{oy} \cos(\omega t + \varphi) \sin\theta \tag{8.55a}$$

$$\mathscr{E}'_y = -E_{ox} \cos(\omega t) \sin\theta + E_{oy} \cos(\omega t + \varphi) \cos\theta \tag{8.55b}$$

By comparing Eqs. (8.54a, b) with Eqs. (8.55a, b), with the help of trigonometry, we obtain

$$\boxed{\tan(2\theta) = \frac{2 E_{ox} E_{oy} \cos\varphi}{(E_{ox})^2 - (E_{oy})^2}} \tag{8.56}$$

For $\varphi = 0$, this relationship is reduced to $\tan\theta = E_{oy}/E_{ox}$, which corresponds to a linearly polarized wave.

Example 8.6 A uniform plane wave with electric field phasor $\tilde{\mathbf{E}} = (E_{ox}\, \mathbf{a}_x + E_{oy}\, e^{j\varphi}\, \mathbf{a}_y)\, e^{-jkz}$ propagates in free space. Determine the time-averaged power density.

Solution

From Eq. (8.21), the magnetic field phasor is

$$\tilde{\mathbf{H}} = \frac{1}{\eta_o}\left(\mathbf{a}_k \times \tilde{\mathbf{E}}\right) = \frac{1}{\eta_o}\mathbf{a}_z \times \left(E_{ox}\, \mathbf{a}_x + E_{oy}\, e^{j\varphi}\, \mathbf{a}_y\right) e^{-jkz}$$

$$= \frac{1}{\eta_o}\left(E_{ox}\,\mathbf{a}_y - E_{oy}\,e^{j\varphi}\,\mathbf{a}_x\right)e^{-jkz}$$

The time-averaged Poynting vector is

$$\langle \mathbf{S}\rangle = \frac{1}{2}\mathrm{Re}\left[\tilde{\mathbf{E}}\times\tilde{\mathbf{H}}^*\right]$$

$$= \frac{1}{2\eta_o}\mathrm{Re}\left[\left(E_{ox}\,\mathbf{a}_x + E_{oy}\,e^{j\varphi}\,\mathbf{a}_y\right)e^{-jkz}\times\left(E_{ox}\,\mathbf{a}_y - E_{oy}\,e^{-j\varphi}\,\mathbf{a}_x\right)e^{jkz}\right]$$

$$= \frac{1}{2\eta_o}\left(E_{ox}^2 + E_{oy}^2\right)\mathbf{a}_z$$

The time-averaged power density is equal to the sum of those of the two linearly polarized component waves.

Exercise 8.13
Sketch the locus of $\tilde{\mathbf{E}}$, as expressed by Eq. (8.51), if $E_{ox} = 2E_{oy}$ and $0 \le \varphi \le 2\pi$.

Ans.

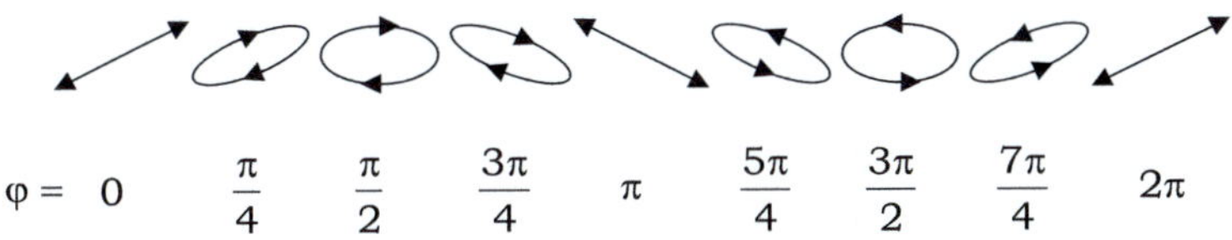

Exercise 8.14
For $\mathcal{E} = 3\,\mathbf{a}_x\sin(\omega t - kz - 30^\circ) + 2\,\mathbf{a}_y\cos(\omega t - kz)$, the tip of $\mathcal{E}$ traces an ellipse in the xy-plane, as shown below. Find (a) the x-coordinate of point p_1, and (b) the difference in time phase between points p_2 and p_3 (surprisingly, it is not $\pi/2$).

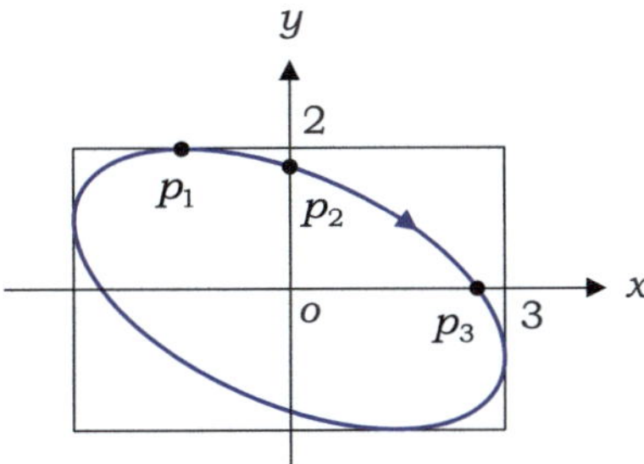

Ans. (a) -1.5, (b) $\omega t - kz = \pi/6$ for p_2, and $\pi/2$ for p_3. $\Delta\varphi = \pi/3$.

Exercise 8.15

For a right-hand elliptically polarized wave with $\tilde{\mathbf{E}} = (2E_o\mathbf{a}_x - jE_o\mathbf{a}_y)e^{-jkz+j\pi/4}$, sketch the electric field vector in the $z = 0$ plane for $\omega t = 0$ and $\omega t = 3\pi/4$.

Ans.

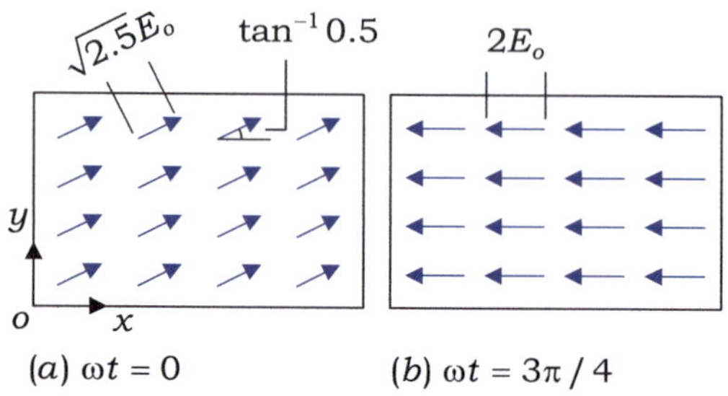

(a) $\omega t = 0$ (b) $\omega t = 3\pi/4$

Review Questions

RQ 8.13   Define wave-polarization.   [(8.51)]

RQ 8.14   How is a uniform plane wave expressed if it exhibits linear or circular polarization?   [(8.42)(8.44)(8.47)]

RQ 8.15   Can a circularly polarized wave be resolved into two linearly polarized waves?   [(8.44)(8.47)]

RQ 8.16   Can an elliptically polarized wave be resolved into two circularly polarized waves?   [(8.44)(8.47)(8.51)]

RQ 8.17   Is the handedness of the elliptical polarization reversed by multiplying the electric phasor by -1?   [(8.51)]

RQ 8.18   Is the angular velocity of rotation of $\mathscr{E}$ of an elliptically polarized wave constant over time?   [Exercise 8.14]

8.2.4 *Wave Propagation in Lossy Medium*

We now turn our attention to the propagation of electromagnetic waves in lossy media. Dielectrics with damping losses and conductors with finite conductivities are examples of lossy media. In this case, electromagnetic waves undergo attenuation as they propagate through the material.

The power-loss mechanism in imperfect dielectrics can be well explained by the damping force and/or finite conductivity. When an electromagnetic wave propagates through a dielectric, its electric field induces electric dipoles in the material, such that the bound charges oscillate in accordance with the time-harmonic electric field. Meanwhile, strong interactions between neighboring atoms tend to hinder the electric dipoles from oscillating in unison with the electric field. The damped oscillation causes the electric dipoles to lag behind the electric field in time phase. Therefore, the electric susceptibility, which is a proportionality constant between the electric polarization and electric field, is now complex in the material. To overcome the damping forces, the incident wave must expend energy and its amplitude decreases as the wave propagates through the material.

If a material medium contains a substantial number of free electrons, the electric field of the incident wave induces a conduction current that leads to ohmic power loss in the material. Power loss is directly related to the rate of decrease of the power density of the wave with distance. Therefore, the wave must undergo attenuation.

The magnetic field of the incident wave also interacts with the lattice atoms through magnetic dipole moments, and the wave may suffer additional power loss if the magnetic dipoles are affected by a damping mechanism. Similarly, the damping effect can be incorporated into Maxwell's equations through complex magnetic susceptibility or complex permeability. However, in most practical cases, the power loss of this type is negligible compared with its electric counterpart; thus, the permeability is assumed to be real.

In this section, we define the complex permittivity in lossy media and establish a relationship between the attenuation of the incident wave and the imaginary part of the complex permittivity.

8.2.4.1 Plane Wave in Dielectric with Damping Loss

In a homogeneous (ε and μ are independent of position), linear (ε and μ are independent of the magnitudes of $\mathscr{E}$ and $\mathscr{H}$), and isotropic (ε and μ are independent of the directions of $\mathscr{E}$ and $\mathscr{H}$) material with damping loss (permittivity is complex), Maxwell's equations are given in phasor form as follows:

$$\nabla \times \tilde{\mathbf{E}} = -j\omega\mu\tilde{\mathbf{H}} \tag{8.57a}$$

$$\begin{aligned}
\nabla \times \tilde{\mathbf{H}} &= j\omega\hat{\varepsilon}\,\tilde{\mathbf{E}} \\
&= j\omega(\varepsilon' - j\varepsilon'')\tilde{\mathbf{E}} \\
&= j\omega\varepsilon_0(\varepsilon_r' - j\varepsilon_r'')\tilde{\mathbf{E}}
\end{aligned} \tag{8.57b}$$

$$\nabla \cdot \tilde{\mathbf{E}} = 0 \tag{8.57c}$$

$$\nabla \cdot \tilde{\mathbf{H}} = 0 \tag{8.57d}$$

where $\hat{\varepsilon}$ is the complex permittivity with the real part ε' and imaginary part ε'', and ε_r' and ε_r'' are the real and imaginary parts of the relative permittivity ε_r, respectively. In these equations, it is assumed that there are no free charges or current. Note that the damping force is accounted for by either ε'' or ε_r''.

Let us start with the vector Helmholtz equation,

$$\nabla^2 \tilde{\mathbf{E}} + \hat{k}^2 \tilde{\mathbf{E}} = 0 \tag{8.58}$$

The ***complex wavenumber*** $\hat{k}$ is defined as

$$\hat{k} = \omega\sqrt{\mu\hat{\varepsilon}} = \omega\sqrt{\mu\varepsilon'}\sqrt{1 - j\left(\frac{\varepsilon''}{\varepsilon'}\right)} \tag{8.59}$$

Because Eq. (8.58) has the same form as Eq. (8.11), except for the complex wavenumber, it immediately follows that a uniform plane wave is the general solution to Eq. (8.58), which is expressed as

$$\tilde{\mathbf{E}} = \hat{\mathbf{E}}_o \, e^{-j\hat{k}\, \mathbf{a}_k \cdot \mathbf{r}} \tag{8.60}$$

where $\mathbf{a}_k$ is a unit vector in the direction of wave propagation. Without loss of generality, we can set $\mathbf{a}_k = \mathbf{a}_z$ for an unbounded medium. Subsequently, the electric field phasor is reduced to

$$\begin{aligned} \tilde{\mathbf{E}} &= \hat{\mathbf{E}}_o \, e^{-j\hat{k}z} \\[6pt] &\equiv \hat{\mathbf{E}}_o \, e^{-\gamma z} \end{aligned} \tag{8.61}$$

where the ***propagation constant*** γ is defined as

$$\begin{aligned} \gamma &= j\hat{k} = j\omega\sqrt{\mu(\varepsilon' - j\varepsilon'')} \\[6pt] &\equiv \alpha + j\beta \end{aligned} \qquad [\text{m}^{-1}] \tag{8.62}$$

The ***attenuation constant*** α is in nepers per meter [Np/m], and the ***phase constant*** β is in radians per meter [rad/m]. After some manipulations, we can obtain

$$\alpha = \omega\sqrt{\frac{\mu\varepsilon'}{2}}\left[\sqrt{1 + \left(\frac{\varepsilon''}{\varepsilon'}\right)^2} - 1\right]^{1/2} \qquad [\text{Np/m}] \tag{8.63}$$

$$\beta = \omega\sqrt{\frac{\mu\varepsilon'}{2}}\left[\sqrt{1 + \left(\frac{\varepsilon''}{\varepsilon'}\right)^2} + 1\right]^{1/2} \qquad [\text{rad/m}] \tag{8.64}$$

It should be noted that the α/β ratio cannot exceed unity.

For lossless or perfect dielectrics with $\varepsilon'' = 0$, the attenuation constant is zero. Otherwise, ε' is always accompanied by ε'' in the expressions for α and β. We find it convenient to define the **loss tangent** as

$$\tan \xi = \frac{\varepsilon''}{\varepsilon'} \qquad\qquad (8.65)$$

where angle ξ (read "xi") is called the **loss angle**. The loss tangent is a measure of the power loss in lossy media. Several methods can be used to obtain the loss angle, which can be calculated from the α/β ratio or the complex intrinsic impedance $\hat{\eta}$, as discussed later.

Next, we examine the behavior of an electromagnetic plane wave propagating in a dielectric with damping loss. With the help of Eq. (8.62), the electric field of the wave can be expressed in phasor and instantaneous forms as follows:

$$\tilde{\mathbf{E}} = \hat{E}_o \mathbf{a}_E \, e^{-\alpha z} e^{-j\beta z} \qquad\qquad (8.66a)$$

$$\mathscr{E} = E_o \mathbf{a}_E \, e^{-\alpha z} \cos(\omega t - \beta z + \varphi) \qquad\qquad (8.66b)$$

where $\mathbf{a}_E$ is a unit vector along the electric field vector, and $\hat{E}_o$ is the complex amplitude with magnitude E_o and phase angle φ. As is implied by Eqs. (8.66a) and (8.66b), the attenuation constant describes how the amplitude decreases with distance, whereas the phase constant shows how the phase varies with position. The attenuation constant must be positive in passive media in which no wave amplification can occur.

When the electric field in Eqs. (8.66a) and (8.66b) has the same phase at two points that are a distance λ apart along the z-axis, it follows that $\beta\lambda = 2\pi$. Thus,

$$\beta = \frac{2\pi}{\lambda} \qquad [\text{rad/m}] \qquad\qquad (8.67)$$

It is evident from Eq. (8.64) that the wavelength λ in lossy dielectrics depends on both ε' and ε''.

The phase velocity of the wave is always given by

$$\upsilon_p = \frac{\omega}{\beta} \qquad [\text{m/s}] \qquad\qquad (8.68)$$

which also depends on ε' and ε'' of the material through β.

Subsequently, the magnetic field phasor is obtained by substituting the electric field phasor expressed in Eq. (8.66a) into Eq. (8.57a), that is,

$$\nabla \times \tilde{\mathbf{E}} = \nabla \times \left(\hat{E}_o \mathbf{a}_E \, e^{-\gamma z} \right) = -\gamma \, \mathbf{a}_z \times \left(\hat{E}_o \mathbf{a}_E \, e^{-\gamma z} \right)$$
$$= -j\omega\mu \tilde{\mathbf{H}}$$

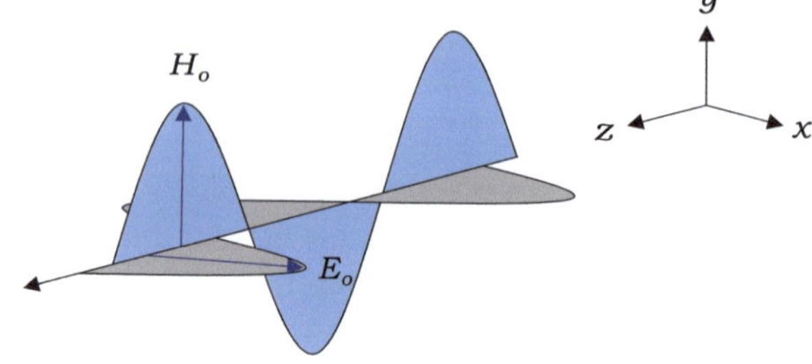

Fig. 8.7 $\tilde{\mathbf{H}}$ always lags behind $\tilde{\mathbf{E}}$ in time phase in lossy media

Thus,

$$\tilde{\mathbf{H}} = \sqrt{\frac{\hat{\varepsilon}}{\mu}}\,\hat{E}_o(\mathbf{a}_z \times \mathbf{a}_E)e^{-\gamma z} = \frac{\hat{E}_o}{\hat{\eta}}(\mathbf{a}_z \times \mathbf{a}_E)e^{-\alpha z}e^{-j\beta z} \qquad (8.69)$$

In a more general form, the magnetic field phasor is expressed as

$$\tilde{\mathbf{H}} = \frac{1}{\hat{\eta}}\mathbf{a}_k \times \tilde{\mathbf{E}} \qquad (8.70)$$

where $\mathbf{a}_k$ is the unit vector of the wavevector. This has the same form as that in Eq. (8.21), except for the ***complex intrinsic impedance***, $\hat{\eta}$, defined as

$$\boxed{\hat{\eta} = \sqrt{\frac{\mu}{\hat{\varepsilon}}} = \sqrt{\frac{\mu}{\varepsilon'(1 - j\varepsilon''/\varepsilon')}}} \quad [\Omega] \qquad (8.71)$$

It is important to remember that $\hat{\eta}$ is the ratio of the complex amplitudes of the electric and magnetic phasors, that is, $\hat{\eta} = \hat{E}_o/\hat{H}_o$. Moreover, Eq. (8.71) indicates that $\tilde{\mathbf{E}}$ and $\tilde{\mathbf{H}}$ are out of phase at all points in lossy media (see Fig. 8.7). The right-hand rule in Eq. (8.25) is also valid for lossy media.

Using the definition $\hat{\varepsilon} = \varepsilon_0(\varepsilon_r' - j\varepsilon_r'')$ and the loss tangent $\tan \xi = \varepsilon''/\varepsilon'$, Eqs. (8.63), (8.64), and (8.71) can be rewritten as

$$\boxed{\alpha = \frac{\omega}{c}\sqrt{\frac{\mu_r \varepsilon_r'}{2}}\sqrt{\sec \xi - 1}} \quad [\text{Np/m}] \qquad (8.72a)$$

$$\boxed{\beta = \frac{\omega}{c}\sqrt{\frac{\mu_r \varepsilon_r'}{2}}\sqrt{\sec \xi + 1}} \quad [\text{rad/m}] \qquad (8.72b)$$

$$\boxed{\hat{\eta} = \eta_o\sqrt{\frac{\mu_r}{\varepsilon_r'}}\sqrt{\cos \xi}\;e^{j\,\xi/2}} \quad [\Omega] \qquad (8.72c)$$

where $\sec\xi = 1/\cos\xi$ from trigonometry.

Example 8.7 For the electric field phasor $\tilde{\mathbf{E}} = \mathbf{a}_x 20\, e^{-0.3z} e^{-j0.5z}$ in a lossy dielectric with $\varepsilon' = 2\varepsilon_0$ and $\mu = \mu_0$, determine (a) λ, (b) $\hat{\eta}$, and (c) $\tilde{\mathbf{H}}$.

Solution

(a) From $\tilde{\mathbf{E}}$ given in the problem, it follows that $\alpha = 0.3$ and $\beta = 0.5$. Thus,

$$\lambda = \frac{2\pi}{\beta} = 4\pi$$

(b) From Eqs. (8.63) and (8.64),

$$\left(\frac{\alpha}{\beta}\right)^2 = \frac{\sqrt{1 + (\varepsilon''/\varepsilon')^2} - 1}{\sqrt{1 + (\varepsilon''/\varepsilon')^2} + 1} \tag{8.73a}$$

Inserting α and β into Eq. (8.73a), the loss tangent is obtained as

$$\frac{\varepsilon''}{\varepsilon'} = 1.88 \tag{8.73b}$$

Substituting Eq. (8.73b) into Eq. (8.71), with $\varepsilon' = 2\varepsilon_0$, we obtain

$$\hat{\eta} = \sqrt{\frac{\mu_0}{2\varepsilon_0}}\,\frac{1}{\sqrt{1 - j(\varepsilon''/\varepsilon')}} = \frac{120\pi}{\sqrt{2}}\,\frac{1}{\sqrt{1 - j1.88}} = 182.7 e^{j0.54}$$

(c) From Eq. (8.69),

$$\tilde{\mathbf{H}} = \mathbf{a}_y 0.11 e^{-j0.54} e^{-0.3z} e^{-j0.5z}$$

The results indicate that $\tilde{\mathbf{H}}$ lags behind $\tilde{\mathbf{E}}$ by 0.54 radians in time phase.

Exercise 8.16
For a wave with $\mathscr{E} = -\mathbf{a}_y 0.5 e^{-0.3z}\sin(10^8 t + 2z)$, determine (a) α, (b) β, (c) υ_p, (d) the direction of propagation, and (e) $\hat{\mathbf{E}}_o$ for cosine reference.

Ans. (a) $\alpha = 0.3\,[\mathrm{Np/m}]$, (b) $\beta = 2\,[\mathrm{rad/m}]$, (c) $\upsilon_p = 5\times10^7\,[\mathrm{m/s}]$, (d) the $-z$-direction, (e) $\hat{\mathbf{E}}_o = 0.5 e^{j\pi/2}\,[\mathrm{V/m}]$.

Exercise 8.17
Starting with Eq. (8.62), express the loss angle in terms of α and β.

Ans. $\xi = \pi - 2\tan^{-1}(\beta/\alpha)$.

8.2.4.2   Plane Wave in Dielectric with Low Conductivity

Conduction current is the dominant source of power loss in lossy media. In most cases, the damping force can be ignored if the material medium has a nonzero conductivity. Under these conditions, Maxwell's equations are given by

$$\nabla \times \tilde{\mathbf{E}} = -j\omega\mu\tilde{\mathbf{H}} \tag{8.74a}$$

$$\nabla \times \tilde{\mathbf{H}} = \tilde{\mathbf{J}} + j\omega\varepsilon\tilde{\mathbf{E}} = (\sigma + j\omega\varepsilon)\tilde{\mathbf{E}} \tag{8.74b}$$

$$\nabla \cdot \tilde{\mathbf{E}} = 0 \tag{8.74c}$$

$$\nabla \cdot \tilde{\mathbf{H}} = 0 \tag{8.74d}$$

Because the damping force is ignored, ε and μ in Eqs. (8.74a) and (8.74b) are both real.

It is more convenient to translate conductivity in Eq. (8.74b) to the complex permittivity such that

$$\begin{aligned} \nabla \times \tilde{\mathbf{H}} &= j\omega\left(\varepsilon - j\frac{\sigma}{\omega}\right)\tilde{\mathbf{E}} \\ &\equiv j\omega\left(\varepsilon' - j\varepsilon''\right)\tilde{\mathbf{E}} \equiv j\omega\hat{\varepsilon}\,\tilde{\mathbf{E}} \end{aligned} \tag{8.75}$$

In the presence of a nonzero conductivity, the complex permittivity $\hat{\varepsilon}$ is newly defined as

$$\boxed{\hat{\varepsilon} \equiv \varepsilon' - j\varepsilon'' = \varepsilon - j\frac{\sigma}{\omega}} \quad [\text{F/m}] \tag{8.76}$$

The real and imaginary parts of $\hat{\varepsilon}$ are directly related to the permittivity ε and conductivity σ of a material, respectively.

The electromagnetic plane wave propagating in a conductive material has the same form as Eq. (8.61):

$$\boxed{\tilde{\mathbf{E}} = \hat{\mathbf{E}}_o e^{-\gamma z}} \tag{8.77}$$

Inserting $\varepsilon' = \varepsilon$ and $\varepsilon'' = \sigma/\omega$ into Eqs. (8.62)–(8.64) leads to the propagation, attenuation, and phase constants in a conductive material as follows:

$$\boxed{\gamma \equiv \alpha + j\beta = j\omega\sqrt{\mu\varepsilon}\sqrt{\left(1 - j\frac{\sigma}{\omega\varepsilon}\right)}} \quad [\text{m}^{-1}] \tag{8.78}$$

$$\alpha = \omega\sqrt{\frac{\mu\varepsilon}{2}}\left[\sqrt{1 + \left(\frac{\sigma}{\omega\varepsilon}\right)^2} - 1\right]^{1/2} \qquad \text{[Np/m]} \tag{8.79}$$

$$\beta = \omega\sqrt{\frac{\mu\varepsilon}{2}}\left[\sqrt{1 + \left(\frac{\sigma}{\omega\varepsilon}\right)^2} + 1\right]^{1/2} \qquad \text{[rad/m]} \tag{8.80}$$

Similarly, from Eq. (8.71), the complex intrinsic impedance of a conductive material is obtained as

$$\hat{\eta} = \sqrt{\frac{\mu}{\varepsilon}}\sqrt{\frac{1}{1 - j\sigma/(\omega\varepsilon)}} \qquad [\Omega] \tag{8.81}$$

In this case, the loss tangent is defined as

$$\tan\xi = \frac{\sigma}{\omega\varepsilon} \tag{8.82}$$

The loss tangent is defined as $\varepsilon''/\varepsilon'$ in the dielectric with damping loss. However, it is also defined as $\sigma/\omega\varepsilon$ in conductive materials. Moreover, conductivity is assumed to be zero in the former case, whereas permittivity is assumed to be real in the latter case.

Low-loss materials include those with $\tan\xi \ll 1$ or $\sigma \ll \omega\varepsilon$. For example, for a low-loss material with $\tan\xi < 0.1$, it follows that the ratio α/β is less than 0.05, and the phase angle of $\hat{\eta}$ is less than 0.05 [rad], which are within the same numerical range as the loss tangent.

Example 8.8 Determine the attenuation and phase constants of a wave with a frequency of 100 [MHz] propagating in a nonmagnetic conductive material with $\hat{\eta} = 200e^{j0.5}$.

Solution

Rewriting Eq. (8.81) in polar form, we have

$$\hat{\eta} = \sqrt{\frac{\mu_0}{\varepsilon_0\varepsilon_r}}\frac{e^{j\xi/2}}{\left[1 + (\tan\xi)^2\right]^{1/4}} = 200e^{j0.5} \tag{8.83}$$

The phase angle in Eq. (8.83) leads to $\xi = 1$; thus, the loss tangent is given by $\tan(1) = 1.56$. Meanwhile, the magnitude in Eq. (8.83) leads to

$$\frac{377}{\sqrt{\varepsilon_r}}\frac{1}{\left[1 + (1.56)^2\right]^{1/4}} = 200$$

This yields $\varepsilon_r = 1.92$. Thus, from Eq. (8.79),

$$
\begin{aligned}
\alpha &= \omega\sqrt{\mu_0\varepsilon_0}\sqrt{\frac{\varepsilon_r}{2}}\left[\sqrt{1 + (\sigma/\omega\varepsilon)^2} - 1\right]^{1/2} \\
&= \frac{2\pi \times 10^8}{3 \times 10^8}\sqrt{\frac{1.92}{2}}\left[\sqrt{1 + (1.56)^2} - 1\right]^{1/2} = 1.90 \ [\text{Np/m}]
\end{aligned}
$$

Similarly, from Eq. (8.80),

$$
\begin{aligned}
\beta &= \omega\sqrt{\mu_0\varepsilon_0}\sqrt{\frac{\varepsilon_r}{2}}\left[\sqrt{1 + (\sigma/\omega\varepsilon)^2} + 1\right]^{1/2} \\
&= \frac{2\pi \times 10^8}{3 \times 10^8}\sqrt{\frac{1.92}{2}}\left[\sqrt{1 + (1.56)^2} + 1\right]^{1/2} = 3.47 \ [\text{rad/m}]
\end{aligned}
$$

The results indicate that $\alpha/\beta = 0.55$, which is much greater than 0.05. Thus, the given material is a high-loss medium.

Example 8.9 The electromagnetic wave in a medium with $\hat{\eta} = 220 + j21 \ [\Omega]$ is characterized by $\tilde{\mathbf{E}} = \mathbf{a}_x 300 e^{-0.21z - j\,2.2z} \ [\text{V/m}]$. Find $\langle \mathbf{S} \rangle$ at the $z = 0.5 \ [\text{m}]$ plane.

Solution

The magnetic field phasor is obtained as

$$
\tilde{\mathbf{H}} = \frac{1}{\hat{\eta}}\left(\mathbf{a}_z \times \tilde{\mathbf{E}}\right) = \mathbf{a}_y \frac{300}{220 + j21} e^{-0.21z - j\,2.2z}
$$

Thus, the time-averaged Poynting vector is

$$
\begin{aligned}
\langle \mathbf{S} \rangle &= \frac{1}{2}\mathrm{Re}\left[\tilde{\mathbf{E}} \times \tilde{\mathbf{H}}^*\right] = \mathbf{a}_z \frac{1}{2}\mathrm{Re}\left[\frac{300^2 e^{-0.42z}}{220 - j21}\right] \\
&= \mathbf{a}_z (202.7) e^{-0.42z}
\end{aligned}
$$

At $z = 0.5 \ [\text{m}]$, the time-averaged power density is

$$
\langle \mathbf{S} \rangle = \mathbf{a}_z (202.7)\, e^{-0.42 \times 0.5} = 164.3\, \mathbf{a}_z \ [\text{W/m}^2]
$$

Exercise 8.18

In a nonmagnetic dielectric, can the following wave or material characteristics be determined from electric field phasor $\tilde{\mathbf{E}} = \mathbf{a}_x 10 e^{-0.3z - j2z}$ alone? (a) α, (b) β, (c) λ, (d) direction of propagation, (e) wave amplitude, (f) wave polarization, (g) loss tangent, (h) ω, (i) ε, (j) υ_p, (k) $\hat{\eta}$, and (l) $\tilde{\mathbf{H}}$.

Ans. (a)–(g), yes; (h)–(l), no.

Exercise 8.19
If one of the unknowns in Exercise 8.18 is set to a value, can the rest be specified?

Ans. Yes.

Exercise 8.20
What phase angle does $\tilde{\mathbf{H}}$ lag behind $\tilde{\mathbf{E}}$ in lossy media?

Ans. $\xi/2$.

Exercise 8.21
What is the maximum possible phase delay of $\tilde{\mathbf{H}}$ with respect to $\tilde{\mathbf{E}}$?

Ans. $45°$.

Review Questions

RQ 8.19	Define attenuation and phase constants.	[(8.63)(8.64)(8.79)(8.80)]
RQ 8.20	Define the loss tangent.	[(8.65)(8.82)]
RQ 8.21	What makes permittivity a complex quantity?	[(8.57b)(8.74b)]
RQ 8.22	Define the intrinsic impedance in lossy media.	[(8.71)(8.81)]
RQ 8.23	Is the phase angle of η in lossy media always positive?	[(8.71)(8.81)]
RQ 8.24	Why should the imaginary part of $\hat{\varepsilon}$ be negative?	[(8.74b)(8.75)]
RQ 8.25	What is the low-loss dielectric material?	[(8.82)]

8.2.4.3 Plane Wave in Good Conductor

The conductivity may not be infinite in **good conductors** but is very large, such that the loss tangent $\sigma/\omega\varepsilon$ is much greater than unity. A material is referred to as a good insulator if its loss tangent is much less than unity. As the loss tangent is inversely proportional to the frequency, a material may be considered a good conductor at low frequencies but may behave as a low-loss dielectric at very high frequencies. In good conductors, the displacement current can be ignored in comparison with the conduction current, as is evident from parentheses in Eq. (8.74b). A good conductor differs from a low-loss dielectric in that the permittivity of a good conductor approximates that of free space, whereas the relative permittivity of a dielectric is typically greater than unity. For good conductors, the propagation constant γ expressed in Eq. (8.78) is reduced to

$$\gamma \cong \frac{1+j}{\sqrt{2}}\sqrt{\omega\mu\sigma} \qquad (8.84)$$

where we have used the relation $\sqrt{-j} = e^{-j\pi/4} = (1-j)/\sqrt{2}$. It follows from Eq. (8.84) that the attenuation and phase constants are equal in magnitude in good conductors, that is,

$$\boxed{\alpha = \beta = \sqrt{\pi f \mu \sigma}}$$ 	(8.85)

where $\omega = 2\pi f$. Note that both α and β increase with frequency and conductivity.

The electric field of an electromagnetic wave penetrating a good conductor can be expressed in phasor and instantaneous forms as follows:

$$\tilde{\mathbf{E}} = \hat{\mathbf{E}}_o e^{-\alpha z} e^{-j\beta z}$$

$$= (E_o \mathbf{a}_E) e^{-z\sqrt{\pi f \mu \sigma}} e^{-jz\sqrt{\pi f \mu \sigma}}$$	(8.86a)

$$\mathscr{E} = (E_o \mathbf{a}_E) e^{-z\sqrt{\pi f \mu \sigma}} \cos(\omega t - z\sqrt{\pi f \mu \sigma})$$	(8.86b)

Here, $e^{-\alpha z}$ indicates that the wave amplitude decreases exponentially with z. In other words, the amplitude is attenuated by a factor of e^{-1} as the wave travels a distance of $1/\alpha$. This distance is denoted by δ and is called the **depth of penetration** or **skin depth** of a conductor:

$$\boxed{\delta = \frac{1}{\sqrt{\pi f \mu \sigma}} = \frac{1}{\alpha} = \frac{1}{\beta}} \quad [\text{m}]$$	(8.87)

where δ is in meters. As the name implies, this is a measure of the depth at which an electromagnetic wave can penetrate a conductor. At microwave frequencies, the skin depth is so small that the electric field and the resulting conduction current can be considered to be confined within a very thin layer of thickness δ on the conductor surface. For example, silver-plated brass waveguides can be used in place of those made of solid silver with little degradation in performance but with significantly reduced material cost.

With the relation $\beta = 2\pi/\lambda$, the skin depth can be expressed as

$$\delta = \frac{\lambda}{2\pi}$$	(8.88)

where λ is the wavelength measured in the conductor. This reveals that only a fraction of a cycle of the incident wave is packed within the skin depth. To show the penetration of an electromagnetic wave through a good conductor, the electric field in the conductor, characterized by $\mathscr{E} = e^{-z/\delta}\cos(\omega t - z/\delta)$, is plotted against the distance from the surface at two different times, $\omega t = 0$ and $\omega t = \pi/2$, as shown in Fig. 8.8.

The phase velocity in a good conductor is always given by ω/β. Thus,

$$\upsilon_p = \frac{\omega}{\beta} = \frac{1}{\sqrt{\mu \varepsilon}} \frac{\sqrt{2}}{\sqrt{\tan \xi}} \quad [\text{m/s}]$$	(8.89)

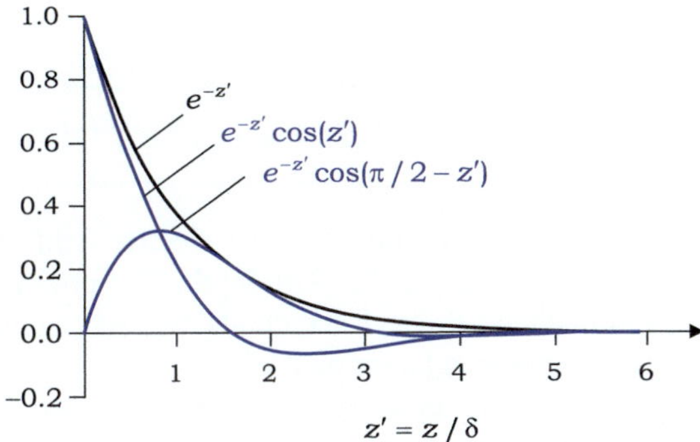

Fig. 8.8 Electric field $\mathscr{E} = e^{-z/\delta}\cos(\omega t - z/\delta)$ in conductor is plotted against the penetration distance from the conductor surface at $\omega t = 0$ and $\omega t = \pi/2$

which depends on the frequency through ξ. The phase velocity is several orders of magnitude lower than the speed of light in free space. For example, the phase velocity in copper is $\upsilon_p = 1.3 \times 10^4$ [m/s] at 1 [GHz].

Because $\sigma/\omega\varepsilon \gg 1$ for good conductors, the intrinsic impedance given in Eq. (8.81) becomes

$$\hat{\eta} \cong \sqrt{\frac{\mu}{\varepsilon}}\sqrt{\frac{j\omega\varepsilon}{\sigma}} = \sqrt{\frac{\mu}{\varepsilon}}\frac{\sqrt{j}}{\sqrt{\tan\xi}} \tag{8.90}$$

which is much smaller in magnitude than that of lossless material. Using the skin depth, the intrinsic impedance of a good conductor can be expressed as

$$\boxed{\hat{\eta} = \frac{(1+j)}{\sigma\delta}} \quad [\Omega] \tag{8.91}$$

Evidently, $\tilde{\mathbf{H}}$ lags behind $\tilde{\mathbf{E}}$ by $\pi/4$ [rad] in time phase.

Next, we focus on the power density of electromagnetic waves in a good conductor. Suppose that the wave is linearly polarized in the x-direction and propagates in the $+z$-direction. Using Eq. (8.87), the electric and magnetic field phasors are expressed as

$$\boxed{\tilde{\mathbf{E}} = \mathbf{a}_x E_o e^{-(1+j)z/\delta}} \tag{8.92a}$$

$$\boxed{\tilde{\mathbf{H}} = \frac{1}{\hat{\eta}}\left(\mathbf{a}_z \times \tilde{\mathbf{E}}\right) = \mathbf{a}_y \frac{\sigma\delta}{1+j}E_o e^{-(1+j)z/\delta}} \tag{8.92b}$$

It was assumed that the wave is transmitted to the conductor through the surface at $z = 0$. Then, the time-averaged Poynting vector is given by

$$\langle \mathbf{S} \rangle = \frac{1}{2}\mathrm{Re}\left[\tilde{\mathbf{E}} \times \tilde{\mathbf{H}}^*\right] = \frac{1}{2}\mathrm{Re}\left[\mathbf{a}_z E_o^2 \frac{\sigma\delta(1+j)}{2}e^{-2z/\delta}\right]$$

or

$$\langle \mathbf{S} \rangle = \mathbf{a}_z \frac{\sigma \delta}{4} E_o^2 e^{-2z/\delta} \qquad [\text{W/m}^2] \qquad (8.93)$$

where E_o is the amplitude of the electric field at the conductor surface, σ is the conductivity, and δ is the skin depth. The power density decreases rapidly as the wave propagates through the conductor. At distance δ from the conductor surface, $\langle \mathbf{S} \rangle$ is reduced by a factor of $1/e^2$.

The skin depth enables us to express the power dissipated in a good conductor in terms of **ac-resistance**. Consider a case in which an electromagnetic plane wave is linearly polarized in the x-direction and propagates in the $+z$-direction in a good conductor of thickness d, as shown in Fig. 8.9. The electric field of the wave induces a conduction current characterized by the current density,

$$\tilde{\mathbf{J}} = \mathbf{a}_x \sigma E_o e^{-(1+j)z/\delta} = \mathbf{a}_x J_o e^{-(1+j)z/\delta} \qquad (8.94)$$

where J_o is the amplitude of the current density on the conductor surface.

The total time-averaged power dissipated in volume $w \times \ell \times d$ [m^3] can be computed as

$$\langle P \rangle = \int_V \frac{1}{2} \text{Re}\left[\tilde{\mathbf{J}} \cdot \tilde{\mathbf{E}}^*\right] dv = \frac{1}{2\sigma} w\ell \text{Re}\left[\int_{z=0}^{z=d} \tilde{\mathbf{J}} \cdot \tilde{\mathbf{J}}^* \, dz\right]$$

$$= \frac{1}{2\sigma} w\ell J_o^2 \int_{z=0}^{z=d} e^{-2z/\delta} dz \qquad (8.95)$$

Assuming a very small skin depth (i.e., $\delta \ll d$), the upper limit of the integration can be replaced with $z = \infty$ to obtain

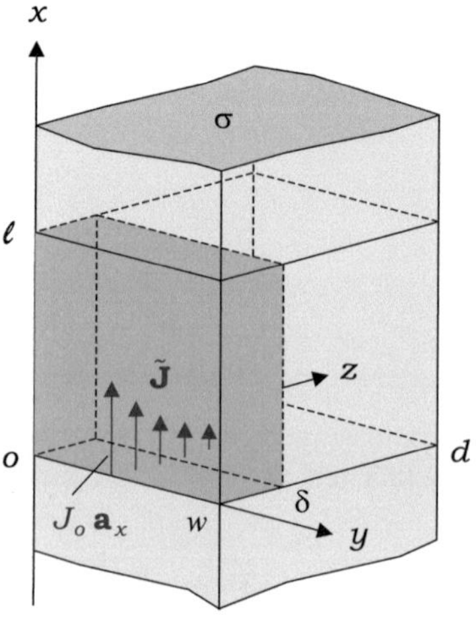

Fig. 8.9 Conduction current due to plane wave transmitted into the conductor

$$\boxed{\langle P \rangle = \frac{1}{4\sigma} w \ell \delta \, J_o^2} \quad [\text{W}] \qquad\qquad (8.96)$$

where $w\ell$ is the area of the entrance surface, and δ is the skin depth. It should be noted that J_o is the amplitude of the volume current density at the entrance surface, and not the surface current density.

Next, the total AC current flowing through the conductor can be obtained by integrating Eq. (8.94) over the cross section at $x = 0$ such that

$$\tilde{I}_T = \int_S \tilde{\mathbf{J}} \cdot d\mathbf{s} = w J_o \int_{z=0}^{z=d} e^{-(1+j)z/\delta} dz$$

$$= \frac{w\delta J_o}{1+j} \ [\text{A}] \qquad\qquad (8.97)$$

where $S = wd$ is the cross-sectional area. From the perspective of electric power, the power in Eq. (8.96) can be considered as the power dissipated in a certain resistor through which the current, as expressed in Eq. (8.97), flows. If the current $\tilde{I}_T$ flows uniformly in a resistor with resistance R_{ac}, the dissipated power is given by

$$\boxed{\langle P \rangle = \frac{1}{2} R_{ac} \,\text{Re}\!\left[\tilde{I}_T \tilde{I}_T^*\right]} \quad [\text{W}] \qquad\qquad (8.98)$$

Comparing Eqs. (8.96) and (8.98), with the help of Eq. (8.97), the so-called *ac-resistance* is defined as

$$\boxed{R_{ac} = \frac{\ell}{\sigma \delta w}} \quad [\Omega] \qquad\qquad (8.99)$$

Surprisingly, R_{ac} corresponds to the resistance of a thin slab with cross section $w\delta$ and length ℓ on the conductor surface. Therefore, the total power dissipated in a conductor with DC resistance $R = \ell/\sigma wd$ [Ω] is equal to the power dissipated in a thin slab of cross section $w\delta$ and length ℓ with *ac-resistance* $R_{ac} = \ell/\sigma w\delta$ [Ω], into which all the currents are uniformly packed.

Example 8.10 A uniform plane wave in air is normally incident on a semi-infinite medium with $\varepsilon = 2\varepsilon_0$, $\mu = \mu_0$, and $\sigma = 5$ [S/m], which fills half-space $z \geq 0$. If the electric field at $z = 0^+$ oscillates along the x-axis at a frequency of 100 [MHz] and has an amplitude of 1 [V/m], determine in region $z \geq 0$, (a) $\mathscr{E}$ and $\mathscr{H}$, (b) phase velocity, (c) wavelength, and (d) skin depth.

Solution

(a) The loss tangent is first evaluated to determine whether the medium is a good conductor:

$$\frac{\sigma}{\omega\varepsilon} = \frac{5}{2\pi \times 10^8 \times 2 \times (1/36\pi) \times 10^{-9}} = 450 \gg 1$$

Because this is a good conductor, from Eq. (8.85),

$$\alpha = \beta = \sqrt{\pi f \mu \sigma} = \sqrt{\pi \times 10^8 \times 4\pi \times 10^{-7} \times 5} = 44.4$$

From Eq. (8.91),

$$\hat{\eta} = \frac{(1+j)}{\sigma\delta} = 12.6\,e^{j\pi/4}$$

Therefore,

$$\mathscr{E} = \mathbf{a}_x e^{-44.4z}\cos(2\pi \times 10^8 t - 44.4z)\ [\text{V/m}]$$

$$\mathscr{H} = \mathbf{a}_y 79.4 e^{-44.4z}\cos(2\pi \times 10^8 t - 44.4z - \pi/4)\ [\text{mA/m}]$$

(b) The phase velocity is

$$v_p = \frac{\omega}{\beta} = \frac{2\pi \times 10^8}{44.4} = 1.42 \times 10^7\ [\text{m/s}]$$

(c) The wavelength is

$$\lambda = \frac{2\pi}{\beta} = \frac{2\pi}{44.4} = 0.14\ [\text{m}]$$

(d) The skin depth is

$$\delta = \frac{1}{\alpha} = \frac{1}{44.4} = 2.23\ [\text{cm}]$$

Example 8.11 Express the skin depth of copper, which is characterized by $\sigma = 5.8 \times 10^7$ [S/m], as a function of frequency.

Solution

From Eq. (8.87), along with $\mu \approx \mu_0$,

$$\delta = \frac{1}{\sqrt{\pi f \mu_0 \sigma}} = \frac{1}{\sqrt{\pi f \times 4\pi \times 10^{-7} \times 5.8 \times 10^7}}$$

Thus, the skin depth of copper is expressed as

$$\delta = \frac{0.066}{\sqrt{f}} \ [\text{m}] \tag{8.100}$$

For example, at a frequency of 1 [GHz], the skin depth of copper is only 2.1 [μm].

Example 8.12 For an electromagnetic wave that is normally incident on a conductor, as shown in Fig. 8.9, compute the time-averaged power transmitted into the conductor through an entrance surface with cross-sectional area $w \times \ell$ [m^2] to show that it is equal to the total dissipated power in the conductor.

Solution

The time-averaged power density at the $z = 0^+$ plane can be obtained using Eq. (8.93) as

$$\langle \mathbf{S} \rangle = \mathbf{a}_z \frac{\sigma \delta}{4} E_o^2$$

The total power flowing through the entrance surface is

$$\langle P \rangle = \int_{x=0}^{\ell} \int_{y=0}^{w} \langle \mathbf{S} \rangle \cdot d\mathbf{s} = \frac{1}{4}\sigma\delta E_o^2 w\ell = \frac{1}{4\sigma}\delta w\ell J_o^2 \tag{8.101}$$

where we have used $J_o = \sigma E_o$ and oriented vector $d\mathbf{s}$ along the direction of $\langle \mathbf{S} \rangle$, not outward from the conductor. The transmitted power in Eq. (8.101) is equal to the power dissipated in the conductor, as given in Eq. (8.96), conforming to the law of energy conservation.

Example 8.13 The coaxial cable is made of cylindrical conductors with conductivity σ. The inner and outer conductors have radii a and b, respectively, and are much thicker than the skin depth. Determine ac-resistance per unit length.

Solution

As the conduction current may be considered uniform within thin layers of thickness δ on the inner conductor and on the inner surface of the outer conductor, the ac-resistance per unit length of the cable is given by

$$R_{ac} = \frac{1}{2\pi\delta\sigma}\left(\frac{1}{a} + \frac{1}{b}\right) \ [\Omega/\text{m}] \tag{8.102}$$

Exercise 8.22
Find the skin depth of seawater with $\sigma = 4$ [S/m] and $\varepsilon = 81\varepsilon_0$ at 100 [kHz].

Ans. 0.8 [m].

Exercise 8.23
At what distance from the conductor surface does the time-averaged power density of the transmitted wave decrease to 1% of its initial value on the surface?

Ans. $2.3 \times \delta$.

Exercise 8.24

Find the magnitude of the intrinsic impedance of copper at 1 [GHz].

Ans. 0.012 [Ω].

Exercise 8.25

As σ cannot be infinite, α and β cannot be strictly equal. Determine the range of the loss tangent for which α and β may be regarded as the same, with an error less than 1%.

Ans. $(\sigma/\omega\varepsilon) > 99.5$.

Review Questions

RQ 8.26	What distinguishes a good conductor from a good dielectric?	[(8.82)]
RQ 8.27	Is it true that α and β are numerically equal in good conductors?	[(8.85)]
RQ 8.28	What is skin depth?	[(8.87)]
RQ 8.29	By what factor does the phase velocity in a highly conductive medium differ from that in a perfect dielectric?	[(8.89)]
RQ 8.30	Is the phase angle of $\hat{\eta}$ always 45° in good conductors?	[(8.91)]
RQ 8.31	By what factor does the intrinsic impedance of a highly conductive medium differ from that of a perfect dielectric?	[(8.90)]
RQ 8.32	Express the time-averaged power density of an electromagnetic wave in a good conductor.	[(8.93)]
RQ 8.33	What is *ac*-resistance?	[(8.99)]

8.3 Plane Waves at the Interface

When an electromagnetic wave impinges on an interface between two different media, part of the wave is reflected from the interface and the remainder is transmitted across the interface into the second medium. The electric and magnetic fields of the three waves must satisfy the boundary conditions imposed by the Maxwell's equations. Because the boundary condition is satisfied at all points on the interface, the phase relationship between the electric fields of the waves remains the same across the interface. This is called the ***phase-matching condition***. In this section, we discuss how the laws of reflection and refraction follow from the phase-matching condition. Next, we examine a uniform plane wave incident normally onto a plane interface and explain the standing wave. Extending our discussion to a uniform plane wave obliquely incident onto a plane interface, Fresnel's equations for the reflection and transmission coefficients are derived. We introduce the reflectance and transmittance and compute the powers of the reflected and transmitted waves.

8.3.1 Normal Incidence of Plane Wave

A uniform plane wave is normally incident onto an infinite-plane interface between
the two lossless dielectrics, as shown in Fig. 8.10. The wave is assumed to be linearly
polarized in the x-direction and propagate in the $+z$-direction in medium 1 before
impinging on medium 2, which extends between $z = 0$ and $z = \infty$. Thus, the electric
and magnetic field phasors of the incident wave are given by

$$\tilde{\mathbf{E}}^i = \mathbf{a}_x E_o^i\, e^{-j\beta_1 z} \tag{8.103a}$$

$$\tilde{\mathbf{H}}^i = \mathbf{a}_y \frac{E_o^i}{\eta_1}\, e^{-j\beta_1 z} \tag{8.103b}$$

where superscript i denotes the incident wave, and subscript 1 denotes medium 1.
The phase constant β_1 and intrinsic impedance η_1 are both real in lossless medium
1. Without loss of generality, the amplitude E_o^i is also assumed to be real.

Under normal incidence, the planar wavefront of the incident wave coincides with
the plane interface. Thus, the phase of $\tilde{\mathbf{E}}^i$ is constant across the interface at any given
time. Under this condition, the reflected wave with $\tilde{\mathbf{E}}^r$ and transmitted wave with $\tilde{\mathbf{E}}^t$
should also have constant phases across the interface to satisfy the phase-matching
condition. This can be accomplished only when the reflected and transmitted waves
are both uniform plane waves, with wavevectors normal to the interface. Thus, the
electric and magnetic field phasors for the reflected wave, propagating in the $-z$-
direction, are expressed as

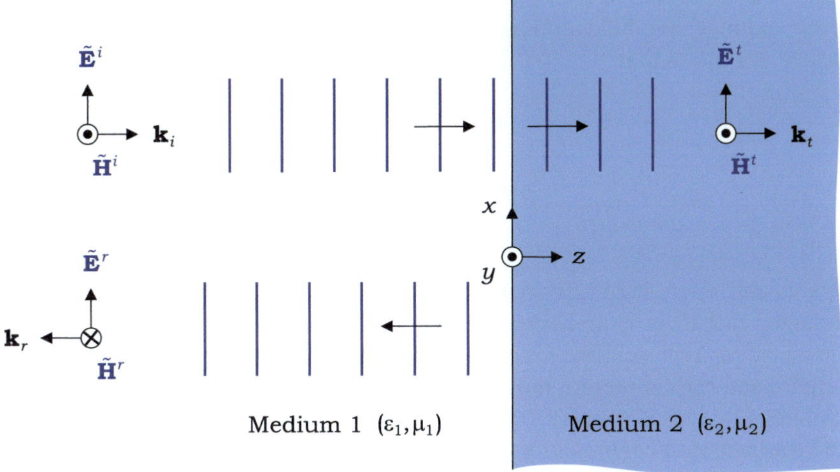

Fig. 8.10 Normal incidence of uniform plane wave on plane interface

$$\tilde{\mathbf{E}}^r = \mathbf{a}_x E_o^r e^{j\beta_1 z} \tag{8.104a}$$

$$\tilde{\mathbf{H}}^r = -\mathbf{a}_y \frac{E_o^r}{\eta_1} e^{j\beta_1 z} \tag{8.104b}$$

where the superscript r denotes the reflected wave. To simplify the notation, $\tilde{\mathbf{E}}^r$ is assumed to be directed along the $+x$-direction, the same as for $\tilde{\mathbf{E}}^i$. Accordingly, $\tilde{\mathbf{H}}^r$ points in the $-y$-direction, following the right-hand rule. Note that the phase constant and intrinsic impedance of the reflected wave are the same as those of the incident wave, because the two waves propagate in the same medium. However, amplitude E_o^r is yet to be specified. Similarly, for a transmitted wave propagating in the $+z$-direction, the electric and magnetic field phasors are expressed as

$$\tilde{\mathbf{E}}^t = \mathbf{a}_x E_o^t e^{-j\beta_2 z} \tag{8.105a}$$

$$\tilde{\mathbf{H}}^t = \mathbf{a}_y \frac{E_o^t}{\eta_2} e^{-j\beta_2 z} \tag{8.105b}$$

where the superscript t denotes the transmitted wave. For notational convenience, $\tilde{\mathbf{E}}^t$ is assumed to be along the $+x$-axis, and not along the $-x$-axis. However, the amplitude E_o^t is yet to be determined. Because medium 2 is different from medium 1, the phase constant β_2 is different from β_1.

We are now ready to apply electromagnetic boundary conditions to the electric and magnetic fields at the interface and relate the unknown terms E_o^r and E_o^t to a known quantity E_o^i set by the wave generator. The boundary condition requires the tangential components of $\tilde{\mathbf{E}}$ and $\tilde{\mathbf{H}}$ to be continuous across the interface. As all the electric and magnetic fields in Eqs. (8.103a)–(8.105b) are already tangential to the interface, it follows that

$$E_o^i + E_o^r = E_o^t \tag{8.106a}$$

$$\frac{E_o^i}{\eta_1} - \frac{E_o^r}{\eta_1} = \frac{E_o^t}{\eta_2} \tag{8.106b}$$

where η_1 and η_2 are the intrinsic impedances of media 1 and 2, respectively, and it is assumed that there is no surface current at the interface. The left-hand sides of these equations represent the total electric and magnetic fields at $z = 0^-$, whereas the right-hand sides represent those at $z = 0^+$.

Solving Eqs. (8.106a) and (8.106b) for E_o^r and E_o^t leads to the **reflection coefficient** Γ and **transmission coefficient** τ for normal incidence:

$$\boxed{\Gamma \equiv \frac{E_o^r}{E_o^i} = \frac{\eta_2 - \eta_1}{\eta_2 + \eta_1}} \tag{8.107}$$

$$\boxed{\tau \equiv \frac{E_o^t}{E_o^i} = \frac{2\eta_2}{\eta_2 + \eta_1}} \tag{8.108}$$

The reflection coefficient becomes negative for $\eta_1 > \eta_2$, implying that the reflected and incident waves are 180° out-of-phase at the interface. In contrast, the transmission coefficient τ is always positive; thus, the transmitted wave is always in phase with the incident wave at the interface. Although τ may be larger than unity for $\eta_2 > \eta_1$, this does not necessarily mean that the wave power increases in the second medium (see Example 8.14).

The expressions for Γ and τ in Eqs. (8.107) and (8.108) remain valid for interfaces that involve lossy media. In such cases, η_1 and η_2 are replaced by their complex counterparts; therefore, the reflection and transmission coefficients are complex quantities.

From Eqs. (8.107) and (8.108), it follows that

$$\boxed{1 + \Gamma = \tau} \tag{8.109}$$

It is important to remember that this relationship is only true for normal incidence.

Example 8.14 For a uniform plane wave normally incident onto an interface between two lossless media with η_1 and η_2, derive the relationship $|\Gamma|^2 + (\eta_1/\eta_2)|\tau|^2 = 1$ from the law of energy conservation.

Solution

The time-averaged power densities of the incident, reflected, and transmitted waves are

$$\langle \mathbf{S}^i \rangle = \frac{1}{2}\mathrm{Re}\left[\tilde{\mathbf{E}}^i \times \tilde{\mathbf{H}}^{i*}\right] = \mathbf{a}_z \frac{1}{2\eta_1}\left|E_o^i\right|^2 \tag{8.110a}$$

$$\langle \mathbf{S}^r \rangle = -\mathbf{a}_z \frac{1}{2\eta_1}\left|E_o^r\right|^2 = -\mathbf{a}_z \frac{1}{2\eta_1}|\Gamma|^2\left|E_o^i\right|^2 \tag{8.110b}$$

$$\langle \mathbf{S}^t \rangle = \mathbf{a}_z \frac{1}{2\eta_2}\left|E_o^t\right|^2 = \mathbf{a}_z \frac{1}{2\eta_2}|\tau|^2\left|E_o^i\right|^2 \tag{8.110c}$$

The law of energy conservation requires that

$$\left|\langle \mathbf{S}^i \rangle\right| = \left|\langle \mathbf{S}^r \rangle\right| + \left|\langle \mathbf{S}^t \rangle\right| \tag{8.110d}$$

Substituting Eqs. (8.110a, b, c) into Eq. (8.110d) yields

$$\boxed{|\Gamma|^2 + \frac{\eta_1}{\eta_2}|\tau|^2 = 1} \tag{8.111}$$

Example 8.15 The $z = 0$ and $z = \ell$ planes separate the three lossless media, as shown in Fig. 8.11. A uniform plane wave with $\tilde{\mathbf{E}}^i = \mathbf{a}_x E_o e^{-j\beta_1 z}$ is normally incident on the interface at $z = 0$ and undergoes multiple reflections in the second medium, which can be incorporated into a forward wave $\tilde{\mathbf{E}}^f$ and backward wave $\tilde{\mathbf{E}}^b$. Find (a) $\tilde{\mathbf{E}}^f$ and $\tilde{\mathbf{E}}^b$, (b) total $\tilde{\mathbf{E}}$ and $\tilde{\mathbf{H}}$ at $z = 0^+$, and effective intrinsic impedance, $\eta_e = \tilde{E}/\tilde{H}$, at $z = 0^+$, and (c) $\tilde{\mathbf{E}}^r$ and $\tilde{\mathbf{E}}^t$.

Solution

All electric fields are assumed to be directed along the x-axis.

(a)   In medium 2, the sum of the odd-numbered waves constitutes the forward wave $\tilde{\mathbf{E}}^f$ and that of the even-numbered waves constitutes the backward wave $\tilde{\mathbf{E}}^b$. Thus, at $z = 0^+$,

$$\tilde{E}^f(z = 0^+) = E_o \tau_1 \left[1 + \Gamma_1 \Gamma_2 e^{-j2\beta_2 \ell} + (\Gamma_1 \Gamma_2 e^{-j2\beta_2 \ell})^2 + \ldots \right]$$

$$= \frac{E_o \tau_1}{1 - \Gamma_1 \Gamma_2 e^{-j2\beta_2 \ell}}$$

$$\equiv E_o^f \tag{8.112a}$$

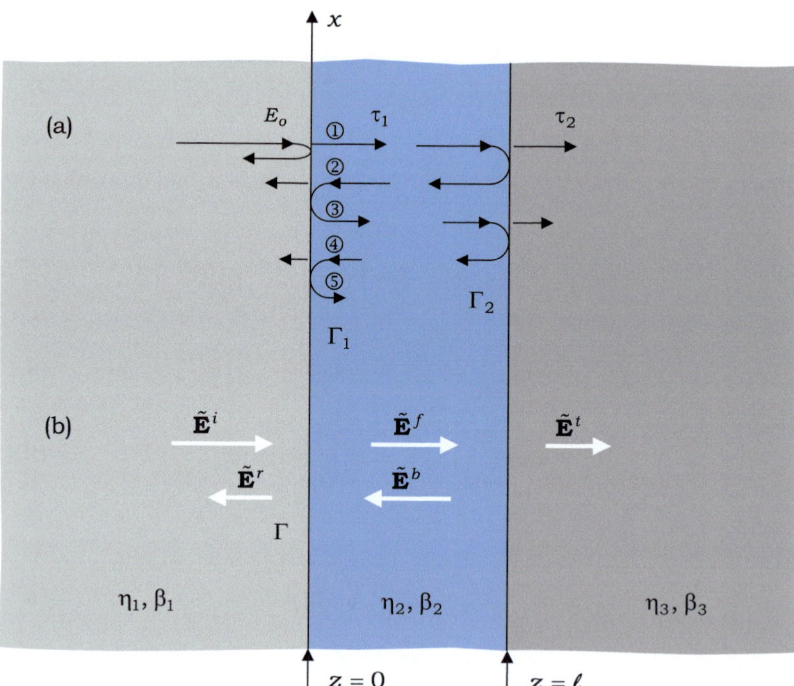

Fig. 8.11 Three contiguous lossless media. **a** Multiple reflections in medium 2. **b** Five representative plane waves in three regions

$$\tilde{E}^b(z = 0^+) = E_o \tau_1 \Gamma_2 e^{-j2\beta_2\ell} \left[1 + \Gamma_1\Gamma_2 e^{-j2\beta_2\ell} + (\Gamma_1\Gamma_2 e^{-j2\beta_2\ell})^2 + \dots\right]$$

$$= \frac{E_o \tau_1 \Gamma_2 e^{-j2\beta_2\ell}}{1 - \Gamma_1\Gamma_2 e^{-j2\beta_2\ell}}$$

$$\equiv E_o^b \tag{8.112b}$$

where β_2 is the phase constant in medium 2, τ_1 is the transmission coefficient at the first interface, and Γ_1 and Γ_2 are the internal reflection coefficients at the first and second interfaces, respectively. From Eqs. (8.107) and (8.108),

$$\tau_1 = \frac{2\eta_2}{\eta_2 + \eta_1}, \ \Gamma_1 = \frac{\eta_1 - \eta_2}{\eta_1 + \eta_2}, \ \text{and} \ \Gamma_2 = \frac{\eta_3 - \eta_2}{\eta_3 + \eta_2} \tag{8.112c}$$

Note that Γ_1 has an opposite sign to Γ, which applies to the opposite side of the interface. Finally, the two waves in medium 2 are expressed as

$$\tilde{\mathbf{E}}^f = \mathbf{a}_x E_o^f e^{-j\beta_2 z}, \ \text{and} \ \tilde{\mathbf{E}}^b = \mathbf{a}_x E_o^b e^{j\beta_2 z}$$

(b)   The total electric field at $z = 0^+$ is obtained from Eqs. (8.112a) and (8.112b) as

$$\tilde{E}(z = 0^+) = E_o^f + E_0^b$$

$$= \frac{E_o \tau_1}{1 - \Gamma_1\Gamma_2 e^{-j2\beta_2\ell}}\left[1 + \Gamma_2 e^{-j2\beta_2\ell}\right]$$

Thus, at $z = 0^+$,

$$\tilde{\mathbf{E}} = \tilde{E}(z = 0^+)\,\mathbf{a}_x$$

Using the relationship $\mathbf{a}_k = \mathbf{a}_E \times \mathbf{a}_H$ and the intrinsic impedance of medium 2, η_2, we obtain

$$\tilde{H}(z = 0^+) = H_o^f + H_o^b = (E_o^f - E_o^b)/\eta_2$$

$$= \frac{1}{\eta_2}\frac{E_o \tau_1}{1 - \Gamma_1\Gamma_2 e^{-j2\beta_2\ell}}\left[1 - \Gamma_2 e^{-j2\beta_2\ell}\right]$$

Thus, at $z = 0^+$,

$$\tilde{\mathbf{H}} = \tilde{H}(z = 0^+)\,\mathbf{a}_y$$

The effective intrinsic impedance at $z = 0^+$ is defined as

$$\boxed{\eta_e \equiv \frac{\tilde{E}(z = 0^+)}{\tilde{H}(z = 0^+)} = \eta_2 \frac{1 + \Gamma_2 e^{-j2\beta_2\ell}}{1 - \Gamma_2 e^{-j2\beta_2\ell}}} \tag{8.112d}$$

(c) The reflection of the incident wave from the first interface is determined by the intrinsic impedance of medium 1, η_1, and the effective intrinsic impedance η_e. Following the same procedure as that used for Eq. (8.112c), the reflection coefficient at the first interface is given by

$$\Gamma = \frac{\eta_e - \eta_1}{\eta_e + \eta_1} \tag{8.112e}$$

Thus, the reflected wave propagating in medium 1 is expressed as

$$\tilde{\mathbf{E}}^r = \mathbf{a}_x \Gamma E_o\, e^{j\beta_1 z} \tag{8.112f}$$

Similarly, following Eq. (8.112c), the transmission coefficient at the second interface is given by

$$\tau_2 = \frac{2\eta_3}{\eta_3 + \eta_2} \tag{8.112g}$$

The wave transmitted into medium 3 is then obtained as

$$\tilde{\mathbf{E}}^t = \mathbf{a}_x \tau_2 [E_o^f\, e^{-j\beta_2 \ell}] e^{-j\beta_3(z-\ell)}$$

$$= \mathbf{a}_x \frac{E_o \tau_1 \tau_2\, e^{-j\beta_2 \ell}}{1 - \Gamma_1 \Gamma_2\, e^{-j2\beta_2 \ell}}\, e^{-j\beta_3(z-\ell)} \qquad (z \geq \ell) \tag{8.112h}$$

where brackets represent the electric field of the forward wave at $z = \ell^-$. It can be shown that the power transmitted into medium 3 is equal to the power of the incident wave in medium 1 multiplied by $1 - |\Gamma|^2$.

If $\eta_e = \eta_1$, there is no reflection of the incident wave at the first interface, even though the transmitted wave still bounces back and forth within medium 2. Under this condition, all the backward component waves in medium 2 completely cancel each other out at $z = 0^+$.

Exercise 8.26

A uniform plane wave in free space ($z < 0$) with $\tilde{\mathbf{E}}^i = \mathbf{a}_x 10 e^{-j\beta_1 z}$ and $f = 2\,[\text{GHz}]$ normally impinges on a dielectric with $\varepsilon = 2\varepsilon_0$ and μ_0. Find $\tilde{\mathbf{E}}^r$ and $\tilde{\mathbf{E}}^t$.

Ans. $\tilde{\mathbf{E}}^r = -\mathbf{a}_x 1.72 e^{j41.9z}$, $\tilde{\mathbf{E}}^t = \mathbf{a}_x 8.28 e^{-j59.2z}$.

Exercise 8.27

When light with $\lambda = 500$ [nm] is normally incident on a semi-infinite glass with refractive index $n = 1.5$, what portion of the incident power is reflected?

Ans. 4%.

Exercise 8.28

If $|\Gamma| = |\tau|$ for the normal incidence of a plane wave upon an interface between two lossless media, what is the ratio of the transmitted to the reflected power?

Ans. 3.

8.3.1.1 Standing-Wave Ratio

When the interface is as shown in Fig. 8.10, the total electric field in medium 1 is equal to the sum of those of the incident and reflected waves. Using Eqs. (8.103a) and (8.104a), along with Eq. (8.107), the total electric field in medium 1 is expressed as

$$\tilde{\mathbf{E}}_1 = \tilde{\mathbf{E}}^i + \tilde{\mathbf{E}}^r = \mathbf{a}_x E_o^i e^{-j\beta_1 z}\left[1 + \Gamma e^{j2\beta_1 z}\right]$$

$$\equiv \mathbf{a}_x \hat{E}_1 \tag{8.113}$$

where Γ is the reflection coefficient at the interface, and $\hat{E}_1$ denotes the magnitude of phasor $\tilde{\mathbf{E}}_1$.

Because the reflection coefficient in Eq. (8.107) can also be applied to an interface between two lossy media if η_1 and η_2 are replaced with complex intrinsic impedances, the reflection coefficient is generally complex. That is,

$$\Gamma = |\Gamma| e^{j\phi} \tag{8.114}$$

where ϕ is the phase angle. Care should be taken when calculating ϕ to ensure that it is within the correct quadrant. The inverse trigonometric function $y = \arctan(x)$ is defined to provide a principal value in the range $-\pi/2 \leq y \leq \pi/2$. For example, most electronic calculators would yield $\arctan(0.5/(-0.5)) = -0.785$ [rad] for $\Gamma = -0.5 + j0.5$, but the correct phase angle for Γ is $\phi = \pi - 0.785$ [rad].

Inserting Eq. (8.114) into Eq. (8.113) gives the magnitude of the electric field phasor in medium 1 as follows:

$$\hat{E}_1 = E_o^i e^{-j\beta_1 z}\left[1 + |\Gamma| e^{j(2\beta_1 z + \phi)}\right] \tag{8.115}$$

Here, $|\hat{E}_1|$ represents the amplitude of the total electric field oscillating at an angular frequency ω, whereas the phase angle of $\hat{E}_1$ specifies the phase delay of the time-harmonic electric field at a given point in space. It is important to note that the amplitude varies with z in medium 1.

The amplitude is the maximum when the phase angle of the second term in brackets in Eq. (8.115) is zero or a multiple of 2π. The maximum value of $|\hat{E}_1|$ occurs at distance $z_{\max}$ from the interface. That is,

$$\boxed{\left|\hat{E}_1\right|_{\max} = E_o^i(1 + |\Gamma|)} \tag{8.116a}$$

$$z_{\max} = -\frac{1}{2\beta_1}(\phi + 2\pi n)$$

$$(n = 0, \pm 1, \pm 2, \ldots) \tag{8.116b}$$

The amplitude maxima are repeated for every $\lambda/2$ [m]. Moreover, $|\hat{E}_1|_{\max}$ depends on the magnitude of Γ, whereas $z_{\max}$ depends on the phase angle of Γ. Because medium 1 extends between $z = 0$ and $-\infty$, $z_{\max}$ must be negative. Therefore, care should be taken to select the proper sign for the integer n.

An amplitude minimum occurs at the midpoint between the two adjacent amplitude maxima. The minimum of $|\hat{E}_1|$ occurs at a distance $z_{\min}$ from the interface, for which the phase angle of the second term in the brackets in Eq. (8.115) is equal to $(2n + 1)\pi$. That is,

$$\left|\hat{E}_1\right|_{\min} = E_o^i(1 - |\Gamma|) \tag{8.117a}$$

$$z_{\min} = -\frac{1}{2\beta_1}(\phi + 2\pi n + \pi)$$

$$(n = 0, \pm 1, \pm 2, \ldots) \tag{8.117b}$$

Amplitude minima occur periodically with $\lambda/2$ [m] spacing. Accordingly, the separation between the minimum and the nearest maximum is $\lambda/4$ [m].

After some manipulations, Eq. (8.115) can be rewritten as

$$\begin{aligned}
\hat{E}_1 &= E_o^i e^{-j\beta_1 z}\left[1 - |\Gamma| + |\Gamma| + |\Gamma| e^{j(2\beta_1 z + \phi)}\right] \\
&= E_o^i e^{-j\beta_1 z}(1 - |\Gamma|) + E_o^i|\Gamma| e^{j\phi/2}[2\cos(\beta_1 z + \phi/2)]
\end{aligned}$$

In instantaneous form,

$$\mathscr{E}_1 = E_o^i(1 - |\Gamma|)\cos(\omega t - \beta_1 z) + 2E_o^i|\Gamma|\cos(\beta_1 z + \phi/2)\cos(\omega t + \phi/2) \tag{8.118}$$

The first term on the right side of Eq. (8.118) represents a traveling wave with amplitude $E_o^i(1 - |\Gamma|)$. The second term represents a **standing wave** that does not travel through the medium. However, the electric field oscillates at a frequency ω with amplitude $2E_o^i|\Gamma|\cos(\beta_1 z + \phi/2)$. A standing wave consists of two plane waves of equal amplitudes propagating in opposite directions. There is no power transfer from one point to another by the standing wave. In view of Eq. (8.118), when the incident wave of amplitude E_o^i reflects off the interface, the reflected wave with amplitude $|\Gamma|E_o^i$ interferes with the incoming wave and produces a standing wave in medium 1. The remainder, with amplitude $(1 - |\Gamma|)E_o^i$, travels in the $+z$-direction without interruption. The amplitude of the traveling wave is constant in the medium, whereas the amplitude of the standing wave varies sinusoidally with the distance as $\cos(\beta_1 z + \phi/2)$.

For example, by setting $|\Gamma| = 0.5$ and $\phi = 0$ in Eq. (8.118), the electric fields of the traveling and standing waves are plotted as functions of $\beta_1 z$ at three different times, $\omega t = 0, 0.3\pi$, and 1.3π, as shown in Fig. 8.12.

Next, the total electric field in medium 1 is plotted as a function of $\beta_1 z$ at five different times, $\omega t = 0, \pi/4, \pi/2, 3\pi/4$, and π, as shown in Fig. 8.13. At $\omega t = 0$, the electric field is represented by a cosine curve. As time passes, the curve apparently moves in the $+z$-direction, but the amplitude decreases with time for $0 < \omega t < \pi/2$ and increases for $\pi/2 < \omega t < \pi$. However, the curve is confined between the two blue curves, representing the amplitude of the total electric field.

The ***standing-wave ratio*** S is defined as the ratio of the maximum to minimum amplitude of the total electric field:

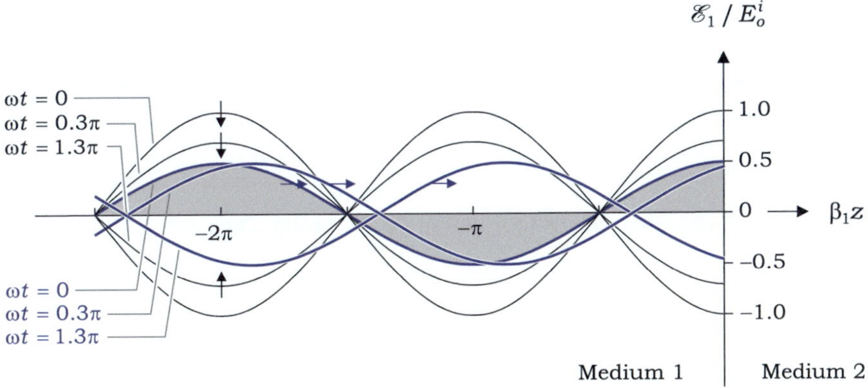

Fig. 8.12 Traveling wave (blue) and standing wave (black) in medium 1

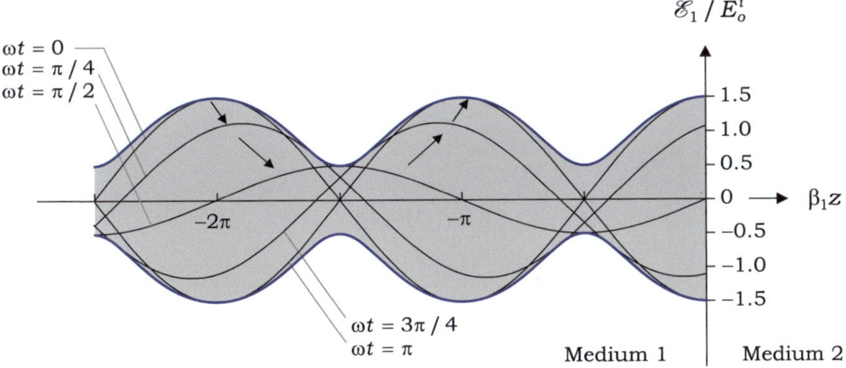

Fig. 8.13 Total electric field in medium 1

$$\boxed{S = \frac{\left|\hat{E}_1\right|_{max}}{\left|\hat{E}_1\right|_{min}} = \frac{1 + |\Gamma|}{1 - |\Gamma|}} \qquad (8.119)$$

where S is a dimensionless quantity. When $|\Gamma|$ is varied from 0 to 1, S varies from 1 to ∞. The standing-wave ratio can also be expressed on a logarithmic scale in decibels, as $20 \log_{10} S$ [dB]. For example, $S = 4$ is equivalent to 12.04 [dB]. The standing-wave ratio S should not be confused with the Poynting vector $\mathbf{S}$ or the unit [S] representing the siemens.

Example 8.16 A standing wave with $S = 4$ is formed in free space ($z < 0$). The first maximum occurs at a distance of 0.2 [m] from the interface at $z = 0$, and the spacing between two adjacent maxima is 0.5 [m]. Determine η_2 of the material in region $z \geq 0$.

Solution

Any two adjacent maxima are half the wavelength apart. Thus, $\lambda = 1$ and

$$\beta_1 = \frac{2\pi}{\lambda} = 2\pi$$

The first maximum corresponds to $n = 0$ in Eq. (8.116b). Thus,

$$z_{max} = -\frac{1}{2\beta_1}\phi = -0.2 \text{ [m]}, \text{ which yields } \phi = 0.8\pi$$

From Eq. (8.119),

$$|\Gamma| = \frac{S - 1}{S + 1} = 0.6$$

Therefore, the reflection coefficient is

$$\Gamma = 0.6e^{j0.8\pi} \qquad (8.120a)$$

Rewriting Eq. (8.107), we have

$$\frac{\eta_2}{\eta_1} = \frac{1 + \Gamma}{1 - \Gamma} \qquad (8.120b)$$

Thus,

$$\eta_2 = 377\frac{1 + 0.6e^{j0.8\pi}}{1 - 0.6e^{j0.8\pi}} = 154e^{j0.83} \text{ [}\Omega\text{]}$$

Example 8.17 A uniform plane wave with $\mathscr{E}^i = \mathbf{a}_x 20\cos(3 \times 10^9 t - 17.3z)$ travels in a lossless medium in region $z < 0$ and impinges normally on a *lossy* medium with $\varepsilon_r = 4$ and $\sigma = 0.5$ [S/m] occupying region $z > 0$. Assuming that both materials are nonmagnetic, determine Γ, S, and the distance of the first maximum from the interface.

Solution

The incident wave has an angular frequency of $\omega = 3 \times 10^9$ [rad/s] and a phase constant of $\beta_1 = 17.3$ [rad/m]. From the relation $\beta_1 = \omega\sqrt{\mu_0\varepsilon_0\varepsilon_r}$, we obtain $\varepsilon_r = (1.73)^2$ and

$$\eta_1 = \sqrt{\mu_0/\varepsilon_0\varepsilon_r} = 217.92 \ [\Omega]$$

For $z > 0$, η_2 is obtained from Eq. (8.81) as

$$\eta_2 = \sqrt{\frac{\mu_0}{\varepsilon}}\sqrt{\frac{1}{1 - j\sigma/(\omega\varepsilon)}} = \frac{377}{2}\sqrt{\frac{1}{1 - j0.5/(3 \times 10^9 \times 4 \times 8.854 \times 10^{-12})}}$$

$$= 85.90\, e^{j0.68} \ [\Omega]$$

By inserting η_1 and η_2 into Eq. (8.107), we obtain

$$\Gamma = \frac{\eta_2 - \eta_1}{\eta_2 + \eta_1} = 0.55 e^{j2.61}$$

From Eq. (8.119),

$$S = \frac{1 + |\Gamma|}{1 - |\Gamma|} = 3.44$$

Using $\phi = 2.61$, $\beta_1 = 17.3$, and $n = 0$ in Eq. (8.116b) yields

$$z_{\max} = -\frac{\phi}{2\beta_1} = -7.5 \ [\text{cm}]$$

Exercise 8.29
If the incident wave $\tilde{\mathbf{E}}^i$ shown in Fig. 8.10 is delayed in time phase, is there any change in the position of the maximum amplitude in region $z < 0$?

Ans. No, see Eqs. (8.116a) and (8.116b).

Exercise 8.30
For the normal incidence of a uniform plane wave on an interface between two lossless media, express S in terms of η_1 and η_2 of the materials.

Ans. $S = \eta_2/\eta_1$ or η_1/η_2, which is greater than unity.

8.3.1.2   Interface Involving Perfect Conductor

A perfect conductor has infinite conductivity and, thus, zero intrinsic impedance. Let us consider the interface between a lossless dielectric in region $z < 0$ and perfect conductor in region $z \geq 0$. By setting $\eta_2 = 0$ in Eq. (8.107), we have $\Gamma = -1$, meaning that the incident wave is totally reflected by the perfect conductor and the reflected wave is $180°$ out of phase with the incident wave at the conductor surface. The expressions for the incident and reflected waves given in Eqs. (8.103a, b) and (8.104a, b) remain valid for this case. The total electric and magnetic fields in the dielectric are

$$\tilde{\mathbf{E}}_1 = \tilde{\mathbf{E}}^i + \tilde{\mathbf{E}}^r = \mathbf{a}_x E_o^i \left[e^{-j\beta_1 z} - e^{j\beta_1 z} \right]$$

$$= -\mathbf{a}_x j 2 E_o^i \sin(\beta_1 z) \tag{8.121a}$$

$$\tilde{\mathbf{H}}_1 = \tilde{\mathbf{H}}^i + \tilde{\mathbf{H}}^r = \mathbf{a}_y \frac{E_o^i}{\eta_1} \left[e^{-j\beta_1 z} + e^{j\beta_1 z} \right]$$

$$= \mathbf{a}_y \frac{2 E_o^i}{\eta_1} \cos(\beta_1 z) \tag{8.121b}$$

In instantaneous form,

$$\mathscr{E} = \mathbf{a}_x 2 E_o^i \sin(\beta_1 z) \sin(\omega t) \tag{8.122a}$$

$$\mathscr{H} = \mathbf{a}_y \frac{2 E_o^i}{\eta_1} \cos(\beta_1 z) \cos(\omega t) \tag{8.122b}$$

In Eqs. (8.122a) and (8.122b), the z- and t-dependences are separated, implying that the incident and reflected waves combine to form a standing wave. The standing wave cannot transmit power from one point to another, which can be verified by inserting Eqs. (8.121a, b) into Eq. (8.38) and noting that $\tilde{\mathbf{E}}_1 \times \tilde{\mathbf{H}}_1^*$ is purely imaginary in this case. Alternatively, substituting $\eta_2 = 0$ into Eq. (8.108) leads to $\tau = 0$, indicating that no wave, and thus, no power is transmitted into the second medium.

The electric and magnetic fields of the standing wave are plotted against $\beta_1 z$ at five different times, as shown in Fig. 8.14. The electric field is zero on the conductor surface at all times, which is a manifestation of the boundary condition for the electric field; that is, the tangential component of the electric field should vanish on the surface of a perfect conductor. Spatial points of zero amplitude, called nodes, occur periodically along the z-axis with a half-wavelength spacing. The nodes of $\mathscr{H}$ also occur periodically with the same spacing as those of $\mathscr{E}$, but are displaced by $\lambda/4$ [m] with respect to those of $\mathscr{E}$. Therefore, the first node of $\mathscr{H}$ occurs at distance $\lambda/4$ from the conductor surface.

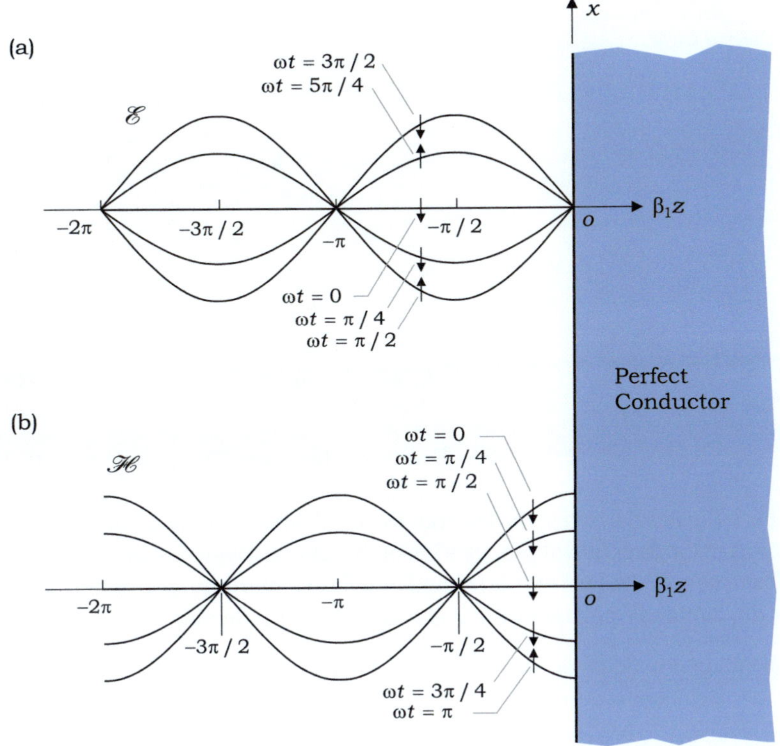

Fig. 8.14 The electric and magnetic fields of a standing wave formed in front of a perfect conductor are plotted at five different times

Example 8.18 A right-hand circularly polarized wave $\tilde{\mathbf{E}}^i = (E_o\,\mathbf{a}_x - jE_o\,\mathbf{a}_y)\,e^{-j\beta z}$ in free space normally impinges on a perfect conductor filling the half-space $z \geq 0$. Find

(a) $\tilde{\mathbf{E}}^r$ and its polarization state,
(b) the current density on the conductor surface, and
(c) the total electric field in free space.

Solution

(a) Because $\Gamma = -1$, the directions of the electric field and wave propagation are reversed for the reflected wave such that

$$\tilde{\mathbf{E}}^r = \left(-E_o\,\mathbf{a}_x + jE_o\,\mathbf{a}_y\right)e^{j\beta z}$$

$$\mathcal{E}^r = -E_o\,\mathbf{a}_x \cos(\omega t + \beta z) - E_o\,\mathbf{a}_y \sin(\omega t + \beta z)$$

At the $z = 0$ plane, we have

Fig. 8.15 Polarization state
of the reflected wave

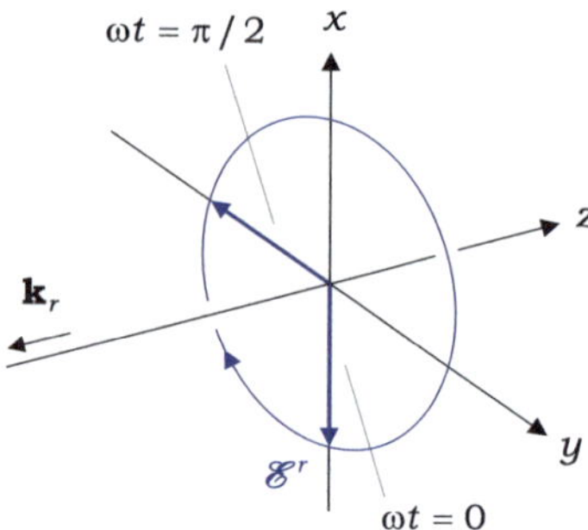

$$\mathscr{E}^r = -E_o\,\mathbf{a}_x \qquad (\omega t = 0) \qquad\qquad (8.123a)$$

$$\mathscr{E}^r = -E_o\,\mathbf{a}_y \qquad (\omega t = \pi/2) \qquad\qquad (8.123b)$$

This indicates that when the left four fingers follow the rotation of $\mathscr{E}^r$, the thumb
points in the direction of propagation of the reflected wave, as shown in Fig. 8.15.
Thus, the reflected wave is left-hand circularly polarized. The polarization state
of the circularly polarized wave is reversed after reflection.

(b)   The magnetic fields of the incident and reflected waves are

$$\tilde{\mathbf{H}}^i = \frac{1}{\eta_o}\left(\mathbf{a}_k \times \tilde{\mathbf{E}}^i\right) = \frac{E_o}{\eta_o}\left(\mathbf{a}_y + j\,\mathbf{a}_x\right)e^{-j\beta z}$$

$$\tilde{\mathbf{H}}^r = \frac{1}{\eta_o}\left(\mathbf{a}_k \times \tilde{\mathbf{E}}^r\right) = \frac{E_o}{\eta_o}\left(\mathbf{a}_y + j\,\mathbf{a}_x\right)e^{j\beta z}$$

The total magnetic fields at the $z = 0$ plane is

$$\tilde{\mathbf{H}}(0) = \tilde{\mathbf{H}}^i(0) + \tilde{\mathbf{H}}^r(0) = \frac{2E_o}{\eta_o}\left(\mathbf{a}_y + j\,\mathbf{a}_x\right)$$

With an outward unit normal to the surface, $\mathbf{a}_n = -\mathbf{a}_z$, the surface current
density on the conductor surface is obtained as

$$\tilde{\mathbf{J}}_s = \mathbf{a}_n \times \tilde{\mathbf{H}}(0) = \frac{2E_o}{\eta_o}\left(\mathbf{a}_x - j\,\mathbf{a}_y\right) \qquad\qquad (8.123c)$$

(c)   The total electric field in free space is

$$\tilde{\mathbf{E}} = \tilde{\mathbf{E}}^i + \tilde{\mathbf{E}}^r = -\mathbf{a}_x 2j E_o \sin(\beta z) - \mathbf{a}_y 2 E_o \sin(\beta z)$$

$$\mathscr{E} = \mathrm{Re}\!\left[\tilde{\mathbf{E}}\,e^{j\omega t}\right] = 2E_o \sin(\beta z)\left[\mathbf{a}_x \sin(\omega t) - \mathbf{a}_y \cos(\omega t)\right]$$

We see from Eq. (8.123c), a surface current is induced on the perfect conductor because of the magnetic field, even if there is no tangential component of the electric field.

Exercise 8.31

A standing wave with $\omega = 5 \times 10^8$ [rad/s] is formed in free space ($z < 0$) because of the perfect conductor in region $z \geq 0$. Where is the first maximum amplitude?

Ans. $z = -0.94$ [m].

Exercise 8.32

By what time phase does $\tilde{\mathbf{H}}$ lag behind $\tilde{\mathbf{E}}$ in the standing wave shown in Fig. 8.14?

Ans. $\pi/2$ [rad].

Review Questions

RQ 8.34	Express Γ and τ for the case of normal incidence.	[(8.107)(8.108)]
RQ 8.35	What boundary conditions are involved in Γ and τ?	[(8.106a,b)]
RQ 8.36	In what way is τ related to Γ for normal incidence?	[(8.109)]
RQ 8.37	Express the law of energy conservation in terms of Γ and τ.	[(8.111)]
RQ 8.38	Define the standing-wave ratio.	[(8.119)]
RQ 8.39	What is the separation between two adjacent amplitude maxima in a standing-wave pattern?	[(8.116a,b)]
RQ 8.40	What is the value of Γ at the surface of the perfect conductor?	[(8.121a,b)]
RQ 8.41	How does an incident electromagnetic wave induce surface current on a perfect conductor?	[(8.123c)]

8.3.2   *Oblique Incidence of Plane Wave*

We now consider the more general case of oblique incidence of a uniform plane wave onto a plane interface. As in the case of normal incidence, the incident wave is partly reflected from and partly transmitted across the interface. In this case, the reflection and transmission strongly depend on the propagation direction and polarization state of the incident wave as well as the intrinsic impedances of the two adjoining materials. Under certain conditions, an incident wave may experience total transmission or no reflection from the interface.

The boundary conditions for the electric and magnetic fields govern the behavior of the reflected and transmitted waves at the interface, leading to the reflection and transmission coefficients that prescribe the amplitudes of the corresponding waves. These coefficients explicitly refer to the *plane of incidence*, which includes the wavevector of the incident wave and normal to the interface. The wavevectors of the incident, reflected, and transmitted waves lie in the plane of incidence, as discussed later in this section. The direction of wave propagation can be conveniently specified by the rotation angle of the wavevector with respect to the normal to the interface. The *angle of incidence*, *angle of reflection*, and *angle of transmission* are all positive and range from zero to 90°. Moreover, the plane of incidence is used to specify the polarization state of the wave, which may be *perpendicular* or *parallel*. As the name suggests, the electric field vector can be decomposed into two mutually orthogonal components that are perpendicular or parallel to the plane of incidence. After determining the reflection of the two component-waves, the reflected wave is obtained by adding the component waves together. A similar procedure can be followed to obtain the transmitted wave.

Let us begin with a plane interface at $z = 0$ that separates the two lossless nonmagnetic dielectrics ($\alpha = 0$ and $\mu = \mu_0$), as shown in Fig. 8.16. The general expressions for the incident, reflected, and transmitted waves are as follows:

$$\tilde{\mathbf{E}}^i = \mathbf{E}_o^i e^{-j\mathbf{k}_i \cdot \mathbf{r}} = \mathbf{E}_o^i e^{-j[(k_i \sin\theta_i)x + (k_i \cos\theta_i)z]} \tag{8.124a}$$

$$\tilde{\mathbf{E}}^r = \mathbf{E}_o^r e^{-j\mathbf{k}_r \cdot \mathbf{r}} = \mathbf{E}_o^r e^{-j[(k_r \sin\theta_r)x - (k_r \cos\theta_r)z]} \tag{8.124b}$$

$$\tilde{\mathbf{E}}^t = \mathbf{E}_o^t e^{-j\mathbf{k}_t \cdot \mathbf{r}} = \mathbf{E}_o^t e^{-j[(k_t \sin\theta_t)x + (k_t \cos\theta_t)z]} \tag{8.124c}$$

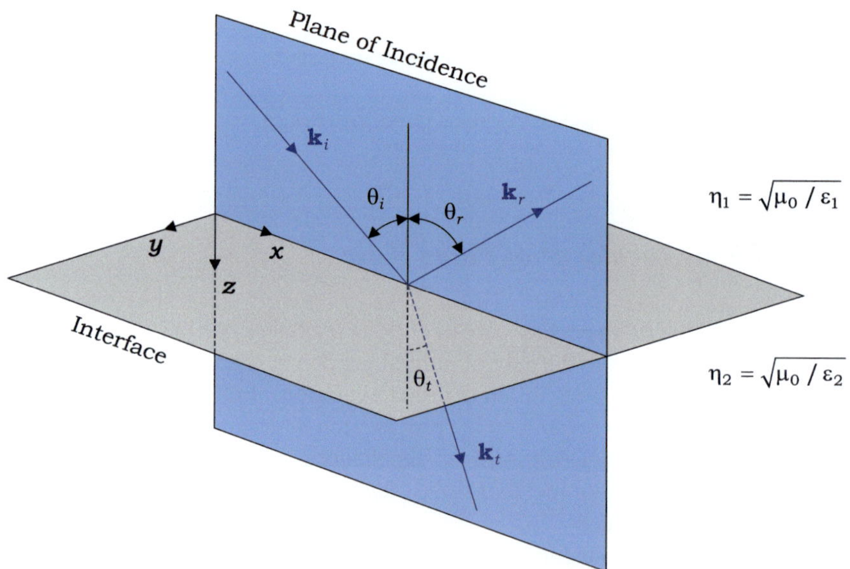

Fig. 8.16 Plane of incidence contains wavevectors of the incident, reflected, and transmitted waves

where i, r, and t denote the incident, reflected, and transmitted waves, respectively, and k is the magnitude of wavevector $\mathbf{k}$. The electric field phasor $\tilde{\mathbf{E}}^i$ should not be confused with amplitude $\mathbf{E}_o^i$, which is a constant vector with complex scalar-components. Our goal is to relate the reflected and transmitted waves with the as-yet-unknown $\mathbf{E}_o^r$, $\mathbf{E}_o^t$, k_r, k_t, θ_r, and θ_t to the incident wave characterized by $\mathbf{E}_o^i$, k_i, and θ_i by applying the boundary conditions for electric and magnetic fields.

The boundary condition requires the tangential component of $\mathbf{E}$ to be continuous across the interface at $z = 0$. Setting $z = 0$ in Eqs. (8.124a, b, c) and applying the boundary condition leads to

$$\left(\mathbf{E}_o^i\right)_{\tan} e^{-j(k_i \sin \theta_i)x} + \left(\mathbf{E}_o^r\right)_{\tan} e^{-j(k_r \sin \theta_r)x} = \left(\mathbf{E}_o^t\right)_{\tan} e^{-j(k_t \sin \theta_t)x} \qquad (8.125)$$

where subscript tan denotes the tangential component. Note that Eq. (8.125) can only be satisfied at all points on the interface when the three exponents are identical, such that

$$\boxed{k_i \sin \theta_i = k_r \sin \theta_r = k_t \sin \theta_t} \qquad (8.126)$$

This is known as the **phase-matching condition**, which states that the tangential components of $\mathbf{k}_i$, $\mathbf{k}_r$, and $\mathbf{k}_t$ are equal.

Here, we digress briefly and introduce the **index of refraction**, or the **refractive index**, of lossless nonmagnetic materials, which is defined as

$$\boxed{n = \sqrt{\frac{\varepsilon}{\varepsilon_0}} = \sqrt{\varepsilon_r}} \qquad (8.127)$$

where ε_r is the relative permittivity or the dielectric constant of the material. In general, materials with high densities exhibit high relative permittivities. With the refractive index, the wavenumber is given by

$$k = \omega\sqrt{\mu_0 \varepsilon} = n\frac{\omega}{c} \qquad (8.128)$$

and the phase velocity is given by

$$\upsilon_p = \frac{c}{n} \qquad (8.129)$$

where c is the speed of light in vacuum. Next, the intrinsic impedance of the material is expressed as

$$\eta = \frac{\eta_o}{n} \qquad (8.130)$$

where $\eta_o = \sqrt{\mu_0/\varepsilon_0}$ is the intrinsic impedance of free space.

As the incident and reflected waves propagate in the same medium, it is certain that $k_i = k_r$. From Eq. (8.126),

$$\boxed{\theta_i = \theta_r} \tag{8.131}$$

This is known as the *law of reflection*.

Next, inserting Eq. (8.128) into Eq. (8.126) leads to ***Snell's law of refraction***:

$$\boxed{n_1 \sin \theta_i = n_2 \sin \theta_t} \tag{8.132}$$

where θ_i and θ_t are the angles of incidence and transmission, respectively, and n_1 and n_2 are the refractive indices of media 1 and 2, respectively.

The laws of reflection and refraction are based on the phase-matching conditions in Eq. (8.126) and hold true regardless of the polarization state of the waves. The two laws and phase-matching conditions are illustrated in Fig. 8.17. The wavefronts of three waves with the same phase meet at a point on the interface. The points of the equal phase are equally spaced along the x-axis. In a denser medium, the transmitted wave with a reduced wavelength bends toward the surface normal so that its wavefronts match those of the incident and reflected waves at the interface.

Example 8.19 A uniform plane wave with wavevector $\mathbf{k}_i = \mathbf{a}_x + \sqrt{3}\,\mathbf{a}_y + 2\sqrt{3}\,\mathbf{a}_z$ propagates in free space and impinges on a lossless dielectric with $\varepsilon_r = 1.69$ filling the $z \geq 0$ region. Determine (a) $\mathbf{k}_r$, and (b) $\mathbf{k}_t$.

Solution

(a) The law of reflection comprises of two parts:
  (I) $\theta_r = \theta_i$, and
  (II) $\mathbf{k}_i$ and $\mathbf{k}_r$ are in the plane of incidence.

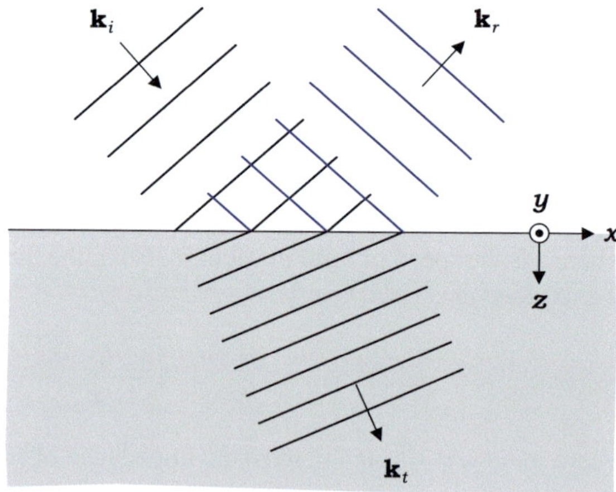

Fig. 8.17 Wavefronts of the incident, reflected, and transmitted waves

Using these, along with $|\mathbf{k}_r| = |\mathbf{k}_i|$, we obtain

$$\mathbf{k}_r = \mathbf{a}_x + \sqrt{3}\,\mathbf{a}_y - 2\sqrt{3}\,\mathbf{a}_z \tag{8.133a}$$

(b) Snell's law consists of two parts:
   (I) $n_i \sin\theta_i = n_t \sin\theta_t$, and
   (II) $\mathbf{k}_i$ and $\mathbf{k}_t$ are in the plane of incidence.

The first condition implies that the tangential components of $\mathbf{k}_i$ and $\mathbf{k}_t$ are equal. With an unknown k_{tz}, we write

$$\mathbf{k}_t = \mathbf{a}_x + \sqrt{3}\,\mathbf{a}_y + k_{tz}\mathbf{a}_z \tag{8.133b}$$

In the second medium, the wavenumber is $k_t = n_t k_i = \sqrt{1.69}\sqrt{1+3+12} = 5.2$. Using this in Eq. (8.133b), we obtain

$$(5.2)^2 = 1 + (\sqrt{3})^2 + (k_{tz})^2, \text{ or } k_{tz} = 4.8$$

Thus,

$$\mathbf{k}_t = \mathbf{a}_x + \sqrt{3}\,\mathbf{a}_y + 4.8\,\mathbf{a}_z \tag{8.133c}$$

Exercise 8.33
What is the maximum angle of transmission for sunlight in the water of a placid lake with a refractive index of 1.33?

Ans. 48.8°.

Exercise 8.34
A uniform plane wave with $\mathbf{k}_i = 3\mathbf{a}_x + 4\mathbf{a}_y + 5\mathbf{a}_z$ in one medium with $n = 1.2$ impinges on another with $n = 2.4$ occupying region $z \geq 0$. Find $\mathbf{k}_r$, and $\mathbf{k}_t$.

Ans. $\mathbf{k}_r = 3\mathbf{a}_x + 4\mathbf{a}_y - 5\mathbf{a}_z$, $\mathbf{k}_t = 3\mathbf{a}_x + 4\mathbf{a}_y + \sqrt{175}\,\mathbf{a}_z$.

8.3.2.1 Perpendicular Polarization

The electric field vector is perpendicular to the plane of incidence for a uniform plane wave with **perpendicular polarization** or **s-polarization**, which is obliquely incident on the interface. The reflected and transmitted waves should also be uniform plane waves with perpendicular polarization; otherwise, the phase-matching and boundary conditions cannot be satisfied across the interface. Accordingly, the three waves can be expressed as follows (see Fig. 8.18):

$$\tilde{\mathbf{E}}^i = \mathbf{a}_y E_o^i\, e^{-j\,\mathbf{k}_i \cdot \mathbf{r}} = \mathbf{a}_y E_o^i\, e^{-j[(k_i \sin\theta_i)x + (k_i \cos\theta_i)z]} \tag{8.134a}$$

$$\tilde{\mathbf{E}}^r = \mathbf{a}_y E_o^r e^{-j\mathbf{k}_r \cdot \mathbf{r}} = \mathbf{a}_y E_o^r e^{-j[(k_r \sin\theta_r)x - (k_r \cos\theta_r)z]} \tag{8.134b}$$

$$\tilde{\mathbf{E}}^t = \mathbf{a}_y E_o^t e^{-j\mathbf{k}_t \cdot \mathbf{r}} = \mathbf{a}_y E_o^t e^{-j[(k_t \sin\theta_t)x + (k_t \cos\theta_t)z]} \tag{8.134c}$$

where E_o^i, E_o^r, and E_o^t are the amplitudes of the incident, reflected, and transmitted waves, respectively. Our goal is to express E_o^r and E_o^t in terms of E_o^i, which is determined by the source of the incident wave. Substituting Eqs. (8.134a, b, c) into Eq. (8.23) leads to the following magnetic field phasors:

$$\tilde{\mathbf{H}}^i = (\mathbf{a}_k^i \times \mathbf{a}_y)\frac{E_o^i}{\eta_1} e^{-j\mathbf{k}_i \cdot \mathbf{r}} = (-\mathbf{a}_x \cos\theta_i + \mathbf{a}_z \sin\theta_i)\frac{E_o^i}{\eta_1} e^{-j\mathbf{k}_i \cdot \mathbf{r}} \tag{8.135a}$$

$$\tilde{\mathbf{H}}^r = (\mathbf{a}_k^r \times \mathbf{a}_y)\frac{E_o^r}{\eta_1} e^{-j\mathbf{k}_r \cdot \mathbf{r}} = (\mathbf{a}_x \cos\theta_r + \mathbf{a}_z \sin\theta_r)\frac{E_o^r}{\eta_1} e^{-j\mathbf{k}_r \cdot \mathbf{r}} \tag{8.135b}$$

$$\tilde{\mathbf{H}}^t = (\mathbf{a}_k^t \times \mathbf{a}_y)\frac{E_o^t}{\eta_2} e^{-j\mathbf{k}_t \cdot \mathbf{r}} = (-\mathbf{a}_x \cos\theta_t + \mathbf{a}_z \sin\theta_t)\frac{E_o^t}{\eta_2} e^{-j\mathbf{k}_t \cdot \mathbf{r}} \tag{8.135c}$$

The unit vectors along the three wavevectors are

$$\mathbf{a}_k^i = \mathbf{a}_x \sin\theta_i + \mathbf{a}_z \cos\theta_i$$

$$\mathbf{a}_k^r = \mathbf{a}_x \sin\theta_r - \mathbf{a}_z \cos\theta_r$$

$$\mathbf{a}_k^t = \mathbf{a}_x \sin\theta_t + \mathbf{a}_z \cos\theta_t$$

Note that all the magnetic field vectors lie in the plane of incidence, as shown in Fig. 8.18.

The boundary conditions for electric and magnetic fields require

$$\tilde{\mathbf{E}}^i\Big|_{z=0,\mathrm{tan}} + \tilde{\mathbf{E}}^r\Big|_{z=0,\mathrm{tan}} = \tilde{\mathbf{E}}^t\Big|_{z=0,\mathrm{tan}} \tag{8.136a}$$

$$\tilde{\mathbf{H}}^i\Big|_{z=0,\mathrm{tan}} + \tilde{\mathbf{H}}^r\Big|_{z=0,\mathrm{tan}} = \tilde{\mathbf{H}}^t\Big|_{z=0,\mathrm{tan}} \tag{8.136b}$$

where subscript tan denotes tangential component. Substituting Eqs. (8.134a, b, c) and (8.135a, b, c) into Eq. (8.136a, b), setting $z = 0$, and invoking the phase-matching condition given in Eq. (8.126), we obtain

$$E_o^i + E_o^r = E_o^t \tag{8.137a}$$

$$-\cos\theta_i \frac{E_o^i}{\eta_1} + \cos\theta_i \frac{E_o^r}{\eta_1} = -\cos\theta_t \frac{E_o^t}{\eta_2} \tag{8.137b}$$

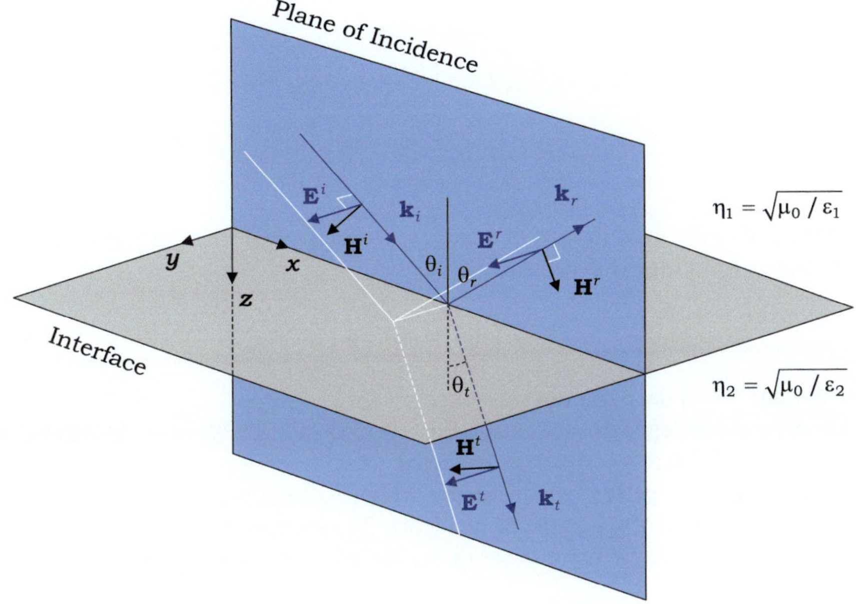

Fig. 8.18 Three uniform plane waves with perpendicular polarization in two adjacent dielectrics. White lines are for visual effects, traced by the tip of **E**

where the law of reflection ($\theta_r = \theta_i$) is used.

Solving these equations for E_o^r and E_o^t yields the **reflection coefficient** $\Gamma_\perp$ and **transmission coefficient** $\tau_\perp$ for perpendicular polarization, as follows:

$$\Gamma_\perp = \frac{E_o^r}{E_o^i} = \frac{\eta_2 \cos \theta_i - \eta_1 \cos \theta_t}{\eta_2 \cos \theta_i + \eta_1 \cos \theta_t} \tag{8.138}$$

$$\tau_\perp = \frac{E_o^t}{E_o^i} = \frac{2\eta_2 \cos \theta_i}{\eta_2 \cos \theta_i + \eta_1 \cos \theta_t} \tag{8.139}$$

These are also called **Fresnel's equations** for perpendicular polarization.

For nonmagnetic dielectrics, using the relation $\eta = \eta_o/n$, Eqs. (8.138) and (8.139) can be rewritten as

$$\Gamma_\perp = \frac{1 - CN}{1 + CN} \tag{8.140}$$

$$\tau_\perp = \frac{2}{1 + CN} \tag{8.141}$$

with

$$C = \frac{\cos \theta_2}{\cos \theta_1} \tag{8.142a}$$

$$N = \frac{n_2}{n_1} \tag{8.142b}$$

where θ_1 and θ_2 are used instead of θ_i and θ_t, respectively, for notational simplicity. It follows from Eqs. (8.140) and (8.141) that

$$\boxed{1 + \Gamma_\perp = \tau_\perp} \tag{8.143}$$

This is valid for all the angles of incidence.

For normal incidence ($\theta_i = 0 = \theta_t$), Eqs. (8.138) and (8.139) are reduced to Eqs. (8.107) and (8.108), respectively, as expected.

If medium 2 is a perfect conductor for which $\eta_2 = 0$, we have $\Gamma_\perp = -1$ and $\tau_\perp = 0$, irrespective of the angle of incidence. In this case, $\tilde{\mathbf{E}}^i + \tilde{\mathbf{E}}^r = 0$ at $z = 0$ and there is no tangential component of the electric field on the perfect conductor, as expected.

Example 8.20 A uniform plane wave with $\mathscr{E}^i = 10\,\mathbf{a}_y \cos(\omega t - \sqrt{3}\,x - z)$ travels in free space and impinges on a lossless dielectric with $\varepsilon_r = 4$ occupying the $z \geq 0$ region. Find

(a) $\mathbf{k}_i$ and θ_i,
(b) $\mathscr{E}^r$, and
(c) the total electric field in free space.

Solution

(a) The phasor form of $\mathscr{E}^i$ is

$$\tilde{\mathbf{E}}^i = 10\,\mathbf{a}_y e^{-j(\sqrt{3}\,x + z)} \tag{8.144a}$$

The wavevector of $\tilde{\mathbf{E}}^i$ is

$$\mathbf{k}_i = \sqrt{3}\,\mathbf{a}_x + \mathbf{a}_z$$

The angle of incidence is the angle between $\mathbf{k}_i$ and surface normal $\mathbf{a}_z$. Thus,

$$\theta_i = \tan^{-1}\frac{\sqrt{3}}{1} = 60^0$$

(b) Setting $\theta_i = 60^0$, $n_1 = 1$, and $n_2 = 2$ in the Snell's law $n_1 \sin \theta_i = n_2 \sin \theta_t$, we obtain

$$\cos \theta_t = \sqrt{1 - (n_1 \sin \theta_i / n_2)^2} = 0.901$$

Setting $C = \cos \theta_2 / \cos \theta_1 = 1.802$ and $N = n_2/n_1 = 2$ in Eq. (8.140) yields

$$\Gamma_\perp = \frac{1 - CN}{1 + CN} = -0.57$$

Substituting $k_r = k_i = 2$ and $\theta_r = \theta_i = 60^o$ into Eq. (8.134b) gives

$$\tilde{\mathbf{E}}^r = \mathbf{a}_y(-5.7)e^{-j(\sqrt{3}x-z)} \tag{8.144b}$$

$$\mathscr{E}^r = \mathbf{a}_y(-5.7)\cos(\omega t - \sqrt{3}\,x + z)$$

(c) In the region $z < 0$, combining Eqs. (8.144a) and (8.144b) yields

$$\tilde{\mathbf{E}} = \tilde{\mathbf{E}}^i + \tilde{\mathbf{E}}^r = \mathbf{a}_y e^{-j\sqrt{3}x}\left[10e^{-jz} - 5.7e^{jz}\right]$$

In instantaneous form, the total electric field is

$$\mathscr{E} = \mathrm{Re}\left[\tilde{\mathbf{E}}e^{j\omega t}\right]$$
$$= \mathbf{a}_y\left[10\cos(\omega t - \sqrt{3}\,x - z) - 5.7\cos(\omega t - \sqrt{3}\,x + z)\right]$$

Exercise 8.35
Under what conditions does $\Gamma_\perp$ become negative at the interface between two lossless dielectrics?

Ans. $n_2 > n_1$.

8.3.2.2 Parallel Polarization

For the oblique incidence of a uniform plane wave with ***parallel polarization*** or ***p-polarization***, the electric field vector is parallel to the plane of incidence. The same procedure used for perpendicular polarization can be followed to obtain the reflection and transmission coefficients. The sign convention for the positive direction of the electric field vector is such that the component of $\mathbf{E}$ tangential to the interface is directed along the tangential component of the wavevector (see Fig. 8.19). Subsequently, the right-hand rule, $\mathbf{a}_k = \mathbf{a}_E \times \mathbf{a}_H$, specifies the direction of the magnetic field vector such that it is perpendicular to the plane of incidence. In this case, the three magnetic field vectors are not all parallel to each other at the interface. It should be noted that some texts first set the magnetic field vectors to be perpendicular to the plane of incidence, parallel to each other, and then define the positive directions of the electric field vectors using the right-hand rule. Specifically, the positive direction

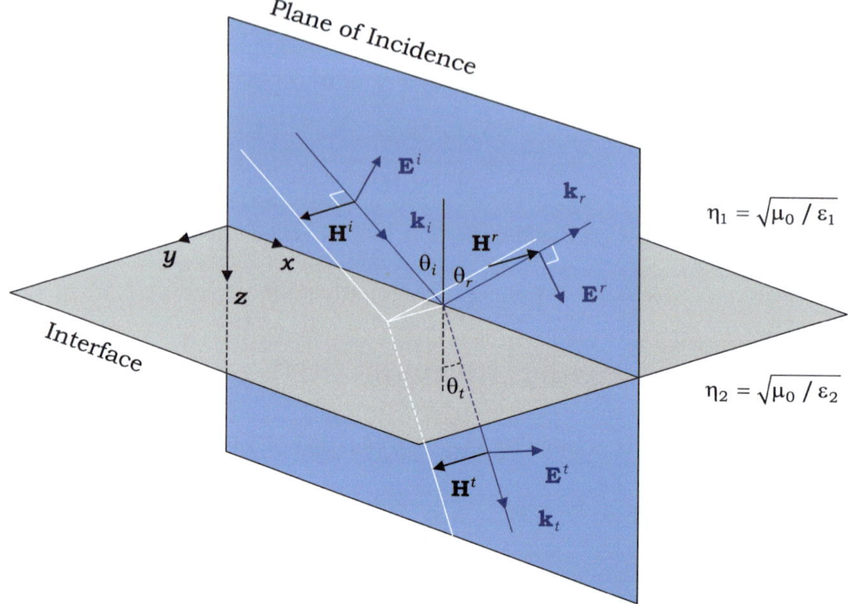

Fig. 8.19 Three uniform plane waves with parallel polarization in two adjacent dielectrics. White lines are for visual effects

of $\tilde{E}^r$ is opposite to that in our case, and thus the reflection coefficient for parallel polarization has a different form.

As before, the reflected and transmitted waves should be uniform plane waves with parallel polarization as the incident wave to satisfy the phase-matching and boundary conditions at the interface. Therefore, these three waves can be expressed as follows (see Fig. 8.19):

$$\tilde{\mathbf{E}}^i = \{\mathbf{a}_x \cos\theta_i - \mathbf{a}_z \sin\theta_i\} E_o^i\, e^{-j[(k_i \sin\theta_i)x + (k_i \cos\theta_i)z]} \tag{8.145a}$$

$$\tilde{\mathbf{E}}^r = \{\mathbf{a}_x \cos\theta_r + \mathbf{a}_z \sin\theta_r\} E_o^r\, e^{-j[(k_r \sin\theta_r)x - (k_r \cos\theta_r)z]} \tag{8.145b}$$

$$\tilde{\mathbf{E}}^t = \{\mathbf{a}_x \cos\theta_t - \mathbf{a}_z \sin\theta_t\} E_o^t\, e^{-j[(k_t \sin\theta_t)x + (k_t \cos\theta_t)z]} \tag{8.145c}$$

where the brackets represent the phase of the wave, which is the same as that for perpendicular polarization, and the braces define the direction of the electric field vector.

The magnetic field phasors can be obtained using Eq. (8.23) as

$$\tilde{\mathbf{H}}^i = \mathbf{a}_y \frac{E_o^i}{\eta_1} e^{-j[(k_i \sin\theta_i)x + (k_i \cos\theta_i)z]} \tag{8.146a}$$

$$\tilde{\mathbf{H}}^r = -\mathbf{a}_y \frac{E_o^r}{\eta_1} e^{-j[(k_r \sin\theta_r)x - (k_r \cos\theta_r)z]} \tag{8.146b}$$

$$\tilde{\mathbf{H}}^t = \mathbf{a}_y \frac{E_o^t}{\eta_2} e^{-j[(k_t \sin\theta_t)x + (k_t \cos\theta_t)z]} \tag{8.146c}$$

The boundary conditions for the electric and magnetic fields require

$$\tilde{\mathbf{E}}^i\Big|_{z=0,\tan} + \tilde{\mathbf{E}}^r\Big|_{z=0,\tan} = \tilde{\mathbf{E}}^t\Big|_{z=0,\tan} \tag{8.147a}$$

$$\tilde{\mathbf{H}}^i\Big|_{z=0,\tan} + \tilde{\mathbf{H}}^r\Big|_{z=0,\tan} = \tilde{\mathbf{H}}^t\Big|_{z=0,\tan} \tag{8.147b}$$

Substituting Eqs. (8.145a, b, c) and (8.146a, b, c) into Eq. (8.147a, b), setting $z = 0$, and invoking the phase-matching condition, we obtain

$$\cos\theta_i \, E_o^i + \cos\theta_r \, E_o^r = \cos\theta_t \, E_o^t \tag{8.148a}$$

$$\frac{E_o^i}{\eta_1} - \frac{E_o^r}{\eta_1} = \frac{E_o^t}{\eta_2} \tag{8.148b}$$

where common exponential factors are omitted.

Solving these equations for E_o^r and E_o^t yields the **reflection coefficient** $\Gamma_\|$ and **transmission coefficient** $\tau_\|$ for parallel polarization, as follows:

$$\boxed{\Gamma_\| = \frac{E_o^r}{E_o^i} = \frac{\eta_2 \cos\theta_t - \eta_1 \cos\theta_i}{\eta_2 \cos\theta_t + \eta_1 \cos\theta_i}} \tag{8.149}$$

$$\boxed{\tau_\| = \frac{E_o^t}{E_o^i} = \frac{2\eta_2 \cos\theta_i}{\eta_2 \cos\theta_t + \eta_1 \cos\theta_i}} \tag{8.150}$$

These are called **Fresnel's equations** for parallel polarization.

For nonmagnetic lossless dielectrics, using the relation $\eta = \eta_o/n$, the above equations can be rewritten as follows:

$$\boxed{\Gamma_\| = \frac{C - N}{C + N}} \tag{8.151}$$

$$\boxed{\tau_\| = \frac{2}{C + N}} \tag{8.152}$$

with

$$C = \frac{\cos \theta_2}{\cos \theta_1}$$

$$N = \frac{n_2}{n_1}$$

as before.

It follows from Eqs. (8.149) and (8.150) that

$$1 + \Gamma_{\parallel} = \tau_{\parallel}\left(\frac{\cos \theta_t}{\cos \theta_i}\right) \qquad (8.153a)$$

or

$$1 + \Gamma_{\parallel} = \tau_{\parallel}\, C \qquad (8.153b)$$

Unlike Eq. (8.143), the right-hand side of these equations contains an additional factor.

For the normal incidence ($\theta_i = 0 = \theta_t$), Eqs. (8.149) and (8.150) are reduced to Eqs. (8.107) and (8.108), respectively.

If medium 2 is a perfect conductor with $\eta_2 = 0$, then $\Gamma_{\parallel} = -1$ and $\tau_{\parallel} = 0$ irrespective of the angle of incidence. In this case, the tangential component (or x-component) of the total electric field vanishes on the surface of the conductor, but its normal component (or z-component) is nonzero.

When the wave in medium 1 reflects off medium 2 with a larger refractive index ($n_2 > n_1$), it is referred to as **external reflection**. The converse is called **internal reflection**. In Fig. 8.20, the reflection and transmission coefficients are plotted as functions of the angle of incidence for external reflection (i.e., $n_1 = 1$ and $n_2 = 1.5$). The coefficients for internal reflection (i.e., $n_1 = 1.5$ and $n_2 = 1$) are plotted as functions of the angle of incidence in Fig. 8.21. The figures show that $|\Gamma_{\perp}|$ increases monotonically from its initial value to unity as θ_i increases. In contrast, $|\Gamma_{\parallel}|$ first decreases from its initial value to zero and then increases to unity. This implies that a wave with perpendicular polarization reflects better than a wave with parallel polarization, for both external and internal reflections.

Example 8.21 A uniform plane wave with parallel polarization $\tilde{\mathbf{E}}^i = 4\mathbf{a}_E\, e^{-j(2x+3z)}$ propagates in free space and impinges obliquely on a lossless dielectric with $\varepsilon_r = 2.25$ that occupies the $z \geq 0$ region. Find (a) $\mathbf{a}_E$, (b) $\tilde{\mathbf{E}}^r$, and (c) $\tilde{\mathbf{E}}^t$.

Solution

(a) From the given wavevector $\mathbf{k}_i = 2\mathbf{a}_x + 3\mathbf{a}_z$, the angle of incidence is obtained as

$$\theta_i = \tan^{-1}\frac{2}{3} \cong 33.69^o$$

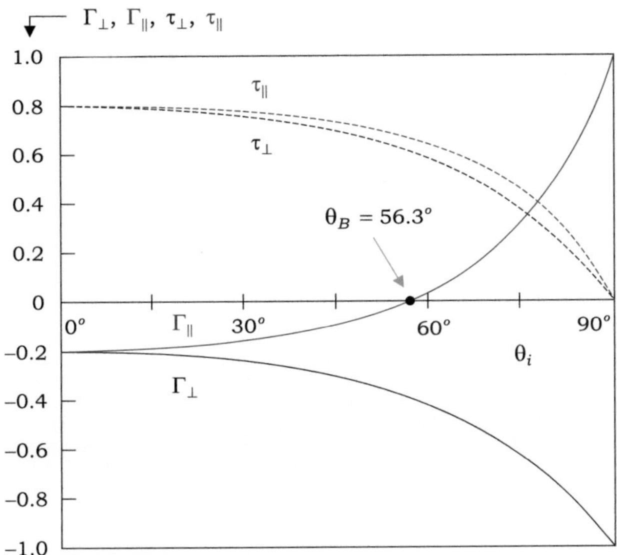

Fig. 8.20   External reflection for $n_1 = 1$ and $n_2 = 1.5$

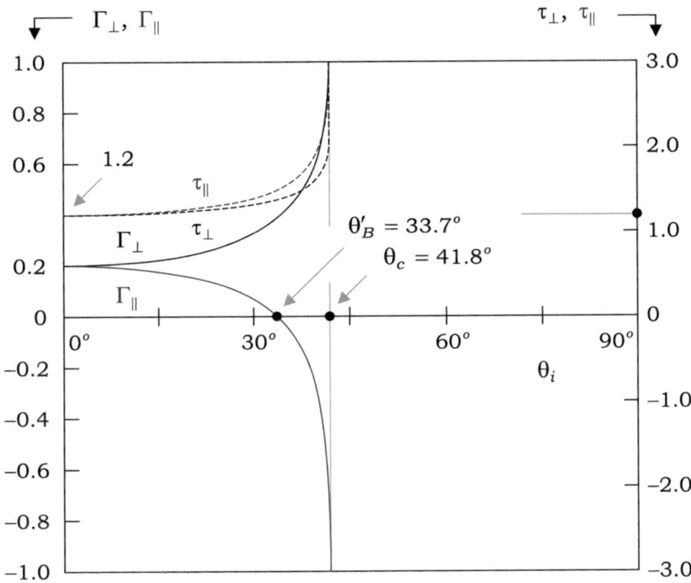

Fig. 8.21   Internal reflection for $n_1 = 1.5$ and $n_2 = 1$

For parallel polarization, we have $(\mathbf{a}_E)_x \| (\mathbf{k}_i)_x$ and $\mathbf{a}_E \cdot \mathbf{k}_i = 0$. Inserting $\mathbf{a}_E = 2\mathbf{a}_x + b\mathbf{a}_z$ into the relation $\mathbf{a}_E \cdot \mathbf{k}_i = 0$ yields

$$\mathbf{a}_E = \frac{1}{\sqrt{13}}(3\mathbf{a}_x - 2\mathbf{a}_z)$$

Thus, the incident wave is given by

$$\tilde{\mathbf{E}}^i = (3.33\,\mathbf{a}_x - 2.22\,\mathbf{a}_z)e^{-j(2x+3z)} \tag{8.154a}$$

(b) Using $n_1 = 1$ and $n_2 = \sqrt{2.25} = 1.5$ in Snell's law of refraction yields

$$\theta_t = \sin^{-1}\left(\frac{1}{1.5}\sin 33.69^o\right) \cong 21.70^o$$

Inserting $C = \cos\theta_2/\cos\theta_1 = 1.117$ and $N = n_2/n_1 = 1.5$ into Eq. (8.151) yields

$$\Gamma_\parallel = \frac{1.117 - 1.50}{1.117 + 1.50} = -0.146$$

By inserting $E_o^i = 4$, $\Gamma_\parallel = -0.146$, $k_r = k_i = \sqrt{13}$, and $\theta_r = \theta_i = 33.69^o$ into Eq. (8.145b), we obtain

$$\tilde{\mathbf{E}}^r = (\mathbf{a}_x \cos\theta_r + \mathbf{a}_z \sin\theta_r)\Gamma_\parallel E_o^i\, e^{-j[(k_r \sin\theta_r)x - (k_r \cos\theta_r)z]}$$
$$= -(0.49\,\mathbf{a}_x + 0.32\,\mathbf{a}_z)\, e^{-j(2x-3z)} \tag{8.154b}$$

(c) From Eq. (8.152),

$$\tau_\parallel = \frac{2}{1.117 + 1.50} = 0.764$$

By inserting $\tau_\parallel = 0.764$, $k_t = k_i\sqrt{\varepsilon_r} = 1.5\sqrt{13}$, and $\theta_t = 21.69^o$ into Eq. (8.145c), we obtain

$$\tilde{\mathbf{E}}^t = (\mathbf{a}_x \cos\theta_t - \mathbf{a}_z \sin\theta_t)\tau_\parallel E_o^i\, e^{-j[(k_t \sin\theta_t)x + (k_t \cos\theta_t)z]}$$
$$= (2.84\,\mathbf{a}_x - 1.13\,\mathbf{a}_z)e^{-j(2x+5z)} \tag{8.154c}$$

The tangential component of $\tilde{\mathbf{E}}$ has a value of $3.33 - 0.49 = 2.84$ and the normal component of $\tilde{\mathbf{D}}$ has a value of $(-2.22 - 0.32) = 2.25 \times (-1.13)$ on either side of the interface. If the floating-point errors are ignored, the boundary conditions are satisfied.

Exercise 8.36

A plane wave with parallel polarization is expressed by $\tilde{\mathbf{E}} = \mathbf{a}_E 10 e^{-j(3x+2y+4z)}$ in free space and is obliquely incident on an interface at $z = 0$. Determine $\mathbf{a}_E$.

Ans. $\mathbf{a}_E = (12\mathbf{a}_x + 8\mathbf{a}_y - 13\mathbf{a}_z)/\sqrt{377}$.

8.3.2.3   Brewster Angle

A uniform plane wave with parallel polarization may undergo no reflection at all if it impinges on a plane interface at an angle called the **Brewster angle**; the wave is totally transmitted into the second medium. However, there is no Brewster angle for the perpendicular polarization. The Brewster angle for external reflection, θ_B, is complementary to that for internal reflection, θ'_B, such that $\theta_B + \theta'_B = 90^o$ (Figs. 8.20 and 8.21). The Brewster angle is also known as the **polarizing angle**. When an unpolarized wave impinges on a surface at a polarizing angle, only the perpendicular component reflects off the surface, causing the surface to behave like a polarizer.

Using Snell's law of refraction in Eqs. (8.138) and (8.149), with $\eta = \eta_o/n$, we can rewrite the reflection coefficients as

$$\boxed{\Gamma_\perp = \frac{\sin(\theta_t - \theta_i)}{\sin(\theta_t + \theta_i)}} \tag{8.155}$$

$$\boxed{\Gamma_\| = \frac{\tan(\theta_t - \theta_i)}{\tan(\theta_t + \theta_i)}} \tag{8.156}$$

It is evident from Eq. (8.156) that $\Gamma_\| = 0$ for $\theta_t + \theta_i = 90^o$, and thus the Brewster angle is given by $\theta_B \equiv \theta_i = 90^o - \theta_t$. In this case, Snell's law of refraction becomes

$$n_1 \sin \theta_B = n_2 \sin \theta_t$$
$$= n_2 \sin(90^o - \theta_B)$$

This gives

$$\boxed{\tan \theta_B = \frac{n_2}{n_1}} \tag{8.157}$$

By changing the roles of 1 and 2 in Eq. (8.157), the Brewster angle for a wave traveling from medium 2 to 1 can be obtained as

$$\tan \theta'_B = \frac{n_1}{n_2} \tag{8.158}$$

It follows from Eqs. (8.157) and (8.158) that the two Brewster angles are complementary:

$$\theta_B + \theta'_B = 90^o \tag{8.159}$$

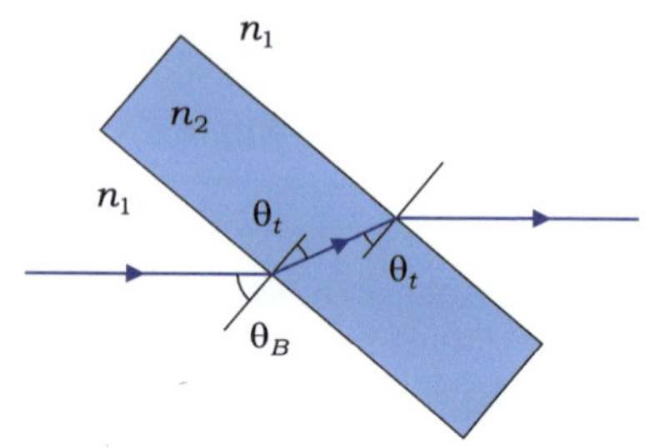

Fig. 8.22 Uniform plane wave incident at the Brewster angle

The reflection coefficient for the perpendicular polarization in Eq. (8.155) cannot be zero if n_1 and n_2 are not equal; thus, there is no Brewster angle for perpendicular polarization.

Example 8.22 A uniform plane wave with parallel polarization is incident onto a dielectric slab at the Brewster angle, as shown in Fig. 8.22. Show that the wave experiences no reflection at the exit surface either.

Solution

Using $\theta_B = 90^o - \theta_t$, Eq. (8.157) can be rewritten as

$$\frac{n_2}{n_1} = \frac{\sin \theta_B}{\cos \theta_B} = \frac{\cos \theta_t}{\sin \theta_t} \tag{8.160}$$

Comparing Eq. (8.160) with Eq. (8.158), θ_t is recognized as θ'_B, meaning that the internal wave is also incident onto the exit surface at the Brewster angle.

Exercise 8.37
Find θ_B and θ'_B for a diamond plate with $n = 2.42$ and $\mu = \mu_0$ placed in free space.

Ans. $\theta_B = 67.5^o$, $\theta'_B = 22.5^o$.

Exercise 8.38
For a uniform plane wave with parallel polarization incident on an interface at Brewster angle, express $\tau_\parallel$ in terms of the indices of refraction.

Ans. $\tau_\parallel = n_1/n_2$.

Exercise 8.39
Find the angle between the wavevectors $\mathbf{k}_r$ and $\mathbf{k}_t$ at the Brewster angle.

Ans. 90°.

Review Questions

RQ 8.42   What is the plane of incidence?   [Fig. 8.16]

RQ 8.43   Define perpendicular and parallel polarizations.   [Fig. 8.18,Fig. 8.19]

RQ 8.44	State the phase-matching condition.	[(8.126),Fig. 8.17]
RQ 8.45	State Snell's law of refraction.	[(8.132)]
RQ 8.46	Write Fresnel's equations.	[(8.138) (8.139) (8.149)(8.150)]
RQ 8.47	Define Brewster angle.	[(8.157)]

8.3.3 Total Internal Reflection

We now turn our attention to the internal reflection of a uniform plane wave at the dielectric-dielectric interface and discuss total internal reflection (TIR). Internal reflection occurs for a wave incident from medium 1 with refractive index n_1 on medium 2 with lower refractive index n_2. In this case, the angle of transmission θ_t in medium 2 is always greater than the angle of incidence θ_i in medium 1. As θ_i gradually increases from $0°$ to a higher value, θ_t increases from $0°$ to $90°$ before θ_i reached $90°$. The angle of incidence corresponding to $\theta_t = 90^o$ is called the **critical angle**, and is denoted as θ_c. If θ_i is further increased past the critical angle, there is no transmission of the wave into medium 2, and the incident wave is completely reflected by the interface, which is known as **total internal reflection**.

Setting $\theta_t = 90^o$ in Snell's law yields the critical angle:

$$\sin \theta_c = \frac{n_2}{n_1} \tag{8.161}$$

where $n_1 > n_2$.

Although there is no transmission of electromagnetic energy into medium 2 in the case of total internal reflection, a residual electromagnetic field should exist in medium 2; otherwise, the boundary conditions are not satisfied at the interface. For $\theta_i > \theta_c$, Snell's law does not allow a real solution for θ_t. To see this, we start with a trigonometric identity $\sin^2 \theta + \cos^2 \theta = 1$ and write

$$\cos \theta_t = \pm\sqrt{1 - \sin^2 \theta_t} = \pm j\sqrt{(n_1/n_2)^2 \sin^2 \theta_i - 1} \tag{8.162}$$

where $j = \sqrt{-1}$, and Snell's law is used to express θ_t in terms of θ_i. To obtain an expression for the electric field in medium 2, Eq. (8.162) is substituted into Eq. (8.134c) or Eq. (8.145c). Consequently,

$$\tilde{\mathbf{E}}^t = \mathbf{a}_E E_o^t\, e^{-j[(k_t \sin \theta_t)x + (k_t \cos \theta_t)z]}$$

$$\equiv \mathbf{a}_E E_o^t\, e^{-\alpha_e z} e^{-j\beta_e x} \tag{8.163}$$

where $\mathbf{a}_E = \mathbf{a}_y$ for the perpendicular polarization, and $\mathbf{a}_E = (\mathbf{a}_x \cos\theta_t - \mathbf{a}_z \sin\theta_t)$ for the parallel polarization. Furthermore,

$$k_t \sin\theta_t = k_t(n_1/n_2)\sin\theta_i \equiv \beta_e \tag{8.164a}$$

$$k_t \cos\theta_t = \pm jk_t\sqrt{(n_1/n_2)^2 \sin^2\theta_i - 1} \equiv -j\alpha_e \tag{8.164b}$$

In Eq. (8.164b), the negative sign is taken for the reason that will become evident shortly. The wave given in Eq. (8.163) is called the **evanescent wave**, for which the amplitude decays exponentially in the z-direction, whereas the phase changes sinusoidally in the x-direction. The evanescent wave propagates along the interface, whereas its amplitude decays exponentially in the direction normal to the interface. This wave is bound to the interface, forming a **surface wave**.

Substitution of Eq. (8.162) into Eqs. (8.138) and (8.149) leads to the reflection coefficients for the case of total internal reflection:

$$\Gamma_\perp = \frac{\cos\theta_i + j\sqrt{\sin^2\theta_i - (n_2/n_1)^2}}{\cos\theta_i - j\sqrt{\sin^2\theta_i - (n_2/n_1)^2}} \tag{8.165a}$$

$$\Gamma_\parallel = \frac{j\sqrt{\sin^2\theta_i - (n_2/n_1)^2} + (n_2/n_1)^2 \cos\theta_i}{j\sqrt{\sin^2\theta_i - (n_2/n_1)^2} - (n_2/n_1)^2 \cos\theta_i} \tag{8.165b}$$

It immediately follows that

$$\left|\Gamma_\perp\right| = \left|\Gamma_\parallel\right| = 1 \tag{8.166}$$

All the energy of the incident wave is reflected back into medium 1, irrespective of the polarization state of the wave and residual electric field in medium 2.

Example 8.23 A dielectric slab waveguide is infinite in extent in the x- and y-directions with refractive indices $n_1 > n_2 > 1$, as shown in Fig. 8.23. Find the maximum angle of incidence θ_m for the total internal reflection at the interface between the core and the cladding.

Solution

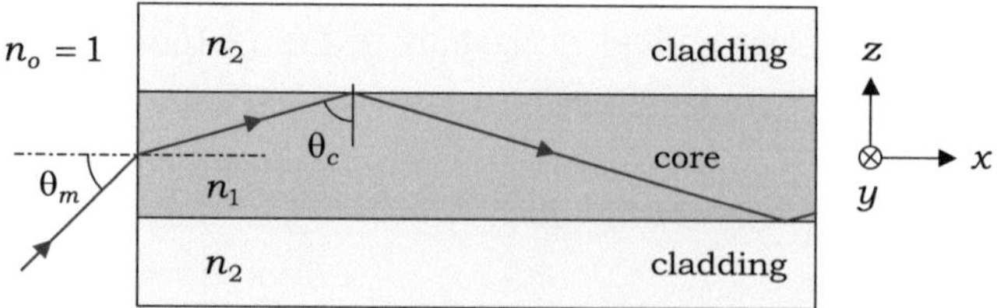

Fig. 8.23 Dielectric slab waveguide lying along the x-axis (see Fig. 10.2)

By applying Snell's law to the entrance surface, we obtain

$$n_o \sin \theta_m = n_1 \sin(90^o - \theta_c)$$
$$= n_1 \cos \theta_c \tag{8.167a}$$

Inserting Eq. (8.161) into Eq. (8.167a) yields

$$n_o \sin \theta_m = n_1 \sqrt{1 - (n_2/n_1)^2}$$

Thus, the maximum angle of incidence is

$$\theta_m = \sin^{-1}\left[\frac{1}{n_o}\sqrt{n_1^2 - n_2^2}\right] \tag{8.167b}$$

If a wave is incident at an angle less than θ_m, it is guided through the slab by successive total internal reflections at the upper and lower interfaces.

Exercise 8.40
Determine θ_c for a diamond plate with $n = 2.42$ residing in free space.

Ans. $\theta_c = 24.4^o$.

Review Questions

RQ 8.48	What is total internal reflection?	[Fig. 8.23]
RQ 8.49	Define the critical angle.	[(8.161)]
RQ 8.50	Express the evanescent wave in the second medium.	[(8.163)]
RQ 8.51	What is the reflection coefficient for TIR?	[(8.165a,b)]

8.3.4 *Reflectance and Transmittance*

The time-averaged power density of an electromagnetic wave is well represented by the time-averaged Poynting vector, $\langle \mathbf{S} \rangle$. In radiometry, **irradiance** I_e is defined as the radiant flux received by the surface of a unit area. Expressed mathematically,

$$I_e \equiv \langle \mathbf{S} \rangle \cdot \mathbf{a}_n \quad [\text{W/m}^2] \tag{8.168}$$

where $\mathbf{a}_n$ is a unit vector normal to a given surface directed along the general direction of the Poynting vector, which is not necessarily directed outward. Meanwhile, the **intensity** I is defined as the time-averaged power through a unit area, where the area is taken in the plane perpendicular to the Poynting vector. In lossless nonmagnetic materials, the intensity is generally given by

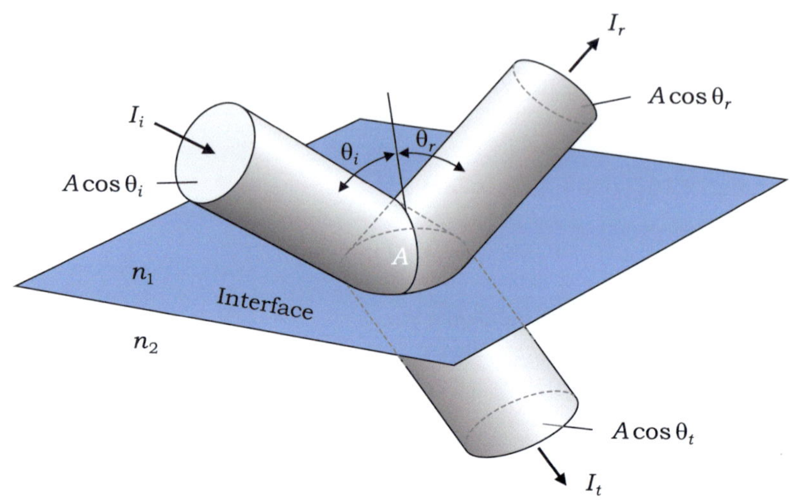

Fig. 8.24 Three waves intersect at the interface

$$\boxed{I = \frac{n}{2\eta_o} E_o^2} \quad [\text{W/m}^2] \tag{8.169}$$

where n is the refractive index of the medium, η_o is the intrinsic impedance of free space, and E_o is the electric field amplitude.

Consider the case in which the incident, reflected, and transmitted waves with finite cross-sectional areas intersect at the interface between two different materials, as shown in Fig. 8.24. If A is the common area shared by the three waves at the interface, the cross-sectional areas of the incident, reflected, and transmitted waves are given by $A \cos \theta_i$, $A \cos \theta_r$, and $A \cos \theta_t$, respectively. When the waves have intensities of I_i, I_r, and I_t, the wave powers are $I_i A \cos \theta_i$, $I_r A \cos \theta_r$, and $I_t A \cos \theta_t$, respectively.

The **reflectance** R is defined as the ratio of the reflected power to the incident power:

$$\boxed{R = \frac{I_r A \cos \theta_r}{I_i A \cos \theta_i} = \left(\frac{E_o^r}{E_o^i} \right)^2} \tag{8.170a}$$

where we have used Eq. (8.169), along with $\theta_r = \theta_i$. Then it follows that

$$\boxed{R = \Gamma^2} \tag{8.170b}$$

The reflectance is equal to the square of the reflection coefficient.

The **transmittance** T is defined as the ratio of the transmitted power to the incident power:

$$T = \frac{I_t A \cos \theta_t}{I_i A \cos \theta_i} = \left(\frac{n_2 \cos \theta_t}{n_1 \cos \theta_i}\right)\tau^2 \qquad (8.171\text{a})$$

or

$$T = \tau^2 C N \qquad (8.171\text{b})$$

where τ is the transmission coefficient, and C and N are given by Eqs. (8.142a) and (8.142b).

The sum of the powers of the reflected and transmitted waves must be equal to that of the incident wave, in accordance with the law of energy conservation. Therefore,

$$I_i A \cos \theta_i = I_r A \cos \theta_r + I_t A \cos \theta_t \qquad (8.172\text{a})$$

This can be rewritten with $\theta_i = \theta_r$ as follows:

$$1 = \frac{I_r}{I_i} + \frac{I_t \cos \theta_t}{I_i \cos \theta_i} \qquad (8.172\text{b})$$

Substituting Eqs. (8.170a, b) and (8.171a, b) into Eq. (8.172b) yields

$$1 = R + T \qquad (8.173)$$

The sum of the reflectance and transmittance is always unity, irrespective of wave polarization. As the reflectance is $R = |\Gamma|^2$, the transmittance is given by $T = 1 - |\Gamma|^2$.

Example 8.24 For the three waves in Example 8.21, which are Eqs. (8.154a, b, c),

(a) determine the wave powers through a unit area of the interface, and
(b) verify the law of energy conservation.

Solution

(a) The incident power through a unit area of the interface is

$$I_i \cos \theta_i = \frac{1}{2\eta_o}(E_o^i)^2 \cos \theta_i = \frac{1}{2 \times 377}\left[(3.33)^2 + (2.22)^2\right]\cos(33.69^o)$$
$$= 17.68 \ [\text{mW}]$$

The reflected power through a unit area of the interface is

$$I_r \cos \theta_r = \frac{1}{2\eta_o}(E_o^r)^2 \cos \theta_r = \frac{1}{2 \times 377}\left[(0.49)^2 + (0.32)^2\right]\cos(33.69^o)$$
$$= 0.38 \ [\text{mW}]$$

The transmitted power through a unit area of the interface is

$$I_t \cos \theta_t = \frac{\sqrt{2.25}}{2 \times 377} (E_o^t)^2 \cos \theta_t$$

$$= \frac{1.5}{2 \times 377} \left[(2.84)^2 + (1.13)^2 \right] \cos(21.69^o)$$

$$= 17.27 \text{ [mW]}$$

(b)   The incident power is 17.68 [mW], whereas the sum of the reflected and transmitted powers is calculated as 17.65 [mW]. When the floating-point errors are ignored, the result conforms to the law of energy conservation.

Exercise 8.41

Starting with $\tau_\| = n_1/n_2$ for a uniform plane wave incident at the Brewster angle, determine the transmittance.

Ans. $T_\| = 1$.

Review Questions

RQ 8.52	Define irradiance.	[(8.168)]
RQ 8.53	Define intensity.	[(8.169)]
RQ 8.54	Define reflectance and transmittance.	[(8.170a,b)(8.171a,b)]
RQ 8.55	Express the law of energy conservation in terms of reflectance and transmittance.	[(8.173)]

8.4   Waves in Dispersive Medium

According to the atomic model of matter, a dielectric can be regarded as an assemblage of discrete atoms placed at equivalent lattice points in free space. When an electromagnetic wave of frequency ω is introduced into a material, its electric field induces electric dipoles oscillating at ω, which in turn generate secondary wavelets of the same frequency, propagating with the same velocity as the primary wave. However, the damping of the electric dipoles results in a phase lag or lead of the secondary wave with respect to the primary wave. Nevertheless, the linear superposition of primary and secondary waves constitutes an internal wave in the material that behaves as a primary wave at a later time and new position. The phase delay between the primary and secondary waves accumulates as the wave propagates through the material. Therefore, the phase velocity in the material is different from that in free space. Because the damping effect depends on the frequency, the phase velocity also depends on the frequency, which is called the **dispersion**. As we see from Eqs. (8.68) and (8.80), the lossy dielectrics are essentially dispersive.

Until now, we have limited our discussion to harmonic waves that extend to infinity. If an electromagnetic wave is used to transport meaningful information, it should be finite in time, as a short digital pulse. A wavepacket is a wave of finite extent in both the space and time dimensions. Fourier analysis shows that a wavepacket consists of a band of frequencies. In dispersive media, different frequency components travel with different phase velocities, resulting in broadening of the wavepacket. The propagation velocity of a wavepacket is called the ***group velocity***, and is equal to the propagation velocity of the energy.

Let us consider a train of wavepackets resulting from the interference between two harmonic waves of equal amplitude but slightly different angular frequencies, such as $\omega^- = \omega_o - \Delta\omega$ and $\omega^+ = \omega_o + \Delta\omega$. The two waves have slightly different phase constants, $\beta^- = \beta_o - \Delta\beta$ and $\beta^+ = \beta_o + \Delta\beta$. Assuming that they are both polarized in the x-direction and propagate in the z-direction, the total electric field is given by

$$\mathscr{E}(z, t) = \mathbf{a}_x E_o \cos(\omega^- t - \beta^- z) + \mathbf{a}_x E_o \cos(\omega^+ t - \beta^+ z)$$
$$= \mathbf{a}_x 2 E_o \cos(\Delta\omega t - \Delta\beta z) \cos(\omega_o t - \beta_o z) \qquad (8.174)$$

where

$$\Delta\omega = \frac{1}{2}(\omega^+ - \omega^-) \qquad (8.175a)$$

$$\Delta\beta = \frac{1}{2}(\beta^+ - \beta^-) \qquad (8.175b)$$

$$\omega_o = \frac{1}{2}(\omega^+ + \omega^-) \qquad (8.175c)$$

$$\beta_o = \frac{1}{2}(\beta^+ + \beta^-) \qquad (8.175d)$$

It can be seen from Eq. (8.174) that the electric field varies rapidly with the space variable z because of β_o, but its envelope changes slowly with z because of $\Delta\beta$.

As in the case of a uniform plane wave, the phase velocity of a wavepacket is the propagation velocity of a point of constant phase (i.e., crest). This is equal to the ratio of the traveled distance, Δz, to the elapsed time, Δt, as measured by an observer moving with the crest. That is,

$$\boxed{v_p = \frac{\omega_o}{\beta_o}} \quad [\text{m/s}] \qquad (8.176)$$

The group velocity is defined as the velocity of the envelope. By following the same procedure as that used for the phase velocity, we obtain the group velocity by taking the ratio of the traveled distance Δz to the elapsed time Δt, as measured by an observer moving with the envelope. In Eq. (8.174), the peak of the envelope corresponds to a point with $(\Delta\omega t - \Delta\beta z) = 0$. By taking the limits as $\Delta\omega \to 0$ and $\Delta\beta \to 0$, the group velocity is obtained as

$$\boxed{v_g = \frac{d\omega}{d\beta}} \quad [\text{m/s}] \tag{8.177}$$

Group velocity represents the velocity of the modulated amplitude of a wavepacket or the velocity of the time-averaged energy of the wave, whereas phase velocity represents the velocity of a wavefront. It is noteworthy that the phase and group velocities are equal for a uniform plane wave with a single frequency.

Although the phase and group velocities depend on the frequency in the dispersive media, they behave differently as the frequency varies. In the region of **normal dispersion**, where the refractive index of the medium increases with the frequency, the group velocity is less than the phase velocity. By contrast, in the region of **anomalous dispersion**, where the refractive index decreases with frequency, the group velocity is always greater than the phase velocity.

To gain a qualitative understanding of the relationship between the group and phase velocities, we consider a train of wavepackets, as shown in Fig. 8.25, which results from the interference between the two waves with slightly different frequencies, ω^+ and ω^-. The three points A, B, and C in Fig. 8.25a represent the three crests of the ω^+ wave at time $t = 0$. As shown in Fig. 8.25b, the crests of the ω^+ and ω^- waves meet at point B, corresponding to the envelope peak. For normal dispersion $(n^+ > n^-)$, the ω^- wave travels faster than the ω^+ wave because of the greater phase velocity. Therefore, at a later time $t = \Delta t$, the crests of the ω^+ and ω^- waves overlap at point A' (see Fig. 8.25c), which is a new position for point A travelling to the right during Δt. For an observer moving with point B, the group velocity appears to be less than the phase velocity (see Fig. 8.25d). The converse is true for anomalous dispersion with $n^+ < n^-$.

The $\omega - \beta$ diagram in Fig. 8.26 shows a plot of the dispersion relation: (a) normal dispersion and (b) anomalous dispersion. The slope of the line drawn from the origin to a point on the curve represents the phase velocity of the wave at that frequency and the tangent to the curve represents the group velocity of the wavepacket with mean frequency ω_o.

Example 8.25 If the wavepacket given in Eq. (8.174) travels in free space, find (a) $\mathcal{H}(z, t)$, and (b) $\langle \mathbf{S} \rangle$.

Solution

(a)   Using the relations $|\mathcal{H}| = |\mathcal{E}|\sqrt{\varepsilon_0/\mu_0}$ and $\mathbf{a}_k = \mathbf{a}_E \times \mathbf{a}_H$, we obtain

$$\mathcal{H}(z, t) = \mathbf{a}_y 2E_o\sqrt{\varepsilon_0/\mu_0}\,\cos(\Delta\omega t - \Delta\beta z)\cos(\omega_o t - \beta_o z) \tag{8.178a}$$

(b)   The Poynting vector is obtained from Eqs. (8.174) and (8.178a) as

$$\mathbf{S} = \mathcal{E} \times \mathcal{H}$$

$$= \mathbf{a}_z 4E_o^2\sqrt{\varepsilon_0/\mu_0}\,\cos^2(\Delta\omega t - \Delta\beta z)\cos^2(\omega_o t - \beta_o z) \tag{8.178b}$$

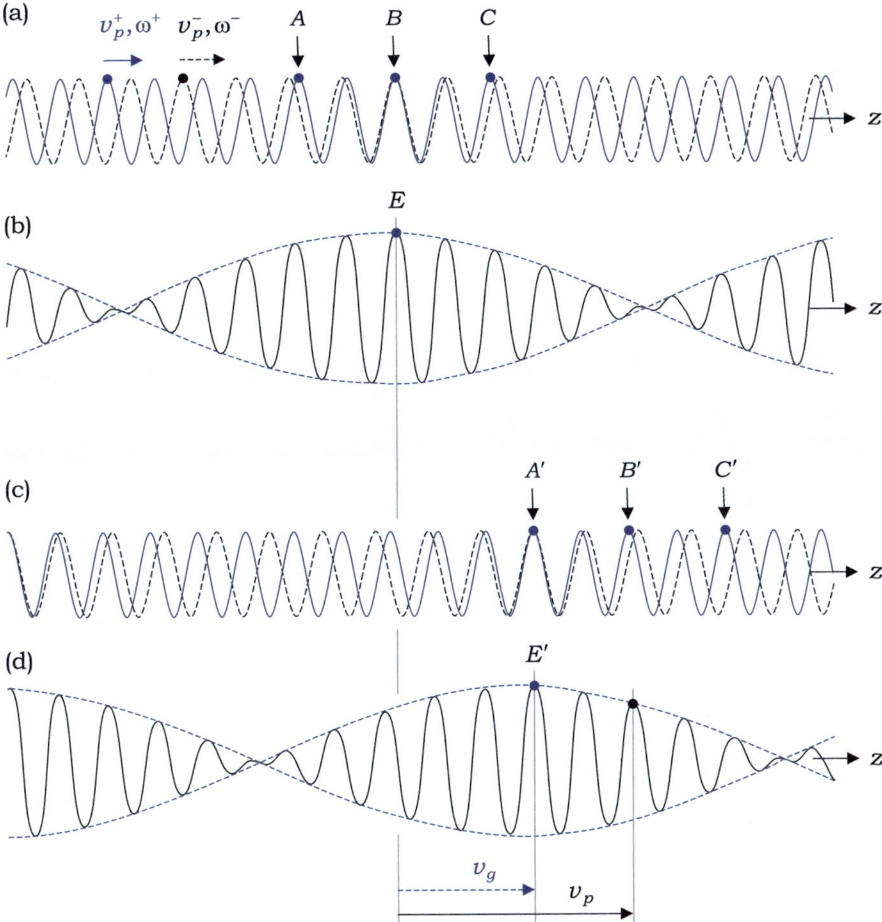

Fig. 8.25 A train of wavepackets due to the interference of two waves with slightly different frequencies. **a** and **b** The envelope peaks at point *B*. **c** and **d** The envelope peaks at point *A'* at a later time in the medium with normal dispersion

The time-averaged power density is defined as

$$\langle \mathbf{S} \rangle = \frac{1}{T} \int_{-T/2}^{T/2} \mathbf{S}\, dt \tag{8.178c}$$

In this equation, $\tau_1 \ll T \ll \tau_2$, where $\tau_2 = 2\pi/\Delta\omega$ is the temporal period of the envelope, and $\tau_1 = 2\pi/\omega_o$ is the temporal period of the phase. Using Eq. (8.178b) in Eq. (8.178c) yields

$$\langle \mathbf{S} \rangle = \mathbf{a}_z 2E_o^2 \sqrt{\varepsilon_0/\mu_0}\, \cos^2(\Delta\omega t - \Delta\beta z) \tag{8.178d}$$

(a)   (b)

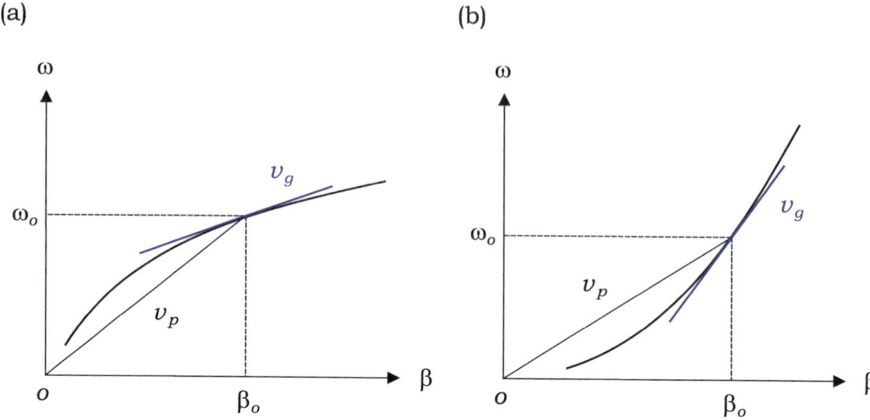

Fig. 8.26 ω–β diagram of dispersive medium: **a** Normal dispersion, and **b** Anomalous dispersion

It is evident from Eq. (8.178d) that the time-averaged power density of the wavepacket propagates at a velocity of $\Delta\omega/\Delta\beta$, or the group velocity.

Exercise 8.42

In a dispersive material with $n = n_o(1 + \omega/\omega_o)$, determine the phase and group velocities for $\omega = \omega_o$. [Hint: $\beta = 2\pi n/\lambda$, $c = f\lambda$.]

Ans. $v_p = c/2n_o$, $v_g = c/3n_o$.

Exercise 8.43

Show that group velocity can be expressed as

$$v_g = \frac{c}{n + \omega(dn/d\omega)}.$$

Review Questions

RQ 8.56   State dispersion in words.   [Fig. 8.26]

RQ 8.57   Define the group velocity.   [(8.177)]

RQ 8.58   Define the phase and group velocities in the ω–β diagram.   [Fig. 8.26]

8.5   Problems

Uniform plane waves in homogeneous medium

8.1   Find the electric and magnetic field phasors of the following waves:

(a) $\mathscr{E} = \mathbf{a}_x E_o\, e^{-\alpha z} \sin(\omega t - \beta z + \pi/4)$ in the material with $\hat{\eta} = \eta_o\, e^{j\pi/4}$.

(b) $\mathscr{E} = -\mathbf{a}_x E_o e^{-\alpha z} \cos(\omega t + \beta z)$ in the material with $\hat{\eta} = \eta_o e^{-j\pi/4}$.

(c) $\mathscr{E} = \mathbf{a}_y E_o \cos(kz) \cos(\omega t)$ in free space.

(d) $\mathscr{E} = (\mathbf{a}_y + 2\mathbf{a}_z) E_o \cos(\omega t - kx)$ in free space.

(e) $\mathscr{E} = \mathbf{a}_z E_o \sin(k_1 x) \cos(\omega t - k_2 y)$ in free space.

8.2 An electromagnetic plane wave in free space has an electric field amplitude of 5.3 [V/m] and propagates in the direction of $(2\mathbf{a}_x - \sqrt{3}\,\mathbf{a}_y)$. If the distance vector $\mathcal{R} = \sqrt{3}\,\mathbf{a}_x + 2\mathbf{a}_y + 3\mathbf{a}_z$ [m] is from one point to its nearest neighbor for a phase change of 2π, find an expression for $\mathscr{E}$.

8.3 A 200 [MHz] uniform plane wave propagates in a lossless medium, for which $\varepsilon_r = 12$ and $\mu_r = 5$. Determine (a) β, (b) λ, (c) υ_p, and (d) η.

8.4 An electromagnetic plane wave propagating in free space is characterized by the electric field phasor $\tilde{\mathbf{E}} = 6\,\mathbf{a}_x e^{-j3y-j4z}$. Determine

(a) the direction of propagation,

(b) the wavelength,

(c) the frequency, and

(d) $\tilde{\mathbf{H}}$.

8.5 An electromagnetic wave with $\tilde{\mathbf{H}} = (10 - j2)\,(3\mathbf{a}_y + j5\mathbf{a}_z)e^{-j20x}$ propagates through a lossless medium, for which $\varepsilon_r = 2.5$ and $\mu_r = 4$. Find $\tilde{\mathbf{E}}$.

8.6 A 300 [MHz] wave has $\tilde{\mathbf{H}} = (2\mathbf{a}_x - \mathbf{a}_z)e^{-j20y}$ [A/m] in a lossless medium. For $|\mathscr{E}| = 700$ [V/m], determine (a) η, (b) ε_r, and (c) μ_r of the medium.

8.7 Starting from the general expression $\tilde{\mathbf{E}} = \mathbf{a}_E E_o e^{-j\mathbf{k}\cdot\mathbf{r}}$ for a uniform plane wave travelling in a lossless medium with ε and μ, derive the following relations using the phasor form of Maxwell's equations:

(a) $\mathbf{k} \times \tilde{\mathbf{E}} = \omega\mu\tilde{\mathbf{H}}$,

(b) $\mathbf{k} \times (\mathbf{k} \times \tilde{\mathbf{E}}) = -\omega^2\mu\varepsilon\,\tilde{\mathbf{E}}$,

(c) $\mathbf{k} \cdot \tilde{\mathbf{E}} = 0$, and

(d) $\mathbf{k} \cdot (\mathbf{k} \times \tilde{\mathbf{E}}) = 0$.

8.8 For phasor $\tilde{\mathbf{E}} = E_o\,\rho^{-1}e^{-jkz}\,\mathbf{a}_\rho$ with $k = \omega\sqrt{\mu_0\varepsilon_0}$ in cylindrical coordinates, verify that $\tilde{\mathbf{E}}$ represents an electromagnetic wave in free space using (a) Maxwell's, and (b) Helmholtz's equations.

Poynting vector

8.9 For the waves stated in Problem 8.1, find the time-averaged power density.

8.10 Given that the electric field of a uniform plane wave, traveling in a lossless nonmagnetic medium, is $\mathscr{E} = 2\,\mathbf{a}_x \cos(6 \times 10^8 t - 5z)$ [V/m], determine

(a) ε_r,

(b) η,

(c) the time-averaged power density, and

(d) the total power entering a sphere of radius 1 [m], centered at the origin.

8.11   In a lossless dielectric with $\varepsilon_r = 1.6$ and an unknown permeability, an electromagnetic plane wave characterized by $\tilde{\mathbf{E}} = (4\mathbf{a}_z - j2\mathbf{a}_y)e^{j15x}$ [V/m] has $\langle \mathbf{S} \rangle = 30.6$ [mW/m^2]. Determine (a) η, (b) μ_r, and (c) $\tilde{\mathbf{H}}$.

8.12   In a lossless nonmagnetic dielectric, an electromagnetic wave has an electric field amplitude of 56 [V/m] and time-average power density of 7.2 [W/m^2]. Determine the phase velocity of the wave.

8.13   A typical laser pointer is rated at 1 [mW], and emits laser light with a radius of 1 [mm]. By assuming that the light is uniform over its cross section, determine its electric field.

8.14   A short current element $I\,d\ell$ placed at the origin in free space generates a spherical wave with $\tilde{\mathbf{E}} = j(30\beta I\,d\ell/R)\sin\theta\,e^{-j\beta R}\,\mathbf{a}_\theta$. Find $\tilde{\mathbf{H}}$, and show that $\langle \mathbf{S} \rangle$ is proportional to ω^2.

Wave polarization

8.15   For a wave in free space with $\tilde{\mathbf{E}} = (E_o\mathbf{a}_x + jE_o\mathbf{a}_y)e^{jkz}$, determine (a) the direction of wave propagation, and (b) the polarization state.

8.16   Resolve a linearly polarized wave with $\tilde{\mathbf{E}} = (E_{ox}\mathbf{a}_x + E_{oy}\mathbf{a}_y)e^{-jkz}$ into two circularly polarized waves.

8.17   A thin quartz plate exhibits **optical activity**, such that a right-hand circularly polarized wave at normal incidence undergoes a phase lead of θ_o [rad] at the exit surface, whereas a left-hand circularly polarized wave undergoes a phase lag of θ_o [rad]. For a linearly polarized wave with $\tilde{\mathbf{E}} = E_o\mathbf{a}_x e^{-jkz}$ incident normally on the plate, determine its polarization state at the exit surface.

8.18   A linearly polarized wave with $\tilde{\mathbf{E}} = E_o\mathbf{a}_y e^{-jkz}$ is normally incident on a quarter-wave plate, whose *slow axis* is at an angle of 30° to the x-axis, as shown in Fig. 8.27. If a wave with $\mathbf{E}$ parallel to the slow axis is delayed by 90° in time phase relative to a wave with $\mathbf{E}$ perpendicular to that axis, determine the polarization state of the transmitted wave.

8.19   Given $\mathscr{E} = 2\,\mathbf{a}_x\sin(\omega t - kz + 30^o) + 3\,\mathbf{a}_y\cos(\omega t - kz)$, (a) sketch the locus of the tip of $\mathscr{E}$, and determine the direction of rotation. (b) What must be changed

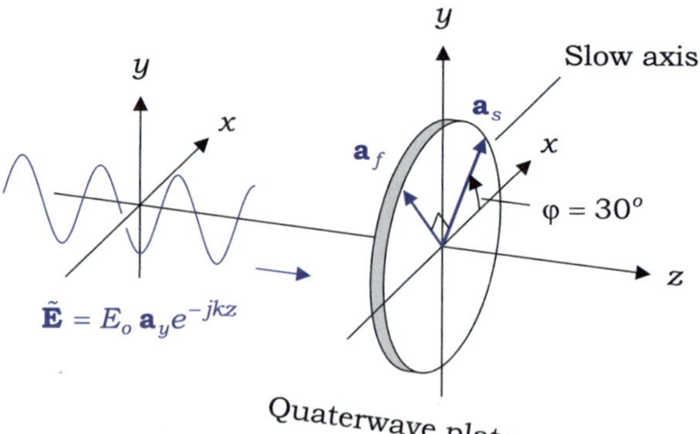

Fig. 8.27 Quarter-wave plate (Problem 8.18)

Fig. 8.28 Elliptical locus traced by $\tilde{\mathbf{E}}$ (Problem 8.20)

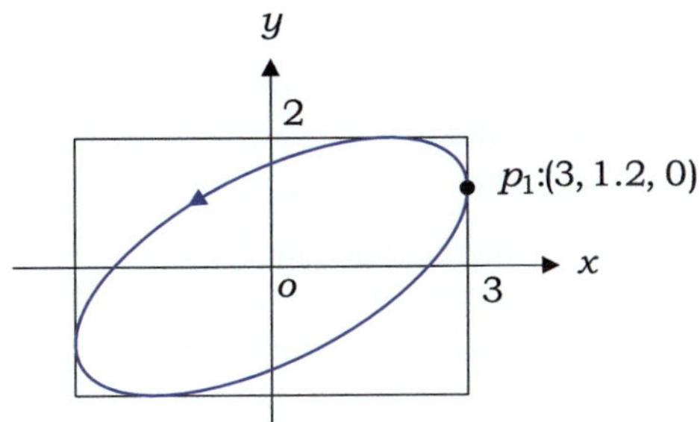

in $\mathscr{E}$ such that the new locus, including the direction of rotation, is a mirror image of the original locus with respect to the y-axis.

8.20 When a right-hand elliptically polarized wave propagates in the $+z$-direction in free space, the tip of its electric field vector traces an ellipse in the $z = 0$ plane, as shown in Fig. 8.28. Find an expression for $\tilde{\mathbf{E}}$.

Dielectrics with damping loss

8.21 For a plane wave of $f = 1$ [GHz] propagating in the $+z$-direction in a low-loss dielectric with $\varepsilon = (2 - j0.3)\varepsilon_0$, $\mu = 1.2\mu_0$, and $\sigma = 0$, compare (a) the amplitudes, and (b) the phases of $\mathscr{E}$ at points p_1: (2, 3, 5) and p_2: (−4, 7, 8).

8.22 A 100 [MHz] uniform plane wave traveling in a nonmagnetic dielectric is described by $\tilde{\mathbf{E}} = 2.4\,\mathbf{a}_x e^{-0.1z} e^{-j6z}$. Determine ε_r', ε_r'', and $\tilde{\mathbf{H}}$.

8.23 An electromagnetic plane wave propagating in a nonmagnetic dielectric is characterized by $\tilde{\mathbf{E}} = (10\mathbf{a}_x + j5\mathbf{a}_y)e^{-3z-j4z}$ for $\omega = 5 \times 10^8$. Find $\tilde{\mathbf{H}}$.

8.24 If the magnetic field of an electromagnetic wave lags behind the electric field by 30° in time phase in a dielectric with $\varepsilon = \varepsilon' - j\varepsilon''$, determine the β/α ratio.

8.25 Check whether the following numerical values are relevant to an electromagnetic wave traveling in a lossy medium: (a) $\gamma = 0.2 + j1.8$ [m^{-1}] and $\hat{\eta} = 400 + j50$ [Ω], and (b) $\gamma = 9.8 + j10$ [m^{-1}] and $\hat{\eta} = 0.102 + j0.1$ [Ω].

8.26 For an electromagnetic wave traveling in the $+z$-direction in a nonmagnetic dielectric with $\varepsilon_r' = 2.53$ and $\tan \xi = 0.03$, the electric field at $z = 0.5$ [m] is given by $\mathscr{E} = 400\,\mathbf{a}_x \sin(2\pi \times 10^8 t - \pi/8)$. Find $\mathscr{E}$ everywhere.

8.27 The time-averaged power carried by a wave of $\omega = 3 \times 10^9$ [rad/s] is reduced by 10 [dB] over a distance of 32.3 [cm] along the direction of wave propagation in a nonmagnetic dielectric. If $\tilde{\mathbf{H}}$ lags behind $\tilde{\mathbf{E}}$ by 0.22 [rad] in time phase, determine $\hat{\eta}$ of the material.

Dielectrics with low conductivity

8.28 Two nonmagnetic materials with $\varepsilon_r = 2.25$ have different conductivities, $\sigma = 0$ and $\sigma = 50$ [S/m]. For a wave with a frequency of 100 [MHz] travelling in the materials individually, compare (a) β, (b) λ, and (c) υ_p.

8.29 In a material with real ε and μ_0, an electromagnetic wave is characterized by an electric field $\mathscr{E}(z, t) = \mathbf{a}_x 967.1 e^{-\alpha z} \cos(1.5 \times 10^9 t - \beta z)$ and a magnetic field $\mathcal{H}(z, t) = \mathbf{a}_y 10 e^{-\alpha z} \cos(1.5 \times 10^9 t - \beta z - 0.7194)$. Determine (a) $\hat{\eta}$, (b) ε_r, (c) σ, (d) α, and (e) β.

8.30   A 1 [GHz] electromagnetic wave traveling in a nonmagnetic medium with a real ε has $\alpha = 20.32$ [Np/m] and $\beta = 38.85$ [rad/m]. Determine

   (a)   the loss tangent,
   (b)   the dielectric constant, and
   (c)   the intrinsic impedance.

8.31   In a lossy dielectric with $\varepsilon_r = 1.5$ and $\mu_r = 1$, the electromagnetic plane wave has $\beta = 120$ [rad/m] at $f = 50$ [MHz]. Determine the (a) loss tangent, and (b) conductivity.

8.32   For an electromagnetic wave with $\tilde{\mathbf{E}} = \mathbf{a}_x 300 e^{-0.21z - j(2.2z)}$ [V/m] travelling in a medium with $\hat{\eta} = 220 + j21$ [Ω], find

   (a) $\langle \mathbf{S} \rangle$,
   (b)   ohmic power-loss per unit volume at $z = 0.5$ [m] using $-\nabla \cdot \langle \mathbf{S} \rangle$, and
   (c)   repeat (b) using $(1/2)\text{Re}[\tilde{\mathbf{E}} \cdot \tilde{\mathbf{J}}^*]$. [Hint: $\sigma = \text{Re}[\gamma/\hat{\eta}]$.]

Good conductors

8.33   The power density of a 30 [MHz] wave is reduced by 10 [dB] over a distance of 0.15 [mm] in a good conductor. Assuming that the conductor is nonmagnetic, determine (a) σ, and (b) $\hat{\eta}$.

8.34   For a 100 [MHz] wave propagating in a good conductor with $\sigma = 10^7$ [S/m] and $\mu = \mu_0$, determine (a) δ, (b) υ_p, and (c) $\hat{\eta}$.

8.35   In a good conductor, an electromagnetic wave with $\lambda = 2$ [mm] propagates at a phase velocity $\upsilon_p = 1.6 \times 10^5$ [m/s]. Determine the conductivity of the material.

8.36   For gold with conductivity $\sigma = 4.1 \times 10^7$ [S/m], determine the frequency at which the skin depth is 1 [μm].

8.37   Given that the skin depth of brass at 200 [MHz] is 9 [μm], determine its conductivity.

8.38   For a copper wire with a diameter of 1 [mm] and length of 100 [m], compare the ac-resistance with the DC resistance at 10 [MHz].

8.39   When a wave with $\tilde{\mathbf{E}} = \mathbf{a}_x E_o e^{-jkz}$ in air is incident normally on a good conductor with ε_0, μ_0, and σ, occupying the region $z \geq 0$, show that the induced current may be considered as a surface current with density $\tilde{\mathbf{J}}_s = 2E_o/\eta_o$. [Hint: Eqs. (8.97) and (8.108).]

8.40   For a wave with an electric field phasor $\tilde{\mathbf{E}} = \mathbf{a}_x E_o e^{-(1+j)z/\delta}$ traveling in a good conductor with skin depth δ and conductivity σ, express (a) $\tilde{\mathbf{H}}$, (b) $\tilde{\mathbf{J}}$, (c) $\langle \mathbf{S} \rangle$, and (d) show that $-d|\langle \mathbf{S} \rangle|/dz = (1/2)\text{Re}[\tilde{\mathbf{E}} \cdot \tilde{\mathbf{J}}^*]$.

Normal incidence

8.41   The uniform plane wave with $\mathcal{E}^i = 10\,\mathbf{a}_x \cos(3 \times 10^{15} t - 10^7 z)$ travelling in region 1 ($z < 0$) normally impinges on a lossless nonmagnetic dielectric in region 2 ($z \geq 0$). If the phase constants are such that $\beta_2 = 1.5\,\beta_1$, determine

    (a)   the reflection coefficient Γ,

    (b)   the standing-wave ratio S,

    (c)   the ratio of the reflected power to the incident power, and

    (d)   find expressions for $\mathscr{E}^r$ and $\mathscr{E}^t$.

8.42   A plane wave in the air is normally incident onto a medium in the region $z \geq 0$ and reflected back into the air, producing a total electric field of $\tilde{\mathbf{E}} = (5.880 + j1.192)\,\mathbf{a}_x e^{-j1.6z} + (0.964 + j7.942)\,\mathbf{a}_x \cos(1.6z + 1.25)$. Determine the reflection coefficient and intrinsic impedance of the material.

8.43   A uniform plane wave is normally incident on the interface between two lossless media. If the power densities of the reflected and transmitted waves are equal, determine (a) the η_2/η_1 ratio, and (b) the standing-wave ratio.

8.44   For the normal incidence of a uniform plane wave onto a lossless dielectric-dielectric interface, verify that even if the transmission coefficient is greater than unity, the power transmitted across the interface is always less than that of the incident wave.

8.45   A uniform plane wave with power density 5 [W/m^2] propagates in air in the $+z$-direction and is reflected from a lossless nonmagnetic dielectric in region $z \geq 0$. If the standing-wave ratio is measured as $S = 4.0$, determine (a) Γ, (b) τ, (c) η_2, and (d) power density of the transmitted wave.

8.46   The expressions for Γ and τ in Eqs. (8.107) and (8.108) are also valid for a wave in air with $\tilde{\mathbf{E}}^i = 2\mathbf{a}_x e^{-j4z}$, which is normally incident on a lossy dielectric ($\varepsilon_r = 2$, $\mu = \mu_0$, and $\tan \xi = 1.732$) in the $z \geq 0$ region. (a) Determine α, β, and η in the material and express $\tilde{\mathbf{E}}^r$ and $\tilde{\mathbf{E}}^t$. (b) Determine $\langle \mathbf{S} \rangle$ for the three waves at the interface. Is the energy conservation satisfied?

8.47   Modify the expression for the law of energy conservation in Eq. (8.111), if the interface is between two lossy materials with complex intrinsic impedances $\hat{\eta}_1$ and $\hat{\eta}_2$.

8.48   Red light with a wavelength of 630 [nm] in air is normally incident on a good conductor with $\sigma = 5 \times 10^7$ [S/m] occupying region $z \geq 0$. Determine (a) $|\Gamma|$, and (b) fraction of incident power transmitted to the conductor.

8.49   A 500 [MHz] wave in air is incident normally on a material with $\varepsilon = 1.5\,\varepsilon_0$, $\mu = \mu_0$, and a relatively low conductivity of $\sigma = 10$ [S/m]. Determine the ratio of the transmitted power to the incident power.

8.50   The standing-wave ratio is $S = 4$ in air because a plane wave of amplitude $E_o = 30$ [V/m] is normally incident on a lossy medium in region $z \geq 0$. Find the total power dissipated in the bar region with a cross-sectional area of 1 [m] $\times$ 1 [m], which extends from $z = 0$ to $z = \infty$.

8.51   The three materials, as shown in Fig. 8.11, are lossless. The thickness of medium 2 is $\ell = \lambda_2/4$, where λ_2 is the wavelength measured in medium 2. Show that $\Gamma = 0$ at the first interface at $z = 0$ if $\eta_2 = (\eta_1 \eta_3)^{1/2}$.

8.52   Referring to Fig. 8.11, under the condition of $\eta_2 = (\eta_1 \eta_3)^{1/2}$, as in Problem 8.51, (a) show that $\Gamma_1 = -\Gamma_2$, and (b) find the standing-wave ratio in medium 2, assuming $\eta_1 < \eta_3$.

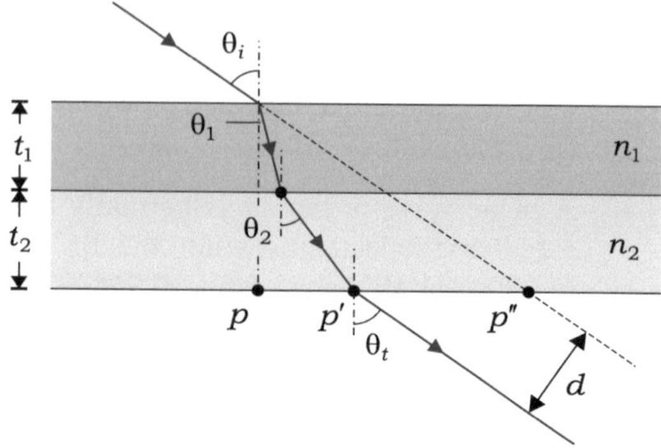

Fig. 8.29 Multiple refractions of a beam in double slabs (Problem 8.54)

8.53 The three media shown in Fig. 8.11 correspond to a thin aluminum foil with $\sigma = 3.82 \times 10^7$ [S/m] and $\ell = \delta$ (skin depth) placed in air. When blue light with $\lambda = 0.5$ [μm] is normally incident on the foil, by assuming that $\varepsilon = \varepsilon_0$ and $\mu = \mu_0$ in the three media, determine the ratio of the transmitted to the incident electric field amplitude.

Oblique incidence

8.54 An infinite slab consists of two dielectrics and is placed in free space, as shown in Fig. 8.29. When a light beam is incident on the slab at an angle θ_i, (a) show that the transmitted beam is parallel to the incident beam, and (b) find the beam separation d.

8.55 A plane wave with $\tilde{\mathbf{E}}^i = -300\,\mathbf{a}_y\,e^{-j(10x+15z)}$ in free space impinges on the surface of the dielectric with $n = 1.30$, filling the $z \geq 0$ region. Find the

    (a) angles of incidence, reflection, and transmission,
    (b) wavevectors of the three waves,
    (c) reflection and transmission coefficients, and
    (d) expressions for $\tilde{\mathbf{E}}^r$ and $\tilde{\mathbf{E}}^t$.

8.56 A wave with parallel polarization $\tilde{\mathbf{E}}^i = 100(-4\mathbf{a}_x + 3\mathbf{a}_z)e^{-j(9x+12z)}$ in free space impinges on a dielectric with $n = 1.30$ in the $z \geq 0$ region, find the

    (a) angles of incidence, reflection, and transmission,
    (b) wavevectors of the three waves,
    (c) wave polarizations of the reflected and transmitted waves,
    (d) reflection and transmission coefficients, and
    (e) expressions for $\tilde{\mathbf{E}}^r$ and $\tilde{\mathbf{E}}^t$.

8.57 For a plane wave with $\tilde{\mathbf{E}}^i = (2\mathbf{a}_x + 2\mathbf{a}_y - 4\mathbf{a}_z)e^{-j(4x+8y+6z)}$ in free space impinging on the $z = 0$ plane, resolve it into two component waves with perpendicular and parallel polarizations.

8.58 If the $z = 0$ plane stated in Problem 8.57 is perfectly conductive, resolve the reflected wave into two component waves with perpendicular and parallel polarizations.

8.59   A parallel-polarized plane wave with $\tilde{\mathbf{E}}^i = (2\mathbf{a}_x - \mathbf{a}_z)30\sqrt{5}\,e^{-j(10x+20z)}$ is obliquely incident from the air on a perfect conductor occupying the region $z \geq 0$. Find

    (a)   expressions for $\tilde{\mathbf{H}}^i$, $\tilde{\mathbf{E}}^r$, and $\tilde{\mathbf{H}}^r$,
    (b)   total $\tilde{\mathbf{E}}$ and $\tilde{\mathbf{H}}$ on the conductor surface, and
    (c) $\tilde{\mathbf{J}}_s$ induced on the conductor surface.

8.60   A plane wave with $\tilde{\mathbf{E}}^i = 4\mathbf{a}_E\,e^{-j(x+\sqrt{3}z)}$ in free space impinges obliquely on a dielectric with $\varepsilon_r = 2.25$ occupying the $z \geq 0$ region. If the amplitude of the perpendicular polarization is 3 [V/m], express (a) $\tilde{\mathbf{E}}^i$, and (b) $\tilde{\mathbf{E}}^r$.

8.61   For a plane wave incident from air on crown glass with $n = 1.52$, determine (a) the Brewster angle, and (b) the reflection coefficient for a wave with perpendicular polarization incident at the Brewster angle.

8.62   A plane wave in air is incident on fused quartz ($n = 1.46$) occupying the $z \geq 0$ region and experiences no reflection. Assuming that the xz-plane is the plane of incidence, find the directions of (a) the propagation of the incident wave, and (b) the polarization vectors of the incident and transmitted waves.

8.63   A left-hand circularly polarized wave in free space is incident at the Brewster angle on a dielectric with $n = 1.54$ occupying the $z \geq 0$ region. Determine the polarization states of the reflected and transmitted waves by taking the xz-plane as the plane of incidence.

8.64   A plane wave with $\hat{\mathbf{E}} = (6\mathbf{a}_x - 8\mathbf{a}_z)e^{-j4\mathbf{a}_x - j3\mathbf{a}_z}$ in air is incident at the Brewster angle on a lossless dielectric with $n = 1.333$ occupying the half-space $z \geq 0$. Show that the incident power is completely transmitted into the dielectric.

Total internal reflection

8.65   A right-angle prism is made of crown glass ($n = 1.52$), as shown in Fig. 8.30. Determine the range of incidence angles over which the internal wave undergoes total internal reflection at the hypotenuse.

8.66   The hypotenuse of a right-angle prism with refractive index n_1 is coated with material with n_2, as shown in Fig. 8.31. Show that if $n_1 > \sqrt{2}$, the beam

Fig. 8.30 Right-angle prism
(Problem 8.65)

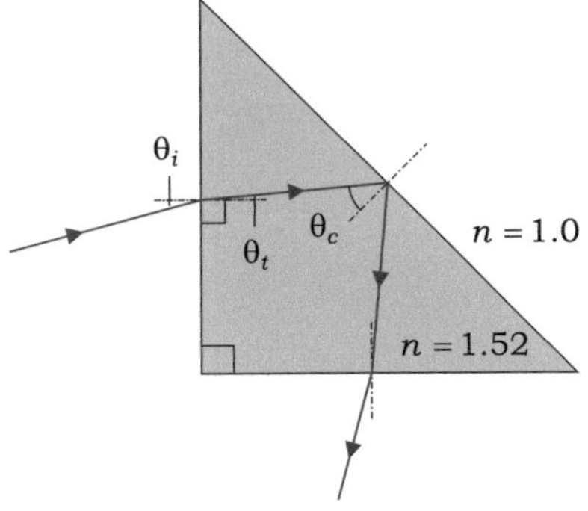

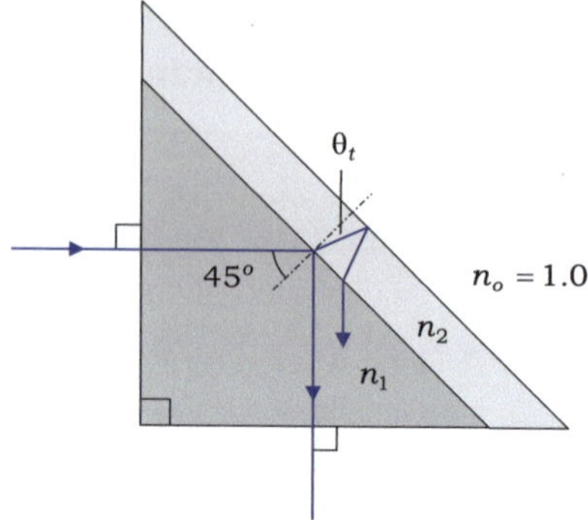

Fig. 8.31 Coating on a right-angle prism (Problem 8.66)

incident at 45° always undergoes total internal reflection, either at the n_1n_2-interface or n_2n_o-interface.

8.67 An evanescent wave with $\tilde{\mathbf{E}}^t = \mathbf{a}_y E_o^t e^{-\alpha_e z} e^{-j\beta_e x}$ is formed in region $z \geq 0$ because of the total internal reflection occurring in the opposite region. (a) Express $\tilde{\mathbf{H}}^t$ using the intrinsic impedance η_o. (b) Show that no power is transmitted to the $z \geq 0$ region.

8.68 Assuming the conductivity is independent of frequency in a nonmagnetic good conductor, find (a) the phase, and (b) the group velocity in the material.

Chapter 9
Transmission Line

Thus far, our discussion has focused on electromagnetic plane waves propagating through unbounded media and those incident on plane interfaces. A uniform plane wave is an unbounded and unguided wave in the sense that its electric field fills all spaces and its electromagnetic energy spreads out over the entire space. Although such unbounded waves may be useful for broadcasting radio and TV signals, the transfer of electromagnetic power and information from one location to another is inefficient.

A transmission line can provide an efficient way of transmitting electromagnetic power from a source to a load, which is fundamentally two parallel conductors connecting the source and the load. Common transmission lines include coaxial cables, parallel-plate lines, and two-wire lines (Fig. 9.1). Transmission lines support *transverse electromagnetic* (TEM) waves, which are characterized by electric and magnetic fields lying in a plane perpendicular to the line. TEM waves on a transmission line exhibit the same propagation characteristics as uniform plane waves propagating in an unbounded medium.

Although the electric and magnetic fields on the transmission line vary over time, their spatial distributions over the transverse plane are identical to those obtained under static conditions. For example, considering the cylindrical and translational symmetries of an infinite coaxial cable along the z-axis, the static electric field of the cable is expressed as $\mathbf{E}_s = E_x(x, y)\,\mathbf{a}_x + E_y(x, y)\,\mathbf{a}_y$. Extending this to the time-harmonic case, the electric field phasor in the cable can be expressed as $\tilde{\mathbf{E}} = \mathbf{E}_s e^{-jkz}$. This can be shown to satisfy the vector Helmholtz equation, and thus constitutes a TEM wave propagating along the cable. The Poynting vector ensures that time-averaged power flows along and is guided by the cable with no energy loss.

When a TEM wave travels on a transmission line that consists of perfectly conducting parallel conductors, the electric field perpendicular to and magnetic field tangential to the conductor surface induce electric surface charges and currents along the conductor, respectively, in accordance with the boundary conditions given in Eqs. (6.51b) and (6.51c). It is convenient to use a cosine curve to describe the spatial variation of these charges and currents on the transmission line, such that the peaks

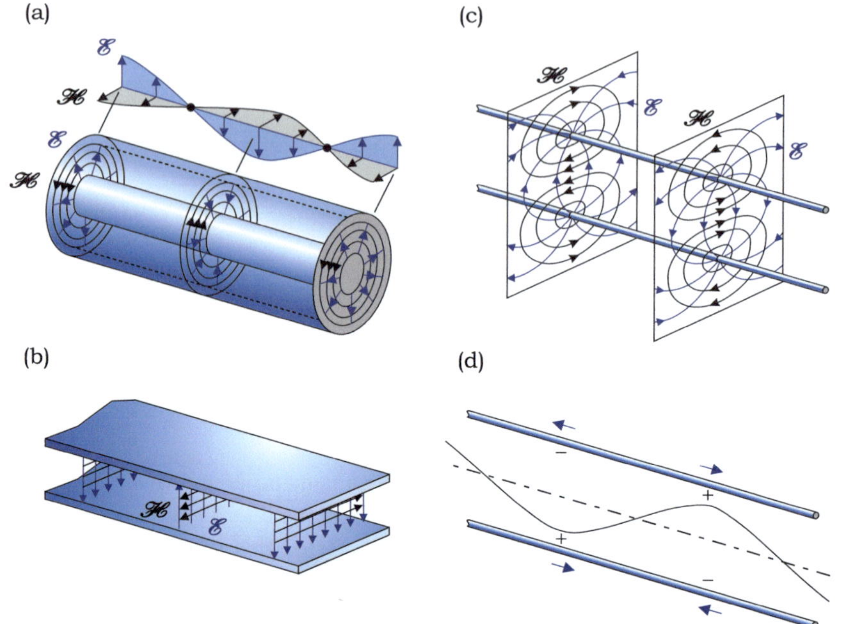

Fig. 9.1 Three common transmission lines: **a** Coaxial line, **b** Parallel-plate line, and **c** Two-wire line. **d** Induced charge and current on a transmission line

represent the maximum density of positive charges or the maximum forward current at their respective positions, whereas the valley represents the maximum density of negative charges or the maximum backward current (Fig. 9.1d). As the wave advances along the line, the induced charge and current constitute two separate waves that travel along the line. As $\mathscr{E}$ and $\mathscr{H}$ are out of phase in unbounded lossy media, the charge and current waves are generally out of phase in lossy transmission lines.

We can use either electromagnetic or electric circuit theory to analyze the wave behavior on a transmission line. Whereas the former is concerned with electric and magnetic fields, the latter deals with scalar quantities, such as voltage and current, which are much easier to work with than vector fields. In circuit theory, the electric field and surface charge on a transmission line are translated into the voltage and charge of the capacitor formed by the two conductors, whereas the magnetic field and surface current are translated into the magnetic flux linkage and current of the inductor formed by the two parallel conductors.

The current wave travelling on a transmission line is distinct from the current flowing in an electric circuit operating at low frequencies. Consider the case of a short positive pulse launched on a transmission line by a generator on the left end. A negative pulse provides contrast to the positive pulse (Fig. 9.2a). More specifically, launching a positive pulse is equivalent to launching equal amounts of positive and negative charges on the upper and lower conductors, respectively. Subsequently, the propagation of a pulse on the line may be considered as a process of charging and

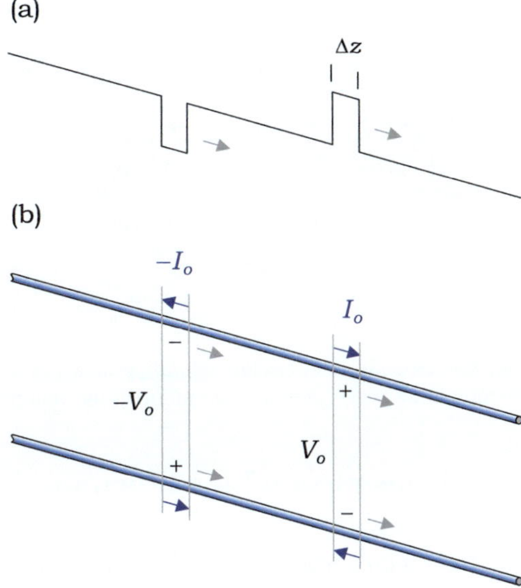

Fig. 9.2 **a** A positive pulse is followed by a negative pulse on the transmission line. **b** + and −, positive and negative charges; Blue arrows, localized currents

discharging successive capacitors of the line by means of the current flowing in successive inductors of the line. In other words, the capacitor near the leading edge of the positive pulse is charged, whereas that near the trailing edge is discharged. Accordingly, positive charges on the upper conductor and negative charges of equal amount on the lower conductor move together to the right, even if the right end of the line is opened (Fig. 9.2b). Moreover, the localized currents flow in opposite directions in the upper and lower conductors because of the opposite polarities of the two charges.

To examine the difference between a transmission line and low-frequency electric circuit, we consider two identical circuits of the same length, $\ell = 15$ [cm], as shown in Fig. 9.3. When they are connected to the same generator voltage but with different frequencies of 100 [kHz] and 2.5 [GHz], the voltage waves are given by $v_1 = V_o \cos(2\pi \times 10^5 t - 2\pi \times 10^{-3} z/3)$ and $v_2 = V_o \cos(5\pi \times 10^9 t - 50\pi z/3)$, respectively, by assuming that the reflection from the load is ignored and the wave velocity is the same as the velocity of light in vacuum. For example, at time $t = 0$, the load voltages are $v_1 = V_o \cos(-2\pi \times 10^{-3} \times 0.15/3) = 0.99999995 V_o$ and $v_2 = 0$, and the generator voltages are V_o in the two circuits. For practical purposes, the voltage may be considered to be constant in a low-frequency circuit. Therefore, the voltage oscillates with the same amplitude and phase at all the points along the circuit. Moreover, a small change in the generator voltage immediately takes effect at the load. In contrast, in a high-frequency circuit, the wavelength is comparable to the length of the circuit. Therefore, the time-harmonic voltages at different points in the circuit are generally out of phase and the circuit should be treated as a transmission line. The distinctive features of a transmission line are the time delay of the time-harmonic voltages at points on the line, and the reflection of the voltage and current waves at the load, which are discussed in the following section.

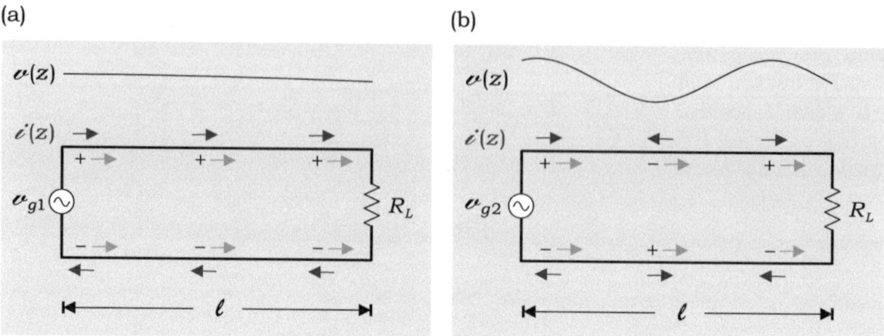

Fig. 9.3 Two identical circuits operating at **a** low frequency and **b** high frequency. $+$ and $-$, charges; Gray arrow, direction of motion of the charge; Blue arrow, localized current

9.1 Transmission Line Equations

We begin with a transmission line consisting of two infinitely long parallel conductors. Recalling from Chaps. 3 and 5, any two conductors form a capacitor and any current-carrying conductor is accompanied by a magnetic flux linkage. Accordingly, a transmission line has a capacitance and inductance that are proportional to the line length. Because the electric and magnetic fields in the transverse plane are the same as if the line is under static conditions, the line capacitance C can be obtained by assuming that the two conductors are maintained at a constant potential and the line inductance L by assuming that the conductors carry a steady current. In general, conductors have finite conductivities. Moreover, the insulating materials between conductors have nonzero conductivities. Therefore, a transmission line may have a series resistance R and shunt conductance G as well as capacitance and inductance, as illustrated in Fig. 9.4a. It should be noted that $G \neq 1/R$.

When a short pulse of width Δz travels along the transmission line, the associated charge and current are confined to a small region of width Δz (Fig. 9.2). The interrelationship between the charge and current is governed by the capacitance and inductance of the small part as well as the resistance and conductance. Therefore, it is most convenient to introduce the ***distributed parameters*** $\overline{R}, \overline{L}, \overline{G}$, and $\overline{C}$, which are measured in units of $[\Omega/\text{m}]$, $[\text{H/m}]$, $[\text{S/m}]$, and $[\text{F/m}]$, respectively. The propagation of a pulse can then be described well in terms of $\overline{R}\Delta z, \overline{L}\Delta z, \overline{G}\Delta z$, and $\overline{C}\Delta z$ (Fig. 9.4b). It should be noted that $\overline{R}\Delta z$ and $\overline{L}\Delta z$ are the total resistance and inductance encountered by the currents flowing in the upper and lower conductors.

Transmission-line equations are a pair of first-order partial differential equations that describe wave behavior on a transmission line. Consider a line segment of length Δz as shown in Fig. 9.5. Hereafter, for simplicity, the bar at the top of the distributed parameters is omitted. In the figure, $v(z, t)$ and $i(z, t)$ represent the instantaneous voltage and current at the left end of the segment, respectively, and $v(z + \Delta z, t)$ and $i(z + \Delta z, t)$ represent those at the right end, respectively. Applying Kirchhoff's voltage law to the segment, we obtain

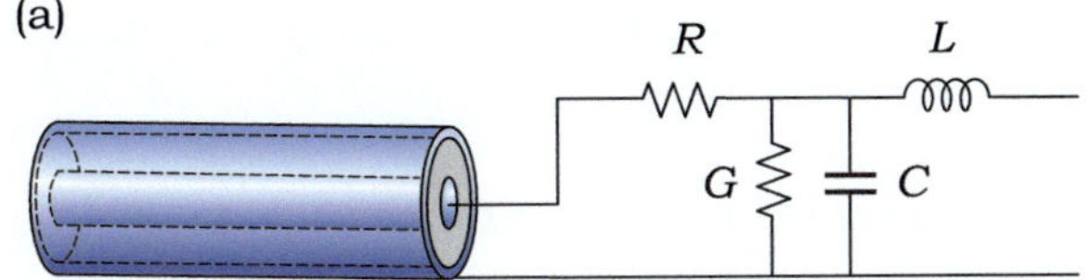

Fig. 9.4 Transmission line parameters. **a** Equivalent lumped-element circuit, and **b** Equivalent distributed-parameter network

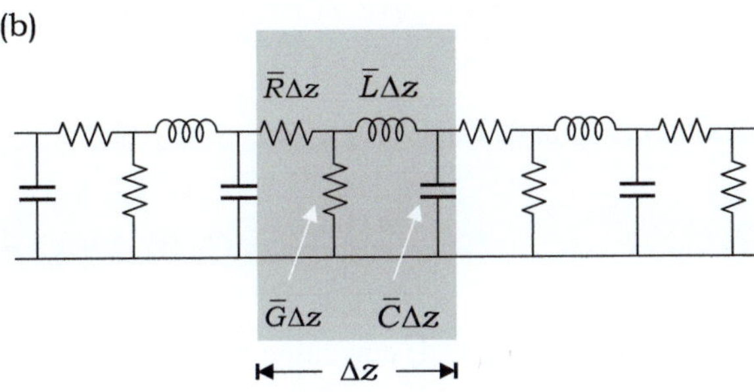

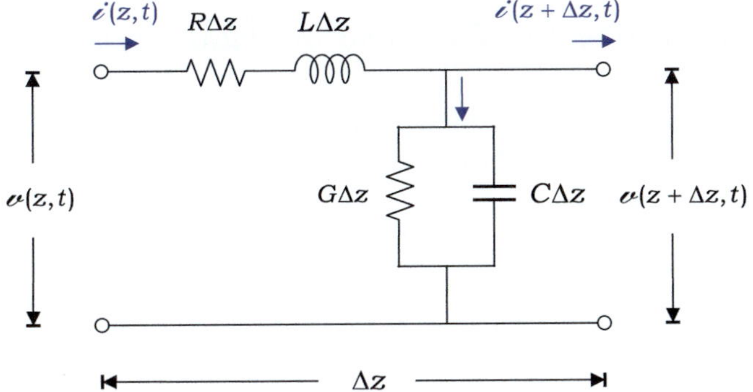

Fig. 9.5 Equivalent lumped-element circuit for a line segment of length Δz

$$v(z, t) = (R\Delta z)\, i(z, t) + (L\Delta z)\frac{\partial i(z, t)}{\partial t} + v(z + \Delta z, t) \qquad (9.1a)$$

This can be rearranged such that

$$-\frac{v(z + \Delta z, t) - v(z, t)}{\Delta z} = R\, i(z, t) + L\frac{\partial i(z, t)}{\partial t} \qquad (9.1b)$$

Taking the limit of Eq. (9.1b) as Δz approaches zero, we obtain

$$\boxed{-\frac{\partial v(z, t)}{\partial z} = R\, i(z, t) + L\frac{\partial i(z, t)}{\partial t}} \qquad (9.2)$$

Next, by applying Kirchhoff's current law to the upper node of the circuit, we obtain

$$i(z, t) = i(z + \Delta z, t) + (G\Delta z)v(z + \Delta z, t) + (C\Delta z)\frac{\partial v(z + \Delta z, t)}{\partial t} \quad (9.3a)$$

This can also be rearranged such that

$$-\frac{i(z + \Delta z, t) - i(z, t)}{\Delta z} = G\,v(z + \Delta z, t) + C\frac{\partial v(z + \Delta z, t)}{\partial t} \quad (9.3b)$$

Taking the limit of Eq. (9.3b) as $\Delta z \to 0$ yields

$$\boxed{-\frac{\partial i(z, t)}{\partial z} = G\,v(z, t) + C\frac{\partial v(z, t)}{\partial t}} \quad (9.4)$$

Equations (9.2) and (9.4) constitute transmission-line equations.

9.1.1 *Transmission Line Equations in Phasor Form*

Under sinusoidal steady-state conditions, a transmission line supports time-harmonic voltage and current, which can be conveniently expressed in phasor form as

$$v(z, t) = \mathrm{Re}\left[\tilde{V}(z)\,e^{j\omega t}\right] \quad (9.5a)$$

$$i(z, t) = \mathrm{Re}\left[\tilde{I}(z)\,e^{j\omega t}\right] \quad (9.5b)$$

where $\tilde{V}(z)$ and $\tilde{I}(z)$ are the voltage and current phasors, respectively, which are generally complex and depend only on space coordinate z. Substituting Eqs. (9.5a, b) into Eqs. (9.2) and (9.4), and noting that the time derivatives of v and i are equivalent to multiplications of the corresponding phasors by $j\omega$, the transmission-line equations can be expressed in phasor form as follows:

$$\boxed{-\frac{\partial \tilde{V}}{\partial z} = (R + j\omega L)\tilde{I}} \quad (9.6a)$$

$$\boxed{-\frac{\partial \tilde{I}}{\partial z} = (G + j\omega C)\tilde{V}} \quad (9.6b)$$

These can be combined to form the following second-order differential equations:

$$\frac{\partial^2 \tilde{V}}{\partial z^2} = (R + j\omega L)(G + j\omega C)\tilde{V} \tag{9.7a}$$

$$\frac{\partial^2 \tilde{I}}{\partial z^2} = (G + j\omega C)(R + j\omega L)\tilde{I} \tag{9.7b}$$

These can be rewritten in a more compact form as

$$\boxed{\frac{\partial^2 \tilde{V}}{\partial z^2} - \gamma^2 \tilde{V} = 0} \tag{9.8a}$$

$$\boxed{\frac{\partial^2 \tilde{I}}{\partial z^2} - \gamma^2 \tilde{I} = 0} \tag{9.8b}$$

where the **propagation constant** γ is defined as

$$\boxed{\begin{aligned} \gamma &= \sqrt{(R + j\omega L)(G + j\omega C)} \\ &\equiv \alpha + j\beta \end{aligned}} \tag{9.9}$$

Because the cosine function is a solution to Eqs. (9.8a) and (9.8b), the voltage and current both take the form of a time-harmonic wave traveling on a transmission line. The propagation characteristics of the wave are determined by the propagation constant γ, whose real and imaginary parts are the **attenuation constant** α [Np/m] and the **phase constant** β [rad/m], respectively. From the analogy between Eqs. (9.8a) and (9.8b) and the Helmholtz equations for $\tilde{\mathbf{E}}$ and $\tilde{\mathbf{H}}$, many relations of $\tilde{\mathbf{E}}$ and $\tilde{\mathbf{H}}$ derived previously in Chap. 8 can also be applied to $\tilde{V}$ and $\tilde{I}$ on a transmission line.

9.1.2 The Relationship Between Parameters

The distributed parameters R, L, G, and C can be combined to obtain the propagation constant γ as expressed in Eq. (9.9). These depend on the conductor geometry and constitutive parameters of the material. Some parameters are related to each other such that one parameter can be determined from another using a simple relation without actually calculating it, which is derived from the conductor geometry and material parameters.

Because the transmission lines are composed of conductors with high conductivities, the series resistance can be ignored in the expression for γ in most cases. Accordingly, the waves on such transmission lines are TEM waves, and their behavior is the same as that of a uniform plane wave travelling in an unbounded dielectric. Under this condition, the propagation constant becomes

$$\gamma = \sqrt{j\omega L(G + j\omega C)} = j\omega\sqrt{LC}\sqrt{1 - j\frac{G}{\omega C}} \qquad \text{(transmission line)} \quad (9.10)$$

Recalling from Eq. (8.78), the propagation constant of a uniform plane wave traveling in an unbounded lossy dielectric is given by

$$\gamma = j\omega\sqrt{\mu\varepsilon}\sqrt{\left(1 - j\frac{\sigma}{\omega\varepsilon}\right)} \qquad \text{(dielectric)} \qquad (9.11)$$

Note that σ in Eq. (9.11) is the conductivity of the dielectric and not that of the conductors.

The ratio of the shunt conductance to the capacitance of the transmission line can be deduced from Eq. (4.61) as

$$\boxed{\frac{G}{C} = \frac{\sigma}{\varepsilon}} \qquad (9.12)$$

where σ and ε are the conductivity and permittivity of the material between conductors, respectively.

Next, comparing Eq. (9.10) with Eq. (9.11) and applying Eq. (9.12), we obtain

$$\boxed{LC = \mu\varepsilon} \qquad (9.13)$$

where ε and μ are the permittivity and permeability of the material between conductors, respectively.

The relations in Eqs. (9.12) and (9.13) depend on the constitutive parameters of the material between the conductors but not on the conductor geometry. Therefore, these can be conveniently used to determine the line parameters. For example, if the capacitance C of a transmission line is known, inductance L can be determined from Eq. (9.13), and vice versa.

Exercise 9.1

Which of the transmission-line parameters (R, L, G, and C) can determine whether the wave on the line is a TEM wave?

Ans. $R = 0$.

Exercise 9.2

Identify the loss tangent in Eq. (9.10).

Ans. $\tan\xi = G/\omega C$.

Exercise 9.3

Does L in Eq. (9.13) involve the internal inductance?

Ans. No. ($\mathbf{E} = 0$ inside conductors for TEM waves).

Review Questions

RQ 9.1	Distinguish between a transmission line and an electric circuit.	[Fig. 9.3]
RQ 9.2	Write the transmission line equations in instantaneous and phasor forms.	[(9.2)(9.4)(9.6a,b)]
RQ 9.3	Write the wave equations for $\tilde{V}$ and $\tilde{I}$ on the transmission line.	[(9.8a,b)]
RQ 9.4	Define the propagation constant on the transmission line.	[(9.9)]
RQ 9.5	Under what conditions does a transmission line support the TEM waves?	[(9.10)]
RQ 9.6	In what ways are the transmission line parameters related to constitutive parameters?	[(9.12)(9.13)]

9.2 Transmission Line Parameters

In the previous chapters, several line parameters for the three common transmission lines were obtained. If necessary, the other parameters can be obtained from the relations given in Eqs. (9.12) and (9.13).

9.2.1 Coaxial Transmission Line

An inner conductor of radius a is concentric with an outer cylinder of inner radius b, as shown in Fig. 9.6a. The space between conductors is filled with a dielectric with ε and μ. The capacitance per unit length is obtained using Eq. (3.152) such that

$$C = \frac{2\pi\varepsilon}{\ln(b/a)} \quad [\text{F/m}] \tag{9.14a}$$

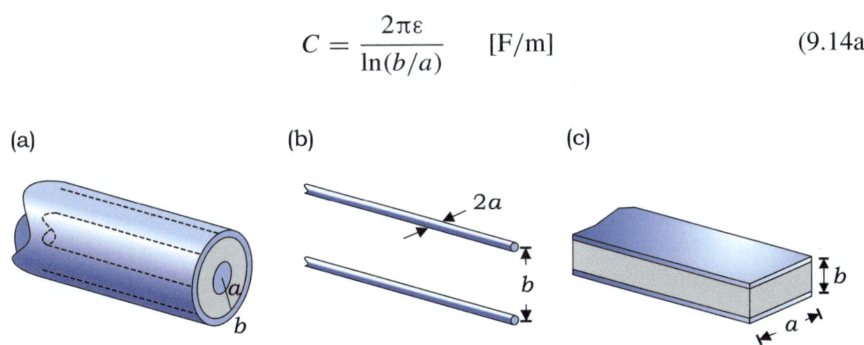

Fig. 9.6 Three most common transmission lines: **a** Coaxial line, **b** Two-wire line, and **c** Parallel-plate line

The external inductance per unit length can be obtained using Eq. (5.110g) or computed using Eqs. (9.13) and (9.14a):

$$L = \frac{\mu}{2\pi} \ln(b/a) \quad [\text{H/m}] \tag{9.14b}$$

The internal inductance can be ignored at high frequencies because the magnetic field hardly penetrates the conductor because of the extremely short skin depth.

Substitution of Eq. (9.14a) into Eq. (9.12) gives the shunt conductance per unit length:

$$G = \frac{2\pi\sigma}{\ln(b/a)} \quad [\text{S/m}] \tag{9.14c}$$

where σ is the conductivity of the dielectric in the space between conductors.

The ac-resistance, expressed by Eq. (8.99), allows us to compute the series resistance per unit length by assuming that the current is uniformly distributed over the skin depth. The cross-sectional areas of the current path are $S_i = 2\pi a\delta$ and $S_o = 2\pi b\delta$ for the inner and outer conductors, respectively. Thus, the series resistance per unit length is

$$R = \frac{1}{\sigma_c S_i} + \frac{1}{\sigma_c S_o} = \frac{1}{2\pi}\left(\frac{1}{a} + \frac{1}{b}\right)\sqrt{\frac{\pi f \mu_c}{\sigma_c}} \quad [\Omega/\text{m}] \tag{9.14d}$$

where σ_c and μ_c are the conductivity and permeability of the conductor, respectively, and are *unrelated* to the dielectric between the conductors. In Eq. (9.14d), it is assumed that the skin depth is much smaller than the radius of the inner conductor and the thickness of the outer cylinder. A nonzero series resistance implies that the conductivity σ_c is not infinite and that the longitudinal component of the electric field does not vanish on the conductor. Strictly speaking, a wave on such a transmission line is not a TEM wave.

9.2.2 Two-Wire Transmission Line

Two parallel conducting wires of radius a are held at a center-to-center distance b apart in a dielectric with ε and μ, as shown in Fig. 9.6b. From Eq. (3.154c), the capacitance per unit length is

$$C = \frac{\pi\,\varepsilon}{\cosh^{-1}(b/2a)} \quad [\text{F/m}] \tag{9.15a}$$

Substitution of Eq. (9.15a) into Eq. (9.13) gives the inductance per unit length:

$$L = \frac{\mu}{\pi} \cosh^{-1}(b/2a) \quad [\text{H/m}] \tag{9.15b}$$

If $b \gg a$, $\cosh^{-1}(b/2a)$ can be approximated as $\ln(b/a)$ and Eq. (9.15b) represents the same external inductance per unit length, as in Eq. (5.111a).

Substituting Eq. (9.15a) into Eq. (9.12) yields the shunt conductance per unit length:

$$G = \frac{\pi \sigma}{\cosh^{-1}(b/2a)} \quad [\text{S/m}] \tag{9.15c}$$

If the current is assumed to be uniformly distributed within the skin depth, then the cross-sectional area of the current path in the wire is $S = 2\pi a \delta$. The series resistance per unit length of the line is obtained from Eq. (8.99) such that

$$R = \frac{2}{2\pi a \delta \sigma_c} = \frac{1}{\pi a}\sqrt{\frac{\pi f \mu_c}{\sigma_c}} \quad [\Omega/\text{m}] \tag{9.15d}$$

where we assume $\delta \ll a$. Note that σ_c and μ_c in Eq. (9.15d) are the constitutive parameters of the conductor, and factor 2 in the numerator accounts for the two conductors.

9.2.3   *Parallel-Plate Transmission Line*

Two parallel conducting plates of equal width a are separated by a dielectric with thickness b, as shown in Fig. 9.6c. The capacitance per unit length is obtained using Eq. (3.142) as

$$C = \frac{\varepsilon a}{b} \quad [\text{F/m}] \tag{9.16a}$$

where the fringing effects at the edges are ignored by assuming that $b \ll a$.

Substituting Eq. (9.16a) into Eq. (9.13) gives the inductance per unit length:

$$L = \frac{\mu b}{a} \quad [\text{H/m}] \tag{9.16b}$$

Next, substituting Eq. (9.16b) into Eq. (9.12) gives the shunt conductance per unit length:

$$G = \frac{\sigma a}{b} \quad [\text{S/m}] \tag{9.16c}$$

By using the same procedure as that used in the previous cases and the fact that the cross-sectional area of the current path is $S = a\delta$, the series resistance per unit length of a parallel-plate line is obtained as

$$R = \frac{2}{a\delta\sigma_c} = \frac{2}{a}\sqrt{\frac{\pi f \mu_c}{\sigma_c}} \quad [\Omega/\text{m}] \tag{9.16d}$$

where we assumed that the skin depth was much less than the thickness of the plate. Factor 2 accounts for both plates.

From Eqs. (9.14a)–(9.16d), we see that the series resistance R depends explicitly on the frequency, and that the internal inductance is ignored at high frequencies. This is because the skin depth is extremely short and no magnetic field can exist in the conductor. In most practical situations, the insulating materials in the transmission line are assumed nonmagnetic ($\mu = \mu_0$).

Exercise 9.4

Two parallel wires of diameter 0.643 [mm] are separated by a center-to-center distance of 1.214 [mm] and insulated from one another by a material with $\varepsilon_r = 2.25$ and $\mu = \mu_0$. Determine C and L per unit length.

Ans. $C = 50$ [pF/m], $L = 500$ [nH/m].

Exercise 9.5

Do the line parameters R, L, G, and C depend on frequency even if the constitutive parameters ε, ε_c, μ, μ_c, σ, and σ_c are independent of frequency?

Ans. $R \sim \sqrt{f}$.

Exercise 9.6

When the C parameter of a transmission line is given, is it possible to determine the R parameter from the relationship $RC = \varepsilon/\sigma$, as given in Eq. (4.61)?

Ans. No.

9.3 Wave Propagation on the Transmission Line

The wave equations in Eqs. (9.8a) and (9.8b) are second-order differential equations, and thus each has two independent solutions. The general solution to the wave equation is given by the superposition of two time-harmonic waves propagating in opposite directions along the transmission line. That is,

$$\tilde{V}(z) = V_o^+ e^{-\gamma z} + V_o^- e^{+\gamma z} \equiv \tilde{V}^+(z) + \tilde{V}^-(z) \tag{9.17a}$$

$$\tilde{I}(z) = I_o^+ e^{-\gamma z} + I_o^- e^{+\gamma z} \equiv \tilde{I}^+(z) + \tilde{I}^-(z) \tag{9.17b}$$

where V_o^+, V_o^-, I_o^+, and I_o^- are the complex amplitudes determined from the boundary conditions at the input and output ends. Voltage phasors $\tilde{V}^+(z)$ and $\tilde{V}^-(z)$ represent voltage waves propagating in the $+z$- and $-z$-directions, respectively. These are referred to as the *forward* and *backward* waves, respectively. Similarly, the current phasors $\tilde{I}^+(z)$ and $\tilde{I}^-(z)$ represent forward and backward current waves, respectively. In contrast, $\tilde{V}(z)$ and $\tilde{I}(z)$ are the phasors of the total voltage and total current on the line, respectively, and they specify the time-harmonic voltage and current on the line, respectively. If a transmission line is infinitely long (the line is actually semi-infinite with a generator at the left end), only the forward waves $\tilde{V}^+(z)$ and $\tilde{I}^+(z)$ can exist on the line because these waves propagate indefinitely in the forward direction without any reflection in the wave path.

The sign conventions for $\tilde{V}$ and $\tilde{I}$ on the transmission line are shown in Fig. 9.7. A cosine curve moving to the right represents either $\tilde{V}^+(z)$ or $\tilde{I}^+(z)$, whereas a cosine curve moving to the left represents either $\tilde{V}^-(z)$ or $\tilde{I}^-(z)$. In the figure, the symbols $+$ and $-$ denote the polarity of the voltage, and the blue arrow denotes the direction of the current flow. The cosine curve relates only to the upper wire, because the current flows in the two wires in opposite directions. For voltage waves, the peak of the cosine curve signifies that the upper wire has a higher potential than the lower wire at a given point, regardless of the wave propagation direction. For the current wave, the peak of the cosine curve indicates that the current flows in the same direction as the current wave in the upper wire. In the sign convection, the cosine curve represents the voltage or current of the upper wire.

By convention, the left end of the transmission line is connected to a generator and the right end is connected to a load. The boundary condition at the load relates $\tilde{V}^+(z = \ell)$ to $\tilde{V}^-(z = \ell)$, where ℓ is the line length. This is discussed in detail in Sect. 9.4. Here, without considering the boundary condition, the relationship between $\tilde{V}^+(z)$ and $\tilde{I}^+(z)$ is derived. By substituting the general solution in Eqs. (9.17a) and (9.17b) into the transmission-line equation in Eq. (9.6a), we obtain

$$\gamma V_o^+ e^{-\gamma z} - \gamma V_o^- e^{+\gamma z} = (R + j\omega L)\left[I_o^+ e^{-\gamma z} + I_o^- e^{+\gamma z} \right] \tag{9.18}$$

Equating the coefficients of $e^{-\gamma z}$ (or $e^{\gamma z}$) on both sides of Eq. (9.18) leads to the *characteristic impedance* of the transmission line, denoted by Z_o,

$$\boxed{\begin{aligned} Z_o &\equiv \frac{V_o^+}{I_o^+} = -\frac{V_o^-}{I_o^-} \\[2mm] &= \sqrt{\frac{R + j\omega L}{G + j\omega C}} \equiv R_o + jX_o \end{aligned}} \quad [\Omega] \tag{9.19}$$

The real part of the characteristic impedance is called the *resistance* and the imaginary part is called the *reactance*. The resistance, R_o, is measured in ohms. It should not be

(a) Forward voltage and current waves

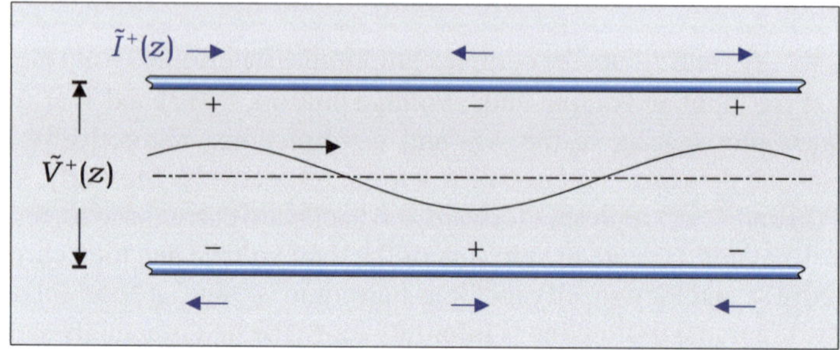

(b) Backward voltage and current waves

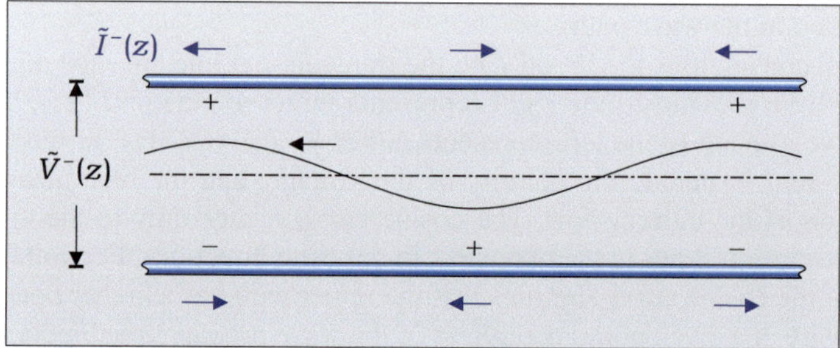

Fig. 9.7 Sign conventions for $\tilde{V}$ and $\tilde{I}$ on the transmission line. A cosine curve is used to describe $\tilde{V}$ or $\tilde{I}$ on the upper conductor

confused with the series resistance R, measured in ohms per meter. The characteristic impedance of the transmission line is analogous to the intrinsic impedance of the medium. Both impedances represent the ratio between the two complex amplitudes. The characteristic impedance is the ratio between the complex amplitudes of the voltage and current, whereas the intrinsic impedance is the ratio between the complex amplitudes of the electric and magnetic fields.

The propagation constant γ and characteristic impedance Z_o are two distinctive constants of the transmission lines. Taken together, they fully describe the time-harmonic behavior of the voltage and current at all points on the transmission line. They depend on the frequency and line parameters R, L, G, and C but not on the line length. The reciprocal of Z_o, denoted as $Y_o = 1/Z_o$, is the ***characteristic admittance***.

Exercise 9.7
For a transmission line with $R = 20\,[\Omega/\text{m}]$, $L = 0.4\,[\mu\text{H/m}]$, $G = 50\,[\mu\text{S/m}]$, and $C = 100\,[\text{pF/m}]$ operating at 10 [MHz], determine γ and Z_o.

Ans. $\gamma = 0.15 + j0.42\,[\text{m}^{-1}]$, $Z_o = 67.6 - j23.3\,[\Omega]$.

Exercise 9.8
Check whether a line with $\gamma = 0.15 + j0.35\,[\text{m}^{-1}]$ and $Z_o = 110 + j80\,[\Omega]$ is physically feasible.

Ans. No. R cannot be negative but $\gamma Z_o = R + j\omega L = -11.5 + j50.5$.

Exercise 9.9
Given $v(t) = 20\cos(10^7 t - 0.13z) + 10\cos(10^7 t + 0.13z)$ [V] on a transmission line with $Z_o = 100$ [Ω], determine the current on the line.

Ans. $i(t) = 0.2\cos(10^7 t - 0.13z) - 0.1\cos(10^7 t + 0.13z)$ [A] .

9.3.1 Lossless Transmission Line

Lossless transmission lines are composed of perfect conductors ($\sigma_c = \infty$) and perfect dielectrics ($\sigma = 0$). The waves can propagate on a lossless line without any energy loss. For a lossless transmission line, we have

$$\boxed{R = 0 = G} \tag{9.20}$$

Other constants are

$$\gamma = j\beta = j\omega\sqrt{LC} = j\omega\sqrt{\mu\varepsilon} \quad \text{(propagation constant)} \tag{9.21a}$$

$$Z_o = R_o = \sqrt{\frac{L}{C}} \quad \text{(characteristic resistance)} \tag{9.21b}$$

$$\upsilon_p = \frac{\omega}{\beta} = \frac{1}{\sqrt{LC}} = \frac{1}{\sqrt{\mu\varepsilon}} \quad \text{(phase velocity)} \tag{9.21c}$$

In these equations, we used Eq. (9.13), and μ and ε are the permeability and permittivity of the insulating material, respectively. The propagation constant is purely imaginary, implying that the attenuation constant is zero. In other words, there is no attenuation of the wave propagating along the lossless line. The characteristic impedance is real, implying that the voltage and current are in phase in the time dimension at every point along the lossless line. From Eq. (9.21c), we can see that the phase velocity of the wave on a lossless line is the same as that of an unguided uniform plane wave in the dielectric of the line.

Example 9.1 When β is 5 [rad/m] on a lossless transmission line with $Z_o = 100$ [Ω] operating at a frequency of 200 [MHz], determine (a) C and L, (b) υ_p, (c) ε_r of the insulating material that is known to be nonmagnetic, and (d) the wavelength.

Solution

(a) From Eq. (9.21a),

$$\beta = \omega\sqrt{LC} = 2\pi \times 2 \times 10^8 \sqrt{LC} = 5 \text{ [rad/m]}$$

From Eq. (9.21b),

$$Z_o = \sqrt{L/C} = 100\,[\Omega]$$

The product of β and Z_o yields

$$4\pi \times 10^8 L = 500$$

Thus

$$L = 0.40\,[\mu\text{H/m}], \text{ and } C = 40\,[\text{pF/m}]$$

(b) From (9.21c),

$$\upsilon_p = \frac{\omega}{\beta} = \frac{4\pi \times 10^8}{5} = 2.51 \times 10^8\,[\text{m/s}]$$

(c) The phase velocity can be obtained as

$$\upsilon_p = \frac{1}{\sqrt{\mu_0 \varepsilon_0 \varepsilon_r}} = 2.51 \times 10^8\,[\text{m/s}]$$

Thus, $\varepsilon_r = 1.43$

(d) From $\beta = 2\pi/\lambda = 5$, we can obtain $\lambda = 1.26\,[\text{m}]$. Alternatively,

$$\lambda = \upsilon_p/f = 1.26\,[\text{m}]$$

Exercise 9.10

For a lossless line with $L = 1\,[\mu\text{H/m}]$ and $C = 100\,[\text{pF/m}]$ operating at 50 [MHz], determine (a) Z_o, (b) β, and (c) υ_p.

Ans. (a) 100 [Ω], (b) π [rad/m], (c) 10^8 [m/s].

Exercise 9.11

A nonmagnetic dielectric with $\varepsilon = 2.25\varepsilon_0$ is used in a lossless line operating at 100 [MHz]. Determine (a) υ_p, and (b) β.

Ans. (a) 2×10^8 [m/s], (b) π [rad/m].

Exercise 9.12

A lossless line consists of two parallel conducting plates of width a separated by distance b in a dielectric with $\varepsilon = \varepsilon_0 \varepsilon_r$. Determine the characteristic impedance.

Ans. $Z_o = 120\pi b/(a\sqrt{\varepsilon_r})$.

9.3.2 Distortionless Transmission Line

As we see from Eq. (9.9), the relationship between the propagation constant and frequency is nonlinear, and so is the relationship between the phase constant and frequency. Subsequently, the phase velocity depends on the frequency; that is, waves with different frequencies travel on a transmission line with different phase velocities. This line is referred to as the ***dispersive line***. From Fourier analysis, a short pulse is composed of time-harmonic waves of different frequencies and thus broadens with the distance travelled on a dispersive line. It would overlap with neighboring pulses at the detector end such that it is impossible to distinguish between the pulses. Pulse broadening is a major factor limiting the data transfer rate.

A lossless line is an ideal transmission line that does not exhibit wave dispersion or attenuation. It is characterized by a zero attenuation constant, constant real characteristic impedance, and constant phase velocity. However, in practical situations, it is impossible to obtain $R = 0 = G$. Nevertheless, we can design a ***distortionless line*** with the same characteristics as a lossless line, except for the nonvanishing attenuation constant. The necessary condition for a distortionless line is

$$\boxed{\frac{R}{L} = \frac{G}{C}} \tag{9.22}$$

Under this condition, the propagation constant is given by

$$\gamma = \sqrt{LC\left(\frac{R}{L} + j\omega\right)\left(\frac{G}{C} + j\omega\right)} = \sqrt{LC}\left(\frac{R}{L} + j\omega\right)$$

$$\equiv \alpha + j\beta \tag{9.23a}$$

Then, it follows that

$$\alpha = R\sqrt{\frac{C}{L}} = \sqrt{RG} \tag{9.23b}$$

$$\beta = \omega\sqrt{LC} = \omega\sqrt{\mu\varepsilon} \tag{9.23c}$$

The other constants are as follows:

$$Z_o = \sqrt{\frac{L(R/L + j\omega)}{C(G/C + j\omega)}} = \sqrt{\frac{L}{C}} \qquad \text{(characteristic impedance)} \tag{9.23d}$$

$$v_p = \frac{\omega}{\beta} = \frac{1}{\sqrt{LC}} = \frac{1}{\sqrt{\mu\varepsilon}} \qquad \text{(phase velocity)} \tag{9.23e}$$

It should be noted that the distortionless line parameters β, Z_o, and v_p in Eqs. (9.23a) – (9.23e) are the same as those of a lossless line, except for the attenuation constant.

Example 9.2 The distortionless line has a characteristic impedance $Z_o = 50\,[\Omega]$ and propagation constant $\gamma = 0.01 + j4.0\,[\mathrm{m}^{-1}]$ at a frequency of 100 [MHz]. Determine (a) line parameters R, L, G, and C, and (b) phase velocity.

Solution

(a) Using Eq. (9.22) in Eqs. (9.23b) and (9.23d) results in $\alpha Z_o = R$ and $\alpha = G Z_o$. Then, using the values given in the problem, we obtain

$$R = 0.50\,[\Omega/\mathrm{m}], \text{ and } G = 0.20\,[\mathrm{mS/m}]$$

From Eqs. (9.23c) and (9.23d),

$$\beta = \omega\sqrt{LC} = 4.0 \text{ and } Z_o = \sqrt{\tfrac{L}{C}} = 50$$

Solving these equations yields

$$L = 318.3\,[\mathrm{nH/m}] \text{ and } C = 127.3\,[\mathrm{pF/m}]$$

(b) The phase velocity is

$$v_p = \frac{\omega}{\beta} = \frac{2\pi \times 10^8}{4} = 1.57 \times 10^8\ [\mathrm{m/s}]$$

Example 9.3 A transmission line exhibits very low losses such that $R \ll \omega L$ and $G \ll \omega C$. Find expressions for α, β, and Z_o, and compare them with those of a distortionless line.

Solution

Rewriting Eq. (9.9),

$$\gamma = \sqrt{(R + j\omega L)(G + j\omega C)} = j\omega\sqrt{LC}\sqrt{\left(1 + \frac{R}{j\omega L}\right)\left(1 + \frac{G}{j\omega C}\right)} \quad (9.24\mathrm{a})$$

Applying the binomial approximation to Eq. (9.24a) yields

$$\gamma \simeq j\omega\sqrt{LC}\left(1 + \frac{R}{j2\omega L} + \frac{R^2}{8\omega^2 L^2}\right)\left(1 + \frac{G}{j2\omega C} + \frac{G^2}{8\omega^2 C^2}\right) \quad (9.24\mathrm{b})$$

By collecting the real and imaginary terms in Eq. (9.24b) and ignoring higher-order terms, we obtain

$$\alpha \simeq \frac{1}{2}\left(R\sqrt{\frac{C}{L}} + G\sqrt{\frac{L}{C}} \right) \qquad (9.24c)$$

$$\beta \simeq \omega\sqrt{LC} \qquad (9.24d)$$

Rewriting Eq. (9.19),

$$Z_o = \sqrt{\frac{R + j\omega L}{G + j\omega C}} = \sqrt{\frac{L}{C}}\sqrt{\frac{1 + R/j\omega L}{1 + G/j\omega C}} \qquad (9.24e)$$

By applying the binomial approximation to Eq. (9.24e), we obtain

$$Z_o \simeq \sqrt{\frac{L}{C}}\left[1 + \frac{j}{2}\left(\frac{G}{\omega C} - \frac{R}{\omega L} \right) \right] \qquad (9.24f)$$

Under conditions $R \ll \omega L$ and $G \ll \omega C$, the second-order correction terms in Eq. (9.24f) can be ignored. Subsequently, β and Z_o are the same as those of the distortionless line, as expressed in Eqs. (9.23c) and (9.23d). *The low-loss lines are distortionless.*

Exercise 9.13
A low-loss coaxial line with $\alpha = 1$ [Np/km] satisfies $R/L = G/C$ at 100 [MHz]. Determine α at 200 [MHz]. [Hint: Eq. (9.24c). Only R changes with f.]

Ans. 1.21 [Np/km].

Exercise 9.14
If the value of ε_r for the dielectric of the line in Exercise 9.13 is doubled, determine α at 100 [MHz].

Ans. 1.06 [Np/km].

9.3.3   *Power Transmission and Attenuation*

The attenuation constant of a wave on a transmission line can be obtained from the power attenuation along that line. The power propagated along the line is equal to the product of the instantaneous voltage and current as follows:

$$\mathcal{P}(z,t) = v(z,t)i(z,t)$$

$$= \frac{1}{2}\left[\tilde{V}^+ e^{j\omega t} + (\tilde{V}^+)^* e^{-j\omega t}\right]\frac{1}{2}\left[\tilde{I}^+ e^{j\omega t} + (\tilde{I}^+)^* e^{-j\omega t}\right]$$

$$= \frac{1}{4}\left[\tilde{V}^+(\tilde{I}^+)^* + (\tilde{V}^+)^*\tilde{I}^+ + \tilde{V}^+\tilde{I}^+ e^{j2\omega t} + (\tilde{V}^+\tilde{I}^+)^* e^{-j2\omega t}\right]$$

$$= \frac{1}{2}\mathrm{Re}\left[\tilde{V}^+\tilde{I}^{+*} + \tilde{V}^+\tilde{I}^+ e^{j2\omega t}\right] \tag{9.25}$$

where $\tilde{V}^+$ and $\tilde{I}^+$ are the phasors of the forward voltage and current waves, respectively. The asterisk $*$ denotes a complex conjugate that reverses the sign of j everywhere, including exponents. It is important to note that $\mathcal{P}$ cannot be expressed as $\mathcal{P} = \mathrm{Re}\left[\tilde{V}^+ e^{j\omega t}\right]\mathrm{Re}\left[\tilde{I}^+ e^{j\omega t}\right]$ because it is a nonlinear function of $v(z,t)$ or $i(z,t)$.

The time-averaged power is defined as

$$\langle\mathcal{P}\rangle = \frac{1}{T}\int_0^T \mathcal{P}(z,t)\,dt \tag{9.26}$$

where T is the temporal period of the time-harmonic function. By inserting Eq. (9.25) into Eq. (9.26), the time-averaged power can be expressed in terms of the voltage and current phasors as

$$\boxed{\langle\mathcal{P}\rangle = \frac{1}{2}\mathrm{Re}\left[\tilde{V}^+\tilde{I}^{+*}\right]} \quad \mathrm{[W]} \tag{9.27}$$

Inserting expressions for $\tilde{V}^+$ and $\tilde{I}^+$ into Eq. (9.27) yields

$$\langle\mathcal{P}\rangle = \frac{1}{2}\mathrm{Re}\left[V_o^+ e^{-(\alpha+j\beta)z} I_o^{+*} e^{-(\alpha-j\beta)z}\right] = \frac{1}{2}\mathrm{Re}\left[Z_o I_o^+ I_o^{+*} e^{-2\alpha z}\right]$$

$$= \frac{1}{2}|I_o^+|^2 R_o e^{-2\alpha z} \tag{9.28}$$

where $V_o^+/I_o^+ = Z_o = R_o + jX_o$. Rewriting Eq. (9.28) gives

$$\boxed{\langle\mathcal{P}(z)\rangle = \langle\mathcal{P}(0)\rangle e^{-2\alpha z}} \quad \mathrm{[W]} \tag{9.29}$$

The time-averaged power decreases exponentially with distance along the line.

In accordance with the principle of energy conservation, the power loss per unit length is equal to the rate of decrease of the time-averaged power with distance. That is,

$$P_L \equiv -\frac{d\langle\mathcal{P}(z)\rangle}{dz} = 2\alpha\langle\mathcal{P}(z)\rangle \tag{9.30}$$

where we have used Eq. (9.29). Therefore, the attenuation constant can be expressed as

$$\alpha = \frac{P_L}{2\langle \mathscr{P}(z)\rangle}$$

(9.31)

where P_L represents the power loss per unit length, and $\langle \mathscr{P}(z)\rangle$ denotes the time-averaged power on the transmission line.

It is convenient to express the power attenuation in decibels (dB) such that

$$\text{Power attenuation in dB} \equiv -10\log_{10}\left[\frac{\mathscr{P}(z)}{\mathscr{P}(0)}\right]$$

(9.32)

Inserting Eq. (9.29) into Eq. (9.32), we have

$$\boxed{\text{Power attenuation in dB} = 8.686\,\alpha\ell}$$

(9.33)

where ℓ is the length of the transmission line. This equation enables us to convert the power attenuation in decibels per unit length into an attenuation constant, and vice versa. For example, 1 [Np/m] corresponds to 8.686 [dB/m].

The decibel is useful for expressing the total power attenuation along a transmission line that consists of different types of lines connected end-to-end. The total power attenuation in decibels is equal to the sum of the power attenuations in decibels on the individual lines.

Example 9.4 When the power attenuation is 1.5 [dB] for every 100 [m] along the transmission line, determine (a) the attenuation constant, and (b) the fraction of the input power reaching a point 1 [km] from the input end.

Solution

(a) From Eq. (9.33),

$$\alpha = \frac{1.5}{8.69 \times 100} = 0.0017\ [\text{Np/m}]$$

(b) From Eq. (9.29),

$$\frac{\langle \mathscr{P}(1000)\rangle}{\langle \mathscr{P}(0)\rangle} = e^{-2\times0.0017\times1000} = 0.033$$

Exercise 9.15
When two lines with power attenuations 2.3 [dB] and 1.8 [dB] are connected, the joint causes an additional attenuation of 3.5 [dB]. Determine the total attenuation.

Ans. 7.6 [dB].

Exercise 9.16
Find the attenuation constant corresponding to a power attenuation 1.737 [dB/m].

Ans. 0.2 [Np/m].

Review Questions

RQ 9.7   What are the general expressions for $\tilde{V}$ and $\tilde{I}$ in transmission lines?   [(9.17a,b)]

RQ 9.8   What is the polarity of the voltage when phasor $\tilde{V}^+$ (or $\tilde{V}^-$) is at its maximum?   [Fig. 9.7]

RQ 9.9   What is the direction of the current flow when phasor $\tilde{I}^+$ (or $\tilde{I}^-$) is at its maximum?   [Fig. 9.7]

RQ 9.10   What is the characteristic impedance of a transmission line?   [(9.19)]

RQ 9.11   Distinguish between line parameter R and characteristic resistance R_o.   [(9.19)]

RQ 9.12   Under what conditions is the transmission line lossless? Describe characteristic parameters of lossless line.   [(9.20) (9.21a,b,c)]

RQ 9.13   Under what conditions is a transmission line distortionless? Describe the characteristic parameters of the distortionless line.   [(9.22) (9.23a,b,c,d,e)]

RQ 9.14   Express the time-averaged power on a transmission line in terms of $\tilde{V}^+$ and $\tilde{I}^+$.   [(9.27)]

RQ 9.15   In what way is α related to the dB power attenuation?   [(9.33)]

9.4   Finite Transmission Line

A finite transmission line typically connects a generator at the left end to a load at the right end as shown in Fig. 9.8. The behaviors of voltage and current on a finite transmission line are governed by the wave equations in Eqs. (9.8a) and (9.8b). The general solutions for the total voltage and total current are expressed as

$$\tilde{V}(z) \equiv \tilde{V}^+(z) + \tilde{V}^-(z) = V_o^+ e^{-\gamma z} + V_o^- e^{\gamma z} \tag{9.34a}$$

$$\tilde{I}(z) \equiv \tilde{I}^+(z) + \tilde{I}^-(z) = \frac{V_o^+}{Z_o} e^{-\gamma z} - \frac{V_o^-}{Z_o} e^{\gamma z} \tag{9.34b}$$

where γ and Z_o are the propagation constant and characteristic impedance, respectively, as given in Eqs. (9.9) and (9.19). V_o^+ and V_o^- are the complex amplitudes of forward and backward waves, respectively. Do not mistake $\tilde{V}^+$ for V_o^+: the former

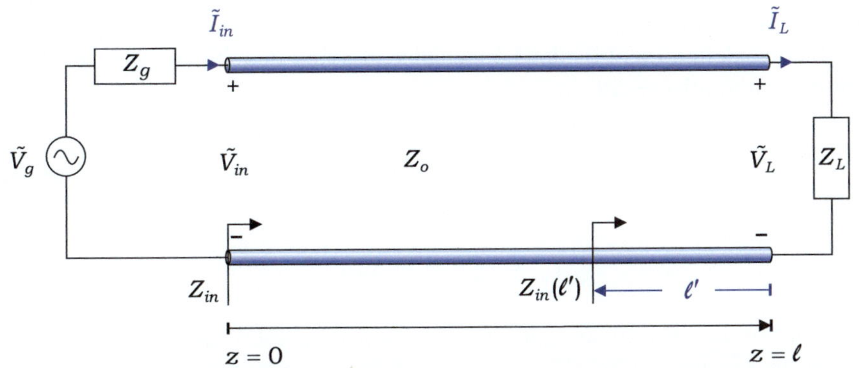

Fig. 9.8 A finite transmission line of length ℓ with characteristic impedance Z_o connects a generator circuit with *internal impedance* Z_g to a load with *load impedance* Z_L. Space variable ℓ' denotes *the distance from the load toward the generator*

is the phasor of the forward wave, which varies with z, and the latter is the complex amplitude, which is constant. Note that phasor $\tilde{V}$ (or $\tilde{I}$) specifies the amplitude of the total voltage (or current) oscillating at frequency ω at the points along the transmission line.

For a transmission line of length ℓ, it is customary to position the generator at $z = 0$ and load at $z = \ell$. The load may be a short circuit, an open circuit, or another transmission line. When a forward wave is launched on a transmission line, its voltage and current are explicitly interrelated through the characteristic impedance of the line. However, the forward wave alone cannot satisfy the voltage-current relationship required at the load if the load impedance is not equal to the characteristic impedance. The impedance mismatch leads to a reflection of the incident wave and initiates a backward wave in such a way that the ratio of the total voltage to the total current at the load is exactly equal to the load impedance.

The wave launched at $z = 0$ may take endless round trips between the internal and load impedances because of impedance mismatches at $z = 0$ and $z = \ell$. To avoid dealing with all the individual reflections, we add all the forward reflections into a forward wave with complex amplitude V_o^+ and all the backward reflections into a backward wave with complex amplitude V_o^-. Therefore, the forward wave represents the net wave incident on the load and the backward wave represents the net wave reflected from the load. The total voltage and current at the load are called the *load voltage* and *load current* and are denoted as $\tilde{V}_L$ and $\tilde{I}_L$, respectively. Setting $z = \ell$ in Eqs. (9.34a) and (9.34b), we have

$$\tilde{V}_L = V_o^+ e^{-\gamma \ell} + V_o^- e^{\gamma \ell} \tag{9.35a}$$

$$\tilde{I}_L = \frac{V_o^+}{Z_o} e^{-\gamma \ell} - \frac{V_o^-}{Z_o} e^{\gamma \ell} \tag{9.35b}$$

The load voltage and current must be related via the relationship $\tilde{V}_L = \tilde{I}_L Z_L$, which is in fact the definition of the load impedance. Solving Eqs. (9.35a) and (9.35b) for complex amplitudes results in

$$V_o^+ = \frac{\tilde{I}_L}{2}(Z_L + Z_o)e^{\gamma\ell} \tag{9.36a}$$

$$V_o^- = \frac{\tilde{I}_L}{2}(Z_L - Z_o)e^{-\gamma\ell} \tag{9.36b}$$

Upon substituting Eq. (9.36a) into Eqs. (9.34a, b), the total voltage and current on the line are obtained as follows:

$$\tilde{V}(z) = \frac{\tilde{I}_L}{2}\left[(Z_L + Z_o)e^{\gamma(\ell-z)} + (Z_L - Z_o)e^{-\gamma(\ell-z)}\right] \text{ [V]} \tag{9.37a}$$

$$\tilde{I}(z) = \frac{\tilde{I}_L}{2Z_o}\left[(Z_L + Z_o)e^{\gamma(\ell-z)} - (Z_L - Z_o)e^{-\gamma(\ell-z)}\right] \text{ [A]} \tag{9.37b}$$

where ℓ is the line length.

Here, we introduce a new space coordinate defined as $\ell' = \ell - z$. It denotes the *distance from the load toward the generator*, or simply the *distance from the load*. Be careful not to confuse space variable ℓ' with a constant line length ℓ. With the variable ℓ', the total voltage and current can be expressed as follows:

$$\boxed{\tilde{V}(\ell') = \frac{\tilde{I}_L}{2}\left[(Z_L + Z_o)e^{\gamma\ell'} + (Z_L - Z_o)e^{-\gamma\ell'}\right]} \tag{9.38a}$$

$$\boxed{\tilde{I}(\ell') = \frac{\tilde{I}_L}{2Z_o}\left[(Z_L + Z_o)e^{\gamma\ell'} - (Z_L - Z_o)e^{-\gamma\ell'}\right]} \tag{9.38b}$$

where $e^{\gamma\ell'}$ represents the forward wave, and $e^{-\gamma\ell'}$ represents the backward wave. In addition, the load position is specified as $\ell' = 0$, whereas the generator position is specified as $\ell' = \ell$. When we deal with the reflection coefficient, standing-wave ratio, and input impedance, it is more convenient to work with Eqs. (9.38a, b) than with Eqs. (9.34a, b).

Example 9.5 For a finite transmission line of length ℓ with a characteristic impedance Z_o, as shown in Fig. 9.8, start with the voltage phasor in Eq. (9.34a) to express the following quantities in terms of V_o^+, V_o^-, Z_o, and γ: (a) *current phasor* $\tilde{I}$, (b) *input voltage* $\tilde{V}_{in}$, (c) *load voltage* $\tilde{V}_L$, (d) *input impedance* Z_{in}, (e) *load impedance* Z_L, and (f) *voltage reflection coefficient* Γ. These quantities are discussed in detail in the following sections.

Solution

(a) $\tilde{I}(z) = \dfrac{V_o^+}{Z_o} e^{-\gamma z} - \dfrac{V_o^-}{Z_o} e^{\gamma z}$ $\hfill$ (9.39a)

(b) $\tilde{V}_{in} = \tilde{V}(z = 0) = V_o^+ + V_o^-$ $\hfill$ (9.39b)

(c) $\tilde{V}_L = \tilde{V}(z = \ell) = V_o^+ e^{-\gamma \ell} + V_o^- e^{\gamma \ell}$ $\hfill$ (9.39c)

(d) $Z_{in} = \dfrac{\tilde{V}_{in}}{\tilde{I}_{in}} = Z_o \dfrac{V_o^+ + V_o^-}{V_o^+ - V_o^-}$ $\hfill$ (9.39d)

(e) $Z_L = \dfrac{\tilde{V}_L}{\tilde{I}_L} = Z_o \dfrac{V_o^+ e^{-\gamma \ell} + V_o^- e^{\gamma \ell}}{V_o^+ e^{-\gamma \ell} - V_o^- e^{\gamma \ell}}$ $\hfill$ (9.39e)

(f) $\Gamma = \dfrac{V_o^- e^{\gamma \ell}}{V_o^+ e^{-\gamma \ell}}$ $\hfill$ (9.39f)

Exercise 9.17
Which of the following quantities are (a) phasors, and (b) constant phasors? [Hint: The symbol $\sim$ is omitted here.]

(1) V^+, (2) V^-, (3) I^+, (4) I^-, (5) V_{in}, (6) I_{in}, (7) V_L, (8) I_L, (9) V_g, (10) V_o^+, (11) V_o^-, (12) I_o^+, (13) I_o^-, (14) Z_o, (15) Z_{in}, (16) Z_L, (17) Z_g, (18) Γ.

Ans. (a) (1)–(9), (b) (5)–(9).

Review Questions

RQ 9.16	Identify the impedances associated with a transmission line.	[Fig. 9.8]
RQ 9.17	Explain how Z_o relates $\tilde{I}$ to $\tilde{V}$.	[(9.34b)]
RQ 9.18	Write general expressions for the voltage phasor $\tilde{V}(z)$	[(9.34a)(9.38a) (9.42a)]
RQ 9.19	What is the distance from the load toward the generator?	[Fig. 9.8]

9.4.1   Reflection Coefficient and Standing-Wave Ratio

A wave traveling on a transmission line with a characteristic impedance Z_o must undergo a reflection at the load if the load impedance Z_L is not equal to Z_o. The ratio between the complex amplitudes of the reflected and incident voltage waves at the load is called the **voltage reflection coefficient**, and is denoted as Γ. By setting $\ell' = 0$ in Eq. (9.38a), the voltage reflection coefficient of the load impedance is expressed as

$$\boxed{\Gamma = \frac{Z_L - Z_o}{Z_L + Z_o} \equiv |\Gamma| e^{j\phi}} \tag{9.40}$$

This is generally complex, with a magnitude less than or equal to unity.

It is evident from Eq. (9.40) that the reflection coefficient vanishes if the characteristic and load impedances are equal, and there is no reflection of the incident wave at the load. In this case, the transmission line is said to be *matched to the load*.

For a future reference, Eq. (9.40) is expressed as

$$\frac{Z_L}{Z_o} = \frac{1 + \Gamma}{1 - \Gamma} \tag{9.41}$$

Upon substituting Eqs. (9.36a, b) and (9.40) into Eqs. (9.37a, b), the voltage and current phasors can be expressed in terms of the reflection coefficient as

$$\boxed{\tilde{V}(z) = V_o^+ e^{-\gamma z}\left[1 + \Gamma e^{-2\gamma\ell'}\right]} \quad \text{[V]} \tag{9.42a}$$

$$\boxed{\tilde{I}(z) = I_o^+ e^{-\gamma z}\left[1 - \Gamma e^{-2\gamma\ell'}\right]} \quad \text{[A]} \tag{9.42b}$$

where $I_o^+ = V_o^+/Z_o$, and $\ell' = \ell - z$. The first term in brackets in Eq. (9.42a) represents the forward wave observed at point z or at a distance ℓ' from the load, whereas the second term represents the backward wave observed at the same point on the line. The same interpretation applies to the current phasor in Eq. (9.42b). The negative sign in front of Γ indicates the reverse direction of the current in the backward wave. Factor 2 in the exponent accounts for the phase delay of the backward wave due to a round trip from a given point on the line to the load.

After some algebraic manipulations, Eq. (9.42a) can be rearranged as

$$\tilde{V}(z) = V_o^+ e^{-\gamma z}\left(1 - |\Gamma| + |\Gamma| + \Gamma e^{-2\gamma\ell'}\right)$$

$$= V_o^+ e^{-\gamma z}(1 - |\Gamma|) + V_o^+ e^{-\gamma z}|\Gamma|\left(1 + e^{j\phi}e^{-2\gamma\ell'}\right)$$

$$= V_o^+ e^{-\gamma z}(1 - |\Gamma|) + V_o^+ |\Gamma| e^{j\phi/2 - \gamma\ell}\left(e^{-j\phi/2}e^{\gamma\ell'} + e^{j\phi/2}e^{-\gamma\ell'}\right) \tag{9.43}$$

where $\Gamma = |\Gamma| e^{j\phi}$. Assuming a lossless line with $\gamma = j\beta$, the instantaneous voltage on a line is obtained as

$$v(z, t) = V_o^+ (1 - |\Gamma|) \cos(\omega t - \beta z)$$
$$+ 2V_o^+ |\Gamma| \cos(\beta\ell' - \phi/2) \cos(\omega t + \phi/2 - \beta\ell) \qquad (9.44)$$

where the first term on the right side represents a wave traveling in the $+z$-direction, and the second term represents a standing wave. Part of the forward wave is reflected from the load and interferes with the incoming wave to form a standing wave, whereas the rest propagates towards the load.

The standing wave causes the amplitude of the time-harmonic voltage to vary with position along the line. By setting $\gamma = j\beta$ in Eq. (9.42a), the voltage phasor on a lossless line is expressed as

$$\tilde{V}(z) = V_o^+ e^{-j\beta z} \left[1 + |\Gamma| e^{j(\phi - 2\beta\ell')} \right] \qquad (9.45)$$

The magnitude of $\tilde{V}$ is determined solely by the term in the brackets in Eq. (9.45). The maximum of $|\tilde{V}|$ occurs at a distance ℓ' from the load where the condition $e^{j(\phi - 2\beta\ell')} = 1$ is satisfied, that is,

$$\phi - 2\beta\ell'_{\text{max}} = 0, \pm 2\pi, \pm 4\pi, \ldots \qquad (9.46a)$$

Thus,

$$\ell'_{\text{max}} = \frac{1}{2\beta}(\phi + 2m\pi) \qquad (9.46b)$$

$$(m = 0, \ 1, \ 2.., \ \ -\pi \leq \phi < \pi)$$

where β is the phase constant of the wave on the line, and ϕ is the phase angle of Γ. Note that m cannot be negative, because ℓ' is always positive. For $\phi > 0$, condition $m = 0$ corresponds to the first voltage maximum from the load. However, for $\phi < 0$, $m = 1$ corresponds to the first voltage maximum. The voltage maxima occur periodically along the line with half-wavelength spacing. By inserting Eq. (9.46b) into Eq. (9.45), the maximum value of $|\tilde{V}|$ is obtained as

$$|\tilde{V}|_{\text{max}} = \left| V_o^+ \right| (1 + |\Gamma|) \qquad (9.46c)$$

A similar procedure can be followed to locate the voltage minimum on the line. The minimum of $|\tilde{V}|$ occurs at a point where the condition $e^{j(\phi - 2\beta\ell')} = -1$ is satisfied, that is,

$$\phi - 2\beta\ell'_{\text{min}} = \pm\pi, \pm 3\pi, \ldots \qquad (9.47a)$$

Thus,

$$\boxed{\ell'_{\min} = \frac{1}{2\beta}[\phi + (2m + 1)\pi]}$$

(9.47b)

$$(m = 0, \ 1, \ 2.., \ -\pi \leq \phi < \pi)$$

As ϕ ranges from $-\pi$ to π, condition $m = 0$ must ensure that $\ell'_{\min}$ is positive so that it corresponds to the first minimum from the load. Evidently, two adjacent minima are separated by the half-wavelength of the wave, and the spacing between the minimum and nearest maximum is a quarter of the wavelength. The minimum voltage is obtained from Eq. (9.45) as

$$|\tilde{V}|_{\min} = |V_o^+|(1 - |\Gamma|)$$

(9.47c)

By following the same procedure, the current maxima and minima can be located on the transmission line. It can be shown that the current maxima occur at the points of the voltage minima.

We are now ready to introduce the **voltage standing-wave ratio** (VSWR, or simply SWR), which is denoted as S and defined as

$$\boxed{S = \frac{|\tilde{V}|_{\max}}{|\tilde{V}|_{\min}} = \frac{1 + |\Gamma|}{1 - |\Gamma|}}$$

(9.48)

Note that $|\tilde{V}|_{\max}$ and $|\tilde{V}|_{\min}$ occur at different points along the line. As $|\Gamma|$ increases from 0 to 1, S increases from 1 to ∞. A higher standing-wave ratio is less desirable because it implies a greater reflection and delivery of less power to the load. The inverse of Eq. (9.48) can be conveniently used in many applications. That is,

$$\boxed{|\Gamma| = \frac{S - 1}{S + 1}}$$

(9.49)

The unknown load impedance can be easily determined by measuring the voltage standing-wave ratio on a slotted line consisting of a lossless coaxial line with a long narrow slit in the outer conductor and small movable probe along the slit. The probe can sample electric fields and measure the maximum and minimum voltage amplitudes, together with their separation. The measured standing-wave ratio provides the magnitude of the reflection coefficient at the load. The separation between the voltage maxima specifies the phase angle of the reflection coefficient. By inserting the reflection coefficient and characteristic impedance of the slotted line into Eq. (9.40), the load impedance can be determined.

Next, we examine the power transmitted from the generator to the load through a *lossless* transmission line. By using the voltage and current phasors given in Eqs.

(9.42a) and (9.42b), the time-averaged power along the line is expressed as

$$
\begin{aligned}
\langle \mathscr{P} \rangle &= \frac{1}{2}\mathrm{Re}\left[\tilde{V}(z)\tilde{I}^{*}(z) \right] \\
&= \frac{1}{2}\mathrm{Re}\left[V_o^{+} e^{-j\beta z}\left(1 + \Gamma e^{-j2\beta \ell'} \right)\left(\frac{V_o^{+}}{R_o} e^{-j\beta z}\left(1 - \Gamma e^{-j2\beta \ell'} \right) \right)^{*} \right] \\
&= \frac{1}{2}\frac{|V_o^{+}|^2}{R_o}\mathrm{Re}\left[1 - |\Gamma|^2 + \left\{ \Gamma e^{-j2\beta \ell'} - \Gamma^{*} e^{j2\beta \ell'} \right\} \right]
\end{aligned}
\tag{9.50}
$$

where the term in the braces is imaginary, and is thus ignored. Consequently, the time-averaged power delivered to the load is

$$
\langle \mathscr{P} \rangle = \frac{1}{2}\frac{|V_o^{+}|^2}{R_o}\left(1 - |\Gamma|^2 \right)
\tag{9.51}
$$

This corresponds to the incident power minus the reflected power, as indicated in the parentheses.

Example 9.6 The 100 [Ω] transmission line is terminated at a load of $Z_L = 120 + j40$ [Ω]. Determine the voltage reflection coefficient and standing-wave ratio.

Solution

From Eq. (9.40),

$$
\Gamma = \frac{Z_L - Z_o}{Z_L + Z_o} = \frac{120 + j40 - 100}{120 + j40 + 100} = 0.20 e^{j0.93}
$$

From Eq. (9.48),

$$
S = \frac{1 + |\Gamma|}{1 - |\Gamma|} = \frac{1 + 0.2}{1 - 0.2} = 1.50
$$

Example 9.7 A 75 [Ω]-lossless line is terminated at an unknown impedance Z_L. Measurements show that $S = 3$ on the line and the first and second voltage maxima are 4 [cm] and 14 [cm] from the load, respectively. Determine (a) Γ, and (b) Z_L.

Solution

(a) The distance between two adjacent maxima is half the wavelength. Thus,

$$
\lambda = 2 \times (14 - 4) = 20 \, [\text{cm}]
$$

$$
\beta = 2\pi/\lambda = 10\pi \, [\text{rad/m}]
$$

Substituting $\ell'_{max} = 0.04\,[\text{m}]$, $m = 0$, and $\beta = 10\pi$ into Eq. (9.46b) yields

$$\phi = 0.8\pi\,[\text{rad}]$$

Letting $S = 3$ in Eq. (9.49),

$$|\Gamma| = \frac{3-1}{3+1} = 0.5$$

The reflection coefficient is thus

$$\Gamma = 0.5e^{j0.8\,\pi}$$

(b) Substituting the results in part (a) into Eq. (9.41) yields

$$Z_L = Z_o\frac{1+\Gamma}{1-\Gamma} = 75\frac{1+0.5e^{j0.8\pi}}{1-0.5e^{j0.8\pi}} = 27.3 + j21.4\,[\Omega]$$

Example 9.8 A 2 [m]-transmission line with $Z_o = 67 - j23\,[\Omega]$ and $\gamma = 0.15 + j0.42\,[\text{m}^{-1}]$ is connected to a generator with $v_g(t) = 20\cos(10^7 t)\,[\text{V}]$ and $Z_g = 0$, but the load is an open circuit ($Z_L = \infty$). Find

(a) the reflection coefficient,
(b) the voltage and current phasors on the line, and
(c) the power dissipated in the line.

Solution

(a) With $Z_L = \infty$ for an open circuit, the reflection coefficient is obtained as

$$\Gamma = \frac{Z_L - Z_o}{Z_L + Z_o} = 1 \qquad\qquad (9.52\text{a})$$

(b) Setting $z = 0$ and $\ell' = 2$ in Eq. (9.42a), the input voltage is obtained as

$$\tilde{V}_{in} = V_o^+\left[1 + e^{-0.6-j1.68}\right] = V_o^+ 1.087e^{-j0.526}\,[\text{V}] \qquad\qquad (9.52\text{b})$$

Setting $\tilde{V}_{in} = \tilde{V}_g$ in Eq. (9.52b) yields

$$V_o^+ = \frac{20}{1.087e^{-j0.526}} = 18.399e^{j0.526}\,[\text{V}]$$

The voltage phasor on the line is, therefore,

$$\tilde{V}(z) = V_o^+ e^{-\gamma z}\left[1 + \Gamma e^{-2\gamma\ell'}\right]$$

$$= 18.399e^{j0.526}e^{-(0.15+j0.42)z}\left[1 + e^{-(0.3+j0.84)\ell'}\right] \tag{9.52c}$$

Using $Z_o = V_o^+/I_o^+ = -V_o^-/I_o^-$ and $Z_o = 70.838e^{-j0.331}$ in Eq. (9.52c) yields

$$\tilde{I}(z) = \frac{V_o^+}{Z_o}e^{-\gamma z}\left[1 - \Gamma e^{-2\gamma\ell'}\right]$$

$$= 0.260e^{j0.857}e^{-(0.15+j0.42)z}\left[1 - e^{-(0.3+j0.84)\ell'}\right] \tag{9.52d}$$

By letting $z = 2$ and $\ell' = 0$ in Eq. (9.52d), it follows that $\tilde{I} = 0$ at the right end, as expected.

(c) Setting $z = 0$ and $\ell' = 2$ in Eqs. (9.52c) and (9.52d), we first obtain $\tilde{V}_{in}$ and $\tilde{I}_{in}$ at the input end and then compute the power delivered by the generator as follows:

$$\tilde{V}_{in} = 18.399e^{j0.526}\left[1 + e^{-(0.6+j1.68)}\right] = 20.0 \tag{9.52e}$$

$$\tilde{I}_{in} = 0.260e^{j0.857}\left[1 - e^{-(0.6+j1.68)}\right] = 0.310e^{j1.332} \tag{9.52f}$$

$$\langle\mathscr{P}_{in}\rangle = \frac{1}{2}\text{Re}\left[\tilde{V}_{in}\tilde{I}_{in}^*\right]$$

$$= \frac{1}{2}\text{Re}\left[4.784e^{-j0.331}\left\{(1 - e^{-1.2}) - e^{-0.6}(e^{j1.68} - e^{-j1.68})\right\}\right]$$

$$= 0.733 \text{ [W]} \tag{9.52g}$$

Because the load is an open circuit, no power is dissipated at the load, and the power in Eq. (9.52g) is dissipated in the transmission line.

The voltage phasor given in Eq. (9.52c) is the result of superposition of the forward and backward voltage waves. The question naturally arises as to whether it is possible to compute the dissipated power by adding the power losses suffered by the forward and backward waves individually during a round trip from the generator to the load end. The answer is no, because the forward and backward waves interfere with each other and give rise to a standing wave along the line. In fact, term $(1 - e^{-1.2})$ in Eq. (9.52g) represents the power loss during a round-trip in the absence of interference. The power dissipated in the line is less than the round-trip power loss.

Exercise 9.18
For a lossless line terminated at a purely reactive load, $Z_L = jX_L$, show that $|\Gamma| = 1$ regardless of the sign of X_L ($X_L > 0$, inductive; $X_L < 0$, capacitive).

Ans. Eq. (9.40).

Exercise 9.19

Locate the voltage maximum nearest to the load if $\Gamma = 0.5e^{-j0.62}$ and $\lambda = 12$ [cm].

Ans. 5.4 [cm].

Exercise 9.20

For a lossless line with R_o terminated at a resistive load $Z_L = R_L$, what is observed at the load (voltage maximum or minimum) if (a) $R_L < R_o$, and (b) $R_L > R_o$.

Ans. (a) Voltage minimum, (b) Voltage maximum.

Exercise 9.21

If a 100 [W] signal is launched on a lossless 50 [Ω]-line that is connected to a load impedance $Z_L = 20 + j100$ [Ω], find the power dissipated in the load.

Ans. 26.9 [W].

Review Questions

RQ 9.20	Define the voltage reflection coefficient.	[(9.40)]
RQ 9.21	Under what condition is a transmission line matched to the load?	[(9.40)]
RQ 9.22	Is the current reflection coefficient the negative of the voltage reflection coefficient?	[(9.19)]
RQ 9.23	What is the distance between the successive voltage maxima?	[(9.46b)]
RQ 9.24	What is the distance between the neighboring voltage maximum and minimum?	[(9.46b)(9.47b)]
RQ 9.25	Does the separation of the successive voltage maxima depend on the terminating load impedance?	[(9.46b)]
RQ 9.26	Define the standing-wave ratio.	[(9.48)]
RQ 9.27	Explain how the SWR and the locations of voltage maxima are used to determine Γ.	[(9.46b)(9.49)]
RQ 9.28	What is the period of the standing-wave envelope?	[(9.46b)]
RQ 9.29	Why is a high SWR on a line undesirable?	[(9.51)]
RQ 9.30	Express the time-averaged power delivered to the load.	[(9.51)]

9.4.2 Input Impedance

The ***input voltage*** $\tilde{V}_{in}$ and ***input current*** $\tilde{I}_{in}$ are the total voltage and current measured at the generator end of the line, respectively. The ***input impedance*** of the transmission line is defined as the ratio of $\tilde{V}_{in}$ to $\tilde{I}_{in}$. That is,

$$Z_{in} \equiv \frac{\tilde{V}_{in}}{\tilde{I}_{in}} = \frac{\tilde{V}(z = 0)}{\tilde{I}(z = 0)} \tag{9.53}$$

where $\tilde{V}$ and $\tilde{I}$ are the voltage and current phasors, respectively. Using Eqs. (9.38a, b) and (9.40) in Eq. (9.53) and setting $\ell' = \ell$, the input impedance can be expressed as

$$Z_{in} = Z_o \frac{1 + \Gamma e^{-2\gamma \ell}}{1 - \Gamma e^{-2\gamma \ell}} \tag{9.54}$$

Using hyperbolic functions, such as $\cosh \theta = (e^{\theta} + e^{-\theta})/2$, $\sinh \theta = (e^{\theta} - e^{-\theta})/2$, and $\tanh \theta = \sinh \theta / \cosh \theta$, the input impedance of a line with length ℓ and characteristic impedance Z_o terminated at load impedance Z_L can be expressed as

$$Z_{in} = Z_o \frac{Z_L + Z_o \tanh \gamma \ell}{Z_o + Z_L \tanh \gamma \ell} \quad [\Omega] \quad \text{(lossy line)} \tag{9.55}$$

where all quantities are complex, except for line length ℓ.

From the standpoint of the generator circuit, the terminated transmission line can be replaced by its input impedance, as shown in Fig. 9.9. An equivalent circuit can provide a useful method for determining input voltage and current.

For a lossless line, the characteristic impedance is real ($Z_o = R_o$) and the propagation constant is imaginary ($\gamma = j\beta$). In this case, $\tanh(j\beta\ell)$ in Eq. (9.55) becomes $j \tan(\beta\ell)$. Thus, the input impedance of a lossless line of length ℓ terminated at load impedance Z_L is given by

$$Z_{in} = R_o \frac{Z_L + j R_o \tan \beta \ell}{R_o + j Z_L \tan \beta \ell} \quad [\Omega] \quad \text{(lossless line)} \tag{9.56}$$

Fig. 9.9 Equivalent circuit for the terminated transmission line

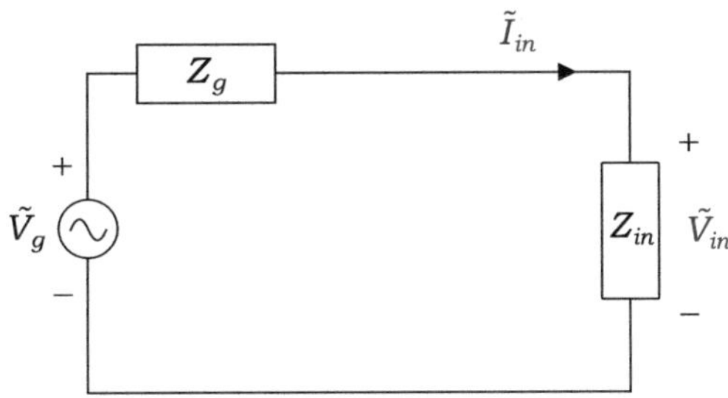

where Z_L and Z_{in} are generally complex. Note that the input impedance varies periodically as a function of the line length ℓ. The term $\beta\ell$ is often referred to as the **electrical length** of the transmission line.

For example, a $\lambda/4$ long lossless line with characteristic impedance R_o terminated at pure resistance R_L has a purely resistive input impedance:

$$\boxed{Z_{in} = \frac{R_o^2}{R_L}} \tag{9.57}$$

A quarter-wave section is often referred to as a **quarter-wave transformer**, and can be used as an impedance inverter.

It is very important to distinguish between the input and characteristic impedances. The former is the ratio of the total voltage to the total current at the input end of the transmission line, whereas the latter is the ratio of the amplitudes of the voltage and current waves traveling in the same direction on the line, that is, $Z_o = V_o^+/I_o^+ = -V_o^-/I_o^-$. Moreover, the input impedance depends on the line length and load impedance, whereas the characteristic impedance is constant and independent of the line length.

Example 9.9 A generator circuit with $v_g(t) = 20\cos(10^8 t)$ [V] and $Z_g = 10\,[\Omega]$ is connected to a 2 [m]-transmission line with $Z_o = 50\,[\Omega]$ and $\gamma = 0.15 + j3.5\,[\text{m}^{-1}]$, which is terminated at a load of $Z_L = 40 + j10\,[\Omega]$. Determine (a) reflection coefficient, (b) input impedance, (c) input voltage, and (d) total voltage along the line.

Solution

(a) From (9.40),

$$\Gamma = \frac{Z_L - Z_o}{Z_L + Z_o} = \frac{40 + j10 - 50}{40 + j10 + 50} = 0.156e^{j2.246} \tag{9.58a}$$

(b) Setting $\gamma\ell = 0.3 + j7.0$, we obtain

$$\tanh(\gamma\ell) = \frac{1 - e^{-0.6 - j14}}{1 + e^{-0.6 - j14}} = 0.482 + j0.749$$

Thus,

$$Z_{in} = Z_o \frac{Z_L + Z_o \tanh\gamma\ell}{Z_o + Z_L \tanh\gamma\ell}$$

$$= 50\frac{(40 + j10) + 50 \times (0.482 + j0.749)}{50 + (40 + j10) \times (0.482 + j0.749)}$$

$$= 55.802 + j6.987 \tag{9.58b}$$

(c) From the equivalent circuit, we obtain

$$\tilde{V}_{in} = \tilde{V}_g \frac{Z_{in}}{Z_g + Z_{in}} = \frac{20 \times (55.802 + j6.987)}{10 + 55.802 + j6.987} = 16.994 + j0.319 \,[\text{V}] \tag{9.58c}$$

(d) Using Eqs. (9.58a, c) in Eq. (9.42a) and letting $z = 0$ and $\ell' = 2$, we obtain

$$16.994 + j0.319 = V_o^+ \left[1 + 0.156 e^{j2.246} e^{-2(0.3 + j7.0)} \right]$$

Thus, $V_o^+ = 16.024 e^{-j0.0399}$

Using Eq. (9.45), the total voltage along the line is obtained as

$$\tilde{V}(z) = 16.024 e^{-j0.0399} e^{-(0.15 + j3.5)z} \left[1 + 0.156 e^{j2.246} e^{-(0.3 + j7.0)(2 - z)} \right]$$

Example 9.10 The lossless line with characteristic resistance R_o and phase constant β is terminated at load Z_L. At a distance ℓ' from the load, *the input impedance looking into the section of length ℓ' is defined as* $Z_{in}(\ell') \equiv \tilde{V}(\ell')/\tilde{I}(\ell')$, and the *reflection coefficient at distance ℓ' from the load is defined as* $\Gamma(\ell') = \tilde{V}^-(\ell')/\tilde{V}^+(\ell')$ (see Fig. 9.10). Find

(a) the general expression for $\Gamma(\ell')$,
(b) $\Gamma(\ell')$ at the locations of $|\tilde{V}|_{\max}$ and $|\tilde{V}|_{\min}$,
(c) $Z_{in}(\ell')$ at the locations of $|\tilde{V}|_{\max}$ and $|\tilde{V}|_{\min}$, and
(d) show that $Z_{in}(\ell')/R_o$ at the location of $|\tilde{V}|_{\max}$ is equal to SWR on the line.

Solution

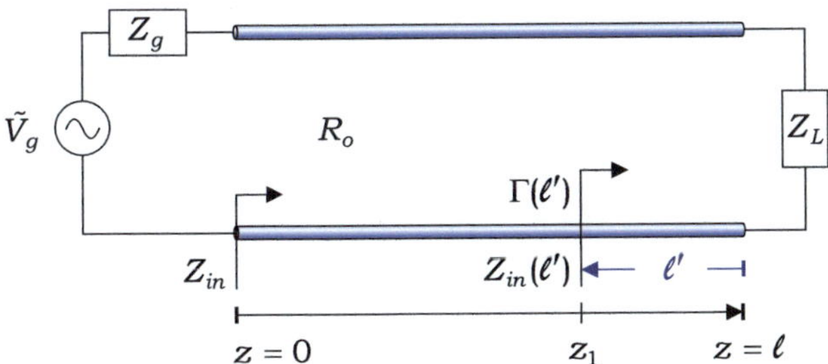

Fig. 9.10 Input impedance and reflection coefficient at distance ℓ' from the load

(a) Letting $\gamma = j\beta$, $z = z_1$, and $\ell' = \ell - z_1$ in Eq. (9.42a) gives

$$\tilde{V}(z_1) = V_o^+ e^{-j\beta z_1}\left[1 + |\Gamma| e^{j(\phi - 2\beta\ell')}\right] \tag{9.59a}$$

We recognize the first and second terms in brackets in Eq. (9.59a) as the forward and backward voltage waves at $z = z_1$. That is,

$$\tilde{V}^+(z_1) = V_o^+ e^{-j\beta z_1}$$

$$\tilde{V}^-(z_1) = V_o^+ e^{-j\beta z_1}|\Gamma| e^{j(\phi - 2\beta\ell')}$$

Thus, the reflection coefficient at a distance ℓ' from the load is given by

$$\Gamma(\ell') = \frac{\tilde{V}^-(z_1)}{\tilde{V}^+(z_1)} = |\Gamma| e^{j(\phi - 2\beta\ell')} \tag{9.59b}$$

(b) From Eq. (9.59a), it follows that $|\tilde{V}|_{\max}$ occurs at a point with $e^{j(\phi - 2\beta\ell')} = 1$ and $|\tilde{V}|_{\min}$ occurs at a point with $e^{j(\phi - 2\beta\ell')} = -1$. Substituting these conditions into Eq. (9.59b) yields

$$\Gamma(\ell') = |\Gamma| \quad : \text{at the location of } |\tilde{V}|_{\max} \tag{9.60a}$$

$$\Gamma(\ell') = -|\Gamma| \quad : \text{at the location of } |\tilde{V}|_{\min} \tag{9.60b}$$

It is worth noting that $\Gamma(\ell')$ is real at the locations of $|\tilde{V}|_{\max}$ and $|\tilde{V}|_{\min}$.

(c) Letting $\gamma = j\beta$ in Eqs. (9.42a) and (9.42b), we obtain *the input impedance looking into the section of length ℓ'* as follows:

$$\boxed{Z_{in}(\ell') = \frac{V_o^+ e^{-j\beta z_1}\left[1 + \Gamma e^{-j2\beta\ell'}\right]}{I_o^+ e^{-j\beta z_1}\left[1 - \Gamma e^{-j2\beta\ell'}\right]} = R_o \frac{1 + |\Gamma| e^{j(\phi - 2\beta\ell')}}{1 - |\Gamma| e^{j(\phi - 2\beta\ell')}}} \tag{9.61}$$

Using the conditions for the voltage maximum and voltage minimum, which are given by $e^{j(\phi - 2\beta\ell')} = \pm 1$, in Eq. (9.61), we have

$$Z_{in}(\ell') = R_o \frac{1 + |\Gamma|}{1 - |\Gamma|} \quad : \text{at the location of } |\tilde{V}|_{\max} \tag{9.62a}$$

$$Z_{in}(\ell') = R_o \frac{1 - |\Gamma|}{1 + |\Gamma|} \quad : \text{at the location of } |\tilde{V}|_{\min} \tag{9.62b}$$

Note that $Z_{in}(\ell')$ is real at the locations of $|\tilde{V}|_{\max}$ and $|\tilde{V}|_{\min}$.

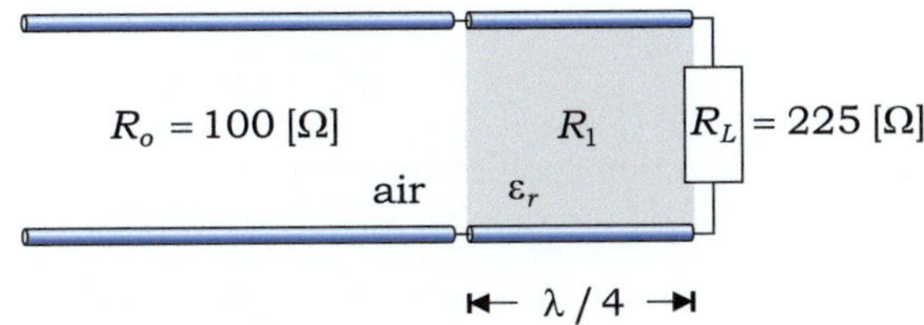

Fig. 9.11 Quarter-wave transformer

(d) Dividing both sides of Eq. (9.62a) by R_o leads to the conclusion that the normalized input impedance at the location of the voltage maximum is numerically equal to the standing-wave ratio on the line. That is,

$$z_{in}(\ell')\big|_{max} = \frac{1 + |\Gamma|}{1 - |\Gamma|} = S \tag{9.63}$$

Example 9.11 A 100 [Ω] lossless air line delivers the maximum possible power to a resistive load of 225 [Ω] through a *quarter-wave transformer* insulated by a material with $\varepsilon_r = 1.44$, as shown in Fig. 9.11. At an operating frequency of 2 [GHz], determine (a) the characteristic impedance and length of the quarter-wave transformer, and (b) the standing-wave ratios on the feedline and quarter-wave lines.

Solution

(a) The input impedance looking into the quarter-wave transformer is obtained using Eq. (9.57) as

$$Z_{in}(\ell' = \lambda/4) = \frac{R_1^2}{R_L}$$

For the maximum power transfer, $Z_{in}(\ell')$ should be equal to $R_o = 100\,[\Omega]$. Thus,

$$\boxed{R_1 = \sqrt{R_o R_L}} \tag{9.64}$$

This gives $R_1 = \sqrt{100 \times 225} = 150\,[\Omega]$. On the quarter-wave line, the phase velocity and phase constant are

$$\upsilon_p = \frac{1}{\sqrt{\mu_0 \varepsilon_0 \varepsilon_r}} = \frac{3 \times 10^8}{1.2} = 2.5 \times 10^8 \text{ [m/s]}$$

$$\beta = \frac{\omega}{\upsilon_p} = \frac{4\pi \times 10^9}{2.5 \times 10^8} = 50.265 \text{ [rad/s]}$$

Therefore, the length of the quarter-wave transformer is

$$\frac{\lambda}{4} = \frac{1}{4}\frac{2\pi}{50.265} = 3.1 \,[\text{cm}]$$

(b) The reflection coefficient at the load is

$$\Gamma = \frac{R_L - R_1}{R_L + R_1} = \frac{225 - 150}{225 + 150} = 0.2$$

The standing-wave ratio on the quarter-wave line is

$$S = \frac{1 + 0.2}{1 - 0.2} = 1.5$$

Because $Z_{in}(\ell') = R_o$, the reflection coefficient at the junction between the two lines is zero. Thus, the standing-wave ratio on the feedline is $S = 1$ and there is no backward wave on the feedline. Meanwhile, $S = 1.5$ on the quarter-wave line and the wave undergoes multiple reflections at the load and junction. However, there is no energy loss in the quarter-wave line. Consequently, the power carried by the forward wave traveling on the feedline is completely delivered to and dissipated in the load.

Exercise 9.22
For a lossless line with $Z_o = R_o$ terminated at Z_L, determine Z_{in} if the line is (a) one-quarter wavelength long, and (b) one-half wavelength long.

Ans. (a) $Z_{in} = R_o^2/Z_L$, (b) $Z_{in} = Z_L$.

Exercise 9.23
When is Z_{in} of a lossless line equal to Z_o?

Ans. The line is matched to a load or infinitely long.

Exercise 9.24
If Z_g is complex, determine Z_{in} for which the maximum power is transferred to the load. [Hint: Start with $Z_g = R_g + jX_g$ and $Z_{in} = R_{in} + jX_{in}$.]

Ans. $Z_{in} = Z_g^*$.

9.4.3 Short-Circuit and Open-Circuit Lines

Transmission lines find applications not only for transmitting power and information but also for fabricating circuit elements with inductive or capacitive reactance.

9.4.3.1 Short-Circuit Line

When a lossless line of length ℓ terminates in a short circuit ($Z_L = 0$), the input impedance at the input terminals can be obtained using Eq. (9.56) as

$$Z_{in}^s = jR_o \tan \beta\ell \tag{9.65}$$

For $0 < \beta\ell < \pi/2$ or $\ell < \lambda/4$, the input impedance of a short-circuited line is purely inductive (a positive imaginary number) and the line behaves like an inductor. In contrast, for $\pi/2 < \beta\ell < \pi$ or $\lambda/4 < \ell < \lambda/2$, the input impedance is purely capacitive (negative imaginary number) and the line behaves like a capacitor. Evidently, $\Gamma = -1$ at the shorted terminal; thus, a voltage minimum and a current maximum occur at the terminal (see Eqs. (9.42a) and (9.42b)). It is interesting to note that the input impedance becomes infinite for $\beta\ell = \pi/2$ or $\ell = \lambda/4$, and the short-circuited line as a whole behaves as an open circuit.

9.4.3.2 Open-Circuit Line

When a lossless line is terminated in an open circuit ($Z_L = \infty$), the input impedance is given by

$$Z_{in}^o = -jR_o \cot \beta\ell \tag{9.66}$$

It is purely capacitive for $0 < \beta\ell < \pi/2$, but purely inductive for $\pi/2 < \beta\ell < \pi$. Substituting $\Gamma = 1$ and $\ell' = 0$ into Eqs. (9.42a) and (9.42b) indicates that a voltage maximum and a current minimum occur at the open terminal. Moreover, the input impedance becomes zero if the line is one-quarter wavelength long ($\ell = \lambda/4$) and the open-circuit line behaves as a short circuit.

9.4.3.3 Shortening and Opening of Lossless Line

Input impedances Z_{in}^s and Z_{in}^o can be measured by shortening and opening the load end of the line, respectively. From these values, we can determine the characteristic impedance and electrical length of the line, as follows:

$$R_o = \sqrt{Z_{in}^s Z_{in}^o} \quad [\Omega] \tag{9.67a}$$

$$\tan^2 \beta\ell = -\frac{Z_{in}^s}{Z_{in}^o} \tag{9.67b}$$

Because the principal value of the arctangent is in the range of $-\pi/2$ to $\pi/2$, care should be taken so that $\beta\ell$ is positive (see Example 9.13).

Example 9.12 Design a capacitor with a capacitance of 2 [pF] by selecting the shortest possible length of a shorted 75 [Ω] lossless line, on which λ is 8 [cm] at 3 [GHz].

Solution

Letting $Z_{in}^s = 1/j\omega C$ in Eq. (9.65), we have

$$\tan\beta\ell = -\frac{1}{\omega C R_o}$$

$$= -\frac{1}{2\pi \times 3 \times 10^9 \times 2 \times 10^{-12} \times 75} = -0.354$$

The tangent function is negative when its argument is within the range of $-\pi/2$ to zero. However, $\beta\ell$ cannot be negative. Thus,

$$\beta\ell = \pi + \tan^{-1}(-0.354) = 2.801$$

The shortest line length is, therefore,

$$\ell = \frac{2.801 \times 0.08}{2\pi} = 3.57\,[\text{cm}]$$

Example 9.13 The lossless line is less than 5 [m] in length. When open circuited, $Z_{in}^o = -j81$ [Ω]. In contrast, when short circuited, $Z_{in}^s = j144$ [Ω] and the first voltage maximum is located 2.5 [m] from the shorted end. Determine (a) R_o, (b) β, and (c) ℓ.

Solution

(a) The characteristic impedance is, from Eq. (9.67a),

$$R_o = \sqrt{Z_{in}^s Z_{in}^o} = \sqrt{(j144)(-j81)} = 108\,[\Omega]$$

(b) A short circuit has $Z_L = 0$ and $\Gamma = -1$. Letting $\ell'_{max} = 2.5$ [m], $\phi = \pi$, and $m = 0$ in Eq. (9.46b) yields

$$\beta = 0.2\pi \qquad\qquad (9.68\text{a})$$

(c) From Eq. (9.67b),

$$\beta\ell = \tan^{-1}\left(\pm\sqrt{144/81}\right) + n\pi = \pm0.93 + n\pi \quad (n = 0, 1, 2..) \qquad (9.68\text{b})$$

Thus,

$$\ell = (\pm 0.93 + n\pi)/(0.2\pi)$$
$$= 1.48\,[\text{m}],\ 3.52\,[\text{m}],\ 6.48\,[\text{m}],\ 8.52\,[\text{m}],\ 11.48\,[\text{m}],\ \cdots$$

Because the line length ℓ must be longer than 2.5 [m], but shorter than 5 [m], the answer is $\ell = 3.52$ [m].

Exercise 9.25

Determine the length of a 100 [Ω] lossless air line terminated in a short circuit if its input impedance is equivalent to an inductor of 0.6 [μH] at 100 [MHz].

Ans. 0.63 [m].

Review Questions

RQ 9.31   Define input impedance Z_{in}.   [(9.53)]

RQ 9.32   What is the Z_{in} value of the matched transmission line?   [(9.55)]

RQ 9.33   What determines the values of $\tilde{V}_{in}$ and $\tilde{I}_{in}$?   [Fig. 9.9]

RQ 9.34   What are the differences between Z_{in} and $Z_{in}(\ell')$?   [Fig. 9.10]

RQ 9.35   What are the differences between Γ and $\Gamma(\ell')$?   [Fig. 9.10]

RQ 9.36   What is a quarter-wave transformer?   [Fig. 9.11]

RQ 9.37   Describe the change in Z_{in} if the short-circuited load end is   [(9.65)(9.66)]
           open.

RQ 9.38   What can be determined from the values of Z_{in}^{s} and Z_{in}^{o}?   [(9.67a,b)]

9.5   The Smith Chart

The Smith chart provides a graphical means of solving various transmission-line problems and allows us to avoid tedious manipulation of complex numbers. The Smith chart begins with a lossless transmission line with a characteristic resistance R_o terminated at a complex load impedance Z_L. If the real part of Z_L is allowed to vary from zero to infinity while its imaginary part is fixed, Γ traces a circle in a complex plane (see Eq. (9.40)). Similarly, if the imaginary part of Z_L varies between $-\infty$ and ∞, whereas the real part of Z_L is fixed, another circle is traced in the complex plane. The intersection of the two circles corresponding to the real and imaginary parts of Z_L determines the reflection coefficient of the load in the complex plane. Conversely, if we can locate a point in the complex plane corresponding to the reflection coefficient of the load, the load impedance can be determined by reading the values of the two circles passing through that point. Interestingly, many transmission-line problems are based on relations that are very similar in form to Eq. (9.41). By following a procedure similar to that used for load impedance, we can determine the standing-wave ratio and input impedance using the Smith chart.

9.5.1   *Relationship Between* Γ *and* Z_L

We consider a lossless transmission line with a characteristic resistance R_o that terminates at a complex load impedance Z_L. The voltage reflection coefficient at the load is given by Eq. (9.40) such that

$$\Gamma = \frac{Z_L - R_o}{Z_L + R_o}$$
$$\equiv \Gamma_R + j\Gamma_I \equiv |\Gamma|e^{j\phi} \tag{9.69}$$

where Γ_R and Γ_I are the real and imaginary parts of the reflection coefficient, respectively, and ϕ is the phase angle. Upon normalizing the load impedance Z_L to R_o, we have

$$z_L \equiv r + jx$$
$$= \frac{Z_L}{R_o} = \frac{R_L + jX_L}{R_o} \tag{9.70}$$

where R_L and X_L are the resistance and reactance of the load impedance, respectively, and r and x are the **normalized load resistance** and **normalized load reactance**, respectively. Subsequently, the relationship between Γ and z_L can be expressed as

$$\Gamma = \frac{z_L - 1}{z_L + 1} \tag{9.71a}$$

$$\boxed{z_L = \frac{1 + \Gamma}{1 - \Gamma} = \frac{1 + |\Gamma|e^{j\phi}}{1 - |\Gamma|e^{j\phi}}} \tag{9.71b}$$

$$r + jx = \frac{1 + \Gamma_R + j\Gamma_I}{1 - \Gamma_R - j\Gamma_I} \tag{9.71c}$$

Because z_L is complex, the right-hand side of Eq. (9.71a) is a point in the complex plane in which the abscissa and ordinate represent the real and imaginary parts of Γ, denoted by Γ_R and Γ_I, respectively. Separating the real and imaginary terms in Eq. (9.71c), we obtain

$$\boxed{\left(\Gamma_R - \frac{r}{1+r}\right)^2 + \Gamma_I^2 = \left(\frac{1}{1+r}\right)^2} \tag{9.72a}$$

$$\boxed{(\Gamma_R - 1)^2 + \left(\Gamma_I - \frac{1}{x}\right)^2 = \left(\frac{1}{x}\right)^2} \tag{9.72b}$$

From a mathematical point of view, Γ_R and Γ_I are orthogonal coordinates in the complex plane, whereas r and x are parameters ranging from 0 to ∞ and from $-\infty$ to ∞, respectively. For a fixed value of r, It is evident from Eq. (9.72a) that all possible combinations of Γ_R and Γ_I form a circle of radius $1/(1+r)$ centered at point $(r/(1+r),\ 0)$ in the complex plane, which is called a constant-r circle. Similarly, for a fixed value of x, we can see from Eq. (9.72b) that all possible combinations of Γ_R and Γ_I form a circle of radius $1/|x|$ centered at $(1,\ 1/x)$, which is called a constant-x circle. All constant-r circles are centered on the Γ_R axis, passing through point $(1, 0)$. Meanwhile, all constant-x circles are centered on the $\Gamma_R = 1$ line, passing through the same point $(1,\ 0)$. These circles are centered on $\Gamma_R = 1$ line in region $\Gamma_I > 0$ for $x > 0$ and in region $\Gamma_I < 0$ for $x < 0$. The $x = 0$ circle has an infinite radius and coincides with the Γ_R axis. In contrast, the $r = 0$ circle has a radius of unity and coincides with the $|\Gamma| = 1$ circle or the perimeter of the Smith chart.

When a load impedance with $r = r'$ and $x = x'$ is given, the reflection coefficient at the load, of course, can be obtained directly from Eq. (9.71a) or by solving Eqs. (9.72a) and (9.72b) for Γ_R and Γ_I. The latter method can be performed graphically. That is, we first locate the intersection of the constant-r' and constant-x' circles in the complex plane and then read its abscissa and ordinate to obtain Γ_R' and Γ_I', respectively, as illustrated in Fig. 9.12. The two circles also intersect at point $(1, 0)$. However, this is a trivial solution because all circles in the complex plane pass through it. Furthermore, it can be shown that a constant-r and a constant-x circle always intersect at a right-angle. Because $|\Gamma|$ has a maximum value of unity, the points within and on the circle of unity radius in the complex plane are meaningful.

The Smith chart contains two families of circles: constant-r and constant-x circles. The constant-x circles with positive x values reside in the upper-half space, whereas those with negative values reside in the lower-half space. If the intersection of the $r = r'$ and $x = x'$ circles is identified in the Smith chart (marked as A' in Fig. 9.12), the magnitude of the reflection coefficient can be determined by measuring the distance from the origin to A' and dividing it by the radius of the $|\Gamma| = 1$ circle, which is the perimeter of the Smith chart. The phase angle of the reflection coefficient can be obtained by measuring the rotation angle of $\overline{OA'}$ with respect to the positive Γ_R-axis.

Example 9.14 A $100\ [\Omega]$ lossless line is terminated at a load of $Z_L = 160 + j100\ [\Omega]$. (a) Determine Γ using a Smith chart. (b) When the line terminates at a new load impedance Z_L', the measurements yield $\Gamma = 0.71\angle -116^\circ$. Determine Z_L'.

Solution

(a) The normalized load impedance is $z_L = Z_L/100 = 1.6 + j1.0$. The intersection of the $r = 1.6$ and $x = 1.0$ circles is marked as A in the Smith chart, as shown in Fig. 9.13. Extend $\overline{OA}$ to point B on the circle of $|\Gamma| = 1$ and read 38° on the "ANGLE OF REFLECTION COEFFICIENT IN DEGREES" scale. Next, the ratio $\overline{OA}/\overline{OB}$ is measured to be $20.5\,[\text{mm}]/49\,[\text{mm}] = 0.42$. Thus,

$$\Gamma = 0.42\angle 38^o \qquad\qquad (9.73a)$$

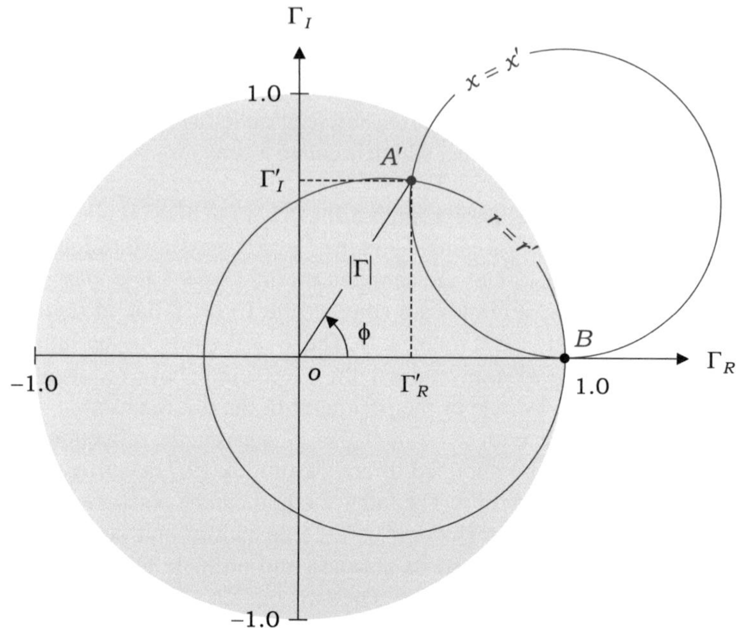

Fig. 9.12 Reflection coefficient $\Gamma = \Gamma_R' + j\Gamma_I' = |\Gamma|e^{j\phi}$ of the normalized load impedance $z_L = r' + jx'$ is denoted by A'

(b) The point with $-116°$ on the "ANGLE OF REFLECTION COEFFICIENT IN DEGREES" scale is marked as D on the circle of $|\Gamma| = 1$ in Fig. 9.13. With $\overline{OD} = 49$ [mm], we obtain $|\Gamma| = 0.71 \times 49 = 34.8$ [mm]. Move a distance of 34.8 [mm] from the center along $\overline{OD}$ to point C, and read $r = 0.225$ and $x = -0.6$. Thus,

$$Z_L = 22.5 - j60 \,[\Omega] \qquad\qquad (9.73b)$$

To verify the result of Eq. (9.73b), we insert Γ given in the problem into Eq. (9.71b) and obtain $Z_L = 23.32 - j60.12$ [Ω], which shows that the error introduced by the Smith chart is less than 4%.

Exercise 9.26

What are the values of r and x for the following points in the Smith chart: (a) center, (b) extreme left, and (c) top?

Ans. (a) $r = 1$ and $x = 0$, (b) $r = 0$ and $x = 0$, (c) $r = 0$ and $x = 1$.

Exercise 9.27

Use the Smith chart to determine Γ of the following normalized load impedances: (a) $z_L = 1$ (matched load), (b) $z_L = 0$ (short circuit), and (c) $z_L = \infty$ (open circuit).

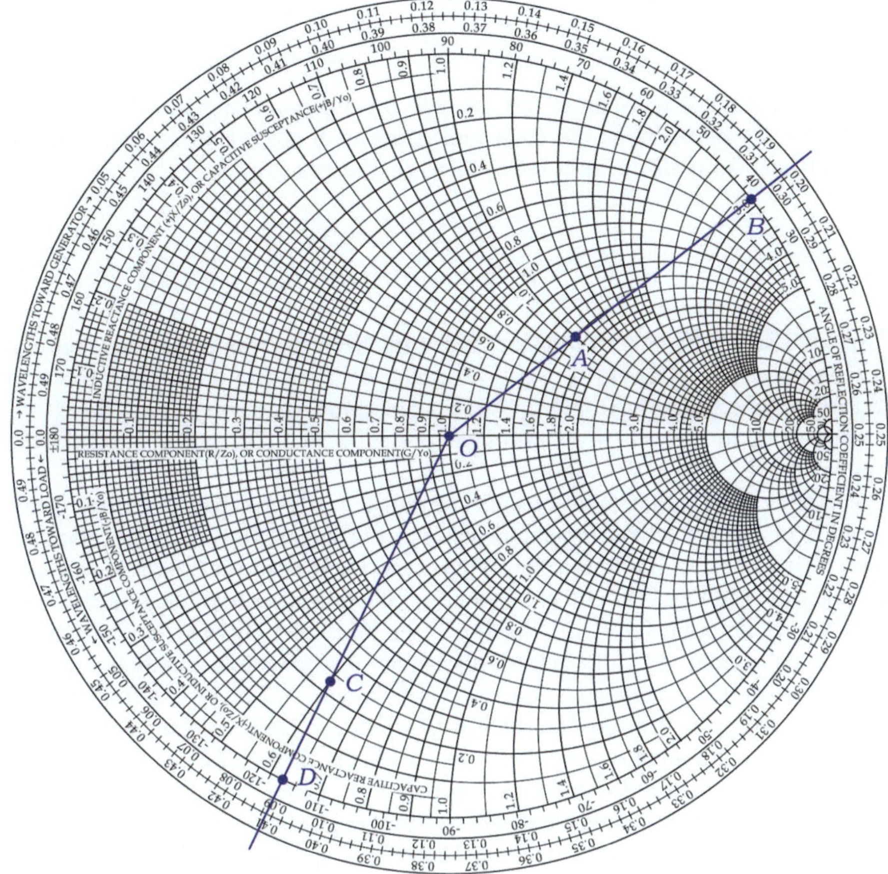

Fig. 9.13 Smith chart for Example 9.14

Ans. (a) $\Gamma = 0$, (b) $\Gamma = -1$, (c) $\Gamma = 1$.

Exercise 9.28
Use the Smith chart to determine z_L, which corresponds to the voltage reflection coefficients, (a) $\Gamma = -0.5$, (b) $\Gamma = j$, and (c) $\Gamma = 1\angle -110^o$.

Ans. (a) $z_L = 0.33$, (b) $z_L = j$, (c) $z_L = -j0.7$.

9.5.2 Relationship Between Γ and Z_{in}

When a lossless line of length ℓ with characteristic resistance R_o is terminated at load impedance Z_L, the input impedance of the line is given by Eq. (9.54):

$$Z_{in} = R_o \frac{1 + |\Gamma|e^{j(\phi - 2\beta\ell)}}{1 - |\Gamma|e^{j(\phi - 2\beta\ell)}} \tag{9.74}$$

where β is the phase constant of the line, and ϕ is the phase angle of the reflection coefficient. By normalizing the input impedance with respect to the characteristic resistance, the normalized input impedance is expressed as

$$\boxed{z_{in} \equiv \frac{Z_{in}}{R_o} = \frac{1 + |\Gamma|e^{j(\phi - 2\beta\ell)}}{1 - |\Gamma|e^{j(\phi - 2\beta\ell)}}} \tag{9.75}$$

This expression is analogous to that of the normalized load impedance given in Eq. (9.71b), where Γ is related to $z_L = r + jx$ through constant-r and constant-x circles in the Smith chart. Similarly, from Eq. (9.75), a new complex quantity $\overline{\Gamma} \equiv |\Gamma|e^{j(\phi - 2\beta\ell)}$ can be related to $z_{in} = r' + jx'$ through constant-r' and constant-x' circles in the Smith chart.

Therefore, for a lossless line of length ℓ terminated at a normalized load impedance $z_L = r + jx$, the normalized input impedance can be obtained by taking the following steps. First, we locate the intersection of the r- and x-circles on the Smith chart, which is marked as A in Fig. 9.14 and corresponds to the reflection coefficient of the load. Next, we locate point for $\overline{\Gamma} \equiv |\Gamma|e^{j(\phi - 2\beta\ell)}$ on the Smith chart by decreasing the phase angle of point A by $2\beta\ell$ [rad]. This can be performed graphically by rotating $\overline{OA}$ by $2\beta\ell$ [rad] in the clockwise direction or by moving the tip of $\overline{OA}$ from point A to point A' along the "circle of constant $|\Gamma|$," which is a circle centered at the origin and passes through point A. At point A', we read new circle parameters, $r = r'$ and $x = x'$, to obtain the normalized input impedance as $z_{in} = r' + jx'$.

Extending the input impedance measured at the generator end of a transmission line to any point on the line, the input impedance looking toward the load at a distance ℓ' from the load is defined as

$$Z_{in}(\ell') = R_o \frac{1 + |\Gamma|e^{j(\phi - 2\beta\ell')}}{1 - |\Gamma|e^{j(\phi - 2\beta\ell')}} \tag{9.76}$$

Note that Eq. (9.76) differs from Eq. (9.74) only in quantity ℓ'; that is, $Z_{in}(\ell')$ is the input impedance as if the transmission line is ℓ' long. Therefore, we can obtain $z_{in}(\ell') = Z_{in}(\ell')/R_o$ in the Smith chart by following the same procedure used for z_{in}. In other words, we locate the point for $\overline{\Gamma} = |\Gamma|e^{j(\phi - 2\beta\ell')}$ on the Smith chart to read the r and x values. More specifically, we enter the Smith chart at point for Γ and move on the circle of constant $|\Gamma|$ to point for $\overline{\Gamma}$ by decreasing the polar angle by $2\beta\ell'$ [rad] and read the values of r and x.

One complete turn on the circle of constant $|\Gamma|$ is equivalent to a change in ℓ' by half the wavelength on the transmission line, as $\ell' = \lambda/2$ for $2\beta\ell' = 2\pi$. To denote the change in ℓ' on a transmission line, a scale called the "wavelengths toward generator" (*wtg*) is constructed around the perimeter of the Smith chart. *wtg* is measured in units of wavelength and increases in the clockwise direction. It is customary to set the zero

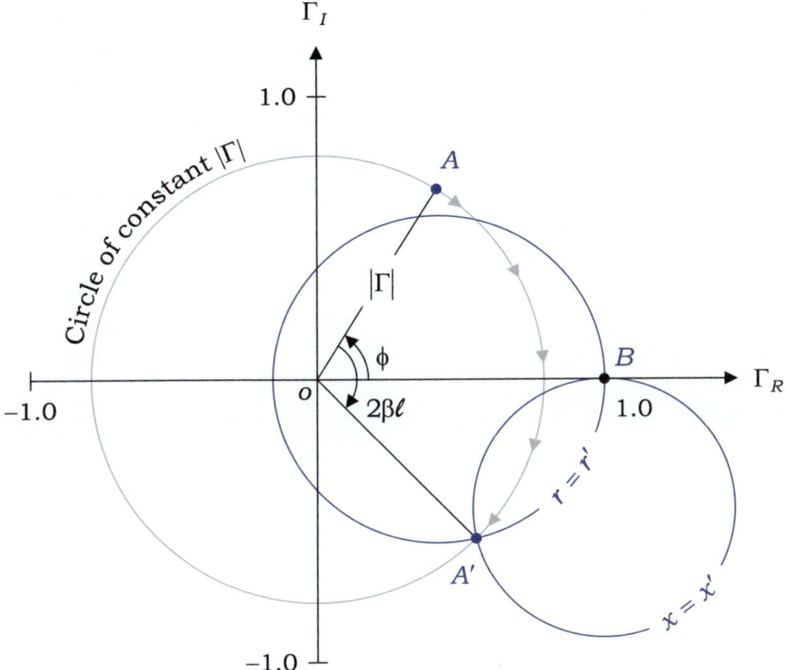

Fig. 9.14 The point for Γ is marked as A and the point for $\overline{\Gamma} \equiv |\Gamma|e^{j(\phi - 2\beta\ell)}$ is denoted as A' on a circle of constant $|\Gamma|$. At A', the reading yields $z_{in} = r' + jx'$

point of the *wtg* scale on the negative Γ_R-axis. If it is necessary to move toward the load, it would be convenient to use another scale called the "wavelengths toward load" constructed just inside the *wtg* scale, which is the exact reverse of the *wtg* scale. The innermost scale is labeled as "angle of reflection coefficient in degrees," which is a protractor. The zero point of this scale is on the positive Γ_R-axis and the angle increases in a counterclockwise direction. As far as moving along a transmission line is concerned, any of these scales will serve this purpose; a change in position on a transmission line is translated into a change in the polar angle on the Smith chart.

Example 9.15 A lossless line of length 0.169λ is terminated with a load impedance that induces $\Gamma = 0.73\angle 55^o$. Determine the normalized input impedance using a Smith chart.

Solution

The phase angle $\phi = 55^o$ is converted into a point on the *wtg* scale:

$$wtg = \frac{\lambda}{2}\frac{(180^o - 55^o)}{360^o} = 0.174\lambda$$

The subtraction in the numerator reflects the fact that ϕ increases in the counter-clockwise direction starting from the positive Γ_R-axis, whereas *wtg* increases in the clockwise direction starting from the negative Γ_R-axis. The factor $\lambda/2$ signifies that a complete turn along the circle of constant $|\Gamma|$ corresponds to a half-wavelength on the transmission line.

The point for Γ must be positioned on the line from the origin to point $\overline{A}$ on the perimeter with $wtg = 0.174\lambda$ and at distance $0.73 \times o\overline{A}$ from the origin. This point is marked A in Fig. 9.15. Next, moving along a circle of constant $|\Gamma|$ and increasing wtg to $0.174\lambda + 0.169\lambda = 0.343\lambda$, we can locate a point for $\overline{\Gamma}$ at point B, where we read $r = 0.49$ and $x = -1.40$. Thus, the normalized input impedance is given by $z_{in} = 0.49 - j1.40$.

Exercise 9.29
If the phase angle of Γ is $-40°$, what is the *wtg* reading?

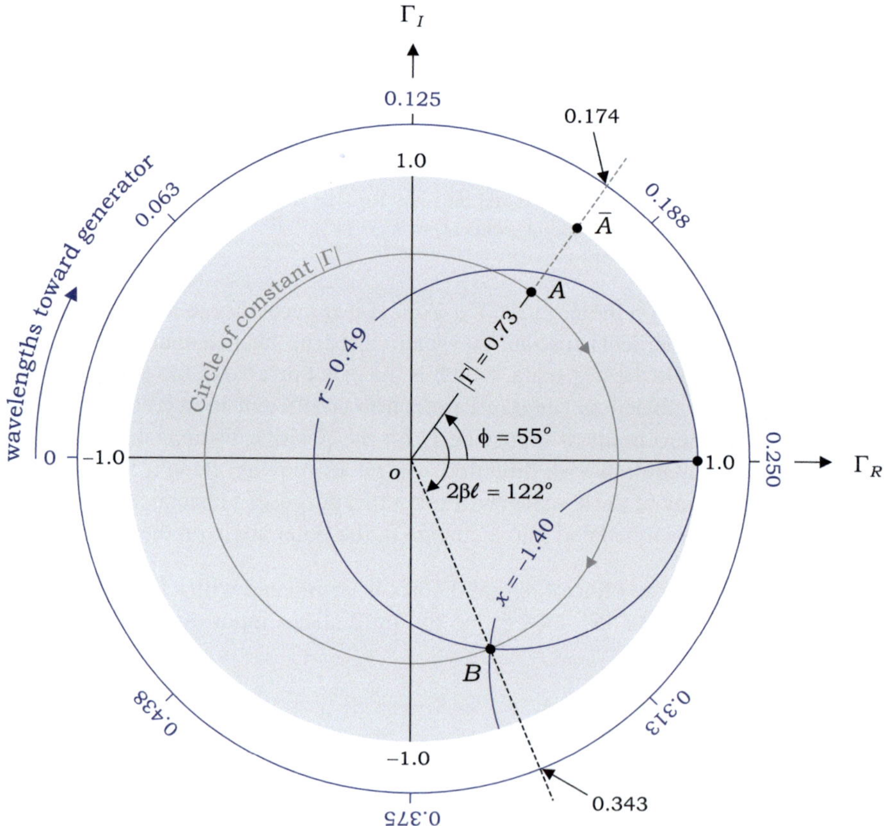

Fig. 9.15 When the reflection coefficient is marked as A, the normalized input impedance is found at point B

Ans. 0.306λ.

Exercise 9.30
For a lossless line of length $\ell = 0.122\lambda$, determine the difference in phase angles between Γ and z_{in} in (a) degrees, and (b) *wtg*.

Ans. (a) $87.8°$, (b) 0.122λ.

Exercise 9.31
If $Z_{in} = j50\,[\Omega]$ for a short-circuited $100\,[\Omega]$-lossless transmission line, determine the line length in wavelengths using a Smith chart.

Ans. $\ell = 0.074\lambda$.

9.5.3   *Relationship Between Γ and Standing-Wave Ratio*

The reflection coefficient at a purely resistive load R_L connected to a lossless line with a characteristic resistance R_o is a positive real number if R_L is greater than R_o, as is evident from the following equation:

$$\Gamma = \frac{R_L - R_o}{R_L + R_o} \tag{9.77}$$

Rewriting Eq. (9.77) with $\Gamma = |\Gamma|$, we obtain

$$\boxed{\frac{R_L}{R_o} = \frac{1 + |\Gamma|}{1 - |\Gamma|} = S} \tag{9.78}$$

which is numerically equal to the standing-wave ratio on the transmission line.

Consider a lossless line with a characteristic resistance R_o that terminates at an arbitrary load impedance Z_L, as shown in Fig. 9.16a. With respect to the input impedance $Z_{in}(\ell')$, moving a distance ℓ' away from the load on the transmission line corresponds to moving on a circle of constant $|\Gamma|$ in the clockwise direction on the Smith chart. On the circle of constant $|\Gamma|$, the point for $|\Gamma|e^{j(\phi - 2\beta\ell')}$ is clearly positioned on the positive Γ_R-axis if $\phi - 2\beta\ell' = 0$. In this case, the corresponding input impedance is purely resistive, such that $Z_{in}(\ell') = R'_{in}$. Next, if the transmission line is cut off at a distance ℓ' from the load and terminated with a load resistance R'_{in}, as shown in Fig. 9.16b, there is no change in the wave traveling on the shortened line. Moreover, from Eq. (9.78), the standing-wave ratio on the shortened line is numerically equal to its normalized *load* resistance, such that $S = R'_{in}/R_o$. In other words, the standing-wave ratio on the original transmission line is numerically equal to the normalized *input* impedance looking toward the load at a distance ℓ' from the load if $z_{in}(\ell')$ is a real number greater than or equal to unity. The standing-wave ratio can be determined at a point on the positive Γ_R-axis of the Smith chart.

(a)

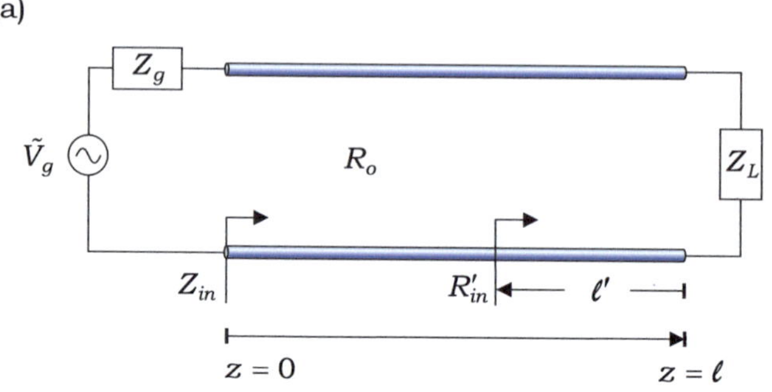

(b)

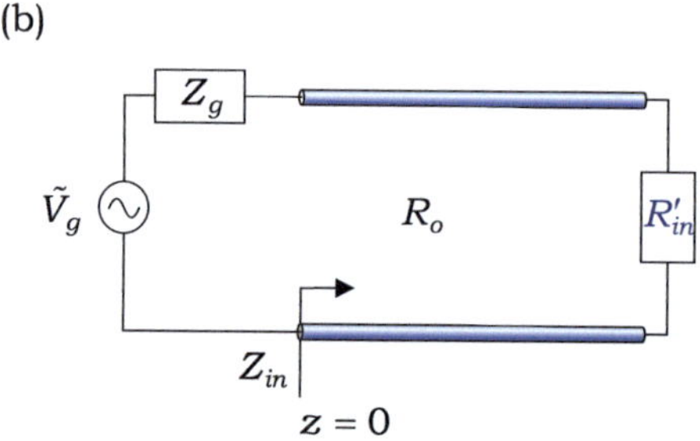

Fig. 9.16 **a** Positive real-valued *input* impedance, R'_{in}, is obtained at distance ℓ' from the load. **b** Shortened line is terminated with *load* impedance R'_{in}, on which $S = R'_{in}/R_o$

Example 9.16 The reflection coefficient is $\Gamma = 0.73\angle 55^\circ$ for an unknown load impedance connected to a lossless line. Determine the standing-wave ratio using a Smith chart.

Solution

The point for Γ is located on the Smith chart and is denoted as A, as shown in Fig. 9.17, such that $\overline{oA}$ makes an angle of 55° with the horizontal axis and has a length as large as seventy-three hundredths of the radius of the Smith chart. Next, we find the intersection of the circle of constant $|\Gamma|$ and the positive Γ_R-axis, and denote it as B. At this point, $r = 6.4$ and $x = 0$. Thus, the standing-wave ratio on the transmission line is $S = 6.4$.

It is evident from Eq. (9.78) that the *circle of constant* $|\Gamma|$ perfectly overlaps the **circle of constant** S (standing-wave ratio). Although the points on the constant-S circle may correspond to different load impedances, they produce the same standing-wave ratio on the transmission line. The value of $|\Gamma|$ is determined by the ratio of the radius of the constant-$|\Gamma|$ circle to that of the Smith chart, whereas S is determined by the value r of the constant-r circle touching the constant-$|\Gamma|$ circle. The constant-S circle intersects the Γ_R-axis at points B and C as shown in Fig. 9.17. From Eqs.

(9.46a) and (9.47a), points B and C correspond to the positions of $|\tilde{V}|_{\max}$ and $|\tilde{V}|_{\min}$ on the transmission line, respectively.

Example 9.17 The 100 [Ω]-lossless line is 0.24λ long. If it is terminated at a load with impedance $Z_L = 140 + j130\,[\Omega]$, use a Smith chart to obtain.

(a)   voltage reflection coefficient,
(b)   input impedance,
(c)   standing-wave ratio, and
(d)   distance of the first voltage maximum from the load.

Solution

(a)   The normalized load impedance is

$$z_L = (140 + j130)/100 = 1.4 + j1.3$$

The intersection of the $r = 1.4$ and $x = 1.3$ circles is marked A in the Smith chart (Fig. 9.18), and we obtain.

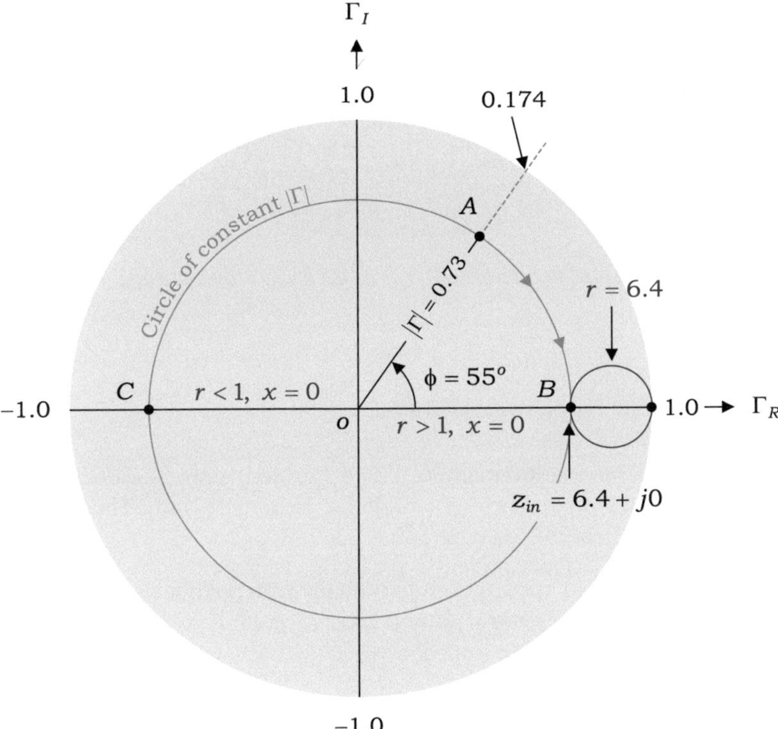

Fig. 9.17 Reflection coefficient marked as A and standing-wave ratio found at point B

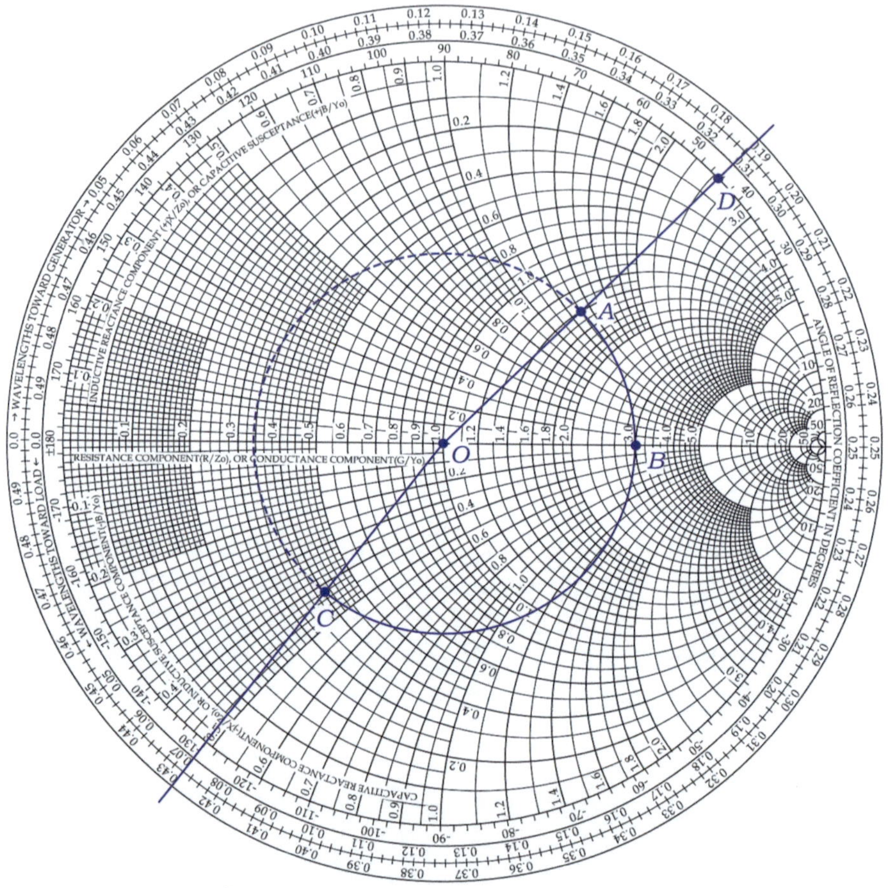

Fig. 9.18 Smith chart is used to determine Γ, z_{in}, S, and the location of $|\tilde{V}|_{\max}$

(1) $\dfrac{\overline{OA}}{\overline{OD}} = 0.5 = |\Gamma|$, and

(2) the *wtg* value for point A is 0.188λ.

The difference in *wtg* between $\overline{OA}$ and the $+\Gamma_R$-axis is 0.062λ, which is converted into a polar angle of $\phi = 0.062 \times 4\pi = 0.78$ [rad]. Thus, the voltage reflection coefficient is $\Gamma = 0.5e^{j0.78}$.

(b) A line length of 0.24λ corresponds to moving from point A to point C along the $|\Gamma| = 0.5$ circle, increasing *wtg* by 0.24λ. At point C, we read $r = 0.40$ and $x = -0.42$. Thus, the input impedance is $Z_{in} = 40 - j42\,[\Omega]$.

(c) The $|\Gamma| = 0.5$ circle crosses the $+\Gamma_R$-axis at point B, where $r = 3.0$ and $x = 0$. Thus, the standing-wave ratio is $S = 3.0$.

(d)   As we see from Eq. (9.45), a voltage maximum occurs at a distance ℓ' from the load, where $e^{j(\phi - 2\beta\ell')} = 1$ is satisfied. Thus, the normalized input impedance $z_{in}(\ell')$ is given by a positive real number (see Eq. (9.61)). Therefore, point B corresponds to the position of the voltage maximum. The difference in *wtg* between points A and B is equal to the distance of the first voltage maximum from the load. Consequently, the answer is $\ell' = 0.062\lambda$.

Exercise 9.32
A lossless line terminates at the load impedance corresponding to (a) point A, and (b) point C in the Smith chart, as shown in Fig. 9.13. Determine the SWR.

Ans. (a) $S = 2.4$, (b) $S = 6.0$.

Exercise 9.33
On a lossless line, the SWR is measured as 2.0, and the first voltage maximum appears at 0.461λ from the load. Determine the normalized load impedance using a Smith chart.

Ans. $z_L = 1.7 - j0.6$.

9.5.4   *Admittance on the Smith Chart*

The admittance is the reciprocal of the impedance. The use of admittances can greatly simplify the solution to certain transmission-line problems. Let us begin by considering a lossless line with characteristic resistance R_o, which is one-quarter wavelength long and terminates at load impedance Z_L. By setting $\beta\ell = \pi/2$ in Eq. (9.56), the input impedance is given by

$$Z_{in} = R_o \frac{Z_L + jR_o \tan(\pi/2)}{R_o + jZ_L \tan(\pi/2)} = \frac{R_o^2}{Z_L}$$

With the admittances $Y_L = 1/Z_L$ and $Y_o = 1/R_o$, this can be rewritten as

$$\boxed{\frac{Z_{in}}{R_o} = \frac{Y_L}{Y_o}} \tag{9.79}$$

where the left-hand side represents the normalized input impedance, denoted by z_{in}, and the right-hand side represents the normalized load admittance, denoted by y_L. Therefore,

$$\boxed{z_{in} = y_L \equiv g + jb} \tag{9.80}$$

where g is the **normalized load conductance**, and b is the **normalized load susceptance**. The normalized input impedance of the quarter-wave line is equal to the

Fig. 9.19 The normalized
input impedance of the
quarter-wave line is
numerically equal to the
normalized *load admittance*

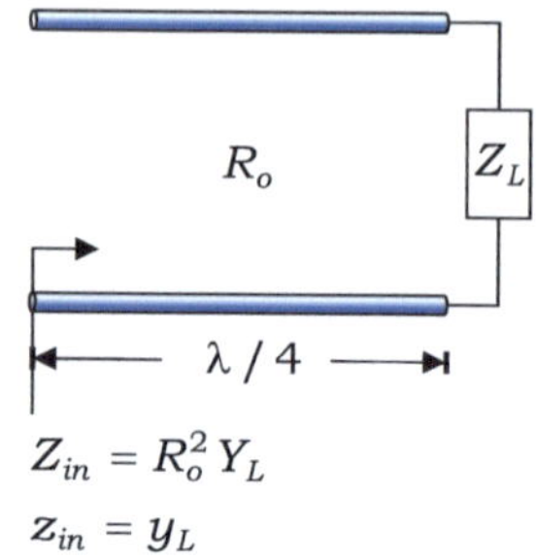

normalized load admittance. Moreover, by definition, the normalized load admittance
is given by

$$\boxed{\; y_L \equiv \frac{Y_L}{Y_o} = R_o Y_L \;}\qquad\qquad (9.81)$$

The relationship in Eq. (9.79) is illustrated in Fig. 9.19.

   The Smith chart may also be used as an admittance chart, for which the constant-r
circles are considered constant-g circles, and the constant-x circles are considered
constant-b circles. With the help of Eq. (9.80), we obtain the load admittance on the
Smith chart as follows: first, given a normalized load impedance $z_L = r' + jx'$, the
Smith chart is entered at the intersection of the $r = r'$ and $x = x'$ circles, marked
as A in Fig. 9.20. Next, the normalized input impedance at distance $\lambda/4$ from the
load is determined by moving to point B on the circle of constant $|\Gamma|$, which can be
accomplished by reducing the phase angle of A by $180°$ or increasing *wtg* by 0.25λ.
Subsequently, we read $r = r''$ and $x = x''$ at point B. Therefore, the normalized input
impedance is $z_{in} = r'' + jx''$ and the normalized load admittance is $y_L = r'' + jx''$.
Finally, the load admittance is given by

$$Y_L = \frac{z_{in}}{R_o} = \frac{r'' + jx''}{R_o}\qquad [\text{S}]\qquad\qquad (9.82)$$

where R_o is the characteristic resistance of the transmission line. The admittance is
measured in units of siemens.

Example 9.18 A 280 [Ω] lossless line of length 0.113λ is terminated at a load with
impedance $Z_L = 168 + j161$ [Ω]. Use a Smith chart to determine (a) load admittance
Y_L, and (b) input admittance Y_{in}.

Solution

The normalized load impedance is $z_L = 0.60 + j0.575$.

The point for z_L is marked A in the Smith chart, as shown in Fig. 9.21, and is located
at 0.104λ on the *wtg* scale. Move along the circle of constant $|\Gamma|$ by a *wtg* of 0.113λ

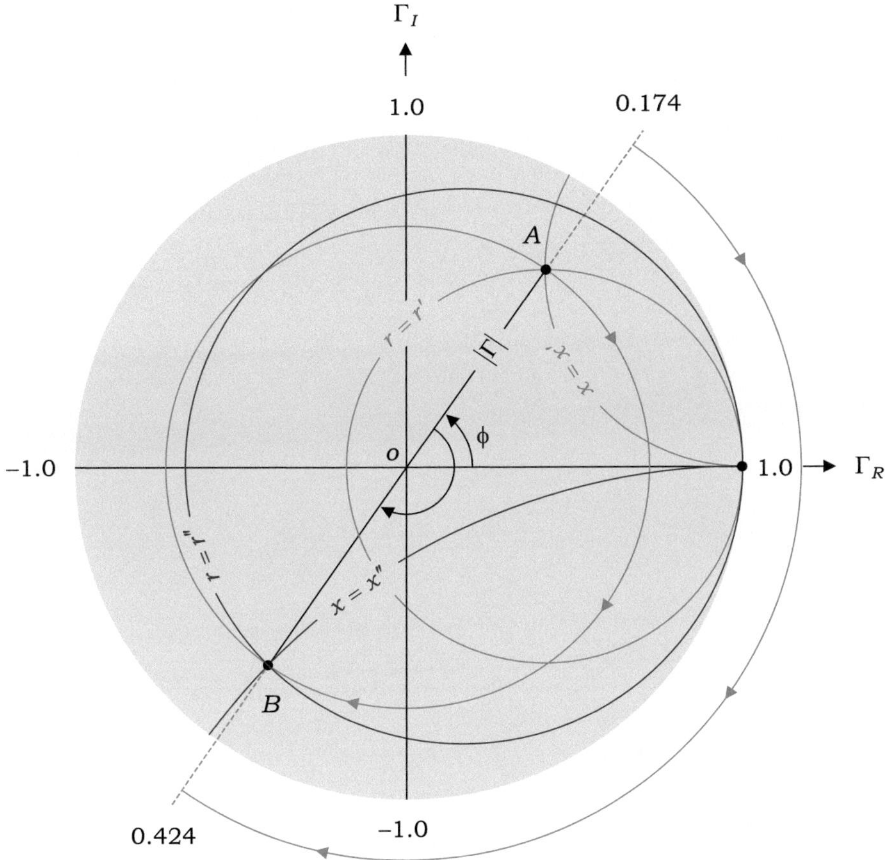

Fig. 9.20 Given z_L, marked as A, y_L is found at point B, which is the diametrical image of point A on the circle of constant $|\Gamma|$

to point B, where we read $r = 2.0$ and $x = 0.8$. Thus, the input impedance is

$$Z_{in} = 280 \times (2.0 + j0.8) = 560 + j224\,[\Omega]$$

Next, the diametrical images of A and B are located on the circle of constant $|\Gamma|$ and are marked as C and D, respectively. At point C, we read $g = 0.87$ and $b = -0.83$ to obtain the load admittance,

$$Y_L = (0.87 - j0.83)/280 = 3.1 - j3.0\,[\text{mS}]$$

At point D, we read $r = 0.43$ and $x = -0.17$ to obtain the input admittance,

$$Y_{in} = (0.43 - j0.17)/280 = 1.5 - j0.6\,[\text{mS}]$$

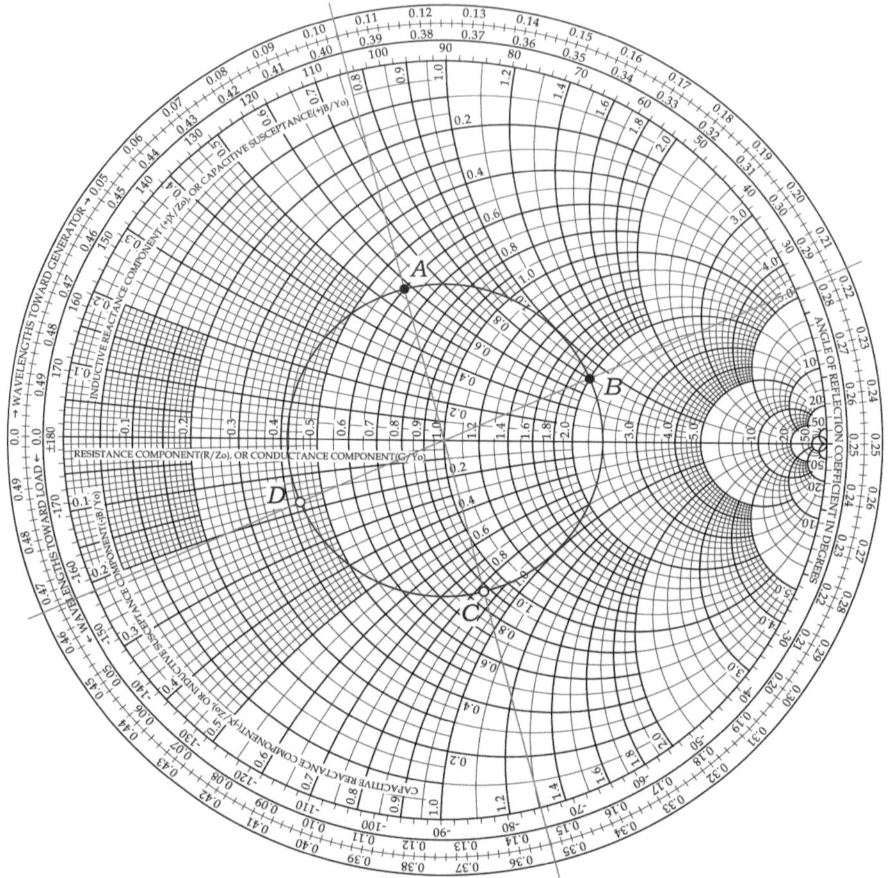

Fig. 9.21 Point A, z_L; Point B, z_{in}; Point C, y_L; Point D, y_{in}. Points C and D are diametrically opposite to points A and B, respectively

Again, the difference in *wtg* between points C and D is equal to the line length.

Exercise 9.34

Solve $\coth(a + jb) = 0.5e^{j0.78}$ for constants a and b using a Smith chart. [Hint: Rewrite it in the form of Eq. (9.71b).]

Ans. $a = 0.32$ and $b = -1.19$.

9.5.5   *Transmission Line Impedance Matching*

A lossless transmission line can transfer maximum possible power to a load if the load impedance is equal to the characteristic resistance of the line. In this case,

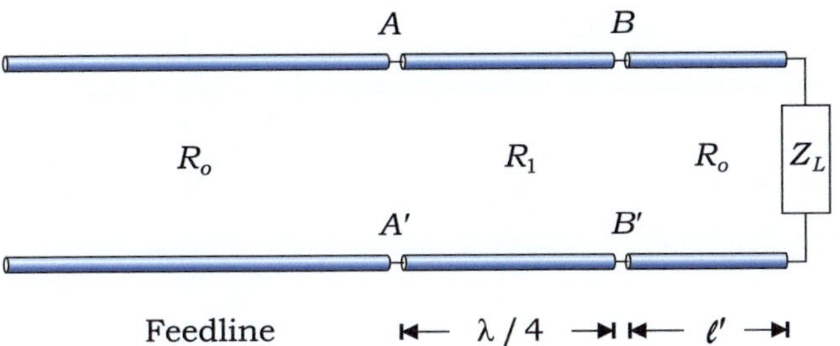

Fig. 9.22 Quarter-wave transformer as the matching network

there is no standing wave on the line because the reflection coefficient of the load is zero. However, in practical situations, it may be impossible to match a load to a transmission line if the load is designed to perform other tasks independently. Alternatively, a matching network may be inserted between the transmission line and load. Transformers and stub-matching networks are commonly used as impedance-matching networks.

9.5.5.1 Quarter-Wave Transformer

A quarter-wave transformer can be used as the matching network, as shown in Fig. 9.22. For a lossless line with characteristic resistance R_o connected to load impedance Z_L, the transmission line is first cut into two pieces at a distance ℓ' from the load such that the input impedance of the ℓ'-section is pure resistance R_{in}. A quarter-wave line with a characteristic resistance of $R_1 = \sqrt{R_o R_{in}}$ is then inserted in between, which ensures that the input impedance at terminal AA' looking into the load is R_o. Consequently, the forward wave traveling on the feedline experiences no reflection at terminal AA', even if multiple reflections can be observed within the quarter-wave line or ℓ'-section. The power delivered to terminal AA' through the feedline is completely dissipated in the load, because there is no power loss in the lossless lines.

9.5.5.2 Single-Stub Method

The input impedance at distance ℓ' from the load can be altered by connecting a short- or open-circuited stub in parallel with the main line. However, an open stub radiates undesirable electromagnetic energy at high frequencies, thereby interfering with neighboring devices. In most practical situations, a short-circuited stub with the same characteristic impedance as the main line is connected parallel to the line for impedance matching. This method is known as the ***single-stub method***. Because we consider parallel connections, it is more convenient to use admittances to match the load to a transmission line.

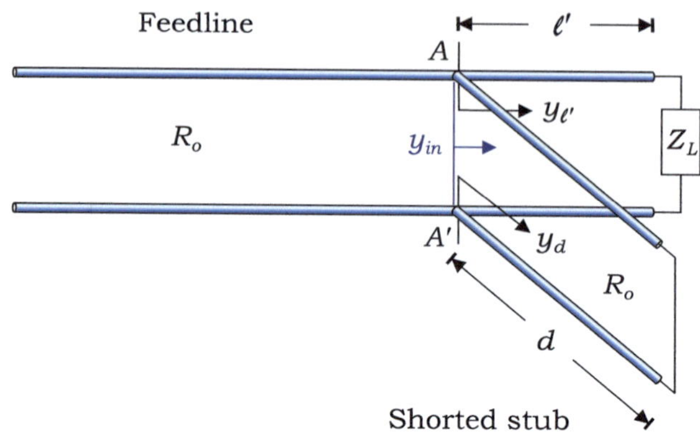

Fig. 9.23 Single-stub method for impedance matching

A typical single-stub matching network is shown in Fig. 9.23, in which a short-circuited stub of length d is connected in parallel with the main line at distance ℓ' from the load. With two degrees of freedom provided by ℓ' and d, we can make the normalized input admittance at junction AA' unity, that is, $y_{in} = 1$. We begin with the condition that the normalized input admittance of a shorted stub is imaginary:

$$y_d = jb_d \tag{9.83}$$

where the susceptance b_d may be either a positive or negative real number. We desire the normalized input admittance of the main line looking toward the load at junction AA' without the stub to be

$$y_{\ell'} = 1 - jb_d \tag{9.84}$$

Thus, the normalized input admittance at the junction with the stub is given by

$$y_{in} = y_{\ell'} + y_d = 1 \tag{9.85}$$

This is called **the impedance matching condition**.

Example 9.19 Load impedance $Z_L = 25 - j75\,[\Omega]$ is connected to a $100\,[\Omega]$-lossless line. Determine location ℓ' and length d of the short-circuited stub connected in parallel for impedance matching if ℓ' and d should be as small as possible.

Solution

The normalized load impedance is $z_L = 0.25 - j0.75$.

The intersection of the $r = 0.25$ and $x = -0.75$ circles is marked A in the Smith chart, as shown in Fig. 9.24, which is located at 0.394λ on the *wtg* scale.

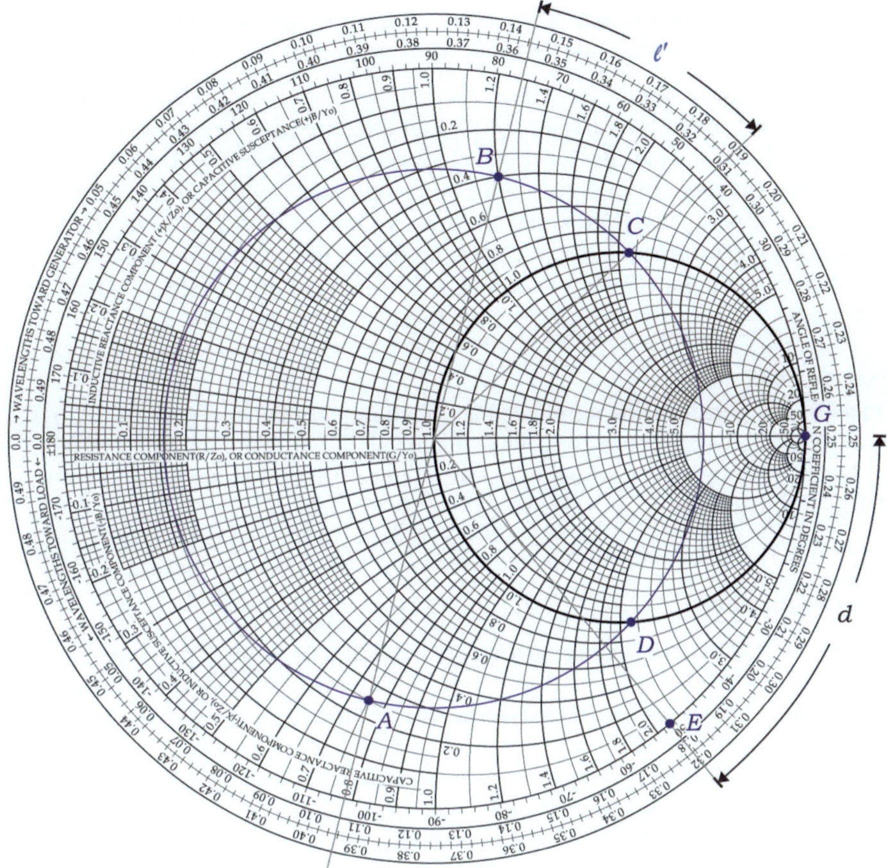

Fig. 9.24 Impedance matching with a single shorted stub

A circle of constant-$|\Gamma|$ is drawn to pass through A. The diametrical image of point A is marked as B, which represents y_L and has a *wtg* value of 0.144λ.

Next, we locate the intersection of the circle of constant-$|\Gamma|$ and the $g = 1$ circle (or $r = 1$ circle) and mark the two intersections with C and D.

At point C with $wtg = 0.19\lambda$, we read $g = 1$ and $b = 2.1$. The difference in *wtg* between points B and C is 0.046λ. Thus, the shorted stub is connected to the main line at a distance $\ell' = 0.046\lambda$ from the load.

The shorted stub has a load impedance $z_L = 0$ that corresponds to the extreme-left point on the Γ_R-axis. The diametrical image is point G (the extreme-right point on the Γ_R-axis). On the circle of constant-$|\Gamma|$ through G or the perimeter of the Smith chart, moving to point E, where the $b = -2.1$ circle intersects the perimeter, *wtg* increases by 0.071λ. Thus, the length of the shorted stub is $d = 0.071\lambda$.

Alternatively, point D may have been used in the single-stub method. In this case, we read $g = 1$ and $b = -2.1$ at point D, and the difference in *wtg* between points B and D is 0.166λ. Thus, the junction is located at a distance $\ell' = 0.166\lambda$ from the load. Next, we move from G to a point on the perimeter where the $b = 2.1$ circle intersects the perimeter. At this intersection, we read 0.179λ on the *wtg* scale. As the moving is performed in the clockwise direction, the *wtg* difference between the intersection and point G is 0.429λ, which is the length of the shorted stub. However, in this case, the single-stub method requires larger values of ℓ' and d.

Exercise 9.35

On a $100\,[\Omega]$-lossless line terminated at a load impedance $Z_L = 160 + j220\,[\Omega]$, the distance between successive voltage maxima is $5\,[\text{cm}]$. For impedance matching, (a) locate a shorted stub closest to the load, and (b) find the shortest length of the stub.

Ans. (a) $2.31\,[\text{cm}]$, (b) $0.81\,[\text{cm}]$.

Review Questions

RQ 9.39	What is the fundamental relationship on which the Smith chart is based?	[(9.71b)]
RQ 9.40	What do the abscissa and ordinate of the Smith chart represent?	[Fig. 9.12]
RQ 9.41	What is the meaning of the perimeter of the Smith chart?	[Fig. 9.12]
RQ 9.42	What are the values of r and x on the horizontal axis of the Smith chart?	[Fig. 9.13]
RQ 9.43	What is meant by the combination of r and x in Smith chart?	[(9.71b)(9.75) (9.78)(9.80)]
RQ 9.44	What is meant by moving on a circle of constant $|\Gamma|$?	[(9.76)]
RQ 9.45	Verify that the circles of constant $|\Gamma|$ and constant S overlap perfectly.	[(9.78)]
RQ 9.46	What is the geometrical relationship between the points for z_L and y_L on the Smith chart?	[Fig. 9.21]
RQ 9.47	Is moving on a circle of constant $|\Gamma|$ for y_{in} performed in the same clockwise direction as that for z_{in}?	[Fig. 9.21]

9.6 Problems

Transmission line parameters

9.1 A coaxial copper transmission line consists of an inner conductor with a radius of 2 [mm] and an outer conductor with an inner radius of 6 [mm]. The space between the conductors is filled with a dielectric of $\varepsilon_r = 2.26$ and $\sigma = 3 \times 10^{-6}$ [S/m]. At an operating frequency of 1 [GHz], determine (a) R, L, G, and C, and (b) α, β, and Z_o.

9.2 A parallel-plate transmission line is composed of two copper strips with a width of 16 [mm] separated by a 2 [mm] thick dielectric slab with $\mu = \mu_0$, $\varepsilon_r = 2.6$, and $\sigma = 10^{-4}$ [S/m]. Determine R, L, G, and C at 100 [MHz].

9.3 A two-wire transmission line is made of two copper wires of radius 1 [mm] separated by a center-to-center distance of 2 [cm] in an insulating material with $\varepsilon_r = 2$ and nonzero σ. When it operates as a distortionless line at 10 [MHz], determine R, L, G, C, and attenuation constant α.

9.4 A 50 [Ω] coaxial lossless line uses a dielectric with $\varepsilon_r = 2.26$ as the insulating material in its gap, which is 4.5 [mm] thick. Determine the radius of the inner conductor.

9.5 A two-wire lossless air line consists of conductors of radius 0.3 [mm] separated by a center-to-center distance of 3 [mm]. If the conductor radius is doubled, how much do Z_o and β change?

9.6 Both the air-filled coaxial line and two-wire copper line exhibit $R \ll \omega L$ and $G = 0$. Compare the attenuation constants for cases (a) $b/a = 3$, and (b) $b/a = 20$. [Hint: Eqs. (9.24a)–(9.24f).]

9.7 The coaxial line operating at a frequency of 1 [GHz] is known to have line parameters $R = 0.79$ [Ω/m], $L = 0.14$ [μH/m], $G = 1.2$ [mS/m], and $C = 360$ [pF/m]. Determine α, β, and Z_o.

9.8 A lossless transmission line has $Z_o = 75$ [Ω] and $\beta = 8.42$ [rad/m] at a frequency of 80 [MHz]. Determine L and C of the line.

9.9 When a transmission line operating at a frequency of 500 [MHz] has $\alpha = 0.02$ [Np/m], $\beta = 30$ [rad/m], and $Z_o = 60$ [Ω], determine R, L, G, and C, and verify that this is a distortionless line.

9.10 If a line with $Z_o = 30$ [Ω], $\alpha = 47$ [mNp/m], and $\upsilon_p = 1.8 \times 10^8$ [m/s] operates as a distortionless line at 100 [MHz], determine its R, L, G, and C.

9.11 A 75 [Ω] coaxial line consisting of an inner copper conductor with radius 1 [mm] and an outer copper conductor with radius 5 [mm] operates as a distortionless line at frequency 1 [GHz]. When the insulating material is nonmagnetic, determine its conductivity and dielectric constant.

9.12 Check that the following statements always hold true in air-spaced transmission lines regardless of the frequency of operation:

(a) The phase angle of γ is in range $\pi/4 < \theta_1 \le \pi/2$.
(b) The phase angle of Z_o is in range $-\pi/4 < \theta_2 \le 0$.
(c) $\theta_1 - \theta_2 = \pi/2$.

9.13 An air-spaced lossless coaxial line with its axis along the z-axis carries a current $i = I_o \cos(\omega t - kz)$.

    (a) Start with $\mathscr{H} = H(z, t)\, \mathbf{a}_\phi$ to obtain magnetic and electric fields.

    (b) Determine the voltage wave from $\mathscr{E}$ using V_o as the amplitude.

    (c) Show that V_o/I_o is equal to Z_o of the coaxial line.

[Hint: Eqs. (9.14) and (9.21b).]

Reflection coefficient

9.14 Measurements on a terminated lossless line indicate that the voltage has a maximum amplitude of 2.4 [V] and a minimum amplitude of 0.8 [V]. Determine the magnitude of the reflection coefficient at the load.

9.15 On a 50 [Ω] transmission line, the standing-wave ratio is measured as 2.7, and the first voltage maximum is observed at a distance of 0.4λ from the load. Determine the voltage reflection coefficient at the load.

9.16 Measurements on a 150 [Ω] lossless line show that the standing-wave ratio is 1.5, and the first voltage minimum and maximum are at distances of 4 [cm] and 10 [cm] from the load, respectively. Determine Γ.

9.17 On a lossless line, the neighboring $|\tilde{V}|_{\max}$ of 1.0 [V] and $|\tilde{V}|_{\min}$ of 0.8 [V] are 3.5 [cm] apart. If a voltage maximum is observed at a distance of 24 [cm] from the load. Determine Γ at the load.

9.18 On a 200 [Ω] air-spaced transmission line terminated at a load impedance $Z_L = 120 - j80\,[\Omega]$, the voltage minima are shifted toward the load by 2.5 [cm] if the load is replaced by a short circuit. Determine the operating frequency.

9.19 An air-spaced transmission line with $Z_o = 100\,[\Omega]$ is terminated at a 75 [Ω] resistor in series with a 25 [pF] capacitor. For an operating frequency of 200 [MHz], determine (a) the reflection coefficient at the load, and (b) the location of the voltage maxima.

9.20 A lossless line with $Z_o = 50\,[\Omega]$ and $\beta = 1.2\pi$ [rad/m] terminates at load impedance $Z_L = 80 + j120\,[\Omega]$. If the amplitude of the forward wave is $V_o^+ = 10\,[V]$, determine

    (a) the reflection coefficient at the load,

    (b) the total voltage on the line, and

    (c) the distance of the first voltage maximum from the load.

9.21 On a 200 [Ω] lossless line terminated at an unknown load impedance, measurements show that the standing-wave ratio is 3.2 and the first voltage minimum is at a distance of 0.4λ from the load. Determine (a) the voltage reflection coefficient at the load, and (b) the load impedance.

9.22 The SWR is measured to be 3.6 on a lossless line terminated at a load impedance $Z_L = 282 + j141\,[\Omega]$. Determine the characteristic resistance of the line, which is less than 200 [Ω].

9.23 A $100\,[\Omega]$ lossless line of length 0.2λ is terminated at a load impedance $Z_L = 120 + j40\,[\Omega]$, where the voltage is measured to be $\tilde{V}_L = 50\,[\mathrm{V}]$. Determine the voltage at the input end of the line.

9.24 If an electric circuit operating at a low frequency is shorted, an excessive current will damage the circuit. However, this is not the case for transmission lines operating at high frequencies. Consider a $100\,[\Omega]$ air-spaced lossless line of $10\,[\mathrm{cm}]$ terminated in a short circuit. Determine the current at the shorted end if the line is directly connected to the following voltage sources: (a) $v_g = 10\cos(3 \times 10^5 t)$, and (b) $v_g = 10\cos(3 \times 10^8 t)$. [Hint: $\Gamma = -1$.]

Input impedance

9.25 A $100\,[\Omega]$ coaxial lossless line uses a dielectric with $\varepsilon = 2.56\varepsilon_0$ and $\mu = \mu_0$ as insulating material and operates at $200\,[\mathrm{MHz}]$. If the line is $1.5\,[\mathrm{m}]$ long and terminated at a load impedance $Z_L = 60 + j50$, determine the input impedance.

9.26 A $100\,[\Omega]$ air-spaced two-wire transmission line of length $0.6\,[\mathrm{m}]$ is terminated at an unknown load impedance Z_L. If the input impedance is measured as $Z_{in} = 100 + j130\,[\Omega]$ at $50\,[\mathrm{MHz}]$, determine Z_L.

9.27 A lossless line with $R_o = 50\,[\Omega]$ terminates at a load with impedance $Z_L = 40 - j50\,[\Omega]$. Find the shortest distance in wavelengths from the load at which the input impedance looking toward the load is real.

9.28 On a $200\,[\Omega]$ lossless line, the successive voltage minima are $15\,[\mathrm{cm}]$ apart, and the SWR is 2.2. Find the input impedance at a distance $\ell' = 10\,[\mathrm{cm}]$ from the load using the fact that the voltage minima are shifted towards the load by $8\,[\mathrm{cm}]$ if the junction at ℓ' is temporarily shorted.

9.29 For a $100\,[\Omega]$ lossless line terminated at load $Z_L = 45 + j54.6\,[\Omega]$, find (a) the location closest to the load at which the normalized input impedance is $z_{in}(\ell') = 1 + jx$, and (b) the value of x.

9.30 A lossless line with $R_o = 100\,[\Omega]$ and length $d_2 = 0.4\lambda$ is terminated by the impedance $Z_2 = 75\,[\Omega]$. It is then connected in parallel to the main line with the same characteristic resistance terminated at $Z_1 = 50\,[\Omega]$, as shown in Fig. 9.25. For $d_0 = 0.8\lambda$ and $d_1 = 0.2\lambda$, determine the input impedance at terminal BB'.

9.31 The generator circuit with $\tilde{V}_g = 15\angle 0^o\,[\mathrm{V}]$ and $Z_g = 20\,[\Omega]$ is connected to a $50\,[\Omega]$ air-spaced lossless line of length $10\,[\mathrm{cm}]$ terminated by load impedance $Z_L = 40 - j30\,[\Omega]$. At an operating frequency of $1.5\,[\mathrm{GHz}]$, determine the (a) input impedance, (b) input voltage, and (c) current through the load.

9.32 A $100\,[\Omega]$ lossless feedline is connected in series to a $50\,[\Omega]$ lossless line that is 0.4λ long and terminated at a load resistance of $100\,[\Omega]$. Determine the standing-wave ratio on each line.

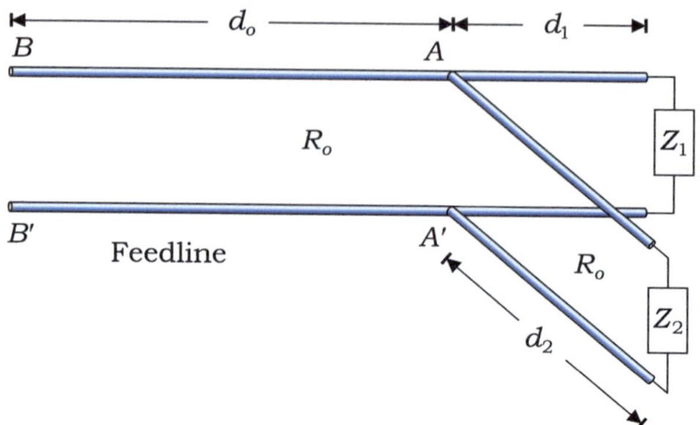

Fig. 9.25 A section of the terminated line is connected parallel to the main line (Problem 9.30)

Impedance matching

9.33   A dipole antenna with input impedance 77 [Ω] was connected to a 500 [MHz] source through a 100 [Ω] coaxial line. To match the antenna to the feedline, a quarter-wave coaxial line with a nonmagnetic dielectric of $\varepsilon_r = 1.22$ was inserted between the antenna and the feedline. Determine the length and ratio of the conductor radii of the quarter-wave coaxial line.

9.34   A complex load impedance Z_L can be matched to a lossless line with characteristic resistance R_o by inserting a shunt resistance R_1 at a distance ℓ_1 from the load where the voltage maximum occurs (Fig. 9.26). Express R_1 in terms of R_o and $|\Gamma|$ of Z_L.

9.35   A load impedance $Z_L = 100 - j50$ [Ω] can be matched to a lossless line with characteristic resistance $R_o = 75$ [Ω] by means of the ℓ_1-section of another lossless line with characteristic resistance R_1, as shown in Fig. 9.27. Express R_1 and ℓ_1 in terms of wavelength, assuming that all lines are air spaced.

Short- and open-circuit lines

9.36   What is the shortest length of a section of a 100 [Ω] air-spaced lossless line terminated in a short circuit for it to have an input impedance equivalent to an inductor of 12 [nH] at 100 [MHz]?

9.37   At the input end of a lossless line with a length of 60 [cm], the input impedances are measured as $Z_{in}^o = -j68.8$ [Ω] and $Z_{in}^s = j36.3$ [Ω] at 200 [MHz] when

Fig. 9.26 Impedance matching using shunt resistance (Problems 9.34 and 9.66)

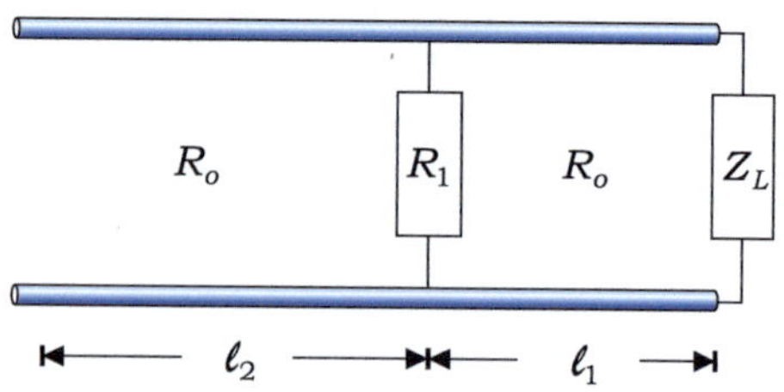

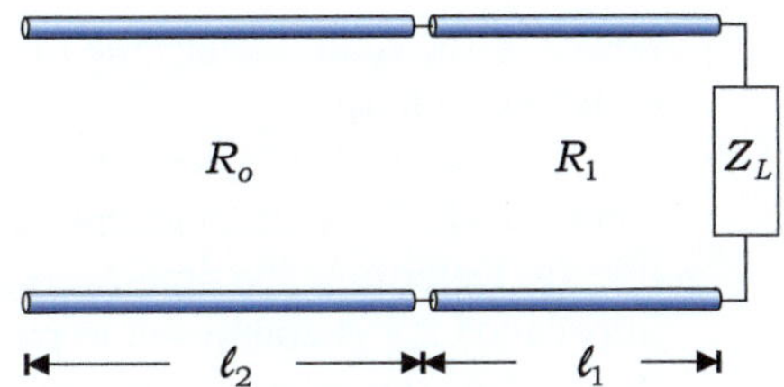

Fig. 9.27 Impedance matching using the ℓ_1-section of the lossless line with R_1 (Problem 9.35)

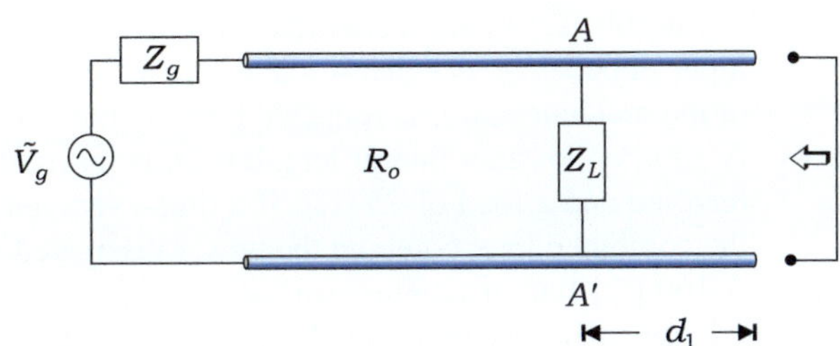

Fig. 9.28 Termination is used as an on/off switch (Problem 9.38)

the line is open- and short-circuited, respectively. Determine (a) R_o of the line, and (b) ε_r of the insulating material. [Hint: $\varepsilon_r \geq 1$.]

9.38 A generator circuit is connected to a lossless line, and impedance Z_L is connected to the line at a distance d_1 from the other end, as shown in Fig. 9.28. By connecting or disconnecting a short circuit to the right end, we wish to switch the power delivered to Z_L, such that the power is on for the short-circuit end and off for the open-circuit end. What is the shortest distance d_1 measured in wavelengths?

Power transfer

9.39 A 0.7λ long $50\,[\Omega]$ lossless line connects load impedance $Z_L = 100\,[\Omega]$ to a generator circuit with $V_g = 15\angle 0^o$ and $Z_g = 20\,[\Omega]$. Determine the

   (a)   input voltage and current,
   (b)   load voltage and current,
   (c)   time-averaged power fed into the line,
   (d)   time-averaged power delivered to the load, and
   (e)   total time-averaged power supplied by the generator.

9.40 The load impedance $Z_L = 75 + j50\,[\Omega]$ is connected to a $100\,[\Omega]$ lossless line, the other end of which is connected to a voltage generator. If the peak voltage at the load is $16\,[V]$, compute the time-averaged powers

   (a)   incident onto the load,
   (b)   dissipated in the load, and
   (c)   reflected by the load.

9.41 The generator circuit with $\tilde{V}_g = 10\angle 0^o\,[V]$ and $Z_g = 20 + j15\,[\Omega]$ is connected to a half-wave lossless line and delivers the maximum possible power to the load. Determine (a) load impedance, and (b) maximum power delivered to load.

9.42 The generator circuit with $\tilde{V}_g = 10\angle 0^o$ [V] and $Z_g = 20 + j15\,[\Omega]$ is connected to the load impedance $Z_L = 100 - j50\,[\Omega]$ via a half-wave lossless line. (a) Determine the time-averaged power delivered to the load. (b) Another impedance Z_A is connected in parallel with Z_L such that the generator feeds the maximum power into the line. Determine Z_A, and the time-averaged power delivered to Z_L.

9.43 For a 200 [Ω] lossless line terminated at an unknown load impedance, the input impedance was measured as $Z_{in} = 169 + j210\,[\Omega]$. What percentage of the incident power is reflected back by the load?

9.44 A 100 [Ω] lossless line of length 0.2λ is inserted between a 50 [Ω] lossless feedline and a load of 75 [Ω]. If a time-averaged power of 4 [W] is fed into the combined line, compute the power dissipated at the load.

9.45 A 100 [Ω]-line of length 2 [m] is connected end-to-end to the 50 [Ω]-line of 1.25 [m] that terminates at $Z_L = 75\,[\Omega]$. These are air-spaced lines, and the amplitudes of the forward voltage waves on these lines are denoted by V_{o1}^+ and V_{o2}^+, respectively. For the input voltage $v_{in} = 10\cos(3\pi \times 10^8 t)$,

   (a) express the voltages on these lines in terms of V_{o1}^+ and V_{o2}^+,

   (b) determine V_{o1}^+ and V_{o2}^+ using the boundary conditions at the joint, and then compute the load voltage and the power delivered to the load, and

   (c) find the input impedance and input power, and then compare the powers obtained in (b) and (c).

Smith chart

9.46 A point on the Smith chart is the intersection of the $r = r'$ and $x = x'$ circles. Express the polar angle of the point in terms of r' and x'.

9.47 A 160 [Ω] lossless line of length 0.268λ is terminated at load impedance $Z_L = 84 + j72\,[\Omega]$. Find (a) reflection coefficient, (b) standing-wave ratio, and (c) input impedance using the Smith chart shown in Fig. 9.29.

9.48 A load impedance of $Z_L = 30 + j44\,[\Omega]$ is connected to a 100 [Ω] air-spaced lossless line of length 16.8 [cm], which operates at 600 [MHz]. By using the Smith chart shown in Fig. 9.30, determine (a) Γ, (b) S, (c) the distance of the first $|\tilde{V}|_{\max}$ from the load, (d) Z_{in}, and (e) Y_{in}.

9.49 On a 200 [Ω] lossless line, the distance between successive voltage minima is 12.5 [cm]. If the first $|\tilde{V}|_{\min}$ is 9.9 [cm] from the load, and the standing-wave ratio is 2.4, determine Z_L using the Smith chart shown in Fig. 9.29.

9.50 The SWR is measured as $S = 2.4$ on a 150 [Ω] lossless line terminated at a purely resistive load. By using the Smith chart shown in Fig. 9.29, determine the load resistance if it is less than the characteristic resistance.

9.51 On a lossless transmission line terminated at normalized load impedance $z_L = 1.0 + j1.5$, the normalized input impedance is $z_{in} = 4.0$. If the line terminates at a purely resistive load, then the SWR is halved. Determine the new z_{in} using the Smith chart shown in Fig. 9.30.

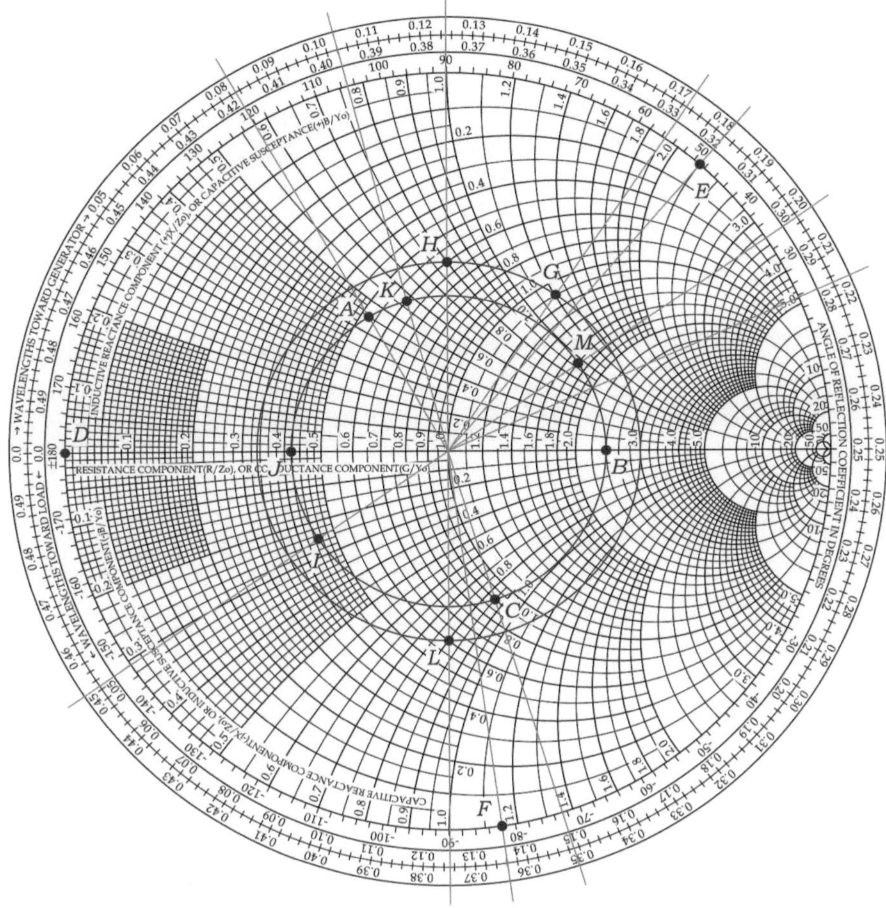

Fig. 9.29 Smith chart for Problems 9.47, 9.49, 9.50, 9.55, 9.58, and 9.61

9.52 A 72.7 [cm] long lossless line with $R_o = 120\,[\Omega]$ is made of a nonmagnetic insulating material with $\varepsilon_r = 2.25$. The line is terminated at the load impedance $Z_L = 360\,[\Omega]$ and operates at 533 [MHz]. Determine the input impedance using the circle of constant $|\Gamma|$, as shown in Fig. 9.18.

9.53 A 50 [Ω] lossless line of length 19.8 [cm] is made of a nonmagnetic insulator with $\varepsilon = 1.44\varepsilon_0$. When the line is terminated at an unknown load impedance Z_L, the input impedance is measured as $Z_{in} = 55 + j85\,[\Omega]$ at an operating frequency of 125 [MHz]. Using the Smith chart shown in Fig. 9.33, determine (a) Z_L, and (b) Γ at the load.

9.54 The input impedance of a 200 [Ω] lossless line terminated in a short circuit is $Z_{in}^s = j250\,[\Omega]$. If the line terminates at an unknown impedance Z_L, the input impedance changes to $Z_{in} = 56 - j68\,[\Omega]$. Use the Smith chart shown in Fig. 9.30 to determine Z_L.

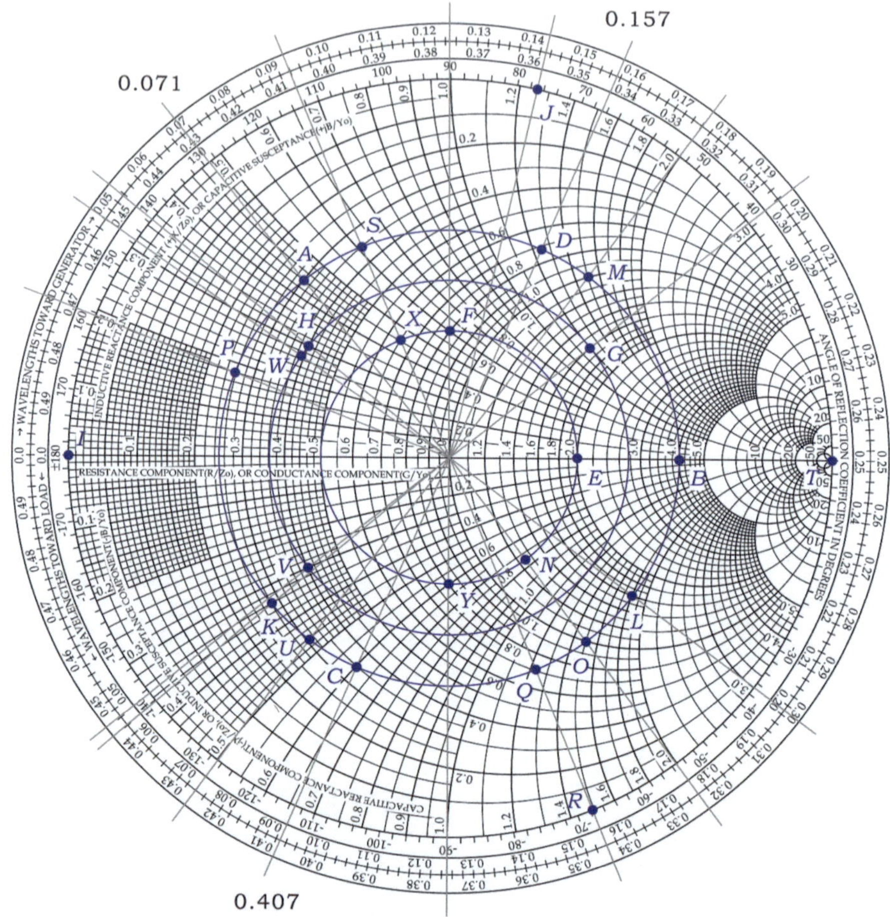

Fig. 9.30 Smith chart for Problems 9.48, 9.51, 9.54, 9.59, 9.62, 9.64, and 9.67

9.55   A 120 [Ω] lossless line of length 1.682λ is terminated in a short circuit. Using the Smith chart shown in Fig. 9.29, (a) determine the input impedance, and (b) repeat (a) if the operating frequency is reduced by half.

9.56   Use the Smith chart in Fig. 9.34 to determine the input impedance of a 120 [Ω] air-spaced lossless line operating at 300 [MHz], if it is (a) 15.6 [cm] long and terminated in a short circuit, and (b) 9.4 [cm] long and terminated in an open circuit.

9.57   For a lossless line terminated at a purely reactive load, using a Smith chart, show that the input impedance is purely reactive.

9.58   Solve equation $1.628e^{j0.829} = \tanh(a + jb)$ for a and b using the Smith chart in Fig. 9.29.

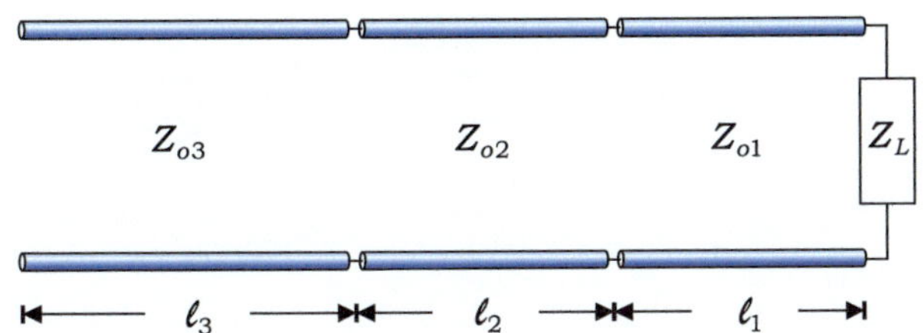

Fig. 9.31 Three-line configuration (Problem 9.60)

9.59 A lossless line with $R_1 = 50\,[\Omega]$ and $\ell_1 = 0.142\lambda$ is terminated at an unknown load impedance Z_L and connected end-to-end to a feedline with $R_o = 100\,[\Omega]$, on which SWR is 2.0 and the first voltage maximum is observed at 0.125λ from the junction between the two lines (Fig. 9.27). Use the Smith chart shown in Fig. 9.30 to determine Z_L.

9.60 Consider a three-line configuration as shown in Fig. 9.31, for which $Z_{o1} = 75\,[\Omega]$, $Z_{o2} = 50\,[\Omega]$, $Z_{o3} = 100\,[\Omega]$, and $Z_L = 30 + j45\,[\Omega]$. The three sections are $\ell_1 = 0.202\lambda$, $\ell_2 = 0.22\lambda$, and $\ell_3 = 0.168\lambda$ in length. Use the Smith chart shown in Fig. 9.32 to determine the input impedance.

9.61 Use the Smith chart shown in Fig. 9.29 to determine the value of y_{in} if (a) $z_{in} = 0.6 + j0.8$, and (b) $z_{in} = 0.45 - j0.25$.

9.62 A lossless line with characteristic admittance $Y_o = 10\,[\text{mS}]$ is 0.206λ long. If the line is terminated at load admittance $Y_L = 10 - j15\,[\text{mS}]$, determine the input admittance using the Smith chart in Fig. 9.30.

9.63 The lossless line with shunt resistance R_1, as shown in Fig. 9.26, has $R_o = 100\,[\Omega]$ and $Z_L = 27 - j41.5\,[\Omega]$. The Smith chart in Fig. 9.33 was used as an admittance chart to determine the normalized input impedance z_{in}, starting from the normalized load impedance z_L. Points A–D represent the intermediate steps taken by someone who knew the values of ℓ_1, ℓ_2, and R_1. Read the Smith chart to determine (a) the smallest possible values of ℓ_1 and ℓ_2, and (b) the value of R_1.

9.64 Use the Smith chart shown in Fig. 9.30 to determine the input impedance of the transmission line, as shown in Fig. 9.25, if $R_o = 100\,[\Omega]$, $Z_1 = 100 + j150\,[\Omega]$, $Z_2 = 160 + j120\,[\Omega]$, $d_1 = 0.104\lambda$, $d_2 = 0.102\lambda$, and $d_o = 0.531\lambda$.

9.65 The 150 $[\Omega]$-line is matched to load $Z_L = 231 + j90\,[\Omega]$ using a shorted stub. Using the Smith chart shown in Fig. 9.32, find the shortest possible (a) distance of the stub from the load, and (b) length of the stub. [Hint: (a) and (b) may not be achieved simultaneously.]

9.66 The lossless line with $R_o = 100\,[\Omega]$ was matched to the load impedance $Z_L = 26 + j18\,[\Omega]$ using a shorted stub (Fig. 9.23). The Smith chart used for impedance matching is shown in Fig. 9.34. From the Smith chart, determine (a) the shortest length of the shorted stub, and (b) the distance of the stub from the load.

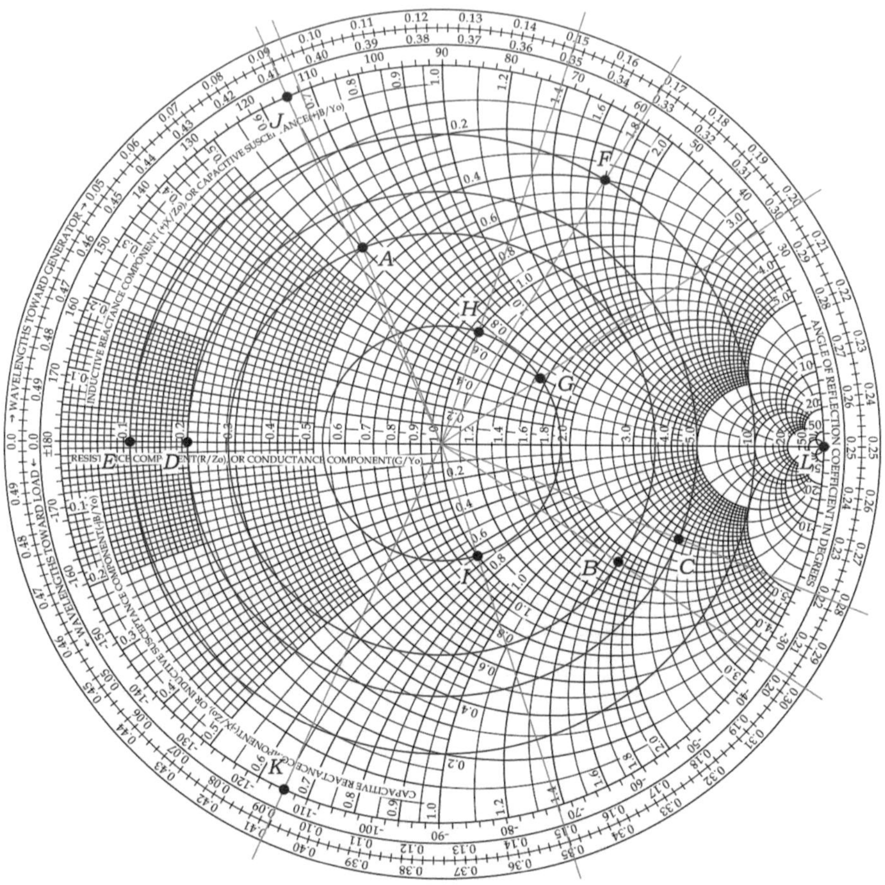

Fig. 9.32 Smith chart for Problems 9.60 and 9.65

9.67 A shorted stub with a length of 0.094λ was used to match the lossless line with $R_o = 100\,[\Omega]$ to an unknown load impedance Z_L (Fig. 9.23). If the position of the stub is $\ell' = 0.082\lambda$, determine Z_L using the Smith chart shown in Fig. 9.30.

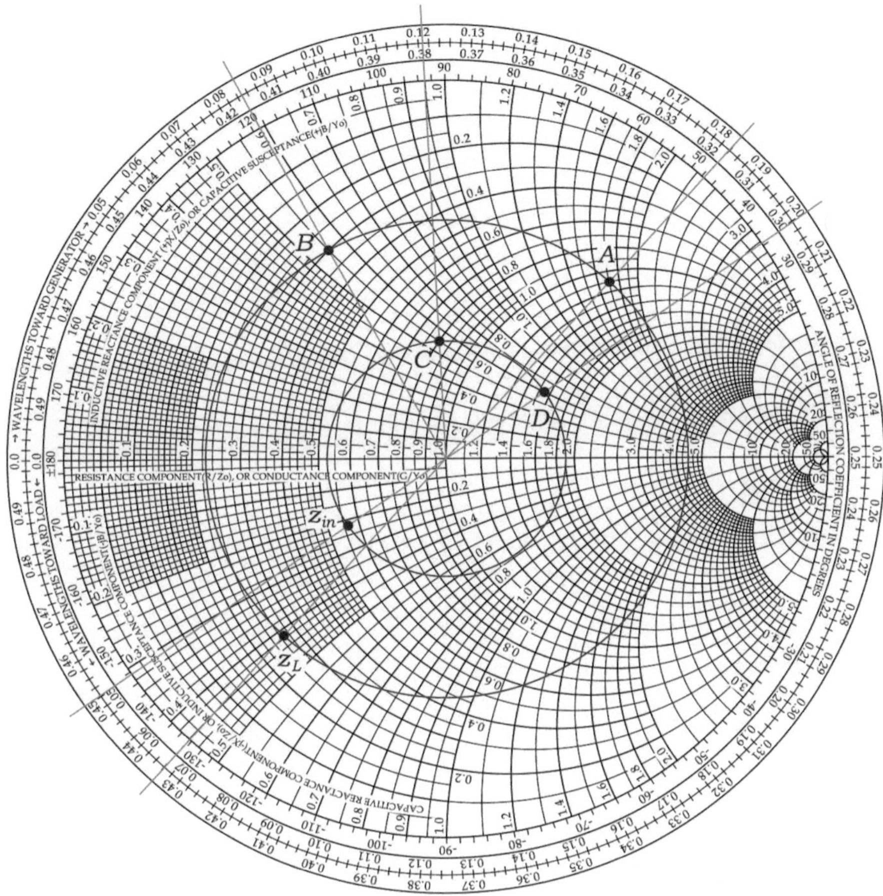

Fig. 9.33 Smith chart for Problems 9.53 and 9.63

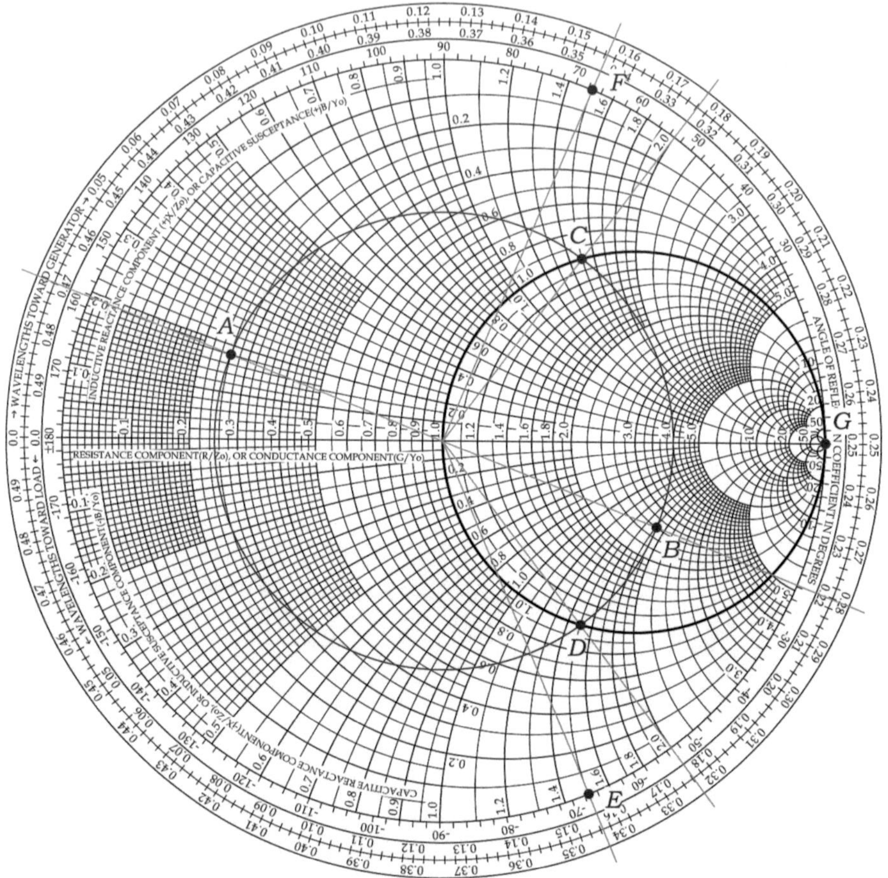

Fig. 9.34 Smith chart for Problems 9.56 and 9.66

Chapter 10
Waveguide

As discussed in the previous chapter, a transmission line can be used to guide a transverse electromagnetic wave from the generator to load. However, at microwave frequencies in the range of 3–300 [GHz], the transmission line exhibits an inefficient transfer of electromagnetic energy because the series resistance of the line is proportional to the square root of the frequency, and the wave traveling along the line has an excessively high attenuation constant. Meanwhile, waveguides are very efficient in transmitting electromagnetic signals from one point to another at such frequencies and are usually hollow metal pipes with uniform rectangular or circular cross sections. The attenuation constant in waveguides is significantly reduced because of the large surface area of conducting walls. A waveguide differs from a transmission line in several ways. Waveguides can support transverse electric (TE) modes, whose electric fields are in the transverse plane of the waveguide but whose magnetic fields are not, and transverse magnetic (TM) modes, whose magnetic fields are in the transverse plane but whose electric fields are not. Hollow metallic waveguides may not support transverse electromagnetic (TEM) waves, whereas transmission lines support only TEM waves. Moreover, a waveguide can operate only above a particular frequency, called the cutoff frequency, whereas a transmission line has no cutoff frequency, even transmitting direct current. A rigorous electromagnetic approach is required for the analysis of waveguides, whereas a simple circuit theory that deals with voltages and currents can be used for the analysis of transmission lines. When the operating frequency approaches the optical frequency of visible or infrared light, the loss due to the skin effect becomes impossibly large in a hollow metallic waveguide. To overcome these difficulties, the air-conductor interface of hollow metallic waveguides has been replaced with a dielectric-dielectric interface in optical fibers and dielectric waveguides. The typical hollow metallic waveguides are shown in Fig. 10.1, and an optical fiber and a dielectric slab waveguide are shown in Fig. 10.2.

© The Author(s), under exclusive license to Springer Nature Switzerland AG 2024

Y. H. Lee, *Introduction to Engineering Electromagnetics*,

https://doi.org/10.1007/978-3-031-28659-9_10

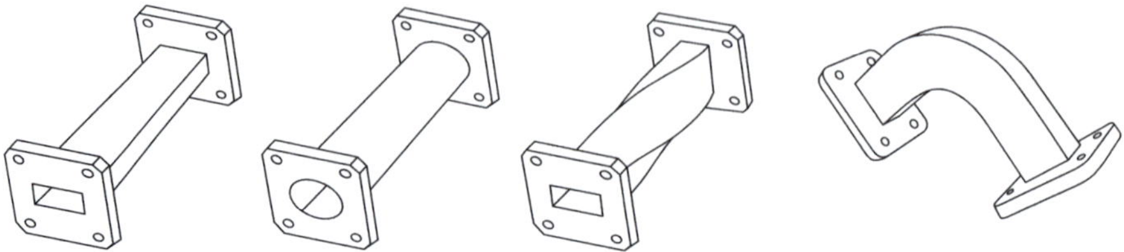

Fig. 10.1 Hollow metallic waveguides

Fig. 10.2 Optical fiber and
dielectric slab waveguide

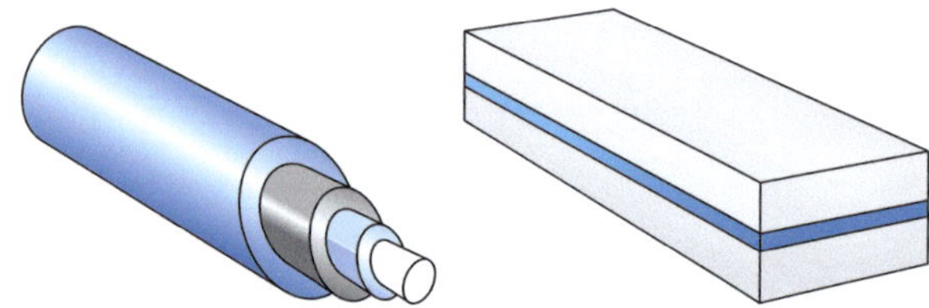

10.1 Parallel-Plate Waveguide

All the metallic waveguides operate according to the same basic principles. To understand how the electromagnetic field propagates in a waveguide, we consider a parallel-plate waveguide consisting of two parallel conducting plates, as shown in Fig. 10.3. Although two parallel plates can be used as transmission lines to support TEM waves, they can also be used as waveguides to support the TE and TM waves. In this case, the electromagnetic field varies spatially in the transverse plane, which is characteristic of waveguides. Therefore, the parallel-plate waveguide is considered the simplest hollow metallic waveguide.

Starting with uniform plane waves propagating in a lossless waveguide made of perfect conductors and dielectrics, and making the electric and magnetic fields satisfy the boundary conditions on the conductor surface, we can find the specific field patterns allowed to propagate in the waveguide.

Fig. 10.3 Parallel-plate
waveguide

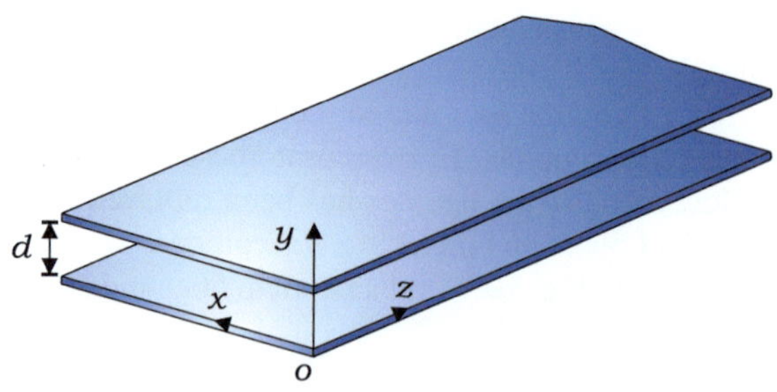

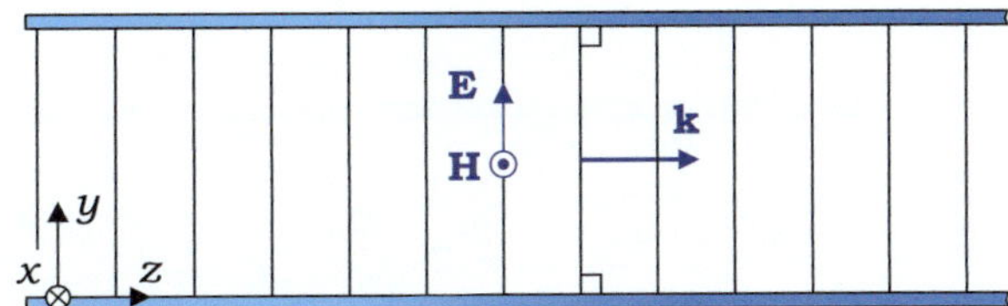

Fig. 10.4 TEM mode in the parallel-plate waveguide

10.1.1 Transverse Electromagnetic (TEM) Wave

Uniform plane waves satisfy the three-dimensional differential wave equation and obey the principle of superposition. Therefore, the sum of the uniform plane waves with the same wavenumber propagating in different directions can be a unique solution in a given region only if the total electric and magnetic fields satisfy the boundary conditions. Accordingly, the behavior of electromagnetic waves in a waveguide can be conveniently analyzed using uniform plane waves.

Consider a parallel-plate waveguide as shown in Fig. 10.3, which is assumed to be lossless. The fringing effect at the edges is ignored by assuming that the separation is much smaller than the plate width. A single uniform plane wave can exist in the space between the plates if its wavevector is directed along the plate orientation and its electric field is perpendicular to the conductor surface, thereby satisfying the boundary condition. In this case, both the electric- and magnetic-field vectors lie in the transverse plane, as shown in Fig. 10.4. This field pattern is known as the *transverse electromagnetic* (TEM) *mode*. The specific field pattern allowed in a waveguide is called the *waveguide mode*. It can propagate in a waveguide without any change in its shape.

10.1.2 Transverse Electric (TE) Wave

The aforementioned TEM mode is not the only mode of propagation for the parallel-plate waveguides. Consider the case in which a uniform plane wave with an electric field directed along the x-axis is incident on the upper plate at an angle θ with respect to the z-axis, as shown in Fig. 10.5. The wave is reflected from the upper plate and then from the lower plate in zigzags. In other words, the incident wave can propagate in the waveguide, bouncing up and down endlessly between the two plates. From the viewpoint of two uniform plane waves, as shown in Fig. 10.5, one with wavefronts represented by solid lines and the other with wavefronts represented by dotted lines propagate simultaneously in the space between the plates. The circle with a cross indicates the positive direction of the electric field vectors. Although the two uniform plane waves may have electric field vectors parallel to the conductor surface, their sum should vanish on the conductor surface to satisfy boundary conditions. From Fig. 10.5, it is evident that the total electric field lies in the transverse plane of the waveguide or the xy-plane, whereas the associated magnetic field has a nonzero

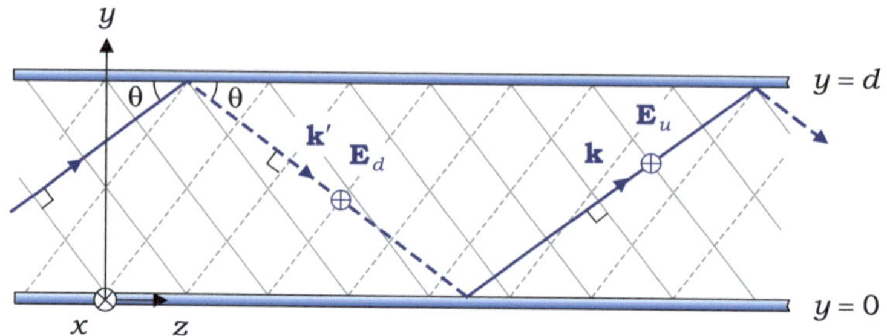

Fig. 10.5 TE mode in the parallel-plate waveguide. Solid lines, wavefronts of wave **k**; Dotted lines, wavefronts of wave **k**′

longitudinal component along the waveguide. The superposition of these two waves constitutes the propagation mode in the waveguide.

Two uniform plane waves should be combined with an appropriate phase relation to form the TE mode in a parallel-plate waveguide. When a uniform plane wave is incident on the upper plate in such a way that the electric field vector is perpendicular to the plane of incidence (the yz-plane) and the wavevector makes an angle θ with the plate, called the *upward wave*, its electric field is expressed as

$$\tilde{\mathbf{E}}_u = \mathbf{a}_x E_o e^{-j(k_y y + k_z z)} \tag{10.1}$$

where E_o is the amplitude, and $\mathbf{a}_x$ is a unit vector in the direction of the electric field. From Fig. 10.5, the wavevector is $\mathbf{k} = k_y \mathbf{a}_y + k_z \mathbf{a}_z = k \sin\theta \, \mathbf{a}_y + k \cos\theta \, \mathbf{a}_z$, and the wavenumber is given by

$$k^2 = k_x^2 + k_y^2 \tag{10.2}$$

Next, in accordance with the law of reflection, the reflected wave has wavevector $\mathbf{k}' = -k \sin\theta \, \mathbf{a}_y + k \cos\theta \, \mathbf{a}_z$, which is the same as $\mathbf{k}$, except for the negative sign of the y-component. Note that $\mathbf{k}$ and $\mathbf{k}'$ have the same magnitude because the two waves propagate in the same medium, that is, air. Therefore, the electric field of the reflected wave is generally expressed as

$$\tilde{\mathbf{E}}_d = \mathbf{a}_x E_d e^{-j(-k_y y + k_z z)} \tag{10.3}$$

where it is assumed that the $+x$-direction is the positive direction of the electric field vector. This wave is also called a *downward wave*.

Fresnel's equation indicates that the reflection coefficient for perpendicular polarization at a perfectly conducting surface is $\Gamma_\perp = -1$. Because the downward wave is considered as a reflected wave at the upper plate, $\tilde{\mathbf{E}}_d$ should have the same magnitude as the upward wave and be multiplied by -1 before adding the two electric fields to obtain the total field in the waveguide. The total electric field in the waveguide is thus expressed as

$$\tilde{\mathbf{E}} = \tilde{\mathbf{E}}_u - \tilde{\mathbf{E}}_d = E_o \mathbf{a}_x \left[e^{-jk_y y} - e^{jk_y y} \right] e^{-jk_z z} \tag{10.4}$$

This can be rearranged as

$$\boxed{\tilde{\mathbf{E}} = -j2E_o\,\mathbf{a}_x\,\sin(k_y y)\,e^{-jk_z z}} \tag{10.5}$$

The total electric field in Eq. (10.5) should satisfy the boundary conditions at the conductor surfaces at $y = 0$ and $y = d$; that is, both $\tilde{\mathbf{E}}(y = 0)$ and $\tilde{\mathbf{E}}(y = d)$ should vanish because otherwise they produce nonzero tangential components at the conductor surface. From the boundary conditions, it follows that

$$\boxed{k_y = \frac{m\pi}{d}} \quad (m = 1, 2, 3, ...) \tag{10.6}$$

where integer m is called the **mode number**.

The electric field in Eq. (10.5) satisfies the differential wave equation and boundary conditions, and thus constitutes an electromagnetic wave propagating in the parallel-plate waveguide with a phase constant k_z. The waveguide mode in Eq. (10.5) has no longitudinal component of the electric field and is thus the **transverse electric** (TE) mode.

The sine term in Eq. (10.5) indicates that a standing wave is formed along the y-axis in the waveguide. The values of k_y in Eq. (10.6) can also be obtained from the transverse resonance condition for upward and downward waves. This quantity is known as the **eigenvalue** of TE mode. If the mode of propagation is considered as a single uniform plane wave bouncing up and down in zigzags between the plates, the total phase shift experienced by the wave during a round trip from the upper plate to the lower plate, and then back to the upper plate, is given by

$$\begin{aligned} \psi &= 2k_y d + 2\phi \\ &= 2\pi m' \end{aligned} \quad (m' = 2, 3, 4, ...) \tag{10.7}$$

where $k_y d$ is the phase shift caused by a travel from the upper to the lower plate, and vice versa, and ϕ is the phase shift caused by a reflection at the perfectly conducting plate. Again, Fresnel's equations provide a reflection coefficient of -1 for perpendicular polarization ($\mathbf{E}$ is perpendicular to the plane of incidence), and thus, $\phi = \pi$ for the TE modes. The round-trip phase shift should be an integer multiple of 2π, as indicated in Eq. (10.7). Otherwise, waves would interfere destructively, and it may not be possible to have a waveguide mode with the same electric field pattern in the transverse plane throughout the waveguide. Because $k_y d$ is positive and ϕ is equal to π, ψ is greater than 2π; thus, integer m' is greater than or equal to 2. Obviously, Eq. (10.7) yields the same eigenvalue as in Eq. (10.6).

It is more convenient to identify the TE mode using k_z rather than k_y. Inserting Eq. (10.6) into Eq. (10.2), we have

$$k_z = \pm\sqrt{k^2 - k_y^2}$$

$$= \pm\sqrt{\left(\frac{n\omega}{c}\right)^2 - \left(\frac{m\pi}{d}\right)^2} \qquad (m = 1, 2, 3, ...) \tag{10.8}$$

$$\equiv \beta_m$$

where $k = n\omega/c$, and n is the refractive index of the dielectric between the plates. Note that the insulating material is assumed to be lossless and nonmagnetic ($\mu = \mu_0$). Here, m is the mode number, c is the speed of light in free space, and d is the separation between the plates. Taking the positive sign in front of the square root of Eq. (10.8), a new constant is defined such that $\beta_m \equiv k_z$, which is called the **phase constant** of mode m. Subsequently, Eq. (10.8) can be rewritten as

$$\boxed{\beta_m = \frac{n\omega}{c}\sqrt{1 - \left(f_{c(m)}/f\right)^2}} \qquad \text{[rad/m]} \tag{10.9}$$

The **cutoff frequency** $f_{c(m)}$ is defined as

$$\boxed{f_{c(m)} = \frac{mc}{2nd}} \qquad \text{[rad/s]} \tag{10.10}$$

Note that k is the phase constant in an unbounded dielectric, whereas β_m is the phase constant in a waveguide oriented along the z-direction.

If the operating frequency f is lower than the cutoff frequency $f_{c(m)}$, the radicand in Eq. (10.9) (or Eq. (10.8)) becomes negative. In this case, taking the negative sign in front of the square root of Eq. (10.8), k_z has a new value, such as $k_z = -j\alpha_m$ with a positive constant α_m. This can be substituted into Eq. (10.5) to exhibit an exponential decay of the electric field along the waveguide. Consequently, *below the cutoff frequency, the corresponding mode does not propagate in the waveguide.*

In contrast, for $f > f_{c(m)}$, mode m propagates in the waveguide with phase constant β_m. It is important to note that although k_y of mode m is independent of the operating frequency, as expressed in Eq. (10.6), k_z, as expressed in Eq. (10.8), depends on the frequency. As wavevector $\mathbf{k}$ makes an angle θ with the conductor surface (Fig. 10.5), the incident angle of mode m, that is, θ_m, can be determined using the relationship $\sin\theta_m = k_y/k$. Furthermore, a comparison of the radicands in Eqs. (10.8) and (10.9) leads to the relationship $k_y/k = f_{c(m)}/f$. It then follows that

$$\sin\theta_m = \frac{k_y}{k} = \frac{f_{c(m)}}{f} \tag{10.11}$$

where $k = n\omega/c$ is the wavenumber in an unbounded dielectric similar to that filling the waveguide. We note from Eq. (10.11) that θ_m becomes $\pi/2$ if f is equal to $f_{c(m)}$, indicating that the wave moves in the transverse direction, and no waves propagate along the waveguide.

The TE mode consists of two plane waves that propagate with the same phase velocity υ_p, but in different directions, as specified by θ_m and $-\theta_m$. Here, $\upsilon_p = \omega/k$ is the phase velocity in an unbounded dielectric. Meanwhile, the phase velocity of mode m in the waveguide is defined as

$$\boxed{\upsilon_{p(m)} = \frac{\omega}{\beta_m} = \frac{c/n}{\sqrt{1 - \left(f_{c(m)}/f\right)^2}}} \qquad (10.12a)$$

The phase constant of the waveguide mode is the z-component of the wavevector in an unbounded dielectric, as expressed by $\beta_m = k \cos\theta_m$ with $k = n\omega/c$. Inserting this into Eq. (10.12a) gives

$$\upsilon_{p(m)} = \frac{c}{n \cos\theta_m} \quad [\text{m/s}] \qquad (10.12b)$$

Although $\upsilon_{p(m)}$ is always greater than the speed of light in an unbounded dielectric c/n, it does not contradict special relativity because $\upsilon_{p(m)}$ is just a phase velocity.

The electromagnetic energy propagates with group velocity in the waveguide. Using Eq. (10.9), the group velocity of mode m is expressed as

$$\boxed{\upsilon_{g(m)} = \frac{d\omega}{d\beta_m} = \frac{1}{d\beta_m/d\omega} = \frac{c}{n}\sqrt{1 - \left(f_{c(m)}/f\right)^2}} \quad [\text{m/s}] \qquad (10.13a)$$

Inserting Eq. (10.11) into Eq. (10.13a) yields

$$\upsilon_{g(m)} = \frac{c}{n} \cos\theta_m \quad [\text{m/s}] \qquad (10.13b)$$

The group velocity of mode m corresponds to the projection of the phase velocity in an unbounded dielectric onto the z-axis, and is always smaller than the aforementioned phase velocity.

The instantaneous electric field of mode m can be obtained by multiplying Eq. (10.5) using the time-harmonic term $e^{j\omega t}$ and taking the real part. Above the cutoff ($f > f_{c(m)}$), the electric field of the TE_m mode is given by

$$\mathscr{E}_{TE} = 2E_o \mathbf{a}_x \sin\left(\frac{m\pi}{d}y\right) \sin(\omega t - \beta_m z) \quad (\omega > \omega_{c(m)}) \qquad (10.14)$$

In contrast, below the cutoff ($f < f_{c(m)}$), the phase constant in Eq. (10.9) becomes imaginary, such that $\beta_m = -j\alpha_m$ with the attenuation constant α_m. In this case, the instantaneous electric field of the TE_m mode is given by

$$\mathscr{E}_{TE} = 2E_o \mathbf{a}_x \sin\left(\frac{m\pi}{d}y\right) e^{-\alpha_m z} \sin(\omega t) \quad (\omega < \omega_{c(m)}) \qquad (10.15)$$

Here, the **attenuation constant** is defined as

$$\alpha_m = \frac{n\,\omega}{c}\sqrt{1 - \left(f/f_{c(m)}\right)^2} \qquad [\text{Np/m}] \tag{10.16}$$

where $\omega_{c(m)} = 2\pi f_{c(m)}$, and $f < f_{c(m)}$. Below the cutoff frequency, the mode becomes nonpropagating in the waveguide and is called the **evanescent mode**.

Exercise 10.1

For a parallel-plate waveguide with an air gap of $d = 5$ [mm], determine the cutoff frequency for the TE mode with $m = 1$.

Ans. 30 [GHz].

Exercise 10.2

What is the mode number of the TE mode as shown in Fig. 10.5?

Ans. $m = 4$ (The phase changes by 4π along the y-axis).

Exercise 10.3

What is the value of the incident angle θ_m at the cutoff frequency?

Ans. 90°.

Exercise 10.4

Find the magnetic field of the TE mode as expressed by Eq. (10.14).

Ans.
$$\mathcal{H}_{TE} = (2E_o\beta_m/\omega\mu_0)\,\mathbf{a}_y\,\sin(m\pi y/d)\,\sin(\omega t - \beta_m z)$$
$$-(2E_o m\pi/d\omega\mu_0)\,\mathbf{a}_z\,\cos(m\pi y/d)\,\cos(\omega t - \beta_m z).$$

10.1.3 *Transverse Magnetic (TM) Wave*

A TM wave may be launched into a parallel-plate waveguide by a uniform plane wave of parallel polarization incident at a certain angle. In this case, the electric field vector lies in the plane of incidence, which corresponds to page, as shown in Fig. 10.6. The positive direction of the electric field vector is set such that its longitudinal component is parallel to that of the wavevector. The incident wave is reflected by the upper and lower plates and progresses along a zigzag path in the waveguide. In other words, the waveguide mode can be resolved into two uniform plane waves: upward and downward waves with electric fields $\tilde{\mathbf{E}}_u$ and $\tilde{\mathbf{E}}_d$, respectively. The upward and downward waves at the upper plate can be considered as the incident and reflected waves, respectively. According to Fresnel's formula, the reflection coefficient for parallel polarization on a perfectly conducting surface is $\Gamma_{\parallel} = -1$. Therefore, before

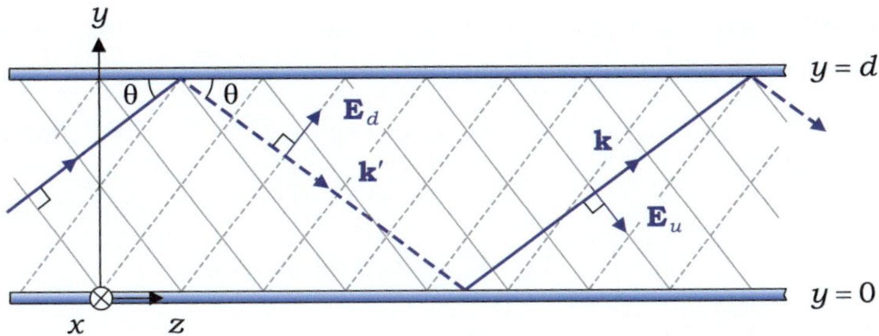

Fig. 10.6 TM mode in the parallel-plate waveguide. Solid lines, wavefronts for wave **k**; Dotted lines, wavefronts for wave **k**$'$

adding the two electric fields, $\tilde{\mathbf{E}}_d$ should be multiplied by -1. The total electric field in the waveguide is thus expressed as

$$
\begin{aligned}
\tilde{\mathbf{E}} &= \tilde{\mathbf{E}}_u - \tilde{\mathbf{E}}_d \\
&= \left(-\mathbf{a}_y \cos\theta + \mathbf{a}_z \sin\theta\right) E_o e^{-j(k_y y + k_z z)} \\
&\quad - \left(\mathbf{a}_y \cos\theta + \mathbf{a}_z \sin\theta\right) E_o e^{-j(-k_y y + k_z z)}
\end{aligned}
\tag{10.17}
$$

The wavevectors are $\mathbf{k} = k\sin\theta\,\mathbf{a}_y + k\cos\theta\,\mathbf{a}_z$ and $\mathbf{k}' = -k\sin\theta\,\mathbf{a}_y + k\cos\theta\,\mathbf{a}_z$ for upward and downward waves, respectively. Equation (10.17) can be rewritten as

$$
\boxed{\tilde{\mathbf{E}} = -2E_o\left[\mathbf{a}_y \cos\theta \cos(k_y y) + \mathbf{a}_z\, j \sin\theta \sin(k_y y)\right] e^{-jk_z z}}
\tag{10.18}
$$

Next, inserting Eq. (10.18) into Eq. (8.9a), the magnetic field associated with the electric field is obtained as

$$
\begin{aligned}
\tilde{\mathbf{H}} &= -\frac{1}{j\omega\mu}\nabla \times \tilde{\mathbf{E}} \\
&= \mathbf{a}_x \frac{2E_o}{\eta}\cos(k_y y) e^{-jk_z z}
\end{aligned}
\tag{10.19}
$$

where $\eta = \sqrt{\mu/\varepsilon}$ is the intrinsic impedance of the dielectric inside the waveguide. The electric and magnetic fields expressed in Eqs. (10.18) and (10.19) constitute a waveguide mode that has no longitudinal component of the magnetic field, and is thus called the ***transverse magnetic*** (TM) mode.

The boundary condition for the electric field requires the tangential component of the electric field to be zero at the conducting surfaces at $y = 0$ and $y = d$. Accordingly, the sine term in Eq. (10.18) should vanish at $y = 0$ and $y = d$, leading to the same eigenvalues as those in Eq. (10.6). Consequently, *the relationships in Eqs. (10.9)–(10.13) and (10.16) pertaining to the TE modes also hold for the TM modes in the parallel-plate waveguides.*

Example 10.1   The parallel-plate waveguide as shown in Fig. 10.3 has an air gap of length d [m].

(a)   Use Eq. (10.6) to verify that the TE modes expressed by Eq. (10.5) are mutually orthogonal, that is,

$$\int_{y=0}^{y=d} \tilde{\mathbf{E}}_m \cdot \tilde{\mathbf{E}}_n^* \, dy = 0 \ \text{ for } m \neq n.$$

(b)   Find the orthonormal set in TE modes by adjusting E_o such that

$$\int_{y=0}^{y=d} \tilde{\mathbf{E}}_m \cdot \tilde{\mathbf{E}}_m^* \, dy = 1.$$

Solution

(a)   With Eqs. (10.5) and (10.6), for $m \neq n$, we obtain

$$\begin{aligned}
\int_{y=0}^{y=d} \tilde{\mathbf{E}}_m \cdot \tilde{\mathbf{E}}_n^* \, dy &= 4E_o^2 \int_{y=0}^{y=d} \sin\left(\frac{m\pi}{d}y\right) \sin\left(\frac{n\pi}{d}y\right) dy \\
&= 2E_o^2 \int_{y=0}^{y=d} \left[\cos\left\{(m-n)\frac{\pi y}{d}\right\} - \cos\left\{(m+n)\frac{\pi y}{d}\right\}\right] dy \\
&= 0
\end{aligned}$$

(10.20)

(b)   Setting $m = n$ in Eq. (10.20) gives

$$\int_{y=0}^{y=d} \tilde{\mathbf{E}}_m \cdot \tilde{\mathbf{E}}_m^* \, dy = 2E_o^2 d$$

With Eq. (10.5) divided by $E_o \sqrt{2d}$, the orthonormal set in the TE modes is given by

$$\boxed{\tilde{\mathbf{E}}_m = -j\sqrt{\frac{2}{d}} \, \mathbf{a}_x \sin\left(\frac{m\pi}{d}y\right) e^{-jk_z z}} \quad (m = 1, 2, 3, \ldots)$$

(10.21)

If the wave in a parallel-plate waveguide has an electric field parallel to the transverse plane, it can be resolved into TE modes with different mode numbers by expanding the electric field in terms of the orthonormal set, as expressed by Eq. (10.21).

Example 10.2 A parallel-plate waveguide with an air gap of 1 [cm] operates at 35 [GHz]. In the waveguide, the electric field is $\tilde{\mathbf{E}} = 10\,\mathbf{a}_x \sin^2(200\pi y)$ in the cross section at $z = 0$. Determine the following quantities for the TE_1, TE_2, and TE_3 modes:

(a) Cutoff frequency.
(b) Electric fields in the $z = 0$ and $z = 2$ [cm] planes.
    [Hint: $(\sin^2 a)\sin b = \frac{1}{2}\sin b - \frac{1}{4}\sin(2a + b) + \frac{1}{4}\sin(2a - b)$.]

Solution

(a) From Eq. (10.10),

$$f_{c(m)} = \frac{mc}{2nd} = \frac{m \times 3 \times 10^8}{2 \times 10^{-2}} = m \times 15\,[\text{GHz}]$$

Thus, the cutoff frequencies for the TE_1, TE_2, and TE_3 modes are

$$f_{c(1)} = 15\,[\text{GHz}], \quad f_{c(2)} = 30\,[\text{GHz}], \quad \text{and} \quad f_{c(3)} = 45\,[\text{GHz}]$$

At an operating frequency of 35 [GHz], the TE_1 and TE_2 modes are above the cutoff frequencies, whereas the TE_3 mode is below the cutoff frequency.

(b) Using Eq. (10.21), the strength of the TE_1 mode in the $z = 0$ plane is obtained as

$$\int_{y=0}^{y=d} \tilde{\mathbf{E}} \cdot \tilde{\mathbf{E}}_1^* \, dy = j100\sqrt{2} \int_{y=0}^{y=0.01} \sin^2(200\pi y)\,\sin(100\pi y)\,dy$$

$$= j0.48 \tag{10.22a}$$

Similarly, the strength of the TE_2 mode in the $z = 0$ plane is

$$\int_{y=0}^{y=d} \tilde{\mathbf{E}} \cdot \tilde{\mathbf{E}}_2^* \, dy = j100\sqrt{2} \int_{y=0}^{y=0.01} \sin^2(200\pi y)\,\sin(200\pi y)\,dy$$

$$= 0 \tag{10.22b}$$

The strength of the TE_3 mode in the $z = 0$ plane is

$$\int_{y=0}^{y=d} \tilde{\mathbf{E}} \cdot \tilde{\mathbf{E}}_3^* \, dy = j100\sqrt{2} \int_{y=0}^{y=0.01} \sin^2(200\pi y)\,\sin(300\pi y)\,dy$$

$$= j0.34 \tag{10.22c}$$

Thus, in the $z = 0$ plane, the electric fields of the three modes are

$$\tilde{\mathbf{E}}_1 = 4.8\sqrt{2}\,\mathbf{a}_x \sin(100\pi y) \tag{10.22d}$$

$$\tilde{\mathbf{E}}_2 = 0 \tag{10.22e}$$

$$\tilde{\mathbf{E}}_3 = 3.4\sqrt{2}\,\mathbf{a}_x \sin(300\pi y) \tag{10.22f}$$

From Eq. (10.9), the phase constant for the TE_1 mode is

$$\beta_1 = \frac{2\pi \times 35 \times 10^9}{3 \times 10^8}\sqrt{1 - \left(\frac{15}{35}\right)^2} = 662.3\,[\mathrm{rad/m}] \tag{10.22g}$$

From Eq. (10.16), the attenuation constant for the TE_3 mode is

$$\alpha_3 = \frac{2\pi \times 45 \times 10^9}{3 \times 10^8}\sqrt{1 - \left(\frac{35}{45}\right)^2} = 592.4\,[\mathrm{Np/m}] \tag{10.22h}$$

Thus, at $z = 2\,[\mathrm{cm}]$, the three modes can be expressed as

$$\tilde{\mathbf{E}}_1 = 4.8\sqrt{2}\,\mathbf{a}_x \sin(100\pi y)e^{-j662.3 \times 0.02} = 6.79\,\mathbf{a}_x \sin(100\pi y)e^{-j0.68} \tag{10.22i}$$

$$\tilde{\mathbf{E}}_2 = 0 \tag{10.22j}$$

$$\tilde{\mathbf{E}}_3 = 3.4\sqrt{2}\,\mathbf{a}_x \sin(300\pi y)e^{-592.4 \times 0.02} = 3.4 \times 10^{-5}\,\mathbf{a}_x \sin(300\pi y) \tag{10.22k}$$

Exercise 10.5

What is the maximum frequency at which a parallel-plate waveguide with an air gap of length 1.5 [cm] is excited in the TEM mode only?

Ans. 10 [GHz].

Exercise 10.6

For a parallel-plate waveguide with a gap of 1 [cm] filled with a dielectric with $n = 2.5$, operating at 15 [GHz], determine the time delay between the TM_1 and TM_2 modes after traveling 1 [cm] in the waveguide.

Ans. 48 [ps].

Review Questions

RQ 10.1	What are the three mode types of parallel-plate waveguides?	[Figs. 10.4–10.6]
RQ 10.2	What is the transverse resonance condition?	[(10.7)]
RQ 10.3	Define α_m, β_m, and the cutoff frequency for the mode of propagation in a parallel-plate waveguide.	[(10.16)(10.9) (10.10)]

RQ 10.4	Does the TEM mode in a parallel-plate waveguide have a cutoff frequency?	[(10.7)]
RQ 10.5	Does the number of guided modes increase with operating frequency in a parallel-plate waveguide?	[(10.10)]
RQ 10.6	Does the incident angle of mode m, θ_m, decrease with operating frequency in a parallel-plate waveguide?	[(10.11)]
RQ 10.7	Distinguish between the phase and group velocities of the guided mode in the parallel-plate waveguide.	[(10.12a,b)] (10.13a,b)]
RQ 10.8	What is the evanescent mode of a parallel-plate waveguide?	[(10.16)]

10.2 Rectangular Waveguide

In the previous section, we gained an understanding of the basic principles of waveguides through a discussion of parallel-plate waveguides. In most practical situations, parallel plates with finite widths produce a fringing effect on the field at the edges and cause loss of electromagnetic energy through the side of the waveguide. To avoid this difficulty, practical waveguides operating at microwave frequencies are made of hollow metal pipes with uniform cross sections. One of the most commonly used waveguides is the rectangular waveguide, which has a rectangular cross section enclosed by four conducting walls (Fig. 10.7). In general, a rectangular waveguide is filled with a dielectric of permeability μ and permittivity ε. Although it may not be easy to visualize the propagation of electromagnetic waves in rectangular waveguides, both the rectangular and parallel-plate waveguides operate based on the same basic principle.

We begin our discussion on rectangular waveguides by assuming that the waveguide is oriented along the z-axis and that the mode of operation is given by a superposition of uniform plane waves having the same frequency and propagation constant along the waveguide. Accordingly, we can write a general expression for the propagation mode in a rectangular waveguide as

Fig. 10.7 Rectangular
waveguide

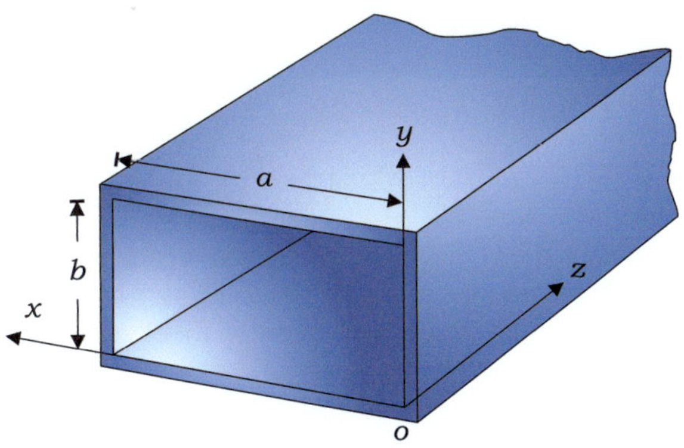

$$\mathscr{E}(\mathbf{r}, t) = \text{Re}\left[\tilde{\mathbf{E}}(\mathbf{r})e^{j\omega t}\right]$$

$$= \text{Re}\left[\left\{E_x(x, y)\,\mathbf{a}_x + E_y(x, y)\,\mathbf{a}_y + E_z(x, y)\,\mathbf{a}_z\right\}e^{-\gamma z}e^{j\omega t}\right] \quad (10.23)$$

where γ is the propagation constant in the z-direction. Note that the space coordinate z is separated from x and y in Eq. (10.23). The three terms, $E_x(x, y)$, $E_y(x, y)$, and $E_z(x, y)$, are responsible for the electric field distribution in the transverse plane, whereas $e^{-\gamma z}$ describes the field variation in the longitudinal direction or along the waveguide. Here, the underlying assumption is that the electric field moves in the z-direction with a propagation constant γ while keeping its transverse distribution unchanged. In phasor notation, the electric field in a rectangular waveguide can be expressed as

$$\boxed{\tilde{\mathbf{E}}(\mathbf{r}) \equiv \left\{E_x(x, y)\,\mathbf{a}_x + E_y(x, y)\,\mathbf{a}_y + E_z(x, y)\,\mathbf{a}_z\right\}e^{-\gamma z}} \quad (10.24)$$

The electric field phasor should satisfy the vector Helmholtz equation in the waveguide, such that

$$\nabla^2\,\tilde{\mathbf{E}} + k^2\,\tilde{\mathbf{E}} = 0 \quad (10.25)$$

where $k = \omega\sqrt{\mu\varepsilon}$ is the wavenumber in the dielectric with permeability μ and permittivity ε, and $\omega = 2\pi f$ is the angular frequency. Inserting Eq. (10.24) into Eq. (10.25) yields

$$\nabla^2_{xy}\left(E_x\mathbf{a}_x + E_y\mathbf{a}_y + E_z\mathbf{a}_z\right) + \left(\gamma^2 + k^2\right)\left(E_x\mathbf{a}_x + E_y\mathbf{a}_y + E_z\mathbf{a}_z\right) = 0 \quad (10.26)$$

where we have used the two-dimensional Laplacian operator, defined as

$$\nabla^2_{xy} \equiv \frac{\partial^2}{\partial x^2} + \frac{\partial^2}{\partial y^2} \quad (10.27)$$

Because the unit vectors in Eq. (10.26) are mutually exclusive, Eq. (10.26) can be separated into three partial differential equations of the same form as follows:

$$\nabla^2_{xy}E_x + \left(\gamma^2 + k^2\right)E_x = 0 \quad (10.28\text{a})$$

$$\nabla^2_{xy}E_y + \left(\gamma^2 + k^2\right)E_y = 0 \quad (10.28\text{b})$$

$$\boxed{\nabla^2_{xy}E_z + \left(\gamma^2 + k^2\right)E_z = 0} \quad (10.28\text{c})$$

where γ is the propagation constant along the waveguide, and k is the wavenumber in an unbounded medium, given by $k = \omega\sqrt{\mu\varepsilon}$. Upon solving Eqs. (10.28a)–(10.28c) and applying the boundary condition for electric fields at the conducting walls of the waveguide, we can uniquely determine the electric field and propagation constant in the waveguide.

Similarly, we can write a general expression for the magnetic field as

$$\mathcal{H}(\mathbf{r}, t) = \text{Re}\left[\tilde{\mathbf{H}}(\mathbf{r})e^{j\omega t}\right]$$
$$= \text{Re}\left[\left\{H_x(x, y)\,\mathbf{a}_x + H_y(x, y)\,\mathbf{a}_y + H_z(x, y)\,\mathbf{a}_z\right\}e^{-\gamma z}e^{j\omega t}\right] \quad (10.29)$$

The magnetic field phasor should also satisfy the Helmholtz equation, such that

$$\nabla_{xy}^2\left(H_x\mathbf{a}_x + H_y\mathbf{a}_y + H_z\mathbf{a}_z\right) + \left(\gamma^2 + k^2\right)\left(H_x\mathbf{a}_x + H_y\mathbf{a}_y + H_z\mathbf{a}_z\right) = 0 \quad (10.30)$$

By separating the three components, we have

$$\nabla_{xy}^2 H_x + \left(\gamma^2 + k^2\right)H_x = 0 \quad (10.31a)$$

$$\nabla_{xy}^2 H_y + \left(\gamma^2 + k^2\right)H_y = 0 \quad (10.31b)$$

$$\boxed{\nabla_{xy}^2 H_z + \left(\gamma^2 + k^2\right)H_z = 0} \quad (10.31c)$$

By solving these equations and applying the boundary conditions for the magnetic fields at the conducting walls, the magnetic field in the waveguide can be uniquely determined.

However, the six field components of **E** and **H** are closely related. If the longitudinal components E_z and H_z are obtained by solving Eqs. (10.28c) and (10.31c), the transverse components of **E** and **H** can be obtained from Maxwell's equations, without solving the partial differential equations in Eqs. (10.28a, b) and (10.31a, b). In the source-free region, for which $\mathbf{J} = 0$, Faraday's law in Maxwell's equations ($\nabla \times \tilde{\mathbf{E}} = -j\omega\mu\tilde{\mathbf{H}}$) can be separated into three parts:

$$\frac{\partial E_z}{\partial y} + \gamma E_y = -j\omega\mu H_x \quad (10.32a)$$

$$-\gamma E_x - \frac{\partial E_z}{\partial x} = -j\omega\mu H_y \quad (10.32b)$$

$$\frac{\partial E_y}{\partial x} - \frac{\partial E_x}{\partial y} = -j\omega\mu H_z \quad (10.32c)$$

Similarly, Ampere's law ($\nabla \times \tilde{\mathbf{H}} = j\omega\varepsilon\tilde{\mathbf{E}}$) can be divided into the following three parts:

$$\frac{\partial H_z}{\partial y} + \gamma H_y = j\omega\varepsilon E_x \quad (10.33a)$$

$$-\gamma H_x - \frac{\partial H_z}{\partial x} = j\omega\varepsilon E_y \tag{10.33b}$$

$$\frac{\partial H_y}{\partial x} - \frac{\partial H_x}{\partial y} = j\omega\varepsilon E_z \tag{10.33c}$$

Note that term $-\gamma$ in the above equations signifies a partial derivative with respect to z. Combining Eqs. (10.32) and (10.33), the transverse components E_x, E_y, H_x, and H_y can be expressed in terms of the longitudinal components E_z and H_z. For example, by eliminating H_x in Eqs. (10.32a) and (10.33b), E_y can be expressed in terms of E_z and H_z. Consequently, the transverse components can be expressed as

$$E_x = -\frac{1}{h^2}\left(\gamma\frac{\partial E_z}{\partial x} + j\omega\mu\frac{\partial H_z}{\partial y}\right) \tag{10.34a}$$

$$E_y = -\frac{1}{h^2}\left(\gamma\frac{\partial E_z}{\partial y} - j\omega\mu\frac{\partial H_z}{\partial x}\right) \tag{10.34b}$$

$$H_x = -\frac{1}{h^2}\left(\gamma\frac{\partial H_z}{\partial x} - j\omega\varepsilon\frac{\partial E_z}{\partial y}\right) \tag{10.34c}$$

$$H_y = -\frac{1}{h^2}\left(\gamma\frac{\partial H_z}{\partial y} + j\omega\varepsilon\frac{\partial E_z}{\partial x}\right) \tag{10.34d}$$

where

$$h^2 = \gamma^2 + k^2 \tag{10.35}$$

It is important to note that Eq. (10.34) is not useful for the case in which both E_z and H_z are zero, because h is also zero in this case (see Exercise 10.7).

A rectangular waveguide can be excited with TE modes containing no E_z-component or with TM modes containing no H_z-component, but not with TEM modes. For example, setting $E_z = 0 = H_z$ in Eqs. (10.24) and (10.29) for the TEM mode and inserting them into two equations of Maxwell, $\nabla \times \tilde{\mathbf{E}} = -j\omega\mu\tilde{\mathbf{H}}$ and $\nabla \cdot \varepsilon\tilde{\mathbf{E}} = 0$, we have

$$\frac{\partial E_y}{\partial x} - \frac{\partial E_x}{\partial y} = 0 \tag{10.36a}$$

$$\frac{\partial E_x}{\partial x} + \frac{\partial E_y}{\partial y} = 0 \tag{10.36b}$$

These equations can be combined to form $\nabla_{xy}^2 E_x = 0$ and $\nabla_{xy}^2 E_y = 0$. Because E_x should satisfy the Helmholtz equation $\nabla^2 E_x + k^2 E_x = \nabla_{xy}^2 E_x + \nabla_z^2 E_x + k^2 E_x = 0$, it follows that E_x varies with z only, meaning that E_x is constant over the transverse plane. However, in this case, E_x may be tangential to the conductor walls in violation of the boundary condition for the electric field. Therefore, E_x should vanish in the waveguide and the same can be said for E_y. *The TEM modes cannot exist in rectangular waveguides.*

Exercise 10.7

If $E_z = 0 = H_z$ in Eqs. (10.24) and (10.29), show that $\gamma = jk$.

Ans. $\nabla^2_{xy} E_x = 0$ from Eq. (10.36). $\gamma^2 + k^2 = 0$ in Eq. (10.28a).

Review Questions

RQ 10.9	What is the underlying assumption regarding the general expression of the waveguide mode in Eq. (10.23)?	[Figs. 10.5, 10.6]
RQ 10.10	Distinguish between the propagation constant γ and wavenumber k in a rectangular waveguide.	[(10.28a,b,c)]
RQ 10.11	What makes the TEM mode nonpropagate in a rectangular waveguide?	[(10.36a,b)]

10.2.1 Transverse Magnetic (TM) Mode

Although the TM mode has no longitudinal component of the magnetic field (i.e., $H_z = 0$) in a rectangular waveguide, it should have a nonzero E_z. Otherwise, it is the TEM mode that is not supported by the waveguide. If E_z is known in the waveguide, the transverse components E_x, E_y, H_x, and H_y can be obtained by inserting E_z into Eq. (10.34).

10.2.1.1 Longitudinal Field Components of TM Mode

In a rectangular waveguide, the longitudinal field component of the TM mode is governed by the Helmholtz equation, given in Eq. (10.28c), which is repeated as follows:

$$\frac{\partial^2 E_z}{\partial x^2} + \frac{\partial^2 E_z}{\partial y^2} + h^2 E_z = 0$$

where $h^2 = \gamma^2 + k^2$. For lossless waveguides, γ is given by jk_z, where k_z is the z-component of the wavevector in an unbounded medium, which is also called the phase constant of the waveguide. Meanwhile, the so-called eigenvalue h corresponds to the projection of the wavevector onto the transverse plane or the xy-plane, that is, $h^2 = k_x^2 + k_y^2$. Again, γ is the propagation constant along the waveguide, and k is the propagation constant in an unbounded dielectric.

We now solve the two-dimensional Helmholtz equation in Eq. (10.28c) using ***the method of separation of variables***. The solution for $E_z(x, y)$ is assumed to be a product of two functions: X depends only on x and Y depends only on y, such that

$$E_z(x, y) = X(x)Y(y) \tag{10.37}$$

If this satisfies the boundary conditions for the electric field at the conducting walls of the waveguide, the uniqueness theorem ensures that it is a unique solution. Inserting Eq. (10.37) into Eq. (10.28c) and rearranging the terms, we have

$$-\frac{1}{X}\frac{d^2X}{dx^2} = \frac{1}{Y}\frac{d^2Y}{dy^2} + h^2 \tag{10.38}$$

Because the left-hand side of Eq. (10.38) depends only on x, and the right side depends only on y, this equation is valid for any values of x and y only if both sides are equal to a constant, called a separation constant. Using the separation constant k_x^2, Eq. (10.38) can be divided into the following two parts:

$$\frac{d^2X}{dx^2} + k_x^2 X = 0 \tag{10.39a}$$

$$\frac{d^2Y}{dy^2} + \left(h^2 - k_x^2\right)Y = 0 \tag{10.39b}$$

which are second-order ordinary differential equations (ODEs).

The general solutions to Eqs. (10.39a) and (10.39b) are given by linear combinations of two independent solutions, that is,

$$X = A\sin(k_x x) + B\cos(k_x x) \tag{10.40a}$$

$$Y = C\sin\left(k_y y\right) + D\cos\left(k_y y\right) \tag{10.40b}$$

with

$$h^2 = k_x^2 + k_y^2 \tag{10.41}$$

The longitudinal electric field E_z vanishes at the conducting walls of the waveguide. Otherwise, an infinite surface current is induced on the walls. According to the boundary condition,

$$E_z(x = 0, y) = 0 \tag{10.42a}$$

$$E_z(x = a, y) = 0 \tag{10.42b}$$

$$E_z(x, y = 0) = 0 \tag{10.42c}$$

$$E_z(x, y = b) = 0 \tag{10.42d}$$

Imposing these conditions on Eq. (10.40) yields

$$B = 0 \text{ and } k_x = \frac{m\pi}{a} \quad (m = 1, 2, 3, ...) \tag{10.43a}$$

$$D = 0 \text{ and } k_y = \frac{n\pi}{b} \quad (n = 1, 2, 3, ...) \tag{10.43b}$$

Consequently, the longitudinal component of the electric field in the TM mode is given by

$$\boxed{E_z(x, y) = E_o \sin\left(\frac{m\pi}{a}x\right)\sin\left(\frac{n\pi}{b}y\right)} \quad (m, n = 1, 2, 3, ...) \tag{10.44}$$

$$\boxed{\mathscr{E}_z(x, y, z, t) = \mathrm{Re}\left[E_o \sin\left(\frac{m\pi}{a}x\right)\sin\left(\frac{n\pi}{b}y\right)e^{-\gamma z}e^{j\omega t}\right]} \quad [\mathrm{V/m}] \tag{10.45}$$

with

$$\boxed{\begin{aligned}\gamma &= j\sqrt{k^2 - h^2} = j\sqrt{\omega^2\mu\varepsilon - \left(\tfrac{m\pi}{a}\right)^2 - \left(\tfrac{n\pi}{b}\right)^2}\\[2mm] &\equiv a_{mn} + j\beta_{mn}\end{aligned}} \quad [\mathrm{m}^{-1}] \tag{10.46}$$

$$\boxed{h^2 = k_x^2 + k_y^2 = \left(\frac{m\pi}{a}\right)^2 + \left(\frac{n\pi}{b}\right)^2} \tag{10.47}$$

$$\boxed{k = \omega\sqrt{\mu\varepsilon}} \quad [\mathrm{rad/m}] \tag{10.48}$$

where the positive integers m and n are mode numbers. The mode number n should not be confused with the refractive index n. A particular field pattern with integers m and n is designated as TM_{mn} mode. The first subscript represents the number of maxima in $\mathscr{E}_z$ along the x-axis, whereas the second subscript represents the number along the y-axis, as can be seen from Eq. (10.45).

If the insulating material has real ε and μ and the operating frequency ω is significantly large, the radicand in Eq. (10.46) is positive, and the propagation constant becomes purely imaginary; that is, $\gamma = j\beta_{mn}$. In this case, the TM_{mn} mode with a longitudinal electric field, as shown in Eq. (10.45), propagates with a phase constant β_{mn} in the waveguide, while forming a standing wave in the transverse plane.

The propagation constant becomes zero at the operating frequency $f_{c(mn)}$, which is called the **cutoff frequency**. With the help of Eq. (10.46), the cutoff frequency for the **mn** mode is obtained as

$$\boxed{f_{c(mn)} = \frac{1}{2\sqrt{\mu\varepsilon}}\sqrt{\left(\frac{m}{a}\right)^2 + \left(\frac{n}{b}\right)^2}} \quad [\mathrm{Hz}] \tag{10.49}$$

At frequencies below the cutoff, the radicand in Eq. (10.46) is negative and γ becomes a positive real number, which is called the attenuation constant denoted by α_{mn}. In this case, the electric and magnetic fields form an evanescent mode that cannot propagate through the waveguide. It should be noted that the attenuation constant in this case is significantly different from that of a lossy transmission line that arises from losses in conductors and dielectrics.

Using the relation $\upsilon_p = 1/\sqrt{\mu\varepsilon} = f\lambda$, the cutoff frequency in Eq. (10.49) can be converted into the **cutoff wavelength** for the **mn** mode, such that

$$\lambda_{c(mn)} = \frac{\upsilon_p}{f_{c(mn)}} = \frac{2}{\sqrt{\left(\frac{m}{a}\right)^2 + \left(\frac{n}{b}\right)^2}} \quad [\text{m}] \tag{10.50}$$

An electromagnetic wave of wavelength $\lambda < \lambda_{c(mn)}$ may propagate in the **mn** mode in the waveguide.

The smallest possible values of the mode numbers **m** and **n** for the TM mode are one. Accordingly, the TM_{11} mode has the lowest cutoff frequency. The TM_{21} mode has the second-lowest cutoff frequency in a rectangular waveguide of dimension $a > b$. For example, an electromagnetic wave with a frequency within the range of $f_{c(11)} < f < f_{c(21)}$ can propagate only in the TM_{11} mode.

At an operating frequency above the cutoff, $f > f_{c(mn)}$, the phase constant for the **mn** mode is given by

$$\boxed{\beta_{mn} = \omega\sqrt{\mu\varepsilon}\sqrt{1 - \left(f_{c(mn)}/f\right)^2}} \quad [\text{rad/m}] \tag{10.51}$$

This was obtained by substituting Eq. (10.49) into Eq. (10.46). It should be noted that β_{mn} is always smaller than the phase constant in an unbounded dielectric given by $k = \omega\sqrt{\mu\varepsilon}$.

Just as the wavelength is related to the wavenumber through $\lambda = 2\pi/k$, the **waveguide wavelength** λ_{mn} is related to the phase constant β_{mn} through

$$\lambda_{mn} = \frac{2\pi}{\beta_{mn}} = \frac{\lambda}{\sqrt{1 - \left(f_{c(mn)}/f\right)^2}} \quad [\text{m}] \tag{10.52}$$

where λ_{mn} is the spatial period of the wave along the waveguide, and λ is the wavelength in an unbounded dielectric. The field pattern of the **mn** mode is repeated at every distance λ_{mn} along the waveguide. Waveguide wavelength λ_{mn} should not be confused with cutoff wavelength $\lambda_{c(mn)}$, which is measured in an unbounded dielectric. Upon using the relation $\upsilon_p = f\lambda = f_{c(mn)}\lambda_{c(mn)}$ in Eq. (10.52), we have

$$\frac{1}{\lambda^2} = \frac{1}{\lambda_{mn}^2} + \frac{1}{\lambda_{c(mn)}^2} \tag{10.53}$$

While the phase velocity in an unbounded dielectric is defined as $\upsilon_p = \omega/k$, the phase velocity in the waveguide is defined as

$$\upsilon_{p(mn)} = \frac{\omega}{\beta_{mn}} \tag{10.54a}$$

This can be rewritten as

$$\boxed{\upsilon_{p(mn)} = \frac{\upsilon_p}{\sqrt{1 - \left(f_{c(mn)}/f\right)^2}}} \tag{10.54b}$$

where $\upsilon_p = 1/\sqrt{\mu\varepsilon}$. Note that $\upsilon_{p(mn)}$ is always larger than υ_p and is dependent on the frequency. Thus, the rectangular waveguides are dispersive.

10.2.1.2 Transverse Field Components of TM Mode

In the previous section, the propagation characteristics of the propagation mode were studied using the longitudinal electric-field component of the TM mode. We now focus on the transverse components of the TM mode, that is, E_x, E_y, H_x, and H_y, which can be obtained using E_z. As stated earlier, a TM wave may be considered a linear combination of uniform plane waves propagating with the same phase constant in the waveguide. Accordingly, we can write the general expressions for the electric and magnetic fields of the TM wave as

$$\mathscr{E}(x, y, z, t) = \mathrm{Re}\left[\tilde{\mathbf{E}}\, e^{j\omega t}\right]$$
$$= \mathrm{Re}\left[\left\{E_x(x, y)\,\mathbf{a}_x + E_y(x, y)\,\mathbf{a}_y + E_z(x, y)\,\mathbf{a}_z\right\}e^{-\gamma z}e^{j\omega t}\right] \tag{10.55a}$$

$$\mathscr{H}(x, y, z, t) = \mathrm{Re}\left[\tilde{\mathbf{H}}\, e^{j\omega t}\right]$$
$$= \mathrm{Re}\left[\left\{H_x(x, y)\,\mathbf{a}_x + H_y(x, y)\,\mathbf{a}_y\right\}e^{-\gamma z}e^{j\omega t}\right] \tag{10.55b}$$

where $E_z(x, y)$ is the longitudinal component of the TM mode as shown in Eq. (10.44), which is repeated below:

$$\boxed{E_z(x, y) = E_o \sin\left(\frac{m\pi}{a}x\right)\sin\left(\frac{n\pi}{b}y\right)} \quad (m, n = 1, 2, 3, ...)$$

Using Eq. (10.44) and setting $H_z = 0$ in Eq. (10.34), the transverse field components of the TM_{mn} mode can be obtained as

$$\boxed{E_x = -\frac{\gamma}{h^2}\left(\frac{m\pi}{a}\right)E_o \cos\left(\frac{m\pi}{a}x\right)\sin\left(\frac{n\pi}{b}y\right)} \tag{10.56a}$$

$$E_y = -\frac{\gamma}{h^2}\left(\frac{n\pi}{b}\right) E_o \sin\left(\frac{m\pi}{a}x\right)\cos\left(\frac{n\pi}{b}y\right) \tag{10.56b}$$

$$H_x = \frac{j\omega\varepsilon}{h^2}\left(\frac{n\pi}{b}\right) E_o \sin\left(\frac{m\pi}{a}x\right)\cos\left(\frac{n\pi}{b}y\right) \tag{10.56c}$$

$$H_y = -\frac{j\omega\varepsilon}{h^2}\left(\frac{m\pi}{a}\right) E_o \cos\left(\frac{m\pi}{a}x\right)\sin\left(\frac{n\pi}{b}y\right)$$

$$(\boldsymbol{m}, \boldsymbol{n} = 1, 2, 3, ...) \tag{10.56d}$$

where $\gamma = j\beta_{mn}$ for lossless waveguides, and eigenvalue h corresponds to the transverse component of the wavevector, that is,

$$h^2 = \gamma^2 + k^2$$
$$= k_x^2 + k_y^2 = \left(\frac{m\pi}{a}\right)^2 + \left(\frac{n\pi}{b}\right)^2 \tag{10.57}$$

where $\boldsymbol{m}$ and $\boldsymbol{n}$ are positive integers for the TM modes, and are not zero because E_z is given by the product of two sine functions, as shown in Eq. (10.44).

The electric and magnetic fields expressed in Eqs. (10.44) and (10.56) comprise the TM_{mn} mode in the rectangular waveguide with dimensions a and b. The instantaneous electric and magnetic fields of the TM_{mn} mode are given by

$$\mathscr{E} = \text{Re}\left[\left(E_x\mathbf{a}_x + E_y\mathbf{a}_y + E_z\mathbf{a}_z\right)e^{-\gamma z}e^{j\omega t}\right]$$

$$\mathscr{H} = \text{Re}\left[\left(H_x\mathbf{a}_x + H_y\mathbf{a}_y\right)e^{-\gamma z}e^{j\omega t}\right]$$

The wave impedance for the TM_{mn} mode in lossless rectangular waveguides is defined as

$$\eta_{\text{TM}} = \frac{E_x}{H_y} = -\frac{E_y}{H_x} \quad [\Omega] \tag{10.58a}$$

Substitution of Eq. (10.56) into Eq. (10.58a) together with $\gamma = j\beta_{mn}$ results in

$$\eta_{\text{TM}} = \frac{\beta_{mn}}{\omega\varepsilon} \tag{10.58b}$$

Next, inserting Eq. (10.51) into Eq. (10.58b), the wave impedance for the TM_{mn} mode is given by

$$\eta_{\text{TM}} = \sqrt{\frac{\mu}{\varepsilon}}\sqrt{1 - \left(f_{c(mn)}/f\right)^2} \quad [\Omega] \tag{10.58c}$$

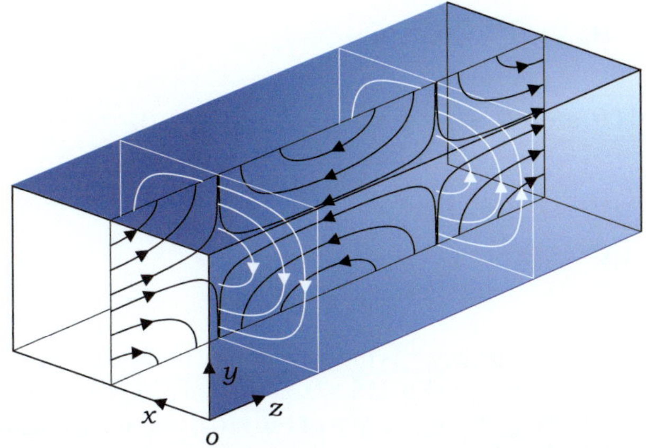

Fig. 10.8 TM_{11} mode in the rectangular waveguide. Black lines, electric field lines; White lines, magnetic field lines

The wave impedance depends on the frequency and is always smaller than the intrinsic impedance of dielectric $\eta = \sqrt{\mu/\varepsilon}$. However, below the cutoff, the wave impedance becomes purely imaginary.

The electric and magnetic field lines of the TM_{11} mode in the rectangular waveguide with $b/a = 3/4$ are shown in Fig. 10.8.

10.2.1.3 Orthonormal Set in TM Modes

The transverse magnetic fields of the TM modes in a rectangular waveguide are mutually orthogonal in the transverse direction, that is,

$$\int_{x=0}^{x=a} \int_{y=0}^{y=b} \tilde{\mathbf{H}}_{mn} \cdot \tilde{\mathbf{H}}^*_{m'n'} \, dx\,dy = 0 \quad (m \neq m' \text{ or } n \neq n') \tag{10.59}$$

This relationship can be readily verified by directly substituting Eq. (10.56). It is more convenient to normalize the transverse **H**-field of the TM_{mn} mode to unity, such that

$$\int_{x=0}^{x=a} \int_{y=0}^{y=b} \overline{\mathbf{H}}_{mn} \cdot \overline{\mathbf{H}}^*_{mn} \, dx\,dy = 1 \tag{10.60}$$

where the bar at the top denotes normalization.

Using Eq. (10.56) together with $\gamma = j\beta_{mn}$, we compute the following:

$$\int_{x=0}^{x=a} \int_{y=0}^{y=b} \tilde{\mathbf{H}}_{mn} \cdot \tilde{\mathbf{H}}_{mn}^* \, dx\,dy$$

$$= \int_0^a \int_0^b \left[(H_x \mathbf{a}_x + H_y \mathbf{a}_y) e^{-j\beta_{mn}z} \right] \cdot \left[(H_x \mathbf{a}_x + H_y \mathbf{a}_y) e^{-j\beta_{mn}z} \right]^* dx\,dy$$

$$= \int_0^a \int_0^b \left(|H_x|^2 + |H_y|^2 \right) dx\,dy$$

$$= \omega^2 \varepsilon^2 \frac{E_o^2}{h^2} \frac{ab}{4}$$

(10.61)

where ω is the angular frequency, ε is the permittivity of the dielectric in the waveguide, h is the eigenvalue given in Eq. (10.57), and a and b are the interior dimensions of the waveguide. By using the square root of Eq. (10.61) as the normalization factor, the normalized **H**-field of the TM$_{mn}$ mode is expressed as

$$\overline{\mathbf{H}}_{mn} = \left(\overline{H}_x \mathbf{a}_x + \overline{H}_y \mathbf{a}_y \right) e^{-j\beta_{mn}z}$$

(10.62)

with

$$\overline{H}_x = \frac{2j}{h\sqrt{ab}} \left(\frac{n\pi}{b} \right) \sin\left(\frac{m\pi}{a} x \right) \cos\left(\frac{n\pi}{b} y \right)$$

(10.63a)

$$\overline{H}_y = -\frac{2j}{h\sqrt{ab}} \left(\frac{m\pi}{a} \right) \cos\left(\frac{m\pi}{a} x \right) \sin\left(\frac{n\pi}{b} y \right)$$

$$(m, n = 1, 2, 3, \ldots)$$

(10.63b)

where $h^2 = (m\pi/a)^2 + (n\pi/b)^2$. The **H**-fields in Eq. (10.63) constitute an orthonormal set in the TM modes in rectangular waveguides.

Example 10.3 A rectangular waveguide with interior dimensions of $a = 1.6\,[\text{cm}]$ and $b = 1\,[\text{cm}]$ is filled with a dielectric with $\varepsilon_r = 1.96$ and $\mu_r = 1$ and is excited with an electromagnetic field of $f = 15\,[\text{GHz}]$. If the wave propagates in TM$_{11}$ mode, determine (a) λ, (b) $f_{c(11)}$, and (c) λ_{11}.

Solution

(a) The wavenumber in an unbounded dielectric is

$$k = \frac{2\pi}{\lambda} = \omega\sqrt{\mu_0 \varepsilon} = \frac{2\pi \times 15 \times 10^9 \times 1.4}{3 \times 10^8}$$

This gives $\lambda = 14.3\,[\text{mm}]$.

(b) From Eq. (10.49), the cutoff frequency is

$$f_{c(11)} = \frac{1}{2 \times 1.4 \times \sqrt{\mu_0 \varepsilon_0}} \sqrt{\left(\frac{1}{1.6 \times 10^{-2}}\right)^2 + \left(\frac{1}{10^{-2}}\right)^2} = 12.6\,[\text{GHz}]$$

(c) From Eq. (10.52), the waveguide wavelength is

$$\lambda_{11} = \frac{\lambda}{\sqrt{1 - \left(f_{c(11)}/f\right)^2}} = \frac{14.3 \times 10^{-3}}{\sqrt{1 - (12.6/15)^2}} = 26.4\,[\text{mm}]$$

Example 10.4 What are the plane-wave components comprising the TM_{21} mode in a rectangular waveguide with interior dimensions a and b?

Solution

Rewriting Eq. (10.44),

$$
\begin{aligned}
E_z &= E_o \sin\left(\frac{2\pi}{a}x\right)\sin\left(\frac{\pi}{b}y\right) \\
&= E_0 \frac{1}{2j}\left(e^{j2\pi x/a} - e^{-j2\pi x/a}\right)\frac{1}{2j}\left(e^{j\pi y/b} - e^{-j\pi y/b}\right) \\
&= -\frac{E_0}{4}\left[e^{j2\pi x/a + j\pi y/b} - e^{j2\pi x/a - j\pi y/b} - e^{-j2\pi x/a + j\pi y/b} + e^{-j2\pi x/a - j\pi y/b}\right]
\end{aligned}
$$

$$\text{(10.64a)}$$

Rewriting Eq. (10.56a),

$$
\begin{aligned}
E_x &= -\frac{j\beta_{21}}{h^2}\left(\frac{2\pi}{a}\right)E_o \cos\left(\frac{2\pi}{a}x\right)\sin\left(\frac{\pi}{b}y\right) \\
&= -\frac{j\beta_{21}}{h^2}\left(\frac{2\pi}{a}\right)E_o \frac{1}{2}\left(e^{j2\pi x/a} + e^{-j2\pi x/a}\right)\frac{1}{2j}\left(e^{j\pi y/b} - e^{-j\pi y/b}\right) \\
&= -\frac{j\beta_{21}}{h^2}\left(\frac{2\pi}{a}\right)\frac{E_o}{4j}\left[e^{j2\pi x/a + j\pi y/b} \right. \\
&\qquad \left. - e^{j2\pi x/a - j\pi y/b} + e^{-j2\pi x/a + j\pi y/b} - e^{-j2\pi x/a - j\pi y/b}\right]
\end{aligned}
$$

$$\text{(10.64b)}$$

Rewriting Eq. (10.56b),

$$
\begin{aligned}
E_y &= -\frac{j\beta_{21}}{h^2}\left(\frac{\pi}{b}\right)E_o \sin\left(\frac{2\pi}{a}x\right)\cos\left(\frac{\pi}{b}y\right) \\
&= -\frac{j\beta_{21}}{h^2}\left(\frac{\pi}{b}\right)E_o \frac{1}{2j}\left(e^{j2\pi x/a} - e^{-j2\pi x/a}\right)\frac{1}{2}\left(e^{j\pi y/b} + e^{-j\pi y/b}\right) \\
&= -\frac{j\beta_{21}}{h^2}\left(\frac{\pi}{b}\right)\frac{E_o}{4j}\left[e^{j2\pi x/a + j\pi y/b} + e^{j2\pi x/a - j\pi y/b} \right. \\
&\qquad \left. - e^{-j2\pi x/a + j\pi y/b} - e^{-j2\pi x/a - j\pi y/b}\right]
\end{aligned}
$$

$$\text{(10.64c)}$$

Each of the four complex exponentials in the brackets of these equations allows us to combine the three electric-field components to form a plane wave of the following form:

$$\tilde{\mathbf{E}} = \left\{ E_x(x, y)\,\mathbf{a}_x + E_y(x, y)\,\mathbf{a}_y + E_z(x, y)\,\mathbf{a}_z \right\} e^{-j\beta_{21} z}$$

In conclusion, the TM_{21} mode consists of four plane waves with spatial variations as specified by

$$e^{j2\pi x/a + j\pi y/b - j\beta_{21} z},\ e^{j2\pi x/a - j\pi y/b - j\beta_{21} z},\ e^{-j2\pi x/a + j\pi y/b - j\beta_{21} z},\ e^{-j2\pi x/a - j\pi y/b - j\beta_{21} z}.$$

Example 10.5 An air-filled rectangular waveguide with dimensions $a = 5\,[\mathrm{cm}]$ and $b = 2.5\,[\mathrm{cm}]$ is excited by electromagnetic waves at a frequency of $8\,[\mathrm{GHz}]$. If the magnetic field is given by $\mathbf{H} = 4\mathbf{a}_x \sin(20\pi x) \sin(60\pi y) + 3\mathbf{a}_y \cos(20\pi x) \cos(60\pi y)\,[\mathrm{A/m}]$ in the $z = 0$ plane, determine

(a)   the mode number for propagating mode,
(b)   the smallest value of α_{mn} for evanescent modes, and
(c)   the magnetic field in the $z = 1\,[\mathrm{m}]$ plane.

Solution

(a)   From Eq. (10.49), the three lowest cutoff frequencies are

$$f_{c(11)} = \frac{1}{2\sqrt{\mu_0 \varepsilon_0}} \sqrt{\left(\frac{1}{0.05}\right)^2 + \left(\frac{1}{0.025}\right)^2} = 6.71\,[\mathrm{GHz}]$$

$$f_{c(21)} = \frac{1}{2\sqrt{\mu_0 \varepsilon_0}} \sqrt{\left(\frac{2}{0.05}\right)^2 + \left(\frac{1}{0.025}\right)^2} = 8.49\,[\mathrm{GHz}]$$

$$f_{c(12)} = \frac{1}{2\sqrt{\mu_0 \varepsilon_0}} \sqrt{\left(\frac{1}{0.05}\right)^2 + \left(\frac{2}{0.025}\right)^2} = 12.37\,[\mathrm{GHz}]$$

Because the operation frequency is greater than $f_{c(11)}$, the TM_{11} mode is the propagation mode, and the TM_{21} and TM_{12} modes are the evanescent modes.

(b)   The TM_{21} mode is an evanescent mode just below the cutoff, whose attenuation constant is obtained from Eq. (10.46) as

$$\alpha_{21} = \pi \sqrt{\left(\frac{2}{0.05}\right)^2 + \left(\frac{1}{0.025}\right)^2 - \left(\frac{2 \times 8 \times 10^9}{3 \times 10^8}\right)^2} = 59.2\,[\mathrm{Np/m}]$$

(c)   In the $z = 0$ plane, from Eq. (10.63),

$$\overline{\mathbf{H}}_{11} = \frac{2j}{h\sqrt{ab}} \left\{ (40\pi)\,\mathbf{a}_x \sin(20\pi x) \cos(40\pi y) - (20\pi)\,\mathbf{a}_y \cos(20\pi x) \sin(40\pi y) \right\}$$

The strength of the TM_{11} mode is computed as

$$\int_{x=0}^{x=0.05} \int_{y=0}^{y=0.25} \tilde{\mathbf{H}} \cdot \overline{\mathbf{H}}_{11}^* \, dx\, dy$$

$$= -\frac{8j}{h\sqrt{ab}}(40\pi) \int_{x=0}^{x=0.05} \sin^2(20\pi x)\, dx \int_{y=0}^{y=0.025} \sin(60\pi y)\cos(40\pi y)\, dy$$

$$+\frac{6j}{h\sqrt{ab}}(20\pi) \int_{x=0}^{x=0.05} \cos^2(20\pi x)\, dx \int_{y=0}^{y=0.025} \cos(60\pi y)\sin(40\pi y)\, dy$$

$$= -\frac{0.3j}{h\sqrt{ab}}$$

$$(10.65a)$$

Multiplying Eq. (10.63) with Eq. (10.65a) and using Eq. (10.57), we obtain

$$H_x = 3.06\sin(20\pi x)\cos(40\pi y) \tag{10.65b}$$

$$H_y = -1.53\cos(20\pi x)\sin(40\pi y) \tag{10.65c}$$

From Eq. (10.51), the phase constant for the TM_{11} mode is

$$\beta_{mn} = k\sqrt{1-\left(f_{c(mn)}/f\right)^2} = \frac{2\pi \times 8 \times 10^9}{3\times 10^8}\sqrt{1-(6.71/8)^2} = 29.04\pi$$

$$(10.65d)$$

Thus, at $z = 1$ [m], the magnetic field phasor is given by

$$\tilde{\mathbf{H}} = \left[3.06\sin(20\pi x)\cos(40\pi y)\,\mathbf{a}_x - 1.53\cos(20\pi x)\sin(40\pi y)\,\mathbf{a}_y\right]e^{-j1.04\pi}$$

where the evanescent mode was ignored because of the large value of α_{21}.

Exercise 10.8
If an air-filled rectangular waveguide was filled with a dielectric with $\varepsilon_r = 2.25$ and $\mu_r = 1$, determine the change in the cutoff frequency for the TM_{mn} mode.

Ans. $f^d_{c(mn)}/f^a_{c(mn)} = 1/1.5$.

Exercise 10.9
Find the group velocity of the propagation mode in rectangular waveguides.

Ans.

$$\boxed{v_{g(mn)} = (d\beta_{mn}/d\omega)^{-1} = v_p\sqrt{1 - (f_{c(mn)}/f)^2}}$$ (10.66)

Exercise 10.10

Show that $v_{p(mn)}v_{g(mn)} = v_p^2$, where v_p is the phase velocity in an unbounded dielectric.

Review Questions

RQ 10.12	Define γ, h, and k for rectangular waveguides.	[(10.46)(10.47) (10.48)]
RQ 10.13	State the boundary condition for E_z in TM mode in rectangular waveguides.	[(10.42)]
RQ 10.14	Can the mode number of TM mode be zero?	[(10.44)]
RQ 10.15	Which TM mode has the lowest cutoff frequency?	[(10.49)]
RQ 10.16	In what way does the phase constant of the propagation mode vary with frequency?	[(10.51)]
RQ 10.17	Is the waveguide wavelength always greater than the wavelength in the corresponding unbounded dielectric?	[(10.52)]
RQ 10.18	Does the phase velocity of the propagation mode decrease with frequency?	[(10.54b)]
RQ 10.19	Define the wave impedance for the propagation mode.	[(10.58a,b,c)]
RQ 10.20	Can one TM mode be converted into another during propagation in a rectangular waveguide?	[(10.59)]

10.2.2 Transverse Electric (TE) Mode

The transverse electric (TE) mode does not have a longitudinal component in the electric field. In other words, a waveguide mode with nonzero H_z must be a TE mode because it is neither a TEM nor a TM mode. The TE mode with propagation constant γ is expressed as

$$\boxed{\mathcal{H}_z(x, y, z, t) = \mathrm{Re}\!\left[H_z(x, y)e^{-\gamma z}e^{j\omega t}\right]} \quad \text{[A/m]} \qquad (10.67)$$

Following the same procedure for the TM mode, we first solve the Helmholtz equation for H_z given in Eq. (10.31c), which is repeated below:

$$\frac{\partial^2 H_z}{\partial x^2} + \frac{\partial^2 H_z}{\partial y^2} + h^2 H_z = 0$$

where h is given by Eq. (10.47). Following the same logic that led to Eq. (10.44), Eq. (10.31c) is first separated into two second-order ordinary differential equations using the method of separation of variables. A general solution for H_z is expressed as

$$H_z(x, y) = [A \sin(k_x x) + B \cos(k_x x)] [C \sin(k_y y) + D \cos(k_y y)] \qquad (10.68)$$

where A, B, C, and D are yet-to-be-determined constants. Next, E_x and E_y are obtained by substituting Eq. (10.68) into Eq. (10.34), setting $E_z = 0$, and applying the following boundary conditions:

$$E_x(x, y = 0) = -\frac{j\omega\mu}{h^2} \frac{\partial H_z}{\partial y}\bigg|_{y=0} = 0 \qquad (10.69a)$$

$$E_x(x, y = b) = -\frac{j\omega\mu}{h^2} \frac{\partial H_z}{\partial y}\bigg|_{y=b} = 0 \qquad (10.69b)$$

$$E_y(x = 0, y) = \frac{j\omega\mu}{h^2} \frac{\partial H_z}{\partial x}\bigg|_{x=0} = 0 \qquad (10.69c)$$

$$E_y(x = a, y) = \frac{j\omega\mu}{h^2} \frac{\partial H_z}{\partial x}\bigg|_{x=a} = 0 \qquad (10.69d)$$

Substitution of Eq. (10.68) into Eq. (10.69) results in $A = C = 0$. Therefore, the longitudinal magnetic field in the TE$_{mn}$ mode is given by

$$\boxed{H_z(x, y) = H_o \cos\left(\frac{m\pi}{a}x\right) \cos\left(\frac{n\pi}{b}y\right)}$$

$$(m, n = 0, 1, 2, \ldots) \qquad (10.70)$$

where H_o was used instead of BD. Upon using Eq. (10.70) in Eq. (10.34) and setting $E_z = 0$, the transverse electric and magnetic fields in the TE$_{mn}$ mode are expressed as

$$\boxed{E_x = \frac{j\omega\mu}{h^2}\left(\frac{n\pi}{b}\right) H_o \cos\left(\frac{m\pi}{a}x\right) \sin\left(\frac{n\pi}{b}y\right)} \qquad (10.71a)$$

$$\boxed{E_y = -\frac{j\omega\mu}{h^2}\left(\frac{m\pi}{a}\right) H_o \sin\left(\frac{m\pi}{a}x\right) \cos\left(\frac{n\pi}{b}y\right)} \qquad (10.71b)$$

$$\boxed{H_x = \frac{\gamma}{h^2}\left(\frac{m\pi}{a}\right) H_o \sin\left(\frac{m\pi}{a}x\right) \cos\left(\frac{n\pi}{b}y\right)} \qquad (10.71c)$$

$$\boxed{H_y = \frac{\gamma}{h^2}\left(\frac{n\pi}{b}\right) H_o \cos\left(\frac{m\pi}{a}x\right) \sin\left(\frac{n\pi}{b}y\right)}$$

$$(m, n = 1, 2, 3, \ldots) \qquad (10.71d)$$

where $\gamma = j\beta_{mn}$ for the lossless waveguide. It is important to note that either m or n may be zero, but not simultaneously.

A comparison of the TE and TM modes reveals that H_z of the TE mode is given by the product of two cosine functions, whereas E_z of the TM mode is given by the product of two sine functions. All transverse components of the TE and TM modes are given by the product of sine and cosine functions.

The electric and magnetic fields in Eqs. (10.70) and (10.71a)–(10.71d) comprise the TE_{mn} mode in the rectangular waveguide with dimensions a and b. The instantaneous electric and magnetic fields of the TE_{mn} mode have general forms such as

$$\mathscr{E} = \mathrm{Re}\left[\left(E_x\,\mathbf{a}_x + E_y\,\mathbf{a}_y\right)e^{-\gamma z}e^{j\omega t}\right]$$

$$\mathscr{H} = \mathrm{Re}\left[\left(H_x\,\mathbf{a}_x + H_y\,\mathbf{a}_y + H_z\,\mathbf{a}_z\right)e^{-\gamma z}e^{j\omega t}\right]$$

The definitions of γ, h, and k in Eqs. (10.46)–(10.48) pertaining to the TM modes still hold true for TE modes. Therefore, the expressions for $f_{c(mn)}$, $\lambda_{c(mn)}$, β_{mn}, λ_{mn}, υ_p, and $\upsilon_{p(mn)}$ given in Eqs. (10.49)–(10.54) remain valid for the TE_{mn} mode. Note that the mode numbers m and n are positive integers for the TM modes, whereas one of them may be zero for the TE modes.

The electric and magnetic field lines of the TE_{11} mode in the waveguide with $b/a = 3/4$ are shown in Fig. 10.9.

The waveguide mode with the lowest possible cutoff frequency is called the **dominant mode**. In rectangular waveguides with $a > b$ (a is generally assigned to the longer dimension), the TE_{10} mode has the lowest cutoff frequency, and is thus the dominant mode. The cutoff frequency of the dominant mode is given by

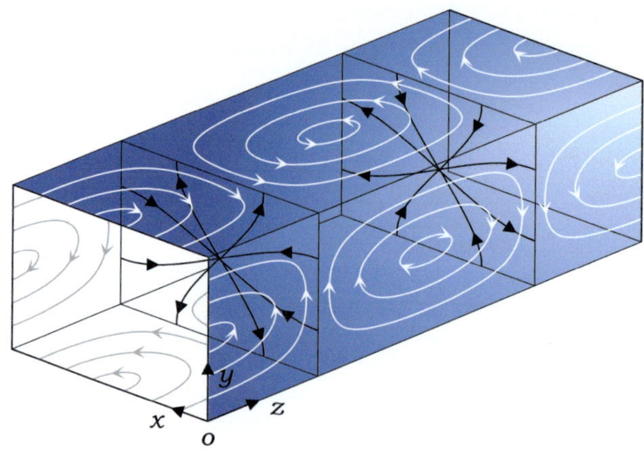

Fig. 10.9 Field patterns of TE_{11} mode. Black lines, **E**-field lines; White lines, **H**-field lines

$$f_{c(10)} = \frac{1}{2a\sqrt{\mu\varepsilon}} = \frac{\upsilon_p}{2a} \ [\text{Hz}] \tag{10.72}$$

Moreover, the TE_{10} mode has the lowest attenuation constant of all modes in the rectangular waveguide.

The wave impedance for the TE_{mn} mode is defined as

$$\eta_{\text{TE}} = \frac{E_x}{H_y} = -\frac{E_y}{H_x} = \frac{\omega\mu}{\beta_{mn}} \tag{10.73a}$$

where we have used Eqs. (10.71a)–(10.71d). Upon substituting Eq. (10.51) into Eq. (10.73a), we have

$$\boxed{\eta_{\text{TE}} = \sqrt{\frac{\mu}{\varepsilon}} \frac{1}{\sqrt{1 - \left(f_{c(mn)}/f\right)^2}}} \ [\Omega] \tag{10.73b}$$

The wave impedances η_{TM} and η_{TE} given in Eqs. (10.58c) and (10.73b) are purely resistive in rectangular waveguides filled with lossless dielectric. Although they depend on the operating frequency, their product is constant such that $\eta_{\text{TM}}\eta_{\text{TE}} = \eta^2$, where $\eta = \sqrt{\mu/\varepsilon}$ is the intrinsic impedance of the dielectric.

As mentioned previously, the phase constants for the TE_{mn} and TM_{mn} modes are the same, as shown in Eq. (10.51). With $\upsilon_p = 1/\sqrt{\mu\varepsilon}$ and $\omega_{c(mn)} = 2\pi f_{c(mn)}$, Eq. (10.51) can be rearranged as

$$\omega = \frac{\beta_{mn}\upsilon_p}{\sqrt{1 - \left(\omega_{c(mn)}/\omega\right)^2}} \tag{10.74}$$

The relation between β and ω is called the dispersion relation, and is well described by the $\omega-\beta$ diagram shown in Fig. 10.10. For a mode with cutoff frequency $\omega_{c(mn)}$, the slope of the tangent to the $\omega-\beta$ curve at point p represents the group velocity of the mode, which propagates with phase constant β_{mn}, whereas the slope of the line joining the origin and point p represents the phase velocity. From Fig. 10.10, it is apparent that $\upsilon_{g(mn)} < \upsilon_{p(mn)}$ and $\upsilon_{g(mn)} < \upsilon_p < \upsilon_{p(mn)}$, which can be verified using Eqs. (10.54b) and (10.66). As the operating frequency increases significantly above the cutoff frequency, both $\upsilon_{p(mn)}$ and $\upsilon_{g(mn)}$ approach υ_p, which is the phase velocity in an unbounded dielectric (see Fig. 10.10).

10.2.2.1 Orthonormal Set in TE Modes

By following the same procedure used for the TM modes, it can be shown that the transverse electric fields of the different TE modes are mutually orthogonal in the transverse plane of a rectangular waveguide. That is,

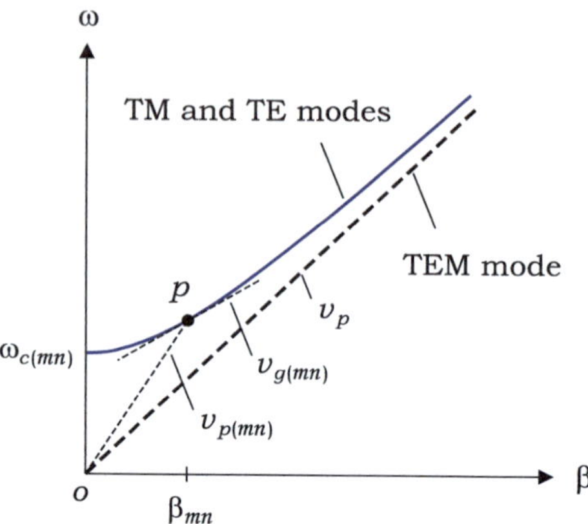

Fig. 10.10 $\omega-\beta$ diagram of the waveguide mode with cutoff frequency $\omega_{c(mn)}$

$$\int_{x=0}^{x=a} \int_{y=0}^{y=b} \tilde{\mathbf{E}}_{mn} \cdot \tilde{\mathbf{E}}_{m'n'}^* \, dx\, dy = 0 \quad (m \neq m' \text{ or } n \neq n') \tag{10.75}$$

This can be verified by directly substituting Eqs. (10.71a) and (10.71b). We normalize the transverse **E**-field of the TE mode to unity such that

$$\int_{x=0}^{x=a} \int_{y=0}^{y=b} \overline{\mathbf{E}}_{mn} \cdot \overline{\mathbf{E}}_{mn}^* \, dx\, dy = 1 \tag{10.76}$$

where the bar at the top denotes normalization.

Using Eqs. (10.71a) and (10.71b) together with $\gamma = j\beta_{mn}$, we compute

$$\int_{x=0}^{x=a} \int_{y=0}^{y=b} \tilde{\mathbf{E}}_{mn} \cdot \tilde{\mathbf{E}}_{mn}^* \, dx\, dy$$

$$= \int_0^a \int_0^b \left[(E_x \mathbf{a}_x + E_y \mathbf{a}_y) e^{-j\beta_{mn}z} \right] \cdot \left[(E_x \mathbf{a}_x + E_y \mathbf{a}_y) e^{-j\beta_{mn}z} \right]^* dx\, dy$$

$$= \int_0^a \int_0^b \left(|E_x|^2 + |E_y|^2 \right) dx\, dy \tag{10.77}$$

$$= \omega^2 \mu^2 \frac{H_o^2}{h^2} \frac{ab}{4} \quad (m, n = 1, 2, \ldots), \text{ or}$$

$$\omega^2 \mu^2 \frac{H_o^2}{h^2} \frac{ab}{2} \quad (m \text{ or } n = 0)$$

where ω is the angular frequency, μ is the permeability of the dielectric in the waveguide, h is the eigenvalue given in Eq. (10.57), and a and b are the interior dimensions of the waveguide. We note that Eq. (10.77) is the same as that in Eq. (10.61) if μ and H_o are replaced with ε and E_o, respectively.

By using the square root of Eq. (10.77) as the normalization factor, the normalized **E**-field of the TE$_{mn}$ mode is expressed as

$$\overline{\mathbf{E}}_{mn} = \left(\overline{E}_x \mathbf{a}_x + \overline{E}_y \mathbf{a}_y\right) e^{-j\beta_{mn}z} \qquad (10.78)$$

with

$$\overline{E}_x = \frac{Aj}{h\sqrt{ab}}\left(\frac{n\pi}{b}\right)\cos\left(\frac{m\pi}{a}x\right)\sin\left(\frac{n\pi}{b}y\right) \qquad (10.79a)$$

$$\overline{E}_y = -\frac{Aj}{h\sqrt{ab}}\left(\frac{m\pi}{a}\right)\sin\left(\frac{m\pi}{a}x\right)\cos\left(\frac{n\pi}{b}y\right)$$

$$(m, n = 0, 1, 2, ...) \qquad (10.79b)$$

where $A = \sqrt{2}$ for the TE$_{10}$ and TE$_{01}$ modes, and $A = 2$ for the other modes. Again, h is the eigenvalue, defined in Eq. (10.57). The **E**-fields expressed in Eq. (10.79) constitute an orthonormal set of TE modes of a rectangular waveguide.

Example 10.6 The TE$_{10}$ mode with a frequency of 3 [GHz] propagates in a rectangular waveguide with dimensions $a = 2b = 4$ [cm] filled with a dielectric with $n = 1.4$. Determine the (a) cutoff frequency, (b) phase constant, (c) phase velocity, and (d) wave impedance.

Solution

(a) From Eq. (10.49), the cutoff frequency is

$$f_{c(10)} = \frac{1}{2\sqrt{\mu_0\varepsilon}}\left(\frac{1}{a}\right) = \frac{3\times10^8}{2\times1.4}\left(\frac{1}{0.04}\right) = 2.68\,[\text{GHz}]$$

(b) From Eq. (10.51), the phase constant is

$$\beta_{mn} = \frac{n\omega}{c}\sqrt{1 - \left(\frac{f_{c(mn)}}{f}\right)^2} = \frac{1.4\times2\pi\times3\times10^9}{3\times10^8}\sqrt{1 - \left(\frac{2.68}{3}\right)^2}$$

$$= 12.58\pi\,[\text{rad/m}]$$

(c) From Eq. (10.54a), the phase velocity is

$$\upsilon_{p(10)} = \frac{\omega}{\beta_{10}} = \frac{2\pi\times3\times10^9}{12.58\pi} = 4.77\times10^8\,[\text{m/s}]$$

(d)   From Eq. (10.73b), the wave impedance is

$$\eta_{TE} = \frac{377}{1.4} \times \frac{1}{\sqrt{1 - (2.68/3)^2}} = 599.2 \,[\Omega]$$

Example 10.7 In a hollow rectangular waveguide with dimensions of $a = 2b = 5$ [cm], the TE mode has $\mathscr{E} = 4(\mathbf{a}_x - \mathbf{a}_y)\sin(20\pi z - 1.8\pi \times 10^{10}t)$ [V/m] at a point with $x = a/8$ and $y = b/4$. Determine (a) cutoff frequency, (b) mode number, and (c) complete expression for $\mathscr{E}$.

Solution

(a)   From $\mathscr{E}$ in the problem, we obtain $\beta_{mn} = 20\pi$ and $\omega = 1.8\pi \times 10^{10}$. Rearranging Eq. (10.51), we have

$$f_{c(mn)} = f\sqrt{1 - \left(\beta_{mn}\frac{c}{\omega}\right)^2} = 9 \times 10^9 \sqrt{1 - \left(\frac{20\pi \times 3 \times 10^8}{1.8\pi \times 10^{10}}\right)^2} = 8.485\,[\text{GHz}]$$

(b)   Setting $x = a/8$ and $y = b/4$ in Eqs. (10.71a) and (10.71b) yields

$$E_x = \frac{j\omega\mu_0}{h^2}\left(\frac{n\pi}{b}\right)H_o \cos\left(\frac{m\pi}{8}\right)\sin\left(\frac{n\pi}{4}\right) \qquad (10.80\text{a})$$

$$E_y = -\frac{j\omega\mu_0}{h^2}\left(\frac{m\pi}{a}\right)H_o \sin\left(\frac{m\pi}{8}\right)\cos\left(\frac{n\pi}{4}\right) \qquad (10.80\text{b})$$

Because these electric field components should lead to an electric field of the form $4(\mathbf{a}_x - \mathbf{a}_y)$, as given in the problem, it is required that

$$n\pi \cos\left(\frac{m\pi}{8}\right)\sin\left(\frac{n\pi}{4}\right) = \frac{m\pi}{2}\sin\left(\frac{m\pi}{8}\right)\cos\left(\frac{n\pi}{4}\right)$$

This can be rearranged in the form of $X \tan X = Y \tan Y$:

$$\left(\frac{n\pi}{4}\right)\tan\left(\frac{n\pi}{4}\right) = \left(\frac{m\pi}{8}\right)\tan\left(\frac{m\pi}{8}\right) \qquad (10.80\text{c})$$

This only holds for $X = Y$. Thus,

$$m = 2n \qquad (10.80\text{d})$$

The cutoff frequency is, therefore,

$$f_{c(mn)} = \frac{1}{2\sqrt{\mu_0\varepsilon_0}}\sqrt{\left(\frac{m}{a}\right)^2 + \left(\frac{n}{b}\right)^2} = 8.485 \times 10^9 \qquad (10.80\text{e})$$

By combining Eqs. (10.80d) and (10.80e) and using $a = 2b$, we obtain

$$m = 2 \quad \text{and} \quad n = 1 \tag{10.80f}$$

(c) Inserting Eq. (10.80f) into Eqs. (10.71a, b) yields

$$\mathscr{E} = A\left[\mathbf{a}_x \cos(40\pi x)\sin(40\pi y) - \mathbf{a}_y \sin(40\pi x)\cos(40\pi y)\right]$$
$$\times \sin\left(20\pi z - 1.8\pi \times 10^{10}t\right)[\text{V/m}] \tag{10.80g}$$

where A is a constant yet to be determined. By enforcing the amplitude of $\mathscr{E}$ to be $4(\mathbf{a}_x - \mathbf{a}_y)$ at the point where $x = a/8$ and $y = b/4$, we obtain $A = 8$.

The instantaneous electric field of the mode of operation is, therefore,

$$\mathscr{E} = 8\left[\mathbf{a}_x \cos(40\pi x)\sin(40\pi y) - \mathbf{a}_y \sin(40\pi x)\cos(40\pi y)\right]$$
$$\times \sin\left(20\pi z - 1.8\pi \times 10^{10}t\right)[\text{V/m}]$$

Example 10.8 Find the plane-wave components of the TE_{10} mode propagating in a rectangular waveguide with dimensions a and b.

Solution

From Eqs. (10.70) and (10.71c), the TE_{10} mode has

$$H_z = H_o \cos\left(\frac{\pi}{a}x\right) = \frac{H_0}{2}\left[e^{j\pi x/a} + e^{-j\pi x/a}\right] \tag{10.81a}$$

$$H_x = \frac{j\beta_{10}}{h^2}\left(\frac{\pi}{a}\right)H_o \sin\left(\frac{\pi}{a}x\right) = \frac{a\beta_{10}}{\pi}\frac{H_o}{2}\left[e^{j\pi x/a} - e^{-j\pi x/a}\right] \tag{10.81b}$$

$$H_y = 0 \tag{10.81c}$$

which can be combined to form

$$\tilde{\mathbf{H}} = \{H_x(x, y)\,\mathbf{a}_x + H_z(x, y)\,\mathbf{a}_z\}e^{-j\beta_{10}z}$$

It follows that the mode consists of two plane-wave components: $e^{j(\pi/a)x - j\beta_{10}z}$ and $e^{-j(\pi/a)x - j\beta_{10}z}$.

Exercise 10.11

Find the frequency range over which an air-filled rectangular waveguide with $a = 2b = 3\,[\text{cm}]$ supports only the dominant mode.

Ans. $5\,[\text{GHz}] < f < 10\,[\text{GHz}]$.

Exercise 10.12
Some TE modes have $\mathbf{E} = 0$ at the center of the rectangular waveguide with dimensions $a = 1.5b$. Determine which mode has the lowest cutoff frequency.

Ans. TE_{11} mode.

Exercise 10.13
An air-filled rectangular waveguide operating at a frequency of 7 [GHz] has a magnetic field $H_z = 10\cos(20\pi x)\cos(40\pi y)$ at $z = 0$. Determine $f_{c(mn)}$ and β_{mn}.

Ans. $f_{c(mn)} = 6.71\,[GHz]$, $\beta_{mn} = 41.8\,[rad/m]$.

Exercise 10.14
What is changed in Eqs. (10.70) and (10.71a)–(10.71d) if the TE_{mn} mode propagates in the $-z$-direction.

Ans. γ is replaced with $-\gamma$.

Review Questions

RQ 10.21	Is the phase velocity of the propagation mode always greater than the group velocity in the rectangular waveguides?	[Fig. 10.10]
RQ 10.22	Do the expressions for $f_{c(mn)}$, β_{mn}, and $\upsilon_{p(mn)}$ hold for both the TM and TE modes?	[(10.49)–(10.54a,b)]
RQ 10.23	Explain the evanescent mode in rectangular waveguides.	[(10.49)]
RQ 10.24	State the boundary condition for H_z of the TE wave.	[(10.69)]
RQ 10.25	Explain the dominant mode of the rectangular waveguide.	[(10.72)]
RQ 10.26	In what ways are η_{TE} and η_{TM} related in a rectangular waveguide?	[(10.58c)(10.73b)]

10.2.3 Power Attenuation

The power of electromagnetic waves may be attenuated in metallic waveguides because of lossy dielectrics and/or imperfect conductor walls. If the attenuation is small, as in most practical situations, the power loss can be conveniently described in terms of the attenuation constant based on the assumption that there is no change in the transverse field pattern.

We begin with the time-averaged power density of an electromagnetic wave, expressed as

$$\langle \mathbf{S} \rangle = \frac{1}{2} \mathrm{Re} \left[\tilde{\mathbf{E}} \times \tilde{\mathbf{H}}^* \right] \tag{10.82}$$

where $\tilde{\mathbf{E}}$ and $\tilde{\mathbf{H}}$ denote electric and magnetic field phasors, respectively. Inserting Eq. (10.24) into Eq. (10.82), with $\gamma = \alpha + j\beta$, the time-averaged power through the cross section of the waveguide is obtained as

$$\langle P \rangle = \int_S \langle \mathbf{S} \rangle \cdot d\mathbf{s} = \int_S \frac{|E_x(x, y)|^2 + |E_y(x, y)|^2}{2\eta_w} e^{-2\alpha z} dx\, dy$$
$$= P_o e^{-2\alpha z} \tag{10.83}$$

where S is the cross-sectional area, and $d\mathbf{s}$ is directed along the z-axis. The wave impedances are $\eta_w = \eta_{\mathrm{TM}}$ and $\eta_w = \eta_{\mathrm{TE}}$ for the TM and TE waves, respectively. It was assumed that there is no change in the transverse field pattern as the wave propagates along the z-axis.

The attenuation constant can be divided into two parts:

$$\alpha = \alpha_d + \alpha_c \tag{10.84}$$

where α_d arises from the lossy dielectric with $\sigma \neq 0$, and α_c is due to the imperfect conductor walls with $\sigma_c \neq \infty$.

All the discussions and expressions developed previously for lossless waveguides are valid for waveguides filled with low-loss dielectrics if complex permittivity is used instead of ε. From Chap. 8, we recall that in low-loss dielectrics, the complex permittivity is defined as

$$\hat{\varepsilon} \equiv \varepsilon' - j\varepsilon'' = \varepsilon - j\frac{\sigma}{\omega} \tag{8.76}$$

where ε and σ are the permittivity and conductivity of the dielectric, respectively. Inserting Eq. (8.76) into Eq. (10.46), the propagation constant for the waveguide mode is obtained as

$$\gamma = \sqrt{\left(\frac{m\pi}{a}\right)^2 + \left(\frac{n\pi}{b}\right)^2 - \omega^2 \mu \left(\varepsilon - j\frac{\sigma}{\omega}\right)}$$
$$\equiv \alpha_{mn} + j\beta_{mn} \tag{10.85}$$

Collecting the real and imaginary parts of Eq. (10.85), we have

$$\alpha_{mn}^2 - \beta_{mn}^2 = \left(\frac{m\pi}{a}\right)^2 + \left(\frac{n\pi}{b}\right)^2 - \omega\mu\varepsilon \tag{10.86a}$$

$$2\alpha_{mn}\beta_{mn} = \omega\mu\sigma \tag{10.86b}$$

Under the condition of a small attenuation $\alpha_{mn} \ll \beta_{mn}$, the left-hand side of Eq. (10.86a) approximates $-\beta_{mn}^2$. The phase constant is thus given by

$$\beta_{mn} = \sqrt{\omega\mu\varepsilon - \left(\frac{m\pi}{a}\right)^2 - \left(\frac{n\pi}{b}\right)^2} = \omega\sqrt{\mu\varepsilon}\sqrt{1 - \left(f_{c(mn)}/f\right)^2} \qquad (10.87)$$

Surprisingly, this is the same as the phase constant in lossless waveguides. Next, inserting Eq. (10.87) into Eq. (10.86), the attenuation constant due to the lossy dielectric can be obtained as

$$\boxed{\alpha_d = \frac{\sigma\sqrt{\mu/\varepsilon}}{2\sqrt{1 - \left(f_{c(mn)}/f\right)^2}}} \qquad [\text{Np/m}] \qquad (10.88)$$

where σ is the conductivity of the dielectric, and $f > f_{c(mn)}$.

It is tedious to analytically determine the attenuation constant α_c, which results from the finite conductivity of conducting walls. In this section, we outline the general procedure to obtain α_c. In the presence of ohmic power losses, the power dissipated in the rectangular waveguide is equal to the total power dissipated in the four conducting walls, such that

$$\langle P_c \rangle = \langle P(y = 0) \rangle + \langle P(y = b) \rangle + \langle P(x = 0) \rangle + \langle P(x = a) \rangle \qquad (10.89)$$

where < > denotes time averaging. From Sect. 8.2, we recall that the time-averaged power dissipated in a conductor with surface area $w \times \ell$ [m^2], conductivity σ, and skin depth δ [m] is given by

$$\langle P_c \rangle = \frac{1}{4\sigma} w\ell\delta J_o^2 \quad [W] \qquad (8.96)$$

where J_o is the amplitude of the volume current density on the conductor surface (not the surface current density) given by $J_o = \sigma E_o$, with the electric field component tangential to the conductor surface E_o. In a rectangular waveguide with dimensions a and b, the dissipated power per unit length of the waveguide can be computed in the following way:

$$\langle \hat{P}_c \rangle = \frac{\sigma\delta}{4} e^{-2\alpha_c z} \int_{x=0}^{x=a} \left[|E_x(y = 0)|^2 + |E_z(y = 0)|^2 \right.$$

$$\left. + |E_x(y = b)|^2 + |E_z(y = b)|^2 \right] dx$$

$$+ \frac{\sigma\delta}{4} e^{-2\alpha_c z} \int_{y=0}^{y=b} \left[|E_y(x = 0)|^2 + |E_z(x = 0)|^2 \right.$$

$$\left. + |E_y(x = a)|^2 + |E_z(x = a)|^2 \right] dy \qquad (10.90)$$

where the hat at the top denotes power per unit length. Again, it was assumed that the transverse field distribution are unaffected by the power loss. Assuming a lossless dielectric in the waveguide and considering the principle of energy conservation, the time-averaged power loss per unit length of the waveguide can be expressed as

$$\langle \hat{P}_c \rangle = -\frac{d}{dz}\left(P_o e^{-2\alpha_c z}\right) = 2\alpha_c P_o e^{-2\alpha_c z} \qquad [\text{W/m}] \qquad (10.91)$$

where P_o is the time-averaged power of the wave at $z = 0$. Finally, equating Eqs. (10.90) and (10.91) leads to an attenuation constant of the wave propagating in the waveguide.

Example 10.9 The air-filled rectangular waveguide is composed of conductors with conductivity σ. Find the attenuation constant for the TE_{10} mode propagating with the phase constant β_{10}.

Solution

The magnetic field of the TE_{10} mode is obtained from Eqs. (10.70) and (10.71c) as

$$\tilde{\mathbf{H}} = \left[\frac{j\beta_{10}\, a}{\pi} H_o \sin(\pi x/a)\, \mathbf{a}_x + H_o \cos(\pi x/a)\, \mathbf{a}_z\right] e^{-(\alpha_c + j\beta_{10})z}$$

Again, it is assumed that the propagation mode is not affected by the ohmic power loss in the conducting walls, except for the variation along the waveguide, expressed by $e^{-\alpha_c z}$. The component of $\mathbf{H}$ tangential to the conductor surface is related to the tangential component of $\mathbf{E}$ through the intrinsic impedance $\hat{\eta} = (1 + j)/\sigma\delta$.

In the $y = 0$ plane, using $\tilde{\mathbf{H}}$, the tangential electric field is obtained as

$$\tilde{\mathbf{E}} = \left[\frac{(1-j)\,\beta_{10}\, a}{\pi\sigma\delta} H_o \sin(\pi x/a)\, \mathbf{a}_z + \frac{(1+j)}{\sigma\delta} H_o \cos(\pi x/a)\, \mathbf{a}_x\right] e^{-(\alpha_c + j\beta_{10})z}$$

$$(10.92a)$$

Similarly, in the $y = b$ plane,

$$\tilde{\mathbf{E}} = \left[\frac{(1-j)\,\beta_{10}\, a}{\pi\sigma\delta} H_o \sin(\pi x/a)\, \mathbf{a}_z - \frac{(1+j)}{\sigma\delta} H_o \cos(\pi x/a)\, \mathbf{a}_x\right] e^{-(\alpha_c + j\beta_{10})z}$$

$$(10.92b)$$

In the $x = 0$ plane,

$$\tilde{\mathbf{E}} = \left[-\frac{(1+j)}{\sigma\delta} H_o\, \mathbf{a}_y\right] e^{-(\alpha_c + j\beta_{10})z} \qquad (10.92c)$$

In the $x = a$ plane,

$$\tilde{\mathbf{E}} = \left[-\frac{(1+j)}{\sigma\delta} H_o\, \mathbf{a}_y \right] e^{-(\alpha_c + j\beta_{10})z} \tag{10.92d}$$

Inserting Eqs. (10.92a)–(10.92d) into Eq. (10.90) yields

$$\langle \widehat{P}_c \rangle = \frac{H_o^2}{\sigma\delta} e^{-2\alpha_c z} \left[b + \frac{a}{2}\left(1 + \frac{a^2\beta_{10}^2}{\pi^2} \right) \right] \tag{10.92e}$$

Next, inserting Eqs. (10.71a, b) and (10.73b) into Eq. (10.83), we obtain

$$\langle P \rangle = \left[\beta_{10}\omega\mu_0 \frac{a^3 b}{4\pi^2} H_o^2 \right] e^{-2\alpha_c z} \equiv P_o e^{-2\alpha_c z} \tag{10.92f}$$

Using Eqs. (10.92e, f) into Eq. (10.91) together with $\delta = 1/\sqrt{\pi f \mu_0 \sigma}$, the attenuation constant for the TE_{10} mode is expressed as

$$\boxed{\alpha_c = \frac{\delta\pi^2}{a^3 b\,\beta_{10}} \left[b + \frac{a}{2}\left(1 + \frac{a^2\beta_{10}^2}{\pi^2} \right) \right]} \quad \text{[Np/m]} \tag{10.93}$$

The attenuation constant for the TE_{01} mode can be obtained by switching the roles of a and b in Eq. (10.93). In Fig. 10.11, the attenuation constants for the TE_{10} and TE_{01} modes in a hollow rectangular copper waveguide with $a = 2b = 2$ [cm] are plotted as functions of the operating frequency.

Example 10.10 Show that the TM_{mn} and TE_{mn} modes can propagate independently in a rectangular waveguide even though they have the same phase constant.

Solution

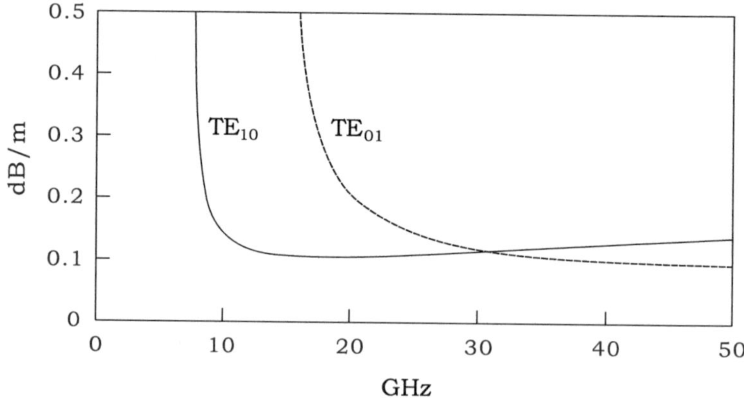

Fig. 10.11 Attenuation constants for TE_{10} and TE_{01} modes in the hollow rectangular copper waveguide with $a = 2b = 2$ [cm]

When the two modes propagate simultaneously in a waveguide, the total electric and magnetic fields are

$$\tilde{\mathbf{E}} = \tilde{\mathbf{E}}_{TM} + \tilde{\mathbf{E}}_{TE} \tag{10.94a}$$

$$\tilde{\mathbf{H}} = \tilde{\mathbf{H}}_{TM} + \tilde{\mathbf{H}}_{TE} \tag{10.94b}$$

The time-averaged power density in the waveguide is

$$\langle P \rangle = \frac{1}{2}\mathrm{Re}\left[\int_{\mathcal{S}} \left(\tilde{\mathbf{E}} \times \tilde{\mathbf{H}}^*\right) \cdot \mathbf{a}_z \, dx\,dy\right]$$

$$= \frac{1}{2}\mathrm{Re}\left[\int_{\mathcal{S}} \left(\tilde{\mathbf{E}}_{TM} \times \tilde{\mathbf{H}}^*_{TM} + \tilde{\mathbf{E}}_{TE} \times \tilde{\mathbf{H}}^*_{TE}\right) \cdot \mathbf{a}_z \, dx\,dy\right]$$

$$+ \frac{1}{2}\mathrm{Re}\left[\int_{\mathcal{S}} \left(\tilde{\mathbf{E}}_{TM} \times \tilde{\mathbf{H}}^*_{TE} + \tilde{\mathbf{E}}_{TE} \times \tilde{\mathbf{H}}^*_{TM}\right) \cdot \mathbf{a}_z \, dx\,dy\right] \tag{10.94c}$$

The second integrand on the right side of Eq. (10.94c) can expressed as

$$\tilde{\mathbf{E}}_{TM} \times \tilde{\mathbf{H}}^*_{TE} + \tilde{\mathbf{E}}_{TE} \times \tilde{\mathbf{H}}^*_{TM} = \left(E_{x,TM}\, H^*_{y,TE} - E_{y,TM}\, H^*_{x,TE}\right)$$
$$+ \left(E_{x,TE}\, H^*_{y,TM} - E_{y,TE}\, H^*_{x,TM}\right) \tag{10.94d}$$

With the help of Eqs. (10.56a)–(10.56d) and (10.71a)–(10.71d), integrating Eq. (10.94d) over the cross section of the waveguide results in a zero. Thus,

$$\langle P \rangle = \frac{1}{2}\mathrm{Re}\left[\int_{\mathcal{S}} \left(\tilde{\mathbf{E}}_{TM} \times \tilde{\mathbf{H}}^*_{TM}\right) \cdot \mathbf{a}_z \, dx\,dy\right] + \frac{1}{2}\mathrm{Re}\left[\int_{\mathcal{S}} \left(\tilde{\mathbf{E}}_{TE} \times \tilde{\mathbf{H}}^*_{TE}\right) \cdot \mathbf{a}_z \, dx\,dy\right] \tag{10.94e}$$

It is evident from Eq. (10.94e) that there is no coupling between the two waveguide modes.

Exercise 10.15
Determine α_c for the TE_{10} mode in a hollow rectangular copper waveguide with $a = 2b = 2$ [cm] operating at a frequency of 11.25 [GHz].

Ans. 0.014 [Np/m].

Exercise 10.16
When a hollow rectangular waveguide with a length of 0.75 [m] has an attenuation of 0.12 [dB/m] and delivers 1.0 [kW] to a matched load, determine the power dissipated in the waveguide walls.

Ans. 20.9 [W].

Review Questions

RQ 10.27	Relate the ohmic power loss in the conductor walls of the waveguide to the attenuation constant.	[(10.91)]
RQ 10.28	Can the energy of one propagation mode be transferred to another in a simple waveguide?	[(10.94e)]
RQ 10.29	What are the differences between α_d and α_c?	[(10.88)(10.93)]
RQ 10.30	Distinguish between attenuation constants of the evanescent and propagation modes.	[(10.16)(10.88) (10.93)]

10.3   Rectangular Cavity Resonator

When the two ends of the rectangular waveguide are terminated by conducting walls, the enclosure becomes a cavity resonator. The electromagnetic fields form standing waves along the three Cartesian axes of the cavity and oscillate at a single resonant frequency. Cavity resonators are primarily used for storing electromagnetic energy. They are also useful in circuits designed for microwave amplification, oscillations, and bandpass filtering. A cavity resonator can be excited with a particular mode through the tip of the inner conductor of the coaxial cable inserted into the cavity and positioned at the maximum electric field (see Fig. 10.12). Another coaxial cable is used as the output probe.

In the previous section, we observed that the TM and TE modes in a rectangular waveguide are identified by the longitudinal field along the propagation direction. Even if there is no unique longitudinal direction in a rectangular resonator, the z-axis is designated as the propagation direction.

Fig. 10.12 Rectangular cavity resonator

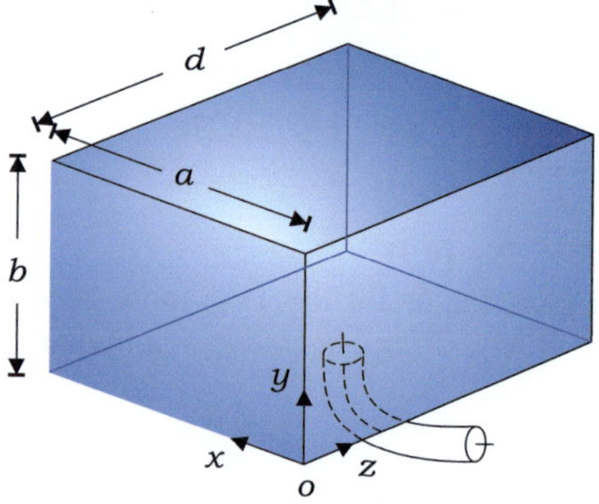

The TE mode in a rectangular waveguide has a longitudinal field, such as

$$H_z = H_o \cos(m\pi x/a)\cos(n\pi y/b)e^{-j\beta_{mn}z} \tag{10.95}$$

The negative sign of the exponent indicates that this mode propagates in the $+z$-direction. Because this wave is reflected by the conducting wall at $z = d$, which is assumed to be perfectly conducting, the two waves propagate in opposite directions in the resonator. The TE wave propagating in the $+z$-direction can be considered a plane wave incident normally on the surface at $z = d$. Considering that the reflection coefficient for perpendicular polarization is $\Gamma = -1$, the total H_z can be expressed as

$$H_z = H_o \cos(m\pi x/a)\cos(n\pi y/b)\left(e^{-j\beta_{mn}z} - e^{j\beta_{mn}z}\right) \tag{10.96}$$

The boundary condition for **B** indicates that its normal component is continuous across the interface. However, time-varying fields cannot exist in perfect conductors; thus, H_z should vanish on conductor walls at $z = 0$ and $z = d$. This requires that $\beta_{mn}d = \pi p$ in Eq. (10.96) for positive integer p. Therefore, the longitudinal magnetic field component of the TE_{mnp} mode in a rectangular resonator is given by

$$\boxed{\tilde{H}_z = H_o \cos(m\pi x/a)\cos(n\pi y/b)\sin(p\pi z/d)} \tag{10.97}$$

where $m = 0, 1, 2, ...$, $n = 0, 1, 2, ...$, and $p = 1, 2, ...$. The mode numbers m and n cannot be zero simultaneously because the TE_{oop} mode cannot exist in a rectangular cavity, just as the TE_{oo} wave cannot exist in a rectangular waveguide.

The transverse field components of the resonator TE_{mnp} mode can be obtained by substituting Eq. (10.97) into Eq. (10.34), setting $E_z = 0$ and replacing term $-\gamma$ with partial differentiation with respect to z. Alternatively, they can be obtained by multiplying Eqs. (10.71a) and (10.71b) with $\sin(p\pi z/d)$, and replacing γ in Eqs. (10.71c) and (10.71d) with $-(p\pi/d)\cos(p\pi z/d)$. The expression for h in Eq. (10.47) still holds for rectangular resonators.

By following a procedure similar to that used to obtain Eq. (10.97), the longitudinal electric field component of the resonator TM_{mnp} mode can be obtained as

$$\boxed{\tilde{E}_z = E_o \sin(m\pi x/a)\sin(n\pi y/b)\cos(p\pi z/d)} \tag{10.98}$$

where $m = 1, 2, ...$, $n = 1, 2, ...$, and $p = 0, 1, 2, ...$. The transverse field components of the TM_{mnp} mode can be obtained by multiplying Eqs. (10.56c) and (10.56d) with $\cos(p\pi z/d)$ and replacing γ in Eqs. (10.56a) and (10.56b) with $(p\pi/d)\sin(p\pi z/d)$. It is evident from Eqs. (10.97) and (10.98) that the integers m, n, and p denote the number of half-waves along the x-, y-, and z-axes, respectively.

10.3.1　Resonant Frequency

The TE_{mnp} mode in Eq. (10.97) can be resolved into eight plane-wave components because $\sin\theta$ and $\cos\theta$ are combinations of $e^{j\theta}$ and $e^{-j\theta}$, each representing a uniform plane wave. Moreover, the arguments for the sinusoidal functions in Eq. (10.97) are recognized as $k_x x$, $k_y y$, and $k_z z$, where k_x, k_y, and k_z are the three Cartesian components of the wavevector with magnitude $k = \omega\sqrt{\mu_0\varepsilon_0}$. In view of these, the resonant frequency for the TE_{mnp} mode can be expressed as

$$\boxed{f_{mnp} = \frac{1}{2\sqrt{\mu_0\varepsilon_0}}\sqrt{\left(\frac{m}{a}\right)^2 + \left(\frac{n}{b}\right)^2 + \left(\frac{p}{d}\right)^2}}\quad \text{[Hz]} \qquad (10.99)$$

where it is assumed that the resonator is air-filled. The resonant frequency increases with the order of the mode.

Based on the same reasoning, Eq. (10.98) leads to the resonant frequency of the TM_{mnp} mode, which is exactly the same as Eq. (10.99). The TM and TE modes with the same mode number have the same resonant frequency and are said to be **degenerate**. Because the mode numbers **m** and **n** cannot be zero simultaneously for the TE modes, TE_{101} is the dominant mode in a rectangular resonator with interior dimensions $a > b < d$.

Exercise 10.17

Determine the resonant frequency of the TE_{110} mode in a hollow rectangular resonator with dimensions $a = 2\,[\text{cm}]$, $b = 2\,[\text{cm}]$, and $d = 3\,[\text{cm}]$.

Ans. 10.6 [GHz].

Exercise 10.18

What are the three dominant degenerate modes for a cubic resonator?

Ans. TE_{101}, TE_{011}, TM_{110}.

10.3.2　Quality Factor of Cavity Resonator

The sum of the electric and magnetic energies of all the resonator modes represents the energy stored in the resonator. In practical situations, the cavity walls have finite conductivity. Thus, power loss in the walls results in a decay in the stored energy. The quality factor Q is a measure of the bandwidth of the cavity resonator and is defined as

$$\boxed{Q = 2\pi \frac{\text{Time-averaged energy stored}}{\text{Energy loss per oscillation cycle}} = \omega\frac{W}{P_L}} \qquad (10.100)$$

where W is the time-averaged energy stored in the cavity, P_L is the time-averaged power loss at the cavity walls, and ω is the resonant angular frequency. In most cases, the energy loss is extremely small; thus, the stored energy is computed using the field patterns for no loss.

As an example, we determine Q of a rectangular resonator for the TE_{101} mode. First, the transverse field components of the mode are

$$H_z = H_o \cos\left(\frac{\pi}{a} x\right) \sin\left(\frac{\pi}{d} z\right) \tag{10.101a}$$

$$H_x = -\left(\frac{a}{d}\right) H_o \sin\left(\frac{\pi}{a} x\right) \cos\left(\frac{\pi}{d} z\right) \tag{10.101b}$$

$$E_y = -j\omega\mu_0 \left(\frac{a}{\pi}\right) H_o \sin\left(\frac{\pi}{a} x\right) \sin\left(\frac{\pi}{d} z\right) \tag{10.101c}$$

The time-averaged electric and magnetic energies of the TE_{101} mode are computed as follows:

$$W_e = \frac{\varepsilon_0}{4} \int_V |E_y|^2 dv = \frac{\varepsilon_0}{4} \left(\frac{\omega\mu_0 a H_o}{\pi}\right)^2 \frac{abd}{4} \tag{10.102a}$$

$$W_m = \frac{\mu_0}{4} \int_V \{|H_x|^2 + |H_z|^2\} dv = \frac{\mu_0}{4} \left\{\left(\frac{a H_o}{d}\right)^2 + H_o^2\right\} \frac{abd}{4} \tag{10.102b}$$

The resonant angular frequency ω is obtained from Eq. (10.99) as

$$\omega = 2\pi f_{101} = \frac{\pi}{\sqrt{\mu_0\varepsilon_0}} \sqrt{\left(\frac{1}{a}\right)^2 + \left(\frac{1}{d}\right)^2} \tag{10.103}$$

Inserting Eq. (10.103) into Eq. (10.102) shows that $W_e = W_m$, and thus,

$$W = W_e + W_m = \frac{\mu_0 H_o^2}{8} \left\{\left(\frac{a}{d}\right)^2 + 1\right\} abd \tag{10.104}$$

To determine P_L, the tangential electric field just inside the conducting wall is first obtained from $\tilde{\mathbf{E}}_t = (\mathbf{a}_n \times \tilde{\mathbf{H}}_t)(1 + j)/\sigma\delta$, where $\mathbf{a}_n$ is the outward unit surface normal, δ is the skin depth, σ is the conductivity, and the subscript t denotes the tangential component as follows:

$$E_y = -H_o \frac{1 + j}{\sigma\delta} \sin\left(\frac{\pi}{d} z\right) \qquad (\text{at } x = 0 \text{ and at } x = a)$$

$$E_x = \pm H_o \frac{1 + j}{\sigma\delta} \cos\left(\frac{\pi}{a} x\right) \sin\left(\frac{\pi}{d} z\right) \quad (\text{at } y = 0 \text{ and at } y = b)$$

$$E_z = \pm\left(\frac{a}{d}\right) H_o \frac{1 + j}{\sigma\delta} \sin\left(\frac{\pi}{a} x\right) \cos\left(\frac{\pi}{d} z\right) \; (\text{at } y = 0 \text{ and at } y = b)$$

$$E_y = -\left(\frac{a}{d}\right) H_o \frac{1+j}{\sigma\delta} \sin\left(\frac{\pi}{a}x\right) \qquad \text{(at } z = 0 \text{ and at } z = d)$$

By substituting $J_o = \sigma E_t$ and using surface integral instead of $w\ell$ in Eq. (8.96), we obtain

$$P_L = \frac{1}{4\sigma\delta d^2}\left[2b(a^3 + d^3) + ad(a^2 + d^2)\right] \tag{10.105}$$

Inserting Eqs. (10.103)–(10.105) into Eq. (10.100) leads to Q in mode TE_{101}:

$$\boxed{Q_{TE_{101}} = \frac{(a^2 + d^2)abd}{\delta\left[2b(a^3 + d^3) + ad(a^2 + d^2)\right]}} \tag{10.106}$$

where $\delta = 1/\sqrt{\pi f_{101}\mu_0\sigma}$ is the skin depth of the cavity wall.

Example 10.11 An air-filled cavity resonator with a square cross section of side 2 [cm] is 2.5 [cm] long and is made of copper. Determine the (a) dominant resonant frequency, and (b) quality factor at that frequency.

Solution

(a)   Setting $a = b = 2$ [cm] and $d = 2.5$ [cm], the TE_{101} mode is the dominant mode of the cavity. The resonant frequency is obtained using Eq. (10.99) as

$$f_{101} = \frac{3 \times 10^8}{2}\sqrt{\left(\frac{1}{0.02}\right)^2 + \left(\frac{1}{0.025}\right)^2} = 9.60\,[\text{GHz}]$$

(b)   The quality factor is obtained using Eq. (10.106) as

$$Q_{TE_{101}} = \frac{(a^2 + d^2)abd}{\delta\left[2b(a^3 + d^3) + ad(a^2 + d^2)\right]} = \frac{(4 + 6.25) \times 10 \times 10^{-2}}{\delta[4(8 + 15.63) + 5(4 + 6.25)]}$$

$$= 7.03 \times 10^{-3}\sqrt{\pi f_{101}\mu_0\sigma} = 7.03 \times 10^{-3}\sqrt{4\pi^2 \times 9.6 \times 10^9 \times 5.8}$$

$$= 10,423$$

Exercise 10.19

If the stored energy W in a cavity with quality factor Q decreases as $e^{-\alpha t}$, then $|dW/dt|$ corresponds to the power dissipated in the walls. Relate α to Q.

Ans. $\alpha = \omega/Q$.

Exercise 10.20

If a cubic cavity resonator is lengthened in the z-direction such that its length is twice as large, what percentages would f_{101} and $Q_{TE_{101}}$ increase?

Ans. f_{101}, -20.9%; $Q_{TE_{101}}$, 7.1%.

Exercise 10.21
How does the quality factor of the TE_{101} cavity mode vary with the dielectric constant of the insulating material?

Ans. $Q \sim (\varepsilon_r)^{-1/4}$.

Review Questions

RQ 10.31	How do field patterns in rectangular resonators differ from those in rectangular waveguides?	[(10.44)(10.70) (10.97)(10.98)]
RQ 10.32	How does the resonant frequency of a rectangular resonator differ from the cutoff frequency of a rectangular waveguide?	[(10.49)(10.99)]
RQ 10.33	Explain the dominant resonant mode and frequency.	[(10.99)]
RQ 10.34	What are the degenerate modes?	[(10.99)]
RQ 10.35	Define the quality factor of the cavity resonators.	[(10.100)]
RQ 10.36	What assumptions are made in the derivation of stored energy in lossy cavity resonators?	[(10.102a,b)]

10.4 Problems

Parallel-plate waveguide

10.1 A parallel-plate waveguide is designed to support only the TEM mode over the frequency range $0 < f < 10\,[\text{GHz}]$. The space between the two plates is filled with a dielectric with $n = 1.5$. Determine maximum separation of plates.

10.2 A parallel-plate waveguide with a gap of length 1.2 [cm] is filled with a dielectric with $n = 2.5$ and operates at 12 [GHz]. When the electromagnetic waves in the waveguide have a nonzero electric field normal to the plates,

(a) determine the three lowest cutoff frequencies,
(b) identify the three modes of propagation, and
(c) compare the group velocities of the modes.

10.3 Starting with the electric field phasor of the TE_m mode in Eq. (10.5), find (a) the magnetic field phasor, and (b) the instantaneous values $\mathscr{E}$ and $\mathscr{H}$ in the parallel-plate waveguide filled with a dielectric with permittivity ε and permeability μ.

10.4 The parallel-plate waveguide is partially filled with a dielectric with permittivity $\varepsilon = 2.25\varepsilon_0$ (Fig. 10.13). The TE_1 mode of frequency 10 [GHz] propagates in the $+z$-direction and impinges on the air-dielectric interface at $z = 0$. For an incident wave with $\tilde{\mathbf{E}} = -j2\,\mathbf{a}_x \sin(50\pi y)\,e^{-j\beta_1 z}$ [V/m], determine (a) $f_{c(1)}$, β_1, and θ in region $z < 0$, and (b) the phase constant and electric field in region $z > 0$.

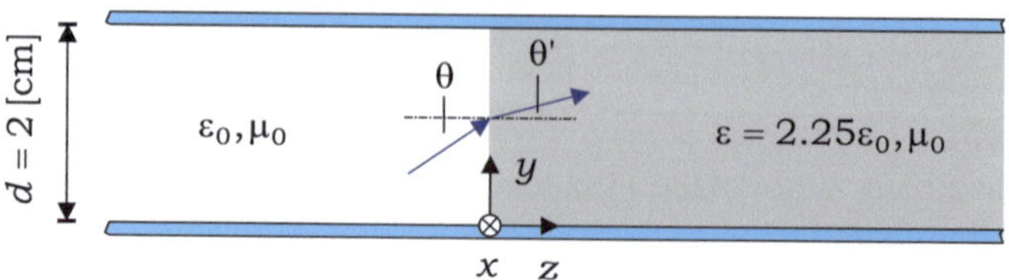

Fig. 10.13 Partially filled parallel-plate waveguide (Problems 10.4 and 10.5)

10.5 The parallel-plate waveguide shown in Fig. 10.13 was filled with new dielectrics of $\varepsilon = 2.25\varepsilon_0$ and $\varepsilon = 1.69\varepsilon_0$ in the $z < 0$ and $z > 0$ regions, respectively. The TE_1 mode propagates in the $+z$-direction in region $z < 0$ and impinges on the interface at $z = 0$. Determine (a) the cutoff frequencies in the two regions, (b) the frequency range over which propagation in the right region is prohibited, and (c) show that the cutoff is related to the total internal reflection of the component plane waves at the interface.

Rectangular waveguide

10.6 For a hollow rectangular waveguide with interior dimensions $a = 2\,[\text{cm}]$ and $b = 1.5\,[\text{cm}]$, determine the frequency range over which the waveguide supports (a) a single mode, and (b) two propagation modes.

10.7 List the modes of propagation in the order of ascending cutoff frequencies for a hollow rectangular waveguide with $a = 2.5\,[\text{cm}]$ and $b = 2\,[\text{cm}]$ operating at 14.6 [GHz] (Let the TE mode come before the TM mode in the list of degenerate modes).

10.8 If the cutoff frequencies for the TM_{12} and TM_{21} modes in an air-filled rectangular waveguide are 30.6 [GHz] and 19.5 [GHz], respectively,

(a) what is the cutoff frequency for the TM_{11} mode, and
(b) determine the waveguide dimensions.

10.9 Design an air-filled rectangular waveguide operating in the dominant mode at a frequency of 6.9 [GHz], which is 15% higher than the cutoff frequency and 8% lower than that of the next higher mode.

10.10 If a hollow rectangular waveguide operates only in the TE_{10} mode over the frequency range of 10 [GHz] to 15[GHz], determine the limits on its interior dimensions a and b.

10.11 The dominant mode for an air-filled rectangular waveguide operates at 30 [GHz], which is 15% higher than the cutoff frequency. Determine the phase constant, phase velocity, and wave impedance.

10.12 Determine the attenuation constant of an electromagnetic wave with an operating frequency 1% lower than the cutoff frequency $f_{c(mn)} = 10\,[\text{GHz}]$ in a hollow rectangular waveguide.

10.13 For the propagation mode $\mathcal{E}_z = 4\sin(40\pi x)\sin(25\pi y)\sin(16\pi \times 10^9 t - \beta z)$ in a hollow rectangular waveguide with interior dimensions $a = 5\,[\text{cm}]$ and

$b = 4\,[\text{cm}]$, (a) identify the mode, and (b) determine the cutoff frequency, phase constant, and wave impedance.

10.14 Charges build up on the conducting walls as the TE_{11} mode propagates in a hollow waveguide with cross section $a \times b$. Find an expression for the instantaneous surface charge density along the edge of the rectangular cross section at $z = 0$.

10.15 Surface currents are also induced on the walls of the rectangular waveguide in Problem 10.14. Find an expression for the instantaneous surface current density along the edge of the rectangular cross section at $z = 0$.

10.16 For the propagation mode with $\mathscr{E}_x = 1.6\cos(50\pi x)\sin(40\pi y)\cos(18\pi \times 10^9 t)$ in a rectangular waveguide filled with a dielectric with $n = 1.2$, determine the (a) cutoff frequency, (b) phase velocity, and (c) group velocity.

10.17 Both the TM and TE modes propagate in an air-filled rectangular waveguide with $a = 2b = 4\,[\text{cm}]$. If the transverse electric field at $z = 0$ is given by $\mathscr{E}_t = 1.5\,\mathbf{a}_x\cos(25\pi x)\sin(50\pi y)\cos(20\pi \times 10^9 t)$, then determine (a) mode number, (b) cutoff frequency, (c) phase constant, and (d) electric field of each mode.

10.18 Show that the magnetic field lines of the TM_{11} mode can be obtained by solving $dy/dx = -(b/a)\tan(\pi y/b)/\tan(\pi x/a)$.

10.19 In an air-filled waveguide with dimensions $a = 2b = 2\,[\text{cm}]$ operating at a frequency of 20 [GHz], determine the difference in arrival times, also called group delays, between the TE_{10} and TE_{01} modes after propagating a distance of 50 [cm].

10.20 The rectangular waveguide is filled with two different dielectrics, as shown in Fig. 10.14. The TE_{10} mode originating from the left impinges on the dielectric-dielectric interface at $z = 0$ and gives rise to reflected and transmitted waves propagating in the TE_{10} modes. By invoking the boundary conditions for $\mathbf{E}$ and $\mathbf{H}$, show that the reflection and transmission coefficients for the TE_{10} mode are the same as those for a uniform plane wave normally incident onto the interface.

10.21 When the rectangular waveguide shown in Fig. 10.14 is filled with dielectrics with $\varepsilon_1 > \varepsilon_2$, (a) express the cutoff frequencies $f_{c(mn)}^L$ and $f_{c(mn)}^R$ for the

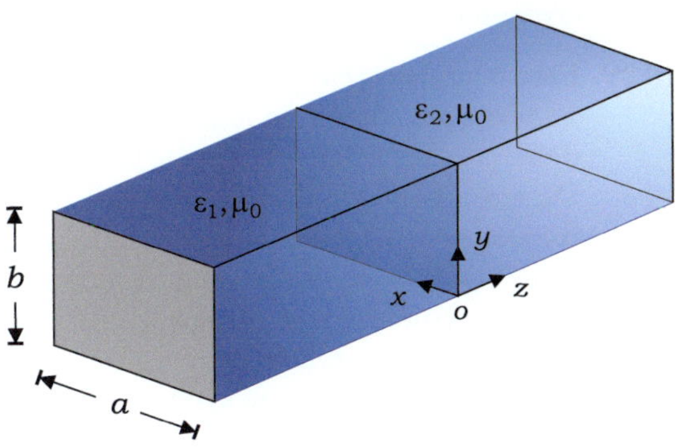

Fig. 10.14 Rectangular waveguide filled with two different dielectrics (Problems 10.20 and 10.21)

TM$_{mn}$ mode in the left and right regions, respectively, and (b) show that the component plane waves in the left region are incident on the interface at the critical angle for the operation frequency $f = f^{R}_{c(mn)}$.

Power and power attenuation

10.22 For the TE$_{10}$ mode propagating in a hollow rectangular waveguide, find the time-averaged (a) power density, $\langle \mathbf{S} \rangle$, (b) power through the cross section of the waveguide, $\langle P \rangle$, (c) electric and magnetic energies per unit length of the waveguide, $\langle \widehat{W}_e \rangle$ and $\langle \widehat{W}_m \rangle$, and (d) show that the ratio of $\langle P \rangle$ to $\langle \widehat{W}_e \rangle + \langle \widehat{W}_m \rangle$ is equal to the group velocity of the mode.

10.23 The dominant mode in an air-filled rectangular waveguide with dimensions $a \times b$ ($a > b$) has an electric field of amplitude E_o. Express the time-averaged power of the mode.

10.24 Show that the dominant mode of a lossless rectangular waveguide can deliver more power if the operating frequency increases above the cutoff, while the maximum electric field intensity is fixed.

10.25 A rectangular waveguide with dimensions $a = 1.5b = 3$ [cm] is filled with a low-loss dielectric with $\varepsilon_r = 2$ and $\sigma = 3 \times 10^{-5}$ [S/m]. For an operating frequency of 5 [GHz], determine the attenuation constant of the mode of operation.

10.26 When the TE$_{10}$ mode with a frequency of 6 [GHz] propagates in a rectangular waveguide with dimensions $a = 2b = 2.4$ [cm], which is filled with polyethylene with $\varepsilon_r = 2.25$, $\mu_r = 1$, and loss tangent $\sigma/\omega\varepsilon = 5.4 \times 10^{-5}$, and coated with gold ($\sigma_c = 4.1 \times 10^7$ [S/m]), determine α_d and α_c.

10.27 Show that the TE$_{mn}$ mode can be decomposed into four plane-wave components and that the mode power is equally distributed among the four components.

10.28 An air-filled rectangular waveguide with dimensions $3a = 4b = 12$ [cm] is 1 [m] long and operates at a frequency of 4 [GHz]. If the dominant mode with $\alpha = 2 \times 10^{-3}$ [Np/m] delivers 1.5 [kW] power to a matched load, find

(a) the time-averaged power dissipated in the waveguide, and
(b) the maximum electric field in the waveguide.

Cavity resonator

10.29 For a 5 [cm] $\times$ 3 [cm] $\times$ 4 [cm] lossless rectangular resonator, if the three mode numbers are allowed to have values of 0 or 1, identify the cavity modes, and determine the resonant frequencies.

10.30 Design a hollow cubic copper resonator operating at a dominant resonant frequency of 12 [GHz] and determine its quality factor.

10.31 If the conductivity of the conducting walls of a cubic resonator is reduced by 20%, determine a new resonator size for which the dominant mode has the same equality factor.

10.32 Express the transverse field components in the TE$_{mnp}$ resonator mode.

10.33 Express the transverse field components in the TM$_{mnp}$ resonator mode.

Chapter 11
Antenna

We studied the propagation of electromagnetic waves in various media in Chap. 8 and the transmission and guiding of waves in Chaps. 9 and 10 without considering the generation of waves. Electromagnetic waves are generated by time-varying currents. Antennas are designed to support specific current distributions and to radiate electromagnetic energy in specific directions. Antennas can also be used to receive electromagnetic energy. It can be shown that the directive pattern for receiving is exactly the same as the directive pattern for transmitting. An antenna array is used to achieve a directivity that is much higher than that obtained with a single antenna element. Furthermore, the beam direction of the antenna array can be steered electronically by controlling the phase of the current fed into antenna elements.

11.1 The Hertzian Dipole

A linear antenna may be regarded as consisting of a large number of very short thin conducting wires carrying time-harmonic currents. In this case, each element is assumed to be infinitesimally small compared to the operating wavelength; thus, the current is considered uniform in an element. The radiation field of a given linear antenna can then be obtained by adding the contributions from the individual elements, taking into account the amplitude and phase of the currents in the elements. In this section, we first discuss the radiation properties of such a current element, called a *Hertzian dipole*, and then extend the results to a half-wave dipole antenna, which is considered a standard antenna for many applications.

We begin with a differential current element of length $d\ell$ oriented along the z-direction at the origin in free space (Fig. 11.1). It carries a time-harmonic current expressed by

$$i(t) = I_o \cos \omega t = \text{Re}\left[I_o\, e^{j\omega t}\right] \tag{11.1}$$

Y. H. Lee, *Introduction to Engineering Electromagnetics*,
https://doi.org/10.1007/978-3-031-28659-9_11

Fig. 11.1 Hertzian dipole

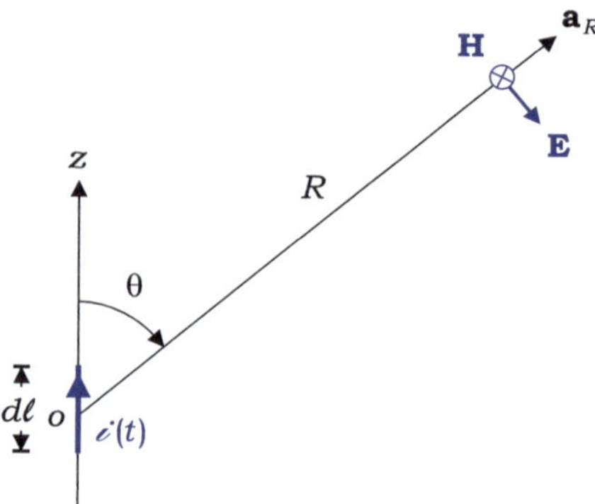

where I_o is the amplitude of the current, which is treated as constant along the element length. The retarded vector magnetic potential of the differential current element can be obtained using Eq. (6.66b) as

$$\tilde{\mathbf{A}} = \frac{\mu_0 I_o d\ell}{4\pi} \left(\frac{e^{-jkR}}{R} \right) \mathbf{a}_z \tag{11.2}$$

where $k = \omega/c = 2\pi/\lambda$ is the phase constant, and R is the radial distance in spherical coordinates. Because the term in parentheses in Eq. (11.2) represents a spherical wave, the current element can be treated as a point source located at the origin.

Evidently, $\tilde{\mathbf{A}}$ can be transformed into spherical coordinates using the relation $\mathbf{a}_z = \cos\theta\, \mathbf{a}_R - \sin\theta\, \mathbf{a}_\theta$. Subsequently, $\tilde{\mathbf{H}}$ can be obtained from $\tilde{\mathbf{A}}$, and $\tilde{\mathbf{E}}$ can be obtained from $\tilde{\mathbf{H}}$ as follows:

$$\tilde{\mathbf{H}} = \frac{1}{\mu_0} \nabla \times \tilde{\mathbf{A}}$$

$$\tilde{\mathbf{E}} = \frac{1}{j\omega\varepsilon_0} \nabla \times \tilde{\mathbf{H}}$$

Therefore,

$$\tilde{H}_\phi = -\left(\frac{I_o k^2 d\ell}{4\pi} \right) \sin\theta \left[\frac{1}{jkR} + \frac{1}{(jkR)^2} \right] e^{-jkR} \tag{11.3a}$$

$$\tilde{E}_R = -\left(\frac{I_o k^2 d\ell}{4\pi} \right) \eta_o 2\cos\theta \left[\frac{1}{(jkR)^2} + \frac{1}{(jkR)^3} \right] e^{-jkR} \tag{11.3b}$$

$$\tilde{E}_\theta = -\left(\frac{I_o k^2 d\ell}{4\pi} \right) \eta_o \sin\theta \left[\frac{1}{jkR} + \frac{1}{(jkR)^2} + \frac{1}{(jkR)^3} \right] e^{-jkR} \tag{11.3c}$$

where $\eta_o = \sqrt{\mu_0/\varepsilon_0} \cong 120\pi\ [\Omega]$ is the intrinsic impedance of free space. The other components, $\tilde{H}_R$, $\tilde{H}_\theta$, and $\tilde{E}_\phi$, are zero everywhere.

The electric field lines of the Hertzian dipole are drawn to scale in Fig. 11.2 by using Eqs. (11.3b) and (11.3c).

In most antenna applications, we are primarily interested in the electromagnetic field at distances far from the source, corresponding to $R \gg \lambda$. This region is called the *far field* or *far zone*, where the terms $1/(kR)^2$ and $1/(kR)^3$ in Eqs. (11.3a)–(11.3c) can be ignored in comparison with the term $1/kR$. The far-zone radiation field of the Hertzian dipole, or simply the ***radiation field*** of the Hertzian dipole, is thus expressed as

$$\tilde{H}_\phi = \frac{jkI_o d\ell}{4\pi} \frac{e^{-jkR}}{R} \sin\theta \qquad (11.4a)$$

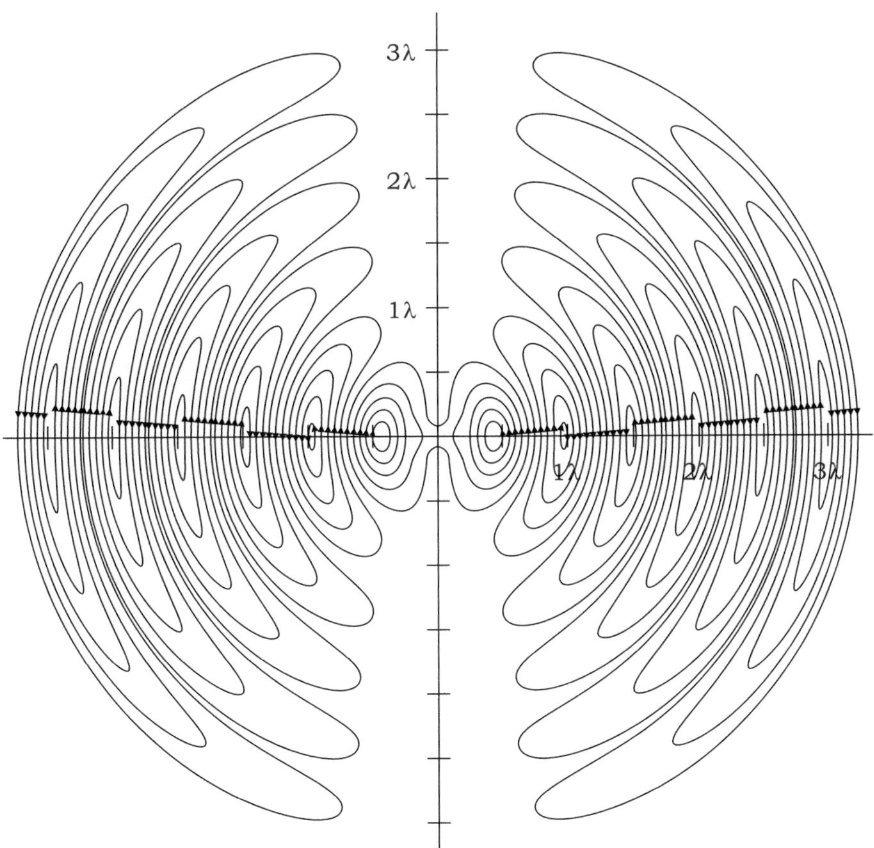

Fig. 11.2 Electric field lines of the Hertzian dipole at one instant of time. Small arrows indicate their directions

$$\boxed{\tilde{E}_\theta = \eta_o \frac{jkI_o d\ell}{4\pi} \frac{e^{-jkR}}{R} \sin\theta}$$
(11.4b)

The radiation field has only H_ϕ and E_θ components and varies inversely with distance to the Hertzian dipole. If $\sin\theta$ is disregarded from Eqs. (11.4a) and (11.4b), the radiation field represents an exact spherical wave.

Review Questions

RQ 11.1   What is a Hertzian dipole?   [Fig. 11.1]

RQ 11.2   Explain the far zone of the antenna.   [(11.3a,b,c)]

RQ 11.3   Define the radiation field of the Hertzian dipole.   [(11.4a,b)]

RQ 11.4   How does the radiation field vary with the distance?   [(11.4a,b)]

RQ 11.5   Under what conditions does the radiation field of the Hertzian   [(11.4a,b)]
          dipole become an exact spherical wave?

11.2   Antenna Characteristics

The electric field of the differential current element is not uniform in space, as shown in Fig. 11.2; no physical antennas radiate equally in any direction. The plot of the relative strength of the far-zone field versus the direction at a fixed radial distance from the antenna is called the **radiation pattern** of the antenna or **antenna pattern**. In general, the radiation patterns vary with θ and ϕ. The difficulty of plotting three-dimensional radiation patterns can be avoided by plotting the strength of the electric field normalized to unity against θ for a fixed ϕ, which is called the **E-plane pattern**, and the strength of the normalized electric field against ϕ for $\theta = \pi/2$, which is called the **H-plane pattern**.

For a Hertzian dipole, the normalized strength of the electric field at a fixed radial distance is, from Eq. (11.4b),

$$F(\theta) = \sin\theta$$
(11.5)

This is known as the *pattern function* of a Hertzian dipole. The E-plane pattern of the Hertzian dipole can be obtained by plotting $F(\theta)$ as a function of θ and is represented by a pair of circles, as shown in Fig. 11.3a. Meanwhile, the H-plane pattern is represented by a circle of unity radius in the xy-plane, as shown in Fig. 11.3b.

The power density of the radiation field of a Hertzian dipole can be obtained by substituting Eqs. (11.4a) and (11.4b) into the time-averaged Poynting vector such that

$$\langle \mathbf{S} \rangle = \frac{1}{2}\mathrm{Re}\left[\tilde{\mathbf{E}} \times \tilde{\mathbf{H}}^*\right] \quad [\mathrm{W/m^2}]$$

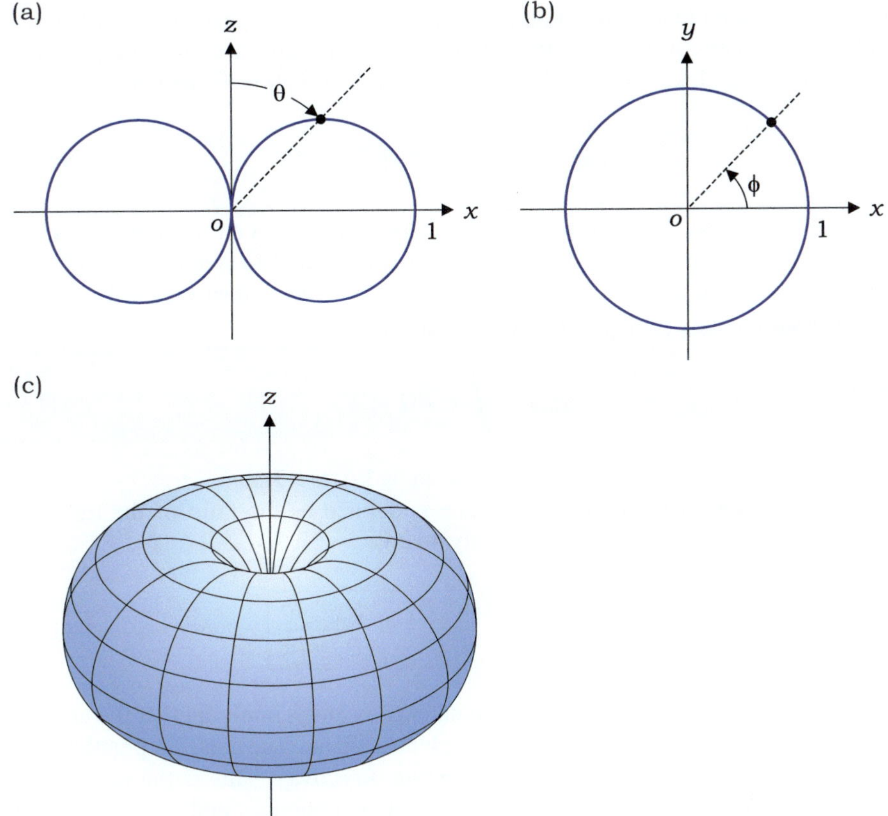

Fig. 11.3 Radiation pattern of the Hertzian dipole. **a** *E*-plane pattern, **b** *H*-plane pattern, and **c** Three-dimensional radiation pattern

where the angle brackets on the left-hand side denote time average. The Poynting vector of the far-zone field is always radially outward, as is evident from Eqs. (11.4a) and (11.4b). The time-averaged power through differential area $d\mathbf{s} = R^2 \sin\theta \, d\theta \, d\phi \, \mathbf{a}_R$ at a distance R from the origin is given by $\langle \Delta\mathscr{P} \rangle = \langle \mathbf{S} \rangle \cdot d\mathbf{s}$. However, in many situations, it is more convenient to work with the ***radiation intensity***, which is defined as the time-averaged power per unit solid angle, measured in watts per steradian [W/sr]. The radiation intensity is thus given by

$$U(\theta, \phi) = \frac{\langle \Delta\mathscr{P} \rangle}{\sin\theta \, d\theta \, d\phi} = |\langle \mathbf{S} \rangle| R^2 \qquad [\text{W/sr}] \qquad (11.6)$$

where $\langle \Delta\mathscr{P} \rangle$ is the time-averaged power through a differential area positioned at point (R, θ, ϕ) on a sphere of radius R, and $|\langle \mathbf{S} \rangle|$ is the magnitude of the power density. It is important to note that the radiation intensity depends only on θ and ϕ and is independent of R even if R^2 is explicit on the rightmost side of Eq. (11.6).

The differential solid angle $d\Omega$ for a differential area ds is defined as

$$d\Omega = \frac{ds}{R^2} = \sin\theta \, d\theta \, d\phi \quad \text{[sr]} \tag{11.7}$$

which is equivalent to the projection of the differential area onto a sphere of unity radius centered at the origin. The solid angle is measured in steradians. While the angular measure of a complete circle is 2π, that of a spherical surface is 4π.

The total radiated power is then expressed as

$$P_{\text{rad}} = \oint U \, d\Omega \quad \text{[W]} \tag{11.8}$$

where $d\Omega$ is the differential solid angle given by Eq. (11.7).

Next, the **directive gain** of an antenna is defined as the ratio of the radiation intensity to the average radiation intensity, that is,

$$G_d(\theta, \phi) = \frac{U(\theta, \phi)}{U_{\text{av}}} = \frac{U(\theta, \phi)}{P_{\text{rad}}/4\pi} \tag{11.9}$$

The directive gain of an omnidirectional antenna is unity, implying that the antenna radiates uniformly in all directions, although no such antenna exists in practice. The maximum directive gain is called the **antenna directivity**. This is the ratio of the maximum radiation intensity to the average radiation intensity and is denoted by D. Thus,

$$D = G_{d,\text{max}} = \frac{U_{\text{max}}}{U_{\text{av}}} \tag{11.10}$$

With the electric field of the radiation field, D can be expressed as

$$D = \frac{4\pi |\tilde{E}_\theta|^2_{\text{max}}}{\oint |\tilde{E}_\theta|^2 d\Omega} \tag{11.11}$$

The directivity is typically expressed in decibels: $D \, [\text{dB}] = 10 \log_{10} D$.

The total radiation power of the antenna, P_{rad}, is less than the input electrical power P_{in} because of ohmic power loss in the antenna structure. The antenna efficiency is conveniently measured in terms of the **power gain** of the antenna, or simply the **gain** of the antenna. The power gain is the ratio of the maximum radiation intensity to the radiation intensity of a lossless omnidirectional antenna at the same input power. That is,

$$G_p = \frac{U_{\max}}{P_{\text{in}}/4\pi} \tag{11.12}$$

where P_{in} is the input power.

Radiation efficiency is defined as the ratio of the radiation power to the input power:

$$\zeta = \frac{P_{\text{rad}}}{P_{\text{in}}} = \frac{G_p}{D} \tag{11.13}$$

It is worth noting that ζ is equal to the ratio of the power gain to directivity. For a lossless antenna, $\zeta = 1$ and $G_p = D$.

The **radiation resistance** is a measure of the radiation power of an antenna. This is equivalent to a resistance dissipating the same power, P_{rad}, as the antenna for the same input current. The radiation resistance R_{rad} is defined as

$$R_{\text{rad}} = \frac{2}{\tilde{I}\tilde{I}^*} P_{\text{rad}} \quad [\Omega] \tag{11.14}$$

where $\tilde{I}$ denotes the current phasor. A high radiation resistance is a desirable feature of an antenna.

For the transmission line connecting the antenna to the generator, the antenna is a load with impedance equal to the input impedance of the antenna. Generally, the input impedance of an antenna is complex such that

$$Z_{\text{in}} = R_{\text{in}} + jX_{\text{in}} \quad [\Omega]$$

The input resistance R_{in} consists of the aforementioned radiation resistance R_{rad} and **loss resistance** R_{loss}. The loss resistance of a wire dipole is equal to the ac-resistance R_{ac} given in Eq. (8.99), that is,

$$R_{\text{loss}} = R_{ac} = \frac{d\ell}{\sigma\delta 2\pi a} \quad [\Omega] \tag{11.15}$$

where σ is the conductivity, δ is the skin depth, and a is the wire radius. With the relation $P_{\text{in}} = (1/2)\tilde{I}\tilde{I}^* R_{\text{in}} = (1/2)\tilde{I}\tilde{I}^*(R_{\text{rad}} + R_{\text{loss}})$, the radiation efficiency is expressed as

$$\zeta = \frac{P_{\text{rad}}}{P_{\text{rad}} + P_{\text{loss}}} = \frac{R_{\text{rad}}}{R_{\text{rad}} + R_{\text{loss}}} \tag{11.16}$$

where R_{rad} is the radiation resistance, and R_{loss} is the loss resistance of the antenna.

Example 11.1 For the Hertzian dipole, determine the following: (a) the magnitude of the time-averaged Poynting vector, (b) radiation intensity, (c) directive gain, (d) directivity, and (e) radiation resistance.

Solution

(a) Using Eqs. (11.4a) and (11.4b), the magnitude of the time-averaged Poynting vector or the power density can be obtained as follows:

$$|\langle \mathbf{S} \rangle| = \frac{1}{2} \mathrm{Re} |\mathbf{E} \times \mathbf{H}^*| = \frac{\eta_o}{8} \left(\frac{I_o d\ell}{\lambda} \right)^2 \frac{\sin^2 \theta}{R^2} \tag{11.17a}$$

The total power radiated by the Hertzian dipole is thus

$$
\begin{aligned}
P_{\mathrm{rad}} &= \int_0^{2\pi} \int_0^{\pi} |\langle \mathbf{S} \rangle|\, R^2 \sin \theta \, d\theta \, d\phi \\
&= \frac{\eta_o}{8} \left(\frac{I_o d\ell}{\lambda} \right)^2 \int_0^{2\pi} \int_0^{\pi} \sin^3 \theta \, d\theta \, d\phi \\
&= \frac{I_o^2}{2} \left[80\pi^2 \left(\frac{d\ell}{\lambda} \right)^2 \right] \text{ [W]}
\end{aligned}
\tag{11.17b}
$$

where we have used $\eta_o = 120\pi$.

(b) Insertion of Eq. (11.17a) into Eq. (11.6) gives the radiation intensity,

$$U(\theta, \phi) = \frac{\eta_o}{8} \left(\frac{I_o d\ell}{\lambda} \right)^2 \sin^2 \theta \quad \text{[W/sr]} \tag{11.17c}$$

(c) Insertion of Eqs. (11.17b, c) into Eq. (11.9) gives the directive gain,

$$G_d(\theta, \phi) = \frac{3}{2} \sin^2 \theta \tag{11.17d}$$

(d) Because $G_d(\theta, \phi)$ has a maximum at $\theta = \pi/2$, the directivity is

$$D = 1.5 \tag{11.17e}$$

(e) Insertion of Eq. (11.17b) into Eq. (11.14) gives the radiation resistance,

$$R_{\mathrm{rad}} = 80\pi^2 \left(\frac{d\ell}{\lambda} \right)^2 \quad [\Omega] \tag{11.17f}$$

It should be noted that Eq. (11.17f) is valid only when condition $d\ell \ll \lambda$ is satisfied.

Example 11.2 A Hertzian dipole is formed by a conducting wire of radius a, length $d\ell$, and conductivity σ. Determine (a) ohmic power loss, and (b) radiation efficiency.

Solution

(a) The loss resistance of the wire dipole is given by Eq. (11.15):

$$R_{\text{loss}} = \frac{d\ell}{\sigma\delta 2\pi a} \quad [\Omega] \tag{11.18a}$$

The ohmic power loss is thus

$$P_{\text{loss}} = \frac{1}{2} I_o^2 R_{\text{loss}} = \frac{I_o^2\, d\ell}{\sigma\delta 4\pi a} \quad [\text{W}] \tag{11.18b}$$

(b) Insertion of Eqs. (11.17f) and (11.18a) into Eq. (11.16) leads to the radiation efficiency of the Hertzian dipole, that is,

$$\zeta = \frac{1}{1 + \frac{1}{160\pi^3\sigma\delta}\left(\frac{\lambda^2}{a\,d\ell}\right)} \tag{11.18c}$$

Because the values of a and $d\ell$ are much smaller than λ, the radiation efficiency is significantly reduced. For example, for a Hertzian dipole with $a = 0.8$ [mm], $d\ell = 1$ [cm], and $\sigma = 5.8 \times 10^7$ [S/m] operating at a frequency of 100 [MHz], the radiation efficiency is only 63%.

Exercise 11.1
For a cone of half-angle $\theta = 45°$ with its vertex at the origin, determine the subtended solid angle.

Ans. $\int_{\phi=0}^{2\pi} \int_{\theta=0}^{45°} \sin\theta\, d\theta\, d\phi = 1.84$ [sr].

Exercise 11.2
If the radiation pattern is in the form of a cone with a half-angle of $\theta = 45°$, for which the normalized pattern function is $F(\theta) = 1$ for $0 \le \theta \le 45°$, and zero otherwise, determine the directivity.

Ans. $D = 6.83$.

Exercise 11.3
The power density varies as $\cos^2\theta$ at a fixed distance R from the power source. Determine the directive gain.

Ans. $G_d(\theta, \phi) = 3\cos^2\theta$.

Exercise 11.4

The maximum power density is 100 [nW/m^2] at a distance of 1 [km] from the Hertzian dipole. If $I_o = 15$ [A], determine the total radiated power and radiation resistance.

Ans. $P_{\text{rad}} = 0.84$ [W], $R_{\text{rad}} = 7.5$ [mΩ].

Review Questions

RQ 11.6	What are the E- and H-plane patterns of the Hertzian dipole?	[Fig. 11.3]
RQ 11.7	What is radiation intensity?	[(11.6)]
RQ 11.8	Define the directive gain and directivity.	[(11.9)(11.10)]
RQ 11.9	Can the directivity of a real antenna be 0 [dB]?	[(11.10)]
RQ 11.10	Define the power gain and radiation efficiency.	[(11.12)(11.13)]
RQ 11.11	What is radiation resistance?	[(11.14)]

11.3 Linear Antenna

A short dipole antenna has low radiation resistance and radiation efficiency; thus, it is an inefficient radiator for electromagnetic power. We now focus on the **linear dipole antenna** and examine its radiation characteristics. A linear dipole antenna is a center-fed thin straight wire with a length comparable to the operating wavelength, and its center terminals are connected to a generator by a transmission line. It is difficult to determine the exact current distribution in a straight wire with finite radius. However, for a very thin straight wire, we can assume a standing wave over the dipole and a sinusoidal current distribution that is symmetrical about the center of the dipole and approaches zero at the ends, as shown in Fig. 11.4. The current can be expressed in phasor form as

$$\tilde{I}(z) = I_o \sin k(\ell/2 - z), \qquad z > 0$$
$$= I_o \sin k(\ell/2 + z), \qquad z < 0 \tag{11.19}$$

where ℓ is the dipole length, $k = 2\pi/\lambda$ is the phase constant, and λ is the wavelength in free space.

The far-zone electric field due to the differential current element $\tilde{I}(z)dz$ is obtained using Eq. (11.4b) as follows:

$$d\tilde{E}_\theta = \eta_o \frac{jk[\tilde{I}(z)\,dz]}{4\pi} \frac{e^{-jkR'}}{R'} \sin\theta \tag{11.20}$$

At large distances from the source ($R \gg \ell$), the following approximation can be used for the phase term in Eq. (11.20):

$$R' \approx R - z\cos\theta \tag{11.21}$$

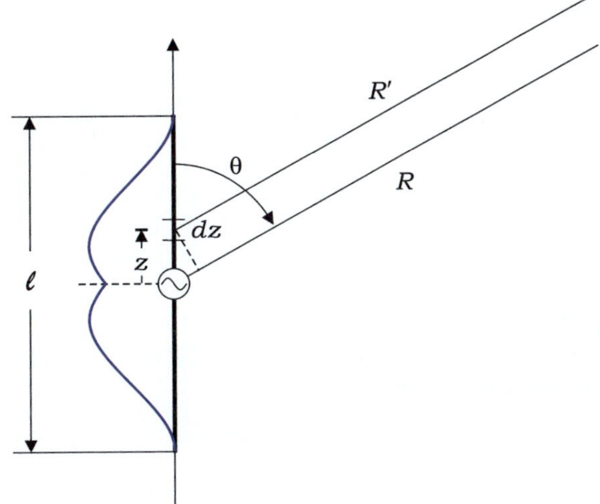

Fig. 11.4 A linear dipole with sinusoidal current distribution

Meanwhile, the term $1/R'$ in the amplitude can be approximated by $1/R$ with minimum error.

The far-zone field of a linear dipole can be obtained by integrating Eq. (11.20) with respect to z along the length of the dipole. That is,

$$\tilde{E}_\theta = \eta_o \frac{jk \sin\theta}{4\pi R} e^{-jkR} I_o \left[\int_0^{\ell/2} \sin k(\ell/2 - z) e^{jkz\cos\theta} dz \right.$$

$$\left. + \int_{-\ell/2}^{0} \sin k(\ell/2 + z) e^{jkz\cos\theta} dz \right] \qquad (11.22)$$

where we have used Eq. (11.19) for $\tilde{I}(z)$. By changing variable, we set $X \equiv -z$ in the second integral of Eq. (11.22) and then use a dummy variable z in place of X such that

$$\tilde{E}_\theta = \eta_o \frac{jk \sin\theta}{4\pi R} e^{-jkR} I_o \int_0^{\ell/2} \sin k(\ell/2 - z) [e^{jkz\cos\theta} + e^{-jkz\cos\theta}] dz$$

This can be rearranged as

$$\tilde{E}_\theta = \eta_o \frac{jk \sin\theta}{4\pi R} e^{-jkR} \left\{ 2I_o \int_0^{\ell/2} \sin k(\ell/2 - z) \cos(kz\cos\theta) dz \right\} \qquad (11.23)$$

The far-zone field of a linear dipole is thus given by

$$\boxed{\tilde{E}_\theta = j60 I_o \frac{e^{-jkR}}{R} F(\theta)} \qquad (11.24)$$

where $F(\theta)$ is the **pattern function** of the linear dipole, and is given by

$$F(\theta) = \frac{\cos\left(\frac{k\ell}{2}\cos\theta\right) - \cos\left(\frac{k\ell}{2}\right)}{\sin\theta}$$

(11.25)

where ℓ is the dipole length, and $k = 2\pi/\lambda$.

Because $|F(\theta)|$ depends on the value of $k\ell/2$ or $\pi\ell/\lambda$, the radiation pattern is significantly different for different antenna lengths. Nevertheless, the E-plane patterns are always symmetrical about the $\theta = \pi/2$ plane. Meanwhile, the H-plane patterns are always circular, because $|F(\theta)|$ is independent of ϕ. The E-plane patterns for the four dipole lengths are obtained using Eq. (11.25) and drawn to the scale, as shown in Fig. 11.5.

Exercise 11.5

A linear dipole antenna radiates a total power of 80 [kW]. If the directivity of the antenna is 6 [dB], determine the maximum electric field at a distance of 20 [km].

Ans. 0.22 [V/m].

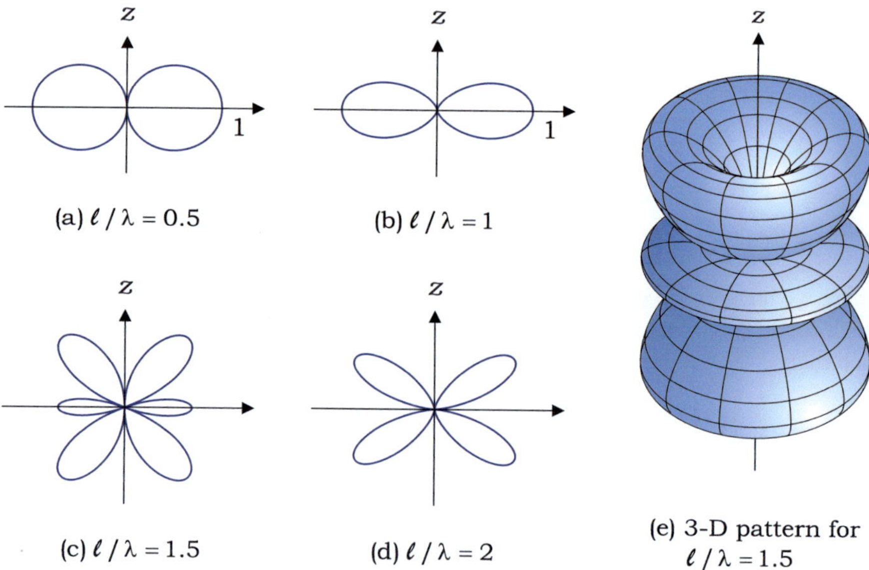

Fig. 11.5 E-plane patterns of center-fed dipole antennas. The 3-D pattern in **e** is obtained by revolving the plane section in **c** with respect to the z-axis

11.3.1 *Half-Wave Dipole*

The half-wave dipole antenna has an overall length of $\ell = \lambda/2$ (half the wavelength). From a practical point of view, it is an important antenna structure because of its highly desirable features such as radiation pattern and radiation resistance. The far-zone field can be easily obtained by considering the half-wave dipole as a series of connected Hertzian dipoles and assuming that the current distribution is given by $\cos(kz)$ along the dipole length. Specifically, this can be obtained by setting $\ell = \lambda/2$ in Eq. (11.19). Subsequently, for the half-wave dipole, the pattern function in Eq. (11.25) is reduced to

$$F(\theta) = \frac{\cos[(\pi/2)\cos\theta]}{\sin\theta} \tag{11.26}$$

The E-plane radiation pattern of the half-wave dipole can be computed from $|F(\theta)|$ and plotted, as shown in Fig. 11.5a. This is similar to that of the Hertzian dipole shown in Fig. 11.3a, except that the plane section is not circular. Nevertheless, the radiation pattern has a maximum value of unity at $\theta = \pi/2$.

The far-zone electric and magnetic fields of a half-wave dipole can be obtained from Eqs. (11.24) and (11.26) as

$$\tilde{E}_\theta = j60I_o \frac{e^{-jkR}}{R} \left\{ \frac{\cos[(\pi/2)\cos\theta]}{\sin\theta} \right\} \tag{11.27a}$$

$$\tilde{H}_\phi = \frac{\tilde{E}_\theta}{\eta_o} \tag{11.27b}$$

where η_o is the intrinsic impedance of free space, and the term in the braces represents the pattern function.

Next, the power density of the far-zone field of the half-wave dipole is computed using Eqs. (11.27a) and (11.27b) such that

$$\langle \mathbf{S} \rangle = \frac{1}{2} \tilde{E}_\theta \tilde{H}_\phi \, \mathbf{a}_R = \frac{15 I_o^2}{\pi R^2} \left\{ \frac{\cos[(\pi/2)\cos\theta]}{\sin\theta} \right\}^2 \mathbf{a}_R \tag{11.28}$$

where we have used $\eta_o = 120\pi$. The total radiated power is then computed as

$$P_{\text{rad}} = \int_0^{2\pi} \int_0^{\pi} |\langle \mathbf{S} \rangle| \, R^2 \sin\theta \, d\theta \, d\phi$$

$$= 30 I_o^2 \int_0^{\pi} \frac{\cos^2[(\pi/2)\cos\theta]}{\sin\theta} \, d\theta$$

$$= 36.6 I_o^2 \; [\text{W}] \tag{11.29}$$

The integration with respect to θ was numerically performed.

Substitution of Eqs. (11.28) and (11.29) into Eqs. (11.6) and (11.9) leads to the following directive gain:

$$G_d(\theta, \phi) = 1.64 \left\{ \frac{\cos[(\pi/2)\cos\theta]}{\sin\theta} \right\}^2 \tag{11.30}$$

Because the maximum $G_d(\theta, \phi)$ occurs at $\theta = \pi/2$, the directivity of the half-wave dipole is

$$D = G_{d,\max} = 1.64 \tag{11.31}$$

which is equivalent to $10 \log_{10}(1.64) = 2.15$ [dB].

The radiation resistance of the half-wave dipole can be obtained using Eq. (11.29) as

$$R_{\text{rad}} = \frac{2 P_{\text{rad}}}{I_o^2} = 73 \, [\Omega] \tag{11.32}$$

Example 11.2 showed that the radiation resistance of the Hertzian dipole is comparable to its loss resistance with a radiation efficiency of only 63%. If the length of the Hertzian dipole is increased to 1.5 [m], keeping in mind that $\lambda = 3$ [m], the dipole becomes a half-wave dipole, and the loss resistance is computed as 0.78 [Ω]. In this case, the radiation efficiency is 99%.

Because the loss resistance is much less than the radiation resistance, the input impedance of the half-wave dipole can be written as

$$Z_{\text{in}} \approx 73 + j X_{\text{in}} \, [\Omega] \tag{11.33}$$

The calculation of reactance is complicated and is beyond the scope of this book. However, X_{in} is approximately 40 [Ω] and is highly sensitive to the dipole length, so that we can make X_{in} zero by slightly reducing the length of the half-wave dipole while leaving the resistance essentially unchanged. Accordingly, it is possible to match a half-wave dipole antenna to a low-attenuation 75 [Ω] coaxial cable without using a matching network.

Example 11.3 A center-fed dipole with a length of 75 [cm] operating at 200 [MHz] is made of an aluminum rod with a diameter of 1 [cm] and conductivity 3.54×10^7 [S/m]. Determine the loss resistance per unit length and radiation efficiency.

Solution

The wavelength is

$$\lambda = \frac{3 \times 10^8}{2 \times 10^8} = 1.5 \, [\text{m}]$$

This is known as a half-wave dipole. Using Eq. (11.15), the loss resistance per unit length is obtained as

$$\frac{R_{\text{loss}}}{d\ell} = \frac{1}{2a}\sqrt{\frac{f\mu_0}{\sigma\pi}} = \frac{1}{10^{-2}}\sqrt{\frac{2 \times 10^8 \times 4\pi \times 10^{-7}}{\pi \times 3.54 \times 10^7}} = 0.15\,[\Omega/\text{m}] \qquad (11.34a)$$

Thus, the ohmic power loss is

$$P_{\text{loss}} = 0.15 \times \frac{1}{2}\int_{-\lambda/4}^{\lambda/4}(I_o^2\cos kz)^2 dz = 0.15 \times \frac{I_o^2}{2} \times \frac{\lambda}{4} = 0.028 I_o^2\,[\text{W}]$$

The total radiated power of the half-wave dipole, P_{rad}, is given by (11.29). Thus, the radiation efficiency is

$$\zeta = \frac{P_{\text{rad}}}{P_{\text{rad}} + P_{\text{loss}}} = \frac{36.6}{36.6 + 0.028} = 99.9\% \qquad (11.34b)$$

Example 11.4 The ***quarter-wave monopole antenna*** is half of a half-wave dipole antenna, which is located vertically on perfectly conducting ground and fed by a source at its base (see Fig. 11.6). Determine radiation resistance and directivity.

Solution

Using the method of images, the conducting ground can be replaced with an image of the antenna, as shown in Fig. 11.6. In this case, the radiation field of the quarter-wave monopole is equal to the upper half of the radiation field of the corresponding half-wave dipole. The power density in Eq. (11.28) applies to region $0 \le \theta \le \pi/2$; thus, the total radiated power of the quarter-wave monopole is one-half of that given in Eq. (11.29):

$$P_{\text{rad}} = 18.3 I_o^2\,[\text{W}] \qquad (11.35a)$$

Fig. 11.6 Quarter-wave monopole on the conducting ground and its image

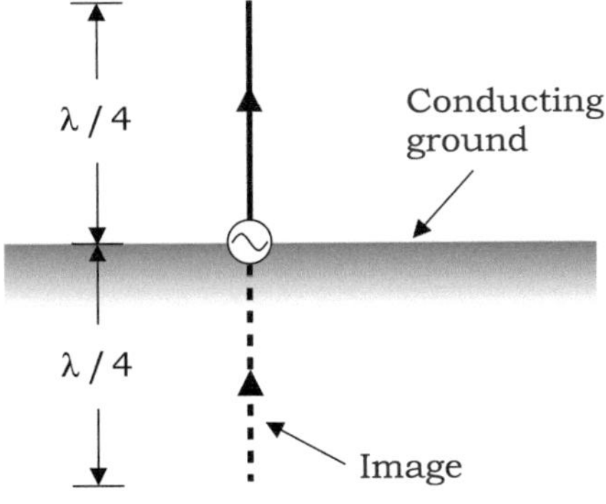

This leads to the radiation resistance of the quarter-wave monopole such as

$$R_{\text{rad}} = \frac{2P_{\text{rad}}}{I_o^2} = 36.6 \ [\Omega] \tag{11.35b}$$

Because of the same power density, U_{max} of the quarter-wave monopole is the same as that of the half-wave dipole. The average radiation intensity is $U_{\text{av}} = P_{\text{rad}}/2\pi$ for the quarter-wave monopole and $U_{\text{av}} = 2P_{\text{rad}}/4\pi$ for the half-wave dipole, where P_{rad} is given by Eq. (11.35a). Consequently, both antennas have the same directivity, that is,

$$D = \frac{U_{\text{max}}}{U_{\text{av}}} = 1.64 \tag{11.35c}$$

11.3.2 Magnetic Dipole Antenna

Example 11.5 A *magnetic dipole antenna* consists of a small circular loop of radius a ($a \ll \lambda$) that carries a uniform current of $I_o \cos \omega t$, as shown in Fig. 11.7. Find the radiation field.

Solution

To determine the radiation field, one may begin with the vector magnetic potential obtained from the loop current. Alternatively, we will use the duality of the electromagnetic fields. Noting that the loop current implies the existence of the ϕ-component of the electric field, the roles of $\tilde{\mathbf{E}}$ and $\tilde{\mathbf{H}}$ in the Hertzian dipole in Eqs. (11.3a)–(11.3c) are interchanged such that

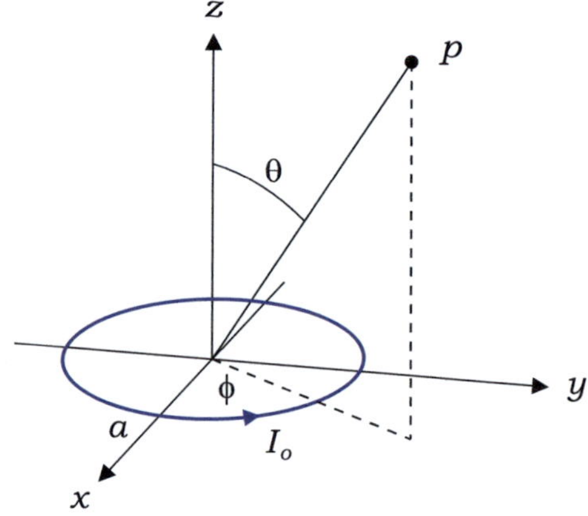

Fig. 11.7 A magnetic dipole antenna

$$\tilde{E}_\phi = -c_o \eta_o \sin\theta \left[\frac{1}{jkR} + \frac{1}{(jkR)^2} \right] e^{-jkR} \tag{11.36a}$$

$$\tilde{H}_R = c_o 2 \cos\theta \left[\frac{1}{(jkR)^2} + \frac{1}{(jkR)^3} \right] e^{-jkR} \tag{11.36b}$$

$$\tilde{H}_\theta = c_o \sin\theta \left[\frac{1}{jkR} + \frac{1}{(jkR)^2} + \frac{1}{(jkR)^3} \right] e^{-jkR} \tag{11.36c}$$

where c_o is a constant for the amplitude of the magnetic field; thus, the amplitude of the electric field is proportional to $-c_o \eta_o$, considering the direction of the Poynting vector. In the near field, only the terms with $1/R^3$ remain effective. Assuming a quasi-static condition, $\tilde{\mathbf{H}}$ in the near field should have the same magnitude as that of the magnetic field of the magnetic dipole, as expressed in Eq. (5.65). This comparison yields $c_o/(jk)^3 = m/4\pi$, where m is the magnetic dipole moment. Inserting c_o into Eqs. (11.36a)–(11.36c), we have

$$\tilde{E}_\phi = -\frac{m(jk)^3}{4\pi} \eta_o \sin\theta \left[\frac{1}{jkR} + \frac{1}{(jkR)^2} \right] e^{-jkR} \tag{11.36d}$$

$$\tilde{H}_R = \frac{m(jk)^3}{4\pi} 2 \cos\theta \left[\frac{1}{(jkR)^2} + \frac{1}{(jkR)^3} \right] e^{-jkR} \tag{11.36e}$$

$$\tilde{H}_\theta = \frac{m(jk)^3}{4\pi} \sin\theta \left[\frac{1}{jkR} + \frac{1}{(jkR)^2} + \frac{1}{(jkR)^3} \right] e^{-jkR} \tag{11.36f}$$

where the magnetic dipole moment is $m = \pi a^2 I_o$. In the far field, the radiation field of the magnetic dipole antenna is given by

$$\boxed{\tilde{E}_\phi = \eta_o \frac{k^2 I_o S}{4\pi} \frac{e^{-jkR}}{R} \sin\theta} \tag{11.37a}$$

$$\boxed{\tilde{H}_\theta = -\frac{k^2 I_o S}{4\pi} \frac{e^{-jkR}}{R} \sin\theta} \tag{11.37b}$$

where S is the loop area. By comparing the radiation fields of the magnetic dipole antenna in Eqs. (11.37a) and (11.37b) with those of the Hertzian dipole in Eqs. (11.4a) and (11.4b), the radiation resistance of the magnetic dipole antenna can be obtained by replacing $d\ell$ with kS in Eq. (11.17f). Therefore,

$$\boxed{R_{\mathrm{rad}} = 20k^4 S^2} \quad [\Omega] \tag{11.38}$$

where $k = 2\pi/\lambda$.

Exercise 11.6
If the maximum electric field at a distance of 200 [km] from a half-wave dipole is 35 [μV/m], determine the total radiated power.

Ans. $P_{rad} = 0.5$ [W].

Exercise 11.7
For a half-wave dipole, what percentage of the maximum radiation intensity occurs in the $\theta = 51°$ direction? What is the half-power beamwidth?

Ans. 50%, $\theta_{1/2} = 78°$.

Review Questions

RQ 11.12	Describe the current distribution along the half-wave dipole.	[(11.19)]
RQ 11.13	What is the essential difference between the Hertzian dipole and half-wave dipole, except for different lengths?	[(11.19)]
RQ 11.14	Derive the radiation field of the half-wave dipole.	[(11.27a,b)]
RQ 11.15	Sketch the E-plane radiation pattern of the half-wave dipole.	[Fig. 11.5]
RQ 11.16	What is the radiation resistance and directivity of the half-wave dipole?	[(11.31)(11.32)]
RQ 11.17	What is a quarter-wave monopole?	[Fig. 11.6]
RQ 11.18	What is the radiation resistance and directivity of a quarter-wave monopole?	[(11.35a,b,c)]

11.4   Antenna Arrays

In the previous section, we observed that single-element linear dipoles radiate electromagnetic energy in such a way that the radiation pattern is symmetrical about the dipole axis. Consequently, they have low directivity. In many antenna applications, it is necessary to design antennas that radiate more power in certain directions and less in other directions. This can be achieved by arranging several antenna elements in a certain configuration with the proper amplitude and phase relations among them. This group of antennas is called an ***antenna array***. This can be used to shape the radiation patterns into multiple beams. Furthermore, the beam direction can be changed electronically by controlling the relative phases of antenna elements. In this section, we focus on the basic principles and characteristics of uniform linear arrays comprising several identical antenna elements that are arranged in a straight line.

11.4.1 *Two-Element Array*

Consider two identical linear dipoles oriented along the z-axis and a distance d apart along the x-axis, as shown in Fig. 11.8. The two elements carry the same current; however, element 1 leads element 0 in phase by ξ radians ($\tilde{I}_0 = I_o$ and $\tilde{I}_1 = I_o e^{j\xi}$). Let observation point p be located in the far zone such that the two elements appear to be very close. Accordingly, it can be assumed that the electric fields of both the elements are directed along $\mathbf{a}_\theta$ at point p.

The total electric field at point p is therefore

$$\tilde{E}_\theta = E_o F(\theta) \left[\frac{e^{-jkR}}{R} + \frac{e^{-jkR_1} e^{j\xi}}{R_1} \right] \tag{11.39}$$

where E_o is a constant, and the pattern function $F(\theta)$ is given by Eq. (11.25). In the far field, R_1 can be approximated as

$$R_1 \approx R - t$$
$$\approx R - d \sin\theta \cos\phi \tag{11.40}$$

where we have used the relation $t = d\,\mathbf{a}_x \cdot \mathbf{a}_R$ in the parallel-ray approximation. Because t is very small compared with R, its effect on the magnitude term in Eq. (11.39) is insignificant, but t should be retained in the phase term because it may represent a significant phase difference. Therefore,

$$\boxed{\tilde{E}_\theta = E_o F(\theta) \frac{e^{-jkR}}{R} \left[1 + e^{j\psi} \right]} \tag{11.41a}$$

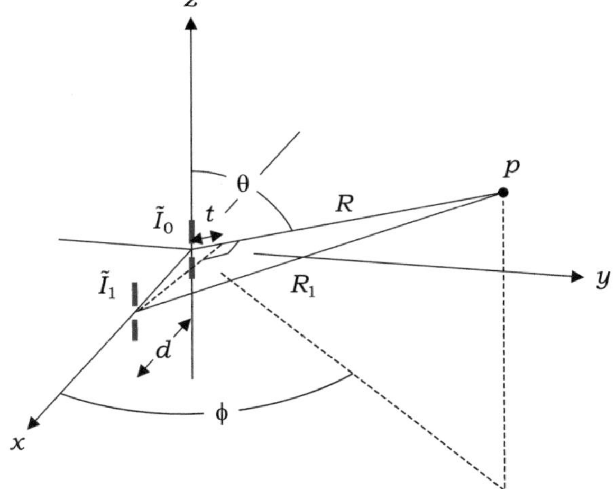

Fig. 11.8 Two-element array

where $F(\theta)$ is as given in Eq. (11.25) and

$$\psi = kd \sin \theta \cos \phi + \xi \qquad (11.41b)$$

Obviously, the magnitude of the radiation field of the two-element array is of the form of

$$\left| \tilde{E}_\theta \right| = \frac{E_o}{R} |F(\theta)||A(\theta, \phi)| \qquad (11.42a)$$

with

$$|A(\theta, \phi)| = \left|1 + e^{j\psi}\right| = \left|2 \cos\left(\frac{\psi}{2}\right)\right| \qquad (11.42b)$$

where $|F(\theta)|$ is the **element factor**, which is the magnitude of the pattern function of the individual antenna elements as expressed in Eq. (11.25), and $|A(\theta, \phi)|$ is known as the **array factor**, which depends on the array configuration, relative amplitude, and phase of the current in the elements. Consequently, the pattern function of an array of identical elements is equal to the product of the element and array factors. This is known as **the principle of pattern multiplication**. It is important to note that no coupling is assumed among the individual elements in the principle of pattern multiplication; that is, the elements induce negligible currents in others. Subsequently, the H-plane pattern is obtained using a normalized array factor. In fact, the main purpose of an antenna array is to modify the H-plane pattern using an array factor.

Example 11.6 Draw H-plane patterns of two-element dipole arrays with (a) $d = \lambda/2$ and $\xi = 0$, and (b) $d = \lambda/4$ and $\xi = -\pi/2$.

Solution

From Eq. (11.42b), the normalized array factor is $\left|\overline{A}(\theta, \phi)\right| = |\cos(\psi/2)|$. Thus, the H-plane pattern in the $\theta = \pi/2$ plane is

$$\left|\overline{A}(\theta = \pi/2, \phi)\right| = \left|\cos\left\{\frac{1}{2}(kd \cos \phi + \xi)\right\}\right| \qquad (11.43a)$$

(a) For $d = \lambda/2$ and $\xi = 0$,

$$\left|\overline{A}(\theta = \pi/2, \phi)\right| = \left|\cos\left(\frac{\pi}{2} \cos \phi\right)\right| \qquad (11.43b)$$

The H-plane pattern reaches a maximum at $\phi = \pi/2$ and $-\pi/2$, or in the broadside direction, as shown in Fig. 11.9a. This array is known as a **broadside array**. The electric fields of the two elements interfere constructively at $\phi = \pi/2$ and $-\pi/2$, and interfere destructively at $\phi = 0$ and π

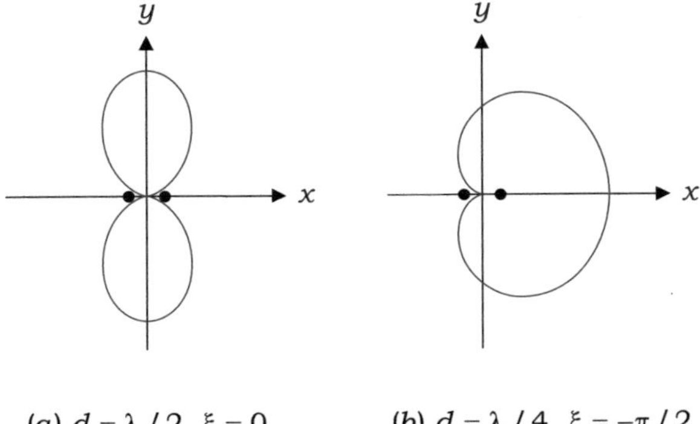

$$(a)\ d = \lambda/2,\ \xi = 0 \qquad\qquad (b)\ d = \lambda/4,\ \xi = -\pi/2$$

Fig. 11.9 *H*-plane pattern of two-element dipole array. The two dots represent the two antenna elements

(b) For $d = \lambda/4$ and $\xi = -\pi/2$,

$$\left|\overline{A}(\theta = \pi/2, \phi)\right| = \left|\cos\left\{\frac{\pi}{4}(\cos\phi - 1)\right\}\right| \tag{11.43c}$$

The H-plane pattern reaches a maximum at $\phi = 0$ and is null at $\phi = \pi$, as shown in Fig. 11.9b. The pattern maximum is now positioned on the line of the array along the $+x$-axis. This array is known as an ***endfire array***. The constructive interference in the $+x$-direction can be explained by the fact that the right-hand element lags in phase by $\pi/2$ compared to the left-hand element, but its electric field arrives at a point on the $+x$-axis by a quarter of a cycle earlier because of the separation $d = \lambda/4$. Similar reasoning can be used to explain the destructive interference occurring in the $-x$-direction.

Example 11.7 The three identical antenna elements are $\lambda/2$ apart and have a current ratio of 1:2:1, as shown in Fig. 11.10. Find the *H*-plane pattern when the three currents are in phase.

Solution

This array is equivalent to two two-element arrays displaced by a distance of $\lambda/2$. For each two-element array, the magnitude of the far field is given by Eq. (11.42a), that is,

$$\left|\tilde{E}_\theta\right| = \frac{E_o}{R}|F(\theta)||A(\theta, \phi)| \tag{11.44a}$$

with

$$\left|A(\theta = \pi/2, \phi)\right| = \left|2\cos\left(\frac{\pi}{2}\cos\phi\right)\right| \tag{11.44b}$$

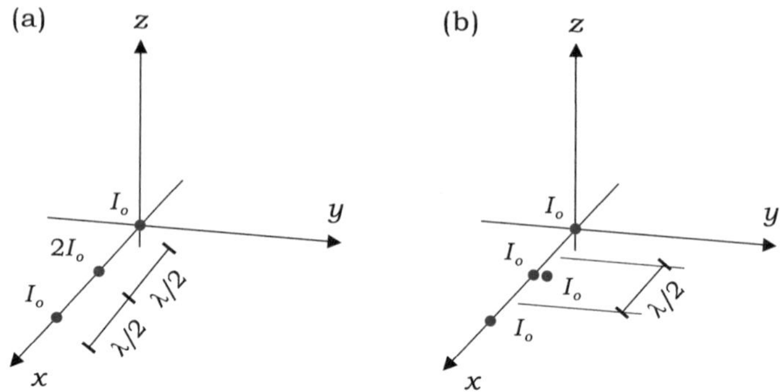

Fig. 11.10 Three-element binomial array. **a** Three elements with a current ratio of 1:2:1. **b** Two equivalent two-element arrays

The two two-element arrays can be considered as identical antenna elements with the same radiation pattern, as given in Eq. (11.44a). The radiation pattern of the three-element array is then given by the product of Eq. (11.44a) and the array factor $|A(\theta, \phi)|$, in accordance with the principle of pattern multiplication, that is,

$$\left|\tilde{E}_\theta\right| = \frac{E_o}{R}|F(\theta)||A(\theta, \phi)|^2 \tag{11.44c}$$

The H-plane radiation pattern can be computed using the normalized array factor $|A(\theta = \pi/2, \phi)|^2/4$, as shown in Fig. 11.11. The pattern is similar to that of the two-element array, except that it has a smaller beamwidth.

If a four-element array is formed by displacing two three-element arrays at a distance $\lambda/2$, the current ratio is 1:3:3:1 and the array factor is given by $|A(\theta, \phi)|^3$. If this process is extended to an N-element array, the array factor is given by $|A(\theta, \phi)|^{N-1} = \left|(1 + e^{j\psi})^{N-1}\right|$, and the current amplitudes vary according to the

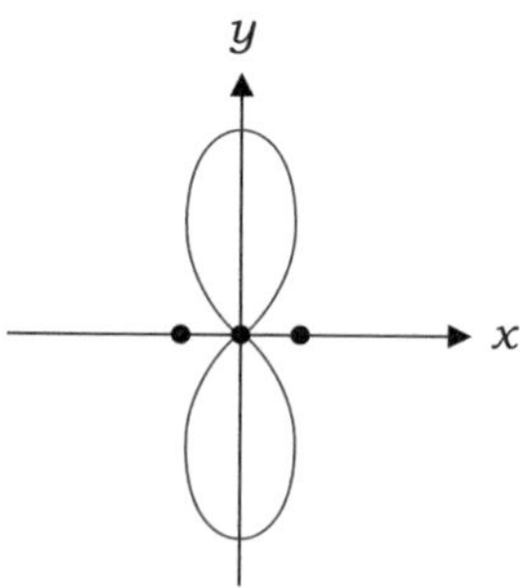

Fig. 11.11 H-plane radiation pattern of a three-element binomial array

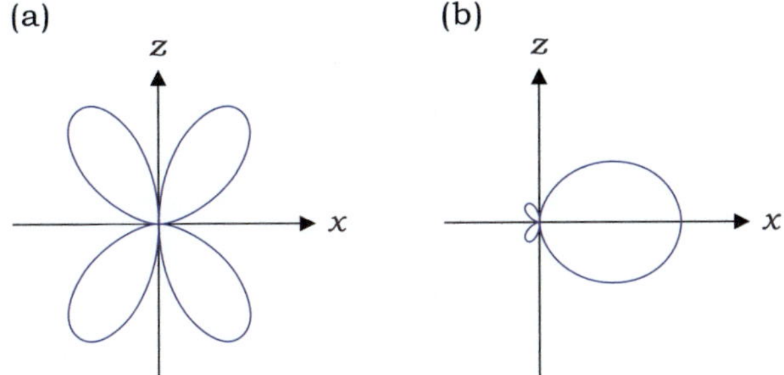

Fig. 11.12 *E*-plane pattern of the two-element array in Exercise 11.8

binomial coefficients $\binom{N-1}{k}$, where $k = 0, 1, .., N - 1$. This array is known as *binomial array*.

Exercise 11.8

If the antenna array in Example 11.6 comprises two identical half-wave dipoles, find an expression for the pattern function in the $\phi = 0°$ plane.

Ans. (a) $\cos[(\pi/2)\cos\theta]\cos[(\pi/2)\sin\theta]/\sin\theta$,
(b) $\cos[(\pi/2)\cos\theta]\cos[(\pi/4)\sin\theta - \pi/4]/\sin\theta$ (See Fig. 11.12).

Exercise 11.9

What is the current ratio of the five-element binomial array?

Ans. 1:4:6:4:1.

Exercise 11.10

For a binomial array that is composed of three identical antennas with directivity D, determine its directivity.

[Hint: $\int_{\phi=0}^{2\pi} [\cos((\pi/2)\cos\phi)]^4 d\phi = 1.57.$]

Ans. $D_{\text{binomial}} = D \times 2\pi/1.57 = 4D$.

11.4.2 Uniform Linear Array

We now consider a ***uniform linear array***, in which more than two identical antennas are positioned at regular intervals along a straight line. The array elements are fed by currents of the same magnitude, but have a progressive phase shift along the line.

Consider a uniform linear array with N identical elements aligned along the x-axis, as shown in Fig. 11.13. Extending Eq. (11.42b) to the array, the array factor can be expressed as

$$|A(\theta, \phi)| = \left|1 + e^{j\psi} + e^{j2\psi} + \cdot + e^{j(N-1)\psi}\right| \tag{11.45}$$

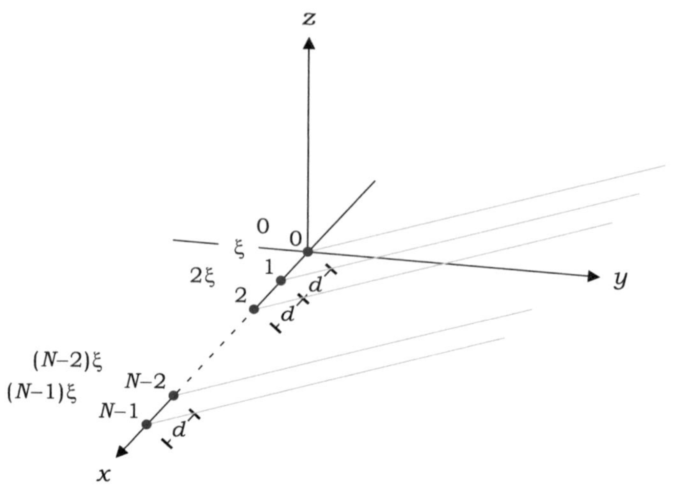

Fig. 11.13 Uniform linear array of N elements under parallel-beam approximation

The geometric progression with a common factor $e^{j\psi}$ in Eq. (11.45) can be rewritten in a closed form, such as

$$|A(\theta, \phi)| = \frac{\left|1 - e^{jN\psi}\right|}{\left|1 - e^{j\psi}\right|} = \frac{\left|\sin\left(\frac{N\psi}{2}\right)\right|}{\left|\sin\left(\frac{\psi}{2}\right)\right|} \qquad (11.46a)$$

with

$$\psi = kd \sin\theta \cos\phi + \xi \qquad (11.46b)$$

Note that $|A(\theta, \phi)|$ has a maximum value of N at $\psi = 0$.

With the help of the principle of pattern multiplication, the magnitude of the far-zone electric field is given by

$$\left|\tilde{E}_\theta(R, \theta, \phi)\right| = \frac{E_o}{R}|F(\theta)||A(\theta, \phi)| \qquad (11.47)$$

where $|F(\theta)|$ is the pattern function given by Eq. (11.25).

For the H-plane radiation pattern in the $\theta = \pi/2$ plane, the phase shift in Eq. (11.46b) is reduced to $\psi = kd \cos\phi + \xi$, which varies from $\xi + kd$ to $\xi - kd$ as ϕ varies from zero to π. To design an endfire array for which maximum radiation occurs along the axis of the array (or the $+x$-axis with $\phi = 0$), it is necessary to have $\psi = kd + \xi = 0$ or $\xi = -kd$, as shown in Eq. (11.46a). Meanwhile, $\psi = \xi = 0$ is required for a broadside array that has the maximum radiation directed normal to the array axis (or along the $+y$-axis with $\phi = \pi/2$). In general, the direction of the main

(a)   (b)

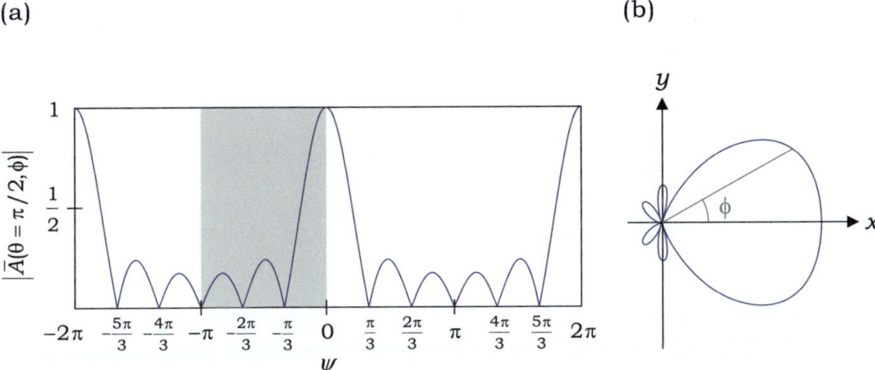

Fig. 11.14 **a** Normalized array factor for the uniform linear array with six elements versus ψ. **b** H-plane radiation pattern for $d = \lambda/4$ and $\xi = -\pi/2$

beam can be changed by changing the progressive phase shift ξ. Antenna arrays with electronic phase shifters are known as ***phased arrays***.

For example, for a uniform linear array with six elements, the normalized array factor $\left|\overline{A}(\theta, \phi)\right|$ is obtained by setting $N = 6$ in Eq. (11.46a), and dividing it by 6. It is plotted as a function of ψ, as shown in Fig. 11.14a. When the inter-element spacing is $d = \lambda/4$ and the inter-element phase shift is $\xi = -\pi/2$, ψ varies from 0 to $-\pi$ as ϕ varies from 0 to π and returns to zero as ϕ increases further towards 2π for a fixed value of $\theta = \pi/2$. In this case, the range of ψ corresponds to the shaded region in Fig. 11.14a. The H-plane radiation pattern of the array is plotted against ϕ, as shown in Fig. 11.14b, where the upper and lower halves of the plot correspond to the shaded region in Fig. 11.14a. Notably, this array can be considered an endfire array.

Figure 11.14b shows that this array is fairly directive such that the radiated power is concentrated mostly in the main lobe. Furthermore, the pattern has four sidelobes, which are usually considered a waste of energy for the antenna operating in the transmitting mode, and the direction of interference for the antenna operating in the receiving mode. The first sidelobes are the ones closest to the main lobe and have the highest level of all the sidelobes in many cases. In general, sidelobes occur when the numerator of Eq. (11.46a) is the maximum, that is,

$$\frac{1}{2}N\psi = \pm(2m + 1)\frac{\pi}{2}, \quad m = 1, 2, 3, \ldots$$

where $m = 0$ corresponds to the maximum value of the main lobe. Accordingly, the first sidelobe can be observed when

$$N\psi = \pm 3\pi \tag{11.48}$$

Next, the level of the first sidelobe relative to the main lobe can be computed by substituting Eq. (11.48) into Eq. (11.46a), and dividing it by N. When N is large, we have

$$\frac{1}{N}\frac{1}{|\sin(3\pi/2N)|} \approx \frac{1}{N}\frac{1}{|3\pi/2N|} = 0.212 \qquad (11.49)$$

The level of the first sidelobe is $20\log_{10}0.212 = -13.5\,[\text{dB}]$ below that of the main lobe, and is independent of N.

Example 11.8 A uniform linear array with $N = 6$ is arranged along the x-axis and spaced $\lambda/2$ apart (Fig. 11.13). Find the half-power beamwidth in the H-plane when the array operates as (a) a broadside array, and (b) an endfire array.

Solution

The normalized array factor was plotted as a function of ψ in Fig. 11.14a. The main lobe extends between $\psi = -\pi/3$ and $\psi = \pi/3$, as is evident from Eq. (11.46a).

(a) For broadside operation, we require the main lobe to be centered at $\phi = \pi/2$, which leads to $\psi = kd\cos(\pi/2) + \xi = 0$ or $\xi = 0$. The half-power point is obtained using the normalized array factor as follows:

$$|\overline{A}(\theta,\phi)| = \frac{1}{6}\left|\frac{\sin(6\psi/2)}{\sin(\psi/2)}\right| = \frac{1}{\sqrt{2}} \qquad (11.50a)$$

This yields $\psi = 0.47$ or $\psi = |kd\cos\phi| = 0.47$, and thus $\phi = 81.4°$ and $98.6°$. The half-power beamwidth is thus obtained as

$$\phi_{\text{hp}} = 98.6° - 81.4° = 17.2° \qquad (11.50b)$$

(b) For the endfire operation, the main lobe must be centered at $\phi = 0$, which leads to $\psi = kd\cos(0°) + \xi = 0$ or $\xi = -kd$. The half-power point is obtained using the following equation:

$$\psi = |kd\cos\phi - kd| = 0.47 \qquad (11.50c)$$

This gives $\phi = \pm 31.7°$. Thus, the half-power beamwidth is

$$\phi_{\text{hp}} = 63.4° \qquad (11.50d)$$

Note that the endfire array has a much broader main lobe than the broadside array.

Exercise 11.11

For a uniform linear array with $N = 6$ and $d = \lambda/4$ (Fig. 11.14), determine the range of positive values of ξ for which the principal lobe of $|\overline{A}(\theta, \phi)|$ becomes uninvolved with the radiation pattern.

Ans. $5\pi/6 \le \xi \le 7\pi/6$.

Exercise 11.12

If the main beam of the uniform linear array is directed along the ϕ_o-direction, then determine the relationship between ξ and d.

Ans. $\xi = -kd \cos \phi_o$.

Review Questions

RQ 11.19	Explain the element and array factors.	[(11.42a,b)]
RQ 11.20	State the principle of pattern multiplication.	[(11.42a,b)]
RQ 11.21	What are broadside and endfire arrays?	[Fig. 11.9]
RQ 11.22	What is a binomial array?	[(11.44c)]
RQ 11.23	What is a uniform linear array?	[(11.47)]

11.5 Antenna in Receiving Mode

We learned that the antennas can be used as electromagnetic energy transmitters. The generator connected to the antenna induces a time-varying current across the antenna, which subsequently produces electromagnetic waves. Antennas can also be used as receivers of incident electromagnetic waves. Receiving antennas convert electromagnetic energy into current and deliver it to a load.

Consider a two-antenna system as shown in Fig. 11.15. When the two antennas operate in transmitting mode and are weakly coupled through transimpedances Z_{1-2} and Z_{2-1}, as shown in Fig. 11.15a, it follows from the equivalent circuits of the antennas that

$$V_1 = Z_{1-1}I_1 + Z_{1-2}I_2 \qquad (11.51a)$$

$$V_2 = Z_{2-2}I_2 + Z_{2-1}I_1 \qquad (11.51b)$$

where Z_{1-1} and Z_{2-2} are the input impedances for isolated antennas 1 and 2, respectively, whose real parts are equal to the radiation resistance if ohmic loss is ignored. The transimpedances Z_{1-2} and Z_{2-1} depend on the separation and orientation of the two antennas as well as the surrounding medium.

Next, we consider the case illustrated in Fig. 11.15b, where antenna 1 operates in transmitting mode and antenna 2 operates in receiving mode. The receiving antenna is represented by a shorted equivalent circuit and its parameters are denoted by a

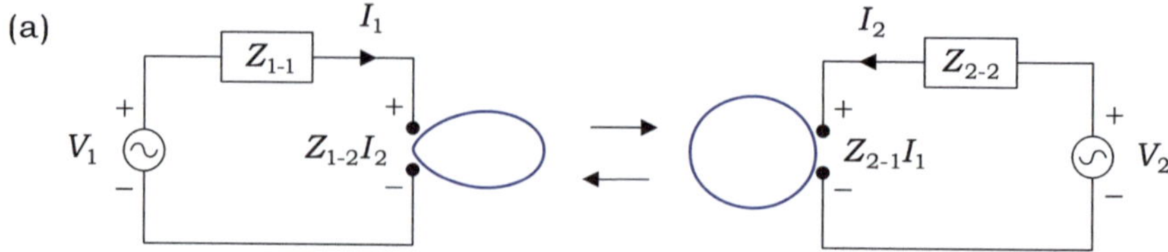

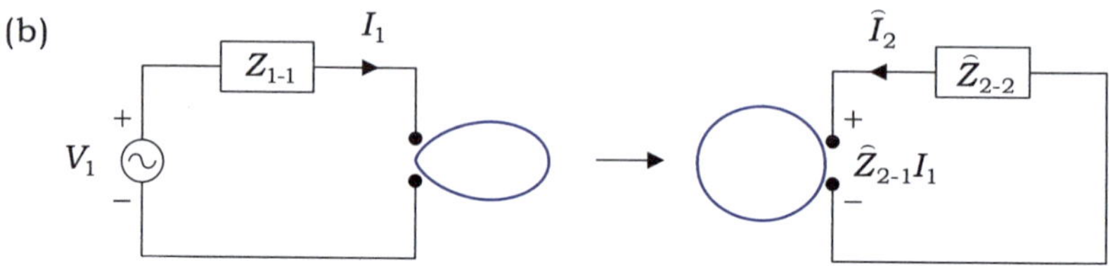

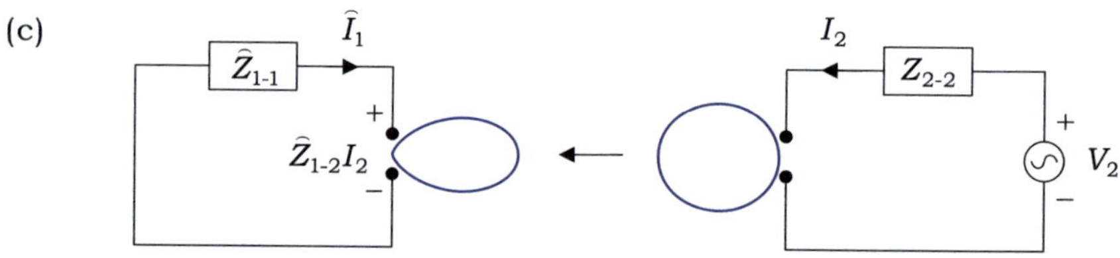

Fig. 11.15 Two-antenna systems. **a** Transmission mode of two antennas. **b** and **c** Transmitter–receiver configurations. The symbol $\frown$ at the top denotes the parameters of receiving antennas

symbol $\frown$ at the top. The two antennas are assumed to be far apart such that the reflection from antenna 2 is ignored, that is, $Z_{1-2} = 0$. Under these conditions, Eqs. (11.51a) and (11.51b) become

$$V_1 = Z_{1-1} I_1 \tag{11.52a}$$

$$0 = \widehat{Z}_{2-2} \widehat{I}_2 + \widehat{Z}_{2-1} I_1 \tag{11.52b}$$

Solving these equations yields

$$\frac{\widehat{I}_2}{V_1} = -\frac{\widehat{Z}_{2-1}}{Z_{1-1}\widehat{Z}_{2-2}} \tag{11.53a}$$

If the roles of antennas 1 and 2 are reversed, as shown in Fig. 11.14c, a similar procedure can be followed to obtain

$$\frac{\widehat{I}_1}{V_2} = -\frac{\widehat{Z}_{1-2}}{\widehat{Z}_{1-1}Z_{2-2}} \tag{11.53b}$$

For the two-antenna system shown in Fig. 11.15b (or 11.15c), the voltage V_1 (or V_2) is the source of the system, and the current $\widehat{I}_2$ (or $\widehat{I}_1$) in the receiver is the output of the system. If the two systems involve linear media such that the reciprocity theorem can be applied, it follows that the current at position 1 due to the voltage at position 2 is identical to the current at position 2 due to the same voltage at position 1, that is,

$$\frac{\widehat{I}_2}{V_1} = \frac{\widehat{I}_1}{V_2} \tag{11.54}$$

Subsequently, inserting Eqs. (11.53a, b) into Eq. (11.54) yields

$$\frac{\widehat{Z}_{2-1}}{Z_{1-1}\widehat{Z}_{2-2}} = \frac{\widehat{Z}_{1-2}}{\widehat{Z}_{1-1}Z_{2-2}} \tag{11.55}$$

Because the transimpedances are independent of the input impedances, and the two input impedances are independent of each other, we arrive at the conclusion that

$$\widehat{Z}_{2-1} = \widehat{Z}_{1-2} \tag{11.56a}$$

$$Z_{1-1} = \widehat{Z}_{1-1} \text{ and } Z_{2-2} = \widehat{Z}_{2-2} \tag{11.56b}$$

The transimpedance is directly related to the directional patterns of the two antenna elements; thus, *the directional pattern of an antenna for reception is identical to the radiation pattern of the antenna.*

In Fig. 11.15b, where $\widehat{I}_2$ is the current generated by the receiver, $\widehat{Z}_{2-2}$ is considered as the internal impedance of the generator circuit. *The generator impedance is equal to the input impedance of the antenna in transmitting mode.*

Next, we consider a transmitter–receiver system, as shown in Fig. 11.16a, where antenna 1 operates in the transmitting mode and antenna 2 operates in the receiving mode. The Thévenin equivalent circuit for antenna 2 is connected to load impedance Z_L, as shown in Fig. 11.16b. Here, V_{oc} is the open-circuit voltage induced in antenna 2, which corresponds to $\widehat{Z}_{2-1}I_1$ in Fig. 11.15b, and Z_g is the internal impedance of the voltage generator in the receiver, which corresponds to the input impedance of the antenna in transmitting mode, Z_{2-2}.

If the surface area of the receiving antenna is sufficiently small to assume that the Poynting vector of the incoming wave is uniform across the antenna surface, the time-averaged power delivered to a matched load is simply $\langle \mathbf{S} \rangle \cdot A \, \mathbf{a}_A$, where $\langle \mathbf{S} \rangle$ is the time-averaged Poynting vector, and A is the surface area of the antenna. As the

(a)   (b)

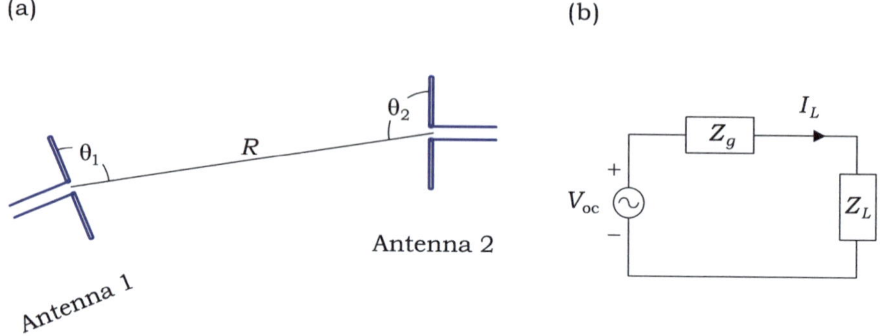

Fig. 11.16 Transmitter–receiver system. **a** Antenna 1 in the transmitting mode and antenna 2 in the receiving mode. **b** Equivalent circuit of the receiving antenna connected to the load

receiving antenna may be oriented in an arbitrary direction, it is convenient to define the **effective area** of the antenna, denoted by A_e, which is the ratio of the power delivered to a matched load to the magnitude of the power density at the antenna, that is,

$$A_e = \frac{P_L}{|\langle \mathbf{S} \rangle|} \quad [\mathrm{m}^2] \tag{11.57}$$

As previously stated, the internal impedance of the voltage generator for the receiving antenna is equal to the input impedance of the antenna in transmitting mode, that is,

$$Z_g = Z_i = R_{\mathrm{rad}} + jX \tag{11.58}$$

where R_{rad} is the radiation resistance when ohmic loss is ignored. Under a matched load condition, $Z_L = Z_g^*$, the power delivered to the load is given by

$$P_L = \frac{1}{2}\left[\frac{|V_{\mathrm{oc}}|}{2R_{\mathrm{rad}}}\right]^2 R_{\mathrm{rad}} \quad [\mathrm{W}] \tag{11.59}$$

where the term in the brackets determines the current flowing through the load.

When the electric field amplitude of the incoming wave is E_o at the site of the receiving antenna, the magnitude of the power density is

$$|\langle \mathbf{S} \rangle| = \frac{E_o^2}{2\eta_o} \quad [\mathrm{W/m}^2] \tag{11.60}$$

where $\eta_o = 120\pi$ in free space. As previously stated, the ratio of P_L to $|\langle \mathbf{S} \rangle|$ determines the effective area of the antenna.

Example 11.9 When a z-directed Hertzian dipole of length $d\ell$ is in the far zone of a radiation source located at the origin, determine the effective area of the dipole.

Solution

Let the radiation field at the dipole site have a magnitude E_o and the propagation direction be an angle θ with the dipole axis. Then,

$$V_{\text{oc}} = E_o d\ell \sin\theta \quad [\text{V}] \tag{11.61a}$$

The radiation resistance of the Hertzian dipole is given by Eq. (11.17f). This, along with Eq. (11.61a), is used in Eq. (11.59) to yield the power delivered to a matched load connected to the antenna:

$$P_L = \frac{1}{640}\left(\frac{E_o \lambda \sin\theta}{\pi}\right)^2 \quad [\text{W}] \tag{11.61b}$$

Inserting Eqs. (11.60) and (11.61b) into Eq. (11.57), the effective area of the Hertzian dipole is obtained as

$$A_e = \frac{3}{8\pi}(\lambda \sin\theta)^2 \tag{11.61c}$$

The effective area can be expressed in terms of the directive gain of a Hertzian dipole such that

$$\boxed{A_e(\theta, \phi) = \frac{\lambda^2}{4\pi}G_d(\theta, \phi)} \tag{11.62}$$

where Eq. (11.17d) is used. This ensures that the directive pattern of the antenna for reception is identical to its radiation pattern. It can be shown that the relationship in Eq. (11.62) holds for all antennas.

We are now ready to derive the power transmission relationship between the transmitting and receiving antennas, which are referred to as antennas 1 and 2, respectively. We start with the total power radiated by antenna 1, P_{rad}, assuming that the antenna has effective area A_{e1} and directive gain G_{d1}. When antenna 2 is at a distance R from antenna 1, the magnitude of the power density at the site of antenna 2 is, from Eqs. (11.6) and (11.9),

$$|\langle \mathbf{S} \rangle| = \frac{P_{\text{rad}}}{4\pi R^2}G_{d1}(\theta, \phi) \quad [\text{W/m}^2] \tag{11.63}$$

If antenna 2 has an effective area A_{e2}, the power delivered to a matched load is

$$P_L = |\langle \mathbf{S} \rangle| A_{e2}(\theta, \phi) \quad [\text{W}] \tag{11.64}$$

Inserting Eqs. (11.62) and (11.63) into Eq. (11.64) leads to the ***Friis transmission formula***:

$$\frac{P_L}{P_{\text{rad}}} = \frac{A_{e1} A_{e2}}{R^2 \lambda^2} \qquad (11.65)$$

Given the total radiated power, the delivered power is directly proportional to the product of the two effective areas, and inversely proportional to the square of the separation distance and wavelength. With Eq. (11.62), the Friis transmission formula can be rewritten as

$$\frac{P_L}{P_{\text{rad}}} = G_{d1} G_{d2} \left(\frac{\lambda}{4\pi R} \right)^2 \qquad (11.66)$$

It should be noted that the relationships in Eqs. (11.65) and (11.66) assume that the lossless antennas are located in the far zone of the other.

Example 11.10 Two half-wave dipoles are separated by a distance of 10 [km] and constitute a transmitter–receiver system, as shown in Fig. 11.16a. For $\theta_1 = 120°$, $\theta_2 = 100°$, and $I_o = 20$ [A] in the transmitting antenna, determine the power received by the receiving antenna at a frequency of 50 [MHz].

Solution

The directive gains are obtained from Eq. (11.30) as

$$G_{d1} = 1.09 \text{ and } G_{d2} = 1.57 \qquad (11.67a)$$

The power radiated by antenna 1 is obtained using Eq. (11.29) as

$$P_{\text{rad}} = 36.6 I_o^2 = 14.6 \,[\text{kW}] \qquad (11.67b)$$

Setting $\lambda = 6$ [m] and $R = 10$ [km] in Eq. (11.66) and using Eq. (11.67a), the received power is computed as follows:

$$P_L = 1.09 \times 1.57 \times \left(\frac{6}{4\pi \times 10^4} \right)^2 \times 14.6 \times 10^3 = 57.0 \,[\mu\text{W}] \qquad (11.67c)$$

Exercise 11.13

Find the maximum effective area of a half-wave dipole operating at 100 [MHz].

Ans. $1.17\,[\text{m}^2]$.

Exercise 11.14

When the half-wave dipole in Exercise 11.13 receives a maximum power of 5 [μW], determine the electric field at the antenna surface.

Ans. 56.7 [mW/m].

Review Questions

RQ 11.24 What is the consequence of the reciprocity theorem for an   [(11.56a,b)]
          antenna that can operate in either the transmitting or
          receiving mode?

RQ 11.25 What is the effective area of the antenna?   [(11.57)]

RQ 11.26 Relate the effective area of the antenna to its directive   [(11.62)]
          gain.

RQ 11.27 Explain Friis transmission formula.   [(11.65)]

11.6 The Radar Equation

A radar uses electromagnetic pulses to detect and locate distant targets. The term
radar is an acronym for **ra**dio **d**etection **an**d **r**anging. When the same antenna is
used for transmitting and receiving pulses, the distance to the target can be measured
by the time delay between the pulse traveling to the target and returning.

The *scattering cross section* or *radar cross section* is used to describe the ability
of a target to reflect the incoming electromagnetic energy. The scattering cross section
is an equivalent area required to intercept the exact amount of incoming power that,
if scattered uniformly in all directions, would produce the same power density as
the real target at the receiver. For a power density incident on the target $|\langle \mathbf{S}_{ti} \rangle|$,
the intercepted power is simply $\sigma |\langle \mathbf{S}_{ti} \rangle|$, where σ is the scattering cross section.
Assuming the intercepted power scatters in all directions, the power density of the
target observed at the receiver site is

$$|\langle \mathbf{S}_{ts} \rangle| = \frac{\sigma |\langle \mathbf{S}_{ti} \rangle|}{4\pi R^2} \quad [\text{W/m}^2] \tag{11.68}$$

where R is the distance between the object and receiver. The subscript t denotes
the target, and the subscripts s and i represent the scattered and incident powers,
respectively. The scattering cross section is thus defined by

$$\boxed{\sigma = 4\pi R^2 \frac{|\langle \mathbf{S}_{ts} \rangle|}{|\langle \mathbf{S}_{ti} \rangle|}} \quad [\text{m}^2] \tag{11.69}$$

Note that σ is a constant, even if R^2 appears explicitly on the right-hand side of
Eq. (11.69).

The same antenna can be used to transmit short electromagnetic pulses and receive
pulses reflected from a target, as shown in Fig. 11.17. In the transmitting mode, the
power density at the target at distance R from the antenna is given by Eq. (11.63),
which is repeated as follows:

$$|\langle \mathbf{S}_{ti} \rangle| = \frac{P_{\text{rad}}}{4\pi R^2} G_d(\theta, \phi) \quad [\text{W/m}^2] \tag{11.70}$$

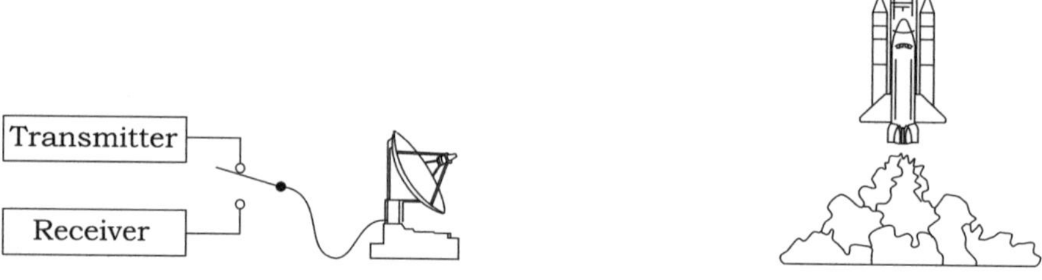

Fig. 11.17 Monostatic radar system using the same antenna for transmitting and receiving short pulses

where P_{rad} is the total radiated power, and G_d is the directive gain. Subsequently, the total power reflected from a target with a scattering cross section σ is $\sigma|\langle \mathbf{S}_{ti}\rangle|$ and is assumed to be uniformly distributed in all directions. Thus, the power density at the receiving antenna is given by $\sigma|\langle \mathbf{S}_{ti}\rangle|/4\pi R^2$. If the effective area of the antenna is A_e, the power delivered to a matched load is

$$P_L = \sigma \frac{|\langle \mathbf{S}_{ti}\rangle|}{4\pi R^2} A_e(\theta, \phi)$$

$$= \sigma \frac{P_{\text{rad}}}{(4\pi R^2)^2} A_e G_d \quad [\text{W}] \tag{11.71}$$

where we have used Eq. (11.70). Inserting Eq. (11.62) into Eq. (11.71) yields

$$\boxed{\frac{P_L}{P_{\text{rad}}} = \frac{\sigma \lambda^2}{(4\pi)^3 R^4} G_d^2(\theta, \phi)} \tag{11.72a}$$

or

$$\boxed{\frac{P_L}{P_{\text{rad}}} = \frac{\sigma/\lambda^2}{4\pi R^4} A_e^2(\theta, \phi)} \tag{11.72b}$$

This equation is known as the **radar equation**. Note that the received power is inversely proportional to R^4, because the radar pulse takes a round trip from the antenna to the target and back to the antenna.

Example 11.11 The radar system operating at 3 [GHz] has a directivity of 40 [dB] and a minimum detectable power of 2 [pW]. Determine the minimum radiated power required to detect a target with scattering cross section 1 [m^2] located 50 [km] from the radar.

Solution

The wavelength is

$$\lambda = \frac{c}{f} = \frac{3 \times 10^8}{3 \times 10^9} = 0.1 \,[\text{m}]$$

From $10 \log_{10} D = 40 \,[\text{dB}]$, the directivity is $D = 10,000$.

From Eq. (11.72a),

$$P_{\text{rad}} = P_L \left[\frac{\sigma \lambda^2}{(4\pi)^3 R^4} G_d^2 \right]^{-1} = 2 \times 10^{-12} \left[\frac{0.01}{(4\pi)^3 (5 \times 10^4)^4} 10^8 \right]^{-1} = 24.8 \,[\text{kW}]$$

Exercise 11.15

For the radar system in Example 11.11, determine the maximum range of the radar for detecting objects with a scattering cross section less than $0.2 \,[\text{m}^2]$.

Ans. 33.4 [km].

Review Questions

RQ 11.28	What is the scattering cross section?	[(11.69)]
RQ 11.29	Explain the radar equation.	[(11.72a,b)]

11.7 Problems

Hertzian dipole

11.1   A Hertzian dipole of length $d\ell = 0.01\lambda$ is fed by a current with $I_o = 2\,[\text{A}]$ at a frequency of 300 [MHz]. Determine the total radiated power and radiation resistance.

11.2   A Hertzian dipole radiates a total power of 2 [kW]. Determine the maximum electric field and radiation intensity at a distance of 10 [km].

11.3   Find the half-power beamwidth of a Hertzian dipole located at the origin. What fraction of the total power is radiated within the beamwidth?

11.4   An oscillating electric dipole located at the origin has electric dipole moment $\mathbf{p} = \mathbf{a}_z q d \cos \omega t$, where d is the separation of the bound charges. In region $d \ll R \ll \lambda$, a quasi-static approximation allows us to determine the electric field using Coulomb's law. Show that the electric field is the same as the near-zone electric field of a Hertzian dipole with a current $I_o = j\omega q$.

11.5   For a short center-fed dipole of length $d\ell$ ($d\ell \ll \lambda$) and current distribution $\tilde{I}(z) = I_o(1 - 2|z|/d\ell)$, whose average is $0.5 I_o$, find the radiation field, and show that it is equal to the radiation field of a Hertzian dipole with a uniform current of $0.5I_o$. By what factors are the radiated power and directivity scaled down with respect to the Hertzian dipole?

11.6 A center-fed dipole antenna is made of copper wire, 1 [m] in length and 2 [mm] in diameter. When operating at 2 [MHz], determine (a) the radiation efficiency, and (b) the antenna gain.

Antenna Characteristics

11.7 When the radiation intensity of an antenna is $U(\theta, \phi) = 30 \sin^2 \theta \cos \phi$ for $0 \le \theta \le \pi$ and $-\pi/2 \le \phi \le \pi/2$, and zero elsewhere, find (a) the directive gain, (b) the **pattern solid angle** for which $U(\theta, \phi)$ is normalized to unity and integrated over a sphere of unity radius, and (c) the half-power beamwidths in the $\theta = \pi/2$, and $\phi = 0$ planes.

11.8 An antenna pattern with a single main lobe has a total power of 72 [W]. When the maximum power density is 60 [μW/m^2] at a distance of 1 [km], compute the pattern solid angle and directivity of the antenna. [Hint: Eqs. (11.6) and (11.8).]

11.9 If an antenna has a radiation efficiency of 95% and directivity of 8 [dB], what is the power gain in decibels?

11.10 When a cone with small half-angle θ_o has its apex at the origin, show that the subtended solid angle can be approximated as $\pi\theta_o^2$.

Linear Antenna

11.11 For a center-fed linear antenna of length $\lambda/4$, determine the current amplitude I_o that produces a maximum electric field of 20 [mV/m] at a distance of 10 [km] in the $\theta = \pi/2$ plane.

11.12 For a linear dipole of length $\ell = 1.5\lambda$, determine the polar angle θ at which the radiation intensity becomes zero. [Hint: Fig. 11.5.]

11.13 Compare the half-power beamwidths of the dipole antennas of lengths 0.5λ and λ, for which $|F(\theta)|_{\max} = 1$ and 2, respectively.

11.14 A 50 [cm] long monopole is made of an aluminum rod with a diameter of 1 [cm] and conductivity 3.54×10^7 [S/m]. When it is vertical over a conducting surface and operates at 3 [MHz], compute the radiation resistance and efficiency.

11.15 A monopole antenna of length λ is vertical over a perfectly conducting surface. What is the current amplitude I_o for the maximum power density 0.5 [μW/m^2] observed at a distance of 4 [km] in the $\theta = 57.6°$ plane?

11.16 For a magnetic dipole antenna with radius a and current $I_o \cos \omega t$, express the total radiated power and directive gain in terms of a, ω, and I_o.

11.17 A loop antenna with a diameter of 10 [cm] operating at 10 [MHz] carries a current with amplitude $I_o = 2$ [A]. Determine the (a) maximum electric field at a distance of 5 [m], and (b) total radiated power.

11.18 A loop antenna with a diameter of 20 [cm] is made of copper wire with a diameter of 2 [mm] and carries a current with an amplitude of 1 [A] at 300 [MHz]. Compute the (a) radiation resistance, (b) radiated power, and (c) radiation efficiency.

Half-wave Dipole

11.19 Given the radiation resistance $R_{\text{rad}} = 73\,[\Omega]$ and directivity $D = 1.64$ of the half-wave dipole, compute (a) the radiated power, and (b) the maximum electric field at a distance of 10 [km] if the current amplitude is $I_o = 2\,[\text{A}]$.

11.20 A 75 [cm] long center-fed thin-wire antenna may operate either as a Hertzian dipole at 2 [MHz] or as a half-wave dipole at 200 [MHz]. When the current amplitude is I_o, compare the radiated powers for the two cases.

11.21 A half-wave dipole is centered at the origin and directed along the z-axis. When it carries a current with current phasor $\tilde{I}(z) = I_o \cos kz$, determine the charge distribution along the antenna. [Hint: Equation of continuity.]

Antenna Array

11.22 For an antenna array with two elements spaced by one wavelength and fed by currents of the same amplitude, determine (a) the array factor, and (b) the progressive phase factor ξ to ensure that the center of the main lobe is at $45°$ with respect to the x-axis in the $\theta = 90°$ plane.

11.23 A two-element antenna array is fed by currents of the same amplitude and phase. If the spacing d varies between λ and 2λ, what is the range of the rotation angle of the main lobe in the first quadrant of the $\theta = 90°$ plane?

11.24 An antenna array consists of two Hertzian dipoles aligned along the z-axis, spaced at distance d, as shown in Fig. 11.18. For the currents $\tilde{I}_0 = I_o$ and $\tilde{I}_1 = I_o e^{j\xi}$, show that the array factor is $|A(\theta, \phi)| = |2\cos[(kd\cos\theta + \xi)/2]|$, and draw the E-plane pattern for (a) $d = \lambda$ and $\xi = 0$, (b) $d = \lambda/2$ and $\xi = 0$, and (c) $d = \lambda/4$ and $\xi = -\pi/2$. [Hint: $R_1 \cong R - d\cos\theta$.]

11.25 Determine the spacing d and progressive phase factor ξ for a three-element uniform linear array so that three main lobes of equal intensity are spaced equally in the H-plane at $\phi = 0°$, $120°$, and $240°$.

11.26 If the number of elements in the antenna array shown in Fig. 11.18 is increased from 2 to 12, which are spaced at a distance $\lambda/2$ and fed by currents of the same amplitude, what should be the progressive phase factor ξ to obtain the main beam at $30°$ above the xy-plane.

Fig. 11.18 A two-element antenna array (Problem 11.24)

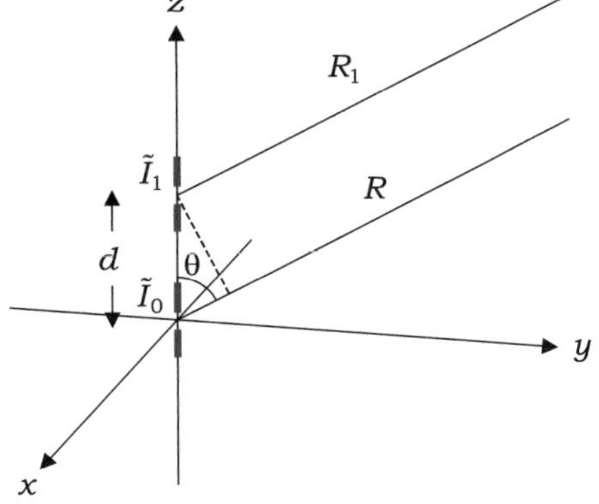

11.27 For a four-element binomial array, (a) write the normalized array factor, and (b) find the half-power beamwidth in the $\theta = \pi/2$ plane. (c) How does the beamwidth change if the array operates as a four-element uniform linear array with the same current amplitude and phase?

11.28 A broadside array consists of five elements spaced $\lambda/2$ apart. If the current amplitude ratio is $1 : 1.5 : 3 : 1.5 : 1$, then find the normalized array factor.

11.29 When a uniform linear array of N elements operates as a broadside array, generating no radiation in the endfire direction in the $\theta = \pi/2$ plane, determine the spacing d.

11.30 An eight-element uniform linear array has two main lobes at $\phi = 60°$ and $\phi = -60°$ in the H-plane, with no side lobes in region $-60° < \phi < 60°$. Determine element spacing d and progressive phase factor ξ.

11.31 Repeat Problem 11.30 if there is no side lobe in the region $60° < \phi < 300°$.

11.32 For an N-element broadside linear array, (a) show that the **null beamwidth** of the main lobe is $\Delta\phi = 2\lambda/Nd$ for a large N, which is the angular width between two nulls on either side of the beam, and (b) determine the half-power beamwidth of the main lobe. [Hint: $\sin x \simeq x - x^3/6$ for $x \ll 1$.]

Effective Area and Friis Formula

11.33 If the electric field amplitude is 30 [mV/m] at the site of an antenna that receives a total power of 2 [μW], determine the effective area of the antenna.

11.34 When an antenna has directivity of 25 [dB] at 3 [GHz], what is the maximum effective area of the antenna?

11.35 For a half-wave dipole, express the (a) maximum effective area, and (b) maximum power delivered to a matched load if the electric field amplitude of the incoming wave is E_o at the site of the antenna. (c) Show that the *effective length* of the dipole, ℓ_e, for which the open-circuit voltage $V_{oc} = E_o \ell_e$ delivers the same power to the load, is given by $\ell_e = \lambda/\pi$.

11.36 When a half-wave dipole with current amplitude I_o operates in transmitting mode, an equivalent Hertzian dipole with length ℓ_e and uniform current I_o generates the same maximum electric field at $\theta = \pi/2$. Determine ℓ_e and compare it with the result of Problem 11.35.

11.37 A Hertzian dipole operates in the receiving mode in the far field, and the incoming wave has an electric field amplitude of E_o. Find the maximum power delivered to a matched load connected to the antenna, starting with (a) the effective area, and (b) the open-circuit voltage $V_{oc} = E_o d\ell$.

11.38 When a half-wave dipole radiates a power of 5 [kW] at 200 [MHz], determine the maximum power received by a second half-wave dipole at a distance of 1 [km], which is parallel to the first.

11.39 In a satellite communication link operating at 15 [GHz], the antennas on Earth and on a satellite in geosynchronous orbit at an altitude of 35,786 [km] have gains of 50 [dB] and 30 [dB], respectively. Determine the power transmitted by the antenna on Earth to deliver a minimum power of 0.2 [nW] to the satellite.

11.40 A monostatic radar with a directive gain of 40 [dB] operates at 4 [GHz] and is used to track a target with a scattering cross section of 0.5 [m^2] in the range of 500 [m]. If the minimum power for detection is 1 [nW], determine the minimum transmitted power.

11.41 Radar antenna with an effective area of 8 [m^2] radiates 250 [kW] at 3 [GHz]. For a target with a scattering cross section of 15 [m^2] located at 100 [km], determine (a) the maximum power density at the target, (b) the power intercepted by the target, and (c) the amount of reflected power received at the antenna.